Arsenic

Arsenic

Environmental Chemistry, Health Threats and Waste Treatment

Edited by

KEVIN HENKE

University of Kentucky Center for Applied Energy Research, USA

A John Wiley and Sons, Ltd., Publication

Library of Congress Cataloging-in-Publication Data

Henke, Kevin R.
    Arsenic : environmental chemistry, health threats, and waste treatment / Kevin Henke.
        p. cm.
    Includes bibliographical references and index.
    ISBN 978-0-470-02758-5 (cloth : alk. paper) 1. Arsenic. 2. Arsenic–Toxicology. 3.
Groundwate–Arsenic content. 4. Arsenic wastes. 5. Environmental chemistry. I. Title.
    TD196.A77H46 2009
    628.5'2–dc22
                                                          2008044505

A catalogue record for this book is available from the British Library

ISBN 978-0470-027585 (H/B)

Typeset in 10/12pt Times by Laserwords Private Limited, Chennai, India
Printed and bound in Great Britain by CPI Antony Rowe, Chippenham, Wiltshire

*To my wife, Yvonne, children Erin and Kyle,
and my parents, Lyle and Lorayne.*

Contents

List of Contributors

David A. Atwood   Department of Chemistry, University of Kentucky, Lexington, KY 40506-0055, USA.

Alan E. Fryar   Department of Earth and Environmental Sciences, University of Kentucky, Lexington, KY 40506-0053, USA.

Kevin R. Henke   University of Kentucky Center for Applied Energy Research, 2540 Research Park Dr., Lexington, KY 40511-8410, USA.

Aaron Hutchison   Department of Science and Mathematics, Cedarville University, 251 N. Main St., Cedarville, OH 45314, USA.

Michael F. Hughes   US Environmental Protection Agency, Office of Research and Development, National Health and Environmental Effects Research Laboratory, Research Triangle Park, NC 27711, USA.

Elaina M. Kenyon   US Environmental Protection Agency, Office of Research and Development, National Health and Environmental Effects Research Laboratory, Research Triangle Park, NC 27711, USA.

Abhijit Mukherjee   Alberta Geological Survey, 4999-98 Avenue, Room 434A, Edmonton, Alberta T6B 2X3, Canada.

Bethany M. O'Shea   Lamont-Doherty Earth Observatory, Columbia University, Geochemistry Division, P.O. Box 1000, 61 Route 9W, Palisades, NY 10964, USA.

David J. Thomas   US Environmental Protection Agency, Office of Research and Development, National Health and Environmental Effects Research Laboratory, Research Triangle Park, NC 27711, USA.

Preface

Arsenic contamination in drinking water aquifers is one of the worst and most widespread environmental problems currently facing humanity. More than 100 million people may be at risk from utilizing arsenic-contaminated groundwater. In recent decades, the tragic plight of millions of people in Bangladesh and West Bengal, India, has been publicized. Arsenic-contaminated groundwaters and other types of arsenic contamination are also serious threats in parts of Argentina, Cambodia, Chile, mainland China, Mexico, Nepal, Pakistan, Taiwan, Vietnam, and the United States (Chapter 6 and Appendix D).

As listed in the bibliographies of our chapters, many excellent books and summary articles have been written on a wide variety of arsenic topics, including: the history of poisoning events and commercial applications, the utilization of chromated copper arsenate wood, the chemistry of arsenic and its compounds, geological occurrences, medical applications and toxicology, analytical and speciation techniques, the environmental impacts of arsenic from coal utilization, groundwater and surface water contamination, and remediation and waste treatment technologies for arsenic in wastes, sediments, soils, flue gas, and water. Our book, *Arsenic: Environmental Chemistry, Health Threats, and Waste Treatment*, can be viewed, to some extent, as an update to these previous publications. More importantly, however, this book will serve as a broad and single resource on the subject.

In its chapters and sections, our book discusses the major historical, geological, chemical, treatment and remediation, and environmental subjects related to arsenic. Although this book is primarily written for chemistry, toxicology, and geology students, the discussions and information would also be useful to scientists and engineers from many different disciplines, medical experts, environmentalists, regulators, waste management personnel, and laypeople. The book also contains overview material, including a glossary in Appendix B, on several fundamental topics in chemistry and geology. Although not every arsenic-related topic can be extensively discussed, the authors of this book have striven to provide a number of key references that contain additional details for our readers.

Our book is divided into seven chapters based on major arsenic-related topics. Chapter 1 provides an introduction to the discussions in the other chapters. In Chapter 2, details on the chemistry and important physical properties of arsenic and its most common naturally occurring compounds (minerals) are reviewed. Background information is also provided on thermodynamics and adsorption isotherms. Chapter 3 reviews the nucleosynthesis of arsenic in massive stars, its distribution in the solar system, important oxidation and reduction reactions, and the distribution and behavior of arsenic in the Earth's natural environments. The toxicology and epidemiology of arsenic are discussed in Chapter 4, which includes summaries of animal tests and the health effects of arsenic inhalation, digestion, and dermal exposure on humans. Chapter 5 discusses the history and commercial use of arsenic in human societies. Examples are also given of criminal and accidental arsenic poisoning events in the nineteenth and twentieth centuries. Chapter 6 concentrates on the catastrophes of arsenic contamination in groundwaters, and discusses examples in Bangladesh, West Bengal (India), the Middle Ganges Plain (India), Chattisgarh (India), the Terai alluvial plain (Nepal), the Indus alluvial system (Pakistan), the Irrawaddy delta (Myanmar), the Mekong plain and delta (Cambodia, Vietnam, and Laos), the Red River delta (Vietnam), the Yellow River plains (Inner Mongolia of China), Taiwan, Ghana, Nigeria, Australia, the United States, Chile, and the Chaco and Pampa plains of Argentina. Finally, Chapter 7 reviews remediation and treatment technologies for arsenic in water, solids, and flue gases. Our book also contains several appendices, which include convenient lists of measurement conversions (Appendix A), a glossary of important terms (Appendix B), tables of thermodynamic data on arsenic and its major compounds and chemical species (Appendix C), maps showing the locations of major sites with arsenic contamination (Appendix D), and a survey of regulations related to arsenic (Appendix E).

The authors acknowledge the support of several individuals during the preparation of this book, including: Ms. Jennifer Cossham, Mr. Richard Davies, Ms. Zoe Mills, Ms. Nicole Elliott, Ms. Gemma Valler and other staff at John Wiley & Sons, Ltd, and Ms. Deepthi Unni, Ms. Kapali Mahalakshmi and Mr. Ashok Kumar L at Laserwords. Ms. Lisa Blue reviewed the contents of some of the chapters. We especially appreciate the support and patience of our spouses and other family members during the preparation of this book. The editor and authors welcome comments, questions, and constructive criticisms from our readers.

Kevin R Henke, editor
Center for Applied Energy Research, The University of Kentucky

1

Introduction

KEVIN R. HENKE

University of Kentucky Center for Applied Energy Research

1.1   Arsenic origin, chemistry, and use

On the periodic table of elements, arsenic (number 33) is in group 15 along with nitrogen, phosphorus, antimony, and bismuth (Chapter 2). The *nucleosynthesis* of the element occurs in massive red giant stars and *supernovas* rather than from the *Big Bang*. After formation, arsenic collects in *nebulae*, which may condense into new stars and planets. In our solar system, arsenic has been detected in the atmospheres of Jupiter and Saturn, and trace amounts occur in *meteorites* and Moon rocks. On Earth, the element is largely concentrated in the *core* and in *clay*- and *sulfide*-rich portions of the *crust*. *Hydrothermal fluids* are important in transporting and concentrating arsenic in crustal rocks (Chapter 3). An appropriate understanding of the chemical properties of arsenic (Chapter 2) and its behavior in natural environments (Chapters 3 and 6) are critical in predicting risks to the environment and human health (Chapter 4), as well as selecting effective and economical technologies for treating arsenic-bearing wastes and remediating arsenic-contaminated sites (Chapter 7).

The only stable and naturally occurring *isotope* of arsenic is ^{75}As, where each atom of this isotope has 33 protons and 42 neutrons. The most common *valence states* of arsenic are -3, 0, $+3$ and $+5$. Arsenic and its compounds include: elemental forms, *organoarsenicals*, *arsenides*, *arsenosulfides*, *arsenites* and *arsenates*. Arsenic forms also partially substitute for sulfide, *sulfate*, and possibly *carbonate* in a variety of *minerals* (Chapter 2). In the presence of surface and near-surface *aerated* water, arsenide and arsenosulfide minerals *oxidize* to more water-soluble arsenates (Chapter 3).

For centuries, humans have utilized arsenic compounds (especially, realgar (As_4S_4), orpiment (As_2S_3), and arsenolite (As_2O_3)) in a wide variety of products, which included: pigments, medicines, alloys,

Arsenic   Edited by Kevin R. Henke
© 2009 John Wiley & Sons, Ltd

pesticides, herbicides, glassware, embalming fluids, and as a depilatory in leather manufacturing. Furthermore, the toxic properties of arsenolite and other arsenic compounds are well known and have been widely used in chemical warfare agents and to commit murder and suicide since ancient times (Chapter 5).

The term *'arsenic'* probably originated from the Persian word *az-zarnikh* or other modifications of its root word, *'zar'*, which referred to yellow or gold orpiment (Azcue and Nriagu, 1994), 3; (Meharg, 2005), 39. In ancient Syria, the term became *zarnika*. The ancient Greeks believed that metals and other substances had masculine or feminine properties. They referred to yellow orpiment pigments as *'αρρενικον'* (arrenikos or arsenikos), which means 'potent' or 'masculine' (Azcue and Nriagu, 1994), 3; (Meharg, 2005), 38–39. From these origins, the Greek name became *'αρσενικόν'* (arsenikon). The term was translated *'arsenicum'* in Latin and ultimately evolved into *'arsenic'* in French, which is also used in modern English-speaking countries. Other modern names for arsenic include: *arsénico* (Spanish), Мышьяк (Russian), 砷 (simplified and traditional Chinese), ヒ素 (Japanese), *Arsen* (German), and *arsenico* (Italian). As European chemists began to distinguish between elements and compounds, the noun *'arsenic'* was eventually restricted to refer to element 33.

1.2 Arsenic environmental impacts

In the twentieth century, arsenic was further utilized in livestock dips and feed supplements, semiconductors, wood preservatives, and medicines. Toward the end of the century, toxicologists and other scientists began to recognize widespread arsenic poisoning in Bangladesh, West Bengal (India), and elsewhere. The arsenic came from their drinking water wells. Additional arsenic-contaminated *groundwater* and other types of arsenic contamination have been identified as serious threats in parts of Argentina, Cambodia, Chile, mainland China, Mexico, Nepal, Pakistan, Taiwan, Vietnam, and the United States (Appendix D). Perhaps, more than 100 million people may be at risk from arsenic-contaminated groundwater (Chapter 6).

Rather than always resulting from industrial spills, the improper disposal of arsenic-bearing wastes, or the excessive application of *arsenical* pesticides, arsenic contamination in groundwater often originates from the mobilization of natural deposits in *rocks*, *sediments*, and *soils*, and sometimes from *geothermal* water (Chapters 3 and 6). The *oxidation* of arsenides and arsenosulfides in mining wastes or natural rock formations can release arsenic into groundwaters and surface waters. Another important source of arsenic in groundwater is the *reductive dissolution* of arsenic-bearing iron and other *(oxy)(hydr)oxides* in *anaerobic* soils, sediments, and rocks (Chapters 3 and 6). Under oxidizing conditions in surface waters and near-surface groundwaters, arsenic is removed from solution by *sorbing* onto or *coprecipitating* with iron and other (oxy)(hydr)oxides. In many areas, organic industrial wastes, as well as soluble organic matter from livestock manure, septic tanks, and landfills infiltrate into the subsurface. These organic compounds are *reductants*, which increase microbial activity and promote reducing conditions in the subsurface. In the absence of abundant sulfide, the reducing conditions convert Fe(III) into water-soluble Fe(II), which dissolves the iron (oxy)(hydr)oxides and releases their arsenic into associated groundwater (Behr and Beane, 2002; Burgess and Pinto, 2005; Stollenwerk and Colman, 2003). The microbial oxidation of organic matter also produces bicarbonate and other carbonate species, which would raise the alkalinity of groundwater and desorb arsenic from mineral surfaces (Burgess and Pinto, 2005; Appelo *et al.*, 2002; García-Sánchez, Moyano and Mayorga, 2005; Anawar, Akai and Sakugawa, 2004). In particular, any orpiment and realgar in *igneous* and *metamorphic* bedrocks would tend to dissolve in the presence of bicarbonate (Kim, Nriagu and Haack, 2000).

In response to the widespread arsenic contamination (Appendix D), many governments have instituted regulations on the disposal of arsenic-bearing wastes and arsenic emissions from ore smelters and *coal*-combustion power plants. The World Health Organization (WHO) also recommended an arsenic limit of 10 $\mu g\,L^{-1}$ (micrograms per liter) for drinking water, and many developed nations have adopted this recommendation as a regulatory standard (Appendix E). Furthermore, several nations have established programs (such as *Superfund* in the United States) to *remediate* arsenic-contaminated areas. In 1999, the *National Priority List* for the Superfund Program of the US Environmental Protection Agency (US EPA) identified 1209 sites in the United States that had serious environmental and human health risks ((US Environmental Protection Agency US EPA, 2002a), 2; Appendix E). After lead, arsenic was the most common inorganic contaminant (568 sites or 47 % of the total; (US EPA, 2002a), 2).

1.3 Arsenic toxicity

Exposure to arsenic can result in a variety of health problems in humans, including various forms of cancer (e.g. skin, lung, and bladder), cardiovascular and peripheral vascular disease, and diabetes (Chapter 4). Overall, both inorganic and organic As(III) forms tend to be more toxic to humans than the As(V) forms. Humans may be exposed to arsenic through inhalation, dermal absorption, and ingestion of food, water, and soil. Inhalation exposure can result from industrial emissions, cigarette smoking, and *flue gas* from coal-combustion power plants and ore smelters. In air, arsenic primarily *sorbs* onto particulate matter. Once arsenic-bearing gases or particles enter the airway and deposit on lung surfaces, the arsenic is absorbed further into the body. Inhalation of arsenic depends on the size of the particles and absorption depends on the solubility of the chemical form of the arsenic (Chapter 4). When compared with ingestion, the risks associated with the dermal absorption of inorganic arsenic are generally low. Like ingestion, any dermal effects would depend on the source of the arsenic (e.g. water, soil, chromated copper arsenate (CCA)-preserved wood). As discussed in Chapter 4, controlled ingestion studies in humans indicate that both As(III) and As(V) are well absorbed from the gastrointestinal tract. Between 45 and 75 % of the dose of various As(III) and As(V) forms are excreted in urine within a few days, which suggests that gastrointestinal absorption is both relatively rapid and extensive (Tam, Charbonneau and Bryce, 1979; Yamauchi and Yamamura, 1979; Buchet, Lauwerys and Roels, 1981a; Buchet, Lauwerys and Roels, 1981b; Lee, 1999).

1.4 Arsenic treatment and remediation

1.4.1 Introduction

Many methods for treating arsenic in water, gases, and solids are utilized in both *waste management* and remediation. Waste management refers to the proper handling, treatment, transportation, and disposal of wastes or other byproducts from mining, utility, agricultural, municipal, industrial, or other operations so that they are not environmental or human health threats. While waste management attempts to prevent environmental contamination, *remediation* deals with sites that have already been contaminated. Remediation refers to the restoration of a site through the treatment of its contaminated soils, sediments, *aquifers*, air, water, previously discarded wastes, and/or other materials so that they no longer pose a threat to the environment or human health.

Inorganic As(V) and As(III) are the dominant forms of arsenic in most natural waters, wastewaters, contaminated soils and sediments, and solid wastes (Chapter 7). The vast majority (approximately 89–98.6 %) of atmospheric arsenic is associated with particulates rather than existing as vapors. Both gaseous and particulate arsenic are inhalation hazards and may also contaminate surface soils, sediments, and waters near their points of origin (Chapter 4; (Shih and Lin, 2003; Chein, Hsu and Aggarwal, 2006; Hedberg, Gidhagen and Johansson, 2005; Martley, Gulson and Pfeifer, 2004)).

1.4.2 Treatment and remediation of water

Considering the millions of people threatened by arsenic-contaminated groundwater, low-cost technologies are desperately needed to effectively treat arsenic in water, especially in developing nations. Artificial sorption and coprecipitation with iron (oxy)(hydr)oxides are some of the more effective and popular technologies for removing inorganic As(V) from water (Chapter 7). Coprecipitation is accomplished by adding salts (such as Fe(III) chlorides or sulfates) to the water to precipitate iron (oxy)(hydr)oxides. Sorbents are commonly placed in treatment columns. Iron (oxy)(hydr)oxides usually have to be imbedded in support materials to maintain the *permeability* of the columns. Popular iron (oxy)(hydr)oxide sorbents include: goethite (α-FeOOH), akaganéite β-FeO(OH), and *ferrihydrites* (variable compositions). Considering that many wastewaters and the vast majority of natural waters have pH values below 9 (Krauskopf and Bird, 1995), 225), any dissolved inorganic As(III) in aqueous solutions primarily exists as unreactive $H_3AsO_3^0$ (Chapter 2). Most water treatment technologies require that $H_3AsO_3^0$ be oxidized to As(V) *oxyanions* (i.e. $H_2AsO_4^-$ and $HAsO_4^{2-}$) before treatment (Chapters 2 and 7). However, zerovalent iron (Fe(0)) is an example of a sorbent that can effectively remove inorganic As(III) from water without preoxidation steps. Zerovalent iron may also be installed in *permeable reactive barriers* (PRBs) to remove arsenic from groundwater. Chapter 7 discusses many other treatment and remediation methods for arsenic-contaminated water.

1.4.3 Treatment and remediation of solid wastes, soils, and sediments

As further discussed in Chapter 7, a variety of technologies are available for treating arsenic-bearing wastes or remediating arsenic in soils and sediments. Many, but not all, of the technologies can be utilized in either waste treatment or remediation. Overall, the main goal of waste treatment and remediation technologies is to either encapsulate the arsenic in an inert matrix that resists *leaching* in natural environments (e.g. *in situ vitrification* and *solidification/stabilization* followed by landfilling) or attempt to isolate the arsenic into a smaller and more manageable volume for disposal (e.g. *pyrometallurgical treatment*).

Waste management and remediation technologies for arsenic in solid materials often utilize heat or electric currents. In pyrometallurgical treatment, incinerators or furnaces volatilize and capture arsenic from soils, sediments, or solid wastes. *Vitrification*, including *in situ vitrification*, refers to the melting of soils, sediments, and solid wastes to primarily incinerate organic contaminants and encapsulate arsenic and other inorganic species into melts. The melts then cool into impermeable and chemically resistant glass. The US EPA considers vitrification to be the *BDAT* (*best demonstrated available technology*) for treating arsenic in soils (US Environmental Protection Agency US EPA, 1999), C.1. Electrokinetic methods refer to in situ and, in some cases, ex situ technologies that remove contaminants from wet soils, sediments, or other solid materials by passing electric currents through them. Unlike in situ vitrification, the currents in electrokinetic methods are too low to melt the materials. Instead, the electric currents cause ions (including As(V) oxyanions) and charged particles in aqueous solutions within contaminated solid materials to migrate toward electrodes, where they may be collected or otherwise treated (Mulligan, Yong and Gibbs, 2001), 193, 199–200.

Solidification/stabilization refers to reducing the mobility of a contaminant in sediments, soils, other solids, or even liquid wastes by mixing them with Portland cement, lime (CaO), cement kiln dust, clays, slags, *polymers*, water treatment *sludges*, iron-rich gypsum ($CaSO_4 \cdot 2H_2O$), coal *flyash*, and/or other *binders* (Mulligan, Yong and Gibbs, 2001), 193; (Leist, Casey and Caridi, 2000), 132; (US Environmental Protection Agency US EPA, 2002b), 4.1; (Mendonça *et al.*, 2006). Arsenic is immobilized by both physical and chemical processes. The binder physically encapsulates the contaminant in an inert matrix that resists leaching (*solidification*). The binder may also create chemical bonds with the arsenic (*stabilization*), such as calcium from the binders reacting with As(V) to form calcium arsenates. To maximize the solidification/stabilization of arsenic-bearing wastes, any As(III) is usually preoxidized (Jing, Liu and Meng, 2005), 1242.

During much of the twentieth century, CCA preservatives were widely used in many nations to protect outdoor wood from microorganisms, fungi, wood-feeding insects, and marine borers (Chapters 5 and 7). *Leaching tests* indicate that arsenic from the wood could contaminate surrounding soils, sediments, and water. The risks of arsenic toxicity from exposure to CCA-treated wood are uncertain. There are reports of children experiencing arsenic poisoning from playing in soils near CCA-treated wood (Nriagu, 2002), 20. However, other studies indicate that arsenic exposure from the wood and associated soils is negligible (Pouschat and Zagury, 2006; Nico *et al.*, 2006). Although the wood is no longer commercially available in many nations, the life expectancy of CCA-treated wood is at least 30 years in terrestrial environments and about 15 years in salt water (Christensen *et al.*, 2004), 228; (Hingston *et al.*, 2001), 54. Thus, the issues dealing with the handling and disposal of CCA-treated wood will persist for decades.

Several nations (e.g. Denmark) or regional and local governments (e.g. Minnesota in the United States) have restricted the landfilling and/or incineration of CCA-treated wood. In some circumstances, CCA-treated wood may be mixed with cement and utilized in construction (Gong, Kamdem and Harcihandran, 2004). CCA-treated wood may also be detoxified with acidic, organic, or other extracting solutions, provided that the use and disposal of the solutions are cost effective (Helsen and Van den Bulck, 2004), 281; (Kakitani, Hata and Katsumata, 2007). Although CCA preservatives are designed to protect wood from fungi and other organisms, several researchers are developing strains that are resistant to CCA so that they could be used to extract arsenic and detoxify wood wastes.

1.4.4 Treatment of flue gases

About 60 % of anthropogenic arsenic emissions to the global atmosphere originate from flue gases emitted by copper ore smelters and coal-combustion facilities (Matschullat *et al.*, 2000), 301. The main method for reducing arsenic emissions to the atmosphere involves capturing As_4O_6 vapors by injecting sorbents into flue gases before they are released into the atmosphere. Potentially effective sorbents for arsenic in flue gases include: hydrated lime ($Ca(OH)_2$, portlandite), lime, calcium carbonate ($CaCO_3$), *limestone*, and *flyash* ((Helsen and Van den Bulck, 2004), 287, 239; (Jadhav and Fan, 2001; Taerakul, Sun and Golightly, 2006; Gupta *et al.*, 2007); Chapter 5). The injection of hydrated lime is especially effective and probably removes volatile As_4O_6 through the formation of calcium arsenates ($Ca_3(AsO_4)_2 \cdot nH_2O$, where $n > 0$), at least over a temperature range of 600–1000 °C ((Mahuli, Agnihotri and Chauk, 1997); Chapter 5).

References

Anawar, H.M., Akai, J. and Sakugawa. H. (2004) Mobilization of arsenic from subsurface sediments by effect of bicarbonate ions in groundwater. *Chemosphere*, **54**(6), 753–62.

Appelo, C.A.J., Van Der Weiden, M.J.J., Tournassat, C. and Charlet, L. (2002) Surface complexation of ferrous iron and carbonate on ferrihydrite and the mobilization of arsenic. *Environmental Science and Technology*, **36**(14), 3096–103.

Azcue, J.M. and Nriagu, J.O. (1994) Arsenic: historical perspectives, in *Arsenic in the Environment: Part I: Cycling and Characterization* (ed. J.O. Nriagu), John Wiley & Sons, Ltd, New York, pp. 1–15.

Behr, R.S. and Beane, J.E. (2002) Arsenic Plumes Where the "Source" Contains no Arsenic. Three Case Studies of Apparent Desorption of Naturally Occurring Arsenic. Arsenic in New England: A Multidisciplinary Scientific

Conference May 29–31, 2002, Manchester, New Hampshire, Sponsored by the National Institute of Environmental Health Sciences, Superfund Basic Research Program.

Buchet, J.P., Lauwerys, R. and Roels, H. (1981a) Comparison of the urinary excretion of arsenic metabolites after a single oral dose of sodium arsenite, monomethylarsonate, or dimethylarsinate in man. *International Archives of Occupational and Environmental Health*, **48**(1), 71–79.

Buchet, J.P., Lauwerys, R. and Roels, H. (1981b) Urinary excretion of inorganic arsenic and its metabolites after repeated ingestion of sodium metaarsenite by volunteers. *International Archives of Occupational and Environmental Health*, **48**(2), 111–18.

Burgess, W.G. and Pinto, L. (2005) Preliminary observations on the release of arsenic to groundwater in the presence of hydrocarbon contaminants in UK aquifers. *Mineralogical Magazine*, **69**(5), 887–96.

Chein, H., Hsu, Y.-D., Aggarwal, S.G. *et al.* (2006) Evaluation of arsenical emission from semiconductor and opto-electronics facilities in Hsinchu, Taiwan. *Atmospheric Environment*, **40**(10), 1901–7.

Christensen, I., Pedersen, A., Ottosen, L. and Riberio, A. (2004) Electrodialytic remediation of CCA-treated wood in larger scale, in *Environmental Impacts of Preservative-Treated Wood*, Florida Center for Environmental Solutions, Conference, February 8–11, Gainesville, FL, Orlando, FL, pp. 227–37.

García-Sánchez, A., Moyano, A. and Mayorga, P. (2005) High arsenic contents in groundwater of central Spain. *Environmental Geology*, **47**(6), 847–54.

Gong, A., Kamdem, D. and Harcihandran, R. (2004) Compression tests on wood-cement particle composites made of CCA-treated wood removed from service, in *Environmental Impacts of Preservative-Treated Wood*, Florida Center for Environmental Solutions, Conference, Gainesville, FL, February 8–11, Orlando, FL, pp. 270–76.

Gupta, H., Thomas, T.J., Park, A.-H.A., Iyer, M.V., Gupta, P., Agnihotri, R., Jadhav, R.A., Walker, H.W., Weavers, L.K., Butalia, T., Fan, L.-S. *et al.* (2007) Pilot-scale demonstration of the OSCAR process for high-temperature multipollutant control of coal combustion flue gas, using carbonated fly ash and mesoporous calcium carbonate. *Industrial and Engineering Chemistry Research*, **46**(14), 5051–60.

Hedberg, E., Gidhagen, L. and Johansson, C. (2005) Source contributions to PM10 and arsenic concentrations in central Chile using positive matrix factorization. *Atmospheric Environment*, **39**(3), 549–61.

Helsen, L. and Van den Bulck, E. (2004) Review of thermochemical conversion processes as disposal technologies for chromated copper arsenate (CCA) treated wood waste, in *Environmental Impacts of Preservative-Treated Wood*, Florida Center for Environmental Solutions, Conference, Gainesville, Florida, February 8–11, Orlando, FL, pp. 277–94.

Hingston, J.A., Collins, C.D., Murphy, R.J. and Lester, J.N. (2001) Leaching of chromated copper arsenate wood preservatives: a review. *Environmental Pollution*, **111**, 53–66.

Jadhav, R.A. and Fan, L.-S. (2001) Capture of gas-phase arsenic oxide by lime: kinetic and mechanistic studies. *Environmental Science and Technology*, **35**(4), 794–99.

Jing, C., Liu, S. and Meng, X. (2005) Arsenic leachability and speciation in cement immobilized water treatment sludge. *Chemosphere*, **59**(9), 1241–47.

Kakitani, T., Hata, T., Katsumata, N. (2007) Chelating extraction for removal of chromium, copper, and arsenic from treated wood with bioxalate. *Environmental Engineering Science*, **24**(8), 1026–37.

Kim, M.-J., Nriagu, J. and Haack, S. (2000) Carbonate ions and arsenic dissolution by groundwater. *Environmental Science and Technology*, **34**(15), 3094–310.

Krauskopf, K.B. and Bird, D.K. (1995) *Introduction to Geochemistry*, 3rd edn, McGraw-Hill, Boston.

Lee, E. (1999) A physiologically based pharmacokinetic model for the ingestion of arsenic in humans, Dissertation in Environmental Toxicology, University of California, Irvine.

Leist, M., Casey, R.J. and Caridi, D. (2000) The management of arsenic wastes: problems and prospects. *Journal of Hazardous Materials*, **76**(1), 125–38.

Mahuli, S., Agnihotri, R., Chauk, S. (1997) Mechanism of arsenic sorption by hydrated lime. *Environmental Science and Technology*, **31**(11), 3226–31.

Martley, E., Gulson B.L. and Pfeifer, H.-R. (2004) Metal concentrations in soils around the copper smelter and surrounding industrial complex of Port Kembla, NSW, Australia. *Science of the Total Environment*, **325**(1–3), 113–27.

Matschullat, J. *et al.* (2000) Arsenic in the geosphere — a review. *Science of the Total Environment*, **249**(1–3), 297–312.

Meharg, A.A. (2005) *Venomous Earth: How Arsenic Caused the World's Worst Mass Poisoning*, Macmillan, New York, 192.

Mendonça, A.A., Brito Galvão, T.C., Lima, D.C. and Soares, E.P. (2006) Stabilization of arsenic-bearing sludges using lime. *Journal of Materials in Civil Engineering*, **18**(2), 135–39.

Mulligan, C.N., Yong, R.N. and Gibbs, B.F. (2001) Remediation technologies for metal-contaminated soils and ground-water: an evaluation. *Engineering Geology*, **60**(1–4), 193–207.

Nico, P.S., Ruby, M.V., Lowney, Y.W. and Holm, S.E. (2006) Chemical speciation and bioaccessibility of arsenic and chromium in chromated copper arsenate-treated wood and soils. *Environmental Science and Technology*, **40**(1), 402–8.

Nriagu, J.O. (2002) Arsenic poisoning through the ages, in *Environmental Chemistry of Arsenic* (ed. W.T. Frankenberger Jr), Marcel Dekker, New York, pp. 1–26.

Pouschat, P. and Zagury, G.J. (2006) In vitro gastrointestinal bioavailability of arsenic in soils collected near CCA-treated utility poles. *Environmental Science and Technology*, **40**(13), 4317–23.

Shih, C.-J. and Lin, C.-F. (2003) Arsenic contaminated site at an abandoned copper smelter plant: Waste characterization and solidification/stabilization treatment. *Chemosphere*, **53**(7), 691–703.

Stollenwerk, K.G. and Colman, J.A. (2003) Natural remediation potential of arsenic-contaminated ground water, in *Arsenic in Ground Water* (eds A.H. Welch and K.G. Stollenwerk), Kluwer Academic Publishers, Boston, pp. 351–80.

Taerakul, P., Sun, P., Golightly, D.W. (2006) Distribution of arsenic and mercury in lime spray dryer ash. *Energy and Fuels*, **20**(4), 1521–27.

Tam, G.K.H., Charbonneau, S.M. and Bryce, F. (1979) Metabolism of inorganic arsenic (74As) in humans following oral ingestion. *Toxicology and Applied Pharmacology*, **50**(2), 319–22.

US Environmental Protection Agency US EPA (1999) Presumptive Remedy for Metals-in-Soils Sites. EPA-540-F-98-054. Office of Solid Wastes and Emergency (5102G).

US Environmental Protection Agency US EPA (2002) Proven Alternatives for Aboveground Treatment of Arsenic in Groundwater, EPA-542-S-02-002. Office of Solid Wastes and Emergency (5102G).

US Environmental Protection Agency US EPA (2002b) Arsenic Treatment Technologies for Soil, Waste, and Water. EPA-542-R-02-004. Office of Solid Wastes and Emergency (5102G).

Yamauchi, H. and Yamamura, Y. (1979) Dynamic change of inorganic arsenic and methylarsenic compounds in human urine after oral intake as arsenic trioxide. *Industrial Health*, **17**(2), 79–83.

2

Arsenic Chemistry

KEVIN R. HENKE[1] **and AARON HUTCHISON**[2]

[1] *University of Kentucky Center for Applied Energy Research*
[2] *Department of Science and Mathematics, Cedarville University*

2.1   Introduction

This chapter provides general background information on the chemistry of arsenic and serves as an introduction for more specific and detailed discussions in later chapters. Chapter 2 reviews: (1) the atomic properties of arsenic, (2) arsenic *valence states* and bonding, (3) some of the prominent chemical and physical properties of elemental arsenic and its solid compounds (including *minerals*), aqueous species, gases, and *organoarsenicals*, (4) arsenic *sorption*, *precipitation*, and *coprecipitation* in water, (5) arsenic *reduction/oxidation* (*redox*), *Eh-pH* diagrams, *methylation*, and *demethylation*, and (6) the *thermodynamic* properties of arsenic and its compounds and aqueous species (also see Appendix C). Chapter 3 discusses the *nucleosynthesis* of arsenic and provides additional details on how the element cycles through terrestrial environments and the oxidation and reduction of inorganic arsenic in nature. Details on the methylation of arsenic in biological organisms and demethylation are presented in Chapter 4. The discussions on the chemistry of arsenic in Chapter 2 are critical in understanding its behavior in terrestrial environments (Chapter 3), the toxic properties of its compounds (Chapter 4), arsenic contamination in groundwater (Chapter 6) and how to treat arsenic in wastes and *remediate* it at contaminated sites (Chapter 7).

2.2   Atomic structure and isotopes of arsenic

Arsenic is a group 15 element on the periodic table along with nitrogen, phosphorus, antimony, and bismuth. The *atomic mass* of arsenic is 74.921 60 *atomic mass units* (amu) and its *atomic number* (Z)

Arsenic   Edited by Kevin R. Henke
© 2009 John Wiley & Sons, Ltd

is 33, which means that 33 protons are located in the nucleus of every arsenic atom. The only stable (nonradioactive) and naturally occurring *isotope* of arsenic is arsenic-75 (^{75}As), where each nucleus of the isotope contains 42 neutrons with the 33 protons, or a *mass number* of 75. Numerous artificial short-lived radioisotopes of arsenic have been produced, including excited-state *isomers* ((Audi *et al.*, 2003); Table 2.1; (Holden, 2007; Lindstrom, Blaauw and Fleming, 2003)). ^{73}As has the longest *half-life*, which is 80.3 days (Holden, 2007). The possible decay modes for radioactive arsenic isotopes are *electron capture* (EC), *electron emission* ($\beta-$), *positron emission* ($\beta+$), *internal transition* (IT), and *neutron emission* (ne) (Audi *et al.*, 2003; Holden, 2007; Lindstrom, Blaauw and Fleming, 2003; Table 2.1).

The electrons in an uncharged arsenic atom (As^0) are located in the s *subshell* of the first *principal quantum number* ($n = 1$), the s and p subshells of principal quantum numbers 2–4 ($n = 2–4$), and the d subshell of the third principal quantum number ($n = 3$). Specifically, the As^0 electron configuration may be written as:

$$1s^2 2s^2 2p^6 3s^2 3p^6 3d^{10} 4s^2 4p^3.$$

The s subshells have one *orbital*, the p subshells have three, and the d subshell has five. Each orbital may contain up to two electrons. For example, the 2p subshell has a total of six electrons, where each of the three 2p orbitals contains two electrons (Faure, 1998), 63–71.

The size of an arsenic atom depends on its valence state and the number of surrounding atoms (its *coordination number*). When *valence electrons* are removed from an atom, the radius of the atom not only decreases because of the removal of the electrons, but also from the protons attracting the remaining electrons closer to the nucleus (Nebergall, Schmidt and Holtzclaw, 1976), 141. An increase in the number of surrounding atoms (coordination number) will deform the electron cloud of an ion and change its ionic radius (Faure, 1998), 91. Table 2.2 lists the radii in ångströms (Å) for arsenic and its ions with their most common coordination numbers.

2.3 Arsenic valence state and bonding

In crystalline substances, arsenic forms *covalent bonds* with itself and most other elements. That is, an arsenic atom in a covalent bond shares its valence electrons with another atom in the bond. However, unless arsenic atoms bond to each other, the valence electrons in a covalent bond are not equally shared between the arsenic atom and the atom of the other element. That is, most of the covalent bonds still have an *ionic character* to them (Faure, 1998), 83–89.

The most common valence states of arsenic are −3, 0, +3, and +5 (Shih, 2005), 86. The −3 valence state forms through the addition of three more electrons to fill the 4p orbital. In the most common form of elemental arsenic (As(0)), which is the rhombohedral or 'gray' form, each arsenic atom equally shares its 4p valence electrons with three neighboring arsenic atoms in a trigonal pyramid structure ((Klein, 2002), 336–337; Figure 2.1). The rhombohedral structure produces two sets of distances between closest arsenic atoms, which are 2.51 and 3.15 Å (Baur and Onishi , 1978), 33-A-2. The +3 valence state results when the three electrons in the 4p orbital become more attracted to bonded nonmetals, which under natural conditions are usually sulfur or oxygen. When the electrons in both the 4s and 4p orbitals tend to be associated more with bonded nonmetals (such as oxygen or sulfur), the arsenic atom has a +5 valence state.

Like sulfide in pyrite, arsenic in arsenic-rich (*arsenian*) pyrite (FeS$_2$) and many *arsenide* and *arseno-sulfide* minerals has a valence state of −1 or 0. These valence states result from arsenic forming covalent bonds with other arsenic atoms or sulfur (Klein, 2002), 340, 369; (Foster, 2003), 35; (O'Day, 2006), 80. In the arsenide niccolite (also called *nickeline, NiAs*), every nickel atom is surrounded by six arsenic atoms, where arsenic has a valence state of −1 and nickel is +1 (Klein, 2002), 360; (Foster, 2003), 35. The

Table 2.1 *Isotopes of arsenic (Audi et al., 2003; Holden, 2007; Lindstrom, Blaauw and Fleming, 2003). ^{75}As is the only stable arsenic isotope. The possible decay modes include electron capture (EC), electron emission ($\beta-$), positron emission ($\beta+$), proton decay (p), internal transition (IT), and neutron emission (ne). Superscripts on some of the arsenic isotope mass numbers designate excited-state isomers. The first (lowest energy) excited state is designated with an 'm' and a second excited state is designated with an 'n'.*

Arsenic isotope	Number of neutrons	Atomic mass	Half-life	Decay mode
60	27	59.993	?	p?
60^m	27	59.993	?	p?
61	28	60.981	?	p?
62	29	61.9732	?	p?
63	30	62.9637	?	p?
64	31	63.9576	40 ms	$\beta+$
65	32	64.9495	170 ms	$\beta+$
66	33	65.944 10	95.8 ms	$\beta+$
66^m	33	65.944 10	1.1 ms	IT
66^n	33	65.944 10	8.2 ms	IT
67	34	66.939 2	42.5 s	$\beta+$
68	35	67.936 8	2.53 min	$\beta+$
68^m	35	67.936 8	111 s	IT
69	36	68.932 28	15.2 min	$\beta+$
70	37	69.930 93	52.6 min	$\beta+$
70^m	37	69.930 93	96 µs	IT
71	38	70.927 114	2.72 d	$\beta+$
72	39	71.926 753	26.0 h	$\beta+$
73	40	72.923 825	80.3 d	EC
74	41	73.923 829	17.77 d	$\beta+$, $\beta-$
75	42	74.921 597	Stable	—
75^m	42	74.921 597	17.62 µs	IT
76	43	75.922 394	1.0778 d	$\beta-$, EC
76^m	43	75.922 394	1.84 µs	?
77	44	76.920 648	38.83 h	$\beta-$
77^m	44	76.920 648	114 µs	IT
78	45	77.921 83	1.512 h	$\beta-$
79	46	78.920 95	9.01 min	$\beta-$
79^m	46	78.920 95	1.21 µs	IT
80	47	79.922 58	15.2 s	$\beta-$
81	48	80.922 13	33.3 s	$\beta-$
82	49	81.924 6	19.1 s	$\beta-$
82^m	49	81.924 6	13.6 s	$\beta-$
83	50	82.925 0	13.4 s	$\beta-$
84	51	83.929 1	4.02 s	$\beta-$, ne
84^m	51	83.929 1	650 µs	$\beta-$
85	52	84.931 8	2.021 s	$\beta-$, ne
86	53	85.936 2	945 µs	$\beta-$, ne
87	54	86.939 6	610 µs	$\beta-$, ne
88	55	87.945	>0.15 µs	$\beta-$?, ne?
89	56	88.949	>0.15 µs	$\beta-$?
90	57	?	>0.15 µs	$\beta-$?
91	58	?	>0.15 µs	$\beta-$?
92	59	?	>0.15 µs	$\beta-$?

Table 2.2   *Radii of arsenic and its ions in angströms (Å).*

Valence state	Coordination number	Radius (Å) and reference(s)
+5	Fourfold	0.34 Pauling (1960), Klein (2002), 67; 0.475 Shannon (1976); Huheey, Keiter and Keiter (1993)
+5	Sixfold	0.46 Pauling (1960), Klein (2002), 67; 0.60 Shannon (1976); Huheey, Keiter and Keiter (1993)
+3	Sixfold	0.58 Pauling (1960), Klein (2002), 67; 0.72 Shannon (1976); Huheey, Keiter and Keiter (1993)
0	Threefold, covalent	1.22 Huheey, Keiter and Keiter (1993)
−3	Sixfold	2.22 Bloss (1971), 209; 2.10 Pauling (1960); Huheey, Keiter and Keiter (1993)

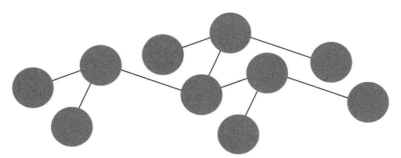

Figure 2.1   *The trigonal pyramidal structure of gray elemental arsenic. Each arsenic atom shares its valence electrons with three other arsenic atoms through covalent bonds.*

crystalline structure of arsenopyrite is based on the structure of marcasite (FeS_2), where one-half of the sulfur atoms are replaced by arsenic (Klein, 2002), 369. That is, arsenopyrite contains As-S^{2-}, where the formal valence state of arsenic is As(I$^-$) (Nesbitt, Uhlig and Szargan, 2002). However, analyses of unaltered arsenopyrite surfaces on a Portuguese sample with an average composition of $Fe_{1.0}As_{0.93}S_{1.12}$ indicated the presence of about 85 % As(I$^-$) and 15 % As(0) (Nesbitt, Muir and Pratt, 1995).

In realgar (AsS or As_4S_4), arsenic has a valence state of +2. The mineral consists of rings of As_4S_4. The rings are analogous to the eight-atom sulfur rings (S_8) in the crystalline structure of yellow elemental sulfur (Klein, 2002), 363.

Arsenic dissolved in natural waters mostly occurs as +3 and +5. As^{3+} and As^{5+} usually bond with oxygen to form inorganic *arsenite* (inorganic As(III)) and *arsenate* (inorganic As(V)), respectively. As(III) mostly exists in low-oxygen *(reducing) groundwaters* and *hydrothermal waters*. Depending on pH, As(III) may mainly exist as $H_3AsO_3^0$, $H_2AsO_3^-$, $HAsO_3^{2-}$, and/or AsO_3^{3-} (Figure 2.2; also see Sections 2.7.3 and 2.7.4). In sulfide-rich and *anoxic* waters, sulfur substitutes for one or more oxygens to form *thioarsenic* species, which could include: $HAs_3S_6^{2-}$, $H_3As_3S_6^0$, $H_2AsO_3S^-$, and $H_2AsS_2O_2^-$ (see Section 2.7.3 and Chapter 3).

As(V) is more common in oxidizing groundwaters and surface waters, and typically occurs as $H_3AsO_4^0$, $H_2AsO_4^-$, $HAsO_4^{2-}$, and/or AsO_4^{3-} depending on pH conditions (Figure 2.2, also see Sections 2.7.3 and 2.7.4). *Thioarsenates* may exist in sulfide-rich and anoxic groundwaters and hydrothermal waters, where sulfur substitutes for one or more oxygens in inorganic As(V). The elements and *ligands* bonded to arsenic strongly control its toxicity (Chapter 4).

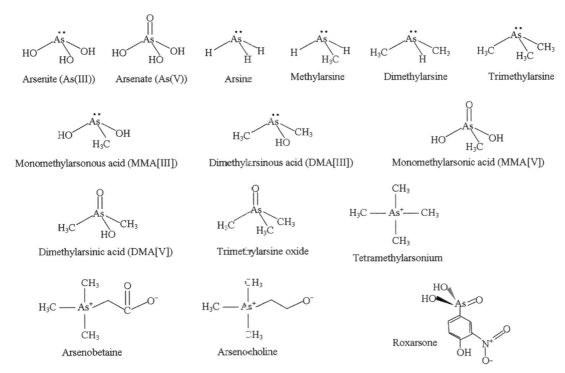

Figure 2.2 *Structures of common arsenic compounds. Many of the structures partially or fully deprotonate under natural conditions (see Chapters 2, 3, and 4).*

2.4 Chemistry of arsenic solids

2.4.1 Elemental arsenic

Arsenic is a *metalloid*. Solid samples of elemental arsenic (As(0)) tend to be brittle, nonductile, and insoluble in water. These properties largely result from arsenic atoms forming strong covalent bonds with each other. Table 2.3 lists the common chemical and physical properties of arsenic, including its *density*, *electronegativity*, and *first ionization potential*.

Although very rare in nature, elemental arsenic may form in hydrothermal deposits at low temperatures (50–200 °C) under very anoxic and low-sulfur conditions (Nordstrom and Archer, 2003), 13. Three solid forms of elemental arsenic are dominant in nature (Table 2.4). The rhombohedral ('gray') form is the more common *polymorph* under ambient temperatures and pressures (Francesconi, Edmonds and Morita, 1994), 191; Figure 2.1. Arsenolamprite (Bmab *space group*) and pararsenolamprite (space group Pmn2$_1$ or P2$_1$nm) are two orthorhombic forms (Matsubara *et al.*, 2001). Specimens of arsenolamprite have been identified in northern Chile, Germany, and the Czech Republic (Clark, 1970; Fettel, 1986; Tucek, 1970). Arsenolamprite has also been found in the roasted *mine tailings* piles of an abandoned South Korean mine. This arsenolamprite is probably not naturally occurring, but condensation from the roasting process (Ahn *et al.*, 2005), 151.

Pararsenolamprite was discovered in a mine in Oita Prefecture, Japan (Matsubara *et al.*, 2001). The Japanese specimens contained antimony and sulfur impurities (As$_{0.96}$Sb$_{0.03}$S$_{0.01}$). In at least the Japanese

Table 2.3 *Selected chemical and physical properties of arsenic. See Table 2.4 for additional information on the elemental polymorphs of arsenic. See Appendix B and the text for explanations of some of the terms and units of measure.*

Property	Value	References
Density (g cm^{-3}) at 25 °C and 1 bar pressure	5.75 (rhombohedral form)	Lide (2007)
Electronegativity, As^{3+}	2.0	Langmuir (1997), 101
Electronegativity, As^{5+}	2.2	Langmuir (1997), 101
First ionization potential (electron volts, eV)	9.7886	Lide (2007)
Luster on fresh surface	Metallic (rhombohedral form)	Nesse (2000)
Tenacity	Brittle (rhombohedral form)	Nesse (2000)

Table 2.4 *Physical and crystalline properties of naturally occurring elemental arsenic polymorphs at 25 °C and 1 bar pressure (Baur and Onishi, 1978; Lide, 2007; Matsubara et al., 2001; O'Neil et al., 2001). See the text and Appendix B for explanations of the terms and units of measure.*

Property	Elemental (rhombohedral form)	Arsenolamprite	Pararsenolamprite
Color	Gray, white, opaque	Gray, white, opaque	Gray, opaque
Density (g cm^{-3}) (varies due to impurities)	5.3–5.75	5.63–5.78	5.88–6.01
Hardness (approximate, Mohs' scale)	3.5	2	2–2.5
Crystal class	Rhombohedral	Orthorhombic	Orthorhombic
Space group	R-3m	Bmab	Pmn2$_1$ or P2$_1$nm
Crystal dimensions:			
a	3.760	3.65	3.633
b	—	4.47	10.196
c	10.548	11.0	10.314
Z (number atoms in unit cell)	6	8	18
V (unit cell volume, Å^3)	129.1	179.5	382.1
As–As bond lengths (Å)	2.51 Å, 3.15 Å	—	—

deposits of Oita Prefecture, pararsenolamprite is more resistant to *weathering* than the more common rhombohedral form of elemental arsenic (Matsubara *et al.*, 2001), 811–812.

A variety of other elemental arsenic forms have been synthesized in laboratories, including amorphous ('black') elemental arsenic, various arsenic clusters, and even larger fullerene-like cage structures (Zhao *et al.*, 2004; Baruah *et al.*, 2004). The compositions of the clusters are As$_n$, where $n = 1 - 5$. As$_4$ or 'yellow arsenic' is the most common form of elemental arsenic in the vapor phase and the bond length between the arsenic atoms in As$_4$ is about 2.435 Å (Zhao *et al.*, 2004; Karttunen, Linnolahti and Pakkanen, 2007), 913.

Arsenic cage structures are As$_n$, where $n = 8, 20, 28, 32, 36$, or 60. However, among the cage structures, only As$_{20}$ was stable enough to avoid dissociation into As$_4$ (Baruah *et al.*, 2004). Because of their symmetry and bonding, arsenic cage structures might have future applications in nanotechnologies (Baruah *et al.*, 2004), 476.

2.4.2 Common arsenic minerals and other solid arsenic compounds

2.4.2.1 Introduction

Geologists define a mineral as a naturally occurring, crystalline, and inorganic solid. Although liquids, gases, synthetic materials, *amorphous* substances, and organic compounds may contain arsenic, they are not minerals. Arsenic minerals include rhombohedral elemental arsenic, arsenolamprite, pararsenolamprite, and over 320 inorganic compounds (Foster, 2003), 39. Chapter 3 discusses the natural occurrences and potential environmental impacts of several of the more common arsenic minerals.

Although over 320 arsenic minerals are known, only about 10 are relatively common in the Earth's *sediments*, *soils*, and crustal rocks (Foster, 2003), 39. Most arsenic minerals and other inorganic solid arsenic substances may be classified into one of five groups, which are: elemental, arsenides, arsenosulfides, arsenites, and arsenates (Table 2.5). Table 2.6 lists the known densities, melting points, and boiling points of various arsenic minerals and other arsenic compounds. In general, arsenide, arsenosulfide, and arsenic-rich sulfide minerals are associated with anoxic hydrothermal ore deposits and *metamorphic and intrusive igneous rocks*. Once, these minerals are exposed to oxygen and water under surface or near-surface conditions, they weather to arsenite and arsenate minerals (Chapter 3; Table 2.5).

Ideal formulas are often listed with minerals. However, in most cases, significant elemental substitutions occur within an individual specimen so that its chemical composition is actually far more complex. In particular, sulfur may partially substitute for arsenic in the crystalline structures of a wide variety of arsenic-bearing minerals (Chapter 3), and substitutions commonly occur between cobalt, nickel, and other metals. Furthermore, some specimens of minerals that normally contain little or no arsenic may be relatively rich in arsenic (i.e. arsenian). Arsenian minerals may contain up to several weight percent (wt %) of arsenic. In these minerals, arsenic partially substitutes into the crystalline structures, usually for sulfur, less commonly for antimony, and sometimes arsenate (AsO_4^{3-}) for phosphate (PO_4^{3-}) (Table 2.7). Specifically, arsenic may attain concentrations of 1 % or higher in various sulfides, including: pyrite, galena, sphalerite, and marcasite (Welch *et al.*, 2000), 597. Most pyrites contain 200–5000 mg kg^{-1} of arsenic, but some may host up to 60 000 mg kg^{-1} (6.0 wt %) in a *solid solution* with sulfur ((Welch *et al.*, 2000), 597; (Reich and Becker, 2006); Table 2.7). Although some pyrite specimens from Nevada, United States, contain up to 19.76 wt % (197 600 mg kg^{-1}) of arsenic, much of the arsenic exists as nanoscale arsenopyrite or other mineral inclusions rather than as a true solid solution in the mineral structure ((Reich and Becker, 2006; Reich *et al.*, 2005), 2784-2786; Chapter 3).

Arsenate commonly substitutes for sulfate and phosphate in a variety of minerals and other solid compounds. Jarosite is an example of a sulfate mineral that may acquire considerable arsenate through substitutions into its crystalline structure. Jarosites often precipitate in *acid mine drainage* and they could be important in controlling the mobility of arsenate in these waters (Chapter 3).

The ideal formula for jarosite is $KFe_3(SO_4)_2(OH)_6$. However, many natural samples are iron deficient and a more realistic composition is $KFe_{2.5}(HSO_4)_{1.5}(SO_4)_{0.5}(OH)_6$ (Savage, Bird and O'Day, 2005), 494. In iron-deficient synthetic jarosites, up to about 30 % of the crystal sites normally filled by sulfate may be occupied by protonated arsenate ($HAsO_4^{2-}$) (Savage, Bird and O'Day, 2005), 495. Extensive As(V) substitutions are also possible in lead-bearing jarosites (Savage, Bird and O'Day, 2005), 475; (Smith *et al.*, 2006).

2.4.2.2 Arsenosulfides and arsenian sulfides

Arsenopyrite (FeAsS), an arsenosulfide, is the most common arsenic mineral on Earth (Welch *et al.*, 2000), 597. The mineral occurs in a variety of hydrothermal deposits and some metamorphic and intrusive igneous rocks Table 2.5; (Klein, 2002), 369. As mentioned earlier in this chapter, the crystalline structure of

Table 2.5 *Relatively common arsenic-bearing minerals (Klein, 2002; O'Day, 2006; Mandal and Suzuki, 2002; Smedley and Kinniburgh, 2002; Mozgova et al., 2005; Utsunomiya et al., 2003; Dunn, Pecor and Newberry, 1980).*

Mineral	Formula	Mineral Group	Origin(s)
Adamite	$Zn_2(OH)(AsO_4)$	Arsenate	Weathering product
Annabergite	$(Ni,Co)_3(AsO_4)_2 \cdot 8H_2O$	Arsenate	Weathering product
Arsenic (rhombohedral)	$As(0)$	Elemental	Hydrothermal deposits
Arsenolite	As_2O_3	Arsenite	Weathering product, fires in coal mines or seams
Arsenolamprite	$As(0)$	Elemental	Hydrothermal deposits
Arsenopyrite	$FeAsS$	Arsenosulfide	Hydrothermal deposits, intrusive igneous rocks, metamorphic rocks, microbial precipitates
Chloanthite	$(Ni,Co)As_{3-x}$	Arsenide	Hydrothermal deposits
Claudetite	As_2O_3	Arsenite	Weathering product
Cobaltite	$(Co,Fe)AsS$	Arsenosulfide	Metamorphic rocks, hydrothermal deposits
Conichalcite	$CaCu(AsO_4)(OH)$	Arsenate	Weathering product
Enargite	Cu_3AsS_4	Arsenosulfide	Hydrothermal deposits
Erythrite	$Co_3(AsO_4)_2 \cdot 8H_2O$	Arsenate	Weathering product
Gersdorffite	$NiAsS$	Arsenosulfide	Metamorphic rocks, hydrothermal deposits
Glaucodot	$(Co,Fe)AsS$	Arsenosulfide	Hydrothermal deposits
Haematolite	$(Mn,Mg)_4Al(AsO_4)(OH)_8$	Arsenate	Weathering product
Hoernesite	$Mg_3(AsO_4)_2 \cdot 8H_2O$	Arsenate	Weathering product
Johnbaumite	$Ca_5(AsO_4)_3(OH)$	Arsenate	Metamorphic rocks
Loellingite	$FeAs_2$	Arsenide	Hydrothermal deposits
Luzonite	Cu_3AsS_4	Arsenosulfide	Hydrothermal deposits
Mansfieldite	$AlAsO_4 \cdot 2H_2O$	Arsenate	Weathering product
Mimetite	$Pb_5(AsO_4)_3Cl$	Arsenate	Weathering product
Niccolite (nickeline)	$NiAs$	Arsenide	Hydrothermal deposits, mafic intrusive igneous rocks
Nickel skutterudite	$(Ni,Co)As_3$	Arsenide	Hydrothermal deposits
Olivenite	Cu_2OHAsO_4	Arsenate	Weathering product
Orpiment	As_2S_3	Arsenosulfide	Hydrothermal deposits, intrusive igneous rocks, volcanic emissions, hot springs, microbial precipitates
Pararealgar	$AsS (As_4S_4)$	Arsenosulfide	Exposure of realgar to sunlight
Pararsenolamprite	$As(0)$	Elemental	Hydrothermal deposits
Pharmacosiderite	$Fe_3(AsO_4)_2(OH)_3 \cdot 5H_2O$	Arsenate	Weathering product
Proustite	Ag_3AsS_3	Arsenosulfide	Hydrothermal deposits
Rammelsbergite	$NiAs_2$	Arsenide	Hydrothermal deposits
Realgar	$AsS (As_4S_4)$	Arsenosulfide	Hydrothermal deposits, intrusive igneous rocks, volcanic emissions, hot springs
Safflorite	$(Co,Fe)As_2$	Arsenide	Hydrothermal deposits
Scorodite	$FeAsO_4 \cdot 2H_2O$	Arsenate	Weathering product
Seligmannite	$PbCuAsS_3$	Arsenosulfide	Hydrothermal deposits
Skutterudite	$(Co,Ni)As_3$	Arsenide	Hydrothermal deposits
Smaltite	$(Co,Ni)As_{3-x}$	Arsenide	Hydrothermal deposits
Sperrylite	$PtAs_2$	Arsenide	Mafic and ultramafic intrusive igneous rocks
Tennantite	$(Cu,Fe)_{12}As_4S_{13}$	Arsenosulfide	Hydrothermal deposits
Westerveldite	$FeAs$	Arsenide	Hydrothermal deposits

Table 2.6 *Physical properties of arsenic and its compounds at 1 bar pressure.*

Substance	Density (g cm⁻³, unless indicated otherwise) at 25 °C	Melting Point (°C)	Boiling point (°C)	Reference(s)
Arsenic (Elemental, rhombohedral form, As(0))	5.75	612 (sublimation)	—	Lide (2007); Dean (1979)
Aluminum arsenide (AlAs)	3.76	1740	—	Lide (2007)
Ammonium dihydrogen arsenate ($NH_4H_2AsO_4$)	2.311	300 (decomposes)	—	Lide (2007)
Arsenic acid hemihydrate ($H_3AsO_4 \cdot 0.5H_2O$)	2.5	35.5	300 (decomposes)	Lide (2007); O'Neil *et al.* (2001)
Arsenic (II) hydride (As_2H_4)	—	—	100	Planer-Friedrich *et al.* (2006)
Arsenic (III) bromide ($AsBr_3$)	3.3972	31	221	Dean (1979)
Arsenic (III) chloride ($AsCl_3$)	2.150	−16	133	Planer-Friedrich *et al.* (2006); Lide (2007)
Arsenic (III) fluoride (AsF_3)	2.7	−5.9	57.8	Planer-Friedrich *et al.* (2006); Lide (2007)
Arsenic (III) iodide (AsI_3)	4.73	141	424	Lide (2007)
Arsenic (V) fluoride (AsF_5)	6.945 g l⁻¹	−79.8	−53 (decomposes)	Planer-Friedrich *et al.* (2006); Lide (2007)
Arsenic (V) oxide (As_2O_5)	4.32	315 (decomposes)	—	Lide (2007); Dean (1979)
Arsenolite (As_2O_3, isometric)	3.86	274	460	Lide (2007)
Arsine (AsH_3)	3.186 g l⁻¹	−116	−62.5	Planer-Friedrich *et al.* (2006); Lide (2007)
Bis(dimethylarsine) oxide ((CH_3)$_2$As)$_2$O	—	−25	120	Dean (1979)
Boron arsenides (BAs)	5.22	1100 (decomposes)	—	Lide (2007)
Cadmium arsenide (CdAs)	6.25	721	—	Lide (2007)
Claudetite (As_2O_3, monoclinic)	3.74	313	460	Lide (2007)
Cobalt arsenide (CoAs)	8.22	1180	—	Lide (2007)

(continued overleaf)

Table 2.6 (continued)

Substance	Density (g cm^{-3}, unless indicated otherwise) at 25 °C	Melting Point (°C)	Boiling point (°C)	Reference(s)
Dimethylarsonous acid (DMA, (CH$_3$)$_2$AsH)	—	—	36	Planer-Friedrich et al. (2006)
Dimethylchloroarsine (CH$_3$)$_2$AsCl	—	—	130	Planer-Friedrich et al. (2006)
Gallium(III) arsenide (GaAs)	5.3176	1238	—	Lide (2007)
Indium arsenide (InAs)	5.67	942	—	Lide (2007)
Iron arsenide (FeAs)	7.85	1030	—	Lide (2007)
Lead hydrogen arsenate (PbHAsO$_4$)	5.943	280 (decomposes)	—	Lide (2007)
Lead(II) arsenate (Pb$_3$(AsO$_4$)$_2$)	5.8	1042 (decomposes)	—	Lide (2007)
Methylarsonous acid (MMA, CH$_3$AsH$_2$)	—	—	~2	Planer-Friedrich et al. (2006)
Nickel arsenide (NiAs)	7.77	967	—	Lide (2007)
Orpiment (As$_2$S$_3$)	3.46	312	707	Lide (2007)
Potassium dihydrogen arsenate (KH$_2$AsO$_4$)	2.87	288	—	Lide (2007)
Realgar (α-As$_4$S$_4$)	3.5	320	565	Lide (2007)
Sodium hydrogen arsenate heptahydrate (Na$_2$HAsO$_4$·7H$_2$O)	1.87	~50 (decomposes)	—	Lide (2007)
Tetramethyldiarsine (CH$_3$)$_2$AsAs(CH$_3$)$_2$	—	−6	165	Dean (1979)
Trimethylarsine ((CH$_3$)$_3$As)	—	—	52	Planer-Friedrich et al. (2006)
Zinc arsenide (Zn$_3$As$_2$)	5.528	1015	—	Lide (2007)

Table 2.7 *Typical arsenic concentrations in selected minerals and other solid substances where arsenic is not a major component. In synthetic and rare natural samples, arsenic concentrations may be much higher (e.g. jarosites in Savage, Bird and O'Day (2005) and calcite in Di Benedetto et al. (2006)).*

Minerals and other compounds	Ideal formula (Klein, 2002)	Typical arsenic concentration ($mg\,kg^{-1}$)	Reference(s)
Amphiboles	$W_{0-1}X_2Y_5Z_8O_{22}(OH,F)_2$, where $W =$ Na^+, K^+; $X = Ca^{2+}$, Fe^{2+}, Mg^{2+}, Mn^{2+}, Na^+; $Y = Al^{3+}$, Fe^{2+}, Fe^{3+}, Mg^{2+}, Mn^{2+}, Ti^{4+}; $Z = Si^{4+}$, Al^{3+}	1.1–2.3	Smedley and Kinniburgh (2002)
Apatite	$Ca_5(PO_4)_3(F,Cl,OH)$	<1–1 000	Smedley and Kinniburgh (2002); Boyle and Jonasson (1973)
Barite	$BaSO_4$	<1–12	Boyle and Jonasson (1973)
Biotite	$K(Mg,Fe)_3(AlSi_3O_{10})(OH)_2$	1.4	Smedley and Kinniburgh (2002)
Calcite	$CaCO_3$	1–8	Boyle and Jonasson (1973)
Calcium sulfates (gypsum and anhydrite)	$CaSO_4\cdot2H_2O$, $CaSO_4$	<1–6	Boyle and Jonasson (1973)
Cassiterite	SnO_2	<5–100	Boyle and Jonasson (1973)
Cerussite	$PbCO_3$	<5	Boyle and Jonasson (1973)
Chalcocite	Cu_2S	95–225	Boyle and Jonasson (1973)
Chalcopyrite	$CuFeS_2$	10–5 000	Smedley and Kinniburgh (2002)
Cinnabar	HgS	<5–12	Boyle and Jonasson (1973)
Dolomite	$CaMg(CO_3)_2$	<3	Boyle and Jonasson (1973)
Fe(III) oxyhydroxides	Varies	⩽76 000	Pichler, Veizer and Hall (1999)
Feldspars	$(NaAlSi_3O_8\text{-}CaAl_2Si_2O_8)$, $KAlSi_3O_8$	<0.1–2.1	Smedley and Kinniburgh (2002)
Fluorite	CaF_2	<2	Boyle and Jonasson (1973)
Galena	PbS	5–10 000	Smedley and Kinniburgh (2002)
Halite	$NaCl$	<3–30	Smedley and Kinniburgh (2002)
Hematite	Fe_2O_3	⩽160	Smedley and Kinniburgh (2002)

(continued overleaf)

Table 2.7 (*continued*)

Minerals and other compounds	Ideal formula (Klein, 2002)	Typical arsenic concentration ($mg\,kg^{-1}$)	Reference(s)
Ilmenite	$FeTiO_3$	<1	Smedley and Kinniburgh (2002)
Jarosite	$KFe_3(SO_4)_2(OH)_6$	34–1 000	Smedley and Kinniburgh (2002)
Magnetite	Fe_3O_4	2.7–41	Smedley and Kinniburgh (2002)
Marcasite	FeS_2	20–126 000	Smedley and Kinniburgh (2002)
Molybdenite	MoS_2	5–500	Boyle and Jonasson (1973)
Olivine	$(Fe,Mg)_2SiO_4$	0.08–0.17	Smedley and Kinniburgh (2002)
Pyroxenes	XYZ_2O_6, where X = Ca, Fe^{2+}, Mg, and/or Na, Y = Al, Fe^{2+}, Fe^{3+}, and/or Mg, and Z = Si,Al	0.05–0.8	Smedley and Kinniburgh (2002)
Pyrite	FeS_2	200–60 000	Welch *et al.* (2000); Reich and Becker (2006)
Pyrochlore	$(Ca,Na)_2(Nb,Ta)_2O_6(O,OH,F)$	<2	Boyle and Jonasson (1973)
Pyrrhotite	$Fe_{1-x}S$, x = 0–0.2, (Klein, 2002), 359	<5–100	Boyle and Jonasson (1973)
Quartz	SiO_2	0.4–1.3	Smedley and Kinniburgh (2002)
Scheelite	$CaWO_4$	<5	Boyle and Jonasson (1973)
Siderite	$FeCO_3$	<3	Boyle and Jonasson (1973)
Sphalerite	ZnS	<5–17 000	Smedley and Kinniburgh (2002); Boyle and Jonasson (1973)
Stibnite	Sb_2S_3	5–500	Boyle and Jonasson (1973)
Tantalite-columbite	$(Fe,Mn)(Ta,Nb)_2O_6$	<2	Boyle and Jonasson (1973)
Wolframite	$(Fe,Mn)WO_4$	<5	Boyle and Jonasson (1973)

arsenopyrite is based on the structure of marcasite (FeS_2), where one-half of the sulfur atoms are replaced by arsenic (Klein, 2002), 369.

In the presence of aerated water, the oxidation of arsenopyrite is relatively rapid and its arsenic oxidizes faster than its sulfide (Foster, 2003), 37. During oxidation, As^{1-} and As^0 on the surfaces of the arsenopyrite initially convert to As^{2+}, As^{3+}, As^{5-}, and possibly As^{1+} (Nesbitt, Muir and Pratt, 1995; Schaufuss *et al.*, 2000). Scorodite ($FeAsO_4 \cdot 2H_2O$) or amorphous Fe(III) arsenates commonly precipitate from the oxidation of arsenopyrite ((Krause and Ettel, 1988), 851; (Williams, 2001), 273; (Craw, Falconer and Youngson, 2003), 73; Chapter 3). If any arsenic on the weathered surfaces of arsenopyrite dissolves into aerated water, the arsenic will exist as As(V) or As(III). Unless As(III) enters reducing environments, it will convert to As(V) over time (Chapter 3).

Arsenosulfides also include realgar, its polymorphs, alacrinite (As_8S_9), and amorphous forms of arsenic sulfide. Realgar and other forms of arsenic sulfide occur in hydrothermal deposits, some intrusive igneous rocks, volcanic emissions, and *hot springs* (Table 2.5). The bacterium, *Pyrobaculum arsenaticum*, also biologically precipitates realgar at temperatures of $68-100\,°C$ (Nordstrom and Archer, 2003), 12.

The formula for realgar is often given as AsS. However, the mineral actually consists of rings of As_4S_4 (Klein, 2002), 363. Realgar also exists as high- and low-temperature polymorphs. At about $240\,°C$, α-realgar converts to high-temperature β-realgar (Naumov, Makreski and Jovanovski, 2007). (Although the older literature is inconsistent and may refer to the low-temperature form as either α- or β-realgar, the more recent literature defines the low-temperature polymorph as α-realgar (Naumov, Makreski and Jovanovski, 2007; Douglass, Shing and Wang, 1992), 1267). The identity of the crystalline structure of β-realgar has been controversial (Douglass, Shing and Wang, 1992), 1267, and has still not been entirely resolved. The chemistry of amorphous arsenic sulfide is also uncertain (Helz *et al.*, 1995), 4602–4603. Nevertheless, As^{2+} is present in the various forms of crystalline and amorphous arsenic sulfide. As^{2+} is unstable in water and oxidizes to As(III) before dissolving from the compounds into aqueous solutions (Lengke and Tempel, 2005), 350.

Pararealgar (As_4S_4, monoclinic, P 2_1/c space group) is a polymorph of realgar. The mineral typically forms from the exposure of α-realgar to oxygen and sunlight (in particular, wavelengths of about $500-670\,nm$). The transformation of α-realgar to pararealgar may initially involve the following reactions (Naumov, Makreski and Jovanovski, 2007):

$$5\ As_4S_4(\alpha\text{-realgar}) + 3\ O_2 \text{ in light } \rightarrow 4\ As_4S_5 + 2\ As_2O_3(\text{arsenolite}) \qquad (2.1)$$

$$As_4S_5 \rightarrow As_4S_4 \text{ (pararealgar) } + S \qquad (2.2)$$

The conversion of α-realgar to pararealgar then becomes cyclic and self-sustaining (i.e. no further exposure to light is required) as the sulfur released by the decomposition of each As_4S_5 can produce additional pararealgar (Naumov, Makreski and Jovanovski, 2007):

$$S + As_4S_4(\alpha\text{-realgar}) \rightarrow As_4S_5 \qquad (2.3)$$

$$As_4S_5 \rightarrow As_4S_4 \text{ (pararealgar) } + S \qquad (2.4)$$

In other words, the sulfur released from Reaction 2.4 is available for Reaction 2.3 to restart the cycle.

Alacrinite (monoclinic, P 2/c space group) is chemically similar to α-realgar (monoclinic, P 2_1/n space group). Its structure contains an equal amount of As_4S_5 and As_4S_4. When compared with realgar, alacrinite is more stable at lower temperatures (Naumov, Makreski and Jovanovski, 2007).

Orpiment (As_2S_3) is a relatively common arsenosulfide mineral that is often associated with realgar in hydrothermal deposits, intrusive igneous rocks, volcanic emissions, hot springs, and microbial precipitates

(Table 2.5). The bright yellow color of orpiment and the scarlet-red color of realgar were once used as pigments and cosmetics in Egypt and Eurasia ((Nriagu, 2002), 4; Chapter 5). The monoclinic crystalline structure of orpiment consists of trigonal pyramids of AsS_3, which form layers. The sulfur and arsenic atoms within the layers are covalently bonded and the layers are held together by weak *van der Waals forces* (Klein, 2002), 363. The oxidation of orpiment and its behavior in aqueous solutions are discussed in Chapter 3.

Several arsenosulfide minerals are valuable ore deposits if their copper, nickel, cobalt, or other metals can be economically recovered without negatively impacting the environment. Relatively common arsenosulfide ore minerals include enargite (Cu_3AsS_4), cobaltite ((Co,Fe)AsS), and gersdorffite (NiAsS) (Table 2.5). In these minerals, some substitutions usually occur between cobalt, nickel, and iron. Substitutions between copper, iron, and zinc are also likely. Unlike most other metallic arsenosulfides, the arsenic in enargite is As^{5+} (Pratt, 2004). That is, enargite is an orthorhombic (Pnm2 space group) thioarsenate, where each arsenic atom is coordinated with four sulfur atoms (AsS_4^{3-}) (Castro and Baltierra, 2005). Below about 300 °C, luzonite, a polymorph of enargite, becomes stable (Lattanzi *et al.*, 2008), 64. Luzonite has tetragonal symmetry with a space group of $\overline{I}42m$ (Klein, 2002), 369.

Jackson *et al.* (2003) studied the oxidation reaction rates of a gersdorffite specimen with the composition of $Ni_{0.68}Fe_{0.19}Co_{0.14}As_{1.08}S_{0.92}$. The oxidation rate of the mineral was more than 10 times greater in aerated water than air, and arsenic was more reactive than sulfur. As^-, As^+, As^{3+}, and As^{5+} were detected in the surface oxidation products in the presence of either air or aerated water. After 5 hours of exposure to air, As^+ and As^{3+} were measured on the surface of the gersdorffite. As^{5+} was finally detected after a total of 10 hours of air exposure (Jackson *et al.*, 2003), 899. In aerated distilled water, As^{3+} and As^{5+} were the dominant arsenic surface oxidation products after only 15 minutes (Jackson *et al.*, 2003), 897.

2.4.2.3 *Arsenides*

The most common arsenides in nature are skutterudite ((Co,Ni)As$_3$), chloanthite ((Ni,Co)As$_{3-x}$, where $x =$ 0.5 − 1.0), smaltite (Co,Ni)As$_{3-x}$ (where $x = 0.5 − 1.0$), niccolite (also called *nickeline, NiAs*), safflorite (Co,Fe)As$_2$, loellingite (also spelled löllingite, FeAs$_2$), westerveldite (FeAs), and rammelsbergite (NiAs$_2$) (Table 2.5). The minerals are most commonly found in hydrothermal deposits. *Mafic* intrusive igneous rocks may also contain significant amounts of niccolite (Klein, 2002), 360. The most common arsenide in platinum ores is sperrylite (PtAs$_2$).

Iron usually substitutes for some nickel and cobalt in skutterudite (Klein, 2002), 369. The arsenic in the crystalline structure of skutterudite occurs as As_4 rings (Cotton *et al.*, 1999), 387. The rings are planar and rectangular with bond lengths of 2.464 ± 0.002 Å and 2.572 ± 0.002 Å at 22 °C (Mandel and Donohue, 1971). In skutterudite, each atom of cobalt or another divalent metal is surrounded by six arsenic atoms in a roughly octahedral formation (Mandel and Donohue, 1971). Chloanthite and smaltite are arsenic-deficient forms of nickel and cobalt skutterudite, respectively (Table 2.5).

Loellingite is an orthorhombic (Pnnm space group) arsenide mineral. Each iron atom in loellingite is coordinated with six arsenic atoms. The arsenic atoms have fourfold coordination, where every arsenic atom is bonded to another arsenic atom and three iron atoms. That is, the arsenic–arsenic bonds in loellingite produce As_2^{2-} units (Nesbitt, Uhlig and Szargan, 2002), 1000.

Nesbitt and Reinke (1999) investigated the oxidation of niccolite (nickeline) in air and water. After 30 hours of oxidation from air, As^{3+} and As^{5+} were present on the surface of the niccolite. On the water-reacted surface, As^{3+} and As^{5+} were detected after 16 hours (Nesbitt and Reinke, 1999), 648.

Gallium(III) arsenide (GaAs) is not naturally occurring, but it is manufactured and extensively used as a semiconductor (Chapter 5). The crystalline structure of GaAs resembles sphalerite (ZnS, isometric, F$\overline{4}$3m space group), where each arsenic atom is coordinated by four gallium atoms (Klein, 2002), 339; (Bloss,

1971), 245. Above absolute zero (0 K), some electrons in GaAs have enough energy to move through the crystal, but not as frequently as in a conductor. GaAs semiconductors have important applications in solar cells, laser windows, and light-emitting diodes (LEDs) ((Brooks, 2007; Azcue and Nriagu, 1994), 13; Chapter 5). Indium (III) arsenide (InAs) and mixed indium-gallium arsenides (InGaAs) have similar applications (Chen and Cho, 1992; Leem *et al.*, 2003; Höglund *et al.*, 2006). These compounds are synthesized with volatile arsenic compounds (Chen and Cho, 1992; Leem *et al.*, 2003; Höglund *et al.*, 2006; Buckley *et al.*, 1990; Ikejiri *et al.*, 2007).

2.4.2.4 *Arsenites*

The two most common naturally occurring arsenites are the polymorphs arsenolite (As_2O_3, isometric) and claudetite (As_2O_3, monoclinic). The structure of arsenolite consists of isometric cages of As_4O_6 (Ballirano and Maras, 2002). In claudetite, each arsenic atom is surrounded by three oxygen atoms that form a trigonal pyramid that resembles the structure of orpiment (Baur and Onishi, 1978), 33-A-4.

Arsenolite and claudetite may weather from elemental arsenic and various arsenide and arsenosulfide minerals, such as arsenopyrite and realgar (Nordstrom and Archer, 2003), 10–11. Coal-fired power plants, ore processing facilities, coal seam and mine fires, or simply burning As(0) in air produce gaseous As_4O_6 ('As_2O_3'), which can condense as arsenolite (Cotton *et al.*, 1999), 400; (Meharg, 2005), 134; Table 2.5. Arsenolite and claudetite are moderately soluble in water (both about $20.5\,g\,l^{-1}$, at $25\,^{\circ}C$; (Lide, 2007)) and form arsenious acid ($H_3AsO_3^0$) and its association anions (see Sections 2.7.3 and 2.7.4).

2.4.2.5 *Arsenates*

Most arsenate minerals result from the extensive oxidation of elemental arsenic, arsenides, or arsenosulfides in the weathering of hydrothermal deposits and metamorphic and intrusive igneous rocks. In particular, skutterudite, smaltite, and chloanthite may oxidize to annabergite (($Ni,Co)_3(AsO_4)_2\cdot 8H_2O$) or erythrite ($Co_3(AsO_4)_2\cdot 8H_2O$) (Table 2.5). Scorodite ($FeAsO_4\cdot 2H_2O$) and amorphous Fe(III) arsenates often precipitate on the surfaces of oxidizing arsenopyrite or from aerated water that has been saturated with As(V) from the weathering and dissolution of arsenopyrite ((Foster, 2003), 40; Chapter 3; (Krause and Ettel, 1988), 851; (Williams, 2001), 273; (Craw, Falconer and Youngson, 2003), 73). The crystalline structure of scorodite is orthorhombic. Each arsenic atom is surrounded by four oxygen atoms, which form four As-O-Fe(III) bridges. Two waters of hydration and four oxygen atoms coordinate around every Fe(III) atom (Xu, Zhou and Zheng, 2007). As discussed in Chapter 3, the formation and dissolution of scorodite may be very important in the controlling the mobility of arsenic in acid mine drainage, hot springs, and other surface and near-surface waters.

As discussed in Chapters 5 and 7, the use of lime to precipitate calcium arsenates is a common method for removing inorganic As(V) from water or *flue gases*. Calcium arsenates were also once extensively used in pesticides (Chapter 5). The compositions of some calcium arsenates, such as johnbaumite ($Ca_5(AsO_4)_3(OH)$; Table 2.5) resemble the very common phosphate mineral, apatite ($Ca_5(PO_4)_3(F,Cl,OH)$), where arsenate replaces phosphate. Some lead arsenates, such as mimetite ($Pb_5(AsO_4)_3Cl$; Table 2.5), also have crystalline structures that are related to apatite. Mimetite may occur in oxidized lead-rich hydrothermal deposits.

Arsenic(V) oxide (As_2O_5) is not very common in nature. However, the compound was often synthesized and once widely used in a variety of manufacturing processes, such as in the production of chromated copper arsenate (CCA) wood preservatives (Chapter 5). The synthesis of As_2O_5 usually involves the oxidation of As_2O_3 or As(0) with nitric acid or other *oxidants* followed by dehydration (Cotton *et al.*, 1999), 400. At $315\,^{\circ}C$ and higher, As_2O_5 decomposes to As_4O_6 vapor and O_2 (Table 2.6). In water,

As_2O_5 can dissolve and form $H_3AsO_4{}^0$, $H_2AsO_4{}^-$, $HAsO_4{}^{2-}$, and/or $AsO_4{}^{3-}$ depending on pH conditions (see Sections 2.7.3 and 2.7.4).

2.4.3   Arsine and other volatile arsenic compounds

Volatile arsenic compounds may be defined as arsenic-bearing inorganic or organic chemicals that have boiling points below 150 °C at atmospheric pressure (Planer-Friedrich *et al.*, 2006). That is, at ambient temperatures and pressures, these compounds are either gases or liquids. Volatile arsenic compounds include: arsine (AsH_3), As(III) chloride ($AsCl_3$), As(III) fluoride (AsF_3), As(V) fluoride (AsF_5), monomethylarsonous acid (MMA(III), $(CH_3)As(OH)_2$), dimethylarsinous acid (DMA(III), $(CH_3)_2As(OH)$), *trimethylarsine* ($(CH_3)_3As$), and *trimethylarsine oxide* ($(CH_3)_3AsO$) (Figure 2.2). Table 2.6 lists the densities, melting points, and boiling points of the more common volatile and less volatile arsenic compounds. Less volatile organoarsenicals are discussed in the next section of this chapter.

One of the more common naturally occurring volatile arsenic compounds is arsine (AsH_3). AsH_3 is also the most volatile arsenic compound with a boiling point of only −62.5 °C ((Planer-Friedrich *et al.*, 2006); Table 2.6). The gas has been detected in landfills, *anaerobic* wastewater treatment facilities, soils, hot springs, and even in the atmospheres of Jupiter and Saturn ((Francesconi and Kuehnelt, 2002), 56; Chapter 3). Bacteria and fungi can produce arsine and methylarsenic species from inorganic arsenic or organoarsenicals (Frankenberger and Arshad, 2002), 367–372. Once arsine forms in sediments and soils, it can volatilize into the atmosphere (Frankenberger and Arshad, 2002), 363.

Volatile methylarsenic compounds occur in *natural gas*, hot springs, soils, sediments, and a wide variety of biological organisms (Chapters 3 and 4). Inorganic arsenic can be methylated to produce MMA(III), DMA(III), and trimethylarsine. Methylarsenates also exist, but these compounds tend to be less volatile (also see Section 2.4.4). The methylation of arsenic in Earth environments is entirely or almost entirely *biotic* (see Section 2.6 and Chapter 4). In laboratory experiments, some volatile arsines have been produced from photochemical (*abiotic*) reactions involving As(III), *carboxylic acids*, and ultraviolet radiation (Guo *et al.*, 2005; McSheehy *et al.*, 2005). In particular, Guo *et al.* (2005) found that formic acid (HCOOH) yielded AsH_3, trimethylarsine resulted from acetic acid (CH_3COOH), triethylarsine came from propionic acid (CH_3CH_2COOH), and tripropylarsine resulted from butyric acid ($C_4H_8O_2$). However, the importance of these photochemical reactions in natural environments is unknown.

In the nineteenth and early twentieth centuries, many people in Europe were poisoned by trimethylarsine vapors (Figure 2.2; Chapter 5). Arsenic was commonly used in green and other wallpaper pigments. The brightly colored and attractive wallpapers were inexpensive and popular. However, mold often grew on the wallpapers and converted the arsenic in the pigments into poisonous and volatile 'Gosio gas', which consists of trimethylarsine (Frankenberger and Arshad, 2002), 370; (Cullen and Reimer, 1989), 714; (Craig, Eng and Jenkins, 2003), 38. By the end of the nineteenth century, arsenic pigments were largely banned in Europe. However, trimethylarsine gas from old wallpapers continued to sicken and even kill Europeans as late as the 1930s ((Meharg, 2005), 69; Chapter 5).

2.4.4   Organoarsenicals

Besides producing volatile arsenic species, biological organisms can also create less volatile and more complex organoarsenicals, such as arsenic sugars and methylarsenates (e.g. dimethylarsinic acid, DMA(V), $(CH_3)_2AsO(OH)$ and monomethylarsonic acid, MMA(V), $(CH_3)AsO(OH)_2$) (Chapters 3 and 4; Figure 2.2).

Arsenic can also replace carbons in *pyridine* or *benzene* rings (O'Day, 2006), 81; (Johansson and Jusélius, 2005).

Organoarsenicals are sometimes detected in water, sediments, soils, and rocks, especially in *geologic materials* that are rich in biological carbon. However, their concentrations are usually higher in biological samples, including algae, plant parts, urine, hair, and carcasses (Francesconi and Kuehnelt, 2002), 53; (Oremland and Stolz, 2003), 939–940. Francesconi and Kuehnelt (2002), Matschullat (2000) and Cullen and Reimer (1989) have extensive lists of organoarsenicals that have been found in different biological species.

The laboratory synthesis of organoarsenicals began when Louis-Claude Cadet de Gassicourt ('Cadet', 1731–1799) produced his combustible fuming liquid in 1757 (Meharg, 2005), 61; (Seyferth, 2001), 1488. During 1837–1843, Robert Wilhelm Bunsen (1811–1899) extensively studied the components of Cadet's fuming liquid, which is a mixture of cacodyl oxide ($(CH_3)_2AsOAs(CH_3)_2$), $(CH_3)_4As_2$, and other arsenic compounds ((Seyferth, 2001), 1494, 1496; Chapter 5). As discussed in Seyferth (2001), the synthesis and analysis of Cadet's fuming liquid led to important developments in organometallic chemistry. Since then, countless organoarsenicals have been synthesized and characterized. A few recent examples are discussed in Qi *et al.* (2007); Sharma, Rai and Singh (2007), and Yang *et al.* (2007). Eventually, some of the recently developed organoarsenicals may have important industrial and medical applications (Yang *et al.*, 2007; Carter *et al.*, 2007; Valkov and Golovchan, 2005).

Some organoarsenicals have been developed as chemical weapons. For example, lewisite is a vesicant (blister) agent ((Bismuth *et al.*, 2004); Chapter 5). It was originally developed in the United States during World War I. Lewisite itself is actually composed of three organoarsenicals. The major component, which is known as *Lewisite I*, is $ClCH{=}CHAsCl_2$. The minor components are Lewisite II, $(ClCH{=}CH)_2AsCl$, and Lewisite III, $(ClCH{=}CH)_3As$ (Muir *et al.*, 2005). Although Lewisite has never been used in actual combat, large stockpiles still existed in 2000 AD (Bismuth *et al.*, 2004) and improper disposal of the agent at the end of World War II has caused significant contamination in the Baltic Sea (Glasby, 1997) and Japan (Kinoshita *et al.*, 2006; Wada, Nagasawa and Hanaoka, 2006).

More benignly, organoarsenicals have significant applications in organic synthesis. The key compound for most of these uses is triphenylarsine, $(C_5H_5)_3As$, which can be successfully utilized in place of its better-known analogue triphenylphosphine, $(C_6H_5)_3P$. Triphenylarsine can substitute for triphenylphosphine as the ligand on palladium *catalysts* for *Stille* (Farina and Krishnan, 1991; Amatore *et al.*, 2003; Lau and Chiu, 2007) and *Suzuki* reactions (Lau *et al.*, 2004). Both the Stille and Suzuki reactions form new carbon to carbon bonds by reacting an *organohalide* or *organotriflate* with either an organotin compound (Stille reaction) or an organoboronic acid (Suzuki reaction). These reactions are key techniques for the formation of synthetic organic compounds. Both reactions are catalyzed by ligated palladium, which is where the triphenylarsine is used. Triphenylarsine is also used in *alkene epoxidation reactions*, which involve the formation of carbon-oxygen three-membered rings (He *et al.*, 2005a). Furthermore, triphenylarsine is the most common starting point for producing arsonium *ylides* (He *et al.*, 2005b). A ylide is a *resonance* compound that can be commonly visualized as alternating between adjacent positive and negative charges or a double bond (X+-Y- ↔ X=Y). In the 1950s, it was discovered that phosphonium ylides were useful in converting *aldehydes* and *ketones* to *alkenes* (the *Wittig reaction*) (Maryanoff and Reitz, 1989). Arsonium ylides are even more reactive in the Wittig reaction than phosponium ylides (He *et al.*, 2005b; Lloyd, Gosney and Ormiston, 1987). Not only can arsonium ylides be used to form traditional alkenes, they can also synthesize carbon–carbon and carbon–oxygen ring structures. These compounds represent powerful synthetic tools for organic chemists.

2.5   Introduction to arsenic oxidation and reduction

2.5.1   Arsenic oxidation

The *oxidation of arsenic* refers to an increase in its valence state to as high as +5 through chemical reactions that cause the arsenic to lose valence electrons. As examples, As(0) may oxidize to As(III), and As(III) to As(V). During the oxidation process, chemical oxidants receive the electrons from the arsenic and are reduced.

By themselves, air and pure oxygen (O_2) only slowly oxidize As(III) in water (Bissen and Frimmel, 2003; Burkitbaev, 2003; Hering and Kneebone, 2002), 173; (Bisceglia *et al.*, 2005). The oxidation of arsenic in natural waters may be considerably enhanced by microorganisms, Fe(III) species, nitrate (NO_3^-), natural organic matter (NOM), or Mn(III,VI) *(oxy)(hydr)oxide* compounds, sometimes even in the absence of O_2 ((Craig, Eng and Jenkins, 2003), 30; (Price and Pichler, 2005; Langner *et al.*, 2001; Evangelou, Seta and Holt, 1998; Schreiber *et al.*, 2003; Redman, Macalady and Ahmann, 2002; Stollenwerk, 2003), 71–72; Chapter 3). The oxidation of As(III) to As(V) further improves if the solid oxidants are poorly crystalline (i.e. have high *surface areas*) and, for at least some Fe(III) oxyhydroxide oxidants, if the reactions are catalyzed by light (Stollenwerk, 2003), 70–71.

The oxidation of arsenopyrite, realgar, orpiment, and other arsenic-bearing minerals in natural environments are discussed in some detail in Chapter 3. Additionally, (Lengke and Tempel, 2005; Lengke and Tempel, 2002; Lengke and Tempel, 2003) concentrated on measuring the oxidation rates of orpiment, realgar, and related amorphous arsenosulfides from O_2. In particular, (Lengke and Tempel, 2002) investigated the oxidation of orpiment in aqueous solutions at 25–40 °C with pH conditions of 6.8–8.2 and dissolved oxygen concentrations of 6.4–17.4 mg L^{-1}. The initial *ionic strength* of the solutions was 0.01 molar (M) as NaCl. The rate of orpiment destruction (R) in moles per square meters per second at 298.15 K (25 °C) may be derived from the following rate law:

$$R = 10^{-11.77(\pm 0.36)}(\text{DO})^{0.36(\pm 0.09)}(\text{H}^+)^{-0.47(\pm 0.05)} \tag{2.5}$$

where:

(DO) = dissolved oxygen concentration in M and

(H^+) = H^+ concentration in M (Lengke and Tempel, 2002).

The following reaction can explain the oxidation of orpiment to inorganic As(III) in the aqueous solutions (Lengke and Tempel, 2002), 3288:

$$\text{As}_2\text{S}_3 + 6\,\text{O}_2 + 6\,\text{H}_2\text{O} \rightarrow 2\,\text{H}_3\text{AsO}_3{}^0 + 3\,\text{SO}_4{}^{2-} + 6\,\text{H}^+ \tag{2.6}$$

Under oxidizing and near neutral pH conditions, inorganic As(III) could then slowly oxidize to inorganic As(V) through the following reaction:

$$2\,\text{H}_3\text{AsO}_3{}^0 + \text{O}_2 \rightarrow \text{H}_2\text{AsO}_4{}^- + \text{HAsO}_4{}^{2-} + 3\,\text{H}^+ \tag{2.7}$$

The oxidation of arsenic in amorphous As_2S_3 complies with the following rate law at 298.15 K (25 °C), pH conditions of 6.9–7.9, and dissolved oxygen concentrations of 6.7–17.3 mg L^{-1} (Lengke and Tempel, 2002), 3289:

$$R = 10^{-16.78}(\text{DO})^{0.42}(\text{H}^+)^{-1.26} \tag{2.8}$$

The oxidation rate of orpiment is slightly lower than the rate for amorphous As_2S_3 at pH conditions of 7–10 and dissolved oxygen concentrations of $1–10\,mg\,L^{-1}$ (Lengke and Tempel, 2002). These results do not consider the possibility of increased oxidation rates under natural conditions due to the presence of bacteria or oxidizing minerals, such as MnO_2. The rate of orpiment oxidation also increases with increasing pH or temperatures (Lengke and Tempel, 2002), 3290–3291 . Under alkaline conditions, As(V) oxyanions are less likely to sorb onto iron (oxy)(hydr)oxides and most other minerals because these minerals are likely to be well above their *zero points of charge* (ZPCs, see Section 2.7.6.2.2). Therefore, arsenic contamination of water could be more problematic if orpiment oxidizes under alkaline conditions (Chapter 3).

In similar studies with realgar and amorphous AsS, (Lengke and Tempel, 2003) derived the following rate laws for the oxidation of arsenic at 298.15 K (25 °C):

$$R_{Realgar} = 10^{-9.63(\pm0.41)}(DO)^{0.51(\pm0.08)}(H^+)^{-0.28(\pm0.05)} \tag{2.9}$$

$$R_{amorphousAsS} = 10^{-13.65(\pm0.82)}(DO)^{0.92(\pm0.08)}(H^+)^{-1.09(\pm0.10)} \tag{2.10}$$

For amorphous AsS, pH conditions ranged from 7.2 to 8.8 and dissolved oxygen concentrations were $6.8–15.8\,mg\,L^{-1}$. The pH conditions for realgar were 7.8–8.8 and dissolved oxygen concentrations were $5.9–16.5\,mg\,L^{-1}$ (Lengke and Tempel, 2003), 864. Under the same conditions, the oxidation rate of amorphous AsS is faster than the rate for realgar by a factor of 2–38 (Lengke and Tempel, 2003), 868.

The valence state of arsenic is a critical factor in effectively treating the contaminant in water. Any As(III) or other reduced forms of arsenic usually require oxidation to As(V) before water treatment technologies can be successfully implemented (Chapter 7). Except at pH conditions above 9.2, dissolved As(III) primarily exists as very unreactive $H_3AsO_3{}^0$ (see Section 2.7.3). By oxidizing dissolved As(III) to As(V) in pH 2–9 waters, As(III) converts into more reactive As(V) oxyanions, such as $H_2AsO_4{}^-$ and $HAsO_4{}^{2-}$. Unlike $H_3AsO_3{}^0$, As(V) oxyanions can be readily removed from water by various *sorbents, ion-exchange* materials, and other treatment technologies (Chapter 7).

Like natural environments, O_2 and air in water treatment systems are slow at oxidizing arsenic. As discussed further in Chapter 7, As(III) oxidation in water may be enhanced with radiation (Hug *et al.*, 2001), electrochemical methods (Arienzo *et al.*, 2002; Licht and Yu, 2005; Maldonado-Reyes, Montero-Ocampo and Solorza-Feria, 2007), a variety of chemical oxidants (Hug and Leupin, 2003; Dodd *et al.*, 2006), and/or bacteria (Ehrlich, 2002; Anderson *et al.*, 2002; De, 2005), 684. Effective chemical oxidants for arsenic in water include: chlorine (Cl_2), ozone (O_3), sodium hypochlorite (NaOCl), chlorine dioxide (ClO_2), potassium permanganate ($KMnO_4$), iron(VI) compounds (such as K_2FeO_4), and manganese (oxy)(hydr)oxides (Foster, 2003), 58; (Bissen and Frimmel, 2003; Licht and Yu, 2005; Dodd *et al.*, 2006; Dutta *et al.*, 2005), 1827; (Tournassat *et al.*, 2002; Tani *et al.*, 2004), 6518; (Manning *et al.*, 2002; Clifford and Ghurye, 2002), 231; (Thirunavukkarasu *et al.*, 2005).

2.5.2   Arsenic reduction

The reduction of arsenic refers to a decrease in its valence state to as low as −3 through chemical reactions that cause the arsenic to gain valence electrons. During the reduction process, *reductants* are oxidized as they donate electrons to arsenic. In general, As(V) converts faster into As(III) in reducing environments than As(III) transforms into As(V) under oxidizing conditions (Stollenwerk, 2003), 71.

Common reductants in natural subsurface environments include hydrogen sulfide (H_2S) and organic carbon with or without the presence of microorganisms (Stollenwerk, 2003), 72. As(V) reduction by H_2S is rapid, especially under acidic conditions (Stollenwerk, 2003), 72. Microorganisms are also important in reducing As(V) to As(III), and As(III) to arsine or dimethylarsine $((CH_3)_2AsH)$ in natural environments ((Craig, Eng and Jenkins, 2003), 29; (Stollenwerk, 2003), 71–72; Figure 2.2). Arsenate-reducing bacteria include *Pseudomonas fluorescens* and *Anabaena oscillaroides* (Cullen and Reimer, 1989), 716–717. Chapter 3 reviews the roles of bacteria, H_2S, and organic carbon in reducing arsenic in various natural environments, and how *reductive dissolution* of iron (oxy)(hydr)oxides can greatly affect the mobility of arsenic in water. Inskeep, McDermott and Fendorf (2002), 191 also discuss in some detail the types of microorganisms that reduce As(V) to As(III) and their possible mechanisms.

2.6   Introduction to arsenic methylation and demethylation

Methylation refers to the addition of one or more methyls (-CH$_3$) onto a chemical species. As mentioned earlier, inorganic arsenic may be methylated into MMA(III), DMA(III), trimethylarsine, trimethylarsine oxide, and a variety of other methyl species (Figure 2.2). In areas contaminated by methylarsenic pesticides, *methylthioarsenates* may also form (Wallschläger and London, 2008). Specifically, Wallschläger and London (2008) confirmed the presence of $(CH_3)AsO_2S^{2-}$, $(CH_3)AsOS_2^{2-}$, $(CH_3)_2AsOS^-$, and $(CH_3)_2AsS_2^-$ in groundwater at an unnamed contaminated site.

The methylation of arsenic is entirely or almost entirely biotic (Frankenberger and Arshad, 2002), 367. Specifically, certain fungi (including yeasts) and bacteria are capable of methylating arsenic ((Bentley and Chasteen, 2002), 257–260; (Cullen and Reimer, 1989), 717–724; Chapter 4). Only limited evidence exists for the chemical (abiotic) methylation of arsenic. As mentioned earlier, some volatile arsines have been produced in the laboratory from photochemical reactions involving As(III), carboxylic acids, and ultraviolet radiation (Guo *et al.*, 2005; McSheehy *et al.*, 2005).

The methylation of arsenic by microorganisms may be described with the *Challenger mechanism* (Bentley and Chasteen, 2002), 254–255; (Dombrowski *et al.*, 2005). The mechanism consists of a series of reduction and oxidative methylation reactions (2.11–2.21, below) that begin with the reduction of inorganic As(V) to inorganic As(III) and end with the formation of trimethylarsine (Figure 2.2). The reductants in natural occurrences of the Challenger mechanism are probably *thiols* and, in particular, *glutathione* and *lipoic acid* (6,8-dithiooctanoic acid) (Bentley and Chasteen, 2002), 255. The methylation of arsenic results from microorganisms transferring methyls (CH$_3$$^+$) from already methylated organic compounds (Cullen and Reimer, 1989), 720. Each methylated species produced in the Challenger mechanism could be excreted by the microorganisms, remain in the organisms, or be converted into the next arsenic species in the sequence.

In inorganic As(V), resonance occurs on the double bond:

$$O^- \!-\! \overset{\overset{\displaystyle OH}{|}}{\underset{\underset{\displaystyle OH}{|}}{As}} \!=\! O \ \rightleftharpoons \ O^- \!-\! \overset{\overset{\displaystyle OH}{|}}{\underset{\underset{\displaystyle OH}{|}}{\overset{+}{As}}} \!-\! O^-$$

$$(2.11)$$

Initially, the resonating inorganic As(V) is biotically or abiotically reduced to inorganic As(III):

$$O^- \!-\! \overset{\overset{\displaystyle OH}{|}}{\underset{\underset{\displaystyle OH}{|}}{\overset{+}{As}}} \!-\! O^- \ +\ 2\,H^+ \ \rightleftharpoons \ \underset{HO}{\overset{\overset{\displaystyle OH}{|}}{As}}\!\!\diagdown_{OH} \ +\ HO^-$$

$$(2.12)$$

The methylation of As(III) by microorganisms produces resonating MMA(V):

$$(2.13)$$

$$(2.14)$$

MMA(V) reduces to MMA(III):

$$(2.15)$$

MMA(III) methylates to resonating DMA(V):

$$(2.16)$$

$$(2.17)$$

DMA(V) reduces to DMA(III):

$$(2.18)$$

DMA(III) methylates to resonating trimethylarsine oxide (TMA(V)):

$$(2.19)$$

$$(2.20)$$

TMA(V) reduces to trimethylarsine (TMA(III)):

$$\underset{\substack{|\\CH_3}}{\overset{\substack{CH_3\\|}}{H_3C-\overset{+}{As}-O^-}} + \ H^+ \ \rightleftharpoons \ \underset{\substack{H_3C\quad\quad CH_3}}{\overset{\substack{CH_3\\|\\As}}{}} + \ HO^-$$

$$(2.21)$$

Biomethylation may also produce more complex *alkyl* arsenic groups. $As(C_2H_5)(CH_3)_2$ has been found in landfill and sewage gas, and probably also exists in natural gas (Bentley and Chasteen, 2002), 251. $As(C_2H_5)_3$ may also occur in landfill gases and probably natural gas (Bentley and Chasteen, 2002), 251. Further details on the reduction and methylation biochemistry of arsenic are discussed in Chapter 4.

Demethylation refers to the removal of methyls from organoarsenicals, which may ultimately transform the organoarsenicals into inorganic arsenic. Although exposure to ultraviolet radiation may demethylate arsenic (Cullen and Reimer, 1989), 741, the role of microorganisms in demethylation is especially important. Under sterile conditions, MMA(V) and DMA(V) are very stable in water (Cullen and Reimer, 1989), 749. However, bacteria can demethylate them and other methylarsenic species into inorganic arsenic (Frankenberger and Arshad, 2002), 364; (Cullen and Reimer, 1989), 749; (Santosa *et al.*, 1996), 703.

Huang, Scherr and Matzner (2007) found that DMA(V) and arsenobetaine $((CH_3)_3As^+CH_2COO^-)$ were the prominent organoarsenicals in samples of organic wetland and forest floor soils from the Fichtelgebirge Mountains of Germany. Incubations at $5\,^\circ C$ of aqueous extracts of the German soils suggest that arsenobetaine rapidly demethylates to an unknown arsenic species perhaps, dimethylarsenoylacetate, $((CH_3)_2As(O)CH_2COO^-)$, which demethylates to DMA(V). Similarly, Khokiattiwong *et al.* (2001) found that arsenobetaine rapidly demethylates to dimethylarsenoylacetate and DMA(V) in seawater (Huang, Scherr and Matzner, 2007). The demethylation of DMA(V) to MMA(V) in the extracts of the German soils is much slower and is followed by the rapid demethylation of MMA(V) to inorganic arsenic (Huang, Scherr and Matzner, 2007).

Lehr *et al.* (2003) found that *Mycobacterium neoaurum* could demethylate MMA(III) and MMA(V) to inorganic arsenic, but not DMA(V) or trimethylarsine oxide. Their results suggest that at least some MMA(V) reductively demethylates to inorganic As(III), which is a reversal of Reaction 2.13 (see above) in the Challenger mechanism (Lehr *et al.*, 2003), 833. Other mechanisms by which microorganisms demethylate arsenic are largely unknown (Lehr *et al.*, 2003). Chapter 4 presents additional information on the demethylation of arsenic.

2.7   Arsenic in water

2.7.1   Introduction

In natural waters, arsenic may exist as one or more dissolved species, whose chemistry would depend on the chemistry of the waters. Over time, arsenic species dissolved in water may: (1) interact with biological organisms and possibly methylate or demethylate (Chapter 4), (2) undergo abiotic or biotic oxidation, reduction, or other reactions, (3) sorb onto solids, often through ion exchange, (4) precipitate, or (5) coprecipitate. This section discusses the dissolution of solid arsenic compounds in water, the chemistry of dissolved arsenic species in aqueous solutions, and how the chemistry of the dissolved species varies with water chemistry and, in particular, pH, redox conditions, and the presence of dissolved sulfides. Discussions also include introductions to sorption, ion exchange, precipitation, and coprecipitation, which have important applications with arsenic in natural environments (Chapters 3 and 6) and water treatment technologies (Chapter 7).

If a reaction in an aqueous solution is at *equilibrium*, an *equilibrium constant* (K_{eq}) may be calculated if the *activities* of all of the products and reactants are known. (However, in practice, K_{eq} is approximated by using molar concentrations instead of activities.) In general, reactions have the following format, where *a moles* of substance A reacts with *b* moles of substance B to form *c* moles of substance C and *d* moles of substance D:

$$aA + bB = cC + dD \tag{2.22}$$

The equilibrium constant then relates the activities of the products and reactants of Reaction 2.22 through the following equation (Faure, 1998). 112:

$$K_{eq} = [C]^c[D]^d/[A]^a[B]^b \tag{2.23}$$

There are several different types of equilibrium constants, including: *solubility product constants* (K_{sp} values) for the dissolution of salts in water, and K_a and K_b *dissociation constants* for the dissolution of *acids* and *bases*, respectively, in water. These constants are discussed in Sections 2.7.2.3 and 2.7.4. MINTEQA2 (Allison, Brown and Novo-Gradac, 1991) and other geochemistry computer models, many of which are available on the internet, use these constants and related thermodynamic data to predict the solubilities of arsenic compounds and other solid substances in water and how dissolved species may react in water. Information on the chemistry and solubility of arsenic in aqueous solutions, including very acidic or highly alkaline mine drainage, is important in predicting the environmental impacts of arsenic on natural environments (Chapter 3).

2.7.2 Aqueous solubility of arsenic compounds and thermodynamics

2.7.2.1 Aqueous solubility

The solubility of a solid in a liquid is the amount of the solid that will dissolve in the liquid to form an equilibrated saturated solution (Faure, 1998), 111. In the literature, the solubility of a particular solid substance in water has been traditionally derived from laboratory measurements of the mass or moles of the solid that dissolves in a given mass or volume of distilled and deionized water. These values are often listed in milligrams per liter, grams per 100 milliliters (Lide, 2007), grams per liter, micrograms per liter ($\mu g\, l^{-1}$), *molal*, molar, or micromolar (μM) (Table 2.8).

Even under ideally controlled laboratory conditions using pure chemicals, the dissolution of a solid compound in water may involve several complex reactions and the formation of numerous dissolved species. As an example, the dissolution of slightly soluble As_2S_3 (orpiment) in water can be investigated in a laboratory *closed system* at 25°C and 1 bar pressure. Nordstrom and Archer (2003), 9 proposed the following reaction to describe the dissolution of orpiment, which initially forms $H_3AsO_3^0$ and H_2S^0:

$$As_2S_3 + 6\,H_2O \rightarrow 2\,H_3AsO_3^0 + 3\,H_2S^0 \tag{2.24}$$

Once formed, H_2S^0 would partially dissociate into HS^- and H^+. A small amount of $H_2AsO_3^-$ and H^+ would form from the dissociation of $H_3AsO_3^0$. Some $H_3AsO_3^0$ and H_2S^0 could also react to produce thioarsenic species, such as $AsS(OH)HS^-$ (see also Section 2.7.3). Depending on the accuracy and completeness of their thermodynamic databases, geochemical computer models may be able to identify the major reactions and estimate the activities of their products.

Table 2.8   *Solubility of some arsenic compounds in water at atmospheric pressure.*

Arsenic compound	Temperature (°C)	Solubility (g/100 g of water, $\pm$ 2 SD)	Reference(s)
Ammonium dihydrogen arsenate ($NH_4H_2AsO_4$)	25	52.7	Lide (2007)
Arsenic (V) oxide (As_2O_5)	20	65.8	Lide (2007)
Arsenolite (As_2O_3)	22	3.0 ± 0.4	Pokrovski *et al.* (1996)
Arsenolite (As_2O_3)	25	2.05	Lide (2007)
Arsenolite (As_2O_3)	50	6.9	Pokrovski *et al.* (1996)
Arsenolite (As_2O_3)	60	7.7 ± 0.8	Pokrovski *et al.* (1996)
Arsenolite (As_2O_3)	90	14 ± 1	Pokrovski *et al.* (1996)
Calcium arsenate ($Ca_3(AsO_4)_2$)	20	0.0036	Lide (2007)
Claudetite (As_2O_3)	22	2.6 ± 0.2	Pokrovski *et al.* (1996)
Claudetite (As_2O_3)	50	5.5 ± 0.4	Pokrovski *et al.* (1996)
Claudetite (As_2O_3)	90	12 ± 1	Pokrovski *et al.* (1996)
Claudetite (As_2O_3)	150	32 ± 2	Pokrovski *et al.* (1996)
Claudetite (As_2O_3)	200	110 ± 0.4	Pokrovski *et al.* (1996)
Claudetite (As_2O_3)	250	313 (+130, −90)	Pokrovski *et al.* (1996)
Claudetite (As_2O_3)	25	2.05	Lide (2007)
Potassium arsenate (K_3AsO_4)	25	125	Lide (2007)
Potassium dihydrogen arsenate (KH_2AsO_4)	6	19	Lide (2007)
Potassium hydrogen arsenate (K_2HAsO_4)	6	18.7	Lide (2007)
Sodium hydrogen arsenate (Na_2HAsO_4)	20	51	Lide (2007)
Sodium hydrogen arsenate heptahydrate ($Na_2HAsO_4 \cdot 7H_2O$)	20	51	Lide (2007)
Zinc arsenate ($Zn_3(AsO_4)_2$)	20	0.000 078	Lide (2007)
Zinc arsenate octahydrate ($Zn_3(AsO_4)_2 \cdot 8H_2O$)	20	0.000 078	Lide (2007)

The dissolution of solid substances under natural conditions is even more complex and may be very different than laboratory results. Specifically, the presence of impurities in a mineral (for example, antimony in orpiment) could noticeably affect its solubility in water. Furthermore, *geothermal* waters, mine drainage, seawater, and other waters in natural environments are sometimes very acidic, highly alkaline, hot, reducing, or contain microorganisms or a lot of dissolved solids (high ionic strength). Any of these properties could substantially affect the solubility of minerals and other solids. Specifically, sulfate-reducing bacteria in a highly reducing water would produce excess H_2S that could partially dissolve in the water. *Le Châtelier's Principle* indicates that an increase in H_2S^0 would reverse Reaction 2.24 to precipitate additional As_2S_3 when compared with the laboratory results. Natural systems also tend to be *open*, which means that a reaction may not be able to attain equilibrium because of periodic changes in temperature, biological activity, or the movement of reactants or products in or out of the system. For example, if H_2S^0 escapes from a highly reducing natural water, Le Châtelier's Principle indicates that Reaction 2.24 would dissolve additional As_2S_3 to replace the lost H_2S^0.

2.7.2.2 Importance of thermodynamic data

Thermodynamic data are critical in developing geochemical models to predict the behavior of arsenic in water and other geologic materials. The data are important in investigating the behavior of arsenic in mineral–water interactions, water quality studies, the formation of ore deposits, Eh-pH diagrams, and environmental remediation projects (Nordstrom and Archer, 2003), 1–2. The reliability of the geochemical models depends on the quality of the thermodynamic data that goes into the models. Unfortunately, published thermodynamic data are not always trustworthy. Many of the data are contradictory and the methods that produce the data are sometimes questionable or have not been thoroughly documented. Nordstrom and Archer (2003) provide a detailed review of the controversies, uncertainties, and problems related to thermodynamic data for arsenic and its compounds. Too often, data in the literature have been passed from reference to reference without critical evaluations. Some of the data have high measurement errors, undefined or poorly defined laboratory conditions, and unrepresentative sampling (Matschullat, 2000), 298; (Nordstrom and Archer, 2003). Furthermore, other questionable data originate from obscure documents or are written in languages that many individuals cannot read and properly interpret. Therefore, thermodynamic results must be accepted with a certain amount of caution. The list of thermodynamic data in Appendix C includes multiple results for the same compounds and other chemical species. Multiple results allow the reader to have some idea of the variability of the values in the literature. However, users are cautioned to avoid mixing data from different literature sources when performing calculations for a reaction. Using data from multiple sources may introduce serious errors (Wagman *et al.*, 1982; Eby, 2004), 474. Once reliable thermodynamic values are obtained, geochemistry models (such as MINTEQA2) may be utilized to model the effects of pH, Eh, temperature, ionic strength, and other factors on the solubility of solid substances under natural conditions.

2.7.2.3 Solubility product constants and thermodynamics

As an alternative to laboratory solubility measurements, solubility product constants (K_{sp}), which are derived from thermodynamic data, can be used to calculate the solubility of solids in water (Table 2.9). Each solubility product constant describes a disassociation of a solid in water and calculates the activities or concentrations of the dissolution products in the saturated solution. The solubility product constant or another equilibrium constant of a reaction may be derived from the *Gibbs free energy* of the reaction ($\Delta G^o{}_R$) as shown in the following equation:

$$\ln K_{sp} = -\Delta G^o{}_R / RT \tag{2.25}$$

where:

K_{sp} = solubility product constant,

$\Delta G^o{}_R$ = Gibbs free energy for the reaction at 298.15 K (25 °C) and 1 bar pressure,

T = Temperature in K, and

R = gas constant, 8.314 41 J mol^{-1}·K^{-1} or 1.987 17 cal mol^{-1}·K^{-1}.

The following equations further define Gibbs free energy and relate it to *enthalpy* and *entropy*:

$$\Delta G^o{}_R = \sum n_i G_i{}^o \text{ (products) } - \sum n_i G_i{}^o \text{(reactants)} \tag{2.26}$$

$$\Delta G^o{}_R = \Delta H^o R - T \Delta S^o{}_R \tag{2.27}$$

Table 2.9 Equilibrium constants (including some solubility product constants, K_{sp} values, from Lide (2007), Krause and Ettel (1988), Bothe and Brown (1999), Davis (2000), Nordstrom and Archer (2003), Langmuir, Mahoney and Rowson (2006), Lee and Nriagu (2007), Zhu et al. (2005), and Zhu et al. (2006)) for reactions with various arsenic compounds in water at 1 bar pressure. $pK = -log_{10} K$.

Arsenic compound	Relevant reaction	Temperature (°C)	Equilibrium pH	Equilibrium constant (pK)	Reference(s)
Ag_3AsO_4	$Ag_3AsO_4 \rightleftharpoons 3\ Ag^+ + AsO_4^{3-}$	25 ± 1	4.92	23.4	Lee and Nriagu (2007)
Ag_3AsO_4	$Ag_3AsO_4 \rightleftharpoons 3\ Ag^+ + AsO_4^{3-}$	25	—	22.0	Lide (2007)
As_2O_3 (arsenolite, cubic)	$As_2O_3 + 3\ H_2O$ (liquid) $\rightleftharpoons 2\ H_3AsO_3^0$	25	—	1.38	Nordstrom and Archer (2003)
As_2O_3 (claudetite, monoclinic)	$As_2O_3 + 3\ H_2O$ (liquid) $\rightleftharpoons 2\ H_3AsO_3^0$	25	—	1.34	Nordstrom and Archer (2003)
As_2S_3(amorphous)	$1.5\ As_2S_3 + 1.5\ H_2S^0 \rightleftharpoons As_3S_4(SH)_2^- + H^+$	25	—	5.5	Nordstrom and Archer (2003)
As_2S_3(amorphous)	$0.5\ As_2S_3 + 0.5\ H_2S^0 + H_2O \rightleftharpoons AsS(OH)(SH)^- + H^+$	25	—	7.9	Nordstrom and Archer (2003)
$As_2S_3(\alpha,\ crystalline)$	$As_2S_3 + 6\ H_2O$ (liquid) $\rightleftharpoons 2\ H_3AsO_3^0 + 3\ HS^- +3\ H^+$	25	—	46.3	Nordstrom and Archer (2003)
$As_2S_3(\alpha,\ crystalline)$	$As_2S_3 + 6\ H_2O$ (liquid) $\rightleftharpoons 2\ H_3AsO_3^0 + 3\ H_2S^0$	25	—	25.3	Nordstrom and Archer (2003)
$Ba_3(AsO_4)_2$	$Ba_3(AsO_4)_2 \rightleftharpoons 3Ba^{2+} + 2\ AsO_4^{3-}$	20	7.0	21.57	Davis (2000)
$Ba_3(AsO_4)_2$	$Ba_3(AsO_4)_2 \rightleftharpoons 3Ba^{2+} + 2\ AsO_4^{3-}$	25	7.88–12.50	23.53	Zhu et al. (2005)
$BaHAsO_4$	$BaHAsO_4 \rightleftharpoons Ba^{2+} + HAsO_4^{2-}$	20	7.0	3.92 ± 0.11	Davis (2000)
$BaHAsO_4 \cdot H_2O$	$BaHAsO_4 \cdot H_2O \rightleftharpoons Ba^{2+} + HAsO_4^{2-} + H_2O$	25	3.63–7.43	5.60	Zhu et al. (2005)
$BiAsO_4$	$BiAsO_4 \rightleftharpoons Bi^{3+} + AsO_4^{3-}$	25	—	9.35	Lide (2007)
$Ca_3(AsO_4)_2 \cdot 2.25H_2O$	$Ca_3(AsO_4)_2 \cdot 2.25H_2O \rightleftharpoons 3\ Ca^{2+} + 2\ AsO_4^{3-} + 2.25\ H_2O$	25	7.14–7.50	21.40	Zhu et al. (2006)

Table 2.9 *(continued)*

Arsenic compound	Relevant reaction	Temperature (°C)	Equilibrium pH	Equilibrium constant (pK)	Reference(s)
$Ca_3(AsO_4)_2 \cdot 3H_2O$	$Ca_3(AsO_4)_2 \cdot 3H_2O \rightleftharpoons 3\,Ca^{2+} + 2\,AsO_4^{3-} + 3\,H_2O$	25	5.54–6.68	21.14	Zhu *et al.* (2006)
$Ca_3(AsO_4)_2 \cdot 3.67H_2O$	$Ca_3(AsO_4)_2 \cdot 3.67H_2O \rightleftharpoons 3\,Ca^{2+} + 2\,AsO_4^{3-} + 3.67\,H_2O$	23 ± 1	11.18	21.00	Bothe and Brown (1999)
$Ca_3(AsO_4)_2 \cdot 4.25H_2O$	$Ca_3(AsO_4)_2 \cdot 4.25H_2O \rightleftharpoons 3\,Ca^{2+} + 2\,AsO_4^{3-} + 4.25\,H_2O$	23 ± 1	7.32–7.55	21.00	Bothe and Brown (1999)
$Ca_4(OH)_2(AsO_4)_2 \cdot 4H_2O$	$Ca_4(OH)_2(AsO_4)_2 \cdot 4H_2O \rightleftharpoons 4\,Ca^{2+} + 2\,AsO_4^{3-} + 2\,OH^- + 4\,H_2O$	23 ± 1	12.14–12.23	29.20	Bothe and Brown (1999)
$Ca_4(OH)_2(AsO_4)_2 \cdot 4H_2O$	$Ca_4(OH)_2(AsO_4)_2 \cdot 4H_2O \rightleftharpoons 4\,Ca^{2+} + 2\,AsO_4^{3-} + 2\,OH^- + 4\,H_2O$	25	9.63–13.40	27.49	Zhu *et al.* (2006)
$Ca_5(AsO_4)_3OH$	$Ca_5(AsO_4)_3OH \rightleftharpoons 5\,Ca^{2+} + 3\,AsO_4^{3-} + OH^-$	23 ± 1	9.54–9.87	38.04	Bothe and Brown (1999)
$Ca_5(AsO_4)_3OH$	$Ca_5(AsO_4)_3OH \rightleftharpoons 5\,Ca^{2+} + 3\,AsO_4^{3-} + OH^-$	25	5.61–13.40	40.12	Zhu *et al.* (2006)
$Ca_5H_2(AsO_4)_4 \cdot 9H_2O$ (Ferrarisite)	$Ca_5H_2(AsO_4)_4 \cdot 9H_2O \rightleftharpoons 5\,Ca^{2+} + 2\,HAsO_4^{2-} + 2\,AsO_4^{3-} + 9\,H_2O$	23 ± 1	6.88	31.49	Bothe and Brown (1999)
$Ca_5H_2(AsO_4)_4 \cdot 9H_2O$ (Guerinite)	$Ca_5H_2(AsO_4)_4 \cdot 9H_2O \rightleftharpoons 5\,Ca^{2+} + 2\,HAsO_4^{2-} + 2\,AsO_4^{3-} + 9\,H_2O$	23 ± 1	~6.91	30.69	Bothe and Brown (1999)
$CaH(AsO_4) \cdot H_2O$	$CaH(AsO_4) \cdot H_2O \rightleftharpoons Ca^{2+} + HAsO_4^{2-} + H_2O$	23 ± 1	6.22	4.79	Bothe and Brown (1999)

(continued overleaf)

Table 2.9 (*continued*)

Arsenic compound	Relevant reaction	Temperature (°C)	Equilibrium pH	Equilibrium constant (pK)	Reference(s)
$Cd_3(AsO_4)_2$	$Cd_3(AsO_4)_2 \rightleftharpoons 3\ Cd^{2+} + 2\ AsO_4^{3-}$	25 ± 1	5.99	33.3	Lee and Nriagu (2007)
$Cd_3(AsO_4)_2$	$Cd_3(AsO_4)_2 \rightleftharpoons 3\ Cd^{2+} + 2\ AsO_4^{3-}$	25	—	32.7	Lide (2007)
$Co_3(AsO_4)_2$ (poorly crystalline)	$Co_3(AsO_4)_2 \rightleftharpoons 3\ Co^{2+} + 2\ AsO_4^{3-}$	25 ± 1	4.58	32.3	Lee and Nriagu (2007)
$Co_3(AsO_4)_2$	$Co_3(AsO_4)_2 \rightleftharpoons 3\ Co^{2+} + 2\ AsO_4^{3-}$	25	—	28.2	Lide (2007)
$CrAsO_4$ (amorphous)	$CrAsO_4$ (amorphous) $\rightleftharpoons Cr^{3+} + AsO_4^{3-}$	25 ± 1	2.68	19.8	Lee and Nriagu (2007)
$Cu_3(AsO_4)_2$	$Cu_3(AsO_4)_2 \rightleftharpoons 3\ Cu^{2+} + 2\ AsO_4^{3-}$	25 ± 1	3.15	35.6	Lee and Nriagu (2007)
$Cu_3(AsO_4)_2$	$Cu_3(AsO_4)_2 \rightleftharpoons 3\ Cu^{2+} + 2\ AsO_4^{3-}$	25	—	35.1	Lide (2007)
$FeAsO_4 \cdot nH_2O$ (amorphous)	$FeAsO_4 \cdot nH_2O + 3\ H^+ \rightleftharpoons Fe^{3+} + H_3AsO_4^0 + X\ H_2O$ (congruent dissolution)	25	<3	23.0 ± 0.3	Langmuir, Mahoney and Rowson (2006)
$FeAsO_4 \cdot 2H_2O$ (Scorodite)	$FeAsO_4 \cdot 2H_2O + 3\ H^+ \rightleftharpoons Fe^{3+} + H_3AsO_4^0 + 2\ H_2O$ (congruent dissolution)	25	<3	25.83 ± 0.07	Langmuir, Mahoney and Rowson (2006)
$FeAsO_4 \cdot 2H_2O$ (Scorodite)	$FeAsO_4 \cdot 2H_2O + 3\ H^+ \rightleftharpoons Fe^{3+} + H_3AsO_4^0 + 2\ H_2O$ (congruent dissolution)	23 ± 1	0.97–2.43	24.41 ± 0.15	Krause and Ettel (1988)

Table 2.9 *(continued)*

Arsenic compound	Relevant reaction	Temperature (°C)	Equilibrium pH	Equilibrium constant (pK)	Reference(s)
$H_3AsO_4^0$	$H_3AsO_4^0 + H_2 \text{ (gas)} \rightleftharpoons H_3AsO_3^0 + H_2O$	25	—	−19.35	Nordstrom and Archer (2003)
$Mg_3(AsO_4)_2$	$Mg_3(AsO_4)_2 \rightleftharpoons 3\,Mg^{2+} + 2\,AsO_4^{3-}$	25 ± 1	6.90	21.1	Lee and Nriagu (2007)
$MnHAsO_4$ (poorly crystalline)	$MnHAsO_4 \rightleftharpoons Mn^{2+} + HAsO_4^{2-}$	25 ± 1	5.72	7.31	Lee and Nriagu (2007)
$NiHAsO_4$ (poorly crystalline)	$NiHAsO_4 \rightleftharpoons Ni^{2+} + HAsO_4^{2-}$	25 ± 1	4.59	6.60	Lee and Nriagu (2007)
$Pb_5(AsO_4)_3OH$	$Pb_5(AsO_4)_3OH \rightleftharpoons 5\,Pb^{2+} + 3\,AsO_4^{3-} + OH^-$	25 ± 1	—	76.1	Lee and Nriagu (2007)
$PbHAsO_4$	$PbHAsO_4 \rightleftharpoons Pb^{2+} + HAsO_4^{2-}$	25 ± 1	4.68	11.8	Lee and Nriagu (2007)
$Sn_3(AsO_4)_2$	$Sn_3(AsO_4)_2 \rightleftharpoons 3\,Sn^{2+} + 2\,AsO_4^{3-}$	25 ± 1	3.09	47.8	Lee and Nriagu (2007)
$Sr_3(AsO_4)_2$	$Sr_3(AsO_4)_2 \rightleftharpoons 3\,Sr^{2+} + 2\,AsO_4^{3-}$	25 ± 1	10.5	17.7	Lee and Nriagu (2007)
$Sr_3(AsO_4)_2$	$Sr_3(AsO_4)_2 \rightleftharpoons 3\,Sr^{2+} + 2\,AsO_4^{3-}$	25	—	18.4	Lide (2007)
$Zn_3(AsO_4)_2$	$Zn_3(AsO_4)_2 \rightleftharpoons 3\,Zn^{2+} + 2\,AsO_4^{3-}$	25 ± 1	5.50	28.9	Lee and Nriagu (2007)
$Zn_3(AsO_4)_2 \cdot 12H_2O$ (Kottigite)	$Zn_3(AsO_4)_2 \cdot 12H_2O \rightleftharpoons 3\,Zn^+ + 2\,AsO_4^{3-} + 12\,H_2O$	25 ± 1	4.87	32.4	Lee and Nriagu (2007)

$$\Delta H^{o}{}_{R} = \sum n_{i} H_{i}{}^{o} \text{ (products)} - \sum n_{i} H_{i}{}^{o} \text{ (reactants)} \tag{2.28}$$

$$\Delta S^{o}{}_{R} = \sum n_{i} S_{i}{}^{o} \text{(products)} - \sum n_{i} S_{i}{}^{o} \text{(reactants)} \tag{2.29}$$

where:

$H_{i}{}^{o}$ = Enthalpies of individual products and reactants at 298.15 K (25 °C) and 1 bar pressure,

ΔH_{R}^{o} = Enthalpy for the reaction at 298.15 K (25 °C) and 1 bar pressure,

n_{i} = number of moles of each product or reactant in the balanced reaction,

$S_{i}{}^{o}$ = Entropies of individual products and reactants at 298.15 K (25 °C) and 1 bar pressure, and

ΔS_{R}^{o} = Entropy for the reaction at 298.15 K (25 °C) and 1 bar pressure.

Although K_{sp} values and other equilibrium constants can be calculated this way, widely used equilibrium constants are often listed in tables, such as Table 2.9. Equilibrium constants are, however, temperature dependent. If a constant is required for a reaction occurring at temperatures other than that listed (usually 25 °C), but between 10 and 40 °C, the *van't Hoff equation* is used to calculate the equilibrium constant at the desired temperature (Langmuir, 1997), 21. The equation states:

$$\ln(K_{eq2}/K_{eq1}) = (-\Delta H_{R}^{o}/R)(1/T_{2} - 1/T_{1}) \tag{2.30}$$

where:

K_{eq1} = reference equilibrium constant, usually at 298.15 K,

K_{eq2} = equilibrium constant at the desired temperature,

$\Delta H^{o}{}_{R}$ = enthalpy of the reaction at 298.15 K or another standard temperature,

R = gas constant (Appendix A),

T_{1} = reference temperature, usually 298.15 K, and

T_{2} = desired temperature.

The equation assumes a constant pressure of 1 bar and that the ΔH^{0} for the reaction is essentially constant from T_{1} to T_{2}, which is usually a reasonable approximation from about 10–40 °C. To obtain accurate equilibrium constants for reactions at higher or lower temperatures, the ΔH would have to be calculated from *heat capacity* measurements at those temperatures (Faure, 1998), 168. Robie, Hemingway and Fisher (1979) also contains thermodynamic data for temperatures as high as 1800 K.

Like pH, solubility product constants are often reported as pK_{sp} values; that is, the negative logs of their K_{sp} values. For example, $Ca_{3}(AsO_{4})_{2} \cdot 4.25H_{2}O$ has a pK_{sp} value of 21.00 at 23 °C ((Bothe and Brown, 1999); Table 2.9) and therefore a K_{sp} of 1.00×10^{-21}. The dissolution reaction for $Ca_{3}(AsO_{4})_{2} \cdot 4.25H_{2}O$ may be written as follows:

$$Ca_{3}(AsO_{4})_{2} \cdot 4.25\ H_{2}O \rightarrow 3\ Ca^{2+} + 2AsO_{4}{}^{3-} + 4.25\ H_{2}O \tag{2.31}$$

Using Equation 2.23 and Reaction 2.29, the following Equation 2.31 may be derived to relate the equilibrium constant of $Ca_{3}(AsO_{4})_{2} \cdot 4.25H_{2}O$ with the activities of its reactants and products:

$$K = [Ca^{2+}]^{3}[AsO_{4}{}^{3-}]^{2}[H_{2}O]^{4.25}/[Ca_{3}(AsO_{4})_{2} \cdot 4.25H_{2}O] = 1.00 \times 10^{-21} \tag{2.32}$$

Water is the solvent for this system and its activity is assumed to be one. Likewise, all solids are assigned an activity of one, which results in Equation 2.30 simplifying into Equation 2.31:

$$K_{sp} = [Ca^{2+}]^2[AsO_4{}^{3-}]^2 = 1.00 \times 10^{-21} \tag{2.33}$$

By definition the reactant for any solubility product constant reaction must be a solid and the solvent must be water, so this is the general form of any K_{sp} and the equation always simplifies down to a nonfraction. From this expression the solubility of $Ca_3(AsO_4)_2 \cdot 4.25H_2O$ is easily calculated. Based on Equation 2.31, the dissolution of 1 mole of the solid will generate three moles of Ca^{2+} and two moles of $AsO_4{}^{3-}$. Therefore:

$$K_{sp} = [3x]^3[2x]^2 = 1.00 \times 10^{-21} \tag{2.34}$$

A little algebra will yield a value of 2.47×10^{-5} for x. In other words, the solubility of $Ca_3(AsO_4)_2 \cdot 4.25H_2O$ in water is 2.47×10^{-5} mol l^{-1} or 11.7 mg L^{-1}. Of course, this is a greatly oversimplified calculation that does not take into account the possibility of some free Ca^{2+} or $AsO_4{}^{3-}$ being initially dissolved in the water or any further reactions that Ca^{2+} and $AsO_4{}^{3-}$ might undergo in a natural system. Due to Le Châtelier's Principle, either situation would affect the real solubility of the salt. This method does provide a ready estimate, however.

2.7.2.4 Activity products

In many cases, reactions in aqueous solutions are not at equilibrium. The failure to achieve equilibrium may be due to slow reaction kinetics, temperature fluctuations, biological activity, or open systems. For a reaction that is not at equilibrium, Equation 2.23 is not true:

$$K_{eq} \neq [C]^c[D]^d/[A]^a[B]^b \tag{2.35}$$

However, an *activity product* (Q) can be calculated from the activities of the reactants and products:

$$Q = [C]^c[D]^d/[A]^a[B]^b \tag{2.36}$$

If $Q/K_{eq} > 1$, the activities of the reaction products (C and D) are higher than expected (*supersaturated*) and Le Châtelier's Principle dictates that the reaction should reverse to regenerate more A and B (reactants) so that equilibrium could be achieved (Faure 1998), 131:

$$aA + bB \leftarrow cC + Dd \tag{2.37}$$

If $Q/K_{eq} < 1$, C and D are undersaturated at equilibrium and Le Châtelier's Principle dictates that to achieve equilibrium, the reaction would need to proceed to the right and yield more products:

$$aA + bB \rightarrow cC + dD \tag{2.38}$$

Computer programs, such as MINTEQA2, can calculate Q values and determine if the products exceed saturation or not.

2.7.2.5 *Incongruent dissolution of scorodite*

Scorodite ($FeAsO_4 \cdot 2H_2O$) is an important mineral in controlling the solubility of As(V) in acid mine drainage (Chapter 3). At pH conditions below about 2.4–3 (Krause and Ettel, 1988; Langmuir, Mahoney and Rowson, 2006), the dissolution of scorodite may be described with the following reaction (Harvey *et al.*, 2006), 6709:

$$FeAsO_4 \cdot 2H_2O + H^+ \rightarrow H_2AsO_4^- + Fe(OH)^{2+} + H_2O \qquad (2.39)$$

In very acidic solutions (pH < 2.4–3) with ionic strengths below 0.1 M and at 25 °C and 1 bar pressure, scorodite has a pK of about 25.83 ± 0.07. The pK of amorphous Fe(III) arsenate is approximately 23.0 ± 0.3 under the same conditions (Langmuir, Mahoney and Rowson, 2006). At higher pH values, scorodite dissolves incongruently, which means that at least one of its dissolution products precipitates as a solid. The incongruent dissolution of scorodite in water leads to the formation of Fe(III) (oxy)(hydr)oxide precipitates; that is, Fe(III) (hydrous) oxides, (hydrous) hydroxides and (hydrous) oxyhydroxides (Chapter 3). During the formation and precipitation of the iron(III) (oxy)(hydr)oxides, As(V) probably coprecipitates with them (Chapter 3; also see Section 2.7.6.3). The dissolution rate of scorodite at 22 °C in pH 2–6 water is slow, around 10^{-9}–10^{-10} mol m^{-2} s^{-1}, which explains its presence in many mining wastes (Harvey *et al.*, 2006).

2.7.3 Dissolved arsenic species

In most natural waters that have detectable arsenic, dissolved inorganic As(III) or As(V) are dominant. Dissolved As(0) and As(-III) species rarely occur in natural waters, and organoarsenicals are often absent or in very low concentrations ((Mandal and Suzuki, 2002), 206; Chapter 3). In local groundwaters and mine drainage waters, inorganic arsenic may form dissolved species with ammonia, cyanide, and fluoride, such as AsO_3F^{2-} and $HAsO_3F^-$ ((Williams, 2001; Ballantyne and Moore, 1988), 478; (Apambire and Hess, 2000; Bundschuh *et al.*, 2000), 29; Chapters 3 and 6). The speciation of dissolved arsenic in water depends on pH, redox conditions, other aqueous chemistry, and biological activity (Shih, 2005), 86–87. Although As(V) is usually the prominent form of arsenic in *oxic* waters, biological activity may result in significant concentrations of metastable As(III) (Chapter 3). In comparison, even mildly reducing conditions in groundwater usually results in more As(III) than As(V) (Francesconi and Kuehnelt, 2002), 56, 64.

In anoxic groundwater and other reducing waters, inorganic arsenite (As(III)) commonly hydrates to arsenious acid, which primarily exists as dissolved $H_3AsO_3^0$ at pH conditions below 9.2 and as its dissociated anions ($H_2AsO_3^-$, $HAsO_3^{2-}$, and AsO_3^{3-}) under more alkaline conditions (Chapter 3; Figure 2.3). As discussed in later chapters in this book, inorganic As(III) is especially toxic and problematic in groundwater. Chapters 3 and 6 provide further details on the occurrences and behavior of dissolved As(III) species in groundwater, seawater, and fresh surface waters. The toxicology of As(III) is discussed in Chapter 4. Methods for removing dissolved As(III) and As(V) from water are reviewed in Chapter 7.

$H_3AsO_3^0$ is the dominant form of As(III) in natural waters at arsenic concentrations up to about one molal, pH conditions of 0–9.2, and at 20–300 °C (Pokrovski *et al.*, 1996). At unusually high As(III) concentrations of 1–2 molal, $H_3AsO_3^0$ will begin to *polymerize* and dehydrate into arsenic (hydr)oxide *oligomers*, such as As_4O_6, $As_3O_3(OH)_3$ or $As_6O_6(OH)_6$ (Pokrovski *et al.*, 1996; Tossell, 1997). However, using thermodynamic data, Tossell (1997), 1615 concluded that AsO_2H and $HAsO(OH)_2$ are unlikely to be stable in aqueous solutions of arsenious acid.

The dominant form of arsenic in oxic natural waters is usually dissolved arsenic acid, which includes $H_3AsO_4^0$ under very acidic (pH < 2) conditions and its associated anions ($H_2AsO_4^-$, $HAsO_4^{2-}$, and/or AsO_4^{3-}) in less acidic, neutral, and alkaline waters (Figure 2.4). In most oxic natural waters that have

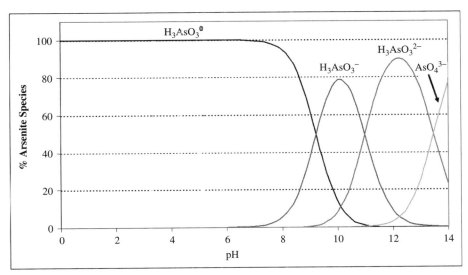

Figure 2.3 Speciation of arsenious acid with pH. Curves calculated from acid dissociation constants listed in Table 2.10.

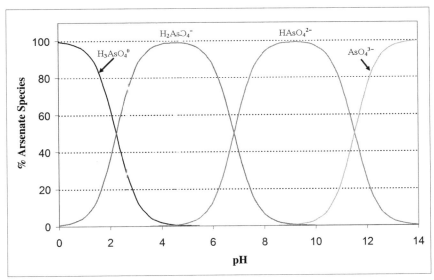

Figure 2.4 Speciation of arsenic acid with pH. Curves calculated from acid dissociation constants listed in Table 2.10.

detectable dissolved arsenic, $H_2AsO_3^-$ and $HAsO_3^{2-}$ are the prominent As(V) species with perhaps metastable $H_3AsO_3^0$.

In sulfide-rich and anoxic waters, dissolved thioarsenic species commonly occur. Thioarsenic species form when sulfur substitutes for one or more of the oxygens in the dissolved forms of arsenious or arsenic acids (e.g. $H_2AsO_3S^-$). Traditionally, aqueous thioarsenic species have been identified as *thioarsenites*

(Chapter 3). However, recent laboratory studies and analyses of geothermal waters from Yellowstone National Park, Wyoming, United States, indicate that thioarsenates rather than thioarsenites form in at least some sulfide-rich and anoxic waters (Stauder, Raue and Sacher, 2005; Wallschläger and Stadey, 2007; Planer-Friedrich *et al.*, 2007). Distinguishing between thioarsenite and thioarsenate aqueous species is often difficult, even with modern and sophisticated analytical methods. Furthermore, sample preservation methods may fail to properly maintain the integrity of the aqueous species until analysis (Planer-Friedrich *et al.*, 2007; Samanta and Clifford, 2006). Effective sample preservation and analytical methods are critically needed before researchers can determine whether thioarsenites, thioarsenates, or both are significantly present in hydrothermal and other high-sulfide and anoxic natural waters.

Planer-Friedrich *et al.* (2007) detected thioarsenates in the Yellowstone geothermal waters over a wide range of pH conditions (2.1–9.3). Overall, thioarsenates are more plentiful in the alkaline waters and represent up to 83 % of the total arsenic. Up to 34 % of the total arsenic in a pH 2.4 sample was monothioarsenate ($H_2AsO_3S^-$) (Planer-Friedrich *et al.*, 2007), 5249. Trithioarsenates ($H_{3-x}AsS_3O^{x-}$, where $x = 0$, 1, 2, or 3) are prominent in the water of the Ojo Caliente hot spring at Yellowstone (51 % of the total arsenic). After discharging from the hot spring, the arsenic is involved in at least three different reactions, which are: (1) the conversion of some trithioarsenates into arsenite, (2) a minor stepwise reaction that involves the replacement of SH^- with OH^- to initially convert trithioarsenates into dithioarsenates ($H_{3-x}AsS_2O_2^{x-}$, where $x = 0$, 1, 2, or 3), followed by the creation of monothioarsenates ($H_{3-x}AsO_3S^{x-}$, where $x = 0$, 1, 2, or 3), and finally the formation of arsenate, and (3) the oxidation of arsenite to arsenate (Planer-Friedrich *et al.*, 2007).

The formation of thioarsenates under sulfide-rich and anoxic conditions is not understood. Stauder, Raue and Sacher (2005) suggested that inorganic As(III) disproportionates into thioarsenates and elemental arsenic, perhaps through the following reaction:

$$5H_3AsO_3^0 + 3H_2S \rightarrow 2As^0 \downarrow + 3H_2AsO_3S^- + 6H_2O + 3H^+ \tag{2.40}$$

However, Wallschläger and Stadey (2007), 3879 disagree. In their mass balance experiments, Wallschläger and Stadey (2007), 3879 did not detect any significant loss of soluble arsenic, which would be expected from the precipitation of elemental arsenic.

The chemical properties of thioarsenates are largely unknown, but are expected to be different than the properties of thioarsenites or other arsenic species (Wallschläger and Stadey, 2007), 3880. If thioarsenates are prominent in sulfide-rich and anoxic environments, measurements of their sorption and other chemical properties are required to understand and predict their mobility in natural environments.

2.7.4 Dissociation of arsenious and arsenic acids

Figure 2.3 shows how the chemistry of dissolved arsenious acid varies with pH. An analogous graph for arsenic acid is in Figure 2.4. As expected, protonated species of the acids are more common under low pH conditions were H^+ is abundant. For both *weak acids*, dissociation constants (K_a values) may be derived to describe their gain or loss of H^+ with changing pH conditions (Table 2.10; (Faure, 1998), 119–120). For example, the following reaction involving the dissociation of $H_3AsO_3^0$ in water at 25 °C and 1 bar pressure has a dissociation constant (K_a) of $10^{-9.2}$ (Wolthers *et al.*, 2005), 3490:

$$H_3AsO_3^0 \rightarrow H_2AsO_3^- + H^+ \tag{2.41}$$

Table 2.10 *Dissociation constants of weak arsenic acids.* $pK_a = -log_{10} K_a$.

Arsenic species	Reaction	Temperature (°C)	pK_a (±2 SD)	Reference(s)
Inorganic As(III)				
	$H_3AsO_3^0 \rightleftharpoons H_2AsO_3^- + H^+$	25	9.23	Drever (1997)
		25	9.17	Nordstrom and Archer (2003)
		25	9.238	Wolthers et al. (2005), 3490
		25	9.22	Raposo et al. (2003)
		25 ± 0.5	9.25 ± 0.05	Zakaznova-Herzog, Seward and Suleimenov (2006)
		50 ± 0.5	8.90 ± 0.05	Zakaznova-Herzog, Seward and Suleimenov (2006)
		100 ± 0.5	8.25 ± 0.05	Zakaznova-Herzog, Seward and Suleimenov (2006)
		150 ± 0.5	7.80 ± 0.09	Zakaznova-Herzog, Seward and Suleimenov (2006)
		200 ± 0.5	7.40 ± 0.17	Zakaznova-Herzog, Seward and Suleimenov (2006)
		250 ± 0.5	7.21 ± 0.23	Zakaznova-Herzog, Seward and Suleimenov (2006)
		300 ± 0.5	7.11 ± 0.14	Zakaznova-Herzog, Seward and Suleimenov (2006)
	$H_2AsO_3^- \rightleftharpoons HAsO_3^{2-} + H^+$	25	12.10	Drever (1997)
		25	10.986	Wolthers et al. (2005), 3490
		25	13.41	Drever (1997)
	$HAsO_3^{2-} \rightleftharpoons AsO_3^{3-} + H^+$	25	13.470	Wolthers et al. (2005), 3490
Inorganic As(V)				
	$H_3AsO_4^0 \rightleftharpoons H_2AsO_4^- + H^+$	25	2.24	Drever (1997)
		25	2.30	Nordstrom and Archer (2003)
		25	2.25	Wolthers et al. (2005), 3490
	$H_2AsO_4^- \rightleftharpoons HAsO_4^{2-} + H^+$	25	6.76	Drever (1997)
		25	6.99	Nordstrom and Archer (2003)
		25	6.83	Wolthers et al. (2005), 3490

(continued overleaf)

Table 2.10 (continued)

Arsenic species	Reaction	Temperature (°C)	pK_a (± 2 SD)	Reference(s)
	$HAsO_4^{2-} \rightleftharpoons AsO_4^{3-} + H^+$	25	11.60	Drever (1997)
		25	11.80	Nordstrom and Archer (2003)
		25	11.520	Wolthers et al. (2005), 3490
H_3AsO_3S				
	$H_3AsO_3S^0 \rightleftharpoons H_2AsO_3S^- + H^+$	25?	3.3	Thilo, Hertzog and Winkler (1970); Schwedt and Rieckhoff (1996)
	$H_2AsO_3S^- \rightleftharpoons HAsO_3S^{2-} + H^+$	25?	7.2	Thilo, Hertzog and Winkler (1970); Schwedt and Rieckhoff (1996)
	$HAsO_3S^{2-} \rightleftharpoons AsO_3S^{3-} + H^+$	25?	11	Thilo, Hertzog and Winkler (1970); Schwedt and Rieckhoff (1996)
$H_3AsO_2S_2$				
	$H_3AsO_2S_2^0 \rightleftharpoons H_2AsO_2S_2^- + H^+$	25?	2.4	Thilo, Hertzog and Winkler (1970); Schwedt and Rieckhoff (1996)
	$H_2AsO_2S_2^- \rightleftharpoons HAsO_2S_2^{2-} + H^+$	25?	7.1	Thilo, Hertzog and Winkler (1970); Schwedt and Rieckhoff (1996)
	$HAsO_2S_2^{2-} \rightleftharpoons AsO_2S_2^{3-} + H^+$	25?	10.9	Thilo, Hertzog and Winkler (1970); Schwedt and Rieckhoff (1996)
MMA (As(V))				
	$H_2AsO_3(CH_3)^0 \rightleftharpoons HAsO_3(CH_3)^- + H^+$	25.0	4.08	Suzuki et al. (2001), 106
		25	4.19	Cox and Ghosh (1994)
	$HAsO_3(CH_3)^- \rightleftharpoons AsO_3(CH_3)^{2-} + H^+$	25.0	8.74	Suzuki et al. (2001), 106
		25	8.77	Cox and Ghosh (1994)
DMA (As(V))				
	$HAsO_2(CH_3)_2^0 \rightleftharpoons AsO_2(CH_3)_2^- + H^+$	25.0	6.12	Suzuki et al. (2001), 106
		25	6.14	Cox and Ghosh (1994)

Following Equation 2.23, the dissociation constant for Reaction 2.41 is further described as follows:

$$K_a = [H_2AsO_3^-][H^+]/[H_3AsO_3^0] = 10^{-9.2} \tag{2.42}$$

Like pH and pK_{sp}, any dissociation constant may be conveniently written as a pK_a value, where $pK_a = -\log_{10} K_a$. That is, the pK_a of Reaction 2.41 at 25 °C and 1 bar pressure is about 9.2 (Table 2.10). Furthermore, like pH, pK_a values are often sensitive to temperature. In hydrothermal waters, for example, the pK_a value of Reaction 2.43 declines with increasing temperature so that the value approaches 7.11 at 300 °C (Zakaznova-Herzog, Seward and Suleimenov, 2006), 1936; Table 2.10.

Dissociation constants for weak acids may be measured by titrating salts of the acids with *strong acids*, such as hydrochloric acid (HCl). By knowing all of the relevant dissociation constants for a weak acid, the fractional concentrations of the various dissolved species of the acid may be determined at any pH as shown in Figures 2.3 and 2.4. To avoid the tedious calculations, MINTEQA2 or other aqueous chemistry programs may be used.

The dissociation constant for Reaction 2.41 indicates that at a pH of 9.2 the activities of $H_3AsO_3^0$ and $H_2AsO_3^-$ are equal. That is, in 25 °C water at 1 bar pressure, the dissociation constant indicates that $H_3AsO_3^0$ is the dominant dissolved form of arsenious acid up to a pH of 9.2 (Figure 2.3). As discussed in Chapter 7 and illustrated in Figure 2.3, the dominance of generally unreactive $H_3AsO_3^0$ in the pH range of most natural waters i.e. pH 4–9; (Krauskopf and Bird, 1995), 30 and many wastewaters explains why As(III) is difficult to remove from water by sorption, ion exchange, and other treatment technologies. In contrast, reactive As(V) oxyanions are common under pH 4–9 conditions and are readily removed from water by a wide variety of sorbents and other chemicals (Figure 2.4; Chapter 7). Thus, most water treatment technologies for arsenic require the preoxidation of any $H_3AsO_3^0$.

2.7.5 Eh-pH diagrams, and their limitations

The negative $\log_{10}$ molar activity of H^+ in an aqueous sample is its pH. Measurements of pH are easily done with electrodes and are usually accurate to at least ± 0.1 unit. In contrast, the *redox potential* or Eh of an aqueous sample is a complex parameter that often involves many separate redox reactions (Drever, 1997), 135–136; (Cherry *et al.*, 1979). Many of these reactions are too sluggish to attain equilibrium in a timely fashion or are hindered from equilibrium by biological activity (Chapter 3). Unless all redox reactions are at equilibrium, a single accurate Eh value cannot be obtained for a water sample. In many cases, the Eh of an aqueous sample can only be estimated by measuring dissolved oxygen, the concentrations of chemical species that are sensitive to reduction/oxidation (for example, Fe(II)), or redox couples, such as Fe(II)/Fe(III), As(III)/As(V), or SO_4^{2-}/HS^- (Yan, Kerrich and Hendry, 2000), 2645. Holm and Curtiss (1989) further argue that the As(V)/As(III) ratio by itself is not a good indicator of the overall Eh of an aqueous sample because of redox disequilibria. As(V)/As(III), Fe(III)/Fe(II), and other redox parameters should be used together to estimate redox conditions. In general, the redox boundary between As(III) and As(V) is approximately +300 mV at pH 4 and −200 mV at pH 9 (O'Day, 2006), 77. However, metastable As(III) and As(V) forms are very common in natural waters (Chapter 3).

Platinum electrodes are sometimes utilized to obtain Eh measurements on water samples. However, they are often ineffective, especially with natural waters (Drever, 1997), 136, 180–182; (Lindberg and Runnells, 1984). Besides problems with chemical disequilibria in many natural waters, platinum electrodes usually fail to quickly respond to the kinetics of the couples that commonly control redox conditions in water samples, including: O_2-H_2O, SO_4^{2-}-H_2S, CO_2-CH_4, NO_3^--N_2, and N_2-NH_4^+ (Drever, 1997), 136. For arsenic, large discrepancies may exist between redox values calculated from the As(V)/As(III) couple and Eh measurements with platinum electrodes (Ryu *et al.*, 2002), 2989–2990. Additionally, when collecting groundwaters or surface waters from reducing environments, extreme care must be taken to prevent air from contaminating and altering the redox chemistry of the samples.

Eh-pH diagrams are sometimes used to predict or describe the major dissolved species and precipitates that should exist at equilibrium in aqueous solutions, including: groundwaters, surface waters, laboratory solutions, and porewaters from soils, sediments, or rocks. However, as previously described, many natural aqueous systems are not at equilibrium and they often contain *metastable* species that are not predicted by Eh-pH diagrams. *Metastable species* refer to compounds, other substances, or ions that are present under redox, pH, pressure, temperature, or other conditions where chemical equilibrium indicates that they should be unstable and absent. Many metastable species (such as As(III) in oxygenated seawater) result from biological activity.

Eh-pH diagrams are usually designed for a specific temperature and pressure (typically 25 °C and 1 bar pressure with O_2 and H_2 gases) and a narrow set of often simplistic chemical conditions. The two diagrams shown in Figure 2.5 represent simple aqueous systems containing only water, 10^{-6} M total arsenic, and ± 0.001 M total sulfide at 25 °C and 1 bar pressure (Vink, 1996). Each field on the diagram lists the arsenic species that should be prominent at equilibrium under those Eh and pH conditions. As seen in Figure 2.5a, the diagram predicts that $H_3AsO_3^0$ is the dominant arsenic species at a pH of 9.0 and an Eh of -0.30 V. Although not quite as abundant as $H_3AsO_3^0$, the discussions in Section 2.7.4 indicate that considerable $H_2AsO_3^-$ should also be present. At a slightly higher pH of 9.2, the activities of $H_3AsO_3^0$ and $H_2AsO_3^-$ should be equal. Although not shown in Figure 2.5b, detailed calculations with MINTEQA2 further indicate that at a pH of 9.0 and an Eh of -0.30 V, trace amounts of other arsenic species should occur (including: $HAsO_3^{2-}$, AsO_3^{3-}, and $H_4AsO_3^+$) along with H_2S, HS^-, and S^{2-}. Thioarsenic species are also expected. However, an absence of thermodynamic data prevents their activities from being estimated.

Variations in temperature, pressure, gas composition, and the addition of new components (such as iron or calcium) would shift the stability fields in the diagrams of Figure 2.5, and create new chemical species and eliminate others. As shown in Figure 2.5, the addition of 0.001 M total sulfide to an aqueous solution with 10^{-6} M total arsenic decreases the size of the $H_3AsO_3^0$ stability field, eliminates As^0, and produces orpiment (As_2S_3) and realgar (AsS) under strong reducing (-Eh) conditions. If the arsenic concentration of the water sample is exceptionally high (above about 0.16 M), arsenolite and claudetite could also precipitate in the $H_3AsO_3^0$ field of Figure 2.5a (Vink, 1996), 25–26.

Faure (1998); Krauskopf and Bird (1995), and other geochemistry textbooks provide examples and other necessary details for preparing Eh-pH diagrams. Software, such as Geochemist's Workbench, is also available for producing the diagrams. However, individuals using Eh-pH diagrams should recognize that the diagrams are likely to oversimplify and possibly inaccurately represent the chemistry of natural waters.

2.7.6   Sorption, ion exchange, precipitation, and coprecipitation of arsenic in water

2.7.6.1   *Introduction*

Once arsenic dissolves in natural water, it may remain in solution for an extended period of time or participate sooner in abiotic or biotic reactions that remove it from solution. Depending upon the pH, redox conditions, temperature, and other properties of an aqueous solution and its associated solids, dissolved arsenic may precipitate or coprecipitate. Arsenic may also sorb onto solid materials, usually through ion exchange. Due to their importance in understanding the behavior of arsenic in natural environments (Chapter 3) and their applications in water treatment (Chapter 7), the sorption, ion exchange, precipitation, and coprecipitation of arsenic have been the subjects of numerous investigations.

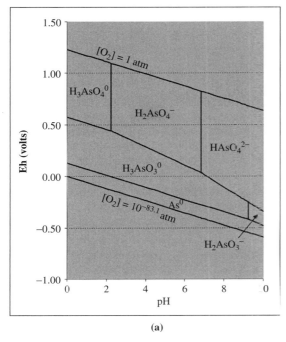
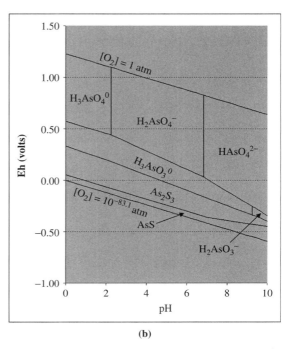

(a)   (b)

Figure 2.5   *Eh-pH diagrams for major arsenic species in water at 25°C and 1 bar pressure containing 10^{-6} M total arsenic in diagram A and 10^{-6} M total arsenic and 0.001 M total sulfide in diagram B. (Reprinted from Vink, B.W., Stability relations of antimony and arsenic compounds in the light of revised and extended Eh-pH diagrams. Chemical Geology,* **130***, 1–2. Copyright 1996 with permission of Elsevier.)*

2.7.6.2   Adsorption, absorption, sorption, and ion exchange

2.7.6.2.1   Introduction   *Adsorption* refers to the removal of ions and other dissolved species from liquids or gases by their accumulation on the surfaces of solid materials, where the adsorbed species is not a major component in the internal chemistry of the solid. The solid materials are often referred to as *adsorbents* or *sorbents*. Sometimes the adsorbed *solute* is called the *adsorbate* (Krauskopf and Bird, 1995), 145. As discussed below, sorbent surfaces often acquire charges that attract and adsorb oppositely charged solutes in associated solutions. Adsorption usually involves ion exchange, where the adsorbing species replaces another species on the surface of the adsorbent (Eby, 2004), 345. In the following example involving an aluminum-bearing surface in an aqueous solution containing arsenic, $H_2AsO_4^-$ displaces OH_2^+ from the aluminum surface in an ion-exchange reaction:

$$\vert Al - OH_2^+ + H_2AsO_4^- \rightarrow \vert Al - H_2AsO_4 + H_2O \qquad (2.43)$$

The displaced species enters the solution as a water *molecule*.

Absorption is the assimilation of a chemical species into the interior of a solid substance. Absorption may include the migration of solutes into the internal pores of a solid material (Fetter, 1993), 117 or the migration or exchange of atoms within the crystalline structure of a mineral (Krauskopf and Bird, 1995), 150. Some researchers use the generic term '*sorption*' to refer to a treatment method where both adsorption and absorption may be involved or if adsorption and absorption cannot be distinguished. Sorption and ion exchange have many important roles in immobilizing arsenic in natural environments (Chapter 3). They also have numerous applications in technologies that treat arsenic in water, gases, sediments, soils, and wastes (Chapter 7).

2.7.6.2.2   Formation of surface charges   Charges often form on the surfaces of solid materials in water and other liquids. The charges may result from substitutions in the crystal lattices of minerals. In particular, *clay minerals* often have negative charges, which result from the substitution of Al^{3+} for Si^{4+} (Krauskopf and Bird, 1995), 137. Chemical reactions between the sorbents and surrounding solutions may also produce surface charges. For example, hydroxides typically form on the surfaces of aluminum, iron(III), and other metal oxides when they come into contact with water (Krauskopf and Bird, 1995), 137, 139. Under either acidic or alkaline conditions, the surfaces may become charged. In acids, excess H^+ may produce positive charges on surface hydroxides through the following reaction (Krauskopf and Bird, 1995), 137:

$$\vdash OH + H^+ \rightarrow \vdash OH_2^+ \tag{2.44}$$

Under more alkaline conditions (Krauskopf and Bird, 1995), 138, hydrogens are removed from the hydroxides and the surfaces become negative:

$$\vdash OH + OH^- \rightarrow \vdash O^- + H_2O \tag{2.45}$$

The overall charge of a sorbent would then be the sum of any permanent structural charges and any charges resulting from surface reactions with the aqueous solution (Krauskopf and Bird, 1995), 140.

The ZPC of a solid is the pH of an aqueous solution in contact with the solid when the solid has a net surface charge of zero. The ZPC depends on the composition of the solid and the concentration and chemistry of the electrolytes in the aqueous solution. In situations where surface charges are only controlled by the adsorption of OH^- or H^+, the ZPC is the *isoelectric point* (Faure, 1998), 218–219; (Langmuir, 1997), 350; (Drever, 1997), 93. Unfortunately, isoelectric points and ZPCs are not consistently defined or utilized in the literature. That is, some authors refer to the '*isoelectric point*' of particular solid sample when they are actually referring to its ZPC as the term is defined in this chapter.

Each type of mineral or other solid substance has its own isoelectric point or range of isoelectric points (Table 2.11). For compounds that have highly variable compositions (e.g. ferrihydrites in Table 2.11 and Chapter 3), isoelectric points differ with individual specimens. ZPCs will also vary with the electrolyte composition of the aqueous solution (Faure, 1998), 219.

At pH values below the isoelectric point or ZPC of a solid, the chemistries of the solid and water result in an overall positively charged surface. The positively charged surface may readily attract and adsorb As(V) oxyanions (Stollenwerk, 2003), 77; Chapters 3 and 7. Above the isoelectric point or ZPC, the net charge on the surface of the solid becomes negative and the adsorption of arsenic oxyanions diminishes (Stollenwerk, 2003), 74. Even if a surface has a net negative charge (i.e. it's above its isoelectric point or ZPC), enough residual positive surface charges might still be present to attract some arsenic oxyanions. Furthermore, because of its neutral charge, $H_3AsO_3^0$ is more readily able to approach a negatively charged surface than an As(V) oxyanion. As shown in the following reaction, $H_3AsO_3^0$ can adsorb:

$$\vdash O^- + H_3AsO_3^0 \rightarrow \vdash H_2AsO_3^0 + OH^- \tag{2.46}$$

A charged surface of a solid in an aqueous solution results in the formation of adsorption complexes, which include the fixed *Stern* layer and the diffuse or *Gouy* layer (Figure 2.6). Ions in the aqueous solution of opposite charge to the surface charge accumulate on the solid surface and create the Stern layer. In Figure 2.6, anions form a Stern layer on the surface of a positively charged mineral. The charged surface

Table 2.11   *Isoelectric points and zero points of charge of various solid materials at about 25°C.*

Material	Isoelectric point	Zero point of charge (ZPC)	Reference(s)
Alumina (γ-Al_2O_3)	—	8.1 (NaNO$_3$ electrolyte)	Cox and Ghosh (1994)
Aluminum (oxy)(hydr)oxides (amorphous)	—	9.4 (NaCl electrolyte)	Goldberg and Johnston (2001)
Carbon black	—	6.4 (NaNO$_3$ electrolyte, 20 ± 1°C)	Borah *et al.* (2008)
Carbon black (modified with sulfuric acid)	—	3.5 (NaNO$_3$ electrolyte, 20 ± 1°C)	Borah *et al.* (2008)
Corundum (Al_2O_3)	9.1	—	Drever (1997), 94
Ferrihydrite ('two-line', approximately $Fe_5HO_8 \cdot 4H_2O$)	—	5.77	Rhoton and Bigham (2005)
Ferrihydrite ('two-line')	—	8.5 (NaCl electrolyte)	Jain, Raven and Loeppert (1999)
Gibbsite (γ-$Al(OH)_3$)	~9	—	Drever (1997), 94
Goethite (α-$FeO(OH)$)	6–7	—	Drever (1997), 94
Goethite (α-$FeO(OH)$)	—	8.5 (NaCl electrolyte)	Gao and Mucci (2001)
Hematite (Fe_2O_3)	Mostly 6–7	—	Drever (1997), 94
Hematite (Fe_2O_3, Nigerian sample)	—	6.6	Ofor (1995), 347
Illite clay	—	2.5 (NaCl electrolytes)	Zhuang and Yu (2002), 624
Iron (oxy)(hydr)oxides (amorphous, coated on activated alumina)	—	6.9 ± 0.3 (1 SD)	Hlavay and Polyák (2005)
Iron (oxy)(hydr)oxides (amorphous)	—	8.0 (NaNO$_3$ electrolyte)	Cox and Ghosh (1994)
Iron (oxy)(hydr)oxides (amorphous)	—	8.5 (NaCl electrolyte)	Goldberg and Johnston (2001)
Iron (oxy)(hydr)oxides (crystalline hematite with hydrated surfaces)	—	9.2 (NaNO$_3$ electrolyte)	Ko *et al.* (2007)
Kaolinite ($Al_2Si_2O_5(OH)_4$)	—	3.7 (NaCl electrolytes)	Zhuang and Yu (2002), 624
Magnetite (Fe_3O_4)	6.5	—	Drever (1997), 94
Montmorillonite clay	—	2.6 (NaCl electrolytes)	Zhuang and Yu (2002), 624
Quartz	2.0	—	Drever (1997), 94
Schwertmannite	7.2	—	Jönsson *et al.* (2005)
Titanium dioxide (anatase)	5.8	5.8	Jing *et al.* (2005)
Titanium dioxide (99% anatase; Hombikat UV 100)	—	6.2 (NaNO$_3$ electrolyte)	Dutta *et al.* (2004), 271
Titanium dioxide (~80% anatase; ~20% rutile; Degussa P25)	—	6.9 (NaNO$_3$ electrolyte)	Dutta *et al.* (2004), 271
Vernadite (δ-MnO_2)	2	—	Lau and Chiu (2007); Drever (1997)

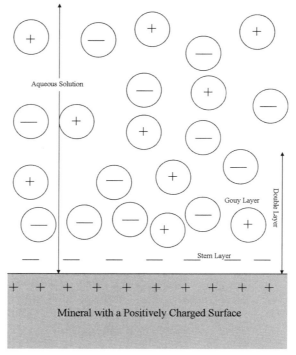

Figure 2.6 *Stern and Gouy layers above a mineral surface in an aqueous solution. Circles in the Gouy layer represent water molecules surrounding the ions.*

also attracts oppositely charged ions in the Gouy layer (Krauskopf and Bird, 1995), 141–144. However, the Gouy layer ions do not come into contact with the adsorbent surface, but are separated from it by the Stern layer. Partial charges on water molecules are also attracted by the ions in the Gouy layer and surround the ions.

In the Gouy layer, the imbalance between cations and anions exponentially decreases with distance away from the solid surface. That is, as shown in Figure 2.6, the number of cations in the Gouy layer increases with distance away from the positively charged mineral surface. Unlike the Stern layer, ions may readily move between the Gouy layer and the surrounding aqueous solution (Krauskopf and Bird, 1995), 141. The size and composition of Stern and Gouy layers are affected by the composition and surface area of the adsorbent and the pH of the aqueous solution (Krauskopf and Bird, 1995), 145. The concentration of electrolytes also affects the size of the Gouy layer. The Gouy layer is much thicker in dilute waters than in saline waters, which contain a lot of electrolytes (Krauskopf and Bird, 1995), 141. Together, the Stern and Gouy layers form a *double layer*. Away from the double layer, the cationic and anionic charges in an aqueous solution are essentially balanced (Figure 2.6).

The Stern layer consists of *outer-* and *inner-sphere complexes* (Figure 2.7). In outer-sphere Stern complexes, the absorbate is indirectly attached to the adsorbent surface through one or more water molecules [Figure 2.7; (Krauskopf and Bird, 1995), 144; (Gräfe and Sparks et al., 2006)). Outer-sphere adsorption is also called *physisorption*. Inner-sphere Stern complexes form from the covalent or ionic bonding of chemical species directly onto the surfaces of solid materials ((Krauskopf and Bird, 1995), 144; Figure 2.7; (Gräfe and Sparks *et al.*, 2006)). The formation of inner-sphere complexes is also called *chemisorption*. In some cases, complexes may grow into three-dimensional surface precipitates (Krauskopf and Bird, 1995), 142.

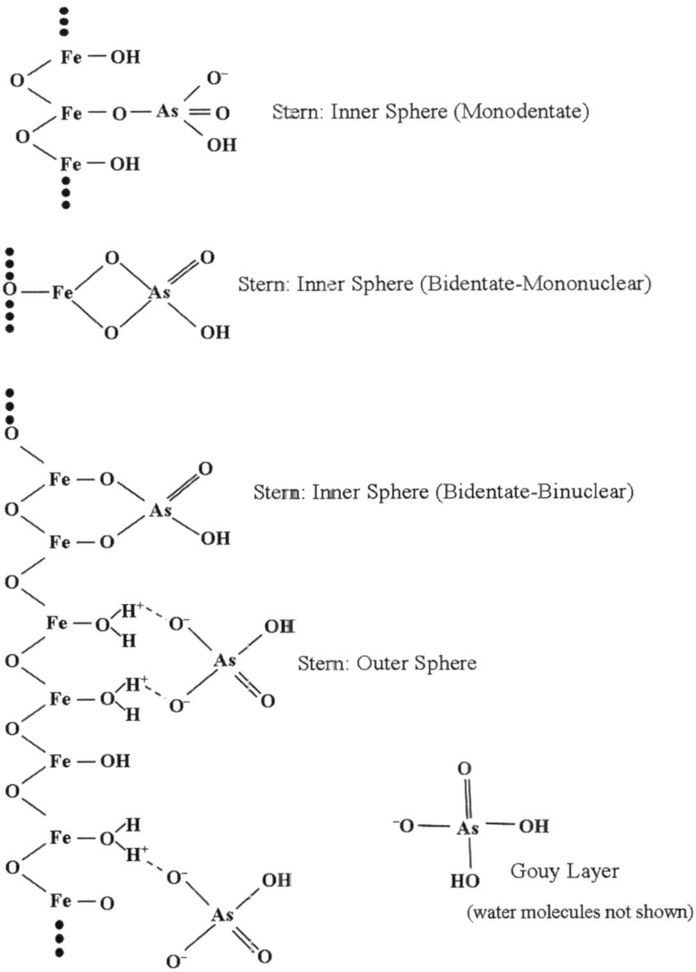

Figure 2.7   *Stern (inner sphere and outer sphere) and Gouy arsenic adsorption complexes associated with the surfaces of iron oxide minerals.*

Stern inner-sphere adsorption complexes are further divided into *monodentate*, *bidentate-mononuclear*, and *bidentate-binuclear* types. In monodentate complexes, each adsorbed species attaches onto only one atom on the adsorbent surface. In Figure 2.7, a single oxygen bridges between a surface iron atom and an absorbed As(V) atom. Bidendate-mononuclear complexes consist of two atoms bridging between an atom of the adsorbed species and an adsorbent metal atom. Figure 2.7 shows two oxygens bridging between iron and As(V) atoms to form a bidendate-mononuclear complex. The bidentate-binuclear complex consists of one adsorbed atom bonding to two separate metal oxides. In Figure 2.7, an As(V) has adsorbed onto two separate pairs of iron and oxygen atoms.

Several possible inner-sphere reactions involving the surface replacement of hydroxides with various arsenic species are possible (Stollenwerk, 2003). 73–74:

$$\models OH + H_3AsO_4^0 \rightarrow \models H_2AsO_4^0 + H_2O \tag{2.47}$$

$$\vdash OH + H_2AsO_4{}^- \rightarrow \vdash HAsO_4{}^- + H_2O \tag{2.48}$$

$$\vdash OH + HAsO_4{}^{2-} \rightarrow \vdash AsO_4{}^{2-} + H_2O \tag{2.49}$$

$$\vdash OH + H_3AsO_3^0 \rightarrow \vdash H_2AsO_3^0 + H_2O \tag{2.50}$$

$$\vdash OH + H_2AsO_3{}^- \rightarrow \vdash HAsO_3{}^- + H_2O \tag{2.51}$$

$$\vdash OH + HAsO_3{}^{2-} \rightarrow \vdash AsO_3{}^{2-} + H_2O \tag{2.52}$$

In the above reactions, $\vdash OH$ represents a solid surface with its hydroxide functional group. Depending upon the pH of the solution and the chemistry of the adsorbent, $\vdash OH_2{}^+$ or $\vdash O^-$ may be significant. The above reactions are also potentially reversible.

Adsorption in Reactions 2.47–2.52 also involves ion exchange (Eby, 2004), 345. During the formation of inner-sphere complexes, adsorbing arsenic commonly replaces hydroxides or other chemical species on the surface of the adsorbent. Complexes in Stern outer-sphere and Gouy layers are also susceptible to ion exchange, especially because they are weakly adsorbed (Krauskopf and Bird, 1995), 150.

In the outer-sphere Stern complexes of Figure 2.7, the partially negatively charged oxygens of water molecules attach onto metal atoms on the adsorbent surface. The positive partial charges on the hydrogens of the waters then electrostatically attract anions from the solution. In other words, the adsorbates form weak hydrogen bonds with the water molecules on the surface of the adsorbent (Figure 2.7; (Prasad, 1994), 143). Unlike inner-sphere, outer-sphere adsorption is weak and usually not associated with a specific crystalline site on a mineral surface (Krauskopf and Bird, 1995), 144.

When compared with inner-sphere complexes, outer-sphere complexes and Gouy layers are more sensitive to the ionic strengths of aqueous solutions (Krauskopf and Bird, 1995), 146–147; (Stollenwerk, 2003), 74. In general, the adsorptive strengths of outer-sphere complexes and Gouy layers decrease as the ionic strength of the solution increases, whereas the relatively strong inner-sphere complexes are either not affected or are actually strengthened by increases in ionic strength (Stollenwerk, 2003), 74; (Goldberg and Johnston, 2001), 205; (Krauskopf and Bird, 1995), 141–148. That is, because inner-sphere complexes adsorb directly onto the surface, they are more resistant to the effects of ionic strength than weakly bound complexes and layers (Stollenwerk, 2003), 74; (Krauskopf and Bird, 1995), 141–148.

By using spectroscopic techniques and monitoring changes in pH and ionic strength, inner- and outer-sphere complexes may be distinguished. For example, Goldberg and Johnston (2001) used *Raman* and *Fourier transform infrared (FTIR) spectroscopy* methods to confirm that As(V) species form inner-sphere surface complexes on both amorphous aluminum and iron 'oxides' ((oxy)hydroxides?). Other studies suggest that inner-sphere complexes are also responsible for the adsorption of As(V) on ferrihydrite and goethite (α-FeO(OH), orthorhombic), methylarsenate on ferrihydrite and aluminum oxide, and As(III) on kaolinite ($Al_2Si_2O_5(OH)_4$), illite (a complex clay mixture), and amorphous aluminum hydroxide (Stollenwerk, 2003), 74. In contrast, As(III) exists as both inner- and outer-sphere complexes on amorphous iron 'oxides' (Goldberg and Johnston, 2001).

Additional information on adsorption mechanisms and models is in Stollenwerk (2003), 93–99 and Prasad (1994). Foster (2003) also discusses in considerable detail how As(III) and As(V) may adsorb and coordinate on the surfaces of various iron, aluminum, and manganese (oxy)(hydr)oxides. In adsorption studies, relevant laboratory parameters include: arsenic and adsorbent concentrations, adsorbent chemistry and surface area, surface site densities, and the equilibrium constants of the relevant reactions (Stollenwerk, 2003), 95. Once laboratory data are available, MINTEQA2 (Allison, Brown and Novo-Gradac, 1991), PHREEQC (Parkhurst and Appelo, 1999), and other geochemical computer programs may be used to derive the adsorption models.

2.7.6.2.3 Adsorption of arsenic Iron, aluminum, and manganese (oxy)(hydr)oxides widely occur as sorbents and coatings on other solid materials in nature. They are often important in adsorbing arsenic from water ((Stollenwerk, 2003), 73; Chapter 3). Below the ZPCs of the (oxy)(hydr)oxides, the presence of abundant $\vdash OH_2^+$ are responsible for the net positive charges. Acidic conditions may also partially dissolve sorbent surfaces and increase surface areas and the number of adsorption sites (Gräfe and Sparks, 2006), 78. High surface areas with net positive charges attract As(V) oxyanions and form strong inner-sphere Stern layers. As pH conditions become more alkaline and rise above the ZPCs of the (oxy)(hydr)oxides, less H^+ are available and surfaces with excess $\vdash O^-$ are more common (Eby, 2004), 341. The pH of a solution associated with an absorbent affects both the surface charges and the charges of the dissolved arsenic species, which would control arsenic adsorption (Stollenwerk, 2003), 77.

As discussed in Section 2.7.3, inorganic As(III) oxyanions only become abundant once the pH values of aqueous solutions rise above 9.2 (Figure 2.3). Most compounds have ZPCs and isoelectric points below 9.2 (Table 2.11). So, their surfaces tend to be negative under very alkaline conditions, which readily repel As(III) oxyanions. As shown in Reaction 2.46, some $H_3AsO_3^0$ adsorbs onto negatively charged surfaces near pH neutral conditions. However, the adsorption of $H_3AsO_3^0$ is not as effective as the adsorption of As(V) oxyanions onto positively charged sorbent surfaces, which is an important factor in natural adsorption and water treatment (Chapters 3 and 7).

Goethite is common in natural environments and water treatment processes (Chapters 3 and 7). Studies in Manning, Fendorf and Goldberg (1998) suggest that As(III) adsorbs onto goethite through the formation of bidentate-binuclear complexes (Figure 2.7). The sorption of As(V) on goethite occurs in two distinct phases, which are an initial fast phase that finishes in less than 5 minutes and a slower phase that lasts for at least several hours (Luengo, Brigante and Avena, 2007). The initial fast phase probably involves the adsorption of As(V) through the formation of inner-sphere complexes ((Zhang and Stanforth, 2005; Fendorf *et al.*, 1997); Figure 2.7). Grossl *et al.* (1997), 324 further argue that the inner-sphere surface complexes are monodentate. Inner-sphere monodentate complexes could rapidly develop as $H_2AsO_4^-$ exchanges for hydroxyl functional groups on the goethite surfaces ((Grossl *et al.*, 1997); e.g. Reaction 2.46).

Several *hypotheses* have been proposed to explain the cause(s) of the second slower phase (Luengo, Brigante and Avena, 2007; Zhang and Stanforth, 2005; Fendorf *et al.*, 1997; Grossl *et al.*, 1997). For example, Luengo, Brigante and Avena (2007), 359 suggested that the slower phase primarily involves the diffusion of As(V) adsorbed during the fast phase into the pores of the goethite grains. They noted that the rate of the second slower phase depends more on how much As(V) is sorbed during the initial fast phase than how much As(V) remains in the water.

2.7.6.2.4 Adsorption isotherms Laboratory adsorption studies measure the partitioning of arsenic between an absorbent and its host aqueous solution. The studies typically consist of mixing a known mass of an adsorbent into several aqueous solutions containing different dissolved concentrations of arsenic. Temperature, pH, and the other properties of the mixtures are usually held constant and efforts are made to prevent the dissolved arsenic from coprecipitating or precipitating in the mixtures. The mixtures are then agitated until they attain equilibrium. After equilibrium is achieved, the water and solids of each mixture are separated and analyzed to determine the amount of arsenic that was adsorbed and the amount that remained dissolved in the water of each mixture. The effectiveness of the absorbent could be evaluated by listing the analytical results in tables. However, the results are usually best visualized with graphs and, if possible, fitting the data distributions to mathematical equations. The adsorbed arsenic concentrations may be plotted on the y axis and the concentrations remaining in the aqueous solutions on the x axis (Figure 2.8). The concentrations may be given in micrograms of arsenic/gram of aqueous solution or adsorbent ($\mu g\,g^{-1}$), but molal (m), molar (M), and micromolar (μM) are more often used. A line and

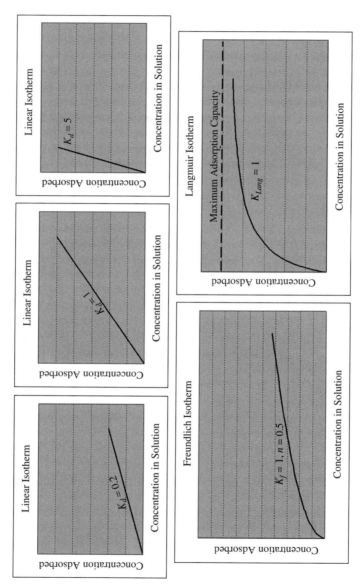

Figure 2.8   Graphs of linear, Freundlich, and Langmuir isotherms.

associated equation may be derived to fit the data distribution on the graph. The line on the graph is often called an *adsorption isotherm* because the line represents data points collected at a constant temperature (most often 25 °C). The simplest model that could fit the data distribution is a straight line, which may be described with this famous geometry equation $y = mx + b$. Obviously, no adsorption of arsenic would occur if the species is absent from an aqueous solution. Therefore, any linear adsorption isotherm would pass through the origin of the graph and $b = 0$. The linear equation then reduces to the following:

$$C_{ads} = K_d C_{soln} \qquad (2.53)$$

where:

C_{ads} = the concentration of adsorbed arsenic (y axis),

K_d = distribution coefficient and

C_{soln} = the arsenic concentration remaining in the solution (x axis).

The slope of this line is the *distribution coefficient* (K_d), which is the ratio of the arsenic concentration on the adsorbent (C_{ads}) to the concentration of the associated remaining arsenic in the aqueous solution (C_{soln}). With each linear adsorption isotherm, K_d has only one value. That is, a linear distribution indicates that the partitioning of arsenic between the adsorbent and the solution is constant over the given range of arsenic concentrations (Eby, 2004), 221. If both concentrations (C_{ads} and C_{soln}) are in the same units (such as molal), K_d is unitless. However, if the adsorbed concentration is given in molal and the dissolved concentration is molar, then K_d has the units of liter/kilogram.

Linear adsorption isotherms may occur with solutions that have very low arsenic concentrations. However, in most cases, the fraction of arsenic that is adsorbed decreases as the arsenic concentration of the host aqueous solution increases (Langmuir, 1997), 354. In such situations, the adsorption isotherm would not fit a straight line. The two most common nonlinear adsorption isotherms are the *Freundlich* and *Langmuir*. With aqueous solutions containing high (e.g. milligrams/liter) concentrations of arsenic, readily available adsorption sites on the surfaces of adsorbents may be quickly occupied (Drever, 1997), 89. Once the readily available surface sites are filled, adsorption will begin to occur on the less numerous sites with weaker binding energies, such as on the corners and edges of the solids. The lack of readily available and high bonding energy adsorption sites for arsenic causes the distribution to curve and may produce a Freundlich isotherm (Figure 2.8). Notice that at very low concentrations, the Freundlich isotherm is almost linear (Figure 2.8). The equation for a Freundlich isotherm is (Eby, 2004), 221:

$$C_{ads} = K_f (C_{soln})^n \qquad (2.54)$$

The Freundlich equation is similar to a linear equation, expect for the presence of the exponent n. For linear distributions, $n = 1$. With Freundlich isotherms usually have $n < 1$, which causes the adsorption isotherm to curve downward at higher concentrations as the readily available adsorption sites are filled and lower proportions of the arsenic from the aqueous solutions are adsorbed (Figure 2.8). The distribution coefficient for a Freundlich isotherm is often written as K_f to stress that the isotherm is not linear (Drever, 1997), 89.

All adsorbents have upper limits to the amount of arsenic that they can adsorb from air, water, or other fluids. That is, there is a finite number of adsorption sites on each gram of adsorbent. The *maximum adsorption capacity*, which is often measured in molal, represents the highest concentration of a solute (such as arsenic) that can be adsorbed by a given mass of a particular adsorbent. The maximum adsorption capacity is routinely obtained from laboratory experiments and measurements, and is closely related to the *cation exchange capacity* (cec) or *anion exchange capacity* (aec) of the materials. The cec or aec provide measurements of the number of negatively or positively charged sites, respectively, on a solid sample. Grinding or using finer particle sizes can often increase the maximum adsorption capacity of a material because surface areas and the number of adsorption corners and edges typically increase with decreasing particle size. There are also laboratory procedures for measuring the surface areas of adsorbents, which are useful in efforts to predict maximum adsorption capacities.

With a Langmuir isotherm, the arsenic concentrations of the aqueous solutions are high enough that they approach the maximum arsenic adsorption capacity of the adsorbent. In Figure 2.8, the dashed line at the top of the Langmuir isotherm graph represents the maximum adsorption capacity (C_{max}) for a theoretical adsorbent. As the arsenic concentrations of the aqueous host solutions are increased, available adsorption sites fill and the Langmuir isotherm converges with the line representing the maximum adsorption capacity. The equation for a Langmuir isotherm may be written as (Drever, 1997), 90:

$$C_{ads} = C_{max}((K_{Lang}C_{soln})/(1 + (K_{Lang}C_{soln})))$$ (2.55)

where:

C_{ads} = adsorbed concentration,

C_{max} = maximum adsorption capacity,

C_{soln} = concentration remaining in the aqueous solution, and

K_{Lang} = distribution coefficient for the Langmuir isotherm.

Langmuir isotherms indicate that there are limits to the amount of arsenic that an adsorbant may adsorb. Knowing these limits are important in developing effective treatment technologies for removing arsenic from water (Chapter 7) and determining the ability of soils, sediments, or other natural materials to remove arsenic from natural waters or acid mine drainage (Chapter 3).

2.7.6.2.5 Interferences with arsenic adsorption and ion exchange

Dissolved organics and anions may interfere with arsenic adsorption and ion exchange in both natural environments and water treatment systems. In some cases, chemical species directly compete with arsenic for adsorption sites. They may also desorb and replace arsenic. Vanadium is one element that could interfere with the adsorption of arsenic onto mineral surfaces. In most cases, vanadium is not abundant in water. However, alkaline (pH 7.0–8.8) groundwaters in the loess aquifers of La Pampa, Argentina contain up to $12\,mg\,L^{-1}$ of vanadium (Smedley *et al.*, 2005). The vanadium readily hinders the sorption of As(V) onto iron (III) (oxy)(hydr)oxides (Chapter 3).

Phosphorus and arsenic are both group 15 elements. The main phosphorus oxyanion, phosphate (PO_4^{3-}), has the same charge and tetrahedral configuration as inorganic As(V), which allow phosphate to out compete As(V) for adsorption sites on clay minerals and various iron and aluminum compounds over a wide range of pH conditions (Stollenwerk, 2003), 85; (Gao and Mucci, 2001); (Su and Puls, 2003), 2582. Phosphate is also capable of desorbing arsenic from *humic acids* and some mineral surfaces (Stollenwerk, 2003), 91; (Mok and Wai, 1994), 111; (Lafferty and Loeppert, 2005), 2120.

Besides phosphate, *silica* is known to commonly compete with As(V) for sorption/ion exchange sites on a wide variety of iron(III) and aluminum compounds (Clifford and Ghurye, 2002), 227; (Su and Puls, 2003), 2582; (Holm, 2002; Smith and Edwards, 2005; Zhang *et al.*, 2004; McNeill, Chen and Edwards, 2002), 146. Silica may directly compete with arsenic for sites or polymerize on adsorbent surfaces and eliminate surface charges that are favorable for arsenic adsorption (Stollenwerk, 2003), 89.

Dissolved organic compounds may compete with arsenic for adsorption and ion-exchange sites on a variety of sorbents (Stollenwerk, 2003), 89. Specifically, *fulvic acid* is known to interfere with As(V)

sorption on aluminum compounds (Stollenwerk, 2003), 89. Over a pH range of 4–8, fulvic acid also hinders the sorption of As(III), As(V), MMA(V), and DMA(V) on iron compounds (Stollenwerk, 2003), 89; (Simeoni, Batts and McRae, 2003). Furthermore, fulvic acids may form soluble complexes with arsenic in water (Mok and Wai, 1994), 110. Carbonate species resulting from the biodegradation of organic matter also often interfere with the sorption of As(III) and may even desorb As(III) from (oxy)(hydr)oxide surfaces (Anawar, Akai and Sakugawa, 2004).

Most natural waters in contact with adsorbents have more sulfate than arsenic. Rather than directly completing with arsenic for surface sites, sulfate or other abundant chemical species may change surface chemistries so that they are less favorable for arsenic adsorption and ion exchange (Stollenwerk, 2003), 85. If the concentrations are high enough, sulfate may accumulate on the surfaces of the adsorbents (such as iron compounds; (Leist, Casey and Caridi, 2000), 131) and dramatically increase the number of negative charges. Greater numbers of negative surface charges would tend to repel arsenic oxyanions and hinder their adsorption or ion exchange.

2.7.6.3 Precipitation and coprecipitation of arsenic

Precipitation refers to dissolved species (such as As(V) oxyanions) in water or other liquids reacting with other dissolved species (such as Ca^{2+}, Fe^{3+}, or manganese cations) to form solid insoluble reaction products. Precipitation may result from evaporation, oxidation, reduction, changes in pH, or the mixing of chemicals into an aqueous solution. For example, As(V) oxyanions in acid mine drainage could flow into a nearby pond and react with Ca^{2+} to precipitate calcium arsenates. The resulting precipitates may settle out of the host liquid, remain suspended, or possibly form *colloids*. Like sorption, precipitation is an important process that affects the movement of arsenic in natural environments and in removing arsenic from contaminated water (Chapters 3 and 7).

In *coprecipitation*, a *minor* or *trace element* (such as arsenic) adsorbs onto or absorbs within the developing or fresh precipitates of other chemical species. Although sometimes difficult to distinguish, sorption involves the incorporation of contaminants onto or within preexisting solids (sorbents), whereas coprecipitation occurs as or shortly after the host solids precipitate from the solution, such as arsenic coprecipitating with iron (oxy)(hydr)oxides in acid mine drainage (Chapter 3). Coprecipitation might also involve arsenic-bearing colloids or other fine-grained particles becoming trapped (absorbed) in the interiors of precipitating compounds (US Environmental Protection Agency (US EPA), 2002), 17; (Yuan *et al.*, 2003). Additionally, arsenic could coprecipitate by substituting into the crystalline structures of precipitating compounds, such as As(V) partially substituting for carbonate in the developing crystalline structures of jarosite (Savage, Bird and O'Day, 2005; Savage, Bird and Ashley, 2000). The roles of the sorption and precipitation/coprecipitation of arsenic in natural environments and water treatment technologies are further discussed in Chapters 3 and 7, respectively.

2.8 Chemistry of gaseous arsenic emissions

Arsenic concentrations in the atmosphere are usually very low, around 0.4–30 ng m^{-3} (Mandal and Suzuki, 2002), 207. Urban areas may have 3–180 ng m^{-3}. Values may increase to about 1000 ng m^{-3} close to industrial facilities (Smedley and Kinniburgh, 2002), 533.

Globally, volcanoes release about 17 150 metric tons (t) of arsenic per year into the atmosphere (1 t equals 1000 kg; (Matschullat, 2000), 300). Other significant natural sources of gaseous arsenic emissions include geothermal vents, wind erosion of soils and sediments, forest and coal seam fires, and sea spray ((Cullen and Reimer, 1989), 740; (Nriagu, 1989); Chapter 3). Under reducing conditions in soils, fungi and other microorganisms may produce gaseous arsine and methylated arsenic vapors (Section 2.4.3; (Mandal and Suzuki, 2002), 205; Chapter 4; (Frankenberger and Arshad, 2002), 363; (Oremland and Stolz, 2003), 939). Perhaps as much as 26 200 t of arsenic may annually volatilize into the atmosphere from microbial activity in soils (Matschullat, 2000), 300–301.

Despite the use of *flue gas* treatment technologies in many countries, coal combustion and ore smelting are still the largest anthropogenic emitters of arsenic to the global atmosphere ((Matschullat, 2000), 301; Chapter 3). Due to its semivolatility, arsenic in smelting ores and combusting coal mostly concentrates in flue gases. Gaseous arsenic in coal combustion and smelting operations primarily exists as arsenic(III) oxide (As_4O_6) (Cotton *et al.*, 1999), 400; (Jadhav and Fan, 2001), 794; (Shih and Lin, 2003), although $AsCl_3$ may be prominent from the combustion of chlorine-rich coals at about 530–930 °C (Urban and Wilcox, 2006). As the flue gases cool after leaving the combustion zone, volatile arsenic largely condenses onto any high surface area, fine-grained *fly ash* particles, especially in coal combustion systems ((Hower *et al.*, 1999; Mukherjee and Kikuchi, 1999), 64; (Guo, Yang and Liu, 2004; Llorens, Fernández-Turiel and Querol, 2001; Galbreath and Zygarlicke, 2004); Chapters 5 and 7).

Arsine gas does not develop under oxidizing conditions in combustors and smelters. However, the gas is used and released by semiconductor and optoelectronics facilities (Chein *et al.*, 2006). Arsine along with As^0, As_4, As_2, and AsO may also result from the production of methane and other combustible gases during *coal gasification* ((Diaz-Somoano, López-Antón and Martínez-Tarazona, 2004), 1241; (López-Antón *et al.*, 2007); Chapter 5). Once in the atmosphere, arsine oxidizes to As_4O_6, which subsequently deposits on atmospheric particles (Chein *et al.*, 2006), 1901–1902.

Other human activities that release arsenic include: herbicide use, automobile traffic, marine vessels, glass manufacturing, steel production, waste incineration, Portland cement manufacturing, and the combustion of CCA-preserved wood ((Shih, 2005), 88; (Matschullat, 2000), 302; (Chein *et al.*, 2006; Wasson *et al.*, 2005; Frey and Zhao, 2004); Chapter 7). Total anthropogenic emissions of arsenic to the atmosphere are about 18 800–25 800 t per year (Shih, 2005), 88.

The vast majority (approximately 89–98.6 %) of atmospheric arsenic, including methyl forms, is sorbed onto *particulates* rather than existing as vapors (Cullen and Reimer, 1989), 741; (Matschullat, 2000), 300. Most particulate arsenic in the atmosphere is inorganic As(V) and As(III) rather than organoarsenicals (Mandal and Suzuki, 2002), 207. The As(III)/As(V) ratio in air is largely controlled by local redox conditions rather than the chemistry of the sources (Cullen and Reimer, 1989), 740-741. Locally, ozone may be an important oxidizer and SO_2 is a significant reductant (Cullen and Reimer, 1989), 740–741. Arsenic on atmospheric particles may be enriched 10–1000 times above values in the continental crust (Cullen and Reimer, 1989), 741. This is probably due to arsenic condensation on high surface area particles. Any arsenic vapors in the atmosphere would mostly consist of As_4O_6 with trace amounts of arsine and methylarsenic species (Cotton *et al.*, 1999), 400; (Francesconi and Kuehnelt, 2002), 86. However, As_4O_6 would quickly sorb onto atmospheric particles or condense as arsenolite. Arsenolite is also moderately soluble in rainwater (about 20.5 g l^{-1} at 25 °C; Lide 2007; Table 2.8) and dissolves to produce arsenious acid ((Cotton *et al.*, 1999), 400). Arsenic in precipitation from unpolluted ocean air averages about 0.019 μg l^{-1} (Hering and Kneebone, 2002), 157 and terrestrial rainwater concentrations (at least over the United States) also have similar averages of around 0.013–0.032 μg l^{-1} ((Smedley and Kinniburgh, 2002), 522; Chapter 3).

In most developed countries, coal combustion and smelting facilities are required to treat gaseous and particulate arsenic in flue gases before the gases are released into the atmosphere (Chapters 5 and 7). Potentially effective sorbents for flue gas arsenic include hydrated lime ($Ca(OH)_2$, portlandite), lime (CaO), calcium carbonate, *limestone*, fly ash, and sometimes *activated carbon* ((Jadhav and Fan, 2001; López-Antón *et al.*, 2007; Helsen and Van den Bulck, 2004), 287, 289; (Taerakul *et al.*, 2006; Gupta *et al.*, 2007); Chapters 5 and 7). The injection of hydrated lime is especially effective and probably removes volatile As_4O_6 through the formation of calcium arsenates at least over a temperature range of 600–1000 °C ((Mahuli *et al.*, 1997); Chapter 5).

Researchers are currently developing combustion and *pyrolysis* technologies to ash CCA-treated wood (Chapter 7). Combustion and pyrolysis converts the wood wastes into a smaller volume that should be more manageable. However, thermal destruction unavoidably volatilizes some of the arsenic and often produces a toxic ash that will require treatment ((Morrell, 2004), 204; (Townsend, Dubey and Solo-Gabriele, 2004), 174; (Rogers *et al.*, 2007); Chapter 7). The fraction of arsenic volatilized during pyrolysis or combustion ranges from 8 to 95 % (Helsen and Van den Bulck, 2004), 285. Volatilized arsenic from the combustion of CCA-treated wood usually consists of arsenic oxides and As(III) and As(V) compounds on particulate matter ((Helsen and Van den Bulck, 2004), 286; Chapter 7). Although arsenic is less volatile during low-temperature pyrolysis than combustion, some arsenic still volatilizes during the process. The volatilization of arsenic during pyrolysis chiefly results from the reduction of As(V) in the wood to As_4O_6 (Helsen and Van den Bulck, 2003; Hata *et al.*, 2003; Helsen *et al.*, 2003).

References

Ahn, J.S., Park, Y.S., Kim, J.-Y. and Kim, K.-W. (2005) Mineralogical and geochemical characterization of arsenic in an abandoned mine tailings of Korea. *Environmental Geochemistry and Health*, **27**(2), 147–57.

Allison, J.D., Brown, D.S. and Novo-Gradac, K.J. (1991) *MINTEQA2/PRODEFA2: A Geochemical Assessment Model for Environmental Systems: Version 3.0 User's Manual*, U.S. Environmental Protection Agency, Washington, DC.

Amatore, C., Bahsoun, A.A., Jutand, A. *et al.* (2003) Mechanism of the stille reaction catalyzed by palladium ligated to arsine ligand: $PhPdl(AsPh_3)(DMF)$ is the species reacting with vinyl stannane in DMF. *Journal of the American Chemical Society*, **125**(14), 4212–22.

Anawar, H.M., Akai, J. and Sakugawa, H. (2004) Mobilization of arsenic from subsurface sediments by effect of bicarbonate ions in groundwater. *Chemosphere*, **54**(6), 753–62.

Anderson, G.L., Ellis, P.J., Khun, P. and Hille, R. (2002) Oxidation of arsenite by Alcaligenes faecalis, in *Environmental Chemistry of Arsenic* (ed. W.T. Frankenberger Jr.), Marcel Dekker, New York, pp. 343–61.

Apambire, W.B. and Hess, J.W. (2000) The aqueous geochemistry of fluoride in the upper regions of Ghana. *Geological Society of America Abstracts with Programs*, **32**(7), 7.

Arienzo, M., Adamo, P., Chiarenzelli, J. *et al.* (2002) Retention of arsenic on hydrous ferric oxides generated by electrochemical peroxidation. *Chemosphere*, **48**(10), 1009–18.

Audi, G., Bersillon, O., Blachot, J. and Wapstra, A.H. (2003) The NUBASE evaluation of nuclear and decay properties. *Nuclear Physics A*, **729**(1), 3–128.

Azcue, J.M. and Nriagu, J.O. (1994) Arsenic: historical perspectives, in *Arsenic in the Environment: Part I: Cycling and Characterization* (ed. J.O. Nriagu), John Wiley & Sons, Inc., New York, pp. 1–15.

Ballantyne, J.M. and Moore, J.N. (1988) Arsenic geochemistry in geothermal systems. *Geochimica et Cosmochimica Acta*, **52**, 475–83.

Ballirano, P. and Maras, A. (2002) Refinement of the crystal structure of arsenolite, As2O3. *Zeitschrift fur Kristallographie: New Crystal Structures*, **217**(2), 177–78.

Baruah, T., Pederson, M.R., Zope, R.R. and Beltra'n, M.R. (2004) Stability of Asn [n = 4, 8, 20, 28, 32, 36, 60] cage structures. *Chemical Physics Letters*, **387**(4–6), 476–80.

Baur, W.H. and Onishi, H. (1978) Arsenic, in *Handbook of Geochemistry* (eds K.H. Wedepohl, C.W. Correns, D.M. Shaw *et al.*), Springer-Verlag, Berlin, pp. II/33–I:33-O.

Bentley, R. and Chasteen, T.G. (2002) Microbial methylation of metalloids: arsenic, antimony, and bismuth. *Microbiology and Molecular Biology Reviews*, **66**(2), 250–71.

Bisceglia, K.J., Rader, K.J., Carbonaro, R.F. *et al.* (2005) Iron(II)-catalyzed oxidation of arsenic(III) in a sediment column. *Environmental Science and Technology*, **39**(23), 9217–22.

Bismuth, C., Borron, S.W., Baud, F.J. and Barriot, P. (2004) Chemical weapons: documented use and compounds on the horizon. *Toxicology Letters*, **149**(1–3), 11–18.

Bissen, M. and Frimmel, F.H. (2003) Arsenic — a review. Part II: oxidation of arsenic and its removal in water treatment. *Acta Hydrochimica et Hydrobiologica*, **31**(2), 97–107.

Bloss, F.D. (1971) *Crystallography and Crystal Chemistry: An Introduction*, Holt, Rinehart and Winston, Inc., New York, p. 545.

Borah, D., Satokawa, S., Kato, S. and Kojima, T. (2008) Surface-modified carbon black for As(V) removal. *Journal of Colloid and Interface Science*, **319**(1), 53–62.

Bothe, J.V. Jr. and Brown, P.W. (1999) The stabilities of calcium arsenates at 23±1 °C. *Journal of Hazardous Materials*, **69**(2), 197–207.

Boyle, R.W. and Jonasson, I.R. (1973) Geochemistry of arsenic and its use as indicator element in geochemical prospecting. *Journal of Geochemical Exploration*, **2**(3), 251–96.

Brooks, W.E. (2007) *Arsenic*, US Geological Survey Mineral Commodity Summaries, January.

Buckley, D.N., Seabury, C.W., Valdes, J.L. *et al.* (1990) Growth of InGaAs structures using in situ electrochemically generated arsine. *Applied Physics Letters*, **57**(16), 1684–86.

Bundschuh, J., Bonorino, G., Viero, A.P. *et al.* (2000) Arsenic and other trace elements in sedimentary aquifers in the Chaco-Pampean plain, in *Arsenic in Groundwater of Sedimentary Aquifers*, Pre-Congress Workshop, 31st International Geological Congress (eds P. Bhattacharya and A.H. Welch), Groundwater Arsenic Research Group, Division of Land and Water Resources, Royal Institute of Technology, Stockholm, pp. 27–32.

Burkitbaev, M. (2003) Radiation-stimulated oxidation reactions of oxo anions in aqueous solutions. *High Energy Chemistry*, **37**(4), 216–19.

Carter, T.G., Vickaryous, W.J., Cangelosi, V.M. and Johnson, D.W. (2007) Supramolecular arsenic coordination chemistry. *Comments on Inorganic Chemistry*, **28**(3–4), 97–122.

Castro, S.H. and Baltierra, L. (2005) Study of the surface properties of enargite as a function of pH. *International Journal of Mineral Processing*, **77**(2), 104–15.

Chen, J.F. and Cho, A.Y. (1992) The effects of GaSb/InAs broken gap on interband tunneling current of a GaSb/InAs/GaSb/AlSb/InAs tunneling structure. *Journal of Applied Physics*, **71**(9), 4432–35.

Chein, H., Hsu, Y.-D., Aggarwal, S.G. *et al.* (2006) Evaluation of arsenical emission from semiconductor and opto-electronics facilities in Hsinchu, Taiwan. *Atmospheric Environment*, **40** (10), 1901–7.

Cherry, J.A., Shaikh, A.U., Tallman, D.E. and Nicholson, R.V. (1979) Arsenic species as an indicator of redox conditions in groundwater. *Journal of Hydrology*, **43**(1–4), 373–92.

Clark, A.H. (1970) Arsenolamprite confirmed from Copiapo area, northern Chile. *Mineralogical Magazine*, **37**(290), 732.

Clifford, D.A. and Ghurye, G.L. (2002) Metal-oxide adsorption, ion exchange, and coagulation-microfiltration for arsenic removal from water, in *Environmental Chemistry of Arsenic* (ed. W.T. Frankenberger Jr.), Marcel Dekker, New York, pp. 217–45.

Cotton, F.A., Wilkinson, G., Murillo, C.A. and Bochmann, M. (1999) *Advanced Inorganic Chemistry*, 6th edn, John Wiley & Sons, Inc., New York.

Cox, C.D. and Ghosh, M.M. (1994) Surface complexation of methylated arsenates by hydrous oxides. *Water Research*, **28**(5), 1181–88.

Craig, P.J., Eng, G. and Jenkins, R.O. (2003) Occurrence and pathways of organometallic compounds in the environment — general considerations, in *Organometallic Compounds in the Environment*, 2nd edn (ed. P.J. Craig), John Wiley & Sons, Ltd, West Sussex, pp. 1–55.

Craw, D., Falconer, D. and Youngson, J.H. (2003) Environmental arsenopyrite stability and dissolution: theory, experiment, and field observations. *Chemical Geology*, **199**(1–2), 71–82.

Cullen, W.R. and Reimer, K.J. (1989) Arsenic speciation in the environment. *Chemical Reviews*, **89**(4), 713–64.

Davis, J.E. (2000) Geochemical controls on arsenic and phosphorus in natural and engineered systems. Master's Thesis, Virginia Polytechnic Institute, Environmental Engineering, Blacksburg, VA.

De, M. (2005) Arsenic — India's health crisis attracting global attention. *Current Science*, **88**(5), 683–84.

Dean, J.A. (ed.) (1979) *Lange's Handbook of Chemistry*, 12th edn, McGraw-Hill, New York.

Di Benedetto, F., Costagliola, P., Benvenuti, M. *et al.* (2006) Arsenic incorporation in natural calcite lattice: evidence from electron spin echo spectroscopy. *Earth and Planetary Science Letters*, **246**(3–4), 458–65.

Diaz-Somoano, M., López-Antón, M.A. and Martínez-Tarazona, M.R. (2004) Retention of arsenic and selenium during hot gas desulfurization using metal oxide sorbents. *Energy and Fuels*, **18**(5), 1238–42.

Dodd, M.C., Vu, N.D., Ammann, A. *et al.* (2006) Kinetics and mechanistic aspects of As(III) oxidation by aqueous chlorine, chloramines, and ozone: relevance to drinking water treatment. *Environmental Science and Technology*, **40**(10), 3285–92.

Dombrowski, P.M., Long, W., Farley, K.J. *et al.* (2005) Thermodynamic analysis of arsenic methylation. *Environmental Science and Technology*, **39**(7), 2169–76.

Douglass, D.L., Shing, C. and Wang, G (1992) The light-induced alteration of realgar to pararealgar. *American Mineralogist*, **77**(11–12), 1266–74.

Drever, J.I. (1997) *The Geochemistry of Natural Waters: Surface and Groundwater Environments*, Prentice Hall, Upper Saddle River, NJ, p. 436.

Dunn, P.J., Pecor, D.R. and Newberry, N. (1980) Johnbaumite, a new member of the apatite group from Franklin, New Jersey. *American Mineralogist*, **65**, 1143–45.

Dutta, P.K., Pehkonen, S.O., Sharma, V.K. and Ray, A.K. (2005) Photocatalytic oxidation of arsenic (III): evidence of hydroxyl radicals. *Environmental Science and Technology*, **39**(6), 1827–34.

Dutta, P.K., Ray, A.K., Sharma, V.K. and Millero, F.J. (2004) Adsorption of arsenate and arsenite on titanium dioxide suspensions. *Journal of Colloid and Interface Science*, **278**(2), 270–75.

Eby, G.N. (2004) *Principles of Environmental Geochemistry*, Brooks/Cole-Thomson Learning, Pacific Grove, CA, p. 514.

Ehrlich, H.L. (2002) Bacterial oxidation of As(III) compounds, in *Environmental Chemistry of Arsenic* (W.T. Frankenberger Jr.), Marcel Dekker, New York. pp. 313–27.

Evangelou, V.P., Seta, A.K. and Holt, A. (1998) Potential role of bicarbonate during pyrite oxidation. *Environmental Science and Technology*, **32**(14), 2084–91.

Farina, V. and Krishnan, B. (1991) Large rate accelerations in the stille reaction with tri-2-furylphosphine and triphenylarsine as palladium ligands: mechanistic and synthetic implications. *Journal of the American Chemical Society*, **113**(25), 9585–95.

Faure, G. (1998) *Principles and Applications of Geochemistry*, 2nd edn, Prentice Hall, Upper Saddle River, NJ, p. 600.

Fendorf, S., Eick, M.J., Grossl, P. and Sparks, D.L. (1997) Arsenate and chromate retention mechanisms on goethite. 1. Surface structure. *Environmental Science and Technology*, **31**(2), 315–20.

Fettel, M. (1986) Arsenolamprit aus dem vorderen Odenwald (In German). *Lapiz*, **11**(11), 25–26.

Fetter, C.W. (1993) *Contaminant Hydrology*, Prentice Hall, Upper Saddle River, NJ, p. 458.

Foster, A.L. (2003) Spectroscopic investigation of arsenic species in solid phases, in *Arsenic in Ground Water* (eds A.H. Welch and K.G. Stollenwerk), Kluwer Academic Publishers, Boston, MA, pp. 27–65.

Francesconi, K.A. and Kuehnelt, D. (2002) Arsenic compounds in the environment, in *Environmental Chemistry of Arsenic* (ed. W.T. Frankenberger Jr.), Marcel Dekker, New York, pp. 51–94.

Frankenberger, W.T., Jr. and Arshad, M. (2002) Volatilization of arsenic, in *Environmental Chemistry of Arsenic* (ed. W.T. Frankenberger Jr.), Marcel Dekker, New York, pp. 363–80.

Francesconi, K.A., Edmonds, J.S. and Morita, M. (1994) Determination of arsenic and arsenic species in marine environmental samples, in *Arsenic in the Environment: Part I: Cycling and Characterization* (ed. J.O. Nriagu), John Wiley & Sons, Inc., New York, pp. 189–219.

Frey, H.C. and Zhao, Y. (2004) Quantification of variability and uncertainty for air toxic emission inventories with censored emission factor data. *Environmental Science and Technology*, **38**(22), 6094–100.

Galbreath, K.C. and Zygarlicke, C.J. (2004) Formation and chemical speciation of arsenic-, chromium-, and nickel-bearing coal combustion PM2.5. *Fuel Processing Technology*, **85**(6–7), 701–26.

Gao, Y. and Mucci, A. (2001) Acid base reaction, phosphate and arsenate complexation, and their competitive adsorption at the surface of goethite in 0.7M NaCl solution. *Geochimica et Cosmochimica Acta*, **65**(14), 2361–78.

Glasby, G.P. (1997) Disposal of chemical weapons in the Baltic Sea. *Science of the Total Environment*, **206**(2–3), 267–73.

Goldberg, S. and Johnston, C.T. (2001) Mechanisms of arsenic adsorption on amorphous oxides evaluated using macroscopic measurements, vibrational spectroscopy, and surface complexation modeling. *Journal of Colloid and Interface Science*, **234**(1), 204–16.

Gräfe, M. and Sparks, D.L. (2006) Solid phase speciation of arsenic, in *Managing Arsenic in the Environment: From Soil to Human Health* (eds R. Naidu, E. Smith, G. Owens *et al.*), CSIRO Publishing, Collingwood, pp. 75–91.

Grossl, P.R., Eick, M., Sparks, D.L. *et al.* (1997) Arsenate and chromate retention mechanisms on goethite. 2. Kinetic evaluation using a pressure-jump relaxation technique. *Environmental Science and Technology*, **31**(2), 321–26.

Guo, X., Sturgeon, R.E., Mester, Z. and Gardner, G.J. (2005) Photochemical alkylation of inorganic arsenic Part 1. Identification of volatile arsenic species. *Journal of Analytical Atomic Spectrometry*, **20**(8), 702–8.

Guo, R., Yang, J. and Liu, Z. (2004) Thermal and chemical stabilities of arsenic in three Chinese coals. *Fuel Processing Technology*, **85**(8–10), 903–12.

Gupta, H., Thomas, T.J., Park, A.-H.A. *et al.* (2007) Pilot-scale demonstration of the OSCAR process for high-temperature multipollutant control of coal combustion flue gas, using carbonated fly ash and mesoporous calcium carbonate. *Industrial and Engineering Chemistry Research*, **46**(14), 5051–60.

Harvey, M.C., Schreiber, M.E., Rimstidt, J.D. and Griffith, M.M. (2006) Scorodite dissolution kinetics: implications for arsenic release. *Environmental Science and Technology*, **40**(21), 6709–14.

Hata, T., Bronsveld, P.M., Vystavel, T. *et al.* (2003) Electron microscopic study on pyrolysis of CCA (chromium, copper and arsenic oxide)-treated wood. *Journal of Analytical and Applied Pyrolysis*, **68–69**, 635–43.

He, H.S., Chung, C.W.Y., But, T.Y.S. and Toy, P.H. (2005b) Arsonium ylides in organic synthesis. *Tetrahedron*, **61**(6), 1385–405.

He, H.S., Zhang, C., Ng, C.K.-W. and Toy, P.H. (2005a) Polystyrene-supported triphenylarsines: useful ligands in palladium-catalyzed aryl halide homocoupling reactions and a catalyst for alkene epoxidation using hydrogen peroxide. *Tetrahedron*, **61**(51), 12053–57.

Helsen, L. and Van den Bulck, E. (2003) Metal retention in the solid residue after low-temperature pyrolysis of chromated copper arsenate (CCA)-treated wood. *Environmental Engineering Science*, **20**(6), 569–80.

Helsen, L. and Van den Bulck, E. (2004) Review of thermochemical conversion processes as disposal technologies for chromated copper arsenate (CCA) treated wood waste, in *Environmental Impacts of Preservative-Treated Wood, Conference*, Orlando, Florida, USA, February 8–11, Florida Center for Environmental Solutions, Gainesville, FL, pp. 277–94.

Helsen, L., Van den Bulck, E., Van Bael, M.K. and Mullens, J. (2003) Arsenic release during pyrolysis of CCA treated wood waste: current state of knowledge. *Journal of Analytical and Applied Pyrolysis*, **68–69**, 613–33.

Helz, G.R., Tossell, J.A., Charnock, J.M. *et al.* (1995) Oligomerization in As(III) sulfide solutions: theoretical constraints and spectroscopic evidence. *Geochimica et Cosmochimica Acta*, **59**(22), 4591–604.

Hering, J.G. and Kneebone, P.E. (2002) Biogeochemical controls on arsenic occurrence and mobility in water supplies, in *Environmental Chemistry of Arsenic* (ed. W.T. Frankenberger Jr.), Marcel Dekker), New York, pp. 155–81.

Hlavay, J. and Polyák, K. (2005) Determination of surface properties of iron hydroxide-coated alumina adsorbent prepared for removal of arsenic from drinking water. *Journal of Colloid and Interface Science*, **284**(1), 71–77.

Höglund, L., Petrini, E., Asplund, C. *et al.* (2006) Optimising uniformity of InAs/(InGaAs)/GaAs quantum dots grown by metal organic vapor phase epitaxy. *Applied Surface Science*, **252**(15), 5525–29.

Holden, N.E. (2007) Table of the isotopes, in *CRC Handbook of Chemistry and Physics*, 88th edn, (ed. D.R. Lide), CRC Press, Boca Raton, FL, pp. 11.50–11.203.

Holm, T.R. (2002) Effects of CO_3^{2-} - /bicarbonate, Si, and PO43 - on arsenic sorption to HFO. *Journal of American Water Works Association*, **94**(4), 174–81.

Holm, T.R. and Curtiss, C.D. (1989) A comparison of oxidation-reduction potentials calculated from the As(V)/As(III) and Fe(III)/Fe(II) couples with measured platinum-electrode potentials in groundwater. *Journal of Contaminant Hydrology*, **5**(1), 67–81.

Hower, J.C., Trimble, A.S., Eble, C.F. *et al.* (1999) Characterization of fly ash from low-sulfur and high-sulfur coal sources: partitioning of carbon and trace elements with particle size. *Energy Sources*, **21**(6), 511–25.

Huang, J.-H., Scherr, F. and Matzner, E. (2007) Demethylation of dimethylarsinic acid and arsenobetaine in different organic soils. *Water, Air, and Soil Pollution*, **182**(1–4), 31–41.

Hug, S.J. and Leupin, O. (2003) Iron-catalyzed oxidation of arsenic(III) by oxygen and by hydrogen peroxide: pH-dependent formation of oxidants in the fenton reaction. *Environmental Science and Technology*, **37**(12), 2734–42.

Hug, S.J., Canonica, L., Wegelin, M. *et al.* (2001) Solar oxidation and removal of arsenic at circumneutral pH in iron containing waters. *Environmental Science and Technology*, **35**(10), 2114–21.

Huheey, J.E., Keiter, E.A. and Keiter, R.L. (1993) *Inorganic Chemistry: Principles of Structure and Reactivity*, Harper-Collins College Publishers, New York, p. 964.

Ikejiri, K., Noborisaka, J., Hara, S. *et al.* (2007) Mechanism of catalyst-free growth of GaAs nanowires by selective area MOVPE. *Journal of Crystal Growth*, **298**(SPEC. ISS), 616–19.

Inskeep, W.P., McDermott, T.R. and Fendorf, S. (2002) Arsenic (V)/(III) cycling in soils and natural waters: chemical and microbiological processes, in *Environmental Chemistry of Arsenic*, (ed. W.T. Frankenberger Jr.), Marcel Dekker, New York, pp. 183–215.

Jackson, D.A., Nesbitt, H.W., Scaini, M.J. *et al.* (2003) Gersdorffite (NiAsS) chemical state properties and reactivity toward air and aerated, distilled water. *American Mineralogist*, **88**(5), 890–900.

Jadhav, R.A. and Fan, L.-S. (2001) Capture of gas-phase arsenic oxide by lime: kinetic and mechanistic studies. *Environmental Science and Technology*, **35**(4), 794–99.

Jing, C., Meng, X., Liu, S. *et al.* (2005) Surface complexation of organic arsenic on nanocrystalline titanium oxide. *Journal of Colloid and Interface Science*, **290**(1), 14–21.

Jain, A., Raven, K.P. and Loeppert, R.H. (1999) Arsenite and arsenate adsorption on ferrihydrite: surface charge reduction and net OH - release stoichiometry. *Environmental Science and Technology*, **33**(8), 1179–84.

Johansson, M.P. and Jusélius, J. (2005) Arsole aromaticity revisited. *Letters in Organic Chemistry*, **2**, 469–74.

Jönsson, J., Persson, P., Sjöberg, S. and Lövgren, L. (2005) Schwertmannite precipitated from acid mine drainage: phase transformation, sulphate release and surface properties. *Applied Geochemistry*, **20**(1), 179–91.

Karttunen, A.J., Linnolahti, M. and Pakkanen, T.A. (2007) Icosahedral and ring-shaped allotropes of arsenic. *ChemPhysChem*, **8**(16), 2373–78.

Khokiattiwong, S., Goessler, W., Pedersen, S.N. *et al.* (2001) Dimethylarsinoylacetate from microbial demethylation of arsenobetaine in seawater. *Applied Organometallic Chemistry*, **15**(6), 481–89.

Kinoshita, K., Shikino, O., Seto, Y. and Kaise, T. (2006) Determination of degradation compounds derived from Lewisite by high performance liquid chromatography/inductively coupled plasma-mass spectrometry. *Applied Organometallic Chemistry*, **20**(9), 591–96.

Klein, C. (2002) *The 22nd Edition of the Manual of Mineral Science (after James D. Dana)*, John Wiley & Sons, Inc., New York, p. 641.

Ko, I., Davis, A.P., Kim, J.-Y. and Kim, K.-W. (2007) Arsenic removal by a colloidal iron oxide coated sand. *Journal of Environmental Engineering*, **133**(9), 891–98.

Krause, E. and Ettel, V.A. (1988) Solubility and stability of scorodite, FeAsO4.2H2O: new data and further discussion. *American Mineralogist*, **73**(7–8), 850–54.

Krauskopf, K.B. and Bird, D.K. (1995) *Introduction to Geochemistry*, 3rd edn, McGraw-Hill, Boston, MA, p. 647.

Lafferty, B.J. and Loeppert, R.H. (2005) Methyl arsenic adsorption and desorption behavior on iron oxides. *Environmental Science and Technology*, **39**(7), 2120–27.

Langmuir, D. (1997) *Aqueous Environmental Geochemistry*, Prentice Hall, Upper Saddle River, NJ, p. 600.

Langner, H.W., Jackson, C.R., Mcdermott, T.R. and Inskeep, W.P. (2001) Rapid oxidation of arsenite in a hot spring ecosystem, Yellowstone National Park. *Environmental Science and Technology*, **35**(16), 3302–9.

Langmuir, D., Mahoney, J. and Rowson, J. (2006) Solubility products of amorphous ferric arsenate and crystalline scorodite (FeAsO$_4$·2H$_2$O) and their application to arsenic behavior in buried mine tailings. *Geochimica et Cosmochimica Acta*, **70**(12), 2942–56.

Lattanzi, P., Da Pelo, S., Musu, E. *et al.* (2008) Enargite oxidation: a review. *Earth-Science Reviews*, **86**(1–4), 62–88.

Lau, K.C.Y. and Chiu, P. (2007) The application of non-cross-linked polystyrene-supported triphenylarsine in Stille coupling reactions. *Tetrahedron Letters*, **48**(10), 1813–16.

Lau, K.C.Y., He, H.S., Chiu, P. and Toy, P.H. (2004) Polystyrene-supported triphenylarsine reagents and their use in Suzuki cross-coupling reactions. *Journal of Combinatorial Chemistry*, **6**(6), 955–60.

Leem, J.-Y., Jeon, M., Lee, J. *et al.* (2003) Influence of GaAs/InAs quasi-monolayer on the structural and optical properties of InAs/GaAs quantum dots. *Journal of Crystal Growth*, **252**(4), 493–98.

Lee, J.S. and Nriagu, J.O. (2007) Stability constants for metal arsenates. *Environmental Chemistry*, **4**(2), 123–33.

Lehr, C.R., Polishchuk, E., Radoja, U. and Cullen, W.R. (2003) Demethylation of methylarsenic species by Mycobacterium neoaurum. *Applied Organometallic Chemistry*, **17**(11), 831–34.

Leist, M., Casey, R.J. and Caridi, D. (2000) The management of arsenic wastes: problems and prospects. *Journal of Hazardous Materials*, **76**(1), 125–38.

Lengke, M.F. and Tempel, R.N. (2002) Reaction rates of natural orpiment oxidation at 25 to 40 °C and pH 6.8 to 8.2 and comparison with amorphous As_2S_3 oxidation. *Geochimica et Cosmochimica Acta*, **66**(18), 3281–91.

Lengke, M.F. and Tempel, R.N. (2003) Natural realgar and amorphous AsS oxidation kinetics. *Geochimica et Cosmochimica Acta*, **67**(5), 859–71.

Lengke, M.F. and Tempel, R.N. (2005) Geochemical modeling of arsenic sulfide oxidation kinetics in a mining environment. *Geochimica et Cosmochimica Acta*, **69**(2), 341–56.

Licht, S. and Yu, X. (2005) Electrochemical alkaline Fe(VI) water purification and remediation. *Environmental Science and Technology*, **39**(20), 8071–76.

Lide, D.R. (ed.) (2007) *CRC Handbook of Chemistry and Physics*, edn88th edn, CRC Press, Boca Raton, FL.

Lindberg, R.D. and Runnells, D.D. (1984) Ground water redox reactions: an analysis of equilibrium state applied to Eh measurements and geochemical modeling. *Science*, **225**(4665), 925–27.

Lindstrom, R.M., Blaauw, M. and Fleming, R.F. (2003) The half-life of 76As. *Journal of Radioanalytical and Nuclear Chemistry*, **257**(3), 489–91.

Llorens, J.F., Fernández-Turiel, J.L. and Querol, X. (2001) The fate of trace elements in a large coal-fired power plant. *Environmental Geology*, **40**(4–5), 409–16.

Lloyd, D., Gosney, I. and Ormiston, R.A. (1987) Arsonium ylides (with some mention also of arsinimines, stibonium and bismuthonium ylides). *Chemical Society Reviews*, **16**, 45–74.

López-Antón, M.A., Díaz-Somoano, M., Fierro, J.L.G. and Martínez-Tarazona, M.R. (2007) Retention of arsenic and selenium compounds present in coal combustion and gasification flue gases using activated carbons. *Fuel Processing Technology*, **88**(8), 799–805.

Luengo, C., Brigante, M. and Avena, M. (2007) Adsorption kinetics of phosphate and arsenate on goethite. A comparative study. *Journal of Colloid and Interface Science*, **311**(2), 354–60.

Mahuli, S., Agnihotri, R., Chauk, S. *et al.* (1997) Mechanism of arsenic sorption by hydrated lime. *Environmental Science and Technology*, **31**(11), 3226–31.

Maldonado-Reyes, A., Montero-Ocampo, C. and Solorza-Feria, O. (2007) Remediation of drinking water contaminated with arsenic by the electro-removal process using different metal electrodes. *Journal of Environmental Monitoring*, **9**(11), 1241–47.

Mandal, B.K. and Suzuki, K.T. (2002) Arsenic round the world: a review. *Talanta*, **58**(1), 201–35.

Mandel, N. and Donohue, J. (1971) The refinement of the crystal structure of skutterudite, CoAs3. *Acta Crystallographica Section B*, **27**, 2288–89.

Manning, B.A., Fendorf, S.E. and Goldberg, S. (1998) Surface structures and stability of arsenic(III) on goethite: spectroscopic evidence for inner-sphere complexes. *Environmental Science and Technology*, **32**(16), 2383–88.

Manning, B.A., Fendorf, S.E., Bostick, B. and Suarez, D.L. (2002) Arsenic(III) oxidation and arsenic(V) adsorption reactions on synthetic birnessite. *Environmental Science and Technology*, **36**(5), 976–81.

Maryanoff, B.E. and Reitz, A.B. (1989) The Wittig olefination reaction and modifications involving phosphoryl-stabilized carbanions. Stereochemistry, mechanism, and selected synthetic aspects. *Chemical Reviews*, **89**(4), 863–927.

Matschullat, J. (2000) Arsenic in the geosphere — a review. *Science of the Total Environment*, **249**(1–3), 297–312.

Matsubara, S., Miyawaki, R., Shimizu, M. and Yamanaka, T. (2001) Pararsenolamprite, a new polymorph of native As, from the Mukuno mine, Oita prefecture, Japan. *Mineralogical Magazine*, **65**(6), 807–12.

McNeill, L.S., Chen, H. and Edwards, M. (2002) Aspects of arsenic chemistry in relation to occurrence, health and treatment, in *Environmental Chemistry of Arsenic* (ed. W.T. Frankenberger Jr.), Marcel Dekker, New York, pp. 141–53.

McSheehy, S., Guo, X.-M., Sturgeon, R.E. and Mester, Z. (2005) Photochemical alkylation of inorganic arsenic Part 2. Identification of aqueous phase organoarsenic species using multidimensional liquid chromatography and electrospray mass spectrometry. *Journal of Analytical Atomic Spectrometry*, **20**(8), 709–16.

Meharg, A.A. (2005) *Venomous Earth: How Arsenic Caused the World's Worst Mass Poisoning*, Macmillan Publishing, New York, p. 192.

Mok, W.M. and Wai, C.M. (1994) Mobilization of arsenic in contaminated river waters, in *Arsenic in the Environment: Part I: Cycling and Characterization* (ed. J.O. Nriagu), John Wiley & Sons, Inc., New York, pp. 99–117.

Morrell, J. (2004) Disposal of treated wood, in *Environmental Impacts of Preservative-Treated Wood, Conference*, February 8–11, Orlando, Florida, USA, Florida Center for Environmental Solutions, Gainesville, FL, pp. 196–209.

Mozgova, N.N., Borodaev, Yu.S., Gablina, I.F. *et al.* (2005) Mineral assemblages as indicators of the maturity of oceanic hydrothermal sulfide mounds. *Lithology and Mineral Resources*, **40**(4), 293–319.

Muir, B., Quick, S., Slater, B.J. *et al.* (2005) Analysis of chemical warfare agents: II. Use of thiols and statistical experimental design for the trace level determination of vesicant compounds in air samples. *Journal of Chromatography A*, **1068**(2), 315–26.

Mukherjee, A.B. and Kikuchi, R. (1999) Coal ash from thermal power plants in Finland, in *Biogeochemistry of Trace Elements in Coal and Coal Combustion Byproducts* (eds K.S. Sajwan, A.K. Alva and R.F. Keefer), Kluwer Academic/Plenum Publishers, New York, pp. 59–76.

Naumov, P., Makreski, P. and Jovanovski, G. (2007) Direct atomic scale observation of linkage isomerization of As_4S_4 clusters during the photoinduced transition of realgar to pararealgar. *Inorganic Chemistry*, **46**(25), 10624–31.

Nebergall, W.H., Schmidt, F.C. and Holtzclaw, H.F., Jr. (1976) *College Chemistry with Qualitative Analysis*, edn5th edn, D. C. Heath and Company, Lexington, MA, p. 1058.

Nesbitt, H.W. and Reinke, M. (1999) Properties of As and S at NiAs, NiS, and Fe (1 - X)S surfaces, and reactivity of niccolite in air and water. *American Mineralogist*, **84**(4), 639–49.

Nesbitt, H.W., Muir, I.J. and Pratt, A.R. (1995) Oxidation of arsenopyrite by air and air-saturated, distilled water, and implications for mechanism of oxidation. *Geochimica et Cosmochimica Acta*, **59**(9), 1773–86.

Nesbitt, H.W., Uhlig, I. and Szargan, R. (2002) Surface reconstruction and As-polymerization at fractured loellingite ($FeAs_2$) surfaces. *American Mineralogist*, **87**(7), 1000–4.

Nesse, W.D. (2000) *Introduction to Mineralogy*, Oxford Press, New York, p. 442.

Nordstrom, D.K. and Archer, D.G. (2003) Arsenic thermodynamic data and environmental geochemistry, in *Arsenic in Ground Water* (eds A.H. Welch and K.G. Stollenwerk), Kluwer Academic Publishers, Boston, MA, pp. 1–25.

Nriagu, J.O. (1989) A global assessment of natural sources of atmospheric trace metals. *Nature*, **338**(6210), 47–49.

Nriagu, J.O. (2002) Arsenic poisoning through the ages, in *Environmental Chemistry of Arsenic* (ed. W.T. Frankenberger Jr.), Marcel Dekker, New York, pp. 1–26.

O'Neil, M.J., Smith, A., Heckelman, P.E. *et al.* (eds) (2001) *The Merck Index: An Encyclopedia of Chemicals, Drugs, and Biologicals*, edn13th edn, Merck & Company, Whitehouse Station, NJ.

O'Day, P.A. (2006) Chemistry and mineralogy of arsenic. *Elements*, **2**(2), 77–83.

Ofor, O. (1995) Oleate adsorption at a Nigerian hematite-water interface: effect of concentration, temperature, and pH on adsorption density. *Journal of Colloid and Interface Science*, **174**, 345–50.

Oremland, R.S. and Stolz, J.F. (2003) The ecology of arsenic. *Science*, **300**(5621), 939–44.

Parkhurst, D.L. and Appelo, C.A.J. (1999) User's Guide to PHREEQC (Version 2): A Computer Program for Speciation, Batch-Reaction, One-Dimensional Transport, and Inverse Geochemical Calculations. Water-Resources Investigations Report 99-4259, U.S. Geological Survey, p. 312.

Pauling, L. (1960) *Nature of the Chemical Bond*, 3rd edn, Cornell University Press, Ithaca, NY.

Pichler, T., Veizer, J. and Hall, G.E.M. (1999) Natural input of arsenic into a coral-reef ecosystem by hydrothermal fluids and its removal by Fe(III) oxyhydroxides. *Environmental Science and Technology*, **33**(9), 1373–78.

Planer-Friedrich, B., Lehr, C., Matschullat, J. *et al.* (2006) Speciation of volatile arsenic at geothermal features in Yellowstone National Park. *Geochimica et Cosmochimica Acta*, **70**(10), 2480–91.

Planer-Friedrich, B., London, J., McCleskey, R.B. *et al.* (2007) Thioarsenates in geothermal waters of Yellowstone National Park: determination, preservation, and geochemical importance. *Environmental Science and Technology*, **41**(15), 5245–51.

Pokrovski, G., Gout, R., Schott, J. *et al.* (1996) Thermodynamic properties and stoichiometry of As(III) hydroxide complexes at hydrothermal conditions. *Geochimica et Cosmochimica Acta*, **60**(5), 737–49.

Prasad, G. (1994) Removal of arsenic(V) from aqueous systems by adsorption onto some geologic materials, in *Arsenic in the Environment: Part I: Cycling and Characterization* (ed. J.O. Nriagu), John Wiley & Sons, Inc., New York, pp. 133–54.

Pratt, A. (2004) Photoelectron core levels for enargite, Cu_3AsS_4. *Surface and Interface Analysis*, **36**(7), 654–57.

Price, R.E. and Pichler, T. (2005) Distribution, speciation and bioavailability of arsenic in a shallow-water submarine hydrothermal system, Tutum Bay, Ambitle Island, PNG. *Chemical Geology*, **224**(1–3), 122–35.

Qi, Y.-F., Li, Y.-G., Wang, E. *et al.* (2007) Hydrothermal synthesis and structures of organic-inorganic hybrid solids based on arsenic-vanadate building blocks. *Journal of Coordination Chemistry*, **60**(13), 1403–18.

Raposo, J.C., Sanz, J., Zuloaga, O. *et al.* (2003) Thermodynamic model of inorganic arsenic species in aqueous solutions. potentiometric study of the hydrolytic equilibrium of arsenious acid. *Journal of Solution Chemistry*, **32**(3), 253–64.

Redman, A.D., Macalady, D.L. and Ahmann, D. (2002) Natural organic matter affects arsenic speciation and sorption onto hematite. *Environmental Science and Technology*, **36**(13), 2889–96.

Reich, M. and Becker, U. (2006) First-principles calculations of the thermodynamic mixing properties of arsenic incorporation into pyrite and marcasite. *Chemical Geology*, **225**(3–4), 278–90.

Reich, M., Kesler, S.E., Utsunomiya, S. *et al.* (2005) Solubility of gold in arsenian pyrite. *Geochimica et Cosmochimica Acta*, **69**(11), 2781–96.

Rhoton, F.E. and Bigham, J.M. (2005) Phosphate adsorption by ferrihydrite-amended soils. *Journal of Environmental Quality*, **34**(3), 890–96.

Robie, R.A., Hemingway, B.S. and Fisher, J.R. (1979) *Thermodynamic Properties of Minerals and Related Substances at 298.15 K and 1 Bar (10^5 Pascals) Pressure and at Higher Temperatures*, Reprinted with corrections. Geological Survey Bulletin 1452, United States Printing Office, Washington, DC, p. 456.

Rogers, J.M., Stewart, M., Petrie, J.G. and Haynes, B.S. (2007) Deportment and management of metals produced during combustion of CCA-treated timbers. *Journal of Hazardous Materials*, **139**(3), 500–5.

Ryu, J.-H., Gao, S., Dahlgren, R.A. and Zierenberg, R.A. (2002) Arsenic distribution, speciation and solubility in shallow groundwater of Owens Dry Lake, California. *Geochimica et Cosmochimica Acta*, **66**(17), 2981–94.

Samanta, G. and Clifford, D.A. (2006) Influence of sulfide (S2 -) on preservation and speciation of inorganic arsenic in drinking water. *Chemosphere*, **65**(5), 847–53.

Santosa, S.J., Mokudai, H., Takahashi, M. and Tanaka, S. (1996) The distribution of arsenic compounds in the ocean: biological activity in the surface zone and removal processes in the deep zone. *Applied Organometallic Chemistry*, **10**(9), 697–705.

Savage, K.S., Bird, D.K. and Ashley, R.P. (2000) Legacy of the California gold rush: environmental geochemistry of arsenic in the southern Mother Lode Gold District. *International Geology Review*, **42**(5), 385–415.

Savage, K.S., Bird, D.K. and O'Day, P.A. (2005) Arsenic speciation in synthetic jarosite. *Chemical Geology*, **215**(1–4), 473–98.

Schaufuss, A.G., Nesbitt, H.W., Scaini, M.J. *et al.* (2000) Reactivity of surface sites on fractured arsenopyrite (FeAsS) toward oxygen. *American Mineralogist*, **85**(11–12), 1754–66.

Schreiber, M.E., Gotkowitz, M.B., Simo, J.A. and Freiberg, P.G. (2003) Mechanisms of arsenic release to water from naturally occurring sources, eastern Wisconsin, in *Arsenic in Ground Water* (eds A.H., Welch and K.G. Stollenwerk), Kluwer Academic Publishers, Boston, MA, pp. 259–80.

Schwedt, G. and Rieckhoff, M. (1996) Separation of thio-and oxothioarsenates by capillary zone electrophoresis and ion chromatography. *Journal of Chromatography A*, **736**(1–2), 341–50.

Seyferth, D. (2001) Cadet's fuming arsenical liquid and the cacodyl compounds of Bunsen. *Organometallics*, **20**, 1488–98.

Shannon, R.D. (1976) Revised effective ionic radii and systematic studies of interatomic distances in halides and chalcogenides. *Acta Crystallographica Section A*, **32**, 751–67.

Sharma, P.K., Rai, A.K. and Singh, Y. (2007) Synthesis and characterization of some novel triphenylarsenic(V) derivatives of monofunctional bidentate 2,2-distributed benzothiazoline ligands. *Heteroatom Chemistry*, **18**(1), 76–80.

Shih, M.-C. (2005) An overview of arsenic removal by pressure-driven membrane processes. *Desalination*, **172**(1), 85–97.

Shih, C.-J. and Lin, C.-F. (2003) Arsenic contaminated site at an abandoned copper smelter plant: waste characterization and solidification/stabilization treatment. *Chemosphere*, **53**(7), 691–703.

Simeoni, M.A., Batts, B.D. and McRae, C. (2003) Effect of groundwater fulvic acid on the adsorption of arsenate by ferrihydrite and gibbsite. *Applied Geochemistry*, **18**(10), 1507–15.

Smedley, P.L. and Kinniburgh, D.G. (2002) A review of the source, behaviour and distribution of arsenic in natural waters. *Applied Geochemistry*, **17**(5), 517–68.

Smedley, P.L., Kinniburgh, D.G., Macdonald, D.M.J. *et al.* (2005) Arsenic associations in sediments from the loess aquifer of La Pampa, Argentina. *Applied Geochemistry*, **20**(5), 989–1016.

Smith, A.M.L., Dubbin, W.E., Wright, K. and Hudson-Edwards, K.A. (2006) Dissolution of lead-and lead-arsenic-jarosites at pH 2 and 8 and 20 °C: Insights from batch experiments. *Chemical Geology*, **229**(4), 344–61.

Smith, S.D. and Edwards, M. (2005) The influence of silica and calcium on arsenate sorption to oxide surfaces. *Journal of Water Supply: Research and Technology - AQUA*, **54**(4), 201–11.

Stauder, S., Raue, B. and Sacher, F. (2005) Thioarsenates in sulfidic waters. *Environmental Science and Technology*, **39**(16), 5933–39.

Stollenwerk, K.G. (2003) Geochemical processes controlling transport of arsenic in groundwater: a review of adsorption, in *Arsenic in Ground Water* (eds A.H. Welch and K.G. Stollenwerk), Kluwer Academic Publishers, Boston, MA, pp. 67–100.

Su, C. and Puls, R.W. (2003) In situ remediation of arsenic in simulated groundwater using zerovalent iron: laboratory column tests on combined effects of phosphate and silicate. *Environmental Science and Technology*, **37**(11), 2582–87.

Suzuki, T.M., Tanco, M.L., Tanaka, D.A.P. *et al.* (2001) Adsorption characteristics and removal of oxo-anions of arsenic and selenium on the porous polymers loaded with monoclinic hydrous zirconium oxide. *Separation Science and Technology*, **36**(1), 103–11.

Taerakul, P., Sun, P., Golightly, D.W. *et al.* (2006) Distribution of arsenic and mercury in lime spray dryer ash. *Energy and Fuels*, **20**(4), 1521–27.

Tani, Y., Miyata, N., Ohashi, M. *et al.* (2004) Interaction of inorganic arsenic with biogenic manganese oxide produced by a Mn-oxidizing fungus, strain KR21-2. *Environmental Science and Technology*, **38**(24), 6618–24.

Thilo, V.E., Hertzog, K. and Winkler, A (1970) Über Vergänge bei der Bildung des Arsen(V)-sulfids beim Ansäuern von Tetrathioarsenatlösungen (In German). *Zeitschrift für Anorganische und Allgemeine Chemie*, **373**(2), 111–21.

Thirunavukkarasu, O.S., Viraraghavan, T., Subramanian, K.S. *et al.* (2005) Arsenic removal in drinking water — impacts and novel removal technologies. *Energy Sources*, **27**(1–2), 209–19.

Tossell, J.A. (1997) Theoretical studies on arsenic oxide and hydroxide species in minerals and in aqueous solution. *Geochimica et Cosmochimica Acta*, **61**(8), 1613–23.

Tournassat, C., Charlet, L., Bosbach, D. and Manceau, A. (2002) Arsenic(III) oxidation by birnessite and precipitation of manganese(II) arsenate. *Environmental Science and Technology*, **36**(3), 493–500.

Townsend, T., Dubey, B. and Solo-Gabriele, H. (2004) Assessing potential waste disposal impact from preservative treated wood products, in *Environmental Impacts of Preservative-Treated Wood, Conference*, February 8–11, Orlando, Florida, Florida Center for Environmental Solutions, Gainesville, FL, pp. 169–88.

Tucek, K. (1970) Nove nalezy nerostu v Ceskoslovensku (new occurrences of minerals in Czechoslovakia) (In Czech). *Casopis Narodniho Muzea. Oddil Prirodovedny*, **137**(3–4), 44–57.

Urban, D.R. and Wilcox, J. (2006) Theoretical study of the kinetics of the reactions Se + O_2 → Se + O and As + HCl → AsCl + H. *Journal of Physical Chemistry A*, **110**(28), 8797–801.

US Environmental Protection Agency (US EPA). (2002) *Proven Alternatives for Aboveground Treatment of Arsenic in Groundwater*, EPA-542-S-02-002, Office of Solid Wastes and Emergency (5102G).

Utsunomiya, S., Peters, S.C., Blum, J.D. and Ewing, R.C. (2003) Nanoscale mineralogy of arsenic in a region of New Hampshire with elevated As-concentrations in the groundwater. *American Mineralogist*, **88**(11–12, PART 2), 1844–52.

Valkov, V.I. and Golovchan, A.V. (2005) Interplay between the spin state of manganese and the stability of the crystal structure of MnAs and MnP compounds. *Low Temperature Physics*, **31**(6), 528–33.

Vink, B.W. (1996) Stability relations of antimony and arsenic compounds in the light of revised and extended Eh-pH diagrams. *Chemical Geology*, **130**(1–2), 21–30.

Wada, T., Nagasawa, E. and Hanaoka, S. (2006) Simultaneous determination of degradation products related to chemical warfare agents by high-performance liquid chromatography/mass spectrometry. *Applied Organometallic Chemistry*, **20**(9), 573–79.

Wagman, D.D., Evans, W.H., Parker, V.B. *et al.* (1982). The NBS tables of chemical thermodynamic properties: selected values for inorganic and C1 and C2 organic substances in SI units. *Journal of Physical and Chemical Reference Data*, **11**(2): complete issue.

Wallschläger, D. and London, J. (2008) Determination of methylated arsenic-sulfur compounds in groundwater. *Environmental Science and Technology*, **42**(1), 228–34.

Wallschläger, D. and Stadey, C.J. (2007) Determination of (oxy)thioarsenates in sulfidic waters. *Analytical Chemistry*, **79**(10), 3873–80.

Wasson, S.J., Linak, W.P., Gullett, B.K. *et al.* (2005) Emissions of chromium, copper, arsenic, and PCDDs/Fs from open burning of CCA-treated wood. *Environmental Science and Technology*, **39**(22), 8865–76.

Welch, A.H., Westjohn, D.B., Helsel, D.R. and Wanty, R.B. (2000) Arsenic in ground water of the United States: occurrence and geochemistry. *Ground Water*, **38**(4), 589–604.

Williams, M. (2001) Arsenic in mine waters: international study. *Environmental Geology*, **40**(3), 267–78.

Wolthers, M., Charlet, L., van Der Weijden, C.H. *et al.* (2005) Arsenic mobility in the ambient sulfidic environment: sorption of arsenic(V) and arsenic(III) onto disordered mackinawite. *Geochimica et Cosmochimica Acta*, **69**(14), 3483–92.

Xu, Y., Zhou, G.-P. and Zheng, X.-F. (2007) Redetermination of iron(III) arsenate dihydrate. *Acta Crystallographica Section E: Structure Reports Online*, **63**(3), i67–i69, doi:10.1107/S1600536807005302.

Yang, Y.X., Jia, R.R., Chen, Y.R. *et al.* (2007) Synthesis and biological activity of the AsI3-urotropine complex. *Russian Journal of Coordination Chemistry*, **33**(9), 698–703.

Yan, X.-P., Kerrich, R. and Hendry, M.J. (2000) Distribution of arsenic(III), arsenic(V) and total inorganic arsenic in porewaters from a thick till and clay-rich aquitard sequence, Saskatchewan, Canada. *Geochimica et Cosmochimica Acta*, **64**(15), 2637–48.

Yuan, T., Luo, Q.-F., Hu, J.-Y. *et al.* (2003) A study on arsenic removal from household drinking water. *Journal of Environmental Science and Health - Part A Toxic/Hazardous Substances and Environmental Engineering*, **38**(9), 1731–44.

Zakaznova-Herzog, V.P., Seward, T.M. and Suleimenov, O.M. (2006) Arsenous acid ionisation in aqueous solutions from 25 to 300 °C. *Geochimica et Cosmochimica Acta*, **70**(8), 1928–38.

Zhang, J. and Stanforth, R. (2005) Slow adsorption reaction between arsenic species and goethite (α -FeOOH): diffusion or heterogeneous surface reaction control. *Langmuir*, **21**(7), 2895–901.

Zhang, W., Singh, P., Paling, E. and Delides, S. (2004) Arsenic removal from contaminated water by natural iron ores. *Minerals Engineering*, **17**(4), 517–24.

Zhao, Y., Xu, W., Li, Q. *et al.* (2004) The arsenic clusters Asn (n = 1–5) and their anions: Structures, thermochemistry, and electron affinities. *Journal of Computational Chemistry*, **25**(7), 907–20.

Zhu, Y., Zhang, X., Xie, Q. *et al.* (2005) Solubility and stability of barium arsenate and barium hydrogen arsenate at 25 °C. *Journal of Hazardous Materials*, **120**(1–3), 37–44.

Zhu, Y.N., Zhang, X.H., Xie, Q.L. *et al.* (2006) Solubility and stability of calcium arsenates at 25 °C. *Water, Air, and Soil Pollution*, **169**(1–4), 221–38.

Zhuang, J. and Yu, G.-R. (2002) Effects of surface coatings on electrochemical properties and contaminant sorption of clay minerals. *Chemosphere*, **49**(6), 619–28.

3

Arsenic in Natural Environments

KEVIN R. HENKE

University of Kentucky Center for Applied Energy Research

3.1 Introduction

Arsenic is the twentieth most plentiful element on the Earth's surface and often has profoundly negative effects on natural environments and human health (Cullen and Reimer, 1989, 713). Chapter 2 provided a general summary of the chemistry of arsenic. Before discussing toxic and environmental impacts in Chapters 4–6, this chapter will provide a general overview of the formation of nonradiogenic arsenic (all as ^{75}As) in red giant stars and *supernovas*, how the element accumulates on Earth, typical arsenic concentrations in various natural environments, and how the element cycles through the Earth's interior, *hydrosphere* (water), rocks, *pedosphere* (soils), and atmosphere over time. The discussions in Chapter 3 also identify and explain the important processes that control the chemical *speciation* and behavior of arsenic in water, air, and *geologic materials* (*sediments*, soils, and *rocks*). These processes include chemical *oxidation–reduction* (*redox* conditions), changes in *pH*, other significant inorganic chemical reactions, *methylation* and *demethylation*, dissolution, *sorption*, *precipitation/coprecipitation*, volatilization, and condensation.

The impacts of arsenic on human health are left to the next chapter, which includes discussions on the toxic effects of arsenic and its biotransformations. Chapter 5 reviews the history of arsenic utilization in human societies, examples of unintentional poisoning events, the role of arsenic in crime, and recent production and market trends. While the discussions in Chapter 3 often have widespread applications in a variety of natural environments, Chapter 6 cites specific field cases, where human activities and natural occurrences of arsenic in water, soils, and sediments have produced exceptionally serious threats to local environments and human populations.

Arsenic Edited by Kevin R. Henke
© 2009 John Wiley & Sons, Ltd

3.2 Nucleosynthesis: the origin of arsenic

3.2.1 The Big Bang

The entire *Universe* came into existence with the initiation of the *Big Bang* about 13.7 billion years ago (Spergel *et al.*, 2003). During the first few seconds after the Big Bang, the Universe expanded from a microscopic singularity. The Universe at that time only consisted of subatomic particles and enormous amounts of energy; that is, no atoms, molecules, or compounds existed. About 13.8 seconds after the beginning of the Big Bang, temperatures were about 3×10^9 K (Faure, 1998, 8). At this point, the Big Bang had cooled enough to allow for limited *nucleosynthesis* or the formation of heavier elements from the *fusion* of the nuclei of lighter elements. During this process, some protons ($^1_1H^+$) and neutrons (^1_0n) fused together to form nuclei of *deuterium* ($^2_1H^+ = D^+$), helium-3 and -4 ($^3_2He^{2+}$; $^4_2He^{2+}$), and trace amounts of lithium-7 ($^7_3Li^{3+}$) (Delsemme, 1998, 22–23). These elements, which formed for about 30 minutes (Faure, 1998, 8), were too hot to have any electrons. They were *plasma* or just bare nuclei. The Big Bang expanded and cooled too quickly to allow for the nucleosynthesis of any elements heavier than $^7_3Li^{3+}$ (Faure, 1998, 8; Delsemme, 1998, 22–23). About 700 000 years later, the temperature of the Universe was 3000 K and was now cool enough for nuclei to pick up electrons (e.g. He^{2+} became He^0; Faure, 1998, 8). Currently, the background temperature of the Universe is about 3 K.

3.2.2 Arsenic formation in stars

After the formation of new elements ceased in the Big Bang, the nucleosynthesis of helium and heavier elements eventually occurred from nuclear fusion reactions in stars and supernovas. Although trace amounts of lithium formed in the Big Bang, the element is easily destroyed in the interiors of stars (Arnett, 1996, 247). Therefore, the first stars to form after the Big Bang only had hydrogen and helium as sources of fusion fuel. At temperatures of about 10×10^6 K and higher in the interior of a star, two protons may fuse and produce a deuterium nucleus, energy (in megaelectron volts, MeV), and subatomic particles (Faure, 1998, 17), as shown in the following reaction:

$$^1_1H^+ + {}^1_1H^+ \rightarrow D^+ + positron(\beta^+) + neutrino(\nu) + 0.422 \text{ MeV} \qquad (3.1)$$

Like nucleosynthesis during the Big Bang, fusing nuclei in stars and supernovas exist as plasma rather than atoms (Faure, 1998, 17).

Additional protons may fuse with deuterium nuclei to produce $^3_2He^{2+}$ (Faure, 1998, 17):

$$D^+ + {}^1_1H^+ \rightarrow {}^3_2He^{2+} + gamma \ ray(\gamma) + 5.493 \text{ MeV} \qquad (3.2)$$

$^3_2He^{2+}$ may undergo further fusion to produce $^4_2He^{2+}$ (Faure, 1998, 17):

$$^3_2He^{2+} + {}^3_2He^{2+} \rightarrow {}^4_2He^{2+} + {}^1_1H^+ + {}^1_1H^+ + 12.859 \text{ MeV} \qquad (3.3)$$

Notice that Reaction 3.3 produces two protons, which may be used in Reaction 3.1.

The *triple-alpha process* is possible in the dense cores of massive stars. The process involves the essentially simultaneous fusion of three $^4_2He^{2+}$ to form $^{12}_6C^{6+}$ (Faure, 1998, 17–18). As the first step:

$$^4_2He^{2+} + {}^4_2He^{2+} \rightarrow {}^8_4Be^{4+}(unstable) \qquad (3.4)$$

$^8_4$Be^{4+} has an extremely short *half-life* (about 10^{-16} seconds) and a $^8_4$Be^{4+} must immediately collide and fuse with another $^4_2$He^{2+} to form $^{12}_6$C^{6+} before it decays:

$$^8_4\text{Be}^{4+} + {}^4_2\text{He}^{2+} \rightarrow {}^{12}_6\text{C}^{6+} + \gamma \tag{3.5}$$

Reaction 3.5 is only possible in the interiors of stars that are at least twice as dense as the Sun (Delsemme, 1998, 44).

In the stable cores of the most massive stars, $^4_2$He^{2+} and other elements may fuse to produce heavier elements up to nonradioactive $^{52}_{26}$Fe^{26+} (Faure, 1998, 13–18; Wallerstein *et al.*, 1997; Burbidge *et al.*, 1957; Delsemme, 1998, 44). Radioactive $^{56}_{28}$Ni^{28+} can form through the following helium fusion reaction:

$$^{52}_{26}\text{Fe}^{26+} + {}^4_2\text{He}^{2+} \rightarrow {}^{56}_{28}\text{Ni}^{28+}(\text{radioactive}) + \gamma \tag{3.6}$$

$^{56}_{28}$Ni^{28+} then decays to radioactive $^{56}_{27}$Co^{27+}, which further decays to stable $^{56}_{26}$Fe^{26+} (Faure, 1998, 18). Even in the interiors of the most massive stable stars, $^{56}_{28}$Ni^{28+} has so many protons (28) that they repel additional $^4_2$He^{2+} and prevent further fusion (Faure, 1998, 18).

Once a massive star begins to run out of fusionable material, it will expand into a red giant. At the end of the red giant phase and shortly before the massive star supernovas, neutron capture reactions begin. These reactions can produce stable elements that are heavier than $^{56}_{26}$Fe, including $^{75}_{33}$As (Faure, 1998, 18–19; Tayler, 1988). The *neutron flux* in red giants is low enough so that short-lived radioactive elements will usually decay before fusing with another neutron (Faure, 1998, 18). This is called the *slow* or *s-process*. Beginning with $^{56}_{26}$Fe^{26+}, $^{75}_{33}$As^{33+} can form through the following series of fusion reactions involving neutron capture and radioactive decay (Faure, 1998, 18–19; Pagel, 1997, 182; Lide, 2007):

$$^{56}_{26}\text{Fe}^{26+} + {}^1_0\text{n} \rightarrow {}^{57}_{26}\text{Fe}^{26+} + \gamma \tag{3.7}$$

$$^{57}_{26}\text{Fe}^{26+} + {}^1_0\text{n} \rightarrow {}^{58}_{26}\text{Fe}^{26+} + \gamma \tag{3.8}$$

$$^{58}_{26}\text{Fe}^{26+} + {}^1_0\text{n} \rightarrow {}^{59}_{26}\text{Fe}^{26+}(\text{radioactive}) + \gamma \tag{3.9}$$

$$^{59}_{26}\text{Fe}^{26+} \rightarrow {}^{59}_{27}\text{Co}^{27+} + \text{electron}(\beta^-) + \bar{v} \text{ (antineutrino)} + 1.565 \text{ MeV} \tag{3.10}$$

$$^{59}_{27}\text{Co}^{27+} + {}^1_0\text{n} \rightarrow {}^{60}_{27}\text{Co}^{27+}(\text{radioactive}) + \gamma \tag{3.11}$$

$$^{60}_{27}\text{Co}^{27+} \rightarrow {}^{60}_{28}\text{Ni}^{28+} + \beta^- + \bar{v} + 2.824 \text{ MeV} \tag{3.12}$$

$$^{60}_{28}\text{Ni}^{28+} + {}^1_0\text{n} \rightarrow {}^{61}_{28}\text{Ni}^{28+} + \gamma \tag{3.13}$$

$$^{61}_{28}\text{Ni}^{28+} + {}^1_0\text{n} \rightarrow {}^{62}_{28}\text{Ni}^{28+} + \gamma \tag{3.14}$$

$$^{62}_{28}\text{Ni}^{28+} + {}^1_0\text{n} \rightarrow {}^{63}_{28}\text{Ni}^{28+}(\text{radioactive}) + \gamma \tag{3.15}$$

$$^{63}_{28}\text{Ni}^{28+} \rightarrow {}^{63}_{29}\text{Cu}^{29+} + \beta^- + \bar{v} + 0.066\,945 \text{ MeV} \tag{3.16}$$

$$^{63}_{29}\text{Cu}^{29+} + {}^1_0\text{n} \rightarrow {}^{64}_{29}\text{Cu}^{29+}(\text{radioactive}) + \gamma \tag{3.17}$$

$^{64}_{29}$Cu^{29+} decays to both stable $^{64}_{30}$Zn^{30-} and $^{64}_{28}$Ni^{28+} as follows (Faure, 1998, 18; Lide, 2007):

$$^{64}_{29}\text{Cu}^{29+} \rightarrow {}^{64}_{30}\text{Zn}^{30+} + \beta^- + \bar{v} + 0.579 \text{ MeV} \tag{3.18}$$

$$^{64}_{29}\text{Cu}^{29+} \rightarrow {}^{64}_{28}\text{Ni}^{28+} + \beta^+ + v + 1.6751 \text{ MeV} \tag{3.19}$$

$^{64}_{30}Zn^{30+}$ converts to radioactive $^{65}_{30}Zn^{30+}$ through another neutron capture reaction and then decays to $^{65}_{29}Cu^{29+}$ (Faure, 1998, 19; Lide, 2007):

$$^{64}_{30}Zn^{30+} + {}^{1}_{0}n \rightarrow {}^{65}_{30}Zn^{30+}(\text{radioactive}) + \gamma \tag{3.20}$$

$$^{65}_{30}Zn^{30+} \rightarrow {}^{65}_{29}Cu^{29+} + \beta^{+} + \nu + 1.3514 \text{ MeV} \tag{3.21}$$

$^{64}_{28}Ni^{28+}$ also produces $^{65}_{29}Cu^{29+}$ through a neutron capture reaction followed by radioactive decay (Faure, 1998, 19; Lide, 2007):

$$^{64}_{28}Ni^{28+} + {}^{1}_{0}n \rightarrow {}^{65}_{28}Ni^{28+}(\text{radioactive}) + \gamma \tag{3.22}$$

$$^{65}_{28}Ni^{28+} \rightarrow {}^{65}_{29}Cu^{29+} + \beta^{-} + \ddot{v} + 2.137 \text{ MeV} \tag{3.23}$$

The series may then continue with the $^{65}_{29}Cu^{29+}$ from Reactions 3.21 and 3.23 (Faure, 1998, 19; Lide, 2007):

$$^{65}_{29}Cu^{29+} + {}^{1}_{0}n \rightarrow {}^{66}_{29}Cu^{29+}(\text{radioactive}) + \gamma \tag{3.24}$$

$$^{66}_{29}Cu^{29+} \rightarrow {}^{66}_{30}Zn^{30+} + \beta^{-} + \ddot{v} + 2.642 \text{ MeV} \tag{3.25}$$

$$^{66}_{30}Zn^{30+} + {}^{1}_{0}n \rightarrow {}^{67}_{30}Zn^{30+} + \gamma \tag{3.26}$$

$$^{67}_{30}Zn^{30+} + {}^{1}_{0}n \rightarrow {}^{68}_{30}Zn^{30+} + \gamma \tag{3.27}$$

$$^{68}_{30}Zn^{30+} + {}^{1}_{0}n \rightarrow {}^{69}_{30}Zn^{30+}(\text{radioactive}) + \gamma \tag{3.28}$$

$$^{69}_{30}Zn^{30+}(\text{radioactive}) \rightarrow {}^{69}_{31}Ga^{31+} + \beta^{-} + \ddot{v} + 0.906 \text{ MeV} \tag{3.29}$$

$$^{69}_{31}Ga^{31+} + {}^{1}_{0}n \rightarrow {}^{70}_{31}Ga^{31+}(\text{radioactive}) + \gamma \tag{3.30}$$

$$^{70}_{31}Ga^{31+}(\text{radioactive}) \rightarrow {}^{70}_{32}Ge^{32+} + \beta^{-} + \ddot{v} + 1.656 \text{ MeV} \tag{3.31}$$

$$^{70}_{32}Ge^{32+} + {}^{1}_{0}n \rightarrow {}^{71}_{32}Ge^{32+}(\text{radioactive}) + \gamma \tag{3.32}$$

$$^{71}_{32}Ge^{32+} \rightarrow {}^{71}_{31}Ga^{31+} + \beta^{+} + \nu + 0.229 \text{ MeV} \tag{3.33}$$

$$^{71}_{31}Ga^{31+} + {}^{1}_{0}n \rightarrow {}^{72}_{31}Ga^{31+}(\text{radioactive}) + \gamma \tag{3.34}$$

$$^{72}_{31}Ga^{31+} \rightarrow {}^{72}_{32}Ge^{32+} + \beta^{-} + \ddot{v} + 4.001 \text{ MeV} \tag{3.35}$$

$$^{72}_{32}Ge^{32+} + {}^{1}_{0}n \rightarrow {}^{73}_{32}Ge^{32+} + \gamma \tag{3.36}$$

$$^{73}_{32}Ge^{32+} + {}^{1}_{0}n \rightarrow {}^{74}_{32}Ge^{32+} + \gamma \tag{3.37}$$

$$^{74}_{32}Ge^{32+} + {}^{1}_{0}n \rightarrow {}^{75}_{32}Ge^{32+}(\text{radioactive}) + \gamma \tag{3.38}$$

$$^{75}_{32}Ge^{32+} \rightarrow {}^{75}_{33}As^{33+} + \beta^{-} + \ddot{v} + 3.39 \text{ MeV} \tag{3.39}$$

$^{75}_{33}As$ is the only stable isotope of arsenic (Chapter 2). Supernovas expel $^{75}_{33}As^{33+}$ from the interiors of stars. They also produce additional arsenic (i.e. perhaps through the 'rapid' or *r-process*; Wallerstein *et al.*, 1997) and heavier elements, such as mercury, gold, and uranium. Not all of the $^{75}_{33}As^{33+}$ will survive the s-process or subsequent supernova. A neutron may enter a ^{75}As nucleus and convert it into radioactive $^{76}_{33}As^{33+}$. ^{76}As has a half-life of 26.3 hours and beta decays to $^{76}_{34}Se^{34+}$ (Faure, 1998, 19).

Once in space, any surviving $^{75}_{33}As^{33+}$ cools and acquires electrons. The arsenic may then accumulate with other elements to form a *nebula*. Based on analyses of meteorites (Section 3.4.2), arsenic within a nebula would tend to associate with iron and/or sulfur. Eventually, the nebula may condense into one or more new stars and possibly planets.

3.3 Arsenic in the universe as a whole

The chemical composition of the Universe is primarily estimated from models of the Big Bang (Delsemme, 1998, 19–42) and spectrographic analyses of nebulas and the atmospheres of stars (Krauskopf and Bird, 1995, 563–564). Currently, hydrogen and helium are the most abundant elements, representing 71 % and 28 %, respectively, of all known matter in the Universe (Delsemme, 1998, 22). Arsenic has been ranked as the 39th most common element in the Universe with an average concentration of 0.008 mg kg^{-1} (Matschullat, 2000, 299; Table 3.1).

3.4 Arsenic chemistry of the solar system

3.4.1 Arsenic in the Sun, Moon, and planets

Debris from earlier supernova(s) condensed into the *Solar Nebula* about 6 billion years ago (Faure, 1998, 22). By about 4.5 billion years ago, the planets had largely condensed from the nebula and the core of the Sun became dense enough to ignite through fusion Reaction 3.1. Based on the chemistry of *chondrite meteorites* (Wasson and Kallemeyn, 1988, 536), the original Solar Nebula had about 6.79 arsenic atoms for every one million atoms of silicon (Table 3.1) and 2.72×10^{10} atoms of hydrogen (Faure, 1998; Anders and Ebihara, 1982, 15).

Information on the current arsenic content of the solar system is largely limited to spectrographic analyses of the Sun, Saturn, and Jupiter; measurements on available Moon rocks and meteorites; and analyses of terrestrial materials. Spectrographic analyses indicate that the arsenic concentration of the Sun is about 0.004 mg kg^{-1} (Matschullat, 2000, 299; Table 3.1). Arsenic is moderately volatile in the vacuum of space (McDonough, 2004, 555) and should be preferentially concentrated on Jupiter and other planets

Table 3.1 Arsenic in extraterrestrial sources (see Table 3.2 for meteorites).

Source	Average number of arsenic atoms per 10^6 silicon atoms ($\pm$ 1 SD)	Average arsenic concentration (mg kg^{-1})	Reference(s)
Universe	–	0.008	Matschullat (2000)
Solar system	6.79 ± 0.75	–	Anders and Ebihara (1982)
Sun	–	0.004	Matschullat (2000)
Lunar highland samples	–	<0.01	Haskin and Warren (1991), 417
Lunar mare basalts	–	<0.1	Haskin and Warren (1991), 417
Lunar regolith and breccias	–	<1	Haskin and Warren (1991), 417

farther from the Sun. Not surprisingly, analyses of the atmospheres of Jupiter and Saturn have detected trace amounts of arsine gas (AsH_3) (Sky and Telescope, 1989; Noll, Larson and Geballe, 1990; Noll and Larson, 1991; Bezard *et al.*, 1989). AsH_3 is the most volatile inorganic arsenic species with a boiling point of about -62.5 °C on the Earth's surface (Planer-Friedrich *et al.*, 2006, 2489). The mole fraction concentration of AsH_3 in the atmosphere of Saturn is about 3 ± 1 ppb, whereas the concentration in the atmosphere of Jupiter is lower, approximately 0.22 ± 0.11 ppb (Noll, Larson and Geballe, 1990; Noll and Larson, 1991, 176). AsH_3 is a relatively heavy gas and its presence in the upper atmospheres of Jupiter and Saturn suggests that the atmospheres are well mixed by heat from the cores of the planets (Sky and Telescope, 1989).

The present lunar surface primarily consists of *anorthosites*, *mafic igneous* rocks (*basalts* and *gabbros*), and *regolith* (impact debris) (Hyndman, 1985, 344–348; Perkins, 1998, 117). Through the Apollo program, the United States collected about 382 kg of Moon rocks (Vaniman *et al.*, 1991, 9). The former Soviet Union obtained about 0.3 kg of lunar samples with their robotic missions. Typically, lunar rocks and regolith contain <1 mg kg^{-1} of arsenic (Haskin and Warren, 1991, 417; Table 3.1). Additionally, chemical analyses indicate that perhaps as many as 26 meteorites originated from the Moon (Korotev *et al.*, 2003). The meteorites were ejected to Earth by other meteorites impacting the lunar surface during the past 10 million years or so (Korotev *et al.*, 2003). Korotev *et al.* (2003) list some chemical results on five aluminum-rich (feldspathic) lunar meteorites. The five meteorites have mean arsenic concentrations that range from <0.2 to 0.79 ± 0.09 (1 SD) mg kg^{-1} (Table 3.2). Although the results are consistent with the low-arsenic concentrations of lunar rocks, they may have been affected by terrestrial contamination and weathering.

No rock samples have been collected from Mercury and Venus, and the arsenic chemistry of their crusts is unknown. Like the Moon, the crustal rocks on Mercury, Venus, and Mars are primarily basalts and other mafic rocks. If the *trace element* chemistry of their basalts is similar to lunar specimens, they should contain <1 mg kg^{-1} of arsenic.

On the basis of their chemistry, a small number of meteorites are believed to have originated from Mars. Warren, Kallemeyn and Kyte (1999, 2107, 2114) had six suspected Martian meteorites analyzed for arsenic. Arsenic was listed as 'not detected' in two of the samples (ALH77005 and Y-793605). Four others (ALH84001, EET70001, EET79001, and QUE94201) had <0.03 to <0.7 mg kg^{-1} of arsenic (Warren, Kallemeyn and Kyte, 1999; Table 3.2). Warren, Kallemeyn and Kyte (1999, 2107, 2114) also admit that the arsenic analyses were imprecise (within about 10 %) and that they could have been influenced by terrestrial weathering.

3.4.2   Arsenic in meteorites and tektites

The chemistry and mineralogy of meteorites suggest that many of them are remnants of condensates from the Solar Nebula or fragments of *asteroids* and *planetesimals* that once inhabited the early solar system (Faure, 1998, 105; Wasson and Kallemeyn, 1988, 536). The mineralogy of meteorites also indicates that some planetesimals were once large and hot enough to differentiate metallic cores and other internal layers (Faure, 1998, 105). A number of meteorites even reveal the existence of liquid water in the interiors of some planetesimals (Chapman, 1999, 341).

Meteorites are generally divided into three broad groups according to their chemistry and mineralogy; that is, *stones*, *stony irons*, and *irons* (Chapman, 1999, 353; Dalrymple, 1991, 264). As the name implies, stony irons are meteorites with intermediate compositions between irons and stones (Dalrymple, 1991, 264). Stones mostly consist of *carbonate minerals*, magnesium- and iron-rich *silicates*, and/or other nonmetallic minerals. They are typically divided into chondrites and *achondrites*. Chondrites are distinguished from achondrites by the presence of abundant tiny silicate spheres (*chondrules*), which are probably quenched liquid droplets that condensed from the Solar Nebula (Dalrymple, 1991, 449). Chondrites are further subdivided by chemistry into several major types, which are identified with letters. The letters are often initials for a prominent member of the type (e.g. CI for Ivuna chondrite) or signify the chemistry or mineralogy of the type (e.g. LL for low iron and low metal, EH for enstatite-high metal iron chondrites; Krot, Keil and Goodrich, 2004; Table 3.2).

Of all meteorites, the volatile-rich CI chondrites are believed to most closely resemble the chemistry of the solar system (Wasson and Kallemeyn, 1988; Palme and Jones, 2004). Chondrites directly condensed from the Solar Nebula (Wasson and Kallemeyn, 1988, 536). Large numbers of chondrites are believed to have agglomerated into the planetesimals, which eventually formed the inner planets of our solar system (Wasson and Kallemeyn, 1988, 536).

Irons largely or entirely consist of nickel–iron alloys. At least some of them are probably the remains of core materials of the planetesimals that once existed in the solar system (Dalrymple, 1991, 274). Based on their chemistry, irons are subdivided into several types, which are usually identified with Roman numerals and letters (IAB, IC, IVA, etc.) Krot, Keil and Goodrich (2004) discusses a common classification system for irons.

On average, iron meteorites with their nickel–iron compounds are relatively rich in arsenic, about 11 mg kg^{-1} (Matschullat, 2000, 299; Table 3.2). Arsenic may partially substitute for sulfur in iron- and sulfide-bearing minerals within the meteorites. Wasson and Richardson (2001) further list the arsenic concentrations of 48 Type IVA irons, which range from 2.00 to 14.5 mg kg^{-1} (Table 3.2). In contrast, stony meteorites only contain about 1.8 mg kg^{-1} of arsenic on average (Matschullat, 2000, 299; Table 3.2). For the various chondrite groups, average arsenic concentrations range from about 1.35 for LL (low metal) chondrites to 3.45 mg kg^{-1} for EH metal iron chondrites (Wasson and Kallemeyn, 1988, 537; Table 3.2). Arsenic concentrations in achondrites typically range from 0.17 to 5.31 mg kg^{-1} (Mittlefehldt, 2004; Table 3.2). The exceptions are the Howardite–Eucrite–Diogenite (HED) achondrites, which are mafic and *ultramafic breccias* that are relatively depleted in arsenic (0.0056–0.6 mg kg^{-1}; Mittlefehldt, 2004, 297, 309). Table 3.2 also lists the arsenic concentrations of a number of individual meteorites.

Tektites are glass spheroids, which cover parts of North America (about 35-million-years old), central Europe (about 15-million-years old), western Africa (about 1.1-million-years old) and Australia–Asia (about 780 000 years old). Most tektites are less than 1 mm in diameter (microtektites), although some are centimeter-sized (Koeberl *et al.*, 1997, 1745). Researchers generally believe that tektites represent terrestrial surface materials that were melted and ejected high into the atmosphere because of large meteorite impacts (Koeberl *et al.*, 1997, 1746).

Koeberl *et al.* (1997) performed arsenic analyses on 11 tektites from Côte d'Ivoire and four microtektites from deep-sea sediments off the coast of western Africa. The tektites and microtektites are believed to have originated from the impact that produced the Bosumtwi crater in Ghana about 1.1 million years ago. The arsenic concentrations of the 11 tektites ranged from 0.17 to 0.72 mg kg^{-1} with an average value of 0.45 ± 0.17 mg kg^{-1} (1 SD; Koeberl *et al.*, 1997, 1750; Table 3.2). The arsenic content of the four microtektites was similar to the tektites, and ranged from 0.29 to 0.58 mg kg^{-1} with an average of 0.42 ± 0.13 mg kg^{-1} (1 SD; Koeberl *et al.*, 1997, 1758; Table 3.2). The overall chemistry of the tektites and microtektites indicates that they mostly consist of melted upper crustal terrestrial rocks (Koeberl *et al.*, 1997, 1764). The low-arsenic (<1 mg kg^{-1}) concentrations in the samples are most likely due to extensive volatilization during the impact and melting event (Koeberl *et al.*, 1997, 1758–1759].

Table 3.2 Arsenic in meteorites and tektites.

Group or origin	Type (name or number of specimen(s))	Arsenic concentration (mg kg^{-1})[a]	Reference(s)
Stone meteorites	Achondrite (Chaunskij)	1.8	Matschullat (2000)
	Achondrite (EET 84302)	2.52	Mittlefehldt (2004)
	Achondrite, HED (Y-74450)	5.31	Mittlefehldt (2004)
		0.0056	Mittlefehldt (2004), 297, 309
	Achondrite (MAC 88177)	0.17	Mittlefehldt (2004)
	Chondrites, CI group	1.84	Wasson and Kallemeyn (1988)
	Chondrites, CI group	1.86	Anders and Grevesse (1989)
	Chondrites, CI group	1.85	McDonough and Sun (1995), 228
	Chondrites, CM group	1.80	Wasson and Kallemeyn (1988)
	Chondrites, CO group	1.95	Wasson and Kallemeyn (1988)
	Chondrites, CV group	1.60	Wasson and Kallemeyn (1988)
	Chondrites, EH group	3.45	Wasson and Kallemeyn (1988)
	Chondrites, EL group	2.20	Wasson and Kallemeyn (1988)
	Chondrites, H group	2.05	Wasson and Kallemeyn (1988)
	Chondrites, L group	1.55	Wasson and Kallemeyn (1988)
	Chondrites, LL group	1.35	Wasson and Kallemeyn (1988)
	Carbonaceous chondrite (Allende, 2 samples)	1.5; 1.7	Buchanan, Zolensky and Reid (1997)
	Carbonaceous chondrite (Oregueil)	1.91	Anders and Ebihara (1982)
	Carbonaceous chondrite (Oregueil)	1.85	Anders and Grevesse (1989)
Iron meteorites		11	Matschullat (2000)
	(Gibeon, 3 specimens)	3.46; 4.63; 4.76	Wasson and Richardson (2001), 952
	IIAB (4 specimens)	4.09; 3.91; 5.23; 7.43	Wasson et al. (1998)
	IIIAB (46 specimens)	3.08–23.9	Wasson et al. (1998); Wasson (1999)
	IVA (48 specimens)	2.00–14.5	Wasson and Richardson (2001), 952
Meteorites of lunar origin	Feldspathic (5 specimens)	0.56 ± 0.05; 0.79 ± 0.09; 0.19 ± 0.08; 0.12 ± 0.08; <0.2	Korotev et al. (2003), 4899
Meteorites of martian origin	(4 specimens)	<0.03; <0.4; <0.7; <0.6	Warren, Kallemeyn and Kyte (1999)
Tektites	(11 specimens from Côte d'Ivoire)	0.17–0.72	Koeberl et al. (1997), 1750
Microtektites	(4 specimens from Atlantic Ocean sediments off of western Africa)	0.29–0.58	Koeberl et al. (1997), 1758

[a]Single values denote average concentrations sometimes with 1 SD. Analyses involving different samples are separated by semicolons. Low and high values are separated by hyphens. Meteorites are susceptible to terrestrial weathering and contamination, which may affect these analytical results.

3.5 Arsenic in the bulk Earth, crusts, and interior

3.5.1 Estimating arsenic concentrations of the bulk Earth and the Earth's core and mantle

Geophysical methods are usually incapable of detecting and quantifying *trace elements* in the deep *mantle* and *core* of the Earth. Geochemists primarily use meteorites, elemental abundances from solar spectra, and the chemical properties of the elements to develop trace element models for the core and mantle. Based on the chemical properties of arsenic (Chapter 2), its moderate volatility, and its tendency to concentrate in iron meteorites rather than stones, McDonough (2004) concluded that during the Earth's formation the element would have been preferentially concentrated in the core and depleted in the silicate portion of the early Earth. The silicate portion of the Earth would have included the primitive mantle and eventually a *crust*. McDonough (2004, 553) estimated the overall arsenic concentration of the silicate portion of the Earth as 0.05 mg kg^{-1} (Table 3.3). Over the past 4.5 billion years, the Earth's primitive mantle has differentiated into a more silicate-rich crust and a 'depleted' mantle. In the process, the small concentrations of arsenic in the primitive mantle would have largely concentrated in the developing crust. As discussed later in this chapter, arsenic is also frequently associated with *hydrothermal fluids* and sulfide minerals

Table 3.3 *Estimated arsenic chemistry of the bulk Earth and its internal layers.*

Material	Arsenic concentration (mg kg^{-1})[a]	Reference
Bulk Earth	1.7	McDonough (2004), 554
Crust: continental	1.0; 1.7; 1.8	Faure (1998), 513; Wedepohl (1995), 1220; Krauskopf and Bird (1995), 590
Crust: continental (Lower)	1.3	Wedepohl (1995), 1219
Crust: continental (Upper)	5.1 ± 1	Sims, Newsom and Gladney (1990), 302
Crust: continental (Upper)	2.0	Wedepohl (1995), 1219
Crust: continental (East China)	3.0	Gao et al. (1998), 1970
Crust: continental (interior North China craton)	4.1	Gao et al. (1998), 1970
Crust: continental (south margin of North China craton)	1.2	Gao et al. (1998), 1970
Crust: continental (North Qinling Belt, China)	3.2	Gao et al. (1998), 1970
Crust: continental (South Qinling Belt, China)	3.9	Gao et al. (1998), 1970
Crust: continental (Yangtze Craton, China)	3.3	Gao et al. (1998), 1970
Core	5	McDonough (2004), 556
Mantle: depleted source of mid-ocean ridge basalts	0.0074 ± 0.0021	Salters and Stracke (2004)
Mantle: primitive	0.066 ± 0.046	Palme and O'Neill (2004), 14
'Silicate' Earth (Earth's mantle and crust, but not the core)	0.05	McDonough (2004), 553

[a]Single values denote average concentrations sometimes with 1 SD. Analyses involving different samples are separated by semicolons.

in various rocks. *Weathering* and biological activity are additionally important in concentrating arsenic in certain sediments and soils.

Using the chemistry of carbonaceous chondrite meteorites and information on element volatility, McDonough (2004, 554) further developed a model to predict the chemistry of the bulk Earth; that is, the Earth as a whole. Based on this model, the arsenic concentration of the bulk Earth is about 1.7 mg kg^{-1} (Table 3.3).

3.5.2 The core

The Earth's core is too deep to be sampled by humans and no samples are known to have erupted onto the Earth's surface. Except for a few facilities with diamond anvils or high-powered guns, pressures in the core are far too high to duplicate under laboratory conditions. Although the Earth's core probably formed under very different conditions than the cores of planetesimals (McDonough, 2004, 555), iron meteorites still provide some insights into the possible chemistry of the terrestrial core. Tungsten (W) and hafnium (Hf) isotopes in terrestrial rocks also suggest that the core differentiated fast and early within the primordial Earth, probably in less than 30 million years after the formation of the solar system (McDonough, 2004, 563). Ratios of phosphorus and neodymium measurements (P/Nd) on basalts and other mafic rocks further indicate that the overall exchange of materials between the core and mantle over the past 3.8 billion years has been <1 % (McDonough, 2004, 565). For most elements, perhaps including arsenic, the Earth's core is essentially a *closed* system. Based on the chemistry of primitive mantle (silicate-rich portion of the Earth) and bulk Earth models, McDonough (2004, 556) estimated an arsenic concentration of 5 mg kg^{-1} for the core (Table 3.3).

3.5.3 The mantle

The Earth's modern mantle extends from the base of the crust to the boundary with the liquid outer core. The mantle consists of several lithologically and physically distinct layers. The uppermost layer is solid and rigid, like the crust. A boundary, the *Mohorovičić Discontinuity* or 'Moho', separates the upper mantle from the overlying crust (Figure 3.1). The velocities of seismic waves dramatically increase as they enter the mantle at this boundary, which indicates that the mantle is ultramafic, whereas the overlying crust is considerably richer in silicate (Press and Siever, 2001, 439). The depth to the Moho from the Earth's surface varies significantly with crustal thickness and ranges from about 5–65 km (Figure 3.1) (Press and Siever, 2001, 439). The *lithosphere* refers to the rigid outer-most layer of the Earth, which includes the crusts and upper mantle. Below the lithosphere is the *asthenosphere*, which is a plastic portion of the mantle (Figure 3.1). In some regions, the asthenosphere is partially molten and contains mafic *magmas*. The asthenosphere extends to depths of about 200 km below the Earth's surface and is underlain by a solid lower mantle (Press and Siever, 2001, 441). The lower portion of the mantle is subdivided into several mineralogically distinct layers and extends down to the boundary with the liquid outer core at a depth of about 2900 km.

The differentiation of the Earth's core about 4.5 billion years ago resulted in the formation of a primitive mantle, which was relatively rich in silicate and depleted in arsenic when compared with the core (McDonough, 2004). Most of the mantle remained 'primitive' until it had undergone enough partial melting to produce a chemically distinct crust (Palme and O'Neill, 2004, 3, 6–7). Partial melting generates mafic magmas, which are still forming at various depths in the mantle (Press and Siever, 2001, 76–78). Once mafic magmas develop and move into the crust, the residual mantle rocks become 'depleted' and less primitive.

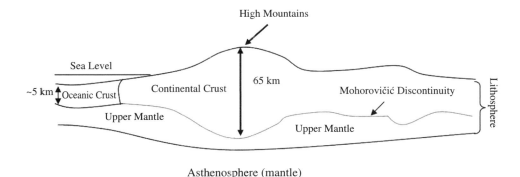

Figure 3.1 *A cross section of the crusts, rigid upper mantle, and plastic mantle asthenosphere of the Earth. The Mohorovičić Discontinuity is the boundary between the crust and upper mantle. The lithosphere includes the crusts and the upper mantle.*

Efforts have been made to determine the compositions of both the primitive mantle and depleted materials in the modern mantle. In the approach of Palme and O'Neill (2004), the estimated chemistry of the primitive mantle was determined by subtracting the likely elemental concentrations of the Earth's core from results on the bulk chemistry of the Earth. The chemistry of the bulk Earth may be derived from chemical data on CI chondrite meteorites, spectrographs of the Sun, and/or analyses of upper mantle rocks. Based on the chemical properties of an element, assumptions can be made on how much of the element was likely to have accumulated in the core. On the basis of this approach, Palme and O'Neill (2004, 14) concluded that the arsenic concentration of the primitive mantle was 0.066 ± 0.046 mg kg^{-1} (Table 3.3).

Occasionally, erupted ultramafic rocks are found in some areas of the ocean floor or as inclusions in basalts (Hyndman, 1985, 184–194; Palme and O'Neill, 2004, 7). Most of these samples are *peridotites* and their overall compositions are consistent with the geophysical properties of the upper mantle. The average arsenic concentration of ultramafic rocks is about 0.7–0.8 mg kg^{-1} (Faure, 1998, 514; Matschullat, 2000, 299; Table 3.4). Mandal and Suzuki (2002, 202) also list a range of 0.3–15.8 mg kg^{-1} of arsenic for various ultramafic samples (peridotites, *dunites*, and *serpentinites*; Table 3.4). In comparison, Salters and Stracke (2004) estimated an arsenic content of only 0.0074 $\pm$ 0.0021 mg kg^{-1} for the depleted mantle source of modern *mid-oceanic ridge basalts* (Table 3.3). Comparisons with the estimated arsenic concentration of 0.066 ± 0.046 mg kg^{-1} for the primitive mantle (Palme and O'Neill, 2004, 14) suggest that arsenic in the present mantle, at least at mid-oceanic ridges, has declined by about an order of magnitude from its original primitive state. Over the past 4.5 billion years, much of the arsenic in the primitive mantle has probably volatilized (Sims, Newsom and Gladney, 1990, 310–311). Hydrothermal fluids and magmas could have transported the volatilized arsenic into the developing crust.

3.5.4 The Earth's crusts

The Earth's mantle is overlain with crust. The crust is divided into two main types: the silicate-rich (*felsic* to *intermediate*) *continental* and the thinner mafic *oceanic crusts*. Oceanic crusts are located in the deeper parts of the ocean floor and have an average thickness of about 5 km. Continents and some shallow seas are typically underlain with continental crust. Continental crusts have an average thickness of about 40 km, but are up to 65 km thick in high mountainous areas (Press and Siever, 2001, 439; Figure 3.1).

Basalts and gabbros are the dominant igneous rocks in oceanic crusts. In rare circumstances, the mafic rocks of the oceanic crust (e.g. at the Mid-Atlantic Ridge) contain various arsenic minerals, including

Table 3.4 *Typical arsenic concentrations in some igneous rocks.*

Igneous rocks	Arsenic concentration (mg kg^{-1})[a]	Reference(s)
Andesite (Bowen Island, British Columbia, Canada)	7	Boyle, Turner and Hall (1998), 205
Andesites (Kanker district, India)	3.8	Pandey *et al.* (2006), 418
Basaltic andesites to andesites (subduction zone, Northeastern Japan, 4 samples)	0.56–7.00	Noll *et al.* (1996)
Basaltic andesites to andesites (subduction zone, Southern Volcanic Zone, Chile, 6 samples)	2.81–7.87	Noll *et al.* (1996)
Basalts (typical average)	2.2	(Faure (1998), 514
Basalts (typical values)	0.18–113	Mandal and Suzuki (2002), 202
Basalts (Iceland, 24 samples)	<0.01–0.24	Arnórsson (2003)
Basalts (Mt. St. Helens, Washington, USA, 2 samples)	0.53; 0.60	Noll *et al.* (1996)
Basalts (typical average in oceanic spreading ridges)	1.0	Matschullat (2000), 299
Basalts (subduction zone, Alaid Volcano, Kamchatka, Russia, 2 samples)	0.85; 0.91	Noll *et al.* (1996)
Basalts (subduction zone, Indian Heaven, Washington, USA, 2 samples)	0.17; 0.30	Noll *et al.* (1996)
Basalts (subduction zone, Klychevskoi Volcano, Kamchatka, Russia)	0.88	Noll *et al.* (1996)
Basalts (subduction zone, Simcoe, Washington, USA, 2 samples)	0.25; 0.38	Noll *et al.* (1996)
Basalts (subduction zone, Tolbachik Volcano, Kamchatka, Russia)	0.89	Noll *et al.* (1996)
Basalts to basaltic andesites (subduction zone, Central America, 19 samples)	0.05–3.86	Noll *et al.* (1996)
Basalts to dacites (subduction zone, Kurile Islands, 13 samples)	0.44–9.19	Noll *et al.* (1996)
Dacites (Iceland, 2 samples)	0.71; 0.75	Arnórsson (2003)
Dacites (Glassy, Julcani District, Peru)	30–70	Noble, Ressel and Connors (1998)
Diabases (typical average)	1.9	Krauskopf and Bird (1995), 590
Felsic volcanics (Kanker district, India)	8.2	Pandey *et al.* (2006), 418
Gabbros (typical values)	0.06–28	Mandal and Suzuki (2002), 202
Gabbros (Kanker district, India)	2.5	Pandey *et al.* (2006), 418
Gabbros and basalts (typical average)	0.7	Matschullat (2000), 299
Granite (Biotite, Fortymile River Watershed, Alaska, USA)	3	Crock *et al.* (1999)
Granite Pegmatites (New Hampshire)	0.15–60	Peters and Blum (2003)
Granites (typical average)	0.5	Krauskopf and Bird (1995), 590
Granites (typical values)	0.18–15	Mandal and Suzuki (2002), 202

Granites (typical average for high calcium varieties)	1.9	Turekian and Wedepohl (1961), Faure (1998), 514
Granites (Kanker district, India)	2.9	Pandey et al. (2006), 418
Granites (typical average for low calcium varieties)	1.5	Turekian and Wedepohl (1961), Faure (1998), 514
Granites and granodiorites (typical average)	3.0	Matschullat (2000), 299
Granodiorite (Mt. Mottarone, Northern Italy)	0.95 ± 0.19	Salvioli-Mariani, Toscani and Venturelli (2001)
Granodiorites (Getchell Mine, Nevada, USA, 4 samples)	63–1500	Davis et al. (2006), 179
Intermediate intrusives: (typical values for diorites, granodiorites, and syenites)	0.09–13.4	Mandal and Suzuki (2002), 202
Intermediate volcanics (Las Espinas Formation: latite, andesite, trachyte): Zimapán, Mexico	10.0–42.0	Armienta et al. (2001), 574
Intermediate volcanics (typical values for latites, andesites, and trachytes)	0.5–5.8	Mandal and Suzuki (2002), 202
Pyroxenite (Fortymile River Watershed, Alaska, USA)	3	Crock et al. (1999)
Quartz diorite (Fortymile River Watershed, Alaska, USA)	5.4	Crock et al. (1999)
Rhyodacite (Betze-Post Pit, Nevada, USA)	34	Noble, Ressel and Connors (1998)
Rhyodacite porphyry (Bowen Island, British Columbia, Canada, 2 samples)	5; 517	Boyle, Turner and Hall (1998)
Rhyolite (Glassy, Deep Star dike, Nevada, USA)	32	Noble, Ressel and Connors (1998)
Rhyolite (Iceland)	1.17	Arnórsson (2003), 1300
Rhyolite pumice (Iceland)	1.28	Arnórsson (2003), 1300
Rhyolites (typical values)	3.2–5.4	Mandal and Suzuki (2002), 202
Rhyolitic ash (La Pampa, Argentina)	7–12	Smedley et al. (2005)
Syenites (typical average)	1.4	Turekian and Wedepohl (1961)
Tuff (Eureka Valley, Eastern California, USA)	17–22	Noble, Ressel and Connors (1998)
Ultramafics (typical averages)	0.8; 0.7	(Faure (1998), 514); Matschullat (2000), 299
Ultramafics (Kanker district, India)	2.0	Pandey et al. (2006), 418
Ultramafics (Minnesota Peak, Alaska, USA, 2 samples)	6.5; 34	Crock et al. (1999)
Ultramafics (peridotite, dunite, and serpentinite)	0.3–15.8	Mandal and Suzuki (2002), 202

[a]Single values denote average concentrations sometimes with one SD. Analyses involving different samples are separated by semicolons. Low and high values are separated by hyphens.

cobaltite (CoAsS), loellingite–safflorite (FeS$_2$(Co,Fe)As$_2$), elemental arsenic (As(0)) intergrowths with elemental gold (Au(0)), luzonite (Cu$_3$AsS$_4$), and tennantite (Cu$_{10}$(Fe,Zn)$_2$As$_4$S$_{13}$) (Mozgova *et al.*, 2005). Faure (1998, 514) estimated the typical arsenic concentration of Earth basalts (which probably includes some continental samples) as 2.2 mg kg^{-1}, whereas Mandal and Suzuki (2002, 202) provided a range of 0.18–113 mg kg^{-1} (Table 3.4). For basalts from oceanic *spreading ridges*, the average arsenic content is around 1.0 mg kg^{-1} (Matschullat, 2000, 299; Table 3.4). According to Mandal and Suzuki (2002, 202), gabbros generally have arsenic concentrations of 0.06–28 mg kg^{-1} (Table 3.4). In comparison, Matschullat (2000, 299) lists an average arsenic content for gabbros and basalts as 0.7 mg kg^{-1} (Table 3.4). Considering all of the results, the average arsenic content of oceanic crusts is probably around 1 mg kg^{-1}.

As a whole, the Earth's continental crust probably contains around 1.0–1.8 mg kg^{-1} of arsenic (Faure, 1998, 513; Krauskopf and Bird, 1995, 590; Wedepohl, 1995, 1220; Table 3.3). However, arsenic measurements on individual continents or parts of continents may significantly deviate from this range. As listed in Table 3.3, Gao *et al.* (1998, 1970) estimated the arsenic concentrations of the continental crusts of eastern China as 1.2–4.1 mg kg^{-1}. The continental crust of eastern China as a whole has an arsenic composition of about 3.0 mg kg^{-1}.

Although the uppermost part of the continental crust is relatively accessible for sampling, it is very heterogeneous and its overall arsenic concentration is difficult to estimate. Sims, Newsom and Gladney (1990, 302) provides an estimate of 5.1 ± 1 mg kg^{-1} (1 SD; Table 3.3). In comparison, Wedepohl (1995) used chemical data from rocks of the Canadian *Precambrian* shield to obtain a somewhat lower arsenic concentration of 2.0 mg kg^{-1} (Table 3.3). *Granites* and *granodiorites*, which are very common in the upper continental crust (Wedepohl, 1995), typically contain around 3 mg kg^{-1} of arsenic (Matschullat, 2000, 299; Table 3.4).

Like the upper continental crust, the lower portion of the continental crust probably consists of heterogeneous distributions of various *metamorphic* and igneous rocks (Sims, Newsom and Gladney, 1990). Available chemical data from rock samples suggest that the lower continental crust is depleted in arsenic when compared with the upper crust (Sims, Newsom and Gladney, 1990, 303–304). Specifically, Wedepohl (1995) estimated the average arsenic concentration of the lower continental crust at 1.3 mg kg^{-1}, which is somewhat lower than his 2.0 mg kg^{-1} of arsenic for the upper crust (Table 3.3). For the continental crust as a whole, Wedepohl (1995, 1220) obtained an arsenic value of 1.7 mg kg^{-1}.

The weathering of surface rocks has had a critical role in the chemical evolution of the continental crust for most of the Earth's history. In the presence of air and water, mafic minerals tend to rapidly weather into iron *(oxy)(hydr)oxides*, *clays*, and other silicate minerals, and at least partially water-soluble salts of alkalis (sodium and potassium) and alkaline earths (calcium and magnesium). In contrast, *quartz* in felsic and intermediate igneous rocks is very stable in the presence of surface air and water, which explains why the mineral readily accumulates in *sands* and other sediments.

Weathering, *erosion*, and *sedimentation* are also important in concentrating arsenic in continental crusts. Togashi *et al.* (2000) even argue that weathering, erosion, and sedimentation are more responsible for enriching arsenic in the upper crust of Japan than contributions from magmas and *lavas*. Additionally, as discussed in the next section, the moderate volatility of arsenic, its sufficient solubility in hot fluids, and its reluctance to enter and accumulate in the mantle should preferentially concentrate the element in the crust.

3.6   Arsenic in hydrothermal and geothermal fluids and their deposits

3.6.1   Introduction

Hydrothermal fluids refer to naturally hot waters, including steam, which originate in the subsurface. The fluids have low (50–150 °C), moderate (150–400 °C), or high (400–600 °C) temperatures (Klein, 2002,

353). They may be associated with magmas, metamorphism, or the *diagenesis* of sediments. That is, hydrothermal fluids may occur in igneous, metamorphic, or *sedimentary rocks*. The moderate volatility of arsenic and its frequent solubility in hot aqueous fluids allows a hydrothermal fluid to extract arsenic from magmas and hot subsurface rocks, and possibly acquire more arsenic by mixing with other hydrothermal fluids (Noll *et al.*, 1996; Webster and Nordstrom, 2003, 106–107; Criaud and Fouillac, 1989, 259–260; Sriwana *et al.*, 1998, 162; Stauffer and Thompson, 1984, 2547; Kubota, Yokota and Ishiyama, 2001). Fluids may then transport arsenic over vast distances through extensive fractures and faults in the crust and eventually deposit it. Once hydrothermal waters cool below 50 °C in the subsurface, they become *groundwaters*. If hydrothermal fluids reach the Earth's surface as *geysers* and *hot springs* (Pentecost, Jones and Renaut, 2003), they are typically identified as *geothermal waters*. Geothermal resources refer to near-surface hydrothermal fluids that are utilized to generate electricity or have other economic benefits.

3.6.2 Origins of hydrothermal fluids and their arsenic

Arsenic-bearing hydrothermal fluids have multiple origins and many of them result from the mixing of two or more waters from very different sources. Low to low-moderate temperature (50–200 °C) hydrothermal fluids in deeply buried sedimentary rocks may be *connate*, which are waters that were trapped during the deposition of the original sediments and remain in the resulting sedimentary rocks (Drever, 1997, 12). In most cases, however, it is impossible to determine whether waters in deep sedimentary rocks are actually connate or groundwater that flowed into them after sediment deposition and burial or even after the sedimentary rocks formed. Due to the ambiguous origins of most waters in deep sedimentary rocks, researchers often refer to any saline solution in them as *formation water* (Drever, 1997, 12).

Other hydrothermal fluids have metamorphic origins. In the deep subsurface from about 200 °C up to the point where rocks begin to melt (about 650 °C and higher; Winkler, 1979, 5, 12; Hyndman, 1985, 511), metamorphic fluids can form from the dehydration of clays and other minerals that contain water in their crystalline structures. Metamorphic hydrothermal waters often result from the *subduction* of *tectonic plates*. In *plate tectonics*, a tectonic plate containing oceanic crust and ocean sediments may collide and subduct underneath another oceanic crustal plate that is younger, warmer, and therefore less dense (Press and Siever, 2001). Hattori and Guillot (2003) argue that metamorphic hydrothermal waters, which contain very soluble inorganic *arsenite* (inorganic As(III); e.g. $H_3AsO_3^0$, $H_2AsO_3^-$, $HAsO_3^{2-}$, and/or AsO_3^{3-}) would volatilize out of the subducting mafic crusts at <400 °C (Figure 3.2). The waters then enter and hydrate mantle peridotites that overlie the subducting plate. The hydration of peridotites produces serpentinites that effectively trap the water at concentrations of >13 wt%. The serpentinites delay the water from entering hotter mantle rocks and immediately causing partial melting (Hattori and Guillot, 2003, 527). As shown in Figure 3.2, the serpentinites also subduct with the plate. Increased heating with depth volatilizes the arsenic-bearing fluids from the subducting serpentinites. Eventually, the serpentinites reach depths of approximately 100 km, where temperatures of about 650 °C completely dehydrate them to *eclogites* (Hattori and Guillot, 2003, 527). Meanwhile, the released arsenic-bearing fluids partially melt surrounding mantle materials (Figure 3.2). The resulting melts erupt to produce *island arc* volcanoes, such as the numerous Aleutian Islands of Alaska and the islands of Japan, Indonesia, and the Philippines. Overall, the volatilization of arsenic from subducting plates returns the element to the crust and prevents it from permanently entering the mantle.

Hydrothermal fluids may also originate from magmas (*magmatic water*) or as *juvenile water* from the deep mantle (Drever, 1997, 12). Magmatic water is an important medium for a variety of *colloidal*, volatile, and dissolved species, including arsenic (almost entirely existing as As(III)). Felsic (granitic) magmas generally contain more magmatic water than mafic melts. As felsic magmas begin to solidify into granites, most of the crystallizing minerals incorporate relatively little (e.g. muscovite) or essentially no

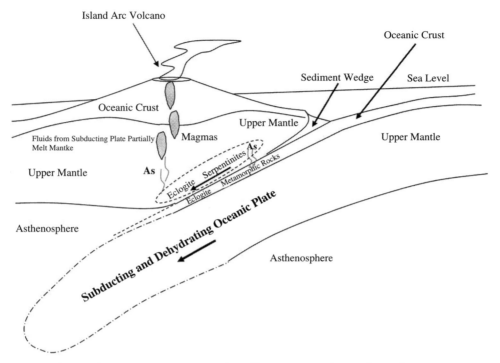

Figure 3.2 *Based on information in (Hattori and Guillot, 2003). Volcanic fronts form as a consequence of serpentinite dehydration in the forearc mantle wedge. Geology 31, (6): 525–528. Hattori and Guillot (2003) argue that arsenic-rich fluids pass from subducting oceanic crusts and their sediments and through mantle serpentinites before partially melting overlying mantle materials and forming magmas.*

(e.g. quartz and potassium feldspar) water into their crystalline structures. Magmatic water may form fluid inclusions in minerals, incorporate into large hydrous minerals in *pegmatites*, or escape through fractures in surrounding rocks.

Atmospheric precipitation (*meteoric water*) or seawater often percolates deep into the subsurface and may come into contact with magmas, hot rocks, or mix with magmatic, formation, and/or metamorphic waters. Common examples are associated with hydrothermal vents, which exist near spreading ridges on the ocean floor. Seawater moving into the subsurface extracts arsenic and other elements from rocks heated by magmas and perhaps from the magmas themselves. The resulting hydrothermal waters then rise and erupt on the ocean floor at temperatures as high as 380 °C (Press and Siever, 2001, 400). As they erupt, the hydrothermal waters deposit iron, copper, and zinc sulfides on the sea floor (Press and Siever, 2001, 533. The waters also react with cold seawater to precipitate plumes of arsenic-bearing iron (oxy)(hydr)oxides (Feely *et al.*, 1991). Specifically, 222 samples of suspended iron (oxy)(hydr)oxide precipitates from the Juan de Fuca spreading ridge in the northeastern Pacific mostly contained $161–459$ mg kg^{-1} of arsenic. The arsenic concentrations of 12 iron (oxy)(hydr)oxides samples from the Mid-Atlantic Ridge were 415 ± 146 mg kg^{-1} (Feely *et al.*, 1991, 619). Extensive hydrothermal ocean floor eruptions would increase the arsenic content of the oceanic crusts over time (Noll *et al.*, 1996, 607). The arsenic is then released into hydrothermal fluids once the crusts subduct and heat (Figure 3.2).

Magmas often exist within a few kilometers or less of the surface in volcanically and recently volcanically active areas of the continents. Meteoric waters may infiltrate deep into the subsurface, become heated, accumulate arsenic from their hot surroundings, and then flow toward the surface to produce geysers and hot springs. The famous continental geysers and hot springs of Yellowstone National Park, Wyoming, USA, often contain high concentrations of arsenic (as much as 3.5 mg kg^{-1}; Table 3.5). In contrast, the geothermal waters and groundwaters in the relatively arsenic-poor mafic rocks of northeastern Iceland typically have 0.010 mg kg^{-1} or less of arsenic (Arnórsson, 2003; Table 3.5).

The amount of arsenic *leaching* from rocks and accumulating in hydrothermal fluids may not always be proportional to the arsenic concentrations of the preleached host rocks (Ballantyne and Moore, 1988, 475). In other words, the hydrothermal fluids with the highest arsenic concentrations may not have had arsenic-rich host rocks and fluids that contain relatively little arsenic may have come into contact with relatively arsenic-rich rocks. Temperature, pH, redox reactions, iron chemistry, rock mineralogy, gas chemistry, pressure, and many other factors are important in controlling whether arsenic would tend to substantially dissolve from host rocks or precipitate/coprecipitate into them. In particular, the concentration of arsenic in hydrothermal fluids is often inversely proportional to the partial pressure of hydrogen sulfide (H_2S) gas (Ballantyne and Moore, 1988, 476). This inverse correlation probably results from the precipitation/coprecipitation of arsenic with sulfides, which occurs in the presence of H_2S gas. Hydrothermal fluids that are dominated by carbon dioxide (CO_2) also tend to carry very little dissolved arsenic (Craw *et al.*, 2002, 150). With hydrothermal fluids containing less than 3000 mg kg^{-1} of chloride, arsenic concentrations often positively correlate with temperature (Ballantyne and Moore, 1988, 476–477). Iron, pH, and redox conditions (*Eh*) may also affect the amount of arsenic in hydrothermal fluids either through the dissolution of arsenic-bearing iron (oxy)(hydr)oxides or sulfide compounds, or the coprecipitation and sorption of arsenic by these compounds (Ballantyne and Moore, 1988, 476).

Field and general chemistry studies cannot always determine the origin of a hydrothermal fluid, especially if the fluid is a complex mixture. Sometimes, geologists use stable (nonradioactive) isotopes and/or chemical ratios to distinguish magmatic and juvenile waters from other hydrothermal fluids. Minerals produced from magmatic or juvenile waters often have carbon, hydrogen, oxygen, and sulfur stable isotope ratios that are distinctive from minerals that form from other fluids, which may involve microbial activity, the leaching of materials from host rocks, diagenesis, metamorphism, or the presence of meteoric or formation waters (Faure, 1998, 311–323; Krauskopf and Bird, 1995, 260–270, 521–523). In particular, quartz forming in magmatic water is often relatively enriched in ^{16}O when compared with ^{18}O, and sulfide minerals produced from the bacterial reduction of sulfate tend to be enriched in ^{32}S when compared with ^{34}S (Krauskopf and Bird, 1995, 261–262).

Kesler, Riciputi and Ye (2005) analyzed sulfur isotopes in minerals from the Betze-Post-Screamer hydrothermal gold deposits of Nevada, USA, and found that the earliest arsenic-rich (*arsenian*) pyrites (FeS_2) primarily originated from magmatic fluids with some contributions from sedimentary rocks. Sulfur isotopes also indicated a magmatic or possibly metamorphic origin for arsenic-bearing hydrothermal deposits in the Sams Creek Granite of South Island, New Zealand (Faure and Brathwaite, 2006). In contrast, Stauffer and Thompson (1984) used Cl/As and Cl/carbonate ratios to conclude that most of the arsenic in the geothermal waters of Yellowstone National Park, Wyoming, USA, had leached from host rocks rather than having a magmatic origin.

3.6.3 Arsenic chemistry of hydrothermal fluids

$H_3AsO_3{}^0$ is the dominant dissolved arsenic species in most hydrothermal fluids because the fluids originate in reducing environments and mostly have pH values of 0–8 (Ballantyne and Moore, 1988, 478; Pokrovski *et al.*, 1996; Chapter 2). The concentrations and compositions of dissolved arsenic species in hydrothermal

Table 3.5 Arsenic concentrations in selected hydrothermal, hot spring, and geothermal waters.

Location of hot spring or well sample (number of samples)	Temperature (°C)	pH	Arsenic concentration (mg kg^{-1})[a]	Reference
Bulgaria: Southwestern: Hotovo	40.0	9.26	0.0098 (84 % As(III)) (0.45 μm filtered)	Criaud and Fouillac (1989)
Bulgaria: Southwestern: Korova	80.1	8.15	0.0078 (100 % As(III)) (0.45 μm filtered)	Criaud and Fouillac (1989)
Bulgaria: Southwestern: Ladjene	56.1	8.64	0.0026 (100 % As(III)) (0.45 μm filtered)	Criaud and Fouillac (1989)
Bulgaria: Southwestern: Sapareva Banja	97.1	8.40	0.0044 (64 % As(III)) (0.45 μm filtered)	Criaud and Fouillac (1989)
Bulgaria: Southwestern: Simitli	62.7	9.17	0.0032 (72 % As(III)) (0.45 μm filtered)	Criaud and Fouillac (1989)
Bulgaria: Southwestern: Sofia	46.6	9.50	0.0050 (92 % As(III)) (0.45 μm filtered)	Criaud and Fouillac (1989)
Bulgaria: Southwestern: Zeleznica	31.1	9.49	0.0040 (60 % As(III)) (0.45 μm filtered)	Criaud and Fouillac (1989)
Chile: El Tatio, spring #226	–	–	47.0	Stauffer and Thompson (1984)
Chile: El Tatio, spring #227	–	–	46.0	Stauffer and Thompson (1984)
Dominica: Valley of Desolation-Boiling Lake: Boiling Spring	97	3.48	0.0055 (98 % As(III)) (0.45 μm filtered)	Criaud and Fouillac (1989)
Dominica: Valley of Desolation-Boiling Lake: Boiling Lake	86	2.86	0.650 (65 % As(III)) (0.45 μm filtered)	Criaud and Fouillac (1989)
Dominica: Wooten Waven: Boiling Spring (3 samples)	95–98	7.70–7.97	0.708–0.775 (0.45 μm filtered)	Criaud and Fouillac (1989)
Dominica: Wooten Waven: Mud Pool (3 samples)	83–91	1.96–3.75	0.0003–0.223 (0.45 μm filtered)	Criaud and Fouillac (1989)
Dominica: Wooten Waven: Sulfate Spring	74	7.05	0.0003 (0.45 μm filtered)	Criaud and Fouillac (1989)
France: Cézallier: Chassolle group: Chassolle drill hole	22.1	6.09	0.300 (0 % As(III)) (0.45 μm filtered)	Criaud and Fouillac (1989)

Arsenic in Natural Environments 87

Table 3.5

France: Cézallier: Chassolle group: La Vessière	10.5	6.60	0.240 (3 % As(III)) (0.45 μm filtered)	Criaud and Fouillac (1989)
France: Cézallier: Chassolle group: Les Gravières 1	11.6	6.31	0.386 (9 % As(III)) (0.45 μm filtered)	Criaud and Fouillac (1989)
France: Cézallier: Chassolle group: Lower Vèze	12.2	6.37	0.328 (25 % As(III)) (0.45 μm filtered)	Criaud and Fouillac (1989)
France: Cézallier: Chassolle group: Ste. Marguerite spring	10.2	6.38	0.360 (5 % As(III)) (0.45 μm filtered)	Criaud and Fouillac (1989)
France: Cézallier: Chassolle group: Upper Vèze	14.9	6.37	0.464 (31 % As(III)) (0.45 μm filtered)	Criaud and Fouillac (1989)
France: Cézallier: Chassolle group: Zagat capture	15.7	6.45	0.509 (5 % As(III)) (0.45 μm filtered)	Criaud and Fouillac (1989)
France: Cézallier: Chassolle group: Zagat slope	13.0	6.68	1.120 (2 % As(III)) (0.45 μm filtered)	Criaud and Fouillac (1989)
France: Chantejail group: Autraguet	15.8	6.35	0.846 (47 % As(III)) (0.45 μm filtered)	Criaud and Fouillac (1989)
France: Chantejail group: Chantejail	10.5	6.50	1.569 (55 % As(III)) (0.45 μm filtered)	Criaud and Fouillac (1989)
France: Chantejail group: Chazelle	11.8	6.39	1.870 (60 % As(III)) (0.45 μm filtered)	Criaud and Fouillac (1989)
France: Chantejail group: La Fage	10.7	6.14	0.816 (29 % As(III)) (0.45 μm filtered)	Criaud and Fouillac (1989)
France: Chantejail group: Le Moulin	11.9	6.34	0.738 (45 % As(III)) (0.45 μm filtered)	Criaud and Fouillac (1989)
France: Chantejail group: Les Trois Sauts	12.0	6.29	1.420 (7 % As(III)) (0.45 μm filtered)	Criaud and Fouillac (1989)
France: Chantejail group: Ouche	11.7	6.20	0.322 (66 % As(III)) (0.45 μm filtered)	Criaud and Fouillac (1989)
Greece: Santorini: Palaea Kameni hydrothermal waters	—	—	0.0065–0.0165	Varnavas and Cronan (1988)
Iceland: Skagafjördur (253 samples)	3.9–89.2	8.14–10.42 (calculated at the temperature)	0.0002–0.010 19	Arnórsson (2003)

Table 3.5 (continued)

Location of hot spring or well sample (number of samples)	Temperature (°C)	pH	Arsenic concentration (mg kg⁻¹)[a]	Reference
Indonesia: West Java: Springs and crater lake (4 samples)	25.8–76.4	0.6–7.0	0.0069–0.279 (0.45 μm filtered)	Sriwana et al. (1998), 169
Italy: Ischia Island (93 samples)	13.5–100.0	5.5–9.1	0.0026–8.345 (filtered)	Aiuppa et al. (2006)
Italy: Ischia Island (37 As samples)	18.0–77.4 (43 samples)	6.31–7.88 (43 samples)	0.001–1.479 (0.45 μm filtered)	Daniele (2004)
Italy: Ischia Island: Thermal springs and shallow groundwaters (73 samples)	11–99.5	4.3–8.8	0.0005–1.558 (filtered)	Lima, Cicchella and Di Francia (2003)
Italy: Phlegrean Field: 3 geothermal wells	337; 250; 250	6.0; 7.5; 6.5, respectively	11; 9.0; 11, respectively (filtered)	Aiuppa et al. (2006)
Italy: Phlegrean Field (64 samples)	14.7–95.1	1.6–8.7	0.0016–6.939 (filtered)	Aiuppa et al. (2006)
Japan: Fushime Geothermal Field, Kyushu (2 samples)	100	7.6; 6.8	1.6; 2.3, respectively	Yokoyama, Takahashi and Tarutani (1993)
Japan: Hachobaru Geothermal Field, Kyushu (8 samples)	91–95	5.5–8.1	3.0–4.4	Yokoyama, Takahashi and Tarutani (1993)
Japan: Makizono Geothermal Field, Kyushu (1 sample)	100	8.5	3.1	Yokoyama, Takahashi and Tarutani (1993)
Japan: Ohtake Geothermal Field, Kyushu (4 samples)	92–97	8.3–8.6	2.2–2.4	Yokoyama, Takahashi and Tarutani (1993)
Japan: Shinji Lowland hot springs	40–85	–	0.010–0.114	Kubota, Yokota and Ishiyama (2001)
Japan: Takigami Geothermal Field, Kyushu (1 sample)	80	8.3	0.7	Yokoyama, Takahashi and Tarutani (1993)
Mexico: Michoacán: 4 Los Azufres geothermal field	--	5.8–7.2	5.1–24 (0.45 μm filtered)	Birkle and Merkel (2000)
New Zealand: Broadlands: Drillhole 8	–	–	5.7	Stauffer and Thompson (1984)
New Zealand: Champagne Pool water, Waiotapu Geothermal System	75.5	5.17	5.3 (0.45 μm filtered)	Phoenix et al. (2005), 325

New Zealand: Waiotapu hydrothermal fluid	220	5.9	5.7	Hedenquist and Henley (1985), 1660
New Zealand: Wairakei: Drillhole 24	–	–	4.5	Stauffer and Thompson (1984)
Papua New Guinea: Tutum Bay: Ambitle Island: Hydrothermal vent	98	6.0	0.950 (all As(III); 0.45 µm filtered)	Price and Pichler (2005)
Russia: Kamchatka: Nalychevskie hot springs	64	6.3	6.4	Nishikawa et al. (2006), 75
Turkey: Drillhole KD1a, Kizildere	–	–	39.0	Stauffer and Thompson (1984)
USA: California: Lassen Peak, Tehama county (3 samples)	–	–	2.20; 11.00; 24.30	Stauffer and Thompson (1984)
USA: Hawaii	–	–	<0.01–0.07	Webster and Nordstrom (2003), 107
USA: Nevada: Steamboat Springs (7 samples)	–	–	2.100–3.310	Stauffer and Thompson (1984)
USA: New Mexico: Jemez Springs and Soda Dam, Valles Caldera: Hot spring	58	6.52	0.460 (0 % As(III) (0.45 µm filtered)	Criaud and Fouillac (1989)
USA: New Mexico: Jemez Springs and Soda Dam, Valles Caldera: Main Jemez	74.3	6.48	0.280 (67 % As(III) (0.45 µm filtered)	Criaud and Fouillac (1989)
USA: New Mexico: Jemez Springs and Soda Dam, Valles Caldera: Soda Dam	47	6.62	1.300 (57 % As(III)) (0.45 µm filtered)	Criaud and Fouillac, 1989)[63]
USA: New Mexico: Jemez Springs and Soda Dam, Valles Caldera: Hidden warm spring	32.5	6.27	0.950 (46 % As(III)) (0.45 µm filtered)	Criaud and Fouillac (1989)
USA: New Mexico: Soda Dam hot spring	–	–	1.77	Reid, Goff and Counce (2003)
USA: New Mexico: Travertine Mound hot spring	–	–	0.830	Reid, Goff and Counce (2003)
USA: New Mexico: Sulfur Springs, Valles Caldera: Women's bathhouse	90	2.50	0.042 (<2 % As(III)) (0.45 µm filtered)	Criaud and Fouillac (1989)
USA: Oregon: Alvord Basin hot springs	30–>90	–	1–5	Koski and Wood (2003)

(continued overleaf)

Table 3.5 (continued)

Location of hot spring or well sample (number of samples)	Temperature (°C)	pH	Arsenic concentration (mg kg^{-1})[a]	Reference
USA: Utah: Roosevelt Hot Springs Thermal Area (14-2 production well)	268 (bottom hole)	5.9	3.0	Christensen, Capuano and Moore (1983)
USA: Yellowstone National Park, Wyoming (53 samples)	–	–	0.160–3.565	Stauffer and Thompson (1984)
USA: Yellowstone, Wyoming: Norris Basin	58–62	3.1	2.5 (mostly As(III))	Langner et al. (2001)
USA: Yellowstone National Park, Wyoming: Hot springs at Norris Geyser Basin	43–53	3.3–3.4	0.28–0.38 (≥90 % As(III); 0.2 μm filtered)	Nordstrom, Ball and McCleskey (2003)
New Zealand: Waiotapu hydrothermal fluid	220	5.9	5.7	Hedenquist and Henley (1985), 1660
New Zealand: Wairakei: Drillhole 24	–	–	4.5	Stauffer and Thompson (1984)
Papua New Guinea: Tutum Bay: Ambitle Island: Hydrothermal vent	98	6.0	0.950 (all As(III); 0.45 μm filtered)	Price and Pichler (2005)
Russia: Kamchatka: Nalychevskie hot springs	64	6.3	6.4	Nishikawa et al. (2006), 75
Turkey: Drillhole KD1a, Kizildere	–	–	39.0	Stauffer and Thompson (1984)
USA: California: Lassen Peak, Tehama county (3 samples)	–	–	2.20; 11.00; 24.30	Stauffer and Thompson (1984)
USA: Hawaii	–	–	<0.01–0.07	Webster and Nordstrom (2003), 107
USA: Nevada: Steamboat Springs (7 samples)	–	–	2.100–3.310	Stauffer and Thompson (1984)
USA: New Mexico: Jemez Springs and Soda Dam, Valles Caldera: Hot spring	58	6.52	0.460 (0 % As(III)) (0.45 μm filtered)	Criaud and Fouillac (1989)

[a]Single values denote average concentrations or single measurements. Analyses involving different samples are separated by semicolons. Low and high values are separated by hyphens. For waters that contain a lot of dissolved solids (including brines and many hydrothermal waters), 1 L of water has a mass greater than 1 kg and concentrations are more accurately reported as milligrams per kilogram for parts per million and micrograms per kilogram for parts per billion, rather than milligrams per liter and micrograms per liter, respectively. Available information on arsenic speciation and micron filtering of the samples is listed.

fluids and the chemistry of any precipitates would depend on redox conditions, pH, temperature, the possible presence of microbes, and the general chemistry of the fluids. Besides hydroxides, dissolved arsenic forms ligands with ammonia, cyanide, fluoride, various organics, and especially sulfide (Ballantyne and Moore, 1988, 478; Williams, 2001).

Under *anoxic* conditions in the subsurface, precipitation/coprecipitation, sorption, and dissolution reactions in hydrothermal fluids commonly involve realgar (AsS or As_4S_4), arsenopyrite (FeAsS), arsenian pyrite (FeS_2), and especially orpiment (As_2S_3). Orpiment dissolves in reducing and low H_2S hydrothermal waters at temperatures up to at least 300 °C as shown in the following reaction (Webster and Nordstrom, 2003, 110):

$$As_2S_3 + 6\,H_2O \rightarrow 2\,H_3AsO_3^{0} + 3\,H_2S \tag{3.40}$$

The dissolution of orpiment also increases under alkaline conditions (Webster and Nordstrom, 2003, 110, 112).

With increasing sulfide concentrations, Reaction 3.40 would initially reverse and precipitate As_2S_3. However, if sulfide concentrations in hydrothermal fluids approach about 0.001 mol kg^{-1}, orpiment could begin to dissolve and produce *thioarsenic* complexes. Traditionally, thioarsenic complexes in sulfide-rich and anoxic waters were identified as *thioarsenites*. For example, Webster and Nordstrom (2003, 111) suggested that $H_2As_3S_6^{-}$ would form from the dissolution of orpiment:

$$3As_2S_3 + 3HS^{-} + H^{+} \rightarrow 2H_2As_3S_6^{-} \tag{3.41}$$

Other possible thioarsenite complexes include neutral or anionic (-1 or -2) forms, such as $HAs_3S_6^{2-}$ or $H_3As_3S_6^{0}$ (Ballantyne and Moore, 1988, 478; Yang and Blum, 1999aa, 167; Helz *et al.*, 1995; Schaufelberger, 1994, 407; Keimowitz *et al.*, 2005; Wilkin and Ford, 2006; Hering and Kneebone, 2002, 165). Although the existence of thioarsenites in hydrothermal and other sulfide-rich and anoxic waters has been widely accepted, recent laboratory studies and analyses of geothermal waters from Yellowstone National Park, Wyoming, USA, indicate that the thioarsenic species in these waters are *thioarsenates* rather than thioarsenites (Stauder, Raue and Sacher, 2005; Planer-Friedrich *et al.*, 2007; see also Chapter 2). Effective sample preservation and analytical methods are needed before researchers can determine whether thioarsenites, thioarsenates, or both are significantly present in hydrothermal and other high-sulfide and anoxic natural waters.

As(III) chloride complexes could exist in some highly acidic and saline fluids. However, in most hydrothermal waters, chloride and inorganic arsenic would not form complexes (Ballantyne and Moore, 1988, 475–476, 478).

3.6.4 Arsenic mineralogy of hydrothermal deposits

Changes in pressure, temperature, and gas *miscibility* in hydrothermal fluids can lead to the precipitation/coprecipitation of arsenic or its sorption onto minerals and other solid surfaces. In particular, as hydrothermal fluids begin to cool or mix with cooler meteoric groundwater as they approach the surface, a variety of arsenic-bearing minerals may precipitate/coprecipitate (Stauffer and Thompson, 1984, 2557). Arsenic-bearing hydrothermal deposits may disseminate through rocks or concentrate in the fractures, faults, and voids of almost any type of sedimentary, metamorphic, or igneous rock, including *dolostones* (Van Moort, Hotchkis and Pwa, 1995; Ayora and Casas, 1986), *limestones* (Turner *et al.*, 1994; Crespo *et al.*, 2000; Percival *et al.*, 1989). *shales* (Crespo *et al.*, 2000; Percival *et al.*, 1989; Panno, Sayre and

Harbottle, 1985), *sandstones* (Scott, 1986; Loredo *et al.*, 2004), basalts (Borba, Figueiredo and Matschullat, 2003), *gneisses* (Skanes, Kerr and Sylvester, 2004), *slates* (Yang and Blum, 1999ab), *schists* (Tingle *et al.*, 1996; Lerouge, Bouchot and Guerrot, 2000; Kerr, Craw and Youngson, 1999; Rasilainen, Nurmi and Bornhorst, 1993), *diorites* (Power *et al.*, 2004), and granites (Faure and Brathwaite, 2006).

Arsenian pyrite is one of the more important host minerals for gold (Reich *et al.*, 2005; Arehart, 1996). Gold probably exists as a *solid solution* (Au^+) in arsenian pyrite or as $Au(0)$ nanoscale inclusions within the minerals (Reich *et al.*, 2005, 2781). Gold-bearing arsenian pyrite may form through the following reaction in hydrothermal fluids (Reich *et al.*, 2005, 2790):

$$Fe^{2+} + 2HAsS_2^0 + 2Au(HS)^0 + 2H_2 \rightarrow Fe(S, As)_2 \cdot Au_2S^0 + 3\,H_2S + 2\,H^+ \tag{3.42}$$

Decreases in H_2S concentrations through boiling and volatilization of hydrothermal fluids would tend to precipitate gold-bearing arsenian pyrite.

The arsenic mineralogy of hydrothermal gold deposits may be quite complex. Although gold and arsenic often coexist in arsenian pyrite, other arsenic- and gold-bearing hydrothermal minerals may not always form at the same time or from the same hydrothermal events even though they may be in close physical proximity to each other. In some cases, stable isotope analyses, outcrop characteristics, radiometric dating, and mineral overgrowths may be used to decipher the origins and age relationships between minerals (e.g. Faure and Brathwaite, 2006; Mehrabi, Yardley and Cann, 1999; Zachar á s *et al.*, 2004; Sidle, Wotten and Murphy, 2001; Sidle, 2002; Bhattacharya *et al.*, 2000). Specifically, the Getchell hydrothermal gold deposits in north-central Nevada, USA, contain arsenopyrite, arsenian pyrite, orpiment, and realgar (Davis *et al.*, 2006; Cline, 2001). Cross-cutting relationships between the minerals and radiometric dating of the host rocks indicate that most of the arsenic minerals formed significantly earlier (e.g. arsenopyrite) or later (orpiment and realgar) than the main gold- and arsenic-bearing pyrites. The mineral successions are so separate in time that they may have actually involved different hydrothermal events (Cline, 2001). In other locations, individual minerals (such as arsenopyrite) may even form during multiple hydrothermal events. For example, the Roudný gold deposits located in the Bohemian Massif of the Czech Republic contain two generations of arsenopyrite and four generations of pyrite with up to 4.5 wt % arsenic (Zachariás *et al.*, 2004).

Beside gold deposits, arsenic-bearing hydrothermal minerals frequently occur in ores containing copper, tin, nickel, lead, uranium, zinc, cobalt, platinum, or colemanite (a boron mineral, $CaB_3O_4(OH)_3 \cdot H_2O$) (Mandal and Suzuki, 2002; Craig, Vaughan and Skinner, 2001, 327; Power *et al.*, 2004; Arehart, 1996; Grip, 1961; Lawrence *et al.*, 1999; Pichler, Hendry and Hall, 2001; Massacci *et al.*, 2003; Çolak, Gemici and Tarcan, 2003). Under reducing conditions, arsenian pyrite and arsenopyrite are the most common arsenic minerals in hydrothermal deposits and hydrothermally altered rocks (Smedley and Kinniburgh, 2002, 528–529). Besides pyrite, arsenic may partially substitute for sulfur in the crystalline structures of several other sulfide minerals, including marcasite (FeS_2), pyrrhotite ($Fe_{0.8-1}S$), sphalerite (ZnS), and stibnite (Sb_2S_3) (Shanks and Lichte, 1996). Natural stibnite may contain more than 0.5 wt% of arsenic (Ashley *et al.*, 2003). Other arsenic-bearing minerals that may be present in hydrothermal deposits include: orpiment, realgar, nickeline ($NiAs$), loellingite ($FeAs_2$), tennantite ($(Cu,Fe)_{12}As_4S_{13}$), and arsenolite (As_2O_3) (Pfeifer *et al.*, 2004, 207; Craig, Vaughan and Skinner, 2001, 353; Mandal and Suzuki, 2002; Van Moort, Hotchkis and Pwa, 1995; Mehrabi, Yardley and Cann, 1999).

Temperature is an important factor in controlling the arsenic mineralogy of hydrothermal deposits. Many mineralogical successions in hydrothermal deposits result from cooling, including the frequent precipitation of arsenopyrite before orpiment and realgar (Cline, 2001; Smedley and Kinniburgh, 2002, 528). Above 250–300 °C, orpiment and realgar will dissolve in aqueous fluids. In the presence of Fe(II) and at temperatures of at least 400–450 °C, dissolved arsenic may precipitate as arsenopyrite (Lerouge, Bouchot and

Guerrot, 2000; Pokrovski, Kara and Roux, 2002, 2361). Arsenian pyrite is also common in hydrothermal deposits at temperatures as low as 150 °C and as high as 250–300 °C (Kesler, Riciputi and Ye, 2005, 132; Reich *et al.*, 2005, 2788; Pokrovski, Kara and Roux, 2002, 2375, 2361). Most pyrites contain only 0.02–0.5 wt % arsenic (Welch *et al.*, 2000, 597). However, arsenian pyrites may host up to 6 wt % arsenic as a solid solution with sulfur (Reich and Becker, 2006). Although pyrites from Nevada, USA, contain as much as 19.76 wt %, much of this arsenic exists as nanoscale arsenopyrite or other mineral inclusions rather than as a true solid solution (Reich *et al.*, 2005; Reich and Becker, 2006, 2784–2786).

At temperatures below about 150–200 °C, arsenic primarily precipitates as orpiment, realgar, or the elemental form (As(0)) (Pokrovski *et al.*, 1996). As(0) forms under stronger reducing conditions than realgar and orpiment, and the element often exists as a solid solution with antimony (i.e. allemontite (SbAs) and 'stibasen' (AsSb); Wretblad, 1941). Very rare intergrowths of As(0) with Au(0) occur in hydrothermal deposits of the Logatchev-2 field at the Mid-Atlantic Ridge between Africa and South America. The intergrowths probably resulted from the cooling and *exsolution* of a high-temperature gold–arsenic alloy (Mozgova *et al.*, 2005, 314).

Arsenic concentrations in most hot spring waters range from 0.1 to 3 mg kg^{-1} (Ballantyne and Moore, 1988, 477). Under reducing conditions, hot spring deposits may include significant orpiment, realgar, and arsenian specimens of marcasite, stibnite, and metastibnite (Sb_2S_3) (Ballantyne and Moore, 1988, 478, 480–481). If inorganic *arsenate* (inorganic As(V); e.g. $H_3AsO_4{}^0$, $H_2AsO_4{}^-$, $HAsO_4{}^{2-}$, and/or $AsO_4{}^{3-}$) is dominant in well-oxidized hot spring waters, a variety of As(V) minerals may precipitate under low water-to-solid ratios and temperatures above 100 °C, including scorodite ($FeAsO_4 \cdot 2H_2O$), annabergite ((Ni_3AsO_4)$_2 \cdot 8H_2O$), pharmakosiderite ($Fe_4(AsO_4)_3(OH)_3 \cdot 6H_2O$), pharmacolite ($CaHAsO_4$), and beudantite ($PbFe_3(AsO_4)(OH)_6$) (Pfeifer *et al.*, 2004, 207). On the other hand, moderate reducing conditions maximize arsenic solubility in near-surface hydrothermal waters (e.g. southern Italy; Aiuppa *et al.*, 2003). The waters are too oxidizing for arsenic sulfides to remain stable, but reducing enough to allow for the release of arsenic by the *reductive dissolution* of iron (oxy)(hydr)oxides (Section 3.11).

3.6.5 Surface and near-surface oxidation of hydrothermal arsenic

As fluids approach the surface and become diluted with aerated groundwater, As(III) will begin to oxidize to As(V). By itself, air is very slow in oxidizing As(III) and considerable As(III) may persist for some time even under well-aerated conditions (Pokrovski *et al.*, 1996, 737; Chapter 7). In surface and near-surface environments, natural chemicals, light, and/or microbial activity can increase the oxidation of As(III) (Yokoyama, Takahashi and Tarutani, 1993; Price and Pichler, 2005; Hering *et al.*, 1999; Mielke, Southam and Nordstrom, 2000, 132). Specifically, manganite (γ-MnOOH) and other naturally occurring manganese (III,IV) (oxy)(hydr)oxides are often effective *oxidants* in water, especially if the compounds are poorly crystalline (high-surface areas) and are in the presence of light and iron (oxy)(hydr)oxides (Stollenwerk, 2003, 70). Once oxidized to As(V), iron and manganese (oxy)(hydr)oxides are capable of sorbing the arsenic (Amirbahman *et al.*, 2006, 534).

Microorganisms may be important in oxidizing As(III) in shallow marine hydrothermal vents (Price and Pichler, 2005) and terrestrial hot springs (Langner *et al.*, 2001). *Thermus aquaticus* and *Thermus thermophilus* are examples of microorganisms that live in hot springs and that are capable of oxidizing As(III) (Gihring, 2001; Inskeep *et al.*, 2004, 3145). Once hot spring waters enter local streams, such as Tantalus Creek at Yellowstone National Park or the Waikato River in New Zealand, most of the arsenic has been oxidized (Webster and Nordstrom, 2003, 119). Nevertheless, periodic increases in total arsenic and As(III) are known to occur in the Waikato River (Webster and Nordstrom, 2003, 120). The increases may be related to temperature-driven overturning in deep lakes at nearby geothermal power plants (Webster and

Nordstrom, 2003, 120). Seasonal changes in the activities of As(V)-reducing bacteria (perhaps *Anabaena oscillaroides*) may also be responsible for As(III) increases in the river (Webster and Nordstrom, 2003, 120).

3.6.6   Arsenic chemistry in hot springs

Once hydrothermal fluids approach the surface, lower pressures cause the liquids to boil. As steam separates from hydrothermal water, arsenic preferentially remains in the liquid phase. Above 200 °C, only about 0.1–0.5% of the arsenic in surface and near-surface hot springs partitions into steam (Ballantyne and Moore, 1988, 477). Table 3.6 lists the arsenic concentrations in condensates of gases from various volcanoes and hot springs.

Amorphous 'As$_2$S$_3$', which has an uncertain composition (Helz *et al.*, 1995, 4602–4603), and orpiment often precipitate in hot springs and geothermal wells. The compounds are currently precipitating in hot

Table 3.6   *Arsenic concentrations in condensates of volcanic and geothermal gases from various locations. Except for three analyses from Nicaragua by Quiseft et al. (1989), all of the data are from Signorelli (1997), 242.*

Location	Temperature (°C)	Arsenic concentration (mg kg^{-1}, in condensates)[a]
Antarctica: Deception Islands (5 samples)	30–97	<0.006–0.80
Chile: Lonquimay	72	0.50
Colombia: Nevada del Ruiz	86	<0.006
Columbia: Galeras (2 samples)	208; 642	8.70; 15.00 (respectively)
Costa Rica: Poàs (2 samples)	117; 118	3.74; 3.75 (respectively)
Italy: Campi Flegrei (8 samples)	157–159	1.00–4.00
Italy: Stromboli	410	<0.006
Italy: Vulcano (7 samples)	90–625	1.20–30.00
Japan: Usu (4 samples)	648–704	1.70–3.90
Japan: Noboribetsu spa (2 samples)	92	1.60; 2.30
Mexico: Colima	168	1.00
Mexico: Chichon	94	<0.006
Mexico: Tacanà	89	1.40
New Zealand: White Island	365	4.00
New Zealand: Whakarewarewa	93	2.50
Nicaragua: Momotombo (3 samples, (Quiseft *et al.*, 1989))	878–886	2.1–7.5
Nicaragua: Momotombo	838	2.00
Portugal: Azores (7 samples)	85–100	3.50–9.50
Russia: Kamchatka: Uzon Caldera	89	4.90
Russia: Kamchatka: Mutnovsky	118	1.50
Russia: Kamchatka: Geyser Valley	97	1.25
Spain: Canary Islands (2 samples)	96; 86	4.40; 3.80 (respectively)
USA: Lassen Peak	94	<0.006
USA: St. Helens	602	<0.006
USA: Kilauea	85; 90	<0.006; 1.00 (respectively)

[a]Single values denote an average or single analysis of the concentration. Analyses involving different samples are separated by semicolons. Low and high values are separated by hyphens.

springs within the Norris Geyser Basin of Yellowstone National Park (Nordstrom, Ball and McCleskey, 2003; Mielke, Southam and Nordstrom, 2000) One orpiment-precipitating spring has a pH of 3.3–3.4 and arsenic ($\geqslant$90% as As(III)) concentrations of 0.28–0.38 mg kg^{-1} (Nordstrom, Ball and McCleskey, 2003). Chemical modeling of the hot spring waters indicates the presence of dissolved thioarsenic species and $H_3AsO_3^0$.

Orpiment and amorphous 'As$_2$S$_3$' are also precipitating with realgar in hot springs at the Kamchatka Peninsula, Russia (Cleverley and Benning, 2003). Like the Norris Basin springs, modeling suggests that $H_3AsO_3^0$ is the primary dissolved arsenic form in the Kamchatka fluids along with possible minor thioarsenic complexes (such as, As$_2$S$_3^0$) (Cleverley and Benning, 2003). Except for amorphous 'As$_2$S$_3$', geochemical modeling indicates that the distribution of arsenic minerals in the springs can be explained by a 15 mg kg^{-1} arsenic fluid cooling from 125 to 25 °C under reducing conditions with only trace amounts of oxygen (Cleverley and Benning, 2003). The low levels of oxygen probably originated from meteoric waters that mixed with the hydrothermal waters. In another study of Kamchatka hot springs, Tazaki *et al.* (2003) concluded that microorganisms have an important role in precipitating arsenic-bearing compounds.

Methylarsenic species occur in the liquids and vapors of some hot springs. Several methylarsenic species are volatile below 150 °C, including methylarsine ((CH$_3$)AsH$_2$: boiling point of -2 °C at atmospheric pressure), dimethylarsine ((CH$_3$)$_2$AsH: boiling point of 36 °C at atmospheric pressure), and trimethylarsine ((CH$_3$)$_3$As: boiling point of 52 °C at atmospheric pressure) (Planer-Friedrich *et al.*, 2006, 2481; Chapter 2). Hirner *et al.* (1998) sampled geothermal waters from Ruapehu (Vulcano Crater Lake), Waimangu (Frying Pan Lake), Waiotapu (Lake Champagne and Lake Ngakoro), and Tokaanu on the North Island of New Zealand. They found that up to 1% of the dissolved arsenic in the waters is methylated and includes methylarsine, dimethylarsine, and trimethylarsine. They also detected arsine (AsH$_3$) in some of the waters at Waimangu, Waiotapu, and Tokaanu.

Planer-Friedrich *et al.* (2006) found several methylarsenic compounds volatilizing from hot springs at Yellowstone National Park. The Yellowstone samples contained total volatile arsenic concentrations of 0.5–200 milligram per cubic meter (mg m^{-3}). Most frequently, the abundance of methylarsenic species in the samples were dimethylchloroarsine ((CH$_3$)$_2$AsCl) > trimethylarsine > dimethylarsenomercaptane (CH$_3$)$_2$AsSCH$_3$ > methyldichloroarsine (CH$_3$AsCl$_2$) (Planer-Friedrich *et al.*, 2006). Although no arsenic-volatilizing microorganisms have been identified in the Yellowstone springs, available chemical and biological evidence suggests that the origin of the volatile methylarsenic compounds is probably biotic rather than abiotic (Planer-Friedrich *et al.*, 2006, 2489).

3.6.7   Arsenic in geothermal power plant scales

In some pipe deposits in geothermal power plants, arsenic is associated with clays or other silicate minerals rather than sulfides or (oxy)(hydr)oxides. Pascua *et al.* (2005) found that about 80% of the arsenic in pipe scales from a Japanese geothermal power plant was associated with Mg-rich *smectite* clays. The arsenic (mostly III) was probably located in the crystalline structures of the clays and/or present as submicron inclusions.

Aqueous As(III) rapidly oxidizes to As(V) in the pipes and drains of some geothermal power plants. Yokoyama, Takahashi and Tarutani (1993, 109–110) speculated that rapid oxidation could be due to iron and manganese (oxy)(hydr)oxides suspended in the water or coating the walls of the pipes. Once oxidized, As(V) may sorb onto the iron (oxy)(hydr)oxides in pipe and drain scales (Webster and Nordstrom, 2003, 116).

3.6.8   Arsenic in volcanic gas emissions

Volcanoes are significant sources of arsenic emissions to the atmosphere (Signorelli, 1997; Nriagu, 1989, 239). Globally, volcanoes release about 17 150 t As year^{-1} (1 t $=$ 1000 kg; Matschullat, 2000, 300). Signorelli (1997) summarizes arsenic measurements of volcanic gases from the literature and results from 62 condensates of volcanic gases (Table 3.6). Unlike some other researchers, Signorelli (1997) did not always find a positive correlation between arsenic concentrations in the condensates and temperature. The properties of magmas and surrounding rocks, as well as volatile chemicals, affect arsenic concentrations in volcanic gases. Specifically, condensing sulfur often scavenges arsenic in cooler ($<$100 °C) gases (Signorelli, 1997, 241).

Kilauea volcano, Hawaii, USA, erupted in January, 1983. During a two-week period including the eruption, atmospheric particulate matter (PM) around the volcano was collected on 0.4 μm air filters (Zoller, Parrington and Phelan Kotra, 1983). Measurements of the atmospheric particles yielded 0.0045–1.600 ng m^{-3} (nanograms of arsenic per cubic meter of air). Before the eruption, the average arsenic concentrations were 0.013–0.039 ng m^{-3} (Zoller, Parrington and Phelan Kotra, 1983).

3.6.9   Environmental impacts of arsenic in hydrothermal and geothermal fluids

Naturally occurring hydrothermal and geothermal solutions often exceed water standards for arsenic (Appendix E). For example, many geothermal waters on Ischia Island, Italy, contain $>$1 mg L^{-1}) of arsenic, whereas the Italian 'intervention limit' for arsenic is only 0.01 mg L^{-1} (Lima, Cicchella and Di Francia, 2003). There are also concerns that arsenic from hot springs or geothermal power plants could contaminate local surface waters and groundwaters. In Montana, USA, natural geothermal arsenic has contaminated the Madison River and local groundwaters. The headwaters of the Madison River are located in Yellowstone National Park. The river exiting the park contains about 0.200 mg L^{-1} of dissolved arsenic (Sonderegger and Sholes, 1989). In Montana, the river waters are used for irrigation and local drinking water *aquifers* have been contaminated with arsenic from infiltration of the irrigation water. Arsenic levels in the blood of local residents have declined since their well water has been treated with *reverse-osmosis* technologies (Sonderegger and Sholes, 1989; Chapter 7).

The recovery and use of geothermal waters by power plants may create significant environmental problems, including surface subsidence, emissions of CO_2, H_2S, and other geothermal gases into the atmosphere, and groundwater contamination from the injection of extremely saline and arsenic-rich waters back into the subsurface (Birkle and Merkel, 2000, 371–373; Webster and Nordstrom, 2003, 123). Until spent waters are returned to the subsurface or otherwise discharged, they are often stored in ponds. The saline and arsenic-rich pond waters can poison wildlife or leak and contaminate surrounding surface waters and groundwaters. By 1994, large numbers of migrating birds had been poisoned by ponded arsenic- and boron-bearing *brines* at the Cerro Prieto geothermal field in northern Mexico (Birkle and Merkel, 2000, 372). More recently, leaking ponds and pipelines at the Los Azufres geothermal field in Michoacán, Mexico, have contaminated surface waters as far away as 10 km from the geothermal facilities with up to 8 mg kg^{-1} of arsenic (Birkle and Merkel, 2000). There are concerns that livestock and crops may come in contact with the contaminated waters and that arsenic and other contaminants might move through the food chain and threaten human health. Birkle and Merkel (2000) made several recommendations to reduce the environmental impacts of the geothermal facilities, including keeping brines away from wildlife and the atmosphere, better sealing of the ponds, and developing procedures that would allow the spent waters to be injected back into the subsurface as soon as possible.

At least in the recent past, the Waikato River and other streams in New Zealand have had excess arsenic (Aggett and Aspell, 1979). For the Waikato River, the arsenic originated from both natural sources and a local geothermal power station. Elsewhere in New Zealand, discharges from the Kawerau geothermal field slightly elevated the typical arsenic concentrations of the Tarawera River from 21 to 29 $\mu g\ kg^{-1}$ (Mroczek, 2005). Most of the arsenic is probably dissolved rather than associated with suspended materials. However, high river flows could increase the transport of arsenic-bearing solids (Mroczek, 2005).

3.7 Oxidation of arsenic-bearing sulfides in geologic materials and mining wastes

3.7.1 Oxidation of sulfide minerals

Arsenopyrite, arsenian pyrite and other Fe(II) sulfides, orpiment, and realgar are the major arsenic-bearing sulfide minerals in the Earth's crust. Arsenic-bearing sulfide minerals occur in sediments, soils, and a variety of rocks, but they are especially common in hydrothermal deposits, *coals*, and felsic igneous rocks. Most sulfide minerals are very stable and insoluble in water if they are left undisturbed under *anaerobic* conditions in the subsurface. However, sulfide minerals may readily decompose if natural processes or mining and other human activities bring them into contact with near ambient temperature (<50 °C) aqueous solutions containing oxidizing bacteria and/or chemical oxidants (such as air and manganese (oxy)(hydr)oxides). Once oxidized, sulfide minerals may release potentially toxic sulfuric acid and soluble trace elements (such as arsenic, lead, chromium, cobalt, nickel, zinc, mercury, copper, and cadmium). In surface environments, atmospheric precipitation may wash the contaminants into nearby streams and lakes, which will often, but not always, acidify the surface waters. Sulfuric acid and trace elements may then harm aquatic wildlife and possibly threaten human health.

3.7.2 Factors influencing the oxidation of arsenic-bearing sulfide minerals

Many factors affect the oxidation rates of sulfide minerals and the chemistry of their oxidation products. A few of the important factors are briefly introduced in this section and discussed in further detail in this and later chapters. As a result of the complex interactions between these different factors, high-arsenic rocks and mining wastes will not automatically produce high-arsenic weathering products and aqueous solutions (Piske, 1990).

3.7.2.1 Mineralogy, trace chemistry, and surface areas of the sulfides

Sulfide compounds will oxidize at different rates depending on their properties. In particular, freshly precipitated sulfides are sometimes amorphous, but crystallize over time in reducing environments (Chaillou *et al.*, 2003, 2997). Crystallization usually increases the stability of sulfide compounds. Specifically, crystalline orpiment tends to oxidize more slowly under aerated surface conditions than amorphous 'As_2S_3' (Helz *et al.*, 1995, 4603). *Polymorphs* may also show substantial differences in oxidation rates. For example, marcasite is less stable in aerated water than pyrite, its polymorph (Klein, 2002, 367). The surface area of a particle is another factor that influences oxidation. Smaller particles generally have higher surface areas, which are more susceptible to oxidation. The presence of trace elements in minerals can also change mineral properties and their stability under oxidizing conditions. In particular, arsenic impurities in pyrite increase the electrical and ionic conductivity of the mineral and its oxidation rate (Savage, Bird and Ashley, 2000, 407).

3.7.2.2 Presence and chemistry of fluids in contact with sulfide minerals

Dry air is an inefficient oxidant of sulfide minerals. In the presence of water, however, oxidants (such as O_2 and Fe(III) compounds) and strong *reductants* (such as H_2S) are important in determining the stability of sulfide minerals. The pH of fluids also affects oxidation rates. Overall, the oxidation rates of arsenic sulfides (orpiment, amorphous As_2S_3, realgar, and amorphous AsS) in water tend to increase with increasing pH (Lengke and Tempel, 2005). In comparison, the oxidation rate of pyrite is greater than orpiment and realgar at pH 2.3–8. However, pyrite oxidation is slower than the oxidation of amorphous As_2S_3 and amorphous AsS at pH 8 (Lengke and Tempel, 2005).

3.7.2.3   *Chemical and physical properties of the materials surrounding the sulfides*

The *permeability*, chemistry, and other properties of the materials surrounding and especially overlying sulfide minerals may greatly affect oxidation. Specifically, fractures and faults in rocks extending into the subsurface may result in oxidizing conditions to depths of 700 m (Arehart, 1996, 389). Coarse sediments and mining wastes have especially high permeabilities, which allow air and aerated water to readily infiltrate and contact buried sulfides (Mascaro *et al.*, 2001; Komnitsas, Xenidis and Adam, 1995). Desiccation cracks or other fractures in otherwise low-permeability materials also provide routes for oxidizing liquids to migrate and react with buried sulfide minerals (Craw *et al.*, 1999).

3.7.2.4   *Mining and other human activities*

Human activities can greatly affect the stability of sulfide minerals by changing redox conditions. In particular, mining exposes underlying sulfide minerals to air and aerated water (Kwong, 2003; Schroth, Parnell and Ketterer, 2000). Furthermore, cyanide solutions or other by-products of mining and ore processing can change the chemistry and oxidizing conditions of water surrounding sulfide minerals. Although many countries have regulations that are designed to minimize water pollution at mining sites, the mining techniques themselves often unavoidably disturb sulfide deposits and promote oxidation. These techniques include the dredging of *placer deposits* in streams, which increases water turbidity and contamination (Gough *et al.*, 1999).

3.7.2.5   *Climate*

Temperature, humidity, precipitation, and evaporation are important factors that contribute to the oxidation of sulfide minerals. In warm and wet climates, excessive precipitation may produce persistently high water tables and extensive biological activity that may create reducing conditions in the shallow subsurface and hinder sulfide oxidation (Seal *et al.*, 2002, 208). At the surface, high humidity and temperatures would promote the oxidation of sulfide minerals (Williams, 2001, 274). Frequent precipitation would also suppress evaporation and the formation of arsenic salt deposits (Seal *et al.*, 2002, 208). Furthermore, precipitation and groundwater, which are controlled by climate, are the major sources of water for the production of arsenic-contaminated runoff from sulfide-bearing rock outcrops.

Climates that are cold or dry tend to have minimal runoff. Cold and dry conditions also suppress the activity of oxidizing bacteria and chemical reaction rates. Although hot arid climates may be very oxidizing, the lack of water hinders the production of contaminated runoff. Extensive evaporation with a lack of running water may allow water-soluble compounds, such as arsenolite, to precipitate in arsenic-rich mining wastes and outcrops.

A lack of water in parts of Argentina explains why sulfide oxidation and *mine drainage* contribute little to the maximum $0.140\,\text{mg}\,\text{L}^{-1}$ of arsenic in local surface waters (Williams, 2001, 274). The sodium chloride-dominated waters are rich in lithium and boron, which suggests that the arsenic is associated with the dissolution of salt deposits rather than sulfide weathering (Williams, 2001, 274). Excessive evaporation in arid climates would initially concentrate arsenic in briny lake water and eventually precipitate it in salt deposits (e.g. Spotts *et al.*, 1997; Johannesson *et al.*, 1997). Without the presence of abundant vegetation to bind and protect soils, winds may easily disperse arsenic-contaminated dust in arid and semiarid climates (Razo *et al.*, 2004, 130).

Arsenic and pH levels in surface runoff from the weathering of sulfide minerals at mines (i.e. mine drainage) may significantly vary with seasonal changes in precipitation and temperature. During the 1993 wet season at the Globe and Phoenix Mine in Zimbabwe, the pH value of the runoff streams was 3.7 and the arsenic concentration was $0.248\,mg\,L^{-1}$ (Williams, 2001, 275). During the subsequent 1993 dry season, water runoff declined by about 95 %. However, the arsenic concentration of the remaining runoff increased to $0.720\,mg\,L^{-1}$ and the pH rose to 9.5. Arsenic in the alkaline runoff probably consisted of water-soluble cyanide complexes. The alkaline cyanides originated from ore processing operations at the former mining site (Williams, 2001, 275).

3.7.2.6   *Bacteria and biological activity*

Bacteria often *catalyze* and considerably enhance the oxidation rates of sulfide minerals (Gleisner and Herbert, 2002, 140; Mihaljevič *et al.*, 2004; Section 3.7.4.1). Sulfide oxidation may also increase because of aquatic organisms burrowing or grazing on the bottom sediments of marine environments, *estuaries*, and lakes. Biologically disruption of surface sediments often exposes underlying sulfides to more oxidizing conditions. Burrowing organisms are also capable of excavating sulfide-rich sediments and entraining the finer-grained particles in oxidizing water.

3.7.2.7   *Time*

Many oxidation reactions require considerable amounts of time to reach equilibrium. In seasonal climates, sufficiently long dry periods may allow stagnant subsurface porewaters to extensively oxidize sulfide grains before flushing from storms. Bacteria also require sufficient time to thoroughly oxidize sulfide surfaces.

3.7.3   Environmental consequences of sulfide and arsenic oxidation

3.7.3.1   *Formation of mine drainage*

Mine drainage refers to surface water or groundwater becoming contaminated with *heavy metals*, arsenic, and/or sulfuric acid as the waters infiltrate into mine shafts, pits, coal piles, ore processing structures, and waste impoundments, such as *mine tailings* piles and disposal ponds. The water accumulates contaminants from oxidizing sulfide minerals in solid mine wastes and the wall rocks of pits and shafts. Water may also mix with cyanide or other liquid wastes at the facilities. In many mined areas, waters have pH values below 2.0 (Krauskopf and Bird, 1995, 148), especially areas with host rocks that are rich in quartz (e.g. most sandstones and felsic igneous rocks) and poor in calcite ($CaCO_3$) and other acid-neutralizing alkali minerals. The resulting mine drainage may then pollute surrounding soils, sediments, surface waters, and groundwaters, which could negatively impact wildlife and human health.

Overall, mine drainage and other arsenic-bearing liquids are expected to disperse more quickly in surface environments than most groundwaters. Although they may rapidly move through subsurface caves and other *karst* features, the vast majority of groundwaters pass through relatively low-permeability soils, sediments, and rocks, and travel more slowly than streams and surface runoff. If iron (oxy)(hydro)oxides and other *sorbents* are not present in host soils, sediments, and rocks, the arsenic may behave 'conservatively' (very nonreactively) and readily migrate in the subsurface.

3.7.3.2 Mine drainage properties and neutralization

Not all mine drainage or natural runoff from rock outcrops are acidic, even when extensive sulfide oxidation is present. Synthetic cyanide solutions, which are often used to extract gold and other metals from ores, can greatly increase the alkalinity of mining wastes and neutralize sulfuric acid (Craw *et al.*, 1999). In other cases, the sulfuric acid is effectively neutralized by alkaline soils, limestones, dolostones, *marbles*, *skarns*, or other carbonate-rich rocks (Pfeifer *et al.*, 2004, 219; Razo *et al.*, 2004; Lee, Lee and Lee, 2001, 491; Mendoza *et al.*, 2006). Reactions between calcium carbonate and sulfuric acid may precipitate gypsum $(CaSO_4 \cdot 2H_2O)$.

While lead, mercury, and other divalent heavy metals tend to be most soluble under acidic conditions, both inorganic As(V) and As(III) may also be very soluble in circumneutral and alkaline waters (Williams, 2001, 267; Pfeifer *et al.*, 2004; Welch *et al.*, 2000, 594; Razo *et al.*, 2004). In some cases, the weathering of pyrite produces iron (oxy)(hydr)oxides that effectively sorb arsenic from mildly acidic to neutral mine drainage (Smedley and Kinniburgh, 2002, 524). Under alkaline conditions and in the presence of undegraded cyanide, however, arsenic may readily form water-soluble complexes (Williams, 2001, 276). Arsenic complexes of cyanide were especially common in discharges from ore processing facilities at the Globe and Phoenix mines in Zimbabwe and Morro Velho in Brazil. At these facilities, the alkaline drainages (pH 8.2–9.0) had arsenic concentrations in excess of $7\,mg\,kg^{-1}$ (Williams, 2001, 270, 272).

Although limestone can neutralize acid in the short term, pH conditions generally must exceed unrealistic values of nine before arsenic appreciably sorbs onto carbonate minerals (Mihaljevič *et al.*, 2004, 61). Neutralizing acid mine drainage with lime or limestone may actually increase the oxidation of pyrite and the mobility of arsenic (Lengke and Tempel, 2005; Nicholson, 1999; Jones, Inskeep and Neuman, 1997). Liming can enhance pyrite oxidation rates by producing Fe(II)–carbonate complexes on the surfaces of the sulfides (Evangelou, Seta and Holt, 1998) and interfere with the ability of iron (oxy)(hydr)oxides to form and effectively sorb arsenic (Jones, Inskeep and Neuman, 1997). Furthermore, the acid-neutralizing effects of carbonate minerals are usually temporary because the minerals tend to rapidly dissolve. The lifetimes of individual carbonate grains under pH conditions of 6–8 generally range from weeks to a few decades, whereas under the same pH conditions, sulfide minerals may persist and release sulfuric acid, heavy metals, and arsenic for centuries to millennia (Lengke and Tempel, 2005, 352).

The highest known arsenic concentrations in mine drainage include an analysis of $850\,mg\,L^{-1}$ at the Richmond mine in Iron Mountain, California, USA, and $400\,mg\,L^{-1}$ in a sample from the Ural Mountains of Russia (Smedley and Kinniburgh, 2002, 525; Nordstrom and Alpers, 1999; Gelova, 1977). The literature contains many other arsenic analyses of mine drainage samples from all over the world, including Mexico (Razo *et al.*, 2004; Mendoza *et al.*, 2006), Montana, USA (Nimick *et al.*, 2003), Arizona, USA (Rösner, 1998), New Mexico, USA (Boulet and Larocque, 1996), South Dakota, USA (May *et al.*, 2001), Colorado, USA (Bednar *et al.*, 2005), Utah, USA (Bednar *et al.*, 2005), Idaho, USA (Bednar *et al.*, 2005), Alaska, USA (Gray and Sanzolone, 1996), Northwest Territories, Canada (Bright, Dodd and Reimer, 1996), British Columbia, Canada (Azcue *et al.*, 1995; Crusius *et al.*, 2003), Ontario, Canada (Azcue and Nriagu, 1995; Ross, Bain and Blowes, 1999), Manitoba, Canada (Moncur *et al.*, 2002), Saskatchewan, Canada (Donahue and Hendry, 2003; Moldovan, Hendry and Jiang, 2001), Spain (Loredo *et al.*, 2004; Loredo *et al.*, 1999; Loredo *et al.*, 2003), France (Migon and Mori, 1999; Bodénan *et al.*, 2004), Río Pilcomayo of Bolivia-Paraguay-Argentina (Hudson-Edwards *et al.*, 2001; Archer *et al.*, 2005), Argentina (Williams, 2001), Tunisia (Jdid *et al.*, 1999), Sweden (Sjöblom, Håkansson and Allard, 2004), Ecuador (Williams, 2001; Appleton *et al.*, 2001), Germany (Zänker *et al.*, 2002; Jakubick, Jenk and Kahnt, 2002), Australia (Ashley *et al.*, 2003), Slovak Republic (Klukanová and Rapant, 1999), Republic of Korea (Lee, Lee and Lee, 2001; Woo and Choi, 2001; Jung, Thornton and Chon, 2002), Poland (Marszałek and Wąsik, 2000),

Malaysia (Williams, 2001), Thailand (Williams, 2001), Philippines (Williams, 2001), Brazil (Williams, 2001; Borba, Figueiredo and Matschullat, 2003; Matschullat *et al.*, 2000), Hungary (Ódor *et al.*, 1998), Tanzania (Bowell *et al.*, 1995), Zimbabwe (Williams, 2001), and Switzerland (Pfeifer *et al.*, 2004). Williams (2001) is an exceptionally broad survey of arsenic in mine drainage samples from 34 sites in Asia, Africa, and South America. The arsenic concentrations of the waters ranged from 0.005 to 72 mg L^{-1}. Like most other mine drainage results, the majority of samples in Williams (2001) contained less than 5 mg L^{-1} of arsenic. The arsenic was also well oxidized. As(V) represented more than 50 % of the dissolved inorganic arsenic at 29 of the 34 sites and at least 18 of the sites had more than 95 % of their dissolved arsenic as As(V) (Williams, 2001, 269–270). The pH values of the samples ranged from 0.52 to 10. Only 3 of the 10 sites with arsenic concentrations above 1 mg L^{-1} had pH values below 4 (Williams, 2001, 269–270).

A limited number of studies have investigated the methylation of arsenic in mine drainage and areas that have been impacted by mining. Bright, Dodd and Reimer (1996) found methylarsenic in sediment porewaters and waters from the subarctic Meg, Keg, Peg, Kam, and Grace Lakes near Yellowknife, Northwest Territories, Canada. The lakes were contaminated by local gold mines (Bright, Dodd and Reimer, 1996). Inorganic forms of arsenic are dominant in the water and sediments. The methylarsenic concentrations are usually less than 10 % of the total dissolved arsenic (Bright, Dodd and Reimer, 1996). The *organoarsenicals* in the lake water primarily consist of dimethylarsinic acid (DMA(V); (CH$_3$)$_2$AsO(OH)) with some monomethylarsonic acid (MMA(V); H$_3$CAsO(OH)$_2$; (Bright, Dodd and Reimer, 1996)). Sediment porewaters contain a greater diversity of monomethylated, dimethylated, and trimethylated arsenic compounds. Sulfate-reducing bacteria are probably responsible for methylating the arsenic (Bright, Dodd and Reimer, 1996).

3.7.3.3 *Long-term environmental impacts of mining wastes*

The current environmental and human health problems associated with mine wastes are not entirely the result of twentieth-century civilizations. At many sites, mining operations and the accumulation of their wastes have been occurring for centuries, such as at Tuscany, Italy since at least the sixth to seventh centuries AD (Mascaro *et al.*, 2001) and the Potosí deposits in Bolivia since AD 1545 (Hudson-Edwards *et al.*, 2001). In some cases, the accumulation of mining and smelting wastes in lakes has left a legible historical record. The lake sediments also provide valuable estimates of premining arsenic and heavy metal levels, which may serve as target goals for remediation efforts.

Local mining and smelting activities have left a record in the sediments of Lake Süßer See, Germany (Becker *et al.*, 2001; Wennrich *et al.*, 2004). Profiles of lead, copper, and zinc concentrations in the upper 400–450 cm of the lake sediments indicate that local copper shales have been mined and smelted in the area for almost 600 years. Before mining and smelting, the lake sediments contained about 10 mg kg^{-1} or less of arsenic (Becker *et al.*, 2001, 208, 215). Sediments at depths of 10–15 cm, which were deposited during the late twentieth century, have arsenic concentrations as high as 1800 mg kg^{-1} (Becker *et al.*, 2001, 208). Another sediment core from Lake Süßer See contained up to 2140 mg kg^{-1} of arsenic, which probably accumulated during the 1950s (Wennrich *et al.*, 2004, 828, 829). Nevertheless, arsenic distributions in sediment cores must be cautiously interpreted (Chaillou *et al.*, 2003). Storms, bioturbation, seasonal changes, groundwater flow, *diffusion*, human activities, and the growth of rooted plants can redistribute arsenic in sediments and complicate historical interpretations of sample cores. In some cases, layered sediments show clear signs of bioturbation, erosion, or other disturbances. ^{210}Pb dating methods may also be used to evaluate postdepositional disturbances in sediment cores (Durant *et al.*, 2004).

3.7.4 Oxidation chemistry of major arsenic-bearing sulfides

3.7.4.1 Oxidation of pyrite and other Fe(II) sulfides

Pyrite is the most common sulfide mineral. It is a major contributor to the formation of mine drainage and sulfate-rich natural runoff. The oxidation of pyrite and other Fe(II) sulfides (e.g. marcasite and pyrrhotite) involves both iron and sulfur, as well as any arsenic impurities. Activation energies suggest that surface reactions dominate the oxidation of pyrite (Lengke and Tempel, 2005). Furthermore, evidence from pyrites in coal and ore deposits suggests that arsenian pyrite is more susceptible to oxidation from weathering than low-arsenic pyrite (Savage *et al.*, 2000, 1239).

The oxidation of pyrite by oxygenated water is often represented by simplified reactions, such as the following example from Krauskopf and Bird (1995, 148):

$$4FeS_2 + 15O_2 + 2H_2O \rightarrow 4Fe^{3+} + 8SO_4^{2-} + 4H^+ \tag{3.43}$$

In reality, the oxidation of pyrite and other Fe(II) sulfides typically involves several intermediate reactions, which may be enhanced by microbial activity or various chemical species, such as bicarbonate (HCO_3^-) (Welch *et al.*, 2000; Evangelou, Seta and Holt, 1998). The exact mechanisms of each intermediate reaction are often very complex and poorly understood (Rimstidt and Vaughan, 2003). Mostly likely, sulfide oxidizes in pyrite before iron. Fe(II) is then released into solution as shown by the following reaction involving oxygen and water (Gleisner and Herbert, 2002, 139–140):

$$2FeS_2 + 7O_2 + 2H_2O \rightarrow 2Fe^{2+} + 4SO_4^{2-} + 4H^+ \tag{3.44}$$

In the presence of aerated water, dissolved Fe(II) readily oxidizes to Fe(III) (Gleisner and Herbert, 2002, 139–140):

$$4Fe^{2+} + O_2 + 4H^+ \rightarrow 4Fe^{3+} + 2H_2O \tag{3.45}$$

Fe(III) is generally soluble in water at pH conditions below about 3.5 (Bednar *et al.*, 2005, 58). At higher pH values, Fe(III) readily reacts with water, oxygen, and/or hydroxides to precipitate Fe(III) (oxy)(hydr)oxides. Fe(III) (oxy)(hydr)oxides include iron oxides (e.g. hematite (Fe_2O_3)), hydrous oxides (certain *ferrihydrites* ($Fe_2O_3 \cdot 2H_2O$), Stollenwerk, 2003, 77), (hydrous) hydroxides (e.g. $Fe(OH)_3$), and (hydrous) oxyhydroxides (e.g. goethite (α-FeOOH) and limonite ($FeO(OH) \cdot nH_2O$, where $n > 0$)). Goethite is an especially common Fe(III) oxyhydroxide that may form from previously precipitated ferrihydrites or directly precipitate in water, possibly through the following reaction (McGregor *et al.*, 1998a, 265):

$$Fe^{3+} + 2H_2O \rightarrow \alpha FeOOH \downarrow +3H^+ \tag{3.46}$$

Bacteria may catalyze and considerably enhance the oxidation of pyrite and Fe(II) in water, especially under acidic conditions (Welch *et al.*, 2000, 597). Many microbial species actually oxidize only specific elements in sulfides. With pyrite, *Acidithiobacillus thiooxidans* is important in the oxidation of sulfur, whereas *Leptospirillum ferrooxidans* and *Acidithiobacillus ferrooxidans* (formerly *Thiobacillus ferrooxidans*) oxidize Fe(II) (Gleisner and Herbert, 2002, 140). *Acidithiobacillus ferrooxidans* obtain energy through Reaction 3.45 (Gleisner and Herbert, 2002, 140). The bacteria are most active at about 30 °C and pH 2–3 (Savage, Bird and Ashley, 2000, 407). *Acidithiobacillus* sp. and *Leptospirillum ferrooxidans* have the ability to increase the oxidation of sulfide minerals by about five orders of magnitude (Welch *et al.*, 2000, 597).

Gleisner and Herbert (2002) investigated the role of bacteria in oxidizing pyrite and producing acid mine drainage. Their experiments used tailings from an ore concentrator at Boliden in northern Sweden. The tailings contained $2960 \pm 60 \, mg \, kg^{-1}$ of arsenic, and included about 0.64 wt % arsenopyrite and approximately 32.4 wt % pyrite (Gleisner and Herbert, 2002, 143). All oxidation rate measurements lasted for 94 days and were performed at room temperature and in the presence of 0.21 atm of O_2 (Gleisner and Herbert, 2002, 141). Bacteria for the microbial samples were collected from an acidic (pH 2.6) tailings pond, which probably contained both *Acidithiobacillus thiooxidans* and *Leptospirillum ferrooxidans*. The pH of the microbial samples was initially maintained at about 2 with hydrochloric acid (HCl), but ranged from 2 to 3 after 9 days. Gleisner and Herbert (2002) obtained a microbial oxidation rate of $\geqslant 2.4 \times 10^{-10}$ mol m^{-2} s^{-1} for pyrite. Under acidic (pH 2–3) and sterile (abiotic) conditions, the oxidation rate was almost an order of magnitude lower or $\geqslant 5.9 \times 10^{-11}$ mol m^{-2} s^{-1}. For untreated samples (no controlling of pH, which was approximately eight, and no addition of bacteria), the oxidation rate was $\geqslant 3.6 \times 10^{-11}$ mol m^{-2} s^{-1} (Gleisner and Herbert, 2002). Gleisner and Herbert (2002) also found that sphalerite and chalcopyrite ($CuFeS_2$) oxidize faster than pyrite.

Aerobic conditions are not always required to oxidize sulfide to sulfate in pyrite and other sulfide minerals (Bednar *et al.*, 2005; Jonas and Gammons, 2000, 58). As shown in the following reaction from Evangelou, Seta and Holt (1998, 2084), dissolved Fe(III) may be an effective oxidant:

$$FeS_2 + 14Fe^{3+} + 8H_2O \rightarrow 15Fe^{2+} + 2SO_4^{2-} + 16H^+ \tag{3.47}$$

Nitrate (NO_3^-) is a common oxidant that is found in fertilizers, livestock manure, and sewage. Like Fe(III), NO_3^- is capable of oxidizing pyrite and other sulfide minerals in surface waters and groundwaters as shown in this reaction (Schreiber *et al.*, 2003, 261):

$$5FeS_2 + 14NO_3^- + 4H^+ \rightarrow 5Fe^{2+} + 10SO_4^{2-} + 7N_2 + 2H_2O \tag{3.48}$$

Arsenic in pyrite primarily exists as As(I–) (Simon *et al.*, 1999, 1076). Once arsenic is released into aqueous solutions by the weathering of arsenian pyrite, it would exist as As(V) or metastable As(III). As shown in the following reaction, Fe(III) is capable of oxidizing inorganic As(III) at pH conditions of about 3.5 or less (Bednar *et al.*, 2005, 58):

$$H_3AsO_3^0 + H_2O + 2Fe^{3+} \rightarrow H_2AsO_4^- + 3H^+ + 2Fe^{2+} \tag{3.49}$$

If substantial potassium and sulfate are present under strongly oxidizing and pH 2–3.5 conditions, As(V) may coprecipitate with jarosite ($KFe_3(SO_4)_2(OH)_6$) (Savage *et al.*, 2000, 1240).

3.7.4.2 Oxidation of arsenopyrite

Arsenopyrite is the most common mineral where arsenic is a major component (Welch *et al.*, 2000, 594, 597). Ideally, unaltered arsenopyrite contains about 85 % As(I–) and 15 % As(0) (Nesbitt, Muir and Pratt, 1995). The oxidation of arsenopyrite is primarily responsible for arsenic-rich acid mine drainage in many gold and other ore mines (Welch *et al.*, 2000, 594, 597).

The oxidation order of iron, sulfur, and arsenic in arsenopyrite has been controversial. Nesbitt and Muir (1998, 140) concluded that iron and arsenic oxidize faster than sulfur, at least when arsenopyrite is exposed to air. In contrast, Craw, Falconer and Youngson (2003, 80) proposed the following reaction to explain arsenopyrite oxidation in water at pH 4–9:

$$FeAsS + 7H_2O \rightarrow Fe^{2+} + H_3AsO_3^0 + 11 \, H^+ + SO_4^{2-} \tag{3.50}$$

Citing Craw, Falconer and Youngson (2003), Walker, Schreiber and Rimstidt (2006, 1674) suggest an alternative oxidation reaction, which is analogous to the oxidation of pyrite by Reaction 3.44:

$$4FeAsS + 11O_2 + 6H_2O \rightarrow 4Fe^{2+} + 4H_3AsO_3^0 + 4SO_4^{2-} \tag{3.51}$$

Both Reactions 3.50 and 3.51 indicate that sulfur oxidizes before iron and that As(I$-$) and As(0) in arsenopyrite oxidize to As(III). Nevertheless, (Craw, Falconer and Youngson (2003), 81) warn that higher than expected pH readings in their experimental data suggest that at least Reaction 3.50 may be a too simplistic description of arsenopyrite oxidation. Some of the arsenic from arsenopyrite may fully oxidize to As(V) rather than existing as $H_3AsO_3^0$ as predicted by Reaction 3.50. Using *X-ray photoelectron spectroscopy* (XPS), Nesbitt and Muir (1998) confirmed that As(III) is not the only arsenic species in surface oxidation products on arsenopyrite. As(V) and even traces of As(I) are also present.

Walker, Schreiber and Rimstidt (2006) list the oxidation rate of arsenopyrite in water as $10^{-10.14\pm0.03}$ mol m^{-2} s^{-1} at 25 °C, pH 6.3$-$6.7, and dissolved O_2 concentrations of 0.3$-$17 mg L^{-1}. Unexpectedly, the dissolved O_2 concentrations had essentially no effect on the oxidation rate. However, in laboratory solutions simulating acid mine drainage (pH = 1.8), the oxidation rate of arsenopyrite was found to increase with increasing temperature (15$-$45 °C) and concentrations of chloride or Fe(III) sulfate ($Fe_2(SO_4)_3$) (Yu, Zhu and Gao, 2004). The oxidation of arsenopyrite primarily released As(III). The oxidation of As(III) to As(V) was slow, but also increased with increasing temperature and chloride or Fe(III) sulfate concentrations (Yu, Zhu and Gao, 2004).

Scorodite, amorphous Fe(III) arsenates, and/or Fe(III) (oxy)(hydr)oxides commonly precipitate once Fe(III) and As(V) are released by the aqueous oxidation of arsenopyrite (Williams, 2001, 273; Krause and Ettel, 1988, 851; Craw, Falconer and Youngson, 2003, 73). Even without O_2, Fe(III) oxidation of arsenopyrite produces scorodite as shown in the following reaction (Dove and Rimstidt, 1985, 838, 842):

$$FeAsS + 14Fe^{3+} + 10H_2O \rightarrow 14Fe^{2+} + SO_4^{2-} + FeAsO_4 \cdot 2H_2O \downarrow + 16H^+ \tag{3.52}$$

Fe(III) can oxidize arsenopyrite about 10 times faster than pyrite and the rates are even more rapid if *Acidithiobacillus ferrooxidans* is present (Welch *et al.*, 2000, 597; Gleisner and Herbert, 2002, 140; Evangelou, Seta and Holt, 1998, 2084). Scorodite, a product of Reaction 3.52, may form *colloids* in water and *natural organic matter* (NOM) could assist in stabilizing the colloids (Buschmann *et al.*, 2006, 6019).

Amorphous Fe(III) arsenates tend to be more soluble in water than crystalline scorodite (Frau *et al.*, 2005, 199). Nevertheless, the exact solubilities of crystalline scorodite and amorphous Fe(III) arsenates are somewhat uncertain and controversial (Welch *et al.*, 2000, 597; Krause and Ettel, 1988; Dove and Rimstidt, 1985; Robins, 1987; Nordstrom and Parks, 1987; Langmuir, Mahoney and Rowson, 2006; Chapter 2). Under at least some poorly understood circumstances, scorodite and other Fe(III) arsenates will precipitate and suppress arsenic solubility in mine drainage (Williams, 2001, 273). The minimum solubility of well-crystallized scorodite probably occurs at pH ~2.5 under ambient temperatures and pressures (Langmuir, Mahoney and Rowson, 2006). At pH $\geqslant 3$, scorodite is likely to decompose into As(V) forms (such as $H_2AsO_4^-$) and iron (oxy)(hydr)oxides (such as goethite) (Krause and Ettel, 1988, 851; Harvey *et al.*, 2006). The As(V) forms may then sorb and/or coprecipitate with goethite or other iron (oxy)(hydr)oxides. Goethite can sorb up to 8 wt% arsenic (Borba, Figueiredo and Matschullat, 2003, 49, 50). The maximum sorption of As(V) on goethite occurs at pH 3$-$6, where the mineral has abundant positive surface charges to attract As(V) oxyanions (Matis *et al.*, 1999; Chapters 2 and 7).

In other situations, such as at mines in Ron Phibun (Thailand) and Globe and Phoenix (Zimbabwe), arsenopyrite may not completely oxidize and water-soluble arsenic species may form instead of less soluble scorodite (Williams, 2001, 274). The Ron Phibun and Globe and Phoenix waters contain up to 5.114 and 7.400 mg L^{-1} of arsenic, respectively (Williams, 2001, 270). About 40% of the total dissolved inorganic arsenic at the two sites is As(III), which would mainly exist as $H_3AsO_3{}^0$ in the usually pH $\leqslant 9$ waters (Williams, 2001, 270; Chapter 2). Water-soluble arsenic–cyanide complexes also occur at the Globe and Phoenix site and the waters of both sites have some As(V) oxyanions.

A variety of bacteria and other microorganisms, such as the archaeum *Ferriplasma acidarmanus*, may be actively involved in the oxidation of arsenopyrite (Gihring *et al.*, 1999; Cruz *et al.*, 2005; Barrett *et al.*, 1993). Specifically, (Gihring *et al.*, 1999) collected *Thiobacillus caldus* and *Ferriplasma acidarmanus* from acid mine drainage at Iron Mountain, California, USA. The mine drainage had a temperature of approximately 42 °C, a pH of 0.7, and contained about $50\,\mathrm{mg\,L^{-1}}$ of arsenic. *T. caldus* growths on the surfaces of arsenopyrite actually hindered the oxidation of the mineral, whereas *F. acidarmanus* was very tolerant of arsenic and accelerated the dissolution of arsenopyrite (Gihring *et al.*, 1999).

Air oxidation layers on arsenopyrite are much thinner than layers produced from oxidizing mine waste-waters (Nesbitt and Muir, 1998, 141). The air-oxidized layers also contain Fe(III) hydroxides, Fe(III) arsenates, and Fe(III) arsenites, which are noticeably different than the Fe(III) oxyhydroxides that form in water-oxidized layers (Nesbitt and Muir, 1998, 141). An arsenopyrite sample from the Halen Mine of Wawa, Ontario, Canada, oxidized in air during 25 years of storage at the University of Western Ontario. The Fe:As ratio of the arsenopyrite surfaces increased from 1:0.93 to 1:12.1 during the 25 years of air oxidation. Nesbitt and Muir (1998, 141–142) concluded that the increased ratio resulted from As(0) diffusing from the interior of the sample to its surfaces and oxidizing to As(III) and As(V).

3.7.4.3 *Oxidation of realgar and orpiment*

The crystalline structure of realgar consists of rings of As_4S_4 (Klein, 2002, 363). That is, arsenic occurs as As(II) in realgar, as well as amorphous AsS. However, As(II) is not stable in aqueous solutions. Instead, As(II) oxidizes to As(III) in the compounds before being released into solution (Lengke and Tempel, 2005, 350).

Besides occurring in hydrothermal deposits and as a sublimation product in volcanic vents and hot springs, orpiment may form as an oxidation product of realgar (Lengke and Tempel, 2002, 3281). In turn, orpiment may eventually oxidize to As(V) in surface and near-surface environments. Rather than immediately oxidizing to As(V) in water, experimental results by Lengke and Tempel (2002, 3288) suggest that at 25–40 °C sulfide first oxidizes in orpiment and that $H_3AsO_3{}^0$ forms as shown in the following reaction:

$$As_2S_3 + 6O_2 + 6H_2O \rightarrow 2H_3AsO_3{}^0 + 3SO_4{}^{2-} + 6H^+ \tag{3.53}$$

Depending on saturation, the $H_3AsO_3{}^0$ may precipitate as arsenolite (Williams, 2001, 274) or remain dissolved in water. As further discussed in Chapters 2 and 7, chemical oxidants (usually Mn(III or IV) or Fe(III) compounds with O_2) and/or microorganisms may eventually transform As(III) into As(V). Studies by Lengke and Tempel (2002) also indicate that the oxidation rates of orpiment tend to increase if pH >8.

The biodegradation of organic matter often creates significant quantities of carbonate species ($H_2CO_3{}^0$, $HCO_3{}^-$, and/or $CO_3{}^{2-}$) in natural waters. Under low-O_2 conditions, carbonates may dissolve orpiment and realgar. This process could produce thioarsenic and perhaps arsenic–carbonate complexes (such as, $As(CO_3)_2{}^-$, $As(CO_3)(OH)_2{}^-$, $AsCO_3{}^+$, $As(CO_3)_2(OH)^{2-}$) in anaerobic groundwater (Kim, Nriagu and Haack, 2000; Lee and Nriagu, 2003). Although $CO_3{}^{2-}$ is more effective than $HCO_3{}^-$ (bicarbonate) in dissolving arsenic from arsenic sulfides, bicarbonate is the dominant carbonate species in circumneutral waters and is probably more responsible for arsenic dissolution (Kim, Nriagu and Haack, 2000, 3097). Although arsenic–carbonate species may decompose into $H_3AsO_3{}^0$ and bicarbonate in groundwater, they may persist under anaerobic and acidic to neutral conditions (Kim, Nriagu and Haack, 2000, 3099).

3.8   Interactions between arsenic and natural organic matter (NOM)

NOM is common in sediments, soils, and near ambient (<50 °C) water. The materials result from the partial decomposition of organisms. They contain a wide variety of organic compounds, including *carboxylic* acids, carbohydrates, phenols, amino acids, and humic substances (Drever, 1997, 107–119; Wang and Mulligan, 2006, 202). Humic substances are especially important in interacting with arsenic. They result from the partial microbial decomposition of aquatic and terrestrial plants. The major components of humic substances are *humin*, *humic acids*, and *fulvic acids*. By definition, humin is insoluble in water. While fulvic acids are water-soluble under all pH conditions, humic acids are only soluble in water at pH >2 (Drever, 1997, 113–114).

Organic matter can affect the mobility of arsenic in water, soils, and sediments through the sorption of dissolved arsenic onto insoluble organic materials, competition between arsenic and organic matter for sorption sites on mineral surfaces, the formation of dissolved complexes between organic materials and arsenic in water, and organic matter affecting the bioavailability of arsenic (Sjöblom, Håkansson and Allard, 2004; Buschmann *et al.*, 2006; Grafe, Eick and Grossl, 2001; Ko, Kim and Kim, 2004; Redman, Macalady and Ahmann, 2002; Lin, Wang and Li, 2002; Warwick, Inam and Evans, 2005). Although humic or fulvic acids will reduce As(V) to As(III) (Palmer, Freudenthal and von Wandruszka, 2006; Tongesayi and Smart, 2006), NOM is often effective in oxidizing As(III) (Redman, Macalady and Ahmann, 2002, 2895). Organic carbon in groundwater, soils, and sediments also frequently promotes the bacterial reduction of sulfate to sulfide and the formation of strong anoxic conditions. Under anoxic conditions, arsenic may precipitate as sulfide compounds, form thioarsenic aqueous species, adsorb onto Fe(II) sulfides, and/or possibly coprecipitate with Fe(II) sulfides (Section 3.12).

Most components of NOM have negative charges under circumneutral and alkaline pH conditions (Wang and Mulligan, 2006, 203). In water, negatively charged organic matter may compete with arsenic oxyanions for sorption sites on (oxy)(hydr)oxides, displace arsenic from mineral surfaces, or hinder arsenic sorption by interfering with the formation of iron (oxy)(hydr)oxides (Wang and Mulligan, 2006, 203, 209; Redman, Macalady and Ahmann, 2002; Ford, Wilkin and Hernandez, 2006). The presence of cations (such as Fe^{2+}, Ca^{2+}, Al^{3+}, or Mn^{3+}) in water with organic matter can reduce charge repulsions and allow for the sorption of arsenic oxyanions (Wang and Mulligan, 2006, 203, 205).

Like other natural organic materials, the overall charges on fulvic and humic acids are usually negative in natural environments (Buschmann *et al.*, 2006, 6018). The negatively charged acids should form more aqueous complexes with $H_3AsO_3^0$ than As(V) oxyanions (Wang and Mulligan, 2006, 203–204). However, for unknown reasons, at least some commercial and natural (Suwannee River, USA) humic acids when dissolved in pH 4.6–8.4 water bind As(V) more strongly than As(III) (Buschmann *et al.*, 2006). Depending on the composition of humic acids and other humic substances, maximum sorption of As(V) is typically around pH 5.5–7 and pH 7–8.5 for As(III) (Wang and Mulligan, 2006, 206; Buschmann *et al.*, 2006, 6017). Overall, humic acids are more effective in sorbing arsenic oxyanions if they are rich in calcium and have high mineral contents (Mok and Wai, 1994, 109). As with other NOM (Wang and Mulligan, 2006, 203, 205), calcium probably reduces charge repulsions between the negatively charged humic acids and the arsenic oxyanions (for some detailed discussions, see Buschmann *et al.*, 2006).

3.9   Sorption and coprecipitation of arsenic with iron and other (oxy)(hydr)oxides

3.9.1   Introduction

As discussed in Chapter 7, many minerals and other solid materials can sorb and/or coprecipitate arsenic in artificial water treatment systems. Iron (oxy)(hydr)oxides are especially important and effective in both natural and artificial systems (Chapters 2 and 7). In nature, iron (oxy)(hydr)oxides sorb and/or coprecipitate arsenic in a wide variety of environments, including mine drainage (Lee, Lee and Lee, 2001; Nesbitt and Muir, 1998), erupting hydrothermal vents at sea floor spreading ridges (Pichler and Veizer, 1999; Canet *et al.*, 2003), hot springs (Le Guern *et al.*, 2003), marine water and sediments (Chaillou *et al.*, 2003; Santosa *et al.*, 1996), estuaries and *fjords* (Mucci *et al.*, 2000; Abdullah, Shiyu and Mosgren, 1995), lakes (Ford, Wilkin and Hernandez, 2006; Viollier *et al.*, 1995; Senn and Hemond, 2002), streams (Fuller and Davis, 1989; Pfeifer *et al.*, 2004; Nimick *et al.*, 2003; Mok and Wai, 1994), and porewaters in aerated sediments and soils (Widerlund and Ingri, 1995; McLaren, Magharaj and Naidu, 2006). Fe(III) oxyhydroxides, in particular, may contain up to 76 000 mg kg^{-1} of arsenic (Pichler, Veizer and Hall, 1999).

3.9.2 Iron, aluminum, and manganese (oxy)(hydr)oxides

Iron (oxy)(hydr)oxides are groups of Fe(III) $\pm$ Fe(II) (hydrous) oxides, (hydrous) hydroxides, and (hydrous) oxyhydroxides. Many of them are poorly crystalline or amorphous. They may also have highly variable compositions, which make them difficult to identify and distinguish. The ferrihydrites are one of the more poorly defined and chemically variable groups of iron (oxy)(hydr)oxide compounds. Some authors, such as Drever (1997, 141) and Langmuir (1997, 254), define 'ferrihydrite' as simply Fe(III) hydroxide, Fe(OH)$_3$. However, actual ferrihydrites are more complex and chemical variable iron hydroxides and hydrous oxides (such as, Fe$_5$HO$_8 \cdot 4$H$_2$O in Stipp *et al.*, 2002, 322; Fe$_2$O$_3 \cdot 2$H$_2$O in Stollenwerk, 2003, 77; Fe$_2$O$_3 \cdot 1.6$H$_2$O in Yu *et al.*, 1999; and Fe$_2$O$_3 \cdot \leqslant 1.8$H$_2$O in Jambor and Dutrizac, 1998, 2581, 2549). Rather than determining the potentially complex chemical formula of every ferrihydrite sample, researchers often classify the compounds by their number of broad powder *X-ray diffraction* (XRD) peaks, such as the poorly crystalline 'two-line' and better crystalline 'six- or seven-line' varieties (Jambor and Dutrizac, 1998; Schwertmann, Friedl and Stanjek, 1999; Janney, Cowley and Buseck, 2000). The chemical compositions of each of the 'line' ferrihydrites are highly variable and inadequately understood (Schwertmann, Friedl and Stanjek, 1999). While Petrunic and Al (2005) listed the composition of their 'two-line ferrihydrite' as Fe$_5$HO$_8 \cdot 4$H$_2$O, the 'two-line ferrihydrite' in Ona-Nguema *et al.* (2005) was identified as 5FeOOH $\cdot 2$H$_2$O. The chemistry of the 'six- or seven-line' ferrihydrites is also controversial and includes at least three possibilities: Fe$_5$HO$_8 \cdot 4$H$_2$O (the same formula as the two-line variety in Petrunic and Al, 2005), Fe$_5$(O$_4$H$_3$)$_3$, and Fe$_2$O$_3 \cdot 2$FeOOH $\cdot 2.6$H$_2$O (Schwertmann, Friedl and Stanjek, 1999, 216).

Over time, 'two-line' ferrihydrite normally transforms into goethite or hematite in laboratory or natural environments (Rancourt *et al.*, 2001, 839). However, extensive sorption of As(V) could delay the transformation (Ford, 2002). The crystallization of arsenic-bearing amorphous iron compounds often releases arsenic from the compounds (Welch *et al.*, 2000, 599). In particular, while aging in seawater from Ambitle Island near Papua New Guinea, 'two-line' ferrihydrites transformed into less arsenic-rich 'six-line' varieties. The arsenic released by the transformation of the ferrihydrites produced distinct crystals of claudetite (As$_2$O$_3$) (Rancourt *et al.*, 2001, 838–839).

Although arsenic-bearing ferrihydrites are mostly amorphous, Rancourt *et al.* (2001, 849) concluded that As(V) is generally tetrahedrally coordinated with Fe(III) in the compounds. The chemical properties of the marine ferrihydrites described in Rancourt *et al.* (2001, 848) more closely resemble synthetic ferrihydrites formed by the coprecipitation of arsenic and Fe(III) rather than compounds that had sorbed arsenic sometime after their precipitation.

As further discussed in Chapters 2 and 7, the sorption of arsenic on iron (oxy)(hydr)oxides is very sensitive to pH and competing anions, such as phosphate and sulfate (Goh and Lim, 2004). In general, the sorption of inorganic As(V) decreases as pH values rise from 3 to 10 (Su and Puls, 2001, 1489). H$_2$AsO$_4^-$ is the dominant As(V) ion at pH 3–6. At pH <6, the surfaces of iron (oxy)(hydr)oxides usually have net positive charges (i.e. they are below their *zero points of charge* (ZPCs) and *isoelectric points*;

Chapter 2). The positively charged surfaces would readily attract and sorb $H_2AsO_4^-$ and other As(V) oxyanions (Su and Puls, 2001, 1489). At pH >6–7, the surface charges of most iron (oxy)(hydr)oxides become dominantly negative and the sorption of As(V) oxyanions diminishes because of charge repulsion.

The sorption of inorganic As(III) on iron (oxy)(hydr)oxides generally increases as pH rises from 3 to 9. Maximum As(III) sorption generally occurs around pH 9 (Su and Puls, 2001, 1489). At this point, As(III) oxyanions start to become prominent (Chapter 2), but yet iron (oxy)(hydr)oxide surfaces still have some residual positive charges.

Aluminum (Lee, Lee and Lee, 2001) and Mn(III, IV) (Wang and Mulligan, 2006; Martin and Pedersen, 2002) (oxy)(hydr)oxides are used in water treatment systems (Chapter 7) and may have locally important roles in naturally removing arsenic from surface water and groundwater, especially in acid mine drainage. The most common crystalline varieties of aluminum (oxy)(hydr)oxides are corundum (α-Al_2O_3), bayerite (α-$Al(OH)_3$, monoclinic; P2$_1$/a space group), gibbsite (γ-$Al(OH)_3$, monoclinic; P 2$_1$/n space group), diaspore (α-$AlO(OH)$, orthorhombic; Pbnm space group), and boehmite (γ-$AlO(OH)$, orthorhombic; Amam space group). Important naturally occurring manganese (oxy)(hydr)oxides include: birnessite (approximately $Na_4Mn_{14}O_{27} \cdot 9H_2O$), cryptomelane ($KMn_8O_{16}$), manganite ($MnOOH$), and pyrolusite ($\beta$-$MnO_2$, tetragonal, P4$_2$/mnm space group).

Although aluminum and iron hydroxides on a molar basis have about equal abilities to remove inorganic As(V) from water at pH <7.5, the iron compounds are more effective at higher pH conditions and with As(III) (Smedley and Kinniburgh, 2002, 534). Iron (oxy)(hydr)oxides also generally have greater capacities to sorb As(V) oxyanions from water than clay minerals or manganese (oxy)hydroxides (Pfeifer *et al.*, 2004; Mucci *et al.*, 2000, 315; Gräfe and Sparks *et al.*, 2006, 83). In most cases, clay minerals and other poor or mediocre sorbents have low-isoelectric points and ZPCs (Chapter 2), which means that except under very acidic conditions their surfaces have excess negative surface charges that repel As(V) oxyanions (Mucci *et al.*, 2000, 315).

3.9.3   Sulfate (oxy)(hydr)oxides and related compounds

Fe(III) may also form complex (oxy)(hydr)oxides with Fe(II) and other chemical species, such as sulfate and carbonate. 'Green rust', which is used in removing arsenic from water (Chapter 7), is a complex assemblage of Fe(II,III) hydroxides with sulfate, chloride, and/or carbonate (for example, $Fe(II)_4Fe(III)_2(OH)_{12}SO_4 \cdot nH_2O$, where $n > 0$) (Refait, Abdelmoula and Génin, 1998; Trolard *et al.*, 1997; Génin *et al.*, 1998; Randall, Sherman and Ragnarsdottir, 2001; Ona-Nguema *et al.*, 2002; Douššová *et al.*, 2005, 32; Stipp *et al.*, 2002, 322).

Sulfate is not always completely flushed from piles of mining wastes and associated brines, sediments, and soils, especially in dry climates. In such cases, a large variety of arsenic-bearing compounds may precipitate from evaporating waters on the surface of the materials or stagnant porewaters within them. The compounds may include: iron (oxy)(hydr)oxides (such as goethite), water-soluble sulfate salts (for example, melanterite, $FeSO_4 \cdot 7H_2O$, (Moncur *et al.*, 2002)), and generally less soluble (oxy)(hydr)oxide sulfates, such as schwertmannite (approximately $Fe_8O_8(OH)_{4.32}(SO_4)_{1.84} \cdot nH_2O$, where $n > 0$; (Yu *et al.*, 1999)) and jarosite (Foster *et al.*, 1998; Baron and Palmer, 1996; Savage, Bird and O'Day, 2005). Schwertmannite is capable of sorbing significant amounts of As(V) and As(III) from water (Carlson *et al.*, 2002; Duquesne *et al.*, 2003). The mineral generally forms in pH 3.0–4.5 waters with sulfate concentrations of 1000–3000 mg L^{-1} (Acero *et al.*, 2006, 4130). Schwertmannite is metastable and over several months converts to goethite with a possible H_3O–jarosite intermediate phase (Acero *et al.*, 2006). The transformation to goethite is slower in schwertmannite specimens that are rich in arsenic and exposed to higher sulfate, lower pH, and lower temperature conditions. The formation of more crystalline goethite from schwertmannite should expel some arsenic, although the expulsion has not been observed in experiments (Acero *et al.*, 2006, 4131).

Jarosites are generally less soluble in acidic waters than under alkaline conditions. Although As(V) is known to partially substitute for sulfate in the mineral (Savage, Bird and Ashley, 2000), controversies exist over whether jarosites preferentially sorb more arsenic than their frequent decomposition product, goethite (Acero *et al.*, 2006, 4137–4138). Examples of exceptionally arsenic-rich jarosites occur in mine tailings at Enguiales, Aveyron, France, where arsenic concentrations are approximately 5.7 wt% (Courtin-Nomade *et al.*, 2003).

In Gunma prefecture, Japan, bacterial mats containing *Gallionella* sp. effectively remove As(V) from mine drainage (initial pH of 4.7) by sorbing and/or coprecipitating the arsenic with iron- and sulfur-bearing compounds, including schwertmannite (Ohnuki *et al.*, 2004). As(V) concentrations in laboratory samples of biologically active drainage water declined from 90 μg L^{-1} (1.2×10^{-6} mol L^{-1}) to 27 μg L^{-1} (3.6×10^{-7} mol L^{-1}) over 17 days. In contrast, iron and As(V) were not removed from sterilized and refrigerated samples of the mine drainage during the 17 day period (Ohnuki *et al.*, 2004, 288). Ohnuki *et al.* (2004) concluded that microorganisms were important in oxidizing Fe(II) and coprecipitating and/or sorbing As(V) in iron and sulfur compounds in mine drainage.

Arsenic and heavy metals in drainage brines, weathering tailings, and ore deposits may precipitate as a number of fairly exotic (oxy)(hydr)oxide sulfates. Specifically, the weathering of outcrops containing arsenic-rich galena (PbS) often produce lead-arsenic jarosites (Smith *et al.*, 2006). Near Golden Point battery in New Zealand weathered mining wastes contain bukovskyite ($Fe_2(AsO_4)(SO_4)(OH) \cdot 7H_2O$) (Mains and Craw, 2005). Pore solutions within waste piles at the Berikul gold mine in the Kemerovo region of Russia precipitated amorphous iron sulfoarsenates and beudantite (Gieré, Sidenko and Lazareva, 2003). Arsenic at this Russian site also coprecipitated with several other iron and sulfate minerals on the surfaces of the piles, including copiapite ($Fe_5(SO_4)_6(OH)_2 \cdot 20H_2O$; 0.27 wt% arsenic), rhomboclase ($HFe(SO_4)_2 \cdot 4H_2O$; 0.87 wt % arsenic), and dietrichite (($Zn,Fe,Mn)Al_2(SO_4)_4 \cdot 22H_2O$; 0.64 wt % arsenic) (Gieré, Sidenko and Lazareva, 2003). Once storms occur, the water-soluble surface salts dissolve and produce highly acidic (pH 1.1) mine drainage with up to 240 mg L^{-1} of arsenic (Gieré, Sidenko and Lazareva, 2003).

In some circumstances, the oxidation of sulfide minerals to sulfate (oxy)(hydr)oxides involves one or more intermediate steps that are related to the properties of the field location. For example, realgar and orpiment in mining wastes at the Kusa mine in Sarawak, Malaysia, initially weather to arsenolite. The arsenolite readily dissolves in sulfate-rich waters in open pits. As the water evaporates, arsenic-rich jarosite precipitates (Williams, 2001, 274).

Arsenic in water may sorb onto colloids or coprecipitate with compounds that form colloids (Buschmann *et al.*, 2006; Hollings, Hendry and Kerrich, 1999, 6019). A variety of iron (sulfate) (oxy)(hydr)oxides, including goethite, can occur as colloids (Buschmann *et al.*, 2006, 6019; Schreiber *et al.*, 2003, 261–262) and they are commonly found in mine drainage. In particular, about 50% of the arsenic (or about 230 mg kg^{-1}) in acid mine drainage at the Himmelfahrt Fundgube mine near Freiberg, Germany, exists as 1.3–5 nm colloids. The colloids are probably mixtures of schwertmannite and hydronium jarosite (Zänker *et al.*, 2002). In hydronium jarosite, As(V) may substitute for sulfate and/or become embedded in the compound as very small scorodite clusters (Zänker *et al.*, 2002, 644). In at least some cases, the development of colloids in mine drainage is an intermediate step in the formation of precipitates (Zänker *et al.*, 2002). However, colloids can also be exceptionally mobile, especially in surface environments, where they and any sorbed contaminants can travel over long distances by wind or water (Loredo *et al.*, 2004; Kimball, Callender and Axtmann, 1995, B67, B71).

3.10   Arsenate inorganic (As(V)) precipitation

Calcium, manganese, aluminum, and Fe(III) may precipitate with inorganic As(V) (arsenate) in oxidizing environments where As(V) is plentiful. As previously mentioned in Section 3.6.4, a large variety of arsenates are known to precipitate at low water-to-solid ratios and temperatures above 100 °C in oxidizing hot springs, including scorodite, annabergite, pharmakosiderite, pharmacolite, and beudantite (Pfeifer *et al.*, 2004, 207). Scorodite and other iron arsenates also form at ambient temperatures (<50 °C) in weathering mining wastes and oxidizing sulfide deposits in ores and coal. In particular, scorodite is a common weathering product of arsenopyrite (Reaction 3.52).

The addition of lime to control acid drainage from mining wastes typically produces calcium arsenates (Pichler, Hendry and Hall, 2001). Bothe and Brown (1999) further concluded that lime precipitates As(V) as a number of hydroxyl and hydrated calcium arsenates ($Ca_4(OH)_2(AsO_4)_2 \cdot 4H_2O$, $Ca_5(AsO_4)_3OH$ (arsenate *apatite*), and/or $Ca_3(AsO_4)_2 \cdot 3H_2O$) rather than anhydrous tricalcium orthoarsenate ($Ca_3(AsO_4)_2$). Calcium arsenates also occur in coal combustion byproducts (Chapter 7). In the *flue gas* treatment systems of coal combustion facilities, volatile arsenic can readily react with calcium to form the arsenates on the surfaces of *flyash* and injected lime (Seames and Wendt, 2000; Yudovich and Ketris, 2005, 175).

Although calcium arsenates may readily precipitate in acid mine drainage and form in flue gas treatment systems, Robins and Tozawa (1982) warn that the compounds may not have long-term stability, which could lead to disposal and environmental problems. Besides dissolving under acidic conditions, the presence of carbonate, bicarbonate, or CO_2 may decompose calcium arsenates. When calcium arsenates react with CO_2, calcium carbonate forms and the arsenic could be released into the environment (Ghimire *et al.*, 2003, 4946; Jing, Korfiatis and Meng, 2003, 5055–5056).

The solubilities of aluminum, manganese, Fe(II), and Fe(III) arsenates in water are very different than the values for calcium arsenates (Chapter 2). While calcium arsenates are less soluble under alkaline conditions, the solubility of the other arsenates generally increases with increasing pH (Sadiq *et al.*, 2002, 306–307). Under pH 2–12 conditions, Mn(II) and Fe(II) arsenates are also less soluble than aluminum and Fe(III) arsenates (Sadiq *et al.*, 2002, 307).

Geochemical modeling is often used to identify the compounds that primarily control the chemistry of arsenic in aqueous solutions. Modeling studies indicate that the arsenic concentrations of Kelly Lake, Ontario, Canada, are controlled by the precipitation and dissolution of Fe(II) arsenates rather than calcium or Fe(III) arsenates (Sadiq *et al.*, 2002). The arsenic in the lake originated from runoff from the nearby Sudbury mining district and airborne particles from local ore smelters (Sadiq *et al.*, 2002).

3.11   Reductive dissolution of iron and manganese (oxy)(hydr)oxides

The reductive dissolution of solid compounds in anaerobic soils, sediments, and waters begins with the reduction of prominent cations within the compounds. Many Fe(III) (oxy)(hydr)oxide compounds are especially susceptible to reductive dissolution. The reduction process converts Fe(III) into more water-soluble Fe(II). The formation of Fe(II) causes the (oxy)(hydr)oxides to decompose in water. In some cases, the Fe(II) rapidly precipitates as new solid compounds, such as siderite ($FeCO_3$) or magnetite (Fe_3O_4).

Extensive reductive dissolution of Fe(III) compounds in water would release arsenic that might have sorbed or coprecipitated with the compounds. However, the arsenic may immediately sorb or coprecipitate with any Fe(II) compounds that might form. As discussed throughout this chapter, reductive dissolution is an important and ubiquitous process that controls the mobility of arsenic in many sediments, soils, wetlands, estuaries, seas, groundwaters, and stagnant lakes.

After organic matter and arsenic-bearing Fe(III), aluminum, or Mn (III or IV) (oxy)(hydr)oxide particles form in the oxygenated surface waters of oceans, seas, estuaries, or lakes, the materials may eventually settle into reducing bottom waters and sediments. By decomposing natural and artificial organic materials, microorganisms have important roles in initiating reducing conditions in stagnant waters, groundwaters, landfills, and buried soils and sediments (Reisinger, Burris and Hering, 2005, 461A; Anawar *et al.*, 2003). Normally, the biodegradation of organic matter first leads to the consumption of O_2 and the release of CO_2, as well as carbonate species in water (Burgess and Pinto, 2005). Unless O_2 is replenished, the soil, sediment, or aqueous environment becomes anoxic. Once dissolved O_2 declines below about 0.5 mg kg^{-1} in water, any NO_3^- converts to nitrite (NO_2^-), ammonium (NH_4^+), or N_2O and N_2 gases (Klinchuch *et al.*, 1999, 10; Smedley and Kinniburgh, 2002, 536). Although the aluminum in gibbsite and other aluminum (oxy)(hydr)oxides is not susceptible to reductive dissolution, microorganisms can desorb arsenic from these compounds (Foster, 2003, 65). Manganese(IV) oxides generally undergo reductive dissolution before Fe(III) oxides (Spliethoff, Mason and Hemond, 1995, 2158; Chaillou *et al.*, 2003, 2996; Klinchuch *et al.*, 1999, 10). Arsenic released by dissolving manganese (oxy)(hydr)oxides may be subsequently captured by iron (oxy)(hydr)oxides (Peterson and Carpenter, 1986, 361). However, dissolved organic matter, phosphate, and other chemicals may hinder arsenic resorption (Dixit and Hering, 2006).

The reductive dissolution of iron (oxy)(hydr)oxides generally begins at Eh values of around 0 mV (Pfeifer *et al.*, 2004). Amorphous iron (oxy)(hydr)oxides also tend to dissolve before crystalline goethite or hematite ((Mucci *et al.*, 2000), 310–311). Dissolution experiments at As/Fe ratios of 0.001–0.005 indicate that adsorbed As(V) is not significantly released from two-line ferrihydrite and goethite until about 50 % of the (oxy)(hydr)oxides have reduced and the surface areas of the iron compounds become too small to retain adsorbed As(V). In contrast, essentially all of the As(V) on lepidocrocite (γ-FeO(OH)) is desorbed once 70 % of the iron compound is reduced ((Pedersen, Postma and Jakobsen, 2006), 4120–4121). Although dissolved Fe(II) released by the reductive dissolution of iron (oxy)(hydr)oxides usually will not hinder the resorption of arsenic onto any remaining (oxy)(hydr)oxides, dissolved organic matter, phosphate, and other chemicals may prevent arsenic resorption (Dixit and Hering, 2006).

The chronological order of the reductions of As(V) and Fe(III) in field and laboratory studies shows some variation. Although both species reduce before sulfate, the reduction of As(V) to As(III) often starts after the initiation of Fe(III) reduction (Smedley and Kinniburgh, 2002), 536; (Pedersen, Postma and Jakobsen, 2006), 4120; (Islam *et al.*, 2004). Inskeep, McDermott and Fendorf (2002), 196–197 further conclude that in most cases As(V) is released into solution by reductive dissolution before being reduced. On the other hand, Campbell *et al.* (2006) found that two strains of the microorganism *Shewanella* sp. reduced As(V) before or simultaneously with Fe(III) reduction in iron (oxy)(hydr)oxides.

Besides *Shewanella* sp., many other microorganisms are directly involved in the reductive dissolution of iron (oxy)(hydr)oxides and/or the reduction of As(V) (García-Sánchez, Moyano and Mayorga, 2005). *Sulfurospirillium barnesii* is able to reduce both dissolved As(V) and As(V) sorbed onto ferrihydrite (Zobrist *et al.*, 2000; Herbel and Fendorf, 2005). In comparison, *Geobacter metallireducens* can reduce Fe(III) in ferrihydrite, but does not substantially reduce or dissolve sorbed As(V). Instead, the As(V) forms nanometer-sized colloids (Tadanier, Schreiber and Roller, 2005). Although the As(V) is not reduced to more toxic As(III), the resulting colloids may be very mobile in natural waters. They could potentially pass through water treatment systems and come into contact with humans (Tadanier, Schreiber and Roller, 2005), 3061.

Geobacter sulfurreducens and *Geothrix fermentans* are also capable of reducing Fe(III) and the resulting Fe(II) may form insoluble precipitates, such as vivianite ($Fe_3(PO_4)_2 \cdot 8H_2O$), siderite, and magnetite (Islam *et al.*, 2005). Instead of mobilizing during the *Geobacter sulfurreducens* or *Geothrix fermentans* reduction of iron (oxy)(hydr)oxides, the arsenic often sorbs or coprecipitates with the Fe(II) compounds (Islam *et al.*, 2005; Coker *et al.*, 2006). Results in (Islam *et al.*, 2005) further demonstrate that the reductive dissolution of iron (oxy)(hydr)oxides by these bacteria cannot explain the mobilization of arsenic in the delta sediments of Bangladesh, West Bengal (India), and elsewhere (Chapter 6; (Chakraborti *et al.*, 2002)). Other microbial species may be responsible for the mobilization of arsenic (Islam *et al.*, 2005).

Iron (oxy)(hydr)oxides may not entirely dissolve if reducing conditions are weak or if they periodically become more oxidizing. Reductive dissolution may also partially or largely convert amorphous Fe(III) oxides or lepidocrocite into insoluble magnetite, which can sorb As(V) (Dixit and Hering, 2003), 4182; (Cummings *et al.*, 2000); (Pedersen, Postma and Jakobsen, 2006), 4121, 4126; (Coker *et al.*, 2006). If sorbed As(V) on persisting iron (oxy)(hydr)oxides is reduced to As(III), the resulting As(III) will not necessarily desorb (Kneebone *et al.*, 2002). That is, As(III) is not always more mobile in water than As(V). (Dixit and Hering, 2003), 4188 found that the relative mobility of As(III) to As(V) depends on solution chemistry (especially pH) and the types of sorbing iron compounds. In their studies at pH 6–9, As(III) sorption was approximately equal to or greater than As(V) sorption on amorphous iron oxides and goethite (Dixit and Hering, 2003), 4188.

Arsenic-bearing iron (oxy)(hydr)oxides often experience cycles of reductive dissolution and reprecipitation in lake, estuary, or marine sediments (Figure 3.3). The cycles usually result from changes in seasons, weather, or long-term climate, which produce periods of alternating oxidizing and reducing conditions in sediments. In lakes, seasonal changes, droughts, or flooding affects biological activity, water levels, and water currents, which alter redox conditions. Along marine coastlines, postglacial rises in sea level over the past 10 000 years have changed redox conditions in sediments of the Kara Sea, Russia (Loring *et al.*, 1998). The reductive dissolution and reprecipitation cycles associated with the postglacial rises have enriched arsenic to greater than 20 mg kg^{-1} in many of the sediments (Loring *et al.*, 1998).

The reductive dissolution of iron and manganese (oxy)(hydr)oxides can release arsenic to groundwaters (Matschullat, 2000), 305; (Bone, Gonneea and Charette, 2006), porewaters in sediments and soils (Widerlund and Ingri, 1995; Bennett and Dudas, 2003), and bottom waters in lakes (Viollier *et al.*, 1995; Seyler and Martin, 1989), fjords (Mucci *et al.*, 2000), and marine environments (Chaillou *et al.*, 2003; Santosa *et al.*, 1996; Loring *et al.*, 1998). Once arsenic and Fe(II) dissolve in sediment porewaters, they may migrate upwards and into more oxidizing sediments where the arsenic could coprecipitate and/or sorb with iron (oxy)(hydr)oxides (Figure 3.3). The reoxidation and rapid precipitation of iron may produce well-defined (oxy)(hydr)oxide layers in sediments, which would delineate the boundary between iron oxidation and reduction (Widerlund and Ingri, 1995).

Arsenic concentrations in the porewaters of a shallow sediment may significantly exceed concentrations in overlying marine or fresh waters. A chemical gradient would then exist at the sediment–water interface, which could allow the arsenic to diffuse upward toward the overlying waters (Figure 3.3; (Azcue and Nriagu, 1995; Martin and Pedersen, 2002; Minami and Kato, 1997; Diamond, 1995)). Although far less common in most waters and sediments than iron compounds, MnO_2 more rapidly oxidizes As(III), especially when the MnO_2 is poorly crystalline and has a high-surface area (Figure 3.3; (Mucci *et al.*, 2000), 315; (Sullivan and Aller, 1996), 1466; (Nicholas *et al.*, 2003), 125; (Stollenwerk, 2003), 70–71). Once oxidized, the As(V) often sorbs onto iron (oxy)(hydr)oxides in bottom sediments or water (Chaillou *et al.*, 2003), 2998, 3001–3002; (Mucci *et al.*, 2000), 315; (Riedel, Sanders and Osman, 1998). Unless divalent cations (such as Ca^{2+}) substantially neutralize negative surface charges, the ZPCs and isoelectric points (Chapter 2) of most manganese compounds are too low for them to be as effective as iron (oxy)(hydr)oxides in sorbing As(V) (Mok and Wai, 1994), 105; (Mucci *et al.*, 2000), 315. After the iron (oxy)(hydr)oxides and their sorbed arsenic are buried and reduced, the arsenic desorption/sorption cycle would restart.

Sometimes clay layers are inserted into mine tailings piles (e.g. Globe and Phoenix Mine in Zimbabwe) or piles are capped or surrounded with clay or plastic liners. The liners and caps are installed to reduce the infiltration of air and oxygenated water, and hinder the formation of mine drainage (Williams, 2001), 275.

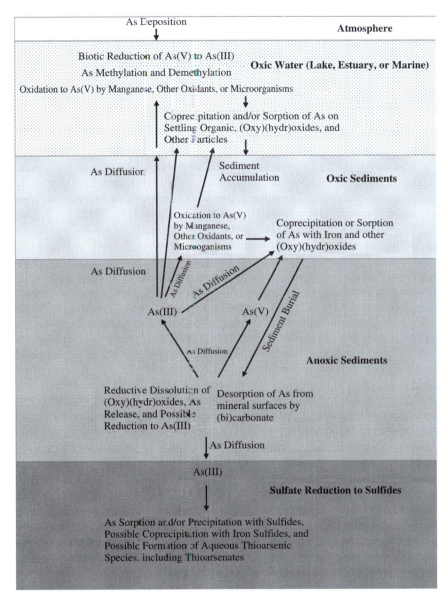

Figure 3.3 *Arsenic transformations in oxidizing and reducing sediments in marine, estuary, or fresh water environments.*

However, clay caps can actually increase arsenic mobility in tailings piles through reductive dissolution of iron (oxy)(hydr)oxides (DeLemos *et al.*, 2006). Specifically, arsenic concentrations in the groundwater of a tailings pile at Snow Lake, Manitoba, Canada, increased to greater than 20 mg L^{-1} (mostly as As(III)) after the installation of a clay and rock cap (Salzsauler, Sidenko and Sherriff, 2005). The cap created reducing conditions in the subsurface tailings, which dissolved iron compounds (i.e. scorodite, jarosite, and amorphous iron sulfoarsenates) that had previously retained arsenic. Similarly, Fe(III) sulfoarsenates, jarosite, and iron oxides in tailings from the Campbell mine at Red Lake, northern Ontario, Canada, became reducing, decomposed, and released their arsenic into porewaters once they were placed in a polyethylene-lined disposal facility (McCreadie, Jambor and Blowes, 1998).

3.12   Arsenic and sulfide at <50 °C

Under strongly reducing conditions (around −150 mV at pH 7; (Beauchemin and Kwong, 2006), 6302) in sediments, soils, and water, bacteria may reduce sulfate to sulfide (Smedley and Kinniburgh, 2002), 534, 536; (Southam and Saunders, 2005). In addition to hydrothermal deposits, arsenic may readily interact with sulfide in anoxic sediments, soils, and waters at temperatures <50 °C (Figure 3.3). In these highly reducing environments, arsenic may precipitate as sulfide compounds, form thioarsenic species, adsorb onto Fe(II) sulfides, and/or possibly coprecipitate with Fe(II) sulfides (Chaillou *et al.*, 2003), 3001; (Reisinger, Burris and Hering, 2005), 459A, 461A; (Nicholas *et al.*, 2003); (Jay *et al.*, 2005).

In a low-iron and sulfide-rich reducing environment, the absence of Fe(II) sulfide precipitation could cause H_2S concentrations to dramatically increase. Eventually, amorphous 'As_2S_3' or orpiment may precipitate, especially under acidic conditions (Cullen and Reimer, 1989), 736; (Helz *et al.*, 1995), (Keimowitz *et al.*, 2005), (Wilkin and Ford, 2006); (O'Day *et al.*, 2004), 13707; (Nordstrom and Archer, 2003), 25. If H_2S concentrations reach about 1 mM in an iron-poor environment with circumneutral to alkaline pH conditions, orpiment will dissolve into thioarsenic species, which would include thioarsenites (such as AsS_3^{3-}, $AsS(SH)_2^-$, and $AsS_2(SH)^{2-}$; (Cullen and Reimer, 1989), 736; (Helz *et al.*, 1995; Keimowitz *et al.*, 2005); (O'Day *et al.*, 2004), 13707) and/or thioarsenates (Stauder, Raue and Sacher, 2005; Planer-Friedrich *et al.*, 2007). Furthermore, as mentioned earlier in Section 3.7.4.3, abundant HCO_3^- would dissolve orpiment to produce additional thioarsenic and perhaps arsenic–carbonate complexes (Kim, Nriagu and Haack, 2000; Lee and Nriagu, 2003).

Thermodynamic data indicate that realgar forms in more reducing environments than orpiment (Nordstrom and Archer, 2003), 13. However, realgar and its polymorphs are rarely seen forming below 100 °C (O'Day *et al.*, 2004), (Nordstrom and Archer, 2003), 11. Similarly, thermodynamics suggest that As(0) could exist under extremely reducing conditions. In reality, however, it probably only forms in hydrothermal fluids (Nordstrom and Archer, 2003), 14, 25.

Although arsenian pyrite is known to form in hydrothermal fluids (Wilkin and Ford, 2006), 170, controversies exist over the coprecipitation of arsenic with Fe(II) sulfides in reducing and low-temperature (<50 °C) soils, sediments, and waters. Many researchers have described the coprecipitation of arsenic with pyrite or FeS (mackinawite and/or greigite) in reducing sediments (e.g. (Wilkin and Ford, 2006; Sracek *et al.*, 2004)). However, unless organic carbon stabilizes the surfaces of FeS, any arsenic-bearing mackinawite and greigite should convert over time to more stable and perhaps less arsenic-rich pyrite (Wilkin and Ford, 2006), 165. Additionally, O'Day *et al.* (2004) were unable to detect arsenian pyrite in a high-iron and -sulfide reducing environment at the location of a former pesticide factory in East Palo Alto, California, USA. Instead, they unexpectedly found extremely small and poorly crystalline realgar on the surfaces of precipitated pyrite. The O'Day *et al.* (2004) study suggests that at least some arsenic-bearing iron sulfides from low-temperature reducing environments could be misidentified as arsenian pyrite.

Bostick and Fendorf (2003), 918 performed detailed laboratory studies and concluded that arsenic could form surface precipitates on FeS (including troilite, pyrrhotite, and mackinawite) under anoxic conditions as shown in the following reaction:

$$3FeS + H_3AsO_3^0 \rightarrow FeS_2 + FeAsS + Fe(OH)_3 \tag{3.54}$$

In further sorption studies at pH 7, Bostick, Chen and Fendorf (2004) concluded that the FeAsS surface precipitates may convert to As_2S_3 over time (also see (Bostick and Fendorf, 2003)).

Under acidic to slightly alkaline conditions with limited reactive Fe(II), As_2S_3 precipitates and dissolved Fe(II) are likely to immediately form through this reaction (Wilkin and Ford, 2006), 169:

$$3FeS + H_3AsO_3{}^0 + 6H^+ \rightarrow As_2S_3 \downarrow + 3Fe^{2+} + 6H_2O \qquad (3.55)$$

Using thermodynamic data, Wilkin and Ford (2006), 169 predict that the molar log activity of Fe(II) must be <-12 at a pH $\sim$8.2 and <-2 at a pH of $\sim$3.4 for Reaction 3.55 to proceed to the right.

As shown in the following reaction, $H_3AsO_3{}^0$ may sorb onto pyrite surfaces under anaerobic conditions, which leads to the formation of FeAsS, Fe(III) hydroxides, and polysulfides, such as FeS_4 (Bostick and Fendorf, 2003):

$$7FeS_2 + 2H_3AsO_3{}^0 \rightarrow 3FeS_4 + 2FeAsS + 2Fe(OH)_3 \qquad (3.56)$$

The sorption of arsenic onto Fe(II) sulfides is most favored at pH $>$5$-$6 (Bostick and Fendorf, 2003).

3.13 Arsenic and its chemistry in mined materials

3.13.1 Environmental issues with arsenic-bearing sulfide minerals in coal and ore deposits

Arsenian pyrite and/or arsenopyrite are the dominant arsenic-bearing minerals in most coals and ore deposits (Smedley and Kinniburgh, 2002), 528–529; (Welch *et al.*, 2000), 592. As further discussed in Chapter 5 and Section 3.7.3.1, the presence of arsenic in ores and coals often creates serious environmental and technical problems during mining, ore processing and smelting, and coal combustion. During ore processing, the release of arsenic from sulfide minerals can kill bacteria and hinder the bioleaching and recovery of valuable metals (Wiertz, Mateo and Escobar, 2006). Arsenic may also leak from heap leaching facilities and contaminate surrounding environments (Kampf, Salazar and Tyler, 2002). Furthermore, coal combustion and many ore smelting operations oxidize sulfide minerals, which lead to the production of volatile arsenic along with acid rain-producing SO_2 and other sulfur gases. Despite the use of flue gas treatment technologies in many countries (Chapters 5 and 7), coal combustion and the smelting of copper ores are still the largest anthropogenic emitters of arsenic to the global atmosphere ((Matschullat, 2000), 301; Section 3.21.1).

Coals are often processed prior to combustion to remove sulfide minerals. Traditionally, recovered waste sulfides have been piled or thrown into ponds at mines and coal combustion facilities, which could lead to the generation of mine drainage. Although raw coal and ores may not always contain very much arsenic, processing techniques and natural weathering often concentrate arsenic in solid wastes and drainage (Boulet and Larocque, 1996). Additionally, arsenic compounds were once used in some ore processing techniques, which further contributed to contamination problems (Rösner, 1998). As shown in the literature, wastes from the mining and utilization of coal and ore deposits frequently contain $>$1 wt % arsenic (Loredo *et al.*, 2004; Moldovan, Hendry and Jiang, 2001; Méndez and Armienta, 2003; Wong *et al.*, 2002).

3.13.2 Behavior of arsenic within mining wastes

As previously discussed in Section 3.7.2, many factors affect the formation of arsenic-bearing drainage and weathering products in mining wastes. In recent years, a number of important studies have been published on the effects of wet and dry cycles on arsenic mobility in mine tailings. In particular, seasonal changes in precipitation often affect the oxidation of sulfide minerals, especially in permeable tailings piles. During relatively dry periods, the water levels in the piles decline and air can enter and contribute to the oxidation of sulfides. Arsenic and other contaminants may also leach into remaining stagnant water. Contaminant pulses occur in tailings piles during the beginning of precipitation events at the end of dry seasons. Precipitation flushes out concentrated salt deposits and any remaining water as drainage (Williams, 2001), 275. With increased precipitation, water levels would rise through the tailings and promote leaching and perhaps dilution. The flushed arsenic may then flow to nearby areas through groundwater or surface water, where it may eventually interact with organisms, dilute, or sorb/coprecipitate with (oxy)(hydr)oxides and other compounds (Section 3.13.3).

Periodic wetting and evaporation may form impermeable layers or 'hardpans' on the surface (Fault Lake tailings impoundment, Ontario, Canada; (McGregor and Blowes, 2002)) or in the shallow subsurface (e.g. depths of 1–1.5 m; (Moncur *et al.*, 2002; Gieré, Sidenko and Lazareva, 2003)) of large tailings deposits. Tailings particles in the hardpans are often cemented with gypsum, goethite, and other minerals that are insoluble or only partially soluble in water (McGregor and Blowes, 2002; McGregor *et al.*, 1998b). Hardpans may also be important in directing the flow of water and controlling the oxidation and leaching of arsenic and other contaminants in the tailings. Although hardpans may cause water to accumulate and concentrate contaminants in the upper oxidizing layers, they may also sorb arsenic and possibly protect underlying sulfides from being oxidized and leached by liquids (McGregor and Blowes, 2002; Gilbert, Cooke and Hollings, 2003).

Morin *et al.* (2003) detected seasonal changes in arsenic speciation and the coprecipitation and sorption of arsenic with Fe(III) compounds in acid drainage at the Carnoulès mine, Gard, France. The drainage had 80–280 mg L^{-1} of As(III). During the wet season, suspended sediments in a nearby stream contained tooeleite (Fe$_6$(AsO$_3$)$_4$(SO$_4$)(OH)$_4 \cdot$ 4H$_2$O) and amorphous As(III)-bearing iron oxyhydroxides. In contrast, amorphous As(V)-bearing Fe(III) oxyhydroxides were prominent during the dry season (Morin *et al.*, 2003). The formation of As(III)-bearing compounds during the wet season may be related to the ability of bacteria (probably *Acidithiobacillus ferrooxidans*) to oxidize Fe(II), but not As(III). The formation of As(V) compounds during the dry season may be due to both abiotic and biotic oxidation of As(III), which could include *Thiomonas* sp. (Morin *et al.*, 2003). Like the wet season, *Acidithiobacillus ferrooxidans* may be responsible for catalyzing the dry season oxidation of Fe(II) (Morin *et al.*, 2003).

Craw, Koons and Chappell (2002) describe the formation of oxidized layers or 'mine soils' on top of pyrite- and arsenopyrite-bearing tailings over a five-year period at the Macraes gold mine, South Island, New Zealand. Wetting and evaporative drying of the oxidized layers produced a surface crust that was enriched in arsenic (about 5 wt %), mostly as scorodite (Craw, Koons and Chappell, 2002), 17–18, 20. The scorodite in the oxidized layers not only results from the *in situ* oxidation of arsenopyrite, but also from precipitation from fluids as they rise through the piles and toward the surface (Craw, Koons and Chappell, 2002), 20. Similar arsenic enrichment also occurred in desiccation cracks. The cracks were up to 2 cm wide and extended to depths of more than 1 m in the tailings (Craw, Koons and Chappell, 2002), 13, 21. No evidence was found of substantial water infiltrating through the cracks and into the deep Macraes tailings. The lack of infiltration may be due to the climate. Evaporation equals or exceeds the annual precipitation of 700 mm (Craw, Koons and Chappell, 2002), 14. Furthermore, the precipitation of gypsum, iron (oxy)(hydr)oxides, and other cements would diminish the permeability of the crack walls (Craw, Koons and Chappell, 2002), 24.

3.13.3 Movement of arsenic from mining wastes and into the environment

Humans and wildlife are exposed to arsenic from mining wastes through food, water, and air (Chapter 4). Arsenic in mine drainage and other waters may also enter plants and aquatic animals. Terrestrial animals could then be exposed to arsenic through food and water. Arsenic may also dissolve into ground-water and contaminate local drinking water and irrigation wells. Additionally, wind can easily transport arsenic-bearing dust from the surfaces of dry tailings piles and create inhalation hazards for local humans and wildlife (Revazyan, 1999).

Streams are another important medium for distributing arsenic-bearing tailings over long distances. Rather than immediately decompose, arsenic-bearing tailings and other solid wastes often wash into nearby streams, where they may be transported over hundreds of kilometers (Hudson-Edwards *et al.*, 2001), 240, 244. As the tailings proceed downstream, they will only partially decompose. Typically, dissolved arsenic concentrations in stream waters impacted by mining wastes consistently decline away from their sources. Surviving heavier particles may accumulate in stream depressions. The rest will eventually be diluted in sediments entering from tributaries and land erosion along the streams. Seas, *wetlands*, and lakes are often the ultimate deposition sites for mine drainage, mine tailings, tailings pond over-flow, acidic seepages from groundwater, and other wastes from mining activities (Martin and Pedersen, 2002), 1516.

The Río Pilcomayo is a major river that originates in Bolivia and forms part of the boundary between Paraguay and Argentina. The water and sediments in the river are substantially contaminated with arsenic from the Potosí mines in Bolivia (Hudson-Edwards *et al.*, 2001), 245. The arsenic content of Río Pilcomayo actually increases from 65 μg L^{-1} at 10.2 km downstream from the mines to 100 μg L^{-1} at about 151 km downstream (Hudson-Edwards *et al.*, 2001), 240, 244. The arsenic increase may be due to a decrease in pH (from 10.3 to 8.42), increased oxidation of the river sediment, and increased activity of sulfide-oxidizing bacteria (Hudson-Edwards *et al.*, 2001), 244–245.

In some cases, mining wastes are catastrophically released from impoundments (Hudson-Edwards *et al.*, 2001; Simón *et al.*, 2002; Carbonell-Barrachina *et al.*, 2004). On April 25, 1998, the retention walls of a tailings pond at Aznalcóllar, Spain, broke and released several metric tons of tailings and mine drainage (Simón *et al.*, 2002; Carbonell-Barrachina *et al.*, 2004). The break impacted an area of about 55 km^2 in the Guadiamar Valley (Simón *et al.*, 2002; Domènech, De Pablo and Ayora, 2002). Local soils were covered with wet tailings layers containing about 73 wt% pyrite and perhaps 1.4 wt% arsenopyrite (Domènech, De Pablo and Ayora, 2002), 341. The average thickness of the tailings was about seven cm (Simón *et al.*, 2002), 322. As the tailings dried, they partially oxidized and produced salts that were rich in sulfates, arsenic, and heavy metals. Through *capillary movement*, the salts concentrated on the surfaces of the tailings. During subsequent storms, the salts dissolved and generated acidic solutions that infiltrated into underlying soils (Simón *et al.*, 2002), 322. The arsenic in the solutions then coprecipitated with Fe(III) to form amorphous and poorly crystalline (sulfate) (oxy)(hydr)oxide compounds, such as schwertmannite and jarosite (Simón *et al.*, 2002), 325–326.

Effective remediation of the Aznalcóllar site would require removing all of the spilled tailings and the reaction zone (upper 10 cm) of the soils, where arsenic and other contaminants from the tailing *leachates* accumulated (Simón *et al.*, 2002). The released tailings were largely excavated from the surface of the contaminated site during the summer of 1998 (Domènech, De Pablo and Ayora, 2002). However, small amounts of sulfide-bearing tailings still remain and could continue to generate toxic drainage (including arsenic) as they oxidize (Domènech, De Pablo and Ayora, 2002), 340.

3.14 Marine waters and sediments

3.14.1 Inorganic arsenic in seawater

Table 3.7 lists some arsenic measurements for various marine waters. Arsenic concentrations in seawater are usually 1.1–1.9 μg L^{-1} with an average of 1.7 μg L^{-1} (Matschullat, 2000), 305. The typical concentration of arsenic at the surface of open ocean water is about 1.5 μg L^{-1} and deep waters in at least the Pacific and

Table 3.7 Arsenic concentrations of various marine and estuary waters.

Location	Arsenic concentration (μg L^{-1})[a]	Reference(s)
Antarctic Ocean	1.045 As(V), 0.003 As(III), 0.023 DMA(V), 0.007 MMA(V), (0.45 μm filtered)	Santosa, Wada and Tanaka (1994), 276
Atlantic Ocean (north, surface)	1.0–1.3 As(V), 0.004–0.32 As(III), (unfiltered)	Middelburg et al. (1988)
Atlantic Ocean (south)	1.58	Van der Weijden et al. (1988)
Australian coast: near Adelaide, South Australia	1.10–1.61 (0.45 μm filtered; 1.2–4.3 % As(III), rest inorganic As(V))	Maher (1985)
Baltic Sea: 17 locations	0.45–1.11 (0.45 μm filtered)	Stoeppler, Burow and Backhaus (1986)
Bay of Biscay: France	1.1 (bottom waters)	Chaillou et al. (2003), 2998
China: various estuaries and coastal seas	0.2–18 ('dissolved')	Xiankun, Jing and Xinian (1994)
Great Astrolabe Lagoon: Fiji: near surface: 12 sites	≤0.5–3.7 (unfiltered?)	Morrison et al. (1997)
Great Astrolabe Lagoon: Fiji: 10 m depth: 12 sites	<0.5–5.7 (unfiltered?)	Morrison et al. (1997)
Indian Ocean (east)	0.452 As(V), 0.232 As(III), 0.050 DMA(V), 0.032 MMA(V), (0.45 μm filtered)	Santosa, Wada and Tanaka (1994), 276
Indonesian Archipelago	0.418 As(V), 0.175 As(III), 0.089 DMA(V), 0.033 MMA(V), (0.45 μm filtered)	Santosa, Wada and Tanaka (1994), 278
Kalix River estuary: Sweden: 30 porewaters from 2 sediment cores	1.32–166 (0.4 μm filtered)	Widerlund and Ingri (1995), 187
Krka estuary: Croatia (former Yugoslavia)	0.13–1.8 (0.4 μm filtering)	Seyler and Martin (1991); Smedley and Kinniburgh (2002), 525
Laurentian Trough estuary: Canada: sediment porewaters: 7 sites	0.8–107.0 (0.4 μm filtered)	Belzile (1988)
Lena River estuary: Russia: 10 samples	0.14–0.78 (0.4 μm filtered)	Martin et al. (1993)
North Sea coast: Nordstrand and Cuxhaven, Germany	2–3 (0.45 μm filtered) (≤15 for unfiltered)	Stoeppler, Burow and Backhaus (1986)

Location	Concentration	Reference
Oslofjord: Bunnefjord, Norway	0.64–2.02 (0.45 μm filtered, total) (0.0097–0.839 as As(III))	Abdullah, Shiyu and Mosgren (1995), 121
Pacific and Atlantic (deep)	1.0–1.8	Cullen and Reimer (1989), 742
Pacific and south Tasman Sea: surface waters	1.0–1.7 (0.45 μm filtered)	Santosa et al. (1996)
Puget Sound: Washington, USA: sediment porewaters: 65 analyses	1.3–110 (0.45 μm filtered, As(III)+As(V))	Peterson and Carpenter (1986)
Ría de Arousa estuary: Galicia, Spain: before and after the sinking of *Prestige* oil tanker	Before: 0.35 ± 0.13 to 1.56 ± 0.19; After: 0.80 ± 0.10 to 1.70 ± 0.09 (probable 0.45 μm filtered)	Peña-Vázquez et al. (2006)
Saanich Inlet: British Columbia, Canada: bottom waters	2.1 2.2	Peterson and Carpenter (1986), 354
Saanich Inlet: Washington, USA: sediment porewaters: 23 analyses	0.6–19 (0.45 μm filtered, As(III)+As(V))	Peterson and Carpenter (1986)
Thames Estuary: United Kingdom: February 1989	7.33 ± 0.43 (0.4 μm filtered, As(III)+As(V))	Millward et al. (1997)
Tutum Bay: Ambitle Island, Papua New Guinea: 32 samples	1.3–8.4 (0.45 μm filtered)	Price and Pichler (2005)
Tyrrhenian Sea: Corsica coast: France: 6 samples	1–10 (unfiltered)	Migon and Mori (1999), 83
Vestfjord: Oslofjord, Norway	0.70–1.57 (0.45 μm filtered, total) (≤0.40 as As(III))	Abdullah, Shiyu and Mosgren (1995), 121
USA: Washington coast: sediment porewaters: 45 analyses	0.97–19 (0.45 μm filtered, As(III)+As(V))	Peterson and Carpenter (1986)
Wright River Estuary: South Carolina, USA	<10–147	Wirth et al. (1996)

[a] Single values denote average concentrations sometimes with 1 SD. Values separated by commas represent duplicate analyses of the same sample. Analyses involving different samples are separated by semicolons. Low and high values are separated by hyphens. Available information on filtering of the samples is listed. As(III) and As(V) denote inorganic arsenite and arsenate, respectively. DMA(V) is dimethylarsinic acid or $(CH_3)_2AsO(OH)$. MMA(V) represents monomethylarsonic acid, $H_3CAsO(OH)_2$.

Atlantic oceans have similar concentrations of around 1.0–1.8 μg L^{-1} (Cullen and Reimer, 1989), 742; (Smedley and Kinniburgh, 2002), 525. Humans may discharge significant amounts of arsenic to water and/or sediments along marine coastlines and in harbors and small seas. Major anthropogenic sources of arsenic include pesticides, herbicides, industrial pollution, emissions from coal utilization, and even atmospheric nuclear fallout from the 1950s and 1960s (Loring *et al.*, 1995). In comparison, most arsenic in open ocean water and sediments originates from submarine volcanism, river water, and atmospheric deposition (Matschullat, 2000), 300.

In open pH 8.2 ocean water, $HAsO_4^{2-}$ and $H_2AsO_4^-$ should be the dominant soluble arsenic species (Smedley and Kinniburgh, 2002), 526. At chemical equilibrium, the ratio of inorganic As(V) to inorganic As(III), As(V)/As(III), in well oxygenated open ocean water is about 10^{26} (Mandal and Suzuki, 2002), 205. However, the actual As(V)/As(III) is often 100 or lower, and in some cases, metastable As(III) accounts for more than 20 % of the arsenic in seawater (Mandal and Suzuki, 2002), 205; (Smedley and Kinniburgh, 2002), 526; (Neff, 1997). As(III) concentrations in seawater are typically around 0.0052 μg L^{-1} and should be highest in anoxic marine bottom waters, such as in the Black Sea, Baltic Sea, and Saanich Inlet, Canada (Smedley and Kinniburgh, 2002), 526; (Sanders, Riedel and Osman, 1994), 296. The As(III) should primarily exist as $H_3AsO_3^0$ ((Smedley and Kinniburgh, 2002), 526; Chapter 2).

Biological reduction by bacteria and phytoplankton is primarily responsible for excess As(III) in aerobic seawater (Francesconi and Kuehnelt, 2002), 64; (Shih, 2005), 88. Eventually, most of the As(III) is either methylated by marine organisms (Francesconi and Kuehnelt, 2002) or is oxidized to inorganic As(V), perhaps by natural MnO_2 or microorganisms (Figure 3.3; (Cullen and Reimer, 1989), 745; (Mucci *et al.*, 2000), 315; (Nicholas *et al.*, 2003)).

3.14.2 Marine arsenic cycle

Matschullat (2000), 300 estimated arsenic inputs to the oceans as 4870 t year^{-1} from submarine volcanism (including hydrothermal fluids), 54 000–61 000 t year^{-1} from river water, and 4300–8200 t year^{-1} from atmospheric deposition. The *residence time* of arsenic in seawater is about 100 000 years (Faure, 1998), 53. Eventually, the vast majority of arsenic in seawater enters sediments and rocks associated with ocean crusts (about 46 400 t year^{-1}). Of the arsenic already in oceanic crustal rocks and sediments, about 38 200 t year^{-1} are subducted (Matschullat, 2000), 300. As previously mentioned in Section 3.6.2, the volatilization of arsenic from subducting oceanic plates and its eruption from hydrothermal vents tend to recycle arsenic into the surface oceanic crust over time (Figure 3.2; (Noll *et al.*, 1996; Hattori and Guillot, 2003)). If hydrothermal fluids erupt from seafloor vents into sufficiently oxygenated seawater and in the presence of oxidizing microorganisms, Fe(II), As(III), and other reduced species from the fluids can oxidize. The oxidation of Fe(II) to Fe(III) precipitates iron (oxy)(hydr)oxides and (Pichler and Veizer, 1999; Metz and Trefry, 1993). Sorption and coprecipitation with iron (oxy)(hydr)oxides are probably the major mechanisms for removing arsenic from erupting hydrothermal fluids on ocean floors (Rancourt *et al.*, 2001), 835. Compared with calcite and most other precipitating minerals, iron (oxy)(hydr)oxides are far more efficient in removing arsenic from water (Le Guern *et al.*, 2003). The arsenic-bearing (oxy)(hydr)oxides will eventually settle into marine sediments or become incorporated into volcanic rocks. If substantial reductive dissolution is avoided, the accumulation of arsenic-bearing iron compounds on the ocean floor would make marine sediments major *reservoirs* for arsenic (Rancourt *et al.*, 2001), 835. The arsenic marine cycle restarts once the oceanic sediments and rocks subduct.

3.14.3   Arsenic methylation in marine environments

The chemical similarities between phosphate and As(V) cause phytoplankton to remove some inorganic As(V) with the phosphate nutrients that they extract from seawater (Smedley and Kinniburgh, 2002), 525; (Cutter and Cutter, 2006). In phytoplankton, As(V) is often reduced to As(III) and possibly methylated ((Cutter and Cutter, 2006); Chapters 2 and 4). The organisms then excrete the inorganic As(III) and methylarsenic into seawater (Figure 3.3; (Neff, 1997)). When compared with inorganic As(III), methylarsenic species tend to be more stable in seawater (Cutter and Cutter, 2006). Nevertheless, methylarsenic eventually decomposes (demethylates) into inorganic forms (Figure 3.3 and Chapters 2 and 4; (Anderson and Bruland, 1991)).

Methylarsenic usually represents less than 10 % of the total arsenic in seawater (Le, 2002), 97. The principal methylarsenic species in seawater are MMA(V) and DMA(V) (Francesconi and Edmonds, 1994), 222. Several other organoarsenicals may also be present. However, analytical methods are not always available to identify every organoarsenical in seawater and other materials. In some circumstances, unidentified organoarsenicals may represent a significant portion of the total arsenic in coastal marine waters, perhaps >25 % (Francesconi and Edmonds, 1994), 224.

Under appropriate conditions, phytoplankton and other marine organisms can excrete considerable quantities of MMA(V) and DMA(V) into seawater (Santosa *et al.*, 1996), 698, 703–704; (Cutter and Cutter, 2006); (Francesconi and Edmonds, 1994). 222–224. Low phosphate concentrations in marine waters actually increase the probabilities of As(V) uptake and methylation by marine algae (e.g. the central north Pacific, near Hawaii, USA; (Santosa *et al.*, 1997)). MMA(V) and DMA(V) concentrations also tend to increase in the *photic zone* of oceans (upper 100 m or so, where photosynthesis is possible), with increasing water temperatures above 25 °C, and along coasts (Santosa *et al.*, 1996), 698, 703–704; (Santosa, Wada and Tanaka, 1994); (Francesconi and Edmonds, 1994), 222–224. Furthermore, MMA(V) and DMA(V) concentrations, especially in marine coastal regions, often change with seasons (Le, 2002), 96–97. All of these changes in methylarsenic concentrations are closely related to biological activity. The ratios of MMA(V) to DMA(V), MMA(V)/DMA(V), may also considerably vary in ocean waters with similar temperatures (Santosa *et al.*, 1996). Santosa *et al.*, (1996), 701 argue that the variations in MMA(V)/DMA(V) demonstrate that the chemical species are being excreted by marine organisms rather than solely resulting from the chemical decomposition of more complex organoarsenicals.

At their sampling sites in the Pacific Ocean, Santosa *et al.* (1997) found that MMA(V) and DMA(V) concentrations were highest at the surface with 0.012–0.016 µg L^{-1} and 0.048–0.185 µg L^{-1}, respectively. The concentrations sharply declined to depths of 200 m. From depths of 200 to at least 5000 m, MMA(V) and DMA(V) concentrations stabilized at about 0.003 µg L^{-1} (Santosa *et al.*, 1997). Santosa *et al.* (1996), 703 argue that the presence of methylarsenic in deep ocean waters is probably not due to diffusion from ocean floor sediments. Instead, the deep water methyl forms may result from the diffusion of methylarsenic-bearing surface waters, the circulation of surface waters to greater depths, and the tendency of methylarsenic not to appreciably sorb onto iron (oxy)(hydr)oxides particles.

3.14.4   Arsenic in marine sediments

Marine, estuary, and fresh water sediments usually have much more arsenic in them than their overlying surface waters (Cullen and Reimer, 1989), 750. Typically, uncontaminated marine sediments contain 5–40 mg kg^{-1} of arsenic (Neff, 1997). Arsenic in marine sediments primarily originates from (1) volcanic and hydrothermal sources, (2) weathering and erosion of continental rocks, and (3) pollution from mining, agricultural runoff, and other sources (Matschullat, 2000; Noll *et al.*, 1996; Hattori and Guillot, 2003; Chaillou *et al.*, 2003; Rancourt *et al.*, 2001; Valette-Silver *et al.*, 1999; De Mora *et al.*, 2004), 76). Like many soils and freshwater sediments, arsenic in oxidizing estuary and marine sediments is often associated with iron and, to a lesser extent, manganese (oxy)(hydr)oxides (Chaillou *et al.*, 2003), 2993; (Santosa *et al.*, 1996), 702. Biological organisms, organic matter, and sulfides are also important 'sinks' for at least temporarily retaining arsenic in sediments and water (Figure 3.3).

Table 3.8 lists the arsenic concentrations of different types of marine and estuary sediments from various locations. Overall, low organic-carbon carbonate muds, oxidizing sands, and coarser-grained sediments have relatively little arsenic. In contrast, reducing marine sediments may contain as much as 3000 mg kg^{-1} of arsenic (Mandal and Suzuki, 2002), 202. Arsenic also tends to be enriched in fine-grained silicate-rich sediments, such as deep-sea clays and marine muds. In most cases, arsenic-rich sediments contain abundant arsenic-accumulating (oxy)(hydr)oxides, organic matter, or sulfides.

Clay minerals are silicates ($Si_4O_{10}^{4-}$) that contain hydroxides of aluminum, magnesium, and/or iron. Ca^{2+}, Na^+, and K^+ may also be present (Klein, 2002), 462–475, 530–534. Although some arsenic oxyanions will adsorb onto the hydroxides, the silicates are not expected to adsorb significant arsenic oxyanions because they have net negative charges under most pH conditions ((Foster, 2003), 55–56; Chapter 2). Aluminum, iron, and other metal (oxy)(hydr)oxide coatings are common on clay minerals and are even more important in sorbing arsenic than the clay minerals themselves (Foster, 2003), 65. The sorption properties of clays under different pH conditions often resemble those of aluminum and iron (oxy)(hydr)oxides because any sorption of arsenic largely occurs on aluminum or iron (oxy)(hydr)oxide coatings and the hydroxide layers within the clay crystalline structures. The maximum sorption of As(V) on kaolinite ($Al_2Si_2O_5(OH)_4$), montmorillonite ($(Al,Mg)_8(Si_4O_{10})_4(OH)_8 \cdot 12H_2O$), illite (variable composition), halloysite ($Al_2Si_2O_5(OH)_4 \cdot 2H_2O$), and chlorite ($(Mg,Fe)_3(Si,Al)_4O_{10}(OH)_2 \cdot (Mg,Fe)_3(OH)_6$) generally occurs under acidic conditions (Stollenwerk, 2003), 83. Like aluminum and iron (oxy)(hydr)oxides, As(III) sorption on clay minerals is less under acidic conditions, but increases and often becomes comparable with As(V) sorption if conditions are alkaline (Stollenwerk, 2003), 83.

3.15   Estuaries

3.15.1   Arsenic in estuaries

Estuaries refer to semienclosed bodies of water (such as bays, lagoons, inlets, and fjords) along marine coasts, where fresh water (usually from rivers) and seawater readily enter and mix. Stream waters entering large bays tend to spread out and drop their sediment load to form *deltas* and wetlands. Changes in pH and salinity from the mixing of marine and fresh stream waters may result in the precipitation, coprecipitation, dissolution, sorption, or desorption of arsenic that enters the estuary. Like marine and lake environments, the oxidation of organic matter, reductive dissolution of iron (oxy)(hydr)oxides, and possible coprecipitation with sulfides may have major roles in controlling the solubility and speciation of arsenic in the pore solutions of reducing estuary sediments (Figure 3.3; (Mucci *et al.*, 2000; Widerlund and Ingri, 1995; Loring *et al.*, 1998; Zhang *et al.*, 2002)).

In contrast to open ocean water, arsenic concentrations in estuary waters often significantly vary with time and location. The arsenic contents of estuary waters and sediments depend on many factors, including (1) the amount of arsenic in the contributing fresh waters (rivers, runoff, groundwater, etc.) and seawater, (2) the types of rocks, sediments, soils, groundwater, and organisms that are present, (3) the quantity and chemistry of any pollution from maritime, industrial, municipal, agricultural, or mining sources, (4) the presence of any hydrothermal waters, and (5) seasonal or other temporal changes in fresh water contributions, suspended sediments, water salinity and temperature, biological activity, precipitation, and water currents (Smedley and Kinniburgh, 2002), 525; (Abdullah, Shiyu and Mosgren, 1995), (Xiankun, Jing and Xinian, 1994), (Sanders, Riedel and Osman, 1994), (Davis, De Curnou and Eary, 1997), (Masuda, Yamatani and Okai, 2005). Pristine estuary surface waters usually contain much less than 4 μg L^{-1} of

Table 3.8 Arsenic concentrations of various marine and estuary sediments.

Location	Arsenic concentration (mg kg^{-1})[a]	Reference(s)
Barataria Basin Estuary: Louisiana, USA	13	Landrum (1994)
Beaufort Sea: Alaskan coast near Prudhoe Bay, USA: 192 samples	4.2–28.4	Trefry et al. (2003)
Beaufort Sea (western): Alaska, USA: 10 surface sediment samples	10.4–43	Valette-Silver et al. (1999)
Carbonate sediments: global average	<1.0	Mandal and Suzuki (2002), 202
Caspian Sea: 105 coastal sediment samples	0.42–22.6	De Mora et al. (2004), 68
Deep ocean clays: global average	13	Turekian and Wedepohl (1961), Faure (1998), 714
Elba-Argentario Basin: southern Tuscany, Italy: 75 samples	4–120	Leoni and Sartori (1997)
Fowey Estuary: Cornwall, United Kingdom: 17 sediment cores	<10–144	Pirrie et al. (2002), 36–37
Great Astrolabe Reef: Fiji: 14 carbonate sediment samples	<0.2–12.4	Morrison et al. (1997)
Great Barrier Reef: 4 sediment samples inside Davies Reef	0.25–1.20	Entsch et al. (1983), 470
Hylebos Waterway: Tacoma, Washington, USA	0.38–1260	Davis, De Curnou and Eary (1997), 1985
Kalix River estuary: Sweden: 30 measurements in 2 cores	9.4–171	Widerlund and Ingri (1995), 187
Kara Sea: Novaya Zemlya Trough, Russia: 86 samples from 8 cores collected in 1965	9.5–240	Galasso, Siegel and Kravitz (2000)
Kara Sea: Russia: 39 samples	10–150 (<2 mm grains)	Loring et al. (1998), 247
La Paz Lagoon sediments: Baja California, Mexico: 81 samples	0.83–44.4	Shumilin et al. (2001)
Marine clays: global averages	4.0–20	Mandal and Suzuki (2002), 202
Marine muds: global averages	3.2–60	Mandal and Suzuki (2002), 202
Ob Estuary: Russia: 22 samples	11–77 (<2 mm grains)	Loring et al. (1998), 247

(continued overleaf)

Table 3.8 (continued)

Location	Arsenic concentration (mg kg^{-1})[a]	Reference(s)
Otsuchi Bay: Japan: 7 surface sediment samples	1.9 ± 1.1 to 23.1 ± 2.9	Takeuchi et al. (2005)
Patos Lagoon (fresh water part): Brazil: 8 samples	3.2–15.6 (<0.5 mm grains)	Mirlean et al. (2003), 1483
Patos Lagoon estuary: Brazil: 29 samples	11.4–49.7 (<0.5 mm grains)	Mirlean et al. (2003), 1483
Pechora Sea: Russia: 16 surface samples	8–308	Loring et al. (1995)
Puget Sound: Washington, USA: 73 analyses in 4 sediment cores	5.2–46	Peterson and Carpenter (1986)
Saanich Inlet: British Columbia, Canada: 22 analyses from 2 sediment cores	1.3–16	Peterson and Carpenter (1986)
Saguenay Fjord: Quebec, Canada: sediment and porewaters from cores, 3 locations	1.4–25.8	Mucci et al. (2000)
Santorini Hydrothermal Field: Santorini, Greece: Metalliferous marine sediments: 25 samples	64–927	Varnavas and Cronan (1988)
United Kingdom: 20 salt marsh sediment samples, including 14 from Essex: range of average values	2–50 (Essex: 2–26, estimated background of 3)	O'Reilly Wiese, Bubb and Lester (1995)
USA: Alaska: marine and coastal sediments from 13 locations along the Pacific, Bering Sea, and Arctic Ocean	1.01–25.7	Valette-Silver et al. (1999)
USA: Southern California coastal marine sediments: 248 samples	1.0–20.4	Schiff and Weisberg (1999), 165
USA: Washington marine coast: 48 analyses on 3 sediment cores	3.7–11	Peterson and Carpenter (1986)
Wright River Estuary: South Carolina, USA	<1.0–82.2	Wirth et al. (1996)
Yenisey Estuary: Russia: 31 samples	6.7–44 (<2 mm grains)	Loring et al. (1998), 247

[a]Single values denote average concentrations sometimes with 1 SD. Low and high values are separated by hyphens. Normally, soil and sediment samples are dried before analysis. Some samples were sieved and only the <0.5 or <2 mm fractions were analyzed.

arsenic, whereas waters impacted by pollution or hydrothermal fluids may have at least 16 µg L^{-1} (Smedley and Kinniburgh, 2002), 523.

Inorganic As(V) is usually the prominent form of arsenic in estuary and other coastal marine surface waters (Table 3.7). However, estuary waters may contain substantial As(III) because of anoxic conditions, surface runoff, or biological activity (Sanders, Riedel and Osman, 1994), 290–291. Although *in situ* biological activity is primarily responsible for the presence of inorganic As(III) in estuary surface waters, at times and in certain locations contributions from surface runoff become more important. Specifically, Abdullah, Shiyu and Mosgren (1995) noticed that As(III) concentrations in the Vestfjord (Olsofjord), Norway, were relatively high (about 5 nM or 0.4 µg L^{-1}) in April 1993 (Table 3.7). During this time, runoff from snow melt was at a maximum and plankton production was low.

Periodic algal blooms and other biological activity may contribute substantial amounts of organoarseni-cals (especially MMA(V) and DMA(V)) to estuary waters and sediments (Sanders, Riedel and Osman, 1994), 296; (Anderson and Bruland, 1991). Although bacteria may demethylate MMA(V) and DMA(V) (Cullen and Reimer, 1989), 749; (Santosa *et al.*, 1996), 703, MMA(V) and DMA(V) are generally more stable in water than inorganic As(III) (Sanders, Riedel and Osman, 1994), 290–291.

3.15.2 Seasonal effects on arsenic in estuaries

During wet seasons, the quantity of arsenic-bearing runoff may be high and rivers often contribute signifi-cant amounts of the contaminant to estuaries. Sediment resuspension from storms or human activities can also release arsenic from porewaters and sediment particles (Xiankun, Jing and Xinian, 1994), 312–313. Besides affecting the quantity of arsenic, seasonal changes influence biological activity, which can affect arsenic speciation.

Chesapeake Bay, USA, is the largest estuary on Earth and almost all of the arsenic entering the headwa-ters is As(V). Although inorganic As(V) is consistently the most abundant arsenic species in the estuary, extensive arsenic reduction and methylation occur during warm months (Sanders, Riedel and Osman, 1994), 295; (Millward *et al.*, 1997), 53. The appearance of As(III) and methylarsenic species correlates well with phytoplankton production. Similar seasonal patterns involving arsenic reduction and methylation are seen in other estuaries (Sanders, Riedel and Osman, 1994), 295.

Any relationships between salinity and arsenic chemistry in estuary waters often vary with location and climate. In some areas, periodic upwelling of high-arsenic and saline bottom waters locally dominates the arsenic chemistry of estuaries (e.g. the Taiwan Strait; (Xiankun, Jing and Xinian, 1994), 332). In other situations, wet season flooding of highly arsenic-contaminated river waters increases the arsenic contents and lowers the salinity of estuaries. In contrast, fairly pristine river waters may dilute both estuary salinity and arsenic concentrations during flooding.

3.15.3 Arsenic in pristine estuaries

The arsenic chemistry of the Krka estuary, Croatia, is very pristine; that is, essentially unaffected by arsenic pollution from humans. Total dissolved arsenic concentrations in the estuary waters vary from 0.13 to 1.8 µg L^{-1} (Table 3.7; (Smedley and Kinniburgh, 2002), 525). The range of concentrations can be readily explained by low-arsenic (0.13 µg L^{-1}) river water simply diluting 1.8 µg L^{-1}-arsenic seawater in various proportions (Smedley and Kinniburgh, 2002), (Seyler and Martin, 1991), 525. In this estuary, arsenic (in particular As(V)) behaves conservatively; that is, the element is relatively unreactive in at least the short term. The arsenic chemistry of most other estuaries, however, is far more complex, especially if the arsenic largely results from pollution (Mucci *et al.*, 2000), 304, 306; (Howard *et al.*, 1988) or is influenced by seasonal changes (Huang *et al.*, 1988). In other words, the arsenic concentrations in most estuary waters simply cannot be predicted by averaging the contributions of the mixed seawater and fresh water in their correct proportions. Nonconservative arsenic behavior may result when the mixing of seawater and fresh water produces very different pH conditions that then favor the sorption and/or coprecipitation of arsenic with Fe(III) compounds (Smedley and Kinniburgh, 2002), 525. In other situations, estuary microorganisms may methylate the arsenic or it may desorb from iron (oxy)(hydr)oxide stream sediments once the sediments enter an estuary and encounter higher salinity (Migon and Mori, 1999), 85; (Howard *et al.*, 1988).

Masuda, Yamatani and Okai (2005) investigated the behavior of arsenic in intertidal sediments on the shore of the Urauchi River estuary, Iriomote Island, Japan. The sediments are remotely located from any human sources of arsenic contamination. In the upper sediments, humic acids sorb onto iron (oxy)(hydr)oxides and inhibit the sorption of arsenic (Masuda, Yamatani and Okai, 2005), 79. At depths of about 1.2–2.4 m, Fe(III)-bearing carboxylic acids decompose, which releases the Fe(III). The released Fe(III) may then form (oxy)(hydr)oxides, which can coprecipitate and/or sorb arsenic from sediment pore-waters (Masuda, Yamatani and Okai, 2005), 79. At approximately 2.4–2.8 m, bacteria are probably involved in the reductive dissolution of iron (oxy)(hydr)oxides and the reduction of sulfate to sulfide. As(V) may be directly reduced to As(III) by the microorganisms or indirectly through chemical reactions with H_2S produced by sulfate-reducing bacteria (Masuda, Yamatani and Okai, 2005), 80. Arsenic then coprecipitates with Fe(II) sulfide compounds (Masuda, Yamatani and Okai, 2005), 79–80. Unless disturbed by humans or natural events, the arsenic will remain immobilized in the sulfides (Masuda, Yamatani and Okai, 2005), 80.

3.15.4 Arsenic in contaminated estuaries

A variety of human activities and products may contaminate estuaries with arsenic, including wood preservatives, pesticides, fertilizer manufacturing, mine drainage and wastes, coal utilization, runoff from slags used as road ballast, and arsenical paints from ships (Landrum, 1994; Davis, De Curnou and Eary, 1997), 370; (Pirrie *et al.*, 2002; Mirlean *et al.*, 2003). In some cases, specific arsenic sources cannot be identified (O'Reilly Wiese, Bubb and Lester, 1995). Nevertheless, the cycling of arsenic between water and sediments in contaminated estuaries is often similar to cycles in pristine estuaries, open marine environments, and lakes.

Sediments in the Hylebos tidal waterway at Tacoma, Washington, USA, are contaminated with up to $1260\,mg\,kg^{-1}$ of arsenic (Table 3.8; (Davis, De Curnou and Eary, 1997), 1985). Several possible sources may have been responsible for the arsenic contamination, including (1) alkaline arsenic-bearing wastewaters discharged from a pesticide manufacturing plant, (2) slags from rock-wool, steel, smelter, and powdered metal plants, (3) runoff from slags used as road ballast, and (4) marinas that stripped arsenical paints from ships (Davis, De Curnou and Eary, 1997), 1985. The feedstocks for the powdered metal plant were especially rich in *detrital* sulfide minerals, including enargite (Cu_3AsS_4) and arsenian pyrite. The feedstocks probably entered the waterway as wind-blown dust or spillage during the unloading of freighters (Davis, De Curnou and Eary, 1997), 1988. The detrital grains are absent in sediments younger than 1954 (when the powdered metal plant closed) and the grains are easily distinguished from *authigenic* sedimentary pyrites, which tend to be *framboidal* and have far less arsenic (Davis, De Curnou and Eary, 1997), 1988–1989. Arsenic, which may have originated from the pesticide plant, is associated with iron (oxy)(hydr)oxides in some of the sediments (Davis, De Curnou and Eary, 1997), 1990. In the sediment porewaters, arsenic solubility is limited by coprecipitation with Fe(II) sulfides (Davis, De Curnou and Eary, 1997), 1990.

3.16   Rivers and other streams

Most unpolluted stream waters contain 0.1–0.8 $\mu g\,L^{-1}$ of dissolved arsenic, although rivers in geothermal areas of New Zealand and the western USA may have 10–370 $\mu g\,L^{-1}$ and arsenic concentrations in streams of the volcanic and geothermal regions of Chile are often >100–1000 $mg\,L^{-1}$ (Table 3.9; (Smedley and Kinniburgh, 2002), 523–524; (Shih, 2005), 88). In most cases, arsenic is far more abundant in stream sediments than their overlying waters (Tables 3.9 and 3.10). Streams also generally transport larger quantities of arsenic through sorption onto suspended and *saltated* particles than through dissolution (Mok and Wai, 1994), 101. The global average of arsenic on stream PM is about 5 $\mu g\,kg^{-1}$ (Matschullat, 2000), 304.

Natural sources of arsenic in streams primarily include geothermal waters, groundwater, wind-blown dust, and drainage from the weathering of sulfides and other arsenic-bearing minerals (Tables 3.9 and 3.10). Arsenic concentrations in streams and other fresh surface waters depend on many factors, including (1) the composition of any underwater sediments and rocks, and any soils, sediments, and rocks surrounding the surface waters; (2) inputs from geothermal sources or high-arsenic groundwaters; (3) any contamination from humans; (4) biological activity, which may reduce and methylate arsenic; and (5) any changes in Eh, pH, temperature, or chemical conditions because of storms, seasonal changes, and human activities. Although streams are usually well oxygenated because of flowing water and relatively shallow depths, algal blooms, sewage discharges, or livestock runoff may substantially consume dissolved O_2 and create reducing conditions in some rivers (e.g. Zala River of Hungary, (Elbaz-Poulichet, Nagy and Cserny, 1997), 272).

Inorganic As(V) is usually the most prominent arsenic species in stream water (Mok and Wai, 1994), 100. In stream sediments, As(V) often sorbs onto clay minerals, iron (oxy)(hydr)oxide particles, and other fine-grained (<63 μm) materials, which tend to have higher surface areas and better sorptive properties than uncoated quartz sands (Smedley and Kinniburgh, 2002), 532; (Mok and Wai, 1994), 108–109; (Sharma, Tobschall and Singh, 2003); (Deacon and Driver, 1999). Dissolved As(V) in some river waters may also precipitate as Fe(III), aluminum, or calcium arsenates (Mok and Wai, 1994), 101. Stream sediments with the highest arsenic contents are usually enriched in high-surface area (oxy)(hydr)oxides, organic matter, and/or detrital sulfides.

If pH, Eh, biological activity, temperature, or other sediment and porewater properties change, arsenic may desorb, dissolve, entrain, or otherwise reenter the overlying water. As mentioned earlier in Section 3.7.2.4, arsenic mobilization is possible if stream sediments are mined, dredged, exposed to mining wastes, or otherwise disturbed by human activities (Hudson-Edwards *et al.*, 2001; Gough *et al.*, 1999). Stream deposits near ore bodies or mines often contain arsenic-rich detrital sulfides (Smedley and Kinniburgh, 2002), 532. Even without human intervention, meandering rivers can periodically bury and then later reexpose the sulfide minerals to additional weathering. Protective weathering rinds sometimes coat the surfaces of the sulfide minerals, which would slow down the decomposition of the minerals and prolong the release of arsenic in streams (Mok and Wai, 1994), 109. The slow weathering of sulfide minerals would allow them to transport farther downstream and contaminate larger sections of the streams over longer periods of time (Hudson-Edwards *et al.*, 2001); (Mok and Wai, 1994), 108–109.

Climate can greatly influence the mobilization of arsenic in streams and other surface waters. Storms and floods may mobilize arsenic through the excavation, oxidation, and dissolution of sulfides in sediments. They may also transport greater than normal quantities of arsenic downstream on suspended and saltated particles (Mok and Wai, 1994), 104. Depending on changes in biological activity and the amount of runoff between seasons, arsenic concentrations in rivers may cycle over time (Neal *et al.*, 2000). In colder water, arsenic is more likely to sorb onto sediments rather than dissolve (Mok and Wai, 1994), 110. During hot and drier months in temperate and arid climates, dissolved arsenic concentrations may considerably increase in streams, lakes, and other surface waters because of water evaporation. In warm water, biological

Table 3.9 Arsenic analyses of water from various streams.

Stream and location	Prominent arsenic source(s)	pH	Arsenic concentration (µg L^{-1})[a]	Reference
Alaknanda River: Devprayag, India	Weathering of geologic materials	7.00–8.50	6.310 (0.2 µm filtered)	Chakrapani (2005), 198
Alamosa River: Colorado, USA	–	–	0.11 (June 1994); 0.14 (September 1994) (0.40 µm filtered)	Taylor et al. (2001)
Amazon River: Brazil	–	–	0.21 (inorganic)	Cutter et al. (2001)
Bhagirathi River: Devprayag, India	Weathering of geologic materials	7.30–8.30	4.630 (0.2 µm filtered)	Chakrapani (2005)
Carmo River, Brazil: 12 samples	Mining	6.9–7.8	7–43 (0.45 µm filtered)	Borba, Figueiredo and Matschullat (2003)
Carson River: California–Nevada, USA	Weathering of geologic materials, hydrothermal sources	7.77–8.5	5–175 (0.4 µm filtered)	Johannesson et al. (1997), 70
Ciwidey River: West Java, Indonesia	Hydrothermal and volcanic activity	3.0–7.1	0.3–3.4 (0.45 µm filtered)	Sriwana et al. (1998), 167
Coeur d' Alene River: Idaho, USA	Mining wastes	7.09–9.07	0.088–1.636 As(III) + As(V), 0.022–1.370 As(III) (0.45 µm filtered)	Mok and Wai (1990)
Columbia River, Washington–Oregon, USA: 23 samples	–	7.46–8.46	0.26–1.4 (unfiltered, total recoverable arsenic using nitric acid)	Johnson and Golding (2002)
Conceição River: Brazil: 12 samples	Mining	5.7–7.5	2–8 (0.45 µm filtered)	Borba, Figueiredo and Matschullat (2003)
Danube River: Vienna, Austria	–	–	<1	Thielen et al. (2004), 423
Danube River: Budapest, Hungary	–	–	1.63 (downstream from city), 1.97 (upstream from city)	Thielen et al. (2004), 423
Davis Creek: California, USA	–	–	2.5	Anderson and Bruland (1991)
Finland (Western, 5 streams)	Geologic materials	4.5–6.2	0.54–0.87 (unfiltered)	Åström and Corin (2000)
Finland: 1160 headwater streams	–	–	<0.20–6.50 (0.45 µm filtered)	Tarvainen, Lahermo and Mannio (1997)
Ganga River: Rishikesh, India	Weathering of geologic materials	7.50–8.30	4.540 (0.2 µm filtered)	Chakrapani (2005), 198

Location	Source	pH	Concentration	Reference
Ghana: near Prestea: 13 samples	Mining	—	150–8250 (0.45 µm)	Serfor-Armah et al. (2006)
Huang He (Yellow River), China: 4 samples	Weathering of loess	7.93–8.38	1.99–2.05 (0.4 µm filtered)	Huang et al. (1988)
Humboldt River (North Fork): Nevada, USA (upstream from mining site)	Mining	7.45–8.43	4–12 (unfiltered)	Earman and Hershey (2004)
Indus River: Pakistan	—	—	620–2536 (<0.46 µm filtered)	Tariq et al. (1996)
Jemez River: New Mexico, USA	Hydrothermal	—	2–300 (unfiltered)	Reid, Goff and Counce (2003)
Kalix River: Sweden (70 samples)	Natural background (unspecified sources)	—	0.16 ± 0.10	Widerlund and Ingri (1995), 190
Lena River: Russia: 2 samples	Natural weathering	—	0.15; 0.16 (0.4 µm filtered)	Martin et al. (1993)
Lisora River: Switzerland	Weathering of geologic materials, mining?	7.9; 8.1	7.9, 12.3, respectively	Pfeifer et al. (2004), 212
Little Arkansas River: Kansas, USA	—	—	2–13	Ziegler et al. (2001)
Little Cottonwood Creek: Utah, USA: 74 samples	Mining and smelting	7.5–8.6	<1–284 (0.45 µm filtered)	Gerner and Waddell (2003)
Lluta River: Chile	—	—	124–305	Yamasaki and Hata (2000)
Madison River: Wyoming and Montana, USA	Hydrothermal from Yellowstone	—	40–200	Sonderegger and Sholes (1989)
Medjerda River: Tunisia–Algeria: during mining operations	Mining	—	250–3000	Jdid et al. (1999)
Medjerda River, Tunisia–Algeria: after mining operations ceased	More toward natural background	—	150–1080	Jdid et al. (1999)
Mississippi River: below Belle Chasse, Louisiana, USA	—	—	1.7 (0.4 µm filtered)	Taylor and Shiller (1995), 1315

(continued overleaf)

Table 3.9 (continued)

Stream and location	Prominent arsenic source(s)	pH	Arsenic concentration (µg L⁻¹)[a]	Reference
Mississippi River: below Arkansas City, Arkansas, USA	–	–	1.0 (0.4 µm filtered)	Taylor and Shiller (1995), 1315
Péroux River: France: Upstream and downstream from mine tailings pile	Mine tailings	7.3 upstream and downstream	9.3 (upstream), 10.4 (downstream) (0.45 µm filtered)	Néel et al. (2003)
Rio Grande: USA/Mexico	–	–	2	Dunbar, Chapin and Chapin (1995)
Rio Grande: Albuquerque, New Mexico, USA: 13 samples	Weathering of geologic materials; hydrothermal water	–	2.0–4.0	Bexfield and Plummer (2003)
Rio Grande: New Mexico, USA	–	–	1.8–3.6	Wilcox (1997)
River Loa: Chile: 5 samples	Volcanic deposits, mining	–	101 300 ± 3000	Pizarro et al. (2003)
Río Loa Basin: Chile	Weathering of geologic materials, hydrothermal, mining? smelter emissions?	–	1400	Romero et al. (2003)
Río Siete: Ecuador	Mining	–	0.1–470 (0.45 µm filtered)	Appleton et al. (2001)
Sacramento River: California, USA	–	–	1.9	Anderson and Bruland (1991)
San Tirso River: Spain	Mining	–	<5 (upstream from mines), 900–13 800 (downstream from mining)	Loredo et al. (2004)
St. Joe River: Idaho, USA	Natural background (unspecified sources)	–	0.276 ± 0.004 (As(III) + As(V)); 0.45 µm filtered)	Mok and Wai (1990)
Tantalus Creek: Yellowstone National Park, Wyoming, USA	Hydrothermal	–	1700 (97 % As(V))	Webster and Nordstrom (2003), 119
Thames River: Howberry Park, United Kingdom	Industry? sewage discharges?	7.82–9.28	0.71–3.95 (0.45 µm filtered), an anomalously high measurement of 40.30	Neal et al. (2000)

Location	Source			Reference
Topolnitza River: Srednogorie, Bulgaria (1987–1990)	Copper smelter	—	750–1500	Rahman, Sengupta and Chowdhury (2006)
Toro River: Chile: 188 samples	Weathering of geologic materials, mining	3.91–7.60	60–3950 (0.45 μm filtered)	Oyarzun et al. (2006)
Tresa River: Switzerland	Weathering of geologic materials, mining?	9.1	3.1	Pfeifer et al. (2004), 213
Truckee River: California–Nevada, USA	Weathering of geologic materials, hydrothermal sources	6.67–8.81	<2–135 (0.4 μm filtered)	Johannesson et al. (1997), 68
Tuolumne River: California, USA	Natural background (unspecified sources)	—	0.2–0.4	Savage, Bird and Ashley (2000), 403
USA: Alabama: northern: impacted by acid mine drainage: 12 sites	Mining	—	<0.2–89 (0.44 μm filtered)	Goldhaber et al. (2001)
USA: Alaska: Cook Inlet Basin streams: 58 sites	Natural weathering of geologic materials	—	<1–9 (0.45 μm filtered); <24 (unfiltered)	Glass and Frenzel (2001)
USA: Kentucky: 59 monitoring sites	—	—	<2.0–3.0	Pope et al. (2004)
USA: Louisiana: 12 monitoring sites	—	—	2.4–5.8	Pope et al. (2004)
USA: Nebraska: 13 monitoring sites	—	—	2.9–6.8	Pope et al. (2004)
USA: Nevada: 59 monitoring sites	—	—	<3.0–251	Pope et al. (2004)
USA: New Jersey: 4 monitoring sites	—	—	1.0–3.0	Pope et al. (2004)
USA: North Dakota: 24 monitoring sites	—	—	1.4–7.9	Pope et al. (2004)
USA: Ohio: 41 monitoring sites	—	—	<2.0–4.0	Pope et al. (2004)

(continued overleaf)

Table 3.9 (continued)

Stream and location	Prominent arsenic source(s)	pH	Arsenic concentration ($\mu g\,L^{-1}$)[a]	Reference
USA: South Dakota: 20 monitoring sites	—	—	<5.0–37.1	Pope et al. (2004)
USA: South Dakota: Black Hills: Spearfish Creek: 66 samples	Mining, natural background	—	<1–50 (total), <1–48 (dissolved)	Driscoll and Hayes (1995)
USA: South Dakota: Black Hills: Spearfish, Whitewood, and Bear Butte creeks watersheds	Mining, natural background	—	0.9–51 (0.45 μm filtered)	May et al. (2001)
USA: South Dakota: Black Hills: Whitewood Creek	Mining	7.9–9.0	<7–110 (dissolved)	Kuwabara, Chang and Pasilis (2003)
USA: 10 southeast coastal rivers: Florida, Georgia, South Carolina, and North Carolina	Weathering of geologic materials; atmospheric deposition	—	0.04–0.74 (0.45 μm filtered)	Waslenchuk (1979)
USA: Tennessee: 70 monitoring sites	—	—	<1.0–6.0	Pope et al. (2004)
USA: Utah: 54 monitoring sites	—	—	<5.0–20.0	Pope et al. (2004)
Velhas River: Brazil: 8 samples	Mining	6.2–7.9	<1–64 (0.45 μm filtered)	Borba, Figueiredo and Matschullat (2003)
Vormböcken River: Sweden: 12 samples	Mining	5.8–7.6	0.9–3.8 (total)	Sjöblom , Hakansson and Allard (2004)
Walker River: California–Nevada, USA	Weathering of geologic materials, hydrothermal sources	6.25–8.69	<2–65 (0.4 μm filtered)	Johannesson et al. (1997), 69
Yakima River: Washington, USA: 24 samples	Pesticides?	6.95–8.56	<0.10–2.7 (unfiltered, total recoverable arsenic using nitric acid)	Johnson and Golding (2002)
Yakima River: Washington, USA: 79 samples	Pesticides	7.1–8.6	<1.0–11 (0.45 μm filtered)	

[a]Single values denote single analyses or average concentrations sometimes with 1 SD. Values separated by commas represent duplicate analyses of the same sample. Analyses involving different samples from the same location are separated by semicolons. Low and high values are separated by hyphens. Available information on arsenic speciation and micron filtering of the samples is listed.

Table 3.10 *Arsenic concentrations in various stream sediments.*

Location	Prominent arsenic source(s)	Arsenic concentration (mg kg^{-1})[a]	Reference
Anoia River: Spain: 20 surface sediments	Agricultural and/or industrial sources?	32.0 ± 6.8 (<63 µm)	Casas *et al.* (2003)
Aril River: India: 5 samples	Manufacturing of agricultural products?	9–16 (<63 µm)	Sharma, Tobschall and Singh (2003)
Blue River Basin: Colorado: USA: 10 stream samples	Mining	13–180 (<63 µm)	Apodaca, Driver and Bails (2000)
Cardener River: Spain: 16 surface sediments	Natural background	25.9 ± 6.1 (<63 µm)	Casas *et al.* (2003)
Carmo River: Brazil: 14 samples	Mining	105–4709 (<63 µm)	Borba, Figueiredo and Matschullat (2003)
Caudal River Basin: Spain: 9 samples	Mining	186 (upstream from mines), 186–41 366 (near mines)	Loredo *et al.* (2004)
Coeur d' Alene River: Idaho: USA: 36 samples	Mining	8.3–179.0	Farag *et al.* (1998)
Coeur d' Alene River: Idaho: USA: 7 samples	Mining	10.68–209.09	Mok and Wai (1990)
Conceição River: Brazil: 9 samples	Mining	29.5–120 (<63 µm)	Borba, Figueiredo and Matschullat (2003)
Danube River: Vienna, Austria	—	17.5	Thielen *et al.* (2004), 423
Danube River: Budapest, Hungary	—	12 (upstream from city), 15 (down stream from city)	Thielen *et al.* (2004), 423
Detroit River: Michigan: USA	—	11.4	Brannon and Patrick (1987), 452
Fortymile River Watershed: Alaska: USA: 25 samples	Natural weathering, mining	3.3–15	Crock *et al.* (1999)
Ganga River: India: 4 samples	—	4–6 (<63 µm)	Sharma, Tobschall and Singh (2003)
Gangan River: India: 7 samples	Brass industry?	4–149 (<63 µm)	Sharma, Tobschall and Singh (2003)
Ghana: near Prestea: 13 samples	Mining	942–10 200	Serfor-Armah *et al.* (2006)

(continued overleaf)

Table 3.10 (continued)

Location	Prominent arsenic source(s)	Arsenic concentration (mg kg^{-1})[a]	Reference
Indus River: Pakistan	—	0.167–7.452	Tariq et al. (1996)
Italy: Tuscany: Fosso dei Noni Creek: 9 samples	Mining	35–421 (total fraction)	Mascaro et al. (2001)
Kalix River: Sweden	Natural background	~5	Widerlund and Ingri (1995), 190
Llobregat River, Spain: 32 surface sediments	Natural background	26.4 ± 8.8 (<63 μm)	Casas et al. (2003)
Mahawan River: India: 4 samples	Industrial and urban wastes?	6–20 (<63 μm)	Sharma, Tobschall and Singh (2003)
Mexico: Baja California: Dry stream sediments: La Paz Lagoon	Natural sources	1.7–28.4	Shumilin et al. (2001)
Mignone River Basin: Central Italy	Mining	2.6–665.5	Spadoni, Voltaggio and Cavarretta (2005)
Pra and Offin rivers: Ghana: 42 samples (21 wet season; 21 dry season)	Gold mining	Wet season: 0.079–0.714 (<64 μm), one outlier: 4.31; dry season: 0.088–0.940 (<64 μm), one outlier: 5.076	Donkor et al. (2005), 485
Ramganga River: India: 22 samples	—	8–645 (<63 μm)	Sharma, Tobschall and Singh (2003)
River Loa: Chile: 5 samples	Volcanism; Mmining	91.2 ± 3.6, 60 % As(III)	Pizarro et al. (2003)
Río Loa Basin: Chile: 60 samples	Weathering of geologic materials, hydrothermal, mining? smelter emissions?	26–2000	Romero et al. (2003)
Sot River: India: 9 samples	Industrial emissions?	4–50 (<63 μm)	Sharma, Tobschall and Singh (2003)
St. Joe River: Idaho: USA: 4 samples	—	2.4 (standard error of the mean: 0.1)	Farag et al. (1998)
Toliman River: Zimapán, Mexico	Mining	96–6575	García, Armienta and Cruz (2001)

Location	Source	Concentration	Reference
Tualatin River: Oregon: USA: 15 sites	–	2.0–16 (<63 μm)	Bonn (1999)
Upper Colorado River Basin: Colorado: USA: 37 streambed sediment samples (8 from nonmined areas)	Natural background (unspecified sources), mining	3.0–180 (3.0–62 from nonmined areas) (<63 μm)	Deacon and Driver (1999)
USA: Alabama: northern: impacted by acid mine drainage: 25 sites	Mining (coal)	5–180 (<80 μm)	Goldhaber et al. (2001)
USA: Alaska: Cook Inlet Basin streams: 47 sites	Natural weathering	1.7–88 (<63 μm)	Frenzel (2002); Glass and Frenzel (2001)
USA: Black Hills: Spearfish, Whitewood, and Bear Butte creeks watersheds	Mining, natural background	10–1951	May et al. (2001)
USA (conterminous) (541 stream bed samples)	–	1–200 (<63 μm)	Rice (1999)
USA: Montana, Idaho, and Washington: northern Rocky Mountains: Clark Fork–Pend Oreille and Spokane River Basins: 16 samples	Mining	2.6–176 (<63 μm)	Maret and Skinner (2000)
Velhas River: Brazil: 5 samples	Mining	68–360 (<63 μm)	Borba, Figueiredo and Matschullat (2003)
Yakima River Basin: Washington: USA: 404 sites	Pesticides, natural weathering	0.7–310 (<63 μm)	Morace et al. (1999)

[a]Single values denote single analyses or average concentrations sometimes with 1 SD. Values separated by commas represent duplicate analyses of the same sample. Analyses involving different samples from the same location are separated by semicolons. Low and high values are separated by hyphens. Normally, soil and sediment samples are dried before analysis. Available information on sieved size fractions is listed.

activity increases, which may mobilize arsenic from sediments through the reduction of arsenic to often more mobile $H_3AsO_3{}^0$ or to methyl forms that are less likely to sorb onto iron (oxy)hydroxides and other compounds (Smedley and Kinniburgh, 2002), 524; (Mok and Wai, 1994), 105, 113; (Seyler and Martin, 1989), 1261.

Biotic methylation in streams often produces volatile arsenic species, such as alkylarsines (e.g. methylarsine), which could escape into the air (Mok and Wai, 1994), 113. In stream water and sediments, DMA(V) and MMA(V) may be directly created by microorganisms or degraded from more complex organoarsenicals. Eventually, methylarsenic in streams tends to degrade to inorganic As(V) (Mok and Wai, 1994), 112.

3.17   Lakes

Lakes are often important sources of municipal water, food, transportation, and recreation for people. Unless they are associated with volcanic deposits, geothermal fluids, or mining sites and other areas impacted by human activity, arsenic concentrations in lakes and their sediments are usually similar to uncontaminated streams; that is, $<1\,\mu g\,L^{-1}$ and $<10\,mg\,kg^{-1}$, respectively (Tables 3.11 and 3.12; (Smedley and Kinniburgh, 2002)). Major sources of arsenic inputs into lakes include runoff from weathered rock outcrops (Tarvainen, Lahermo and Mannio, 1997; Islam *et al.*, 2000; Skjelkvåle *et al.*, 2001), groundwaters (Ford, Wilkin and Hernandez, 2006), arsenic-bearing pesticides and herbicides (Islam *et al.*, 2000; Chen *et al.*, 2000), mine drainage and other mining wastes (Sadiq *et al.*, 2002), deposition of atmospheric emissions from coal-fired power plants and ore smelters (Sadiq *et al.*, 2002; Rognerud and Fjeld, 2001), and streams with geothermal waters (Kneebone and Hering, 2000; Oremland, Stolz and Hollibaugh, 2004) (Tables 3.11 and 3.12). Severe storms, extensive bioturbation, and human activities can also disturb buried contaminants in bottom sediments and remobilize arsenic into overlying lake waters.

The density of liquid water changes with temperature and fresh water reaches its maximum density at 4 °C. Differences in temperature and density with depth in large lakes often lead to the formation of distinct water layers. The *epilimnion* refers to the well-aerated surface layer of lakes (Figure 3.4; (Drever, 1997), 166). If the lake is clear of ice, waves and currents usually produce an epilimnion that has fairly uniform temperature and density. In shallow lakes, the epilimnion may extend all the way to the bottom; that is, the lake is unstratified (Figure 3.3). Nevertheless, even if stratification is absent in a lake, arsenic concentrations and speciation may be significantly different in the photic zone when compared with deeper waters (Cullen and Reimer, 1989), 744. Although inorganic As(V) is usually the prominent arsenic form in the epilimnion, biological activity from seasonally warm water and abundant nutrients in the photic zone may produce considerable inorganic As(III) and methylarsenic (Figures 3.3 and 3.4; (Anderson and Bruland, 1991; Elbaz-Poulichet, Nagy and Cserny, 1997)). The production of organoarsenicals by algae and subsequent deposition and burial in bottom sediments may be a significant process in removing arsenic from lake water (Roberts, Herbert and Louchouarn, 2001). Arsenic demethylation into inorganic forms may also become prominent from winter until biological activity increases in the spring (Figures 3.3 and 3.4; (Anderson and Bruland, 1991)).

One or more *metalimnions* may exist below the epilimnion (Figure 3.4). The temperature of the metalimnion rapidly declines with depth. Eventually, temperature stabilizes in the underlying *hypolimnion*, where the waters are cold, anoxic, and relatively dense (Figure 3.4; (Drever, 1997), 166). The size of the hypolimnion depends on the amount of organic matter produced by summer biological activity in the epilimnion. Once abundant organic matter settles into the hypolimnion, considerable decomposition of the matter would increase the severity and extent of anoxic conditions. On the other hand, a lack of biotic organic matter would thin or possibly eliminate the hypolimnion, which would allow oxidizing conditions

Table 3.11 *Arsenic analyses of water from various lakes, reservoirs, wetlands, and ponds*

Location	Arsenic source(s) favored by the authors	pH	Arsenic concentration ($\mu g\,L^{-1}$)[a]	Reference(s)
Anderson Lake, Oregon, USA	Hydrothermal	8.6	600–700	Finkelstein *et al.* (2004)
Bangladesh: Andulia: 1 surface water	Weathering of geologic materials and irrigation runoff	—	7.79 (0.45 µm filtered); 14 (unfiltered)	Islam *et al.* (2000)
Bangladesh: Mainamoti: 1 surface water	Weathering of geologic materials and irrigation runoff	—	5.74 (0.45 µm filtered); 5.23 (unfiltered)	Islam *et al.* (2000)
Bangladesh: Rajarampur: 3 surface waters	Weathering of geologic materials and irrigation runoff	—	97.05 (unfiltered)	Islam *et al.* (2000)
Bangladesh: Shamta: 2 surface waters	Weathering of geologic materials and irrigation runoff	—	13.71 (0.45 µm filtered); 15.36 (unfiltered)	Islam *et al.* (2000)
Canada: Quebec and Ontario: 16 lakes	Mining and smelting	3.99–8.36	0.10–7.30 (0.2 µm filtered)	Belzile and Tessier (1990)
Canada: near Yellowknife, Northwest Territories: Grace, Meg, Keg, Kam, and Peg lakes	Mining	—	11–530 (~1.5 µm filtered)	Bright, Dodd and Reimer (1996)
Davis Creek Reservoir, California, USA	—	—	1.8	Anderson and Bruland (1991)
Egypt: 3 Coastal lakes near Nile River	—	—	1.2–18.2 (dissolved), 1.2–8.7 (particulate)	Abdel-Moati (1990)
Finland: 152 lakes	Weathering of geologic materials	—	0.08–5.20 (0.45 µm filtered)	Tarvainen, Lahermo and Mannio (1997)
Finland: 36 lakes in Lapland	Weathering of geologic materials, some air deposition?	5.7 (median)	0.17 (median)	Mannio *et al.* (1995)
Ghana: Amansie West: 37 surface waters	—	—	5–2900 (0.45 µm filtered)	Duker, Carranza and Hale (2005)
India: Kanker district: surface waters	Weathering of geologic materials	6.5–7.3	<1–200.0	Pandey *et al.* (2006), 414

(continued overleaf)

Table 3.11 (continued)

Location	Arsenic source(s) favored by the authors	pH	Arsenic concentration ($\mu g\,L^{-1}$)[a]	Reference(s)
Kelly Lake, Ontario, Canada: 65 samples	Mining and deposition of smelter air emissions	6.72–7.95	1–8 (0.45 µm filtered)	Sadiq et al. (2002)
Lake Biwa, Japan	—	6.5–9.0	0.17–12.7	Sohrin et al. (1997)
Lake Coeur d'Alene (southern), Idaho, USA	—	—	0.357 ± 0.005 (As(III) + As(V))	Mok and Wai (1990), 104
Lake Coeur d'Alene, Idaho, USA: 145 samples	—	—	< 1–1	Harrington, Fendorf and Rosenzweig (1998)
Lake Corpus Christi, Texas, USA	Weathered rhyolitic volcanics	—	2–15	Parker et al. (2001)
Lake Ontario, Great Lakes, North America	—	—	0.91 (mean)	Traversy et al. (1975)
Lake Superior, Great Lakes, North America	—	—	0.23 (mean)	Traversy et al. (1975)
Lake Tahoe, California, USA	—	—	1.1	Anderson and Bruland (1991)
Lake Washington, Washington, USA: Bottom water	Natural background	—	1.0	Peterson and Carpenter (1986)
Mono Lake, California, USA: 0–14 m surface waters	Hydrothermal	9.8	14 700	Hollibaugh et al. (2005), 1932
Mono, Big Soda, Walker, Crowley, and Searles desert alkaline lakes: California–Nevada, USA	Hydrothermal and weathering of volcanic rocks	8.5–9.8	60 (Crowley) –220 000 (Searles)	Oremland, Stolz and Hollibaugh (2004)
Moon Lake, Coahoma county, Mississippi, USA	Natural background	—	4.48 ± 2.46 to 5.99 ± 3.95	Cooper and Gillespie (2001)
Nacharam Lake, Hyderabad, India: 9 samples	Unspecified anthropogenic	—	7.94–12.14 (filtered, unknown size)	Govil, Reddy and Gnaneswara Rao (1999)
North Haiwee Reservoir, California, USA: 2 sites	Hydrothermal	7.8 (northwest site), 8.2 (northeast site)	Northwest: 5.7 (total, unfiltered), 1.3 (unfiltered As(III)); Northeast: 5.7 (total unfiltered), 1.5 (unfiltered As(III))	Kneebone et al. (2002)

Location	Source	pH	As concentration	Reference
Pyramid Lake, California–Nevada, USA: 3 samples	Weathering of geologic materials, hydrothermal sources	9.04; 9.25; 9.23	99.6; 99.6; 117.9, respectively	Johannesson et al. (1997), 68
San Joaquin Valley, California, USA: Inlet water for artificial wetland	Agricultural drainage	7.7	80–220	Fox and Doner (2003), 2429
San Joaquin Valley, California, USA: Kern Water Bank	weathering of volcanic rocks	–	<1–211	Swartz and Thyne (1995)
Salton Sea, California, USA	–	–	13	Anderson and Bruland (1991)
Thailand: Mae Moh lignite Basin: surface waters	Mining	–	1.2–325	Bashkin and Wongyai (2002)
Tuskegee Lake, Alabama, USA	–	6.98–7.63 (sampling #1), 7.06–7.68 (sampling #2)	0.06 ± 0.23 (sampling #1); 0.6 ± 2.2 (sampling #2)	Ikem, Egiebor and Nyavor (2003)
USA: New England and New York: 17 lakes	Agriculture	–	0.022–0.587 (0.45 μm filtered; unweighted averages)	Chen et al. (2000)
USA: Florida: ponds and wetland surface waters near a golf course	–	–	≤31	Lewis et al. (2002)
Walker Lake: California–Nevada, USA: 3 samples	Weathering of geologic materials, hydrothermal sources	9.52; 9.46; 9.44	1029; 1399; 989, respectively	Johannesson et al. (1997), 69

[a] Single values denote single analyses or average concentrations sometimes with 1 SD. Values separated by commas denote duplicate analyses of the same sample. Analyses involving different samples are separated by semicolons. Low and high values are separated by hyphens. Available filtering information is listed.

Table 3.12 *Arsenic concentrations in various lake, reservoir, pond, and wetland soils and sediments*

Location	Arsenic source(s) favored by the authors	Arsenic concentration (mg kg^{-1})[a]	Reference(s)
Antarctica: King George Island: Fildes Peninsula: lake sediments: 1500 year record: 31 measurements	Uncertain natural	0.05–0.17	Yin et al. (2006)
Canada: Labrador: 578 lake sediments	Weathering of geologic materials	<0.5–126	McConnell (1999)
Canada: near Yellowknife, Northwest Territories: Grace, Meg, Keg, Kam, and Peg lakes	Mining	14–3090	Bright, Dodd and Reimer (1996)
Crowley Lake: California, USA	Hydrothermal	4–80	Kneebone and Hering (2000)
Hall's Brook Holding Area pond, Woburn, Massachusetts, USA: As contaminated surface sediments	Industrial	20–2100	Wilkin and Ford (2006), 160
Harlan County Lake (Reservoir), Nebraska, USA: 15 sediment core samples	Weathered shales and other rocks; agricultural irrigation drainage?	5.7–9.0	Christensen and Juracek (2001)
Japan: Niigata Plain: Lake deposited organic clay	–	25–64	Kubota, Ishiyama and Yokota (2000)
Japan: Niigata Plain: swamp sediments	–	2–14	Kubota, Ishiyama and Yokota (2000)
Kirwin Reservoir, Kansas, USA: 24 sediment core samples	Weathered shales and other rocks; agricultural irrigation drainage?	4.6–10.0	Christensen and Juracek (2001)
Lake Brompton: Quebec, Canada: sediments	Weathering of geologic materials; mining	12	Belzile, Lecomte and Tessier (1989)
Lake Coeur d'Alene, Idaho, USA, 206 samples	Mining	2–568	Harrington et al. (1998), 652
Lake Eire, Great Lakes, North America	–	3.20	Traversy et al. (1975)
Lake Joannes: Quebec, Canada: sediments	Weathering of geologic materials; mining	14	Belzile, Lecomte and Tessier (1989)
Lake Superior shoreline, Superior, Wisconsin, USA: 3 samples of red lake clay	Natural background	12	Helmke et al. (1977)

Location	Source	Concentration	Reference
Lake Superior, Duluth, Minnesota and Superior, Wisconsin harbors, USA	Unspecified anthropogenic	5.4 ± 0.5	Helmke *et al.* (1977)
Lake Superior, Great Lakes, North America	—	2.03	Traversy *et al.* (1975)
Lake Süßer See, Germany: sediment core	Mining and smelting	2–1800	Becker *et al.* (2001), 208]
Lake Waco (Reservoir) sediments, Texas, USA: 48 samples	Weathered rocks	2.0–16	Abraham (1998)
Lake Washington, Washington, USA: 16 analyses from a sediment core	Copper smelter operated 1890–1985	10–98	Peterson and Carpenter (1986)
Michigan City Harbor, Lake Michigan, Indiana, USA	—	2.3	Brannon and Patrick (1987), 452
Milford Lake (Reservoir), Kansas, USA: 20 sediment core samples	Weathered rocks; agricultural irrigation drainage?	6.1–9.9	Christensen and Juracek (2001)
Milltown Reservoir: Montana, USA	Mining	22–56 (<63 µm)	Nichols (2003)
Milwaukee Harbor, Lake Michigan, Wisconsin, USA	—	1.1	Brannon and Patrick (1987), 452
Moira Lake, Ontario, Canada	Mining	408–1051	Azcue and Nriagu (1995)
Moon Lake and surrounding wetlands, Coahoma county, Mississippi	Natural background	5.614 (lake sediment mean), 6.746 (wetland sediment mean)	Cooper and Gillespie (2001)
New Orleans, Louisiana, USA: 26 soil and sediment samples	—	1.74–24.15	Presley *et al.* (2006)
New Orleans, Louisiana, USA: 6 flood sediments from Hurricane Katrina (2005)	Urban arsenic-bearing herbicides and leaching of CCA wood?	5.51–22.2	Cobb *et al.* (2006)
North Bull Island salt marsh, Dublin Bay, Ireland: top 20 cm of soil core samples	—	2 ± 0.2 to 36 ± 9.9	Doyle and Otte (1997), 4
Norway: surface sediments in 210 lakes	Air emissions from coal and refuse combustion	12.8 (median)	Rognerud and Fjeld (2001), 14

(continued overleaf)

Table 3.12 (continued)

Location	Arsenic source(s) favored by the authors	Arsenic concentration (mg kg^{-1})[a]	Reference(s)
Owens dry lake bed sediment crusts: California, USA: 14 samples	Evaporative concentration, including mine drainage	7 ± 4 to 38 ± 5	Gill et al. (2002)
Owens dry lake bed sediment: California, USA	Evaporative concentration	4.0–105.6	Ryu et al. (2002)
Pescadero salt marsh, California, USA	Natural background	~14	Bostick, Chen and Fendorf (2004), 3299
San Joaquin Valley, California, USA: Preflooded sediments in artificial wetlands	Agricultural irrigation drainage	13.2	Fox and Doner (2003), 2429
Spy Lake, Massachusetts, USA: 68 surface sediments	Likely herbicides	1–2600; background: 10–40	Durant et al. (2004)
Swanson Lake (Reservoir), Nebraska, USA: 20 samples	Weathered shales and other rocks; agricultural irrigation drainage?	5.2–9.3	Christensen and Juracek (2001)
Tulare Lake bed: California, USA	Weathering of marine sedimentary rocks	24	Gao et al. (2006)
Turbio ancient lake sediments (about 9700 years old): Chile: 14 samples	Weathering of geologic materials	119–2344 (<60 μm)	Oyarzun et al. (2004)
USA: Idaho: 8 lake bottom sediments	Mining	40–322	Rabbi (1994)
USA: Iowa, Montana, Nebraska, North Dakota, South Dakota: Surface sediments from 13 pothole and riverine wetlands	Mostly natural background	1.4–9.3 (pothole wetlands); 0.7–6.1 (riverine wetlands)	Martin and Hartman (1984)
USA: Pennsylvania: Presque Isle: Pond and lake sediments	Uncertain natural and/or anthropogenic causes	2.7–393.6	Murnock (2002)
Waconda Lake (Reservoir), Kansas, USA: 24 sediment core samples	Weathered rocks; agricultural irrigation drainage?	5.4–13.1	Christensen and Juracek (2001)
Webster Reservoir, Kansas, USA: 24 sediment core samples	Weathered shales and other rocks; agricultural irrigation drainage?	8–15.1	Christensen and Juracek (2001)

[a]Single values denote single analyses or average concentrations sometimes with 1 SD. Values separated by commas represent duplicate analyses of the same sample. Analyses involving different samples from the same location are separated by semicolons. Low and high values are separated by hyphens. Normally, soil and sediment samples are dried before analysis. For some samples, sieved fractions were analyzed.

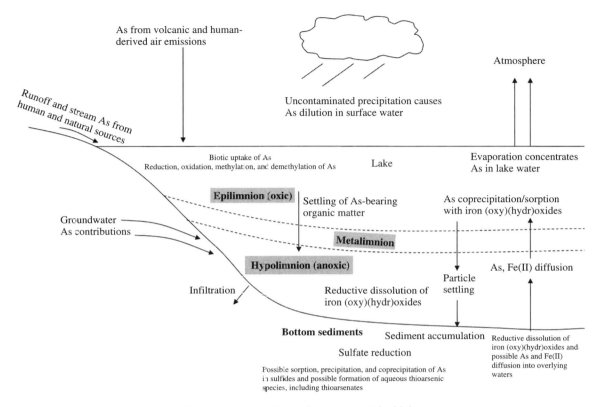

Figure 3.4 *Arsenic cycling in a stratified lake.*

to approach the bottom sediments (Durant *et al.*, 2004), 2999. Partially oxidized bottom sediments with abundant iron (oxy)(hydr)oxides would increase the capacity of the sediments to sorb arsenic.

In lakes that contain very anoxic bottom waters, Fe(II) may diffuse into overlying oxygenated waters and precipitate iron (oxy)(hydr)oxides (Figure 3.4). As(V) in the oxygenated waters may coprecipitate with the iron (oxy)(hydr)oxides and/or sorb onto them (Figures 3.3 and 3.4; (Ford, Wilkin and Hernandez, 2006), 138). Although dissolved organic carbon in lake waters may not interfere with the oxidation of Fe(II), it could hinder the precipitation of iron (hydr)oxides and the sorption of arsenic (Ford, Wilkin and Hernandez, 2006). On the other hand, organic-rich waters may contain largely undecomposed biological materials that are coated with arsenic-sorbing iron (oxy)(hydr)oxides (Jackson and Bistricki, 1995).

Some iron (oxy)(hydr)oxides will reductively dissolve and release their arsenic as they pass through the hypolimnion (Figure 3.4). Although bacteria reduce As(V) to As(III) (Harrington, Fendorf and Rosenzweig, 1998), As(V) reduction in the hypolimnion and underlying bottom sediments is often sluggish and significant metastable As(V) may survive for some time under anoxic conditions (Cullen and Reimer, 1989), 743; (Seyler and Martin, 1989), 1258. As in estuaries and marine environments, reductive dissolution, sorption, precipitation, and possible coprecipitation with sulfides are potentially important in controlling the mobility of arsenic in lake sediments (Harrington, Fendorf and Rosenzweig, 1998). Arsenic and Fe(II) released by the reductive dissolution of iron (oxy)(hydr)oxides in bottom sediments would migrate into sediment porewaters and perhaps into overlying lake water. The burial and compaction of fine-grained sediment would accelerate the expulsion of dissolved arsenic and Fe(II) (Savage, Bird and Ashley, 2000),

410. Like estuary and marine sediments, arsenic may also migrate into deeper sediments where bacteria reduce sulfate to sulfide (Figure 3.4). In such highly reducing environments, arsenic might be incorporated into relatively stable sulfide minerals (Ford, Wilkin and Hernandez, 2006).

A hypolimnion of a lake may be permanent or seasonal. In the fall and/or spring, layered lakes may experience turnovers, which usually redistribute arsenic and other contaminants in the waters. In fall turnovers, the epilimnion waters cool and approach the water temperatures of the underlying layers. Temperature and density differences between the layers disappear. Currents, waves, and storms can then readily mix the top and bottom layers. In the spring, ice thaws on temperate-climate lakes and the waters warm to 4 °C. The dense surface waters sink and mix into the underlying waters, which create a spring turnover.

Pavin Lake, France, is an example of a fresh water body with a permanent epilimnion and hypolimnion. The epilimnion extends from the surface to depths of about 60 m, whereas the hypolimnion exists at depths of 60–90 m (Viollier *et al.*, 1995). Dissolved arsenic concentrations (usually >88 % As(V)) in the epilimnion are generally very low (4.7–6.6 nmol L^{-1} or 0.35–0.49 μg L^{-1}). Coprecipitation and sorption involving manganese and iron (oxy)(hydr)oxide particles are probably responsible for the low arsenic concentrations. In contrast, the dissolved arsenic concentrations dramatically increase to 23.2–132.2 nmol L^{-1} (1.74–9.90 μg L^{-1}) in the hypolimnion (Viollier *et al.*, 1995), 67. (Oxy)(hydr)oxide compounds settling down from epilimnion undergo reductive dissolution in the hypolimnion and release their arsenic. $H_3AsO_3^0$ is the dominant arsenic species in the hypolimnion rather than As(V) or thioarsenic species (Viollier *et al.*, 1995), 68–69. The reduction of As(V) to As(III) is probably due to organic compounds rather than Fe(II) (Viollier *et al.*, 1995), 68–69.

The discharge of organic pollutants into lakes or declines in the concentrations of copper, zinc, and other heavy metal toxins may promote the growth of phytoplankton (e.g. 'algal blooms'). Greater biological activity may then increase anoxic conditions in lake bottoms, which stimulate the reductive dissolution of (oxy)(hydr)oxides and increase the mobilization of arsenic. In particular, Martin and Pedersen (2002) concluded that reduced discharges of copper, zinc, and nickel to Balmer Lake, Ontario, Canada, increased phytoplankton production and arsenic mobility in the lake.

Phosphate and inorganic As(V) may be intimately involved in lake as well as marine and estuary algal blooms. In most cases, phosphate in lakes, estuaries, and other surface waters originates from fertilizer runoff from agricultural fields, golf courses, and lawns (Johannesson *et al.*, 1997; Campos, 2002). Fertilizers may also contain traces of arsenic and other impurities (Campos, 2002). Once phosphate promotes algal blooms in a lake epilimnion or other surface waters, algae may extensively convert As(V) into inorganic As(III) and methylarsenic. Dissolved arsenic concentrations may also considerably increase in the waters if abundant phosphate displaces or outcompetes As(V) during coprecipitation and sorption with a limited supply of manganese and iron (oxy)(hydr)oxides (Pfeifer *et al.*, 2004), 231; (Kneebone and Hering, 2000), 4312.

In some cases, both phosphate and As(V) in surface waters originate from geothermal sources rather than fertilizers. As an example, geothermal arsenic and phosphate in Hot Creek drains into Crowley Lake, California, USA, which provides drinking water for the city of Los Angeles (Kneebone and Hering, 2000). Like many lakes, the arsenic in the epilimnion of Crowley Lake is mostly As(V). The uptake of As(V) by algae results in the production of As(III) and methylation to DMA(V) (Kneebone and Hering, 2000), 4307, 4312. Unlike other aqueous environments, MMA(V) was not detected in water samples from the lake (Kneebone and Hering, 2000), 4312.

As(III) also increases at the expense of As(V) in the deeper and less oxygenated waters of Crowley Lake (Kneebone and Hering, 2000). Unlike many other lakes, only a limited amount of arsenic enters the sediments of Crowley Lake. Kneebone and Hering (2000) also found no evidence of the lake sediments releasing substantial arsenic into the overlying waters. The bottom sediments appear to be permanently reducing and their arsenic (about $4-80\,mg\,kg^{-1}$ dry mass, Table 3.12) is probably effectively sequestered by sulfides (Kneebone and Hering, 2000), 4312.

As discussed in Chapter 6, many groundwaters in Bangladesh are contaminated with arsenic that threatens the health and lives of millions of people. Some individuals have advocated using Bangladeshi lakes and other surface waters as alternatives to contaminated groundwaters. The surface waters at the town of Mainamoti in eastern Bangladesh, for example, have relatively low arsenic concentrations ($5-6\,\mu g\,L^{-1}$) despite the presence of arsenic-rich soils (mean of $18.8\,mg\,kg^{-1}$ for six samples). Nevertheless, Islam *et al.* (2000) have shown that many other Bangladeshi surface waters are high in arsenic (up to $97\,\mu g\,L^{-1}$ for unfiltered samples from a site in Rajarampur; Table 3.11). The surface waters were probably contaminated by local arsenic-rich rocks and soils, and runoff from irrigating fields with arsenic-rich groundwaters (Islam *et al.*, 2000), 1088. In the surface waters at Mainamoti, the arsenic may be diluted by the excessive rainfall (Islam *et al.*, 2000), 1088.

3.18   Wetlands

Wetlands refer to low-lying inland or coastal areas that are normally damp or flooded with water. Swamps, marshes, bogs, sloughs, and fens are examples of wetlands. Like other surface environments, arsenic in wetland water, sediments, and soils (including *peats*) may originate from a variety of sources, including seawater (Dellwig *et al.*, 2002), natural weathering, mining wastes, pesticide runoff, the leaching of chromated copper arsenate (CCA)-treated wood ((Cobb *et al.*, 2006); Chapter 5), geothermal waters (Chagué-Goff, Rosen and Eser, 1999), groundwater (Wilkin and Ford, 2006), and air emissions from coal combustion facilities ((Graney and Eriksen, 2004; Shotyk *et al.*, 2003); Tables 3.11 and 3.12). Peats and other wetland soils often contain a variety of organoarsenicals, including methylarsenic, arsenobetaine ($(CH_3)_3As^+CH_2CO_2^-$), arsenocholine ($(CH_3)_3As^+CH_2CH_2OH$), and arsenosugars, all of which result from biological activity (Huang and Matzner, 2006). Peats are also capable of absorbing significant amounts of arsenic from water. In particular, González *et al.* (2006) found that an intermittent stream entering the Gola di Lago peatland in Ticino canton, Switzerland, had up to $408\,\mu g\,L^{-1}$ of arsenic. However, the stream exiting the peatland contained $<2\,\mu g\,L^{-1}$. Peats near the site of the entering stream contained up to $350\,mg\,kg^{-1}$ of arsenic. As discussed in Chapter 7, artificial and natural wetlands can be used to treat arsenic in aqueous wastes (Fox and Doner, 2003; Buddhawong *et al.*, 2005).

Wetlands are rich in plants, other organisms, and their debris. Arsenic may readily accumulate in wetland sediments, organic-rich soils, and plants and other biological organisms (Chagué-Goff, Rosen and Eser, 1999). Plant roots, in particular, are often coated with iron plaque that contains substantial arsenic (Doyle and Otte, 1997; Foster *et al.*, 2005). Like other environments, plant debris and other organic matter in wetland soils are especially important in affecting the mobility of arsenic. The matter promotes microbial activity, which creates reducing conditions that dissolve arsenic-bearing iron and manganese (oxy)(hydr)oxides (Huang and Matzner, 2006), 2024; (Wilkin and Ford, 2006), (Fox and Doner, 2003). Under strong reducing conditions, arsenic may sorb, precipitate, and/or coprecipitate with sulfide compounds. Microorganisms may also reduce arsenic and methylate it ((Huang and Matzner, 2006); Chapter 4). As an example, microbiological activity in organic soils of a German acidic fen produced As(III) and a variety of organoarsenicals in porewaters, including MMA(V), DMA(V), trimethylarsine oxide (TMAO; $(CH_3)_3AsO$), tetramethylarsonium ions (TETRA; $(CH_3)_4As^+$), arsenobetaine, and unknown species (Huang and Matzner, 2006). The reductive dissolution of iron (oxy)(hydr)oxides was also more important in releasing arsenic to porewater solutions than the decomposition of aluminum or manganese (oxy)(hydr)oxides (Huang and Matzner, 2006), 2031. In temperate-climate wetlands, reductive dissolution, As(V) reduction, and arsenic methylation are often seasonal and decline with decreasing biological activity during cool and cold months.

Arsenic may associate with a variety of minerals and other compounds in wetland soils and sediments. In northwestern Germany, arsenic largely occurs in biotic pyrite and/or other sulfides in coastal peats (Dellwig *et al.*, 2002). Arsenic is associated with FeS, Fe(II)-bearing carbonate minerals, and/or hydroxides in sediments at a contaminated wetland near Woburn, Massachusetts, USA (Wilkin and Ford, 2006). Arsenic associations in wetland soils and sediments may also change with seasonal water levels and temperatures. Using chemical extractions and *X-ray absorption near edge structure* (XANES) *spectra*, La Force, Hansel and Fendorf (2000) concluded that arsenic in the Coeur d'Alene wetlands of Idaho, USA, was associated with carbonate minerals in the summer, iron (hydr)oxides in the fall and winter, and silicates in the spring. Silicate sorbents included high-surface area volcanic ash.

Coastal estuary wetlands are susceptible to arsenic contamination because of hurricanes or other storms. Sediments deposited in 2005 by Hurricane Katrina in New Orleans, Louisiana, USA, were often contaminated with arsenic that exceeded human health standards (Cobb *et al.*, 2006). The arsenic may have originated from runoff containing lawn herbicide residues and the leachates of wood treated with CCA preservatives (Cobb *et al.*, 2006), 4576.

3.19 Groundwater

3.19.1 Subsurface water and groundwater

Subsurface water refers to any liquid, solid, or gaseous water below the Earth's surface. In the subsurface, liquid water is located in *hydrologic* zones, which includes the *unsaturated zone* (Figure 3.5). The unsaturated zone refers to any rocks, sediments, or soils in the shallow subsurface whose pores and other openings contain air and perhaps some water. While gases in the unsaturated zone are generally at atmospheric pressure, any water is below atmospheric pressure.

Another hydrologic zone, the *capillary fringe*, may lie underneath the unsaturated zone (Figure 3.5). The pores in the capillary fringe are saturated with water. However, like the unsaturated zone, the water is at less than atmospheric pressure (Freeze and Cherry, 1979), 44. The boundary between the capillary fringe and the overlying unsaturated zone is typically very uneven (Figure 3.5). Although the capillary fringe may be quite thick in fine-grained sediments, it is usually thin or nonexistent in coarser-grained materials. In clays and other fine-grained sediments, surface tension allows water droplets to adhere between the closely spaced particles, accumulate, and saturate the narrow pores.

Pores, fractures, and other openings in the geologic materials of the *saturated zone* are usually entirely filled with water that is above atmospheric pressure (Freeze and Cherry, 1979), 44. In some cases, organic liquids or various gases may be trapped in interstices of the saturated zone. Depending on their proximity to the surface, biological activity, the permeability of their geologic materials, climate, and many other factors, saturated zones may have reducing or oxidizing conditions.

The *water table* refers to the surface (i.e. two dimensional) between the saturated zone and the overlying capillary fringe or unsaturated zone. In Figure 3.5, the water table passes through a homogeneous and permeable sand layer. If a water table is under atmospheric influence, as shown in Figure 3.5, its water is at atmospheric pressure (Freeze and Cherry, 1979), 39.

Groundwater refers to the liquid water in the saturated zone. Most shallow groundwater is meteoric. Except for waters in caves and other large openings in the subsurface, groundwaters tend to flow more slowly (sometimes less than one mm/year), have fewer microorganisms, contain less O_2, and have temperatures that are less affected by seasonal changes when compared with surface waters. In some cases, groundwater intersects the surface and creates springs in lakes, rivers, wetlands, and other locations. Fresh, brackish, and marine surface waters may also readily infiltrate into the subsurface and impact groundwater. Along marine coastlines, a mixing zone exists between fresh terrestrial groundwater and denser saline

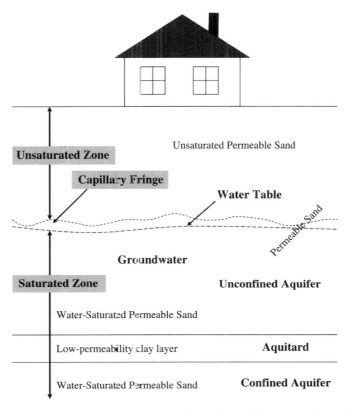

Figure 3.5   *Major subsurface hydrologic zones at a hypothetical location with sand and clay layers.*

groundwater that results from seawater seeping into the subsurface (Bone, Gonneea and Charette, 2006). As mentioned earlier in Section 3.6.2, deep groundwaters are usually very saline and may include meteoric, connate, hydrothermal, and/or other types of water. At great depths, pressure and precipitating cements close most pores, fractures, and other openings in rocks, which stop the downward flow of groundwater.

Rocks, sediments, and soils in the saturated zone include aquifers and *aquitards*. An aquifer is a water-saturated soil, sediment, or rock that is capable of transmitting significant or 'useful' quantities of groundwater. Whether something is an aquifer or not, often depends on human demand. Specifically, a sandstone may provide enough water (or qualify as an aquifer) for a small family farm, but not meet the demands of a large factory. In contrast, an aquitard is a water-saturated rock, sediment, or soil that does not transmit significant quantities of water (Figure 3.5). Aquitards have low permeabilities, which include many shales and clay layers (Freeze and Cherry, 1979), 29, 47. If a well is placed in an aquitard, it will rapidly run dry because groundwater cannot move fast enough through the aquitard to keep the well casing filled.

Aquifers may be classified as unconfined or confined. Although *unconfined aquifers* usually have sediments or soils between them and the surface, the geologic materials are sufficiently permeable so that unconfined aquifers are rapidly influenced by atmospheric conditions, including pressure and precipitation events (Figure 3.5). The upper boundary of an unconfined aquifer is the water table. A *confined aquifer* is located between two aquitards ((Freeze and Cherry, 1979), 48). At least one aquitard occurs between a confined aquifer and the surface. That is, a confined aquifer is substantially insulated from conditions on the Earth's surface.

3.19.2   Impacts of arsenic contamination in shallow (<1 km deep) groundwaters

Unlike surface waters, well water is often not treated before human consumption. Arsenic contamination of well water is one of the most serious and widespread threats to human health and the environment (Chapters 4 and 6). Worldwide, perhaps as many as 150 million people, including about 50 million in Bangladesh, are threatened by arsenic in their water (Nriagu, 2002), 22. The World Health Organization (WHO) has set a recommended limit of 10 μg L^{-1} for arsenic in drinking water. In response, the United States and several other nations have recently lowered their drinking water standards for arsenic from 50 to 10 μg L^{-1} (Appendix E). Nevertheless, many individuals simply cannot afford to treat their water, buy bottled water, or find other alternatives. Even households that have access to bottled water for drinking and cooking are still at risk when they use high-arsenic well water for bathing, washing, and brushing their teeth. Children are especially susceptible to arsenic exposure through prolonged play in bathtubs and the swallowing of bath and shower water (Paulu *et al.*, 2002).

Several important challenges for the twenty-first century will include identifying arsenic-contaminated aquifers, implementing cost-effective technologies to mitigate the contamination, and dealing with the widespread human health problems from arsenic poisoning. Effective mitigation strategies will require a multidisciplinary approach involving scientists, engineers, technicians, medical personnel, government officials, translators, social workers, sociologists, economists, bankers, and others.

3.19.3   'Dissolved' and particulate arsenic in groundwater

In groundwater, arsenic exists as dissolved species and often in PM, including colloids. 'Dissolved' arsenic is usually defined as truly dissolved species (e.g. $H_3AsO_3^0$) and any colloids that are small enough to pass through a 0.45 μm filter. The arsenic that collects on the filters would be identified as particulate (Gong *et al.*, 2006). Colloids in water samples may consist of arsenic-sorbing iron (oxy)(hydr)oxides, clay minerals, various organic complexes, and some microorganisms. They are much larger than truly dissolved species ($\sim$0.01–10 μm; (Eby, 2004), 343) and, unlike dissolved species, colloids have less predictable effects on the chemical and physical properties of their host solutions (Krauskopf and Bird, 1995), 157.

The arsenic concentrations in the PM of a water sample may be determined by analyzing and comparing filtered and unfiltered aliquots of the sample (Kim, Nriagu and Haack, 2002). PM may also be collected on filters and analyzed directly for arsenic (Gong *et al.*, 2006). Identifying the amount of particulate and dissolved arsenic species in groundwater samples has practical applications in water treatment because if most of the arsenic in the samples is in PM, filtering may be an easy and inexpensive treatment process. The ratio of particulate to dissolved arsenic in groundwaters is often very site specific. Groundwater samples from Inner Mongolia, China, for example, may contain six times more particulate than dissolved arsenic (Gong *et al.*, 2006). In contrast, other groundwaters have negligible particulate arsenic, such as the southeastern Michigan, USA, samples described in Kim, Nriagu and Haack (2002).

3.19.4   Arsenic mobility in groundwater

Flow and *diffusion* transport dissolved and mobile particulate arsenic in groundwater. The flow velocity (speed and direction) of groundwater is largely controlled by changes in the elevation of the water table with lateral distance, water pressure and density, and the permeability and other properties of the aquifer. In some circumstances, temperature gradients may also affect groundwater flow (Freeze and Cherry, 1979), 25.

Diffusion involves the movement of dissolved chemical species in liquids or gases because of a concentration gradient. That is, the dissolved species will migrate from areas of high concentration to areas of lower concentration (Fetter, 1993). Diffusion results from the thermal movement of water molecules and occurs even in surface water or groundwater that experiences no flow. Diffusion is commonly seen when a drop of dye is carefully added to a glass of still water.

Arsenic distributions may be fairly uniform in the groundwaters of homogeneous and essentially horizontal aquifers. However, many aquifers are very heterogeneous and consist of fractured bedrocks with localized concentrations of arsenic-bearing minerals or irregular distributions of impermeable clay layers in permeable sands. Such heterogeneities could cause arsenic concentrations in groundwaters to vary by orders of magnitude over a few meters in either lateral or vertical directions. These irregular distributions are difficult to model and predict.

As mentioned earlier in Section 3.6.2, flow in groundwaters, especially connate waters, may be so slow that the waters remain more or less stationary in the subsurface for thousands or even millions of years. Estimating the ages of groundwaters is important in determining their origins, long-term velocities, and potential threats to nearby drinking water wells. Naturally occurring radioactive isotopes are often used to date groundwaters. Carbon-14 (^{14}C) has potential applications with groundwaters that are 150–35 000 years old (Faure, 1998), 294; (Freeze and Cherry, 1979), 135, 290–292. Groundwaters that are more than one million years old are sometimes dated with chlorine-36 (^{36}Cl) (Drever, 1997), 323. Tritium (^{3}H) and krypton-85 (^{85}Kr) may be used to date groundwaters that formed within the past few decades (Freeze and Cherry, 1979), 136–137; (Sidle and Fischer, 2003).

3.19.5   Sources of arsenic contamination in groundwater

Groundwaters containing more than 10 $\mu g\,L^{-1}$ of arsenic are very common (Table 3.13; Appendix D). The arsenic may originate from one or more anthropogenic or natural sources, such as: (1) the improper manufacturing, use, and disposal of products containing arsenic; (2) extensive applications of arsenic-bearing pesticides and phosphate fertilizers; (3) mine drainage and smelter emissions; (4) the percolation into the subsurface of evaporative brines or runoff from weathering outcrops and irrigation; (5) oxidation of sulfide minerals in expanding unsaturated zones resulting from declining water tables; (6) geothermal waters and discharges from power plants; (7) reductive dissolution of arsenic-bearing iron and manganese (oxy)(hydr)oxides, and (8) bacterial degradation of natural or artificial organic materials, production of carbonate species, and subsequent desorption of arsenic from mineral surfaces (Welch *et al.*, 1998; Jacks and Bhattacharya, 2000; Appelo *et al.*, 2002). Even if contamination does not directly originate from human-generated wastes, human activities (such as irrigation, excavations, and application of agrichemicals) may change conditions in subsurface geologic materials and cause them to release substantial quantities of arsenic into groundwaters. Furthermore, groundwaters with significant concentrations of arsenic (>10 $\mu g\,L^{-1}$) derived from natural sources are not always associated with arsenic-rich aquifer materials (Erickson and Barnes, 2005). Reductive dissolution, desorption with bicarbonate-rich groundwaters, evaporative concentration, and other natural processes are capable of producing high-arsenic solutions from relatively low-arsenic sediments and soils. As an example, the Duero River Basin of Spain has bicarbonate groundwaters with up to 260 $\mu g\,L^{-1}$ of arsenic, yet the arsenic concentrations of the basin sediments are mostly <0.2–16 mg kg^{-1} ((García-Sánchez, Moyano and Mayorga, 2005); Tables 3.13 and 3.14).

Unfortunately, it is not always possible to precisely identify the cause(s) of arsenic contamination in groundwater (Chapter 6). Determining the cause(s) requires detailed field and laboratory studies of the geological, chemical, biological, and physical properties of the subsurface materials, which would include identifying groundwater flow patterns and changes in the distribution of arsenic over the study area.

Arsenic has been used in a wide variety of products and applications over the years, including pesticides, livestock dips and feed additives, wood preservatives, semiconductors, dyes, paints, alloys, and medicines

Table 3.13 Areas with elevated arsenic (>10 µg L⁻¹) concentrations in groundwater and spring water

Location	Arsenic source(s) favored by the authors	pH	Concentration ($\mu g\,L^{-1}$)[a]	Reference(s)
Argentina	Loess; high alkalinity; hydrothermal springs	—	< 1–9900	Nordstrom (2002)
Argentina: La Pampa (province)	Desorption under alkaline conditions; weathering of loess and other geologic materials; arid climate and slow groundwater flow	7.0–8.7	< 4–5300 (As(V) dominant)	Smedley et al. (2002)
Argentina: Pampa and Chaco plains	Loess; weathering of geologic materials	usually 6.5–9	<4–5280 As(V)	Bundschuh, Bonorino and Viero (2000)
Argentina: Tucumán: Los Pereyras; 58 wells	Desorption from Fe(III) (oxy)(hydr)oxides	6.64–8.63	0.6–758 (0.45 µm filtered)	Warren, Burgess and Garcia (2005)
Austrailia: Perth: 40 water analyses from various depths in 12 auger holes	Sulfide oxidation, reductive dissolution	2.6–6.9	< 1–7300	Appleyard, Angeloni and Watkins (2006)
Bangladesh	Uncertain, probably multiple causes	—	< 10–1670	Hossain (2006)
Bangladesh: 112 samples nationwide	Reductive dissolution of natural materials, ion exchange	—	<0.7–640 (unfiltered)	Frisbie et al. (2002)
Bangladesh: 9640 samples	—	—	3–4720	Samanta et al. (1999)
Bangladesh: Brahmanbaria district: 35 wells	Reductive dissolution	6.4–7.6	1.80–439	Bhattacharya et al. (2006)
Bangladesh: Manikganj District: Singair Thana: 26 samples	Reductive dissolution of Fe (oxy)(hydr)oxides	—	< 1–240	Mohammad et al. (2003)
Bangladesh: Noakhali District	Multiple unknown	—	2700	Chowdhury et al. (1999)
Bangladesh: Ramganj: police station	Multiple unknown	—	1354	Chowdhury et al. (1999)
Bangladesh: Sonargaon thana: 35 samples	Reductive dissolution of natural materials	6.53–7.59	30–750 (0.7 µm filtered)	Anawar et al. (2003), 115
Brazil: Jundiaí, São Paulo	Phosphate fertilizers	—	130–170	Campos (2002)
Cambodia: 1072 samples	—	—	< 10–1700 (<0.45 µm filtered)	Polya et al. (2005)

Location	Process	pH	Concentration	Reference
Canada: British Columbia: Bowen Island: 32 samples	Ion exchange with clay minerals, sulfide oxidation, desorption from Al, Fe (oxy)(hydr)oxides	7.40 8.86	0.5–580	Boyle, Turner and Hall (1998), 207
Canada: Saskatchewan: 61 wells	–	–	< 1–117	Thompson et al. (1999)
Chile	Hydrothermal; enclosed evaporitic basins; mining	–	100–1000	Nordstrom (2002)
Chile: Lluta River Basin	–	–	5–45	Yamasaki and Hata (2000)
China: Datong Basin: shallow (5–80 m) groundwater	Sulfide oxidation from coal mining?	6.16–9.14	<0.1–1530.1 (0.45 µm filtered)	Guo and Wang (2005)
China: Inner Mongolia	Natural desorption from high alkalinity	–	< 1–2400	Nordstrom (2002)
China: Inner Mongolia: 51 wells	Decomposition of organic matter	6.89–8.37	24–1120	Xiaoying (2001)
China: Inner Mongolia: Ba Men region	–	–	< 1–1800	Gong et al. (2006)
China: Inner Mongolia: Ba Men region (6 samples)	–	–	362.9–671.2 (including dissolved and particulate arsenic)	Gong et al. (2006)
China: Inner Mongolia: Hetao Region: 97 samples	Mining; sulfide oxidation	–	1–969	Zhang (2004)
China: Inner Mongolia: Huhhot Basin	Reductive dissolution; Arsenic desorption from minerals; slow flowing groundwater	–	< 1–1480	Smedley et al. (2003)
China: Xinjiang, Shanxi	Natural	–	40–750	Nordstrom (2002)
Germany	Natural	–	< 10–150	Nordstrom (2002)
Germany: Bavaria: Keuper Sandstone: 160 wells	Leaching of clay and silt layers in the sandstone, saline groundwater under the sandstone and leaching of mineral deposits	6.7–7.8	10–150 (0.45 µm)	Heinrichs and Udluft (1999)
Ghana: Amansie West: 39 groundwater samples	–	–	10–1200 (0.45 µm filtered)	Duker, Carranza and Hale (2005)

(continued overleaf)

Table 3.13 (continued)

Location	Arsenic source(s) favored by the authors	pH	Concentration (μg L^{-1})[a]	Reference(s)
Ghana: Bolgatanga area	Sulfide oxidation	6.0–7.4	<1–141	Smedley (1996)
Ghana: Obuasi area	Sulfide oxidation	3.9–6.8	<1–64	Smedley (1996)
Hungary (1941–1983)	Natural	–	60–4000	Hossain (2006); Rahman, Sengupta and Chowdhury (2006)
India: Behala, Calcutta: 19 wells	Chemical factory	–	100–38 000	Rahman, Sengupta and Chowdhury (2006)
India: Bihar: Semria Ojha Patti	–	–	<50–1654	Rahman, Sengupta and Chowdhury (2006)
India: Chattisgarh: Rajnandgaon district: 146 groundwater samples	Weathering of geologic materials	–	<10–880	Rahman, Sengupta and Chowdhury (2006)
India: Hyderabad, Andhra Pradesh: 5 samples	–	–	8.84–40.25	Govil, Reddy and Gnaneswara Rao (1999)
India: Kanker district	Weathering of geologic materials	5.1–7.3	<1–900.0	Pandey et al. (2006), 412
India: Raichur district, Karnataka: 68 groundwater samples	Oxidation of sulfide minerals	7.12–8.27	0.1–41.5 (0.45 μm)	Sahoo and Pandalai (2000)
India: West Bengal	Reductive dissolution?	–	1–880	Gault et al. (2005)
India: West Bengal: 47 000 samples	–	–	3–3882	Samanta et al. (1999)
Japan: Osaka: Spring waters	Weathering of geologic materials	–	2–54	Ito, Masuda and Kusakabe (2003)
Mexico: Baja California Sur: San Antonio-El Triunfo area	Mining	–	<10–2000	Carrillo-Chavez, Drever and Martinez (2000)
Mexico: Guanajuato mining district	Mining	–	9–500	Morton, Carrillo and Hernandez (2000)
Mexico: Guanajuato: Mineral de Pozos mining site	Mining; weathering of geologic materials	–	11–90	Carrillo-Chavez, Gonzalez-Partida and Morton-Bermea (2002)

Arsenic in Natural Environments 153

Location	Source/Process	pH	Concentration	Reference
Mexico: Lagunera	—		8–624	Rahman, Sengupta and Chowdhury (2006)
Mexico: Lerma-Chapala Basin			1–100	Ortega-Guerrero (2003)
Mexico: Rioverde Basin: 53 wells	Evaporative concentration; other natural		<5–53.5	Planer-Friedrich, Armienta and Merkel (2001)
Mexico: Zimapán: 60 samples	Weathering of geologic materials, mining	5.83–8.97	<14–1000 (0.45 μm filtered)	Armienta et al. (2001)
Nepal: Terai Alluvial Plain	—		<10–2620	Rahman, Sengupta and Chowdhury (2006)
Pakistan: Punjab: Muzaffargarh District	Various natural causes; agricultural and industrial pollution are possible		<10–906	Nickson et al. (2005)
Saudi Arabia: City of Jabail	Seepage of irrigation water and landfill leachate?	6.75–8.48 (104 samples)	0.05–11.14 (0.45 μm filtered; 103 samples)	Sadiq and Alam (1997)
Spain: Caudal River Basin: near Los Rueldos and La Soterraña mines: 10 samples	Mining	—	<1–57 500	Loredo et al. (2004)
Spain: Deuro River Basin: 28 wells	Bicarbonate desorption, microbial reductive dissolution	7.6–9.5	20–260 (0.45 μm filtered)	García-Sánchez, Moyano and Mayorga (2005)
Spain: Madrid Aquifer: 61 samples	pH controlled ion exchange with aquifer materials	6.85–9.05	<10–91 (0.45 μm filtered)	Hernández-García and Custodio (2004)
Switzerland	Hydrothermal; Sulfide oxidation	—	20–2800	Pfeifer and Zobrist (2002)
Switzerland: Sottoceneri area: springs	Weathering of rocks	—	20–60	Temgoua and Pfeifer (2002)
Taiwan	Natural	—	10–1820	Nordstrom (2002)
Taiwan: Putai: 54 samples	Release of As from marine clays	8.1 ± 0.4	470–897 (<0.45 μm)	Chen et al. (1994)
Turkey: Igdeköy-Emet, Kütahya: 10 samples	Oxidation of orpiment and realgar	7.25–8.40	70–7754	Çolak, Gemici and Tarcan (2003)
Ukraine: Donets Basin	Mining	—	<2–1200	Panov et al. (1999)
United Kingdom: England: Chalk aquifer contaminated by gasoline and other hydrocarbons: 12 samples	Enhanced reductive dissolution of Fe and Mn (oxy)(hydr)oxides from hydrocarbons	6.59–7.05	<0.1–7.1 (0.45 μm)	Burgess and Pinto (2005), 890)

(continued overleaf)

Table 3.13 *(continued)*

Location	Arsenic source(s) favored by the authors	pH	Concentration (μg L^{-1})[a]	Reference(s)
United Kingdom: England: Sandstone aquifer contaminated by petroleum hydrocarbons and petrochemicals: 22 samples	Arsenic desorption by increased bicarbonate alkalinity?	5.82–10.05	0.1–157	Burgess and Pinto (2005), 891)
United Kingdom: England: Thames gravel deposit contaminated by aviation fuel and other hydrocarbons: 6 samples	Enhanced reductive dissolution of (oxy)(hydr)oxides from hydrocarbons	6.8–7.5	0.1–70 (0.45 μm)	Burgess and Pinto (2005), 890)
United Kingdom: Southwest England	Mining	–	< 1–80	Nordstrom (2002)
USA: Massachusetts (central): Near 5 landfills	Reductive dissolution	4.88–6.70	< 30–700, sometimes >5000	Hon, Mayo and Brandon (2003)
USA: Michigan (southeastern): 73 wells, >15 m deep	Reductive dissolution; carbonate leaching of arsenic from sulfides	6.9–8.5	0.5–278 (unfiltered)	Kim, Nriagu and Haack (2000); Kim, Nriagu and Haack (2002)
USA: Alaska: Cook Inlet Basin: 220 wells	Natural weathering of geologic materials	–	< 1–150	Pope et al. (2004)
USA: Alaska: Ester Dome near Fairbanks: 30 wells	–	6.36–7.87	< 3–1160 (0.45 μm filtered); < 3–2270 (unfiltered)	Verplanck et al. (2003)
USA: Alaska: Fairbanks area	Reductive dissolution	6.00–7.86	< 3–1670 (unfiltered) (up to 10,000 dissolved in previous studies)	Mueller, Goldfarb and Verplanck (2001), Mueller et al. (2001)
USA: Arizona: Cerbat Mountains: 8 wells	Mining	5.8–7.5	10–1000	Rösner (1998)
USA: Arizona: Verde Valley: 41 samples	Supai and Verde formations	–	10–210	Foust et al. (2004)

Location	Process/comments	pH	As concentration	Reference
USA: California: Hanford (San Joaquin Valley)	—	—	9–75 (20 µm filtered)	Hering and Chiu (2000)
USA: California: Kerr River Alluvial Fan (San Joaquin Valley)	Desorption at high pH	<5.7 to ≥9.1	1–211	Klinchuch et al. (1999)
USA: California: Lassen county	Natural	—	50–1400	Rahman, Sengupta and Chowdhury (2006)
USA: California: Owens dry lake: 30 sites	Evaporative concentration; other?	9.0–9.7	100–96 000 (<0.2 µm)	Ryu et al. (2002)
USA: Connecticut	Arsenic desorption at high pH; possible anthropogenic sources?	>7.7 for 5 wells with highest As	<0.18–24 (dissolved)	Brown and Chute (2001)
USA: Connecticut: 176 analyses	Weathering of geologic materials	—	<5–33	Ayotte et al. (2006), 3581
USA: Idaho: Coeur d'Alene Mining District: Bunker Hill Mine	Mining	2.7–6.4	0.1–1336.1 (As(III) + As(V); 0.15 µm filtered)	Mok, Riley and Wai (1988)
USA: Illinois, Indiana, Ohio, Michigan and Southeastern Wisconsin: glacial deposit aquifers	Reductive dissolution and sulfide oxidation	5.9–9.5	<1–84	Thomas (2003)
USA: Illinois (central): Mahomet and Glasford glacial sand and gravel aquifers	Reductive dissolution	—	<10–266	Kelly et al. (2005)
USA: Illinois: East-central: 47 samples		6.61–7.83	<5–290 (As(V)), <5–146 (As(III)) (0.2 µm filtered)	Holm and Curtiss (1989)
USA: Illinois: Mahomet Aquifer	Bacterial reduction in absence of sulfide	—	<1–266	Holm et al. (2004), 8
USA: Kentucky: statewide database of 4402 measurements at 930 sites	Sulfide oxidation in coals and black shales, possibly other causes	—	<10–265	Fisher (2002)
USA: Maine	—	—	5–400 (0.45 µm filtered)	Culbertson et al. (2002)

(continued overleaf)

Table 3.13 (continued)

Location	Arsenic source(s) favored by the authors	pH	Concentration ($\mu g\,L^{-1}$)[a]	Reference(s)
USA: Maine: 471 analyses	Weathering of geologic materials	–	<5–217	Ayotte et al. (2006), 3581
USA: Maine: Goose River Basin: 39 samples (including 3 springs)	Sulfide oxidation	6.23–8.18	<1–46	Sidle, Wotten and Murphy (2001)
USA: Maine: Goose River Basin: 96 wells and 4 springs	Sulfide oxidation	–	<1–368	Sidle and Fischer (2003); Sidle (2002)
USA: Maine: Northport	Sulfide oxidation	–	<1–5500	Reeve et al. (2001)
USA: Massachusetts (north central): Landfill sites	Reductive dissolution of natural Fe and Mn (oxy)(hydr)oxides	4.88–6.70	5.4–4200	Mayo et al. (2003)
USA: Massachusetts: 473 samples	Weathering of geologic materials	–	<5–1100	Ayotte et al. (2006), 3581
USA: Massachusetts: Woburn: 18 sites	Tannery pollutants	5.6–8.5	1–2850	Davis et al. (1994)
USA: Michigan: 11 samples	–	–	1–1310	Meliker et al. (2002)
USA: Michigan: Huron and Lapeer counties: 2 wells: groundwater at various depths	–	7.1–7.6	<1–171 (unfiltered)	Kim, Nriagu and Haack (2003)
USA: Michigan: Oakland county	Uncertain natural causes, perhaps including sulfide oxidation	generally 7.0–7.5	<10–221	Aichele (2004); Kolker, Haack and Cannon (2003)
USA: Minnesota	Reductive dissolution	–	3–157	Kanivetsky (2000)
USA: Minnesota: 954 wells	Glacial sediments; other possible sources	–	<1–91	Minnesota Pollution Control Agency (1998)
USA: Missouri: Kansas City: 8 wells	Bacterially mediated reductive dissolution	–	1–167	Shahnewaz et al. (2003)
USA: Montana: Milltown: 4 community wells	Mining	–	220–550	Moore and Woessner (2003)
USA: Montana–Wyoming: Madison River valley	Reductive dissolution; irrigation	–	16–176	Nimick (1998)
USA: Montana–Wyoming: Madison River valley: 13 samples	–	–	26–150	Sonderegger and Ohguchi (1988)
USA: Nevada: baseline groundwater near Getchell Mine: 358 As analyses	Mining	6.1–9	<1–7700	Davis et al. (2006)

Location	Source	pH	As (µg l⁻¹)	Reference
USA: Nevada: Getchell Main Pit: 6 groundwater sites	Mining	7.44–8.09	4–1800	Davis et al. (2006)
USA: Nevada: Lone Tree Mine	Sulfide oxidation	–	100–1000	Nicholson et al. (1996)
USA: New Hampshire: 30 domestic bedrock wells	Sulfide oxidation, especially in pegmatites	5.25–8.26	<0.02–399 (<0.45 μm)	Peters and Blum (2003)
USA: New Hampshire: 937 analyses	Weathering of geologic materials	–	<5–300	Ayotte et al. (2006), 3581
USA: New Mexico: Albuquerque Basin: Santa Fe Group sediments:	Natural desorption from mineral surfaces	8.3–9.9	<2–42	Stanton et al. (2001)
USA: New Mexico: Middle Rio Grande Basin: deep briny groundwaters: 4 samples	Cooled hydrothermal waters	6.2–7.0	33.0–1500	Bexfield and Plummer (2003), 320
USA: New Mexico: Socorro Basin: 74 samples	Hydrothermal	–	<2–43	Brandvold (2001)
USA: Ohio (northeastern)	Wood preservatives	–	<5–36	Eshete and Chyi (2003)
USA: Ohio (western): Lockport Dolomite aquifer	–	–	<1–29	Dumouchelle (1998)
USA: Oklahoma: Norman	Desorption from aquifer materials	5.0–9.6	<0.5–232	Smith (2005)
USA: Oregon: Lane County	Natural	–	50–1700	Rahman, Sengupta and Chowdhury (2006)
USA: Oregon: Williamette Basin (10 sites)	Weathering of volcanic materials	–	16–1200 (unfiltered); 16–1100 (0.10 μm filtered)	Hinkle and Polette (1999), 12
USA: Ozark Mountains (Arkansas, Missouri, Oklahoma and Kansas): 215 analyses	–	5.2–8.8	<1–15	Lee and Goldhaber (2001)

(continued overleaf)

Table 3.13 (continued)

Location	Arsenic source(s) favored by the authors	pH	Concentration (μg L^{-1})[a]	Reference(s)
USA: Rhode Island: 144 samples	Weathering of geologic materials	–	<5–46	Ayotte et al. (2006), 3581
USA: South Dakota	Pesticides	–	3–110	Welch et al. (2000), 591
USA: South Dakota: Whitewood Creek Basin: 1986–1987	Mining	6.6–7.1 (5 analyses)	181–336 (9 analyses)	Fuller and Davis (2003)
USA: Utah: Millard County	Natural	–	180–210	Rahman, Sengupta and Chowdhury (2006)
USA: Utah: Salt Lake valley: 30 wells	Sulfide oxidation; mining	6.8–7.8	<1–19.6	Thiros (2003)
USA: Vermont: 269 analyses	Weathering of geologic materials	–	<5–156	Ayotte et al. (2006), 3581
USA: Wisconsin: Brown county: Algoma township	Sulfide oxidation, unknown	–	2–12 000	Schreiber et al. (2003)
USA: Wisconsin: Fox River Valley: 5 wells	Sulfide oxidation	3.8–7.2	0.3–166 (0.2 μm filtered)	Schreiber, Simo and Freiberg (2000)
USA: Wisconsin: Winnebago and Outagamie counties	–	–	<10–15 000	Johnson (2001)
Vietnam: Red River Basin	Reductive dissolution in geologic materials?	–	<1–3050	Berg et al. (2001)

[a]Single values denote single analyses or average concentrations sometimes with 1 SD. Values separated by commas represent duplicate analyses of the same sample. Analyses involving different samples from the same location are separated by semicolons. Low and high values are separated by hyphens. Available information on micron filtering of the samples is listed.

Table 3.14   *Arsenic concentrations in aquifers and associated aquitards*

Location	Concentration (mg kg^{-1})^a	Reference(s)
Bangladesh: Samta village: aquitard sediments and peaty samples	0.7−261.5	Matsumoto and Hosoda (2000)
India: West Bengal: Nadia District	20−133	Bhattacharya and Jacks (2000), 19−20
Spain: Deuro River Basin sediments	<0.2−16	García-Sánchez, Moyano and Mayorga (2005)
USA: California: Kerr River Alluvial Fan (San Joaquin Valley): aquifer soils	1.3−8.2	Klinchuch *et al.* (1999)
USA: Illinois: Mahomet Aquifer (glacial till, sand, and gravel)	3−16	Kirk *et al.* (2004)
USA: Massachusetts: Waquoit Bay: Coastal aquifer: 3 cores	0.361−7.470	Bone, Gonneea and Charette (2006)
USA: Michigan: Huron county well core	0.8−70.7	Kim, Nriagu and Haack (2003)
USA: New Mexico: Albuquerque Basin: Santa Fe Group sediments:	< 2−25	Stanton *et al.* (2001)
Vietnam: Red River Valley: 5 sediment cores	< 1−43	Berg *et al.* (2001)

^{a}Low and high values are separated by hyphens. Normally, soil and sediment samples are dried before analysis.

((Azcue and Nriagu, 1994); Chapter 5). At the old manufacturing sites of many of these products, local groundwaters are heavily contaminated with arsenic (e.g. Behala, Calcutta, India, (Rahman, Sengupta and Chowdhury, 2006); Woburn, Massachusetts, USA; (Davis *et al.*, 1994); Table 3.13). Mine drainage and smelting emissions are other major anthropogenic sources of arsenic, which may contaminate both surface waters and groundwaters. Arsenic-contaminated groundwaters from mining activities occur in the western United States (including Alaska), the United Kingdom, Spain, Mexico, Thailand, China, and elsewhere (Table 3.13; (Hering and Kneebone, 2002; Loredo *et al.*, 2004; Welch *et al.*, 2000; Nordstrom, 2002; Carrillo-Chavez, Gonzalez-Partida and Morton-Bermea, 2002), 156; (Zhang, 2004)). Groundwaters around cemeteries may also contain significant arsenic. Arsenic and mercury were used as embalming fluids in Europe since about 650 AD to the early twentieth century (Spongberg and Becks, 2000), 314. In the United States, arsenic was a popular embalming fluid from the Civil War (1860−1865) to about 1905 (Fetter, 1993), 274−275; (Spongberg and Becks, 2000), 314.

As discussed earlier in Section 3.17, the excessive application of arsenic-bearing pesticides and phosphate fertilizers on agricultural lands, golf courses, and lawns may locally contaminate surface waters and groundwaters (Welch *et al.*, 2000), (Lewis *et al.*, 2002), 590. Phosphates desorb arsenic from mineral surfaces and readily interfere with the sorption and coprecipitation of arsenic onto iron (oxy)(hydr)oxides (Campos, 2002). Commercial phosphate fertilizers also frequently contain >13 mg kg^{-1} of arsenic impurities (Campos, 2002), which may further contribute to groundwater contamination.

Extensive irrigation and droughts lower water tables, which allow air and aerated water to migrate farther into the subsurface through expanded unsaturated zones and abundant well casings (Foster, 2003), 62. Arsenic-bearing sulfides that were previously stable under reducing conditions below the water table are now exposed to oxidation. Weathered Fe(II) sulfides produce iron (oxy)(hydr)oxides, which may largely capture the arsenic released from the sulfides. During wet periods, water tables rise and immerse the arsenic-bearing (oxy)(hydr)oxides. The iron (oxy)(hydr)oxides may then undergo reductive dissolution and release their arsenic into groundwater (Hossain, 2006), 4.

The oxidation of pyrite and other sulfides are responsible for high arsenic concentrations ($>1\,mg\,L^{-1}$) in groundwaters at many locations, including the upper portion of the St. Peter Sandstone in eastern Wisconsin, USA (Table 3.13; (Schreiber *et al.*, 2003; Burkel and Stoll, 1999)). If deeper uncontaminated aquifers are available, the overlying sandstone or other contaminated aquifers should be carefully sealed off from the well casing so that their groundwaters do not enter the well (Burkel and Stoll, 1999). Furthermore, pumping rates on the lower aquifer must be carefully monitored to make sure that the withdrawal of water does not cause arsenic-rich groundwaters from overlying aquifers and aquitards to flow downward and enter the well (McTigue, Stein and Brandon, 2001).

In arid climates, such as the southwestern USA, northern Chile, and northern China, evaporation may concentrate arsenic in alkaline lake brines (Hering and Kneebone, 2002), p. 159. Desert regions also frequently have alkaline and well-oxidized soils and sediments (Holm *et al.*, 2004), 62. Irrigation runoff may leach arsenic and other chemicals from the sediments and soils, which would further contribute to the development of arsenic-rich brines. The brines may then seep into the subsurface and contaminate shallow groundwaters (Welch *et al.*, 2000), 592–593, 596. In some regions, such as the Datong Basin, Shanxi, China, underground coal beds assist in the reduction of As(V) to generally more toxic As(III) in infiltrating brines (Wang, Zhao and Wu, 1998). Under alkaline conditions, iron (oxy)(hydr)oxides in the sediments and soils are usually negatively charged (i.e. above their ZPCs and isoelectric points; Chapter 2) and are not ideal sorbents of arsenic oxyanions.

Arsenic may also accumulate in the groundwaters of areas with wet climates. Smedley (1996) proposed a model to explain the development of high-arsenic groundwaters in the wet and tropical climate of Ghana. High temperatures and extensive flushing of the subsurface by oxygenated rainwater result in the oxidation of arsenian sulfides in shallow subsurface rocks. In the shallow, acidic, and oxidizing subsurface environments, arsenic largely precipitates as Fe(III) arsenates and sorbs and/or coprecipitates with iron (oxy)(hydr)oxides. Heavy precipitation would periodically wash weathering rinds from the sulfide minerals and allow further oxidation. In the deeper subsurface ($>40\,m$) where weathering is less intense, the dissolution of silicates and trace carbonate minerals raises the pH of the groundwaters and releases bicarbonate. Oxygen as O_2 tends to be readily consumed by weathering sulfides, which produces mildly reducing conditions. Under mildly alkaline and reducing conditions, $H_3AsO_3{}^0$ becomes dominant in the groundwater and only partially sorbs or coprecipitates with iron (oxy)(hydr)oxides (Smedley, 1996), 468. Below about $70\,m$, O_2 is essentially absent, sulfide minerals are mostly stable, and arsenic concentrations are low in the groundwaters (Smedley, 1996), 467–468.

Aqueous discharges from geothermal power plants sometimes contaminate groundwaters and surface waters with arsenic. Examples occur in New Zealand (Aggett and Aspell, 1979) and Michoacán, Mexico (Birkle and Merkel, 2000). Other groundwaters become contaminated by natural hydrothermal sources of arsenic (e.g. Idaho and Nevada, USA, (Welch *et al.*, 2000), 593). As mentioned earlier in Section 3.6.9, geothermal waters from Yellowstone National Park contaminated the Madison River of Montana. Groundwaters surrounding the river contained $16-176\,\mu g\,L^{-1}$ of geothermal arsenic (Hering and Kneebone, 2002), (Nimick, 1998), 159; (Sonderegger and Sholes, 1989).

Arsenic is a 'contaminant of concern' in the groundwaters of about 30 % of the 1191 US *Superfund* sites (Appendix E) even though there is no evidence of the generation or disposal of significant arsenic wastes at most of the sites (Welch *et al.*, 2000), 590. Organic compounds at Superfund sites as well as soluble organic matter in livestock manure, septic tanks, and landfills often infiltrate into the subsurface. These organic compounds are reductants, which increase microbial activity, enhance the reductive dissolution of iron and manganese (oxy)(hydr)oxides, and lead to the release of arsenic into groundwater (Burgess and Pinto, 2005; Behr and Beane, 2002; Stollenwerk and Colman, 2003). The microbial oxidation of organic matter also produces bicarbonate and other carbonate species, which would raise the alkalinity of groundwater and desorb arsenic from mineral surfaces (Burgess and Pinto, 2005; García-Sánchez, Moyano and Mayorga, 2005; Appelo *et al.*, 2002; Anawar, Akai and Sakugawa, 2004). Any orpiment and realgar in igneous and metamorphic bedrocks would also tend to dissolve in the presence of bicarbonate (Kim, Nriagu and Haack, 2000).

Clay caps and liners are used to hinder the flow of water into landfills. While they prevent the formation of many toxic leachates, caps and liners also restrict the flow of air and aerated water into landfills, which allow subsurface microorganisms to intensify reducing conditions, promote reductive dissolution, and increase the release of arsenic into groundwaters (Section 3.11; (DeLemos *et al.*, 2006), 70). Stollenwerk and Colman (2003) argue that removing organic wastes from landfills and creating oxidizing conditions in the subsurface might reverse reductive dissolution and remove arsenic from contaminated groundwater. Oxidation would also convert As(III) to As(V), which may then sorb or coprecipitate with iron (oxy)(hydr)oxides. However, reversing the effects of reduction and returning groundwater arsenic concentrations to prelandfill levels may take at least several years (Stollenwerk and Colman, 2003), 379.

High-arsenic peats and other natural organic materials are usually stable in the subsurface unless water tables decline or if they are disturbed by human excavations and other construction activities (Appleyard, Angeloni and Watkins, 2006; Alaerts and Khouri, 2004; Huisman *et al.*, 1997). Specifically, at Perth, Australia, urban construction and the destruction of peat deposits led to the mobilization of arsenic in groundwater (Appleyard, Angeloni and Watkins, 2006). At Perth and other locations, excavations and declining water tables exposed previously reducing and stable peat to O_2 and microbial activity. The partial biodegradation of peat produces dissolved organic compounds, bicarbonate, and ammonium. Ammonium further dissolves organic compounds and promotes reductive dissolution (McArthur *et al.*, 2001). The partial biodegradation of peats may also explain the high alkalinity and ammonium concentrations in some arsenic-contaminated groundwaters near Hanoi, Vietnam (Berg *et al.*, 2001), 2624.

3.19.6   Arsenic chemistry in groundwater

Except in the very shallow subsurface, most groundwaters have less dissolved O_2 than surface waters and As(III) is usually more common than As(V). The majority of groundwaters also have pH values of 6–9 and $H_3AsO_3^0$ should be the dominant arsenic form under reducing conditions ((Clifford and Ghurye, 2002), 217–218; Chapter 2). In some cases, however, considerable metastable As(V) may occur in anaerobic groundwaters.

Theoretically, $H_2AsO_4^-$ and $HAs_2O_4^{2-}$ should dominate under oxidizing conditions in the subsurface. However, As(III) oxidation is often sluggish and the species may persist in oxidizing groundwaters (Boyle, Turner and Hall, 1998), 210. A greater sorption of As(V) oxyanions than $H_3AsO_3^0$ on aquifer solids could also contribute to increasing the ratio of dissolved As(III)/As(V) in groundwaters (Boyle, Turner and Hall, 1998), 206).

Although relatively few analyses are available of methylated arsenic in groundwater, the generally low populations of microorganisms in aquifers suggest that methylated forms should be nearly absent except in areas that have been contaminated by human-generated wastes or infiltrating surface waters that already have abundant methylarsenic (Mandal and Suzuki, 2002), 206–207; (Boyle, Turner and Hall, 1998), 206; (Welch *et al.*, 2000), 590. In a rare case, exceptionally high methylarsenic groundwaters were associated with an old tannery at Woburn, Massachusetts USA. The groundwaters contained up to 420 $\mu g\,L^{-1}$ MMA(V) and 180 $\mu g\,L^{-1}$ DMA(V) or about 26 % of the total arsenic (Davis *et al.*, 1994). Tannery wastes at the site released dissolved organic carbon and arsenic into the groundwaters, which resulted in reducing conditions and arsenic methylation (Davis *et al.*, 1994).

The arsenic chemistry of groundwaters is largely controlled through reactions with their aquifer solids, which may involve sorption, *ion exchange*, coprecipitation, and precipitation (Chapter 7). In most aquifers, arsenic concentrations are controlled by the sorption of arsenic oxyanions on iron (oxy)(hydr)oxides and to a lesser extent on aluminum and manganese (oxy)(hydr)oxides (Foster, 2003), 64. The optimal pH for As(V) sorption on kaolinite and smectite clays is about 5.0, while the optimal value for As(III) is approximately 9.0 (Boyle, Turner and Hall, 1998), 206, 209. For most clay minerals and (oxy)(hydr)oxides in aquifers and other geologic materials, the sorption of inorganic As(III) generally increases with increasing pH, whereas with increasing pH inorganic As(V) sorption generally decreases ((Stollenwerk, 2003), 67; Chapter 7).

Many of the chemical reactions in surface waters are also present in groundwaters. However, aquifers may create arsenic-rich groundwaters that contain other chemicals that are rarely abundant in surface waters (such as boron or fluoride). Such groundwaters are most likely to form in aquifers that consist of pegmatites or other geologic materials with unusual minerals. As examples, groundwaters in the upper regions of Ghana and the Pampa and Chaco plains of Argentina contain unusual dissolved arsenic-fluoride species, including AsO_3F^{2-} and $HAsO_3F^-$ (Bundschuh, Bonorino and Viero, 2000), (Apambire and Hess, 2000), 29. In upper Ghana, more than 90 % of the dissolved arsenic in the local groundwater is AsO_3F^{2-} (Apambire and Hess, 2000). The possible toxic effects of these unusual species are often unknown.

Oxygenated surface waters are sometimes pumped into aquifers to restore and enhance groundwater supplies. In Florida, artificial recharging mobilized up to 130 $\mu g\,L^{-1}$ of arsenic from a limestone aquifer (Price and Pichler, 2006). The injected water probably oxidized arsenian pyrite and/or organic matter (Price and Pichler, 2006; Arthur, Cowart and Dabous, 2000). Although arsenic may sorb onto iron (oxy)(hydr)oxides resulting from the oxidation of pyrite, the arsenic-bearing (oxy)(hydr)oxides may occur as very mobile colloids and pose a health threat to groundwater consumers. Furthermore, once the O_2 is consumed through oxidation reactions in the aquifer, anaerobic conditions would return and the arsenic may be released from the (oxy)(hydr)oxides through reductive dissolution (Price and Pichler, 2006), 53.

3.20   Glacial ice and related sediments

If areas of snow accumulate to thicknesses of several tens of meters or more, and if they persist through the summers, they may be transformed into ice that can slide and plastically flow as a *glacier*. In the process of moving, glaciers can erode, grind, and redistribute large amounts of rocks and sediments. Glacially deposited sediments commonly include poorly *sorted tills* and better sorted sands and gravels that result from flowing meltwaters.

Glaciers are divided into continental and alpine varieties. *Continental glaciers* may be more than 3 km thick (Press and Siever, 2001), 331. They currently cover most of Greenland and Antarctica. *Alpine glaciers* form in mountainous areas, including elevations above 4000–5000 m in tropical regions (Benn and Evans, 1998), 39. About 21 glacial advances or 'ice ages' have occurred during the past 1.8 million years or so (Benn and Evans, 1998), 43, 48. These advances have left extensive continental glacial deposits in North America and Eurasia.

Arsenic has been detected in enriched layers within the cores of glacial ice. The layers correlate with historic volcanic eruptions (Keskin, 1995). Overall, however, tills and other glacial sediments tend to have much higher arsenic concentrations than glacial ice. Specifically, analyses of glacial ice from Sweden yielded arsenic concentrations of only 2.0–3.8 $\mu g\,kg^{-1}$ (Table 3.15; (Weiss and Bertine, 1973)). In contrast, glacial sediments are derived from the weathering and erosion of regional bedrocks, which may include hydrothermal minerals, sulfide-bearing sedimentary rocks, volcanic deposits, and other arsenic-enriched materials (Kelly *et al.*, 2005), 502; (Thomas, 2003; Kettles and Shilts, 1994; Klassen, 2004); Table 3.15).

Groundwater in glacial deposits often contains high concentrations of arsenic (more than 10 $\mu g\,L^{-1}$; Table 3.13). The presence of abundant fine-grained materials and organic carbon in glacial sediment

Table 3.15   *Arsenic in glacial ice, glacial sediments, and loess*

Location	Concentration (mg kg^{-1})^a	Reference(s)
Clay (glacial): Lower Illinois River Basin, Illinois, USA	7.2−8.3	Warner (2001)
Glacial Ice: Sweden	0.0020−0.0038	Weiss and Bertine (1973)
Loess along Huang He (Yellow) River, China	10.4 (average); 25 (maximum)	Huang *et al.* (1988), 83
Loess: La Pampa province, Argentina (45 samples)	3−13	Smedley *et al.* (2005)
Loess: Tafi del Valle, Tucumán, Argentina	43	Warren, Burgess and Garcia (2005), 884
Sands and gravel (glacial): Lower Illinois River Basin, Illinois, USA	3−5	Warner (2001)
Till: British Columbia, Canada	1.9−170	Smedley and Kinniburgh (2002), 531
Till (calcareous): Near Edmonton, Alberta, Canada	10.5	Muloin and Dudas (2005)
Till: Malcantone watershed, Switzerland: 2 samples	76; 339	Pfeifer *et al.* (2004)
Till: Near Edmonton, Alberta, Canada	10.3 ± 0.25	Bennett and Dudas (2003)
Till: southern Kootenay terrane, British Columbia, Canada (2 samples)	5.9; 28.4	Lett *et al.* (1998)

^{a}Single values denote single analyses or average concentrations sometimes with 1 SD. Analyses involving different samples from the same location are separated by semicolons. Low and high values are separated by hyphens. Normally, soil and sediment samples are dried before analysis.

aquifers promotes the extensive reductive dissolution of iron (oxy)(hydr)oxides. In particular, reductive dissolution in the Mahomet and Glasford glacial sediment aquifers of Illinois, USA, has released up to 266 $\mu g\,L^{-1}$ of arsenic into groundwater (Kelly *et al.*, 2005). Reducing conditions were initially intense enough to precipitate iron sulfides and perhaps remove some of the arsenic from the groundwaters. However, once the sulfides entirely precipitated, dissolved arsenic concentrations were able to reaccumulate in the waters (Kelly *et al.*, 2005), 507.

3.21   Arsenic in air and wind-blown sediments

3.21.1   Arsenic emission sources

Arsenic may occur in the open atmosphere, as well as in the air of buildings, soils, and sediments. The vast majority (approximately 89–98.6%) of atmospheric arsenic is sorbed onto particulates rather than existing as vapors (Matschullat, 2000), 300. At any one time, the Earth's atmosphere has about 800–1740 t of arsenic with approximately 85% of it in the Northern Hemisphere due to its greater land area and industrialization (Matschullat, 2000), 299–300. The residence time of arsenic in the atmosphere is about 7–10 days (Matschullat, 2000), 300.

Volcanic eruptions are the major natural source of atmospheric arsenic (about 17 150 t year^{-1}). Considerable arsenic is also released by burning vegetation (approximately 125–3345 t year^{-1}), especially forest fires (Matschullat, 2000); (Galasso, Siegel and Kravitz, 2000), 849. Microbial activity in soils is another likely source of extensive arsenic emissions to the atmosphere with perhaps as much as 26 200 t year^{-1} (Matschullat, 2000), 300–301.

Copper smelting and coal combustion are the largest anthropogenic sources of arsenic emissions to the global atmosphere (Chapter 5). They are estimated at 12 800 and 6240 t As/year, respectively (Matschullat, 2000), 301. Other human activities that release arsenic include herbicide use, semiconductor manufacturing, automobile traffic, marine vessels, glass manufacturing, steel production, waste incineration, Portland cement manufacturing, and the combustion of CCA-preserved wood ((Shih, 2005), 88; (Matschullat, 2000; Chein *et al.*, 2006), 302; (Wasson *et al.*, 2005; Frey and Zhao, 2004); Chapter 7). Total anthropogenic emissions of arsenic to the atmosphere are about 18 800–25 800 t year^{-1} (Shih, 2005), 88.

Arsenic concentrations in outdoor surface air range from about 0.008–1 ng m^{-3} in remote areas to 15 ng m^{-3} in heavily polluted cities ((Matschullat, 2000), 301; Table 3.16). Typically, arsenic measurements of air are obtained by collecting PM of <2.5 or <10 μm on filters and then analyzing the materials. The literature contains numerous surveys and discussions of efforts to identify possible emission sources of arsenic in urban and regional air, including the Arctic (Akeredolu *et al.*, 1994; Landsberger *et al.*, 1992), Toronto, Canada (Lee *et al.*, 2003), Central Chile (Hedberg, Gidhagen and Johansson, 2005), Hong Kong, China (Lee and Hills, 2003), Frankfurt am Main, Germany (Zereini *et al.*, 2005), Mumbai, India (Bhanarkar *et al.*, 2005), Raipur, India (Deb *et al.*, 2002), Naples, Italy (Giordano *et al.*, 2005), Lisbon, Portugal (Freitas, Pacheco and Ventura, 2004), several other locations in Portugal (Freitas and Pacheco, 2007), Edinburgh, UK (Heal *et al.*, 2005), Oxford, UK (Lai *et al.*, 2004), the conterminous USA (Malm and Sisler, 2000), Birmingham, Alabama, USA (Hidy *et al.*, 2000), Alaska, USA (Shaw, 1991), Jacksonville, Florida, USA (Zhao and Frey, 2004), Houston, Texas, USA (Frey and Zhao, 2004), Minnesota, USA (Pratt *et al.*, 2000), Vermont, USA (Polissar, Hopke and Poirot, 2001), and the US Southwest (Eatough *et al.*, 1996). Besides coal combustion and ore smelting, significant arsenic air pollutants in urban areas may result from traffic, including fuel soot and wear on brakes and tires (Matschullat, 2000), 302.

Lichens (such as: *Xanthoria parietina*, *Hypogymnia physodes*, and *Pseudevernia furfuracea*) and mosses (e.g. *Sphagnum capillifolium*, *Hypnum cupressiforme*, *Hylocomium splendens*, *Brachytechium salebrosum*, and *Brachytechium rutabulum*) are often used to monitor arsenic and other contaminants in outdoor air (Giordano *et al.*, 2005; Cuny *et al.*, 2004; Helena, Franc and Cvetka, 2004; Farinha *et al.*, 2004; Lucaciu *et al.*, 2004). These organisms lack roots and largely obtain their nutrients and contaminants from wet (precipitation) and dry air deposition, which makes them suitable for monitoring air quality. In one study, lichens (*Pseudevernia furfuracea*) and mosses (*Sphagnum capillifolium*) were collected from a relatively unpolluted area at Forcella Laverdet in northeastern Italy (Giordano *et al.*, 2005), 433. The original arsenic concentrations (dry weight) of the mosses and lichens were 0.11 and 0.15 mg kg^{-1}, respectively. They were then transported to random locations within Naples, Italy (2001 population of about one million), which is located near the dormant volcano, Mt. Vesuvius (Giordano *et al.*, 2005). After four months of air exposure during both wet and dry seasons, mosses from 22 locations within the city contained 0.6–3.9 mg kg^{-1} of arsenic (dry weight). The lichens from 10 locations had 0.4–2.0 mg kg^{-1} of arsenic (dry weight). Arsenic load factors were calculated for the organisms, where the load factor = (exposed concentration — original concentration)/original concentration. Overall, the arsenic load factors for the mosses were significantly higher (4.5–34.2) than for the lichens (1.8–12.5). The arsenic could have originated from anthropogenic sources, sea salts, and/or volcanic-rich sediments, soils, and weathered rocks (Giordano *et al.*, 2005).

Indoor air in homes and public buildings may contain significant and sometimes dangerous levels of arsenic from coal combustion (Section 3.24.2.3, (Finkelman *et al.*, 2002)), tobacco smoke (Fowles and Dybing, 2003; Chang *et al.*, 2005), or dust (Clayton *et al.*, 1999; Schieweck *et al.*, 2005).

Table 3.16 *Arsenic in outdoor air samples*

Location	Arsenic concentration (ng m^{-3}) with particle size distributions (diameter in microns, μm)[a]	Reference(s)
Canada: Toronto	<0.042−6.0 (≤2.5 μm)	Lee *et al.* (2003)
Chile: Linares: 62 measurements	<0.69−49.9	Hedberg, Gidhagen and Johansson (2005)
Chile: Quillota: 81 measurements	1.7−19€ (≤10 μm)	Hedberg, Gidhagen and Johansson (2005)
Germany: Frankfurt am Main (60 samples)	0.2−3.8 (≤22 μm)	Zereini *et al.* (2005)
Russia: Kola Peninsula	28 ± 12	Kelley *et al.* (1995)
United Kingdom: Oxford (30 residential outdoor samples)	4.0 ± 2.7 (geometric SD; ≤2.5 μm)	Lai *et al.* (2004)
USA (lower 48 states)	0.02−55.36 (≤2.5 μm)	Malm and Sisler (2000)
USA: Barrow, Alaska	0.19 ± 0.150 (≤8 μm)	Shaw (1991)
USA: Birmingham, Alabama: 3 samples at each of 9 residences; Summer, 1997	5 (≤2.5 μm; average)	Hidy *et al.* (2000)
USA: Birmingham, Alabama: 3 samples at each of 9 residences; Winter, 1998	15 (≤2.5 μm; average)	Hidy *et al.* (2000)
USA: College Park: University of Maryland: 4 runs	5.1−12	Aras *et al.* (1973)
USA: Great Lakes states (Illinois, Indiana, Ohio, Michigan, Minnesota, and Wisconsin): 85 measurements	0.74 (mean; ≤50 μm)	Clayton *et al.* (1999)
USA: Minnesota: 717 samples at 25 sites	<5−15 (≤10 μm)	Pratt *et al.* (2000)
USA: Poker Flat, Alaska	0.213 ± 0.278 (≤8 μm; 1984−1985 AD); 0.210 ± 0.180 (≤8 μm; 1985−1987 AD)	Shaw (1991)
USA: Silver Spring, Maryland: 2 analyses	5.0; 5.2	Aras *et al.* (1973)
USA: Smoky Mountains	2.2 (≤2.5 μm)	Hidy *et al.* (2000)
USA: Washington DC: 24 samples	20	Aras *et al.* (1973)

[a]Single values denote single analyses or average concentrations sometimes with 1 SD. Analyses involving different samples from the same location are separated by semicolons. Low and high values are separated by hyphens.

Museums and industrial facilities, especially often have exotic minerals, pigments, and chemicals that release arsenic-bearing dust and vapors. In particular, mold growing on arsenic pigments in old wallpaper has been a potentially dangerous source of trimethylarsine gas in the past (Chapter 5).

3.21.2 Arsenic atmospheric chemistry

Most particulate arsenic in the atmosphere is inorganic As(V) and As(III) rather than organoarsenicals (Mandal and Suzuki, 2002), 207. Any arsenic vapors would mostly consist of As_4O_6 with trace amounts of arsine and methylarsines (Francesconi and Kuehnelt, 2002), 86; (Cotton *et al.* 1999), 400. As$_4$O$_6$ often originates from the combustion of arsenic-bearing sulfides in coal (Zheng *et al.*, 1999), 125. Microorganisms in soils may release volatile arsines (Section 3.21.6).

Both gaseous and particulate arsenic are potential inhalation hazards and may also contaminate surface soils, sediments, organisms, and waters near their points of origin ((Leoni and Sartori, 1997; Chein *et al.*, 2006; Hedberg, Gidhagen and Johansson, 2005; Shih and Lin, 2003; Martley, Gulson and Pfeifer, 2004; Klumpp *et al.*, 2003); Chapter 4). In particular, the upper 20 cm of soils within 15 km of a copper smelter and industrial complex at Port Kembla, New South Wales, Australia, contain up to 26 mg kg^{-1} of arsenic. The soils normally have maximum arsenic concentrations of 5.9 mg kg^{-1} (Martley, Gulson and Pfeifer, 2004).

3.21.3 Arsenic in precipitation

Eventually, atmospheric arsenic falls on the Earth's surface through wet and dry deposition. Sweet, Weiss and Vermette (1998) evaluated the deposition of arsenic over lakes Erie, Michigan, and Superior in the Great Lakes of North America. Overall, the wet and dry deposition rates for the lakes were very similar. Specifically, they were 72–94 and 66–91 μg m^{-2} year^{-1}, respectively.

Arsenic in precipitation from unpolluted ocean air averages about 0.019 μg L^{-1} (Hering and Kneebone, 2002), 157 and terrestrial rainwater concentrations (at least over the USA) also have similar averages of around 0.013–0.032 μg L^{-1} ((Smedley and Kinniburgh, 2002), 522; Table 3.17). As the precipitation infiltrates into the subsurface, its chemistry changes as it reacts with sediments, soils, and rocks. Therefore, the arsenic chemistry of the groundwater of an area may be very different than its precipitation chemistry.

Unless contaminated by coal combustion facilities, ore smelters, or other arsenic emitters, melted snow tends to have much <1 μg L^{-1} of arsenic (Table 3.17). The arsenic concentrations in the precipitation of an area may also change over time. Specifically, snowpacks in Colorado and New Mexico, USA, had less arsenic in 1999–2000 (<0.01–0.02 μg L^{-1} in meltwater) than averages from nearby sampling stations in 1993–1999 (0.05–0.14 μg L^{-1} in meltwater) ((Ingersoll, 2000); Table 3.17). The origin(s) of the arsenic is unknown, but may be related to emissions from nearby coal-fired power plants (Ingersoll, 2000), 2.

3.21.4 Arsenic in atmospheric dust

Winds are rarely capable of lifting and transporting sediment particles larger than coarse-grained *silt* or fine-grained sand (about 60 μm in diameter) (Press and Siever, 2001), 309. Furthermore, very fine-grained (<4 μm) clay minerals tend to adhere together into heavier clumps. Therefore, nonadhering and intermediate diameter (4–63 μm) silt grains are generally best suited for wind transport.

Airborne particles with *aerodynamic diameters* of larger than about 10 μm are usually caught in the nose and mouth before entering human lungs (Vedal, 1995), 1. However, smaller particles may penetrate into the lungs and cause serious health problems. PM with aerodynamic diameters of ⩽2.5 μm (designated as PM$_{2.5}$) is especially capable of entering the deep lungs (Vedal, 1995), 1. The particles become even more problematic if they contain arsenic or other toxins.

The lack of soil-anchoring vegetation and the frequent presence of strong winds in deserts may produce large quantities of arsenic-bearing wind-blown dust. At the Owens *playa* in California, USA, air samples contain as much as 400 ng m^{-3} of arsenic and the <10 μm fraction (PM$_{10}$) of the wind-blown dusts may have up to 50 mg kg^{-1} of the element (Ryu *et al.*, 2002), 2981. Overall, the Owens dry lake bed is a major source of wind-blown dust in the western United States (Gill *et al.*, 2002); (Ryu *et al.*, 2002), 2981. The arsenic-bearing dust originates from salts that precipitate from the capillary movement of shallow groundwater to the surface followed by evaporation. The production of wind-blown dust might

Table 3.17 Arsenic in atmospheric precipitation and dust.

Location	Arsenic Concentration[a]	Reference(s)
Dust: Sahara Desert, Mali	~17 mg kg^{-1}	Holmes and Miller (2002)
Dust: wind-blown from Sahara Desert?, St. John, US Virgin Islands	38 mg kg^{-1}	Holmes and Miller (2002)
Meltwater from snowpacks: Colorado and New Mexico, USA	<0.01–0.14 µg L^{-1} (0.45 µm filtered)	Ingersoll (2000)
Meltwater from snowpacks: north-central Arizona, USA, 8 samples	0.017 ± 0.008 to 0.422 ± 0.022 µg L^{-1}	Barbaris and Betterton (1996)
Meltwater from snowpacks: Rural New Brunswick, Canada: 15 samples	< 3.0 µg L^{-1}	Pilgrim and Schroeder (1997)
Meltwater from snowpacks: Urban New Brunswick, Canada: 15 samples	< 3.0–5.4 µg l^{-1}	Pilgrim and Schroeder (1997)
Precipitation: Rural New Brunswick, Canada: 15 samples	< 1.00–1.96 µg L^{-1}	Pilgrim and Schroeder (1997)
Precipitation: Urban New Brunswick, Canada: 1 sample	1.38 µg L^{-1}	Pilgrim and Schroeder (1997)
Rainwater: Lake Erie, North America	0.1 ± 0.1 µg L^{-1}	Sweet, Weiss and Vermette (1998), 428
Rainwater: Lake Michigan, North America	0.1 ± 0.1 µg L^{-1}	Sweet, Weiss and Vermette (1998), 428
Rainwater: Lake Superior, North America	0.1 ± 0.1 µg L^{-1}	Sweet, Weiss and Vermette (1998), 428
Rainwater: Maritime areas	0.02 µg L^{-1}	Smedley and Kinniburgh (2002)
Rainwater: Mid-Atlantic coast, USA	<0.005–1.1 µg L^{-1}	Smedley and Kinniburgh (2002)
Rainwater: Savannah Georgia, USA	0.2 µg L^{-1}	Waslenchuk (1979), 323
Rainwater: Terrestrial, USA	0.013–0.032 µg L^{-1}	Smedley and Kinniburgh (2002)

[a]Concentrations are in milligrams per kilogram or micrograms per liter. Single values denote single analyses or average concentrations sometimes with 1 SD. Low and high values are separated by hyphens. Available micron filtering information is listed.

be minimized by constructing sand fences, which would allow sand dunes to accumulate and cover the surface of the playa (Ryu *et al.*, 2002), 2991.

Some very fine-grained atmospheric dust is capable of crossing oceans and accumulating on different continents. Dust samples from the Sahara Desert of North Africa (Mali) are known to contain at least 17 mg kg^{-1} of arsenic and may be responsible for contaminating cisterns as far west as the Caribbean and eastern USA ((Holmes and Miller, 2002; Shinn, 2001); Table 3.17). Dust in one cistern from St. John, US Virgin Islands, contained about 38 mg kg^{-1} of arsenic ((Holmes and Miller, 2002); Table 3.17). Arsenic from Sahara Desert dust may also accumulate in soils, sediments, water, and plants in Florida and surrounding areas (Holmes and Miller, 2002).

3.21.5   Arsenic in wind-blown sediment deposits (loess)

Loess refers to deposits of fine-grained wind-blown (*eolian*) sediments, which may accumulate up to 100 m thick. The material mostly consists of silt grains with perhaps 5–30 % clay-sized particles and 5–10 % sand (Benn and Evans, 1998), 291. Like other sediments, loess may contain considerable arsenic depending on the chemistry of its source rocks (Table 3.15).

Loess is especially common in parts of China and US Midwestern states. Chinese loess primarily forms from the wind erosion of the uplifting Himalaya Mountains under arid and semiarid conditions. In the US Midwest, loess accumulated as winds scoured deposits left by retreating glaciers about 10 000 years ago (Press and Siever, 2001), 318–319.

Groundwaters in the loess aquifers of La Pampa, Argentina, may contain $\geq$ 5 mg L^{-1} of arsenic ((Smedley *et al.*, 2005; Smedley *et al.*, 2002); Table 3.13). The aquifers are very oxidizing and As(V) is the dominant arsenic species. The alkalinity of the groundwaters (pH 7.0–8.8) and the unusual presence of vanadium probably hinders the sorption of As(V) onto iron and manganese (oxy)(hydr)oxides. The alkalinity and vanadium may also desorb arsenic from the minerals (Smedley *et al.*, 2005).

3.21.6   Arsenic in soil and sediment gases

Perhaps as much as 26 200 t of arsenic may annually volatilize into the atmosphere from soils (Matschullat, 2000), 300–301. Much of this volatilization is due to microbial activity (Frankenberger and Arshad, 2002), 363–364. Under reducing conditions in soils, fungi and other microorganisms may produce gaseous arsine and methylated arsines, such as: methylarsine, dimethylarsine and trimethylarsine ((Mandal and Suzuki, 2002), 205; (Frankenberger and Arshad, 2002), 363; (Oremland and Stolz, 2003), 939; Chapter 4).

3.22   Petroleum

Coal and *petroleum* (crude *oil* and *natural gas*) are called *fossil fuels* because they formed from the organic remains of ancient plants and animals. After death, most organisms rapidly decay to CO_2 and other products. However, a minority of them die in reducing environments that allow the organisms to be partially preserved. Specifically, plant debris may convert into peat in wetlands. Peat transforms into coal through diagenetic processes, which involve deep burial, compaction, and heating to temperatures >40 °C (Section 3.24.1). Under similar diagenetic conditions, plankton and other marine microorganisms may convert into solid, liquid, and gaseous petroleum. Petroleum consists of *hydrocarbons* that include natural gas (which is mostly methane (CH_4)), solids (asphalts, bitumen and tars), and liquids (crude oil) (North, 1990), 9, 54–55.

Crude oils have an average arsenic concentration of about 15 µg kg^{-1} (Cullen and Reimer, 1989), 758. A typical range of arsenic in crude oils is 2.4–1630 µg kg^{-1} ((Matschullat, 2000), 299; Table 3.18). Although oil usually has relatively low concentrations of arsenic, the microbial degradation of petroleum contaminants in subsurface soils, sediments, and aquifers may promote the reductive dissolution of arsenic-bearing manganese and iron (oxy)(hydr)oxides in the geologic materials ((Burgess and Pinto, 2005; Klinchuch *et al.*, 1999); Section 3.11). The biodegradation of petroleum may also produce bicarbonate that can desorb or dissolve arsenic from minerals (Burgess and Pinto, 2005). In particular, arsenic concentrations up to 70 µg L^{-1} in the groundwaters of an English gravel aquifer were associated with the biodegradation of petroleum contaminants and the subsequent reductive dissolution of iron and manganese (oxy)(hydr)oxides. Bicarbonate released from the biodegradation of petroleum residues was likely responsible for high-arsenic concentrations (up to about 160 µg L^{-1}) in the groundwaters of an English sandstone aquifer ((Burgess and Pinto, 2005); Table 3.13).

Besides occurring in oil deposits, economically valuable quantities of natural gas may be associated with subsurface coal seams (McBeth, Reddy and Skinner, 2003). Natural gas often contains trace amounts of volatile arsine compounds, especially trimethylarsine and to lesser extents: dimethylethylarsine ((CH_3)$_2$(CH_3CH_2)As), methyldiethylarsine (CH_3(CH_3CH_2)$_2$As), and triethylarsine ((CH_3CH_2)$_3$As) (Irgolic *et al.*, 1991; Delgado-Morales, Zingaro and Mohan, 1994). These potentially toxic arsenic compounds must be removed before the gas is used, which may be accomplished with alumina or other sorbents (Nédez *et al.*, 1996). Arsenic-bearing formation waters (Section 3.6.2) also commonly occur in subsurface natural gas and crude oil deposits, where they are unavoidably extracted with the petroleum (Table 3.18; (Hanor, 1998)). In particular, one formation water from the Activo Luna Oilfield, Mexico, had 550 µg kg^{-1} of arsenic ((Birkle, Cid Vázquez and Fong Aguilar, 2005); Table 3.18).

Petroleum facilities may accidentally release significant quantities of arsenic into soils, sediments, and waters. Soils surrounding a crude oil storage facility in Los Angeles, California, USA, normally contain <0.5–8.0 mg kg^{-1}. However, the soils were contaminated with 30–2300 mg kg^{-1} of arsenic. Most of the arsenic probably originated from corrosion inhibitors that were used in production wells and possibly also from arsenic-bearing biocides that were applied to tanks (Wellman, Reid and Ulery, 1999).

3.23   Soils

3.23.1   Distinguishing between soils and sediments

Collectively, the world's soils are known as the *pedosphere*. Although similar in appearance, soils and sediments often have very different origins and distinctive properties. A number of definitions exist to distinguish soils from sediments. Sediments may be defined as weathering by-products that have been removed (eroded) from their place of origin, transported, and then deposited at another location by liquid water, wind, or ice. Although soils may contain sediments, most of the layered and fine-grained organic and inorganic materials in soils form more or less in-place by biological activity, the downward migration of fine-grained particles by infiltrating water, and the weathering of underlying bedrock and sediments.

Biological, chemical, and physical processes produce distinctive layers, or *horizons*, in soils. There are numerous types of soil horizons (Birkeland, 1984). However, the most common from the top of a soil profile to the bottom are the O, A, B, C, and R (Table 3.19). Not all of the horizons are present in every soil. Furthermore, the specific characteristics of each horizon are affected by climate, bedrock composition, the types of vegetation, pH, redox conditions, and time (Faure, 1998), 354–355.

The O horizon includes leaf litter and other partially undecayed biological matter on the surface (Birkeland, 1984), 7. O horizons are usually thin, especially under oxidizing, warm, and wet conditions, where

Table 3.18   Arsenic in petroleum, petroleum-associated formation waters, and petroleum-impacted surface waters

Sample	Arsenic concentration	Reference(s)
Crude oil: Casmalia crude, California, USA: 10 analyses	142 ± 21 µg kg^{-1}	Veal (1966)
Crude oil: Kansas, USA: 2 samples	31 µg kg^{-1}; 56 µg kg^{-1}	Veal (1966)
Crude oil: Middle East: 2 samples	30 µg kg^{-1}; 21 µg kg^{-1}	Veal (1966)
Crude oil: Mississippi, USA	10 µg kg^{-1}	Veal (1966)
Crude oil: Texas, USA	5 µg kg^{-1}	Veal (1966)
Crude oil: Venezuela	92 µg kg^{-1}	Veal (1966)
Crude oils	2.4–1630 µg kg^{-1}	Matschullat (2000)
Crude oils: Elk Hills and Greeley oil fields, Kern River alluvial fan, California, USA: 2 samples	< 1000 µg kg^{-1}	Klinchuch et al. (1999)
Distillate oil	100–200 µg kg^{-1}	Edward Aul and Associates, Incorporated and E. H. Pechan and Associates (1993)
Formation water, oil well: Alberta, Canada	230 µg kg^{-1}	Smedley and Kinniburgh (2002)
Formation water: Activo Luna Oilfield, Mexico	550 µg kg^{-1}	Birkle, Cid Vázquez and Fong Aguilar (2005)
Natural gas	0.01–63 µg dm^{-3}	Irgolic et al. (1991)
Natural gas: Abo gas field, New Mexico, USA	0.2–2.5 µg L^{-1} arsines	Delgado-Morales, Zingaro and Mohan (1994)
Petroleum (10 samples)	46.4–1112.4 µg kg^{-1}	Boyle and Jonasson (1973)
Residual (No. 6 fuel) oil	20–2000 µg kg^{-1}	(Edward Aul and Associates, Incorporated and E. H. Pechan and Associates (1993)
Shale oil: Green River Formation, USA	31 000 µg kg^{-1}	Fruchter et al. (1980)
Surface waters at Prestige oil tanker spill site, Galicia, Spain	$\leqslant 1.70 \pm 0.09$ µg L^{-1}	Peña-Vázquez et al. (2006)
Used oil	< 100–100 000 µg kg^{-1}	Mazur et al. (2004)
Used oil	1000–100 000 µg kg^{-1}	Edward Aul and Associates, Incorporated and E. H. Pechan and Associates (1993)

[a]Single values denote single analyses or average concentrations sometimes with 1 SD. Analyses involving different samples from the same location are separated by semicolons. Low and high values are separated by hyphens. Available information on arsenic speciation is listed. Concentrations are in micrograms per kilogram, micrograms per liter, and micrograms per cubic decimeter, where 1 dm^3 = 1 l.

the materials can readily decay in the presence of organisms and air. Once the materials largely decay, they form a mostly organic A horizon. The A horizon contains critical nutrients for plant growth, including crops. Fertile soils tend to have thick A horizons. In dry and tropical climates, the A horizon is usually thin. Vegetation is sparse in dry climates. In tropical climates, abundant water washes away organic matter or organisms rapidly consume it. Arsenic often accumulates in O and A horizons and recycles through them over time. The cycle includes surface deposition of arsenic-bearing materials by wind and water, the washing of arsenic into the A horizon by infiltrating precipitation, arsenic uptake by plant roots, release by

Table 3.19 Major soil horizons from top to bottom

Horizon	General description
O (top)	Leaf litter and other plant debris. Usually very thin
A	Organic-rich horizon mostly developed from the decay of plant remains. Contains important nutrients for plant growth. Relatively thin in dry and tropical climates. Dry climates lack plant debris. Biological activity and heavy precipitation often destroy organic matter in tropical climates
B	Mineral-rich horizon. Abundant clay minerals, iron (oxy)(hydr)oxides, and/or calcite
C	Weathered bedrock and sediments
R (bottom)	Unweathered bedrock

decaying vegetation, arsenic accumulation in mineral and organic matter, and the restarting of the cycle by wind and water erosion removing arsenic-bearing materials (Baroni *et al.*, 2004), 2.

The B soil horizon largely consists of clay minerals, iron (oxy)(hydr)oxides, and/or calcite. The C horizon, which is composed of sediments and weathered bedrock, usually occurs below the B. Arsenic in oxidizing soils readily sorbs and/or coprecipitates with iron and other (oxy)(hydr)oxides in B and C horizons (Reynolds, Naylor and Fendorf, 1999; Lund and Fobian, 1991). Below the C is the R horizon, which is unweathered bedrock.

Soils may also develop on the surfaces of weathering tailings piles. Specifically, soils with weak A and C horizons formed on the La Petite Faye tailings pile, France, within 35 years (Néel *et al.*, 2003). Soil development chiefly resulted from some plant growth and the oxidation of sulfide minerals, including arsenopyrite converting to scorodite (Section 3.7.4.2). Unlike other mine tailing soils, no gypsum layers were present (Néel *et al.*, 2003).

3.23.2 Arsenic chemistry in soils

Uncontaminated soils typically contain $5-10$ mg kg^{-1} of arsenic. Many of the processes that affect arsenic chemistry in water, mining wastes, and sediments are also present in soils, especially microbial activity, climate, contamination from human activities, reactions with organic matter, sorption onto clay minerals and (oxy)(hydr)oxides, precipitation/coprecipitation, reductive dissolution, and the oxidation of sulfide minerals. While sandy quartz-rich soils derived from felsic rocks tend to have the least amount of arsenic (Mandal and Suzuki, 2002), 203, hydrothermal deposits or any igneous, metamorphic, or sedimentary bedrock that is rich in sulfide minerals could be important sources of arsenic in soils. Arsenic also accumulates in NOM (Section 3.8 and Chapter 6; (Matschullat, 2000), 302; (Mandal and Suzuki, 2002), 203). Therefore, peats and other organic-rich soils, which average about 13 mg kg^{-1} of arsenic, usually have somewhat higher arsenic concentrations than uncontaminated inorganic soils ((Smedley and Kinniburgh, 2002), 533; Table 3.20). Soil texture is another important characteristic that affects arsenic chemistry. Fine-grained clay minerals tend to sorb and trap arsenic more effectively than clean quartz sand (Yan-Chu, 1994), 25; (Scazzola *et al.*, 2004). Specifically, clays in soils of the coastal plains of New Jersey may contain >40 mg kg^{-1} of arsenic, whereas sands usually contain <10 mg kg^{-1} (Barringer, Szabo and Barringer, 1998).

The pedosphere contains around $0.6-1.7 \times 10^9$ t of arsenic. The residence time of arsenic in soils in climates with moderate temperatures and precipitation is about $1000-3000$ years (Matschullat, 2000), 303. Uncontaminated soils derived from the *in situ* weathering of bedrock usually inherent arsenic from the rock (Yan-Chu, 1994), 18. However, soils that develop from wind-blown volcanic ash or sediments brought in over long distances by glaciers or streams may have very different arsenic concentrations than underlying bedrock. Additional natural arsenic contributions may result decaying biological organisms (Section 3.8).

The anthropogenic input of arsenic to the pedosphere is about $28\,400-94\,000$ t year^{-1} (Matschullat, 2000), 303. Major artificial sources of arsenic in soils include mining wastes, sewage sludges, animal manure, CCA- and ammoniacal copper zinc arsenate-treated wood, municipal solid wastes, chemical warfare agents, applications of arsenic-bearing fertilizers and pesticides, and deposition from coal combustion and smelter emissions ((Cullen and Reimer, 1989; Jain, Kim and Townsend, 2005), 749–750; (Yan-Chu, 1994), 19–21; (Welch et al., 2000; Maki et al., 2006; Morrell, Keefe and Baileys, 2003; Chirenje et al., 2003; Morrell and Huffman, 2004; Zagury, Samson and Deshênes, 2003; Bhumbla and Keefer, 1994; Bhattacharyya et al., 2003; Folkes, Kuehster and Litle, 2001; Nriagu and Pacyna, 1988); Chapter 5). Specifically, arsenic concentrations in soils contaminated by mining and smelting activities may easily exceed $10\,000$ mg kg^{-1} or 1 wt% (Table 3.20). Although most artificial accumulations of arsenic in soils are very localized, they may persist in the soils for years (Matschullat, 2000), 303; (Jain, Kim and Townsend, 2005).

Approximately $44\,000$ t of arsenic are annually removed from soils (Matschullat, 2000), 303. Major processes that eliminate arsenic from soils include microbial volatilization (up to 26,200 t year^{-1}; (Matschullat, 2000), 300–301), plant uptake, wind and water erosion, and leaching into precipitation, irrigation water, and groundwater (Matschullat, 2000; Bar-Yosef, Chang and Page, 2005).

Inorganic arsenic (As(III) and As(V)) is more prevalent than organoarsenicals in most soils. Typically, As(V) is prominent at Eh values above 200 mV and pH conditions of 5–8 (Akter and Naidu, 2006), 61. Microorganisms commonly reduce, oxidize, methylate, and demethylate arsenic in soils ((Mandal and Suzuki, 2002), 204; (Yan-Chu, 1994); (Craig, Eng and Jenkins, 2003), 22; (Kuehnelt and Goessler, 2003), 229, 230; Chapters 2 and 4). In particular, Inskeep, McDermott and Fendorf (2002), 204–209 discusses the microbial oxidation of As(III) in soils associated with the Madison River in Montana, USA. As mentioned earlier (Section 3.6.9), the As(III) originated from geothermal waters flowing out of Yellowstone National Park (Inskeep, McDermott and Fendorf, 2002), 205–206.

In many soils and sediments, As(V) mobility is limited by sorption onto clay minerals, organic matter, or iron (oxy)(hydr)oxides, and less commonly onto aluminum or manganese (oxy)(hydr)oxides (Inskeep, McDermott and Fendorf, 2002), 187. The sorption of As(III) onto iron (oxy)(hydr)oxides is generally most effective under slightly alkaline conditions (Section 3.9 and Chapter 7). For As(V) oxyanions, maximum sorption onto clay minerals and iron and aluminum (oxy)(hydr)oxides usually occurs in acidic rather than alkaline soils (Baroni et al., 2004; Arai and Sparks, 2002). Under acidic conditions, many iron minerals are below their ZPCs and isoelectric points (Chapter 2), which results in positive surface charges that readily attract more As(V) oxyanions than uncharged $H_3AsO_3^0$. Manganese (oxy)(hydr)oxides may also sorb As(V), but only under very acidic (approximately below pH 2) conditions as determined by their isoelectric points and ZPCs values ((Sadiq, 1997); Chapter 2).

Although iron, manganese, magnesium, calcium, and aluminum arsenates are usually too water soluble to control arsenic mobility in soils (Inskeep, McDermott and Fendorf, 2002), 187, iron, aluminum, or manganese arsenates occur in some acidic soils. In particular, scorodite may form from the partial weathering of arsenian pyrite or arsenopyrite (Inskeep, McDermott and Fendorf, 2002), 187. Calcium arsenates may be present in alkaline calcium-rich soils (Matschullat, 2000), 303; (Mandal and Suzuki, 2002), 204.

Like sediments, colloids are often important in sorbing and transporting arsenic in soils (Sadiq, 1997; Waychunas, Kim and Banfield, 2005). Colloids may consist of clay minerals, organic matter, calcium carbonate, and various aluminum, manganese, and iron (oxy)(hydr)oxides (Sadiq, 1997). Important iron (oxy)(hydr)oxides include goethite, akaganéite (β-FeO(OH)), hematite, ferrihydrites, and schwertman-

Table 3.20   *Arsenic concentrations in soils and nonmarine sediments*

Soil and location	Arsenic concentration (mg kg^{-1})^a	Reference(s)
Australia: Port Kembla, New South Wales: soils near copper smelter and industrial complex	0.91–19 (0–5 cm depth, upper background value: 5.8); 0.51–26 (5–20 cm, upper background value: 5.9)	Martley, Gulson and Pfeifer (2004)
Bangladesh: agricultural soils, 0–15 cm depth, uncontaminated: 234 samples	1.13–12.60	Hossain (2006)
Bangladesh: Brahmanbaria: soils	< 1–16 (orange band: ~300)	Breit *et al.* (2000)
Bangladesh: Chapainawabganj Sadar thana soils	1.27–56.68	Alam and Sattar (2000)
Bangladesh: Deuli: soils	20–111 (peat and peaty clay); 4–18 (clay rich); (3–7) sandy	Yamazaki *et al.* (2003)
Belgium: Flanders: clay soils: 24 samples	2.6–12.4	De Temmerman *et al.* (2003)
Belgium: Flanders: loam soils: 115 samples	3.6–21.9	De Temmerman *et al.* (2003)
Belgium: Flanders: sandy loam soils: 274 samples	2.2–19.6	De Temmerman *et al.* (2003)
Belgium: Flanders: sandy soils: 225 samples	0.93–34.4	De Temmerman *et al.* (2003)
Belgium: Reppel: soils near arsenic refinery: 16 samples	400–50 900 (<2 mm)	Van Herreweghe *et al.* (2003)
Belgium: Reppel-Bocholt: soils near arsenic refinery	100–36 000	Cappuyns *et al.* (2002)
Brazil: Iron Quadrangle mining region soils	200–860	Matschullat *et al.* (2000)
Canada: Bowen Island, British Columbia: clay gouge in faults and clay from weathering in fractures: 8 samples	44–4940	Boyle, Turner and Hall (1998)
Canada: Clyde Forks area, eastern Ontario: A horizon (40 samples); B horizon (40 samples); and C horizon (39 samples)	A: 2–15; B: 2–100; C: 2–65	Boyle and Jonasson (1973)
Canada: Kejimkujik Park, Nova Scotia: wetland soils	0.4–15.1 (37 O horizon samples, <0.2 mm); 0.7–34 (47 A horizons, <0.2 mm); 0.3–74.9 (50 C horizons, <0.2 mm); 0.067–206.5 (49 C horizons, <63 μm)	Rencz *et al.* (2003)
Canada: Near Blueberry Mountain, Alberta: acid sulfate soil from weathering pyrite in shales, B horizon	37.9 ± 0.3	Bennett and Dudas (2003)
Canada: New Brunswick: garden soils	9.0 ± 1.5 (18 samples, East Saint John); 12.7 ± 2.4 (4 samples, West Saint John); 10.7 ± 0.4 (2 samples, rural Fredericton)	Pilgrim and Schroeder (1997)

(*continued overleaf*)

Table 3.20 *(continued)*

Soil and location	Arsenic concentration (mg kg^{-1})^a	Reference(s)
Canada: Quebec: Montreal area: soils around CCA-treated utility poles	37.4 ± 2.5 to 251 ± 12 (< 300 μm)	Pouschat and Zagury (2006)
Chile: northern, soils	17.2 ± 1.1 (<125 μm)	Pizarro *et al.* (2003)
China: Lanmuchang area, southwest Guizhou	157–504 (10 samples from mining area); 37–89 (5 undisturbed soils); 4.4–7.5 (3 background soils lacking sulfide mineralization)	Xiao *et al.* (2004)
China: South: O, A and B horizons of forest soils	2.8–98.4	Hansen *et al.* (2001)
China: various locations: 792 samples	0.86–95.0	Yan-Chu (1994), 23–24
China: western Hunan: 96 samples	8.8–496 (<1 mm)	Lu and Zhang (2005)
Croatia: Krk Island: Soils	27.1 (<0.7 mm)	Kutle *et al.* (2004)
Czech Republic: soils of Mokrsko, Roudný, and Kašperské Hory gold deposits, Bohemian Massif: 17 sites	10–3489 (<2 mm)	Filippi, Goliáš and Pertold (2004)
France: soils near Douai	6.2–29.0 (near smelters, <2 mm); 3.1–10.7 (reference soils, <2 mm)	Sterckeman *et al.* (2002)
Germany: Fichtelgebirge Mountains: wetland (fen) soil	3.3–4.2	Huang and Matzner (2006)
Ghana: Southwest: acidic soils exposed to mining wastes	40.5–1290	Bowell, Morley and Din (1994)
Ghana: Southwest: aerobic soils on weathered gold–arsenic deposits	189–1025	Bowell, Morley and Din (1994)
Ghana: Southwest: uncontaminated soils from unmineralized bedrock	12.5–20.2	Bowell, Morley and Din (1994)
India: Behala: South Calcutta: Pesticide factory: 2 samples	20 100; 35 500	Chatterjee and Mukherjee (1999)
India: Chhattisgarh: 30 soil samples	9–390	Patel *et al.* (2005)
India: Kanker district: soils	8.8–49.0	Pandey *et al.* (2006)
India: West Bengal: Murshidabad district: 172 soil samples from 86 fallow sites	2.68–-6.79	Roychowdhury *et al.* (2002)
Italy: Mt. Mottarone: soil on weathered granodiorite and mica schists	12.7 ± 0.89	Salvioli-Mariani, Toscani and Venturelli (2001)
Italy: Near Porto Marghera industrial zone	12 ± 8 (4 undisturbed soils, <2 mm); 17 ± 5 (14 agricultural soils, <2 mm); 19 ± 12 (6 urban soils, <2 mm); 8 ± 5 (4 background soils, <2 mm)	Scazzola *et al.* (2004)
Italy: Tuscany: soils near hydrothermal deposits and mining wastes	5.3–2035.3 (<2 mm)	Baroni *et al.* (2004)
Japan: Geothermal Onikobe Basin and Ogachi region: 97 samples of soil A horizons	7.0	Shiikawa (1983)

Table 3.20   *(continued)*

Soil and location	Arsenic concentration (mg kg^{-1})^a	Reference(s)
Japan: Ichinokawa mine, Ehime: 4 soil samples (depth 0−12 cm)	1160−1960	Mitsunobu, Harada and Takahashi (2006), S1
Japan: Ohkunoshima Island: 5 soils contaminated by chemical warfare agents	7−125 000	Maki *et al.* (2006)
Malaysia: 241 soils (0−15 cm depth)	0.28−280	Zarcinas *et al.* (2004)
Mexico: Santa Maria de la Paz mining district, Villa de la Paz-Matehuala, San Luis Potosí: 282 soils	7−17 384	Razo *et al.* (2004)
Netherlands: Graft-De Rijp: soils contaminated with municipal waste and coal ash	0.5−24 (background soils); 15−90 (contaminated soils)	Walraven *et al.* (1997)
New Zealand: Waiotapu Geothermal System: sulfur mounds siliceous pyrite mud	80	Hedenquist and Henley (1985), 1646
New Zealand: Waiotapu Geothermal System: sulfur mounds yellow-orange mud	131 000	Hedenquist and Henley (1985), 1646
Peaty and other organic-rich soils: 14 samples	2−36	Boyle and Jonasson (1973)
Portugal: Cova dos Mouros mine: 286 soils	18−19 950 (<74 μm)	Reis *et al.* (2005)
Russia: Uranium processing site in Southeastern Siberia, Chita region: 4 sites at 0−10 cm depth (including a control site)	8.4 (control); 25; 13; 21	Gongalsky (2003)
Russia: uranium processing site in Southeastern Siberia, Chita region: 4 sites at 20−30 depth (including a control site)	9.7 (control); 10; 11; 18	Gongalsky (2003)
Russia: West Siberian plain soils	15−35	Temerev and Yu Kondakova (2006)
Soils (typical values)	<0.1−97	Mandal and Suzuki (2002), 202
Soils (typical values, 327 samples)	0.1−55	Boyle and Jonasson (1973)
Soils: surface, baseline: USA: Florida: 448 samples	0.02−7.01	Chen, Ma and Harris (1999)
South Korea: agricultural soils	<0.03−7.20	Jo and Koh (2004)
South Korea: vicinity of Dalsung mine: control site surface (0−15 cm depth) soils (15 samples)	5.1−25.3 (< 2 mm)	Jung, Thornton and Chon (2002)
South Korea: vicinity of Dalsung mine: mine dump surface (0−15 cm depth) soils (28 samples)	539−9380 (< 2 mm)	Jung, Thornton and Chon (2002)
Spain: carbonate soils associated with Aznalcóllar mine tailings spill	15.1−1480.5 (0−101 cm depth)	Simón *et al.* (2002)
Spain: La Peña-El Terronal mercury mining site near Mieres, Asturias (52 samples)	5.0−7287	Loredo *et al.* (1999)
Spain: Soil near Barruecopardo tungsten mine near Salamanca	1327 (< 2 mm)	García-Sanchez, Alvarez-Ayuso and Rodriguez-Martin (2002)

(continued overleaf)

Table 3.20   (continued)

Soil and location	Arsenic concentration (mg kg^{-1})^a	Reference(s)
Spain: soil near Terrubias tungsten mine near Salamanca	1215 (< 2 mm)	García-Sanchez, Alvarez-Ayuso and Rodriguez-Martin (2002)
Spain: Tarragona, Catalonia: soils around municipal waste incinerator before and after modernization of flue gas treatment system in 1997	4.30 ± 1.45 (1994 AD); 4.66 ± 1.83 (1997 AD); 5.56 ± 3.18 (< 2 mm; 1999 AD)	Llobet, Schuhmacher and Domingo (2002)
Sweden: Uppsala, urban soils: 75 samples	1.41−15.0 (0−5 cm depth); 1.41−16.2 (5−10 cm); 1.33−23.7 (10−20 cm)	Ljung, Otabbong and Selinus (2006)
Switzerland: Malcantone watershed: 7 soils	72−800	Pfeifer *et al.* (2004)
Taiwan: calcareous soils: 2 samples	23.7; 12.9	Lin, Wang and Li (2002)
Taiwan: road dust	11.7 ± 5.6 (11 samples downtown Hsinchu); 15.7 ± 7.0 (5 samples from Hsinchu industrial park); 13.0 ± 1.7 (25 samples from freeway tunnels)	Wang *et al.* (2005)
Thailand: 318 soils (0−15 cm depth)	0.08−124	Zarcinas *et al.* (2004)
Thailand: Ron Phibun mining district: 23 soil samples	< 10−2123.63	Williams *et al.* (1998)
Turkey: Igdeköy-Emet, Kütahya: soils near colemanite mine	<0.01−7.11	Çolak, Gemici and Tarcan (2003)
United Kingdom: Cornwall: soil: near arsenic processing facility	200−3325	Camm *et al.* (2003)
United Kingdom: northern England (818 top soil samples)	3.0−101 (<2 mm)	Rawlins, Lister and Mackenzie (2002)
USA: Alum Cave, Great Smoky Mountains National Park, Tennessee: soil	52	Hammarstrom *et al.* (2003)
USA: Arizona: Maricopa and Pima counties	3.1−24	Parker, Fossum and Ingersoll (2000)
USA: Central Massachusetts: 283 samples	4−770	Doherty and Hon (2003)
USA: Florida: Gainesville	0.21 − ~660	Chirenje *et al.* (2003)
USA: Florida: Miami	0.32−112	Chirenje *et al.* (2003)
USA: Florida: Shooting facility	0.50−107	Chen, Ma and Harris (2001)
USA: Florida: soils underneath 9 CCA-treated wooden structures: 65 samples	1.18−217; 1.36 (average of control soil)	Townsend *et al.* (2003)
USA: Hanford, Washington: Orchard soils: 31 samples	2.9−270	Yokel and Delistraty (2003)
USA: Mississippi: Nonamended soils: 6 samples	0.59−4.5	Han *et al.* (2004)
USA: Mississippi: soils amended with poultry wastes: 66 samples	1.7−15.2	Han *et al.* (2004)
USA: New Orleans, Louisiana: Post-Katrina flooding: 37 soils	<0.23−28	Cobb *et al.* (2006)

Table 3.20 *(continued)*

Soil and location	Arsenic concentration (mg kg^{-1})^a	Reference(s)
USA: New York: 13 orchard soils	1.60–141	Merwin *et al.* (1994)
USA: New York: Cornell Orchard: high lime sewage sludge-amended silty clay loam soil	51.9 ± 7.4 (12 amended soil samples); 0.6 (average of 3 control soil samples); 14 (sludge sample)	McBride *et al.* (1999)
USA: Pueblo, Colorado: soils from 33 sites	1.8–66.5	Diawara *et al.* (2006)
USA: Roosevelt Hot Springs, Utah: soil near Opal Mound	17	Christensen, Capuano and Moore (1983)
USA: Roosevelt Hot Springs, Utah: soils: nearly 550 samples	< 1–41	Capuano and Moore (1980)
USA: Virginia, New York, and Florida: soils within distances of 300 mm from 19 utility poles coated with ammoniacal copper zinc arsenate preservatives	< 1 to 10.3 ± 11.3	Morrell, Keefe and Baileys (2003)

^{a}Single values denote single analyses or average concentrations sometimes with one SD. Analyses involving different samples from the same location are separated by semicolons. Low and high values are separated by hyphens. Normally, soil and sediment samples are dried before analysis. Available information on sample depths, sieved fractions, number of samples, year of sampling, sample composition, and soil horizons is listed.

nite (Waychunas, Kim and Banfield, 2005). Precipitation and groundwater flow are capable of moving arsenic-bearing colloids in soils.

Some of the soils in the western portion of Hunan province, China, are especially rich in arsenic. The main soil of the province has arsenic concentrations of 8.8–22.8 mg kg^{-1}, although contaminated soils may contain as much as 496 mg kg^{-1} ((Lu and Zhang, 2005); Table 3.20). Some of the contamination is due to weathering of arsenic-rich rock outcrops, especially slates (Lu and Zhang, 2005), 317. However, mining and smelting activities also contribute significant arsenic (Lu and Zhang, 2005).

Substantial arsenic may leach into adjacent soils from wooden structures with CCA or other arsenic-bearing preservatives. Townsend *et al.* (2003) has shown that 65 surface soil samples from underneath nine CCA-treated wood structures in Florida, USA, contain 1.18–217 mg kg^{-1} of arsenic when compared with control soils that average only 1.36 mg kg^{-1} (Table 3.20). In a study involving ammoniacal copper zinc arsenate wood preservatives, arsenic concentrations generally fell to background levels within 150–300 mm away from 19 utility poles in Virginia, New York, and Florida, USA ((Morrell, Keefe and Baileys, 2003); Table 3.20). The environmental effects of arsenic-bearing wood preservatives are further discussed in Chapters 5 and 7.

Extensive use of lead arsenate (schulterite, PbHAsO$_4$) and other arsenic-bearing pesticides has dramatically increased arsenic concentrations in some agricultural and industrial soils (Welch *et al.*, 2000; Chatterjee and Mukherjee, 1999; Peryea and Kammereck, 1997). In particular, some soils on the coastal plains of New Jersey contain 20–270 mg kg^{-1} of arsenic. Nevertheless, the arsenic at these sites largely remains immobilized in the A horizon. That is, arsenic concentrations in the groundwaters rarely exceed 1 μg L^{-1} (Barringer, Szabo and Barringer, 1998). On the other hand, the application of phosphate fertilizers could mobilize pesticide-derived arsenic in agricultural soils (Peryea and Kammereck, 1997). As in aquifers and sediments, phosphate competes with inorganic As(V) for sorption sites on minerals in soils (Dixit and

Hering, 2006; Campos, 2002; Reynolds, Naylor and Fendorf, 1999). Once the As(V) is mobilized, it may migrate deeper into the subsurface and contaminate groundwater.

The chemistry of arsenic contaminants in soils often changes over time. In the soils at an abandoned pesticide factory near Auzon, France, lead arsenate contaminants have mostly degraded. As(V) released by the lead arsenates largely sorbed onto amorphous iron (oxy)(hydr)oxides in the soils, often at concentrations above $7000 \, mg \, kg^{-1}$ (Cancès *et al.*, 2005). Similarly, lead arsenates (including schultenite) were released into soils over a 96-year period at an agrichemical manufacturing facility in the northeastern United States (Arai *et al.*, 2006). In an oxidized soil with $284 \, mg \, kg^{-1}$ of arsenic, substantial (about 29 %) lead arsenate remains. The rest largely occurs as As(V) sorbed onto amorphous iron (oxy)(hydr)oxides. For a more reducing soil with $76 \, mg \, kg^{-1}$ of arsenic, lead arsenates entirely decomposed into As(V) and As(III) forms. The arsenic in the more reducing soil sorbed onto amorphous iron (oxy)(hydr)oxides, precipitated as amorphous 'As_2S_3', and coprecipitated with calcium compounds (Arai *et al.*, 2006).

3.23.3   Soil porewater chemistry

Soil porewaters, which are analogous to sediment porewaters, refer to water trapped in the interstices of soils, especially unsaturated soils. Biological activity and a low volume of water in contact with a large soil surface area often result in porewaters with arsenic concentrations that are very different than the soil matrix or any nearby surface waters. Like groundwaters and sediment porewaters, pH, pollutants, redox reactions, sulfide concentrations, (oxy)(hydr)oxide chemistry, biological activity, and the general chemistry of the water have profound affects on the As(V), As(III), and organoarsenical concentrations of soil porewaters ((Inskeep, McDermott and Fendorf, 2002); Table 3.21). To illustrate how these factors may influence the arsenic chemistry of soil porewaters, Bowell, Morley and Din (1994) performed an extensive study at the Ashanti gold mine in southwestern Ghana. Aerobic soils on weathered bedrock containing gold–arsenic mineralization have $189–1025 \, mg \, kg^{-1}$ of arsenic (Table 3.20). The arsenic concentrations of the soil porewaters are $86.2–557 \, \mu g \, L^{-1}$, which largely consists of inorganic As(V) (Table 3.21). Inorganic As(III), MMA(V), and DMA(V) are also sometimes present. Uncontaminated soils on unmineralized bedrock only contain $12.5–20.2 \, mg \, kg^{-1}$ of arsenic with porewater concentrations of $11.2–20 \, \mu g \, L^{-1}$. Inorganic As(V) is also dominant in the porewaters and no methylarsenic was detected. Anaerobic and aerobic acidic soils on unmineralized bedrock, but associated with mine tailings, have $40.5–1290 \, mg \, kg^{-1}$ of arsenic. The arsenic concentrations of the porewaters in the anaerobic soils are $70.2–110 \, \mu g \, L^{-1}$, whereas porewaters in the aerated soils contain $70.8–599 \, \mu g \, L^{-1}$ (Tables 3.21). Inorganic As(V) and methylarsenic were present in the aerated soils. The anaerobic soils only contained As(III) and As(V) (Bowell, Morley and Din, 1994).

3.24   Sedimentary rocks

3.24.1   Diagenesis and sedimentary rocks

Sedimentary rocks commonly include coals, sandstones, shales, salt deposits, *conglomerates*, and limestones. They form from the deep burial and diagenesis of sediments and, in some cases, soils. Diagenesis involves any chemical reaction or physical change resulting from groundwater, heat, or pressure that occurs during the transformation (lithification) of a buried sediment or soil into a sedimentary rock. In general, diagenesis occurs to depths of roughly 8 km below the Earth's surface and at temperatures of around $40–200 \, °C$. An important diagenetic process involving groundwater is the precipitation of calcite, silica, hematite, or other cements between grains in lithifying sediments and soils. Furthermore, clay minerals may form from feldspar ($KAlSi_3O_8$, $NaAlSi_3O_8$, and $CaAl_2Si_2O_8$) grains during diagenesis. Amorphous silica typically crystallizes into quartz over time. Under subsurface heat, gypsum dewaters to form anhydrite ($CaSO_4$) in salt deposits. Heat and pressure from diagenesis also dehydrate peats and convert them into

Table 3.21 *Arsenic concentrations in soil and sediment porewaters*

Location and description	Source(s) of arsenic	pH	Arsenic concentration ($\mu g\,L^{-1}$)[a]	Reference(s)
Belgium: Reppel-Bocholt: soils near arsenic refinery	Arsenic refinery	–	606–549 000 (0.45 μm filtered)	Cappuyns *et al.* (2002)
Canada: near Birsay, Saskatchewan: till and clay aquitard	Desorption from mineral surfaces	7.06–9.42	3.2–98 (0.45 μm filtered)	Yan, Kerrich and Hendry (2000)
Canada: near Yellowknife, Northwest Territories: Grace, Meg, Keg, Kam, and Peg lakes: Sediments	Reductive dissolution; mining	–	15–5170 (0.22 μm filtered)	Bright, Dodd and Reimer (1996)
Germany: Fichtelgebirge Mountains: wetland (fen) soil	Natural background	4.5–6.0	0.4–2.5	Huang and Matzner (2006)
Ghana: Southwest: acidic soils exposed to mining wastes	Mining wastes	–	70.2–599 (0.45 μm, substantial As(III))	Bowell, Morley and Din (1994)
Ghana: Southwest: aerobic soils on weathered gold–arsenic deposits	Weathered ores	–	86.2–557 (0.45 μm filtered, mostly As(V))	Bowell, Morley and Din (1994)
Ghana: Southwest: uncontaminated soils from unmineralized bedrock	Natural background	–	11.2–20 (0.45 μm filtered, mostly As(V))	Bowell, Morley and Din (1994)
Ireland: North Bull Island salt marsh, Dublin Bay: wetland soils	–	–	0.004–0.036 (0.45 μm filtered)	Doyle and Otte (1997)
New Zealand: Lake Ohakuri: sediments	–	–	570–2320, 520–2100 As(III), (0.45 μm filtered)	Aggett and O'Brien (1985)
United Kingdom: Cornwall	Mining and smelting	6.4–6.5	< 1–129 (mostly As(V))	Hutton *et al.* (2005)
USA: Lake Coeur d'Alene, Idaho: 12 samples	Reductive dissolution	–	40–350 (0.2 μm filtered)	Harrington, Fendorf and Rosenzweig (1998)
USA: Lake Washington, Washington: 24 analyses	Reductive dissolution	–	6.1–60 (0.45 μm filtered, As(III)+As(V))	Peterson and Carpenter (1986)
USA: Tulare Lake bed, California	Desorption from sediments	8.5	588 (total arsenic), 567 (As(V)), 21 (As(III))	Gao *et al.* (2004)

[a] Single values denote single analyses or average concentrations. Values separated by commas denote duplicate analyses of the same sample. Low and high values are separated by hyphens. Available information on arsenic speciation and micron filtering is listed.

coals. During the diagenesis of carbonate rocks, calcite may be transformed into *dolomite* $(CaMg(CO_3)_2)$ as magnesium-rich brines from partially evaporated seawater percolate into the subsurface and react with calcite-rich sediments or limestones (Blatt, Middleton and Murray, 1980), 514, 519–523.

Sedimentary rocks with the highest arsenic concentrations largely consist of materials that readily sorb or contain arsenic, such as organic matter, iron (oxy)(hydr)oxides, clay minerals, and sulfide compounds. Arsenian pyrite and arsenic-sorbing organic matter are especially common in coals and shales. *Ironstones* and *iron formations* are mainly composed of hematite and other iron (oxy)(hydr)oxides that readily sorb or coprecipitate arsenic. Iron compounds also occur as cements in some sandstones. Although almost any type of sedimentary rock could contain arsenic-rich minerals precipitated by subsurface fluids (Section 3.6.4), many sandstones and *carbonates* consist almost entirely of minerals that by themselves retain very little arsenic; namely, quartz in sandstones and dolomite and calcite in limestones.

3.24.2   Coal

3.24.2.1   Introduction

Coals are sedimentary and sometimes metamorphic rocks that contain at least 50 wt % and 70 vol % combustible organic materials, which are often valuable sources of energy (Craig, Vaughan and Skinner, 2001), 503. Unlike many other sedimentary and metamorphic rocks, coals often contain substantial pyrite and other sulfide minerals. The organic matter and sulfide minerals in coals frequently have strong affinities for arsenic.

Like sulfide-bearing ore deposits, the combustion of sulfide minerals in coals is a major source of atmospheric arsenic (Section 3.21.1) and coal mining wastes often produce arsenic-bearing mine drainage (Section 3.7.3.1). Economic, technical, regulatory, and environmental considerations determine whether an arsenic-rich coal is cleaned and utilized or if alternative sources of energy are used. To make these decisions, extensive and accurate arsenic analyses are needed. In situations where pyrite is the major source of arsenic, its removal through coal cleaning can significantly reduce atmospheric arsenic emissions. On the other hand, if arsenic is primarily associated with organic materials in a coal, pyrite removal would have little or no effect on reducing arsenic emissions.

Evaluating the potential health and environmental impacts of arsenic in coal not only must consider the arsenic concentration and mineralogy, but also the characteristics of the combustion and flue gas treatment systems, and how local climate, hydrology, and subsurface geology might affect the leaching of arsenic from any buried ash and other coal utilization byproducts (CUBs) ((Finkelman and Gross, 1999); Chapter 5). In other words, all economic, environmental, and human health impacts must be evaluated at all stages of coal utilization, including mining, transport, combustion, and byproduct utilization and disposal.

3.24.2.2   Coal formation

Although some coals are derived from algal remains and other fine-grained organic materials ((Ren *et al.*, 1999); Sections 3.14.3 and 3.17), most originate from the deep burial of peats that are composed of terrestrial plant debris. As mentioned in Section 3.18, peats may form in marine, estuary, and freshwater wetlands. Reducing conditions in wetlands allow the accumulation and burial of plant debris to exceed oxidation and decay. Once peats are buried to depths of at least 1–2 km, diagenetic temperatures exceed 40 °C and *coalification* begins. That is, the peat dehydrates and transforms into *lignite* coal. Lignites generally contain no more than 75 % water and have *heat contents* of about 15 000–19 000 kilojoule per kilogram (kJ kg^{-1}) (Craig, Vaughan and Skinner, 2001), 143; (Boggs, 1995), 277. With increasing diagenetic temperatures in the subsurface, the coals continue to lose water, carbon is concentrated, and the coals pass through increasing *ranks* (i.e. lignite, *subbituminous*, and then *bituminous*) (North, 1990), 55. Bituminous coals usually have no more than 10 % water and heat contents that range from 24 000 to 33 000 kJ kg^{-1} (Craig, Vaughan and Skinner, 2001), 143; (Boggs, 1995), 277. Besides rank, the amount of water, and heat contents, coals are usually described by their types of organic materials and concentrations of inorganic matter (ash contents) (Boggs, 1995), 277.

Above about 200 °C, bituminous coal is metamorphosed to *anthracite* (Boggs, 1995), 279. Some geologists refer to highly metamorphosed anthracites as *meta-anthracites*. If metamorphic temperatures continue to rise, remaining volatiles are released, which only leaves graphite, a mineral that is nearly pure carbon. Uplift, perhaps from the movement of tectonic plates, and erosion over geologic time may result in coals leaving their deep diagenetic or metamorphic environments to reside in near-surface locations, where they can be mined.

3.24.2.3   *Arsenic in coal*

Arsenic concentrations in coal are highly variable and range from <1 to >30 000 mg kg^{-1} (Table 3.22). US National Committee for Geochemistry, Panel on the Trace Element Geochemistry of Coal Resource Development Related to Health (1980) lists a global average arsenic concentration of 5 mg kg^{-1} for coal. Arsenic speciation and mineralogy also varies with different deposits. Pyrite is the major arsenic-bearing mineral in many coals (Welch *et al.*, 2000), 592, including the bituminous deposits of the US Appalachians and Illinois Basin (Huggins and Huffman, 1996; Huggins *et al.*, 1997; Kolker, 2000; Goldhaber, Lee and Hatch, 2003), samples from the Gunnedah Basin, New South Wales, Australia (Ward *et al.*, 1999), many European coals (Yudovich and Ketris, 2005), 152, and high-arsenic Chinese coals (Shraim *et al.*, 2003), 41. Unlike ore deposits, arsenopyrite is uncommon in at least US deposits (Huggins and Huffman, 1996), 47.

Arsenic may accumulate at any stage of coal development, including (1) depositing in ancestral plants and peats, (2) originating from groundwater during coalification, and (3) resulting from hydrothermal fluids or metamorphism after coal formation. Although often insignificant when compared with other sources of arsenic, some arsenic in coals originated from water that entered the original peat-forming plants. As in other environments, plants in wetlands often acquire inorganic As(V) along with phosphate and methylate it (Ruppert *et al.*, 1992). The methylarsenic (such as MMA(V) and DMA(V)) may then accumulate in the cell walls of the plants. After death and burial under reducing conditions, some microbes convert methylarsenic compounds into volatile (methyl)arsines. If sulfides and Fe(II) are also presence, arsenic may be incorporated into Fe(II) sulfide compounds that eventually transform into arsenian pyrite (Ruppert *et al.*, 1992). In rare cases, arsenian pyrite occurs as replacement materials in cell walls and other parts of fossilized plants (Ruppert *et al.*, 1992).

Detrital iron (oxy)(hydr)oxides, organic matter, and other arsenic-bearing materials in sediments may be transported by water or wind into wetlands and contribute arsenic to peats. Once buried, reductive dissolution releases sorbed arsenic from iron (oxy)(hydr)oxides. Under sulfate-reducing conditions, the arsenic coprecipitates in sulfide minerals or organic matter. During diagenesis, additional arsenic may be released from the organic matter and coprecipitate in sulfide minerals (Eskenazy, 1995), 253.

The ancestral peats of some coals acquired arsenic from natural drainage that resulted from the weathering and erosion of nearby outcrops (Yudovich and Ketris, 2005), 169. Specifically, lignites in northern Greece often have high-arsenic concentrations (up to 207 mg kg^{-1}) and are primarily associated with igneous rocks (Papanicolaou *et al.*, 2004), 157. The arsenic probably originated from the weathering of sulfide minerals in the nearby Eastern Rhodopian Massif or from hydrothermal waters related to volcanism roughly 25–40 million years ago (Yudovich and Ketris, 2005), 169; (Papanicolaou *et al.*, 2004), 157–158. Modern examples of arsenic-bearing drainage entering peat deposits occur in Finland and Indonesia. Finnish peats located near sulfide deposits contain up to 340 mg kg^{-1} of arsenic. Peats on

Table 3.22   Arsenic in whole coals

Location	Arsenic concentration (mg kg^{-1})[a]	Reference(s)
Typical global value	5	US National Committee for Geochemistry, Panel on the Trace Element Geochemistry of Coal Resource Development Related to Health (1980)
Australia: Liddell coal bed, Sydney Basin	0.69 +/± 11--20 %	Palmer et al. (1990)
Australia: New South Wales: Gunnedah Basin: 35 samples	5.29 (average), 38.10 (maximum)	Ward et al. (1999)
Bulgaria	2.0 ± 0.8 to 58 ± 85	Eskenazy (1995)
Bulgaria: Meden buk deposit: 2 localities	140; 1050	Eskenazy (1995)
Canada: Alberta: Mt. Allan: 3 samples	<0.3--0.8	Goodarzi and Swaine (1993)
Canada: Alberta–British Columbia: 22 subbituminous and bituminous samples	<0.3--13.0	Goodarzi and Swaine (1993)
Canada: British Columbia: Byron Creek bituminous: 30 samples	0.4--1.5	Goodarzi and Swaine (1993)
Canada: British Columbia: Fording Mine bituminous: 22 samples	0.2--2.6	Goodarzi and Swaine (1993)
Canada: British Columbia: Tulameen bituminous: 27 samples	1.3--5.3	Goodarzi and Swaine (1993)
Canada: British Columbia: Vancouver Island: Nanaimo bituminous: 10 samples	2.0--6.9	Goodarzi and Swaine (1993)
Canada: Nova Scotia: Joggins Formation: 8 samples	127--6159	Hower et al. (2000)
Canada: Nova Scotia: Phalen seam (Sydney Basin) and Foord seam (Stellarton Basin): 15 samples	0.90--194.5	Mukhopadhyay et al. (1998)
Canada: Nova Scotia: Sydney Basin: 3 seams	30--150	Mukhopadhyay et al. (1999)
Canada: Saskatchewan: 8 lignites	<0.2--5.9	Goodarzi and Swaine (1993)
Canada: Saskatchewan: Estevan lignite	0.5	Beaton, Goodarzi and Potter (1991)
Canada: subbituminous and bituminous feed coals at 7 power stations	1.6--84.4	Goodarzi and Huggins (2005)
Chile: Southern: Pupunahue, Pecket, and Victoria de Lebu mines: 22 samples	<0.04--17	Helle et al. (2000)

Location / samples	Value	Reference
China: East (Xuzhou, Huaibei, and Huainan): 11 samples	66.39 ± 4.11 to 98.32 ± 2.64	He, Liang and Jiang (2002)
China: Guizhou: 32 samples from 19 coal mines	0.23–21.65	Feng et al. (2000)
China: Guizhou: Liupanshui mining district: 22 samples	3–19	Zhuang et al. (2000)
China: Guizhou: Xingren county: Haizhi township	32 000	Ren et al. (1999)
China: Liaoning province: Shenbei lignite	9.9 (average) ± 16	Ren, Xu and Zhao (2004)
China: National: 132 coals	0.21–32 000	Ren et al. (1999)
China: National: 3390 samples	3.75 (average)	Dai, Zeng and Sun (2006), 223
China: North (Shuozhou, Huozhou, Yima, and Pingdingshan): 12 samples	61.74 ± 2.64 to 156.72 ± 7.55	He, Liang and Jiang (2002)
China: Northeast (Fuxin, Beipiao, Tiefa, and Fushun): 10 samples	55.71 ± 2.32 to 129.43 ± 9.36	He, Liang and Jiang (2002)
China: southwest Guizhou: Qianxi Fault Depression Area: 44 samples	0.2–238	Zhang et al. (2004)
China: Western Guizhou: Longtan Formation: /1 whole-seam samples	<0.3–11	Dai et al. (2005)
China: Western Guizhou: Xingren county: 9 anthracite seam samples	2.3–2226	Dai, Zeng and Sun (2006)
China: Yunnan province: Fuyuan county: Laochang mining area	0.5–12.5 (12 nonhydrothermal coals); 201.1–1504.2 (6 hydrothermally affected coals near faults)	Zheng et al. (1999)
Czech Republic: 9 lignite samples	1.59–14.3 (Velenje, Trbovlje, and Sokolov Basins); 142 (Sokolov Basin)	Šlejkovec and Kanduč (2005)
Czech Republic: North Bohemian Basin: 9172 samples	0.1–757.0	Bouška and Pešek (1999)
France: feed coals	<18	Yudovich and Ketris (2005), 147
Greece: Thrace, Eastern Greek Macedonia, Western Greek Macedonia, Epirus, Central Greece, Peloponese, Crete basins	3–207	Papanicolaou et al. (2004)

(continued overleaf)

Table 3.22 (continued)

Location	Arsenic concentration (mg kg^{-1})^a	Reference(s)
India: Jharkhand: North Karanpura coalfield	<0.01–0.49	Priyadarshi (2004)
India: Makum coalfield: 4 Assam coals	0.04–0.24	Mukherjee and Srivastava (2005)
Japan: 271 samples from 12 coal fields	3–325	Coleman and Bragg (1990)
Russia (Former USSR)	25	Yudovich and Ketris (2005), 146
South Africa: Highveld coals: 5 mines	1–5.6	Wagner and Hlatshwayo (2005)
Thailand: Lampang province: Mae Moh lignite: 107 samples from 3 seams	3.07–515 (15.23–21.81 in feed coal of Mae Moh power plant)	Bashkin and Wongyai (2002)
Turkey: Gokler coalfield, Gediz: 45 samples	170–3854	Karayigit, Spears and Booth (2000aa)
Turkey: Kangal lignites, Sivas: 67 seam and borehole core samples	< 21–181	Karayigit et al. (2001)
Turkey: Lignites	100 ± 137	Ölmez et al. (2004)
Turkey: Sorgun Basin: 35 samples from 5 seam sections	2.1–183	Karayigit, Spears and Booth (2000ab)
Turkey: Zeolite-rich Beypazari lignite, central Anatolia: 12 samples	32–148	Querol et al. (1997)
Ukraine: Donetsk Basin	80	Yudovich and Ketris (2005), 147
Ukraine: East Donetsk Basin anthracites	33	Yudovich and Ketris (2005), 147
Ukraine–Russia: Donets Basin	32.7	Kizilshtein and Kholodkov (1999)
United Kingdom: England: Beeston coal bed	8.34 ± <5 %	Palmer et al. (1990)
United Kingdom: England: Shallow Wood coal bed	2.95 ± <5 %	Palmer et al. (1990)
United Kingdom: Feed coals from 13 fields	1.9–63	Yudovich and Ketris (2005), 147
United Kingdom: South Wales Variscan Foreland Basin: 26 samples	2.8–783.6	Gayer et al. (1999)
USA: Alaska: 164 samples	0.36–27	U.S. Geological Survey Coal Quality Database (2006)
USA: Appalachian: 4385 samples	0.08–2200 (average: 35)	U.S. Geological Survey Coal Quality Database (2006)
USA: Appalachian: Alabama: Pottsville Formation: American seam	2170	Huggins et al. (1993)

Location	Value	Reference
USA: Appalachian: Alabama: Pottsville Formation: Pratt seam	600	Huggins *et al.* (1993)
USA: Appalachian: Eastern Kentucky: Knox county: Dean (fire clay) seam	365	Mardon and Hower (2004)
USA: Appalachian: Kentucky	<0.6–98 (mostly < 100 μm)	Tuttle *et al.* (2002)
USA: Appalachian: Kentucky: Leslie county: Breathitt Formation: Fire Clay coal bed: 11 samples	1–418	Ruppert, Hower and Eble (2005)
USA: Appalachian: Pennsylvania: Pittsburgh coal: duplicate analyses	8.4, 4.5	Huggins (2002)
USA: Appalachian: Pennsylvania: Pittsburgh No. 8	69	Huggins *et al.* (1993)
USA: Appalachian: Pennsylvania: Upper Freeport	670	Huggins *et al.* (1993)
USA: Appalachian: Pennsylvania: Upper Freeport: duplicate analyses	11, 17	Huggins (2002)
USA: Appalachian: West Virginia: duplicate analyses	6.2, 6.9	Huggins (2002)
USA: Appalachians: Alabama: Warrior Basin: Upper Cobb coal bed	6.25 ± <5 %	Palmer *et al.* (1990)
USA: Appalachians: Central: Southern West Virginia, Virginia, Kentucky, and northeastern Tennessee: 1769 samples	0.08–680	U.S. Geological Survey Coal Quality Database (2006)
USA: Appalachians: northern (Pennsylvania, Ohio, Maryland, and northwestern West Virginia);1630 samples	0.31–410	U.S. Geological Survey Coal Quality Database (2006)
USA: Appalachians: Pennsylvania: Greene county: Pittsburgh coal	8	Huggins and Huffman (1996)
USA: Appalachians: Pennsylvania: Upper Freeport, Indiana county	17	Huggins and Huffman (1996)
USA: Appalachians: South: Alabama, Georgia, and Southeastern Tennessee: 986 samples	0.21–2200	U.S. Geological Survey Coal Quality Database (2006)
USA: Appalachians: Virginia: Buchanan county: Pocahontas #3: Duplicate samples	10, 6.9	Huggins (2002)

(continued overleaf)

Table 3.22 *(continued)*

Location	Arsenic concentration (mg kg^{-1})[a]	Reference(s)
USA: Appalachians: West Virginia: Raymond (Pittsburgh) coal bed	30.6 ± <5 %	Palmer *et al.* (1990)
USA: Arizona: Black Mesa field: subbituminous–bituminous: 11 samples	0.5–10	U.S. Geological Survey Coal Quality Database (2006)
USA: Gulf Coast: Texas, Arkansas, and Alabama lignites: 141 samples	0.96–22	U.S. Geological Survey Coal Quality Database (2006)
USA: Illinois Basin: Illinois, Indiana, and Western Kentucky: 298 samples	0.40–180	U.S. Geological Survey Coal Quality Database (2006)
USA: Illinois Basin: Indiana: Dugger Formation: Danville Member: 4 samples	0.2–141.0	Mastalerz *et al.* (2004)
USA: Illinois Basin: Indiana: Petersburg Formation: Springfield Coal Member: 5 samples	14.7–130.0	Mastalerz *et al.* (2004)
USA: Illinois Basin: St. Clair county, Illinois #6 coal	5	Huggins and Huffman (1996)
USA: Illinois: Illinois No. 6: bituminous	10	Huggins *et al.* (1993)
USA: Illinois: Illinois No. 6: bituminous: duplicate analysis	4.7, 3.2	Huggins (2002)
USA: Indiana: Dugger Formation: Danville Member: 2 bituminous samples	10.2; 17.4	Mastalerz and Padgett (1999)
USA: Michigan Basin: Michigan: 3 bituminous coals	5; 8; 40	U.S. Geological Survey Coal Quality Database (2006)
USA: Midwest (Western Interior): Arkansas, Iowa, Kansas, Missouri, Nebraska, Oklahoma mostly bituminous: 311 samples	0.72–240	U.S. Geological Survey Coal Quality Database (2006)
USA: Midwest (Western Interior): Kansas bituminous: 32 samples	6.8–60	Brady and Hatch (1997)
USA: National Average	24	U.S. Geological Survey Coal Quality Database (2006)
USA: New Mexico–Colorado: Raton Mesa subbituminous and bituminous: 40 samples	0.17–8.9	U.S. Geological Survey Coal Quality Database (2006)

Location	As (mg/kg)	Reference
USA: New Mexico–Colorado: San Juan River mostly subbituminous and bituminous: 187 samples	0.06–32	U.S. Geological Survey Coal Quality Database (2006)
USA: North Dakota: Hagel seam	11	Huggins *et al.* (2002)
USA: North Dakota: Beulah seam	9	Huggins *et al.* (1993)
USA: North Dakota: Beulah-Zap Seam: duplicate analyses	2.6, 3.3	Huggins (2002)
USA: North Dakota: Beulah-Zap Seam	3.3 ± 0.1; 2.63 ± 0.19	Schobert (1995), Palmer (1990)
USA: North Dakota: Mercer county: Beulah-Zap lignite	3	Huggins and Huffman (1996)
USA: North Dakota–Montana: Fort Union: 280 lignite samples	0.80–560	U.S. Geological Survey Coal Quality Database (2006)
USA: Ohio: Blend of Nos. 5, 6, and 7 seams	19	Huggins *et al.* (2002)
USA: Pennsylvania: Anthracites: 52 samples	0.7–140	U.S. Geological Survey Coal Quality Database (2006)
USA: Rhode Island: Meta-anthracites: 12 samples	3.3–45	U.S. Geological Survey Coal Quality Database (2006)
USA: Rocky Mountain subbituminous: Wyoming: Campbell county: Wyodak-Anderson coal	4	Huggins and Huffman (1996)
USA: Rocky Mountain subbituminous: Wyoming: Wyodak-Anderson: duplicate analyses	3.6, 3.5	Huggins (2002)
USA: Rocky Mountain: Denver Basin subbituminous and lignites: 40 samples	0.36–4	U.S. Geological Survey Coal Quality Database (2006)
USA: Rocky Mountain: Green River mostly subbituminous and bituminous: 416 samples	0.10–130	U.S. Geological Survey Coal Quality Database (2006)
USA: Rocky Mountain: north-central Montana bituminous: 7 samples	8.3–52	U.S. Geological Survey Coal Quality Database (2006)

(continued overleaf)

Table 3.22 *(continued)*

Location	Arsenic concentration (mg kg^{-1})[a]	Reference(s)
USA: Rocky Mountain: Wind River subbituminous and bituminous: 41 samples	0.43–40	U.S. Geological Survey Coal Quality Database (2006)
USA: Texas and Gulf Coast: Lignite, mostly Wilcox Group:	3.5–4.6	Crowley et al. (1998)
USA: Texas: Jackson Group: Manning Formation: Lignite: 11 samples	1.6–18.7	Warwick et al. (1997)
USA: Texas: Near Bremond: Wilcox Group: 2 Shaley lignites	2.0; 8.1	Crowley et al. (1997)
USA: Texas: Near Bremond: Wilcox Group: 7 lignite samples	0.3–3.0	Crowley et al. (1997)
USA: Utah: Blind Canyon: duplicate analyses	0.5, 1.2	Huggins (2002)
USA: Utah: Southwest bituminous: 42 samples	0.47–27	U.S. Geological Survey Coal Quality Database (2006)
USA: Utah–Colorado: Uinta Basin mostly subbituminous and bituminous: 250 samples	0.07–20	U.S. Geological Survey Coal Quality Database (2006)
USA: Utah–Wyoming: Hams Fork subbituminous–bituminous: 29 samples	0.70–21	U.S. Geological Survey Coal Quality Database (2006)
USA: West Virginia: Kanawha county: Lewiston-Stockton	6	Huggins and Huffman (1996)
USA: Wyoming–Montana: Powder River Basin mostly subbituminous: 630 samples	0.20–420	U.S. Geological Survey Coal Quality Database (2006)

[a]Single values denote single analyses or average concentrations sometimes with 1 SD. Values separated by commas denote duplicate analyses of the same sample. Analyses involving different samples are separated by semicolons. Low and high values are separated by hyphens. Available information on micron sieving is listed.

Sumatra in Indonesia have up to $2200 \, mg \, kg^{-1}$ of arsenic and probably originate from river drainage (Yudovich and Ketris, 2005), 168.

In the subsurface, arsenic may readily accumulate in coal. Specifically, groundwater percolating through volcanic deposits may leach arsenic and transfer it to underlying coal beds during or after coalification. *Contact metamorphism* and any associated hydrothermal fluids are other processes that sometimes locally enrich arsenic in coal seams. At a site in British Columbia, a *Cretaceous* bituminous coal seam was contact metamorphosed by a mafic *dike*. The contact temperatures were about 700–900 °C, which could volatilize arsenic and/or dissolve it into associated hydrothermal fluids. The arsenic was then transported and deposited into cooler sections of the seam (Yudovich and Ketris, 2005), 163.

Arsenic in the Beyparzari lignite, Turkey, has a very unusual mineral association. The arsenic mostly occurs in sulfide minerals and *zeolites* (Querol *et al.*, 1997). The zeolites in the Turkish lignites probably originated from alkaline diagenetic solutions altering buried volcanic glass that had been deposited in the ancestral peat bogs (Querol *et al.*, 1997), 268.

Arsenian pyrite in many arsenic-rich coals originated from regional hydrothermal fluids that entered the coals during or after their formation. In the subsurface, hydrothermal fluids may readily flow through faults and enrich adjacent coals in arsenic (e.g. Yunnan province, China, (Ren *et al.*, 1999); Gokler coalfield, Turkey, (Karayigit, Spears and Booth, 2000a)). Hydrothermal arsenian pyrites also occur in the coals of Guizhou province, China (Dai *et al.*, 2005), 135 and in the south (Warrior Basin, (Goldhaber, Lee and Hatch, 2003; Goldhaber *et al.*, 1997; Diehl, Goldhaber and Hatch, 2004)) and central (southeastern Kentucky, (Tuttle *et al.*, 2002)) Appalachian Basin, USA.

In the bituminous coals of the US Illinois and Appalachian basins, arsenic primarily occurs in pyrite. The arsenian pyrite probably originated from subsurface fluids that existed about 270 million years ago during the formation of the Ouachita and Appalachian mountains (Goldhaber, Lee and Hatch, 2003). The arsenic-bearing fluids in the midcontinent Illinois Basin were primarily brines derived from surrounding sedimentary basins that were also responsible for the formation of the Mississippi Valley lead–zinc deposits. In contrast, the fluids that were responsible for the arsenian pyrites in the Appalachians (especially in the coals of the Warrior Basin of Alabama) were metamorphic and not as saline as those in the midcontinent (Goldhaber, Lee and Hatch, 2003).

The infamous arsenic poisoning of more than 3000 people in Guizhou province, China, is primarily associated with the combustion of local high-arsenic coals ((Finkelman *et al.*, 2002), 428; (Zheng *et al.*, 1999); Chapter 4). The population in the region often uses coal-burning stoves for heating and cooking. Peppers, which are a major part of the local diet (Finkelman and Gross, 1999), are dried on the stoves. The drying process may raise the arsenic concentrations of the peppers from $<1 \, mg \, kg^{-1}$ to $500 \, mg \, kg^{-1}$. The combustion of arsenic-rich coal produces As_4O_6 ('As_2O_3') gas at temperatures above 215 °C. As_4O_6 may be inhaled or condense on peppers, other food, and water (Zheng *et al.*, 1999), 125. Arsenic poisoning may also result from the indoor inhalation of coal ash particles ((Finkelman *et al.*, 2002), 428; Section 3.21). Further discussions of the environmental and human health impacts of coal utilization and its byproducts are in Chapter 5.

Some whole coals in southwest Guizhou province, China (Xingren county) have up to $32\,000 \, mg \, kg^{-1}$ of arsenic (Dai *et al.*, 2005; Belkin, 1998; Ren *et al.*, 1999). Although these coals are primarily responsible for the arsenic poisoning in the province, they are very localized (Dai *et al.*, 2005), 128; (Dai, Zeng and Sun, 2006). Dai *et al.* (2005) actually found that 71 whole-seam coals from the western portion of the province contained no more than $11 \, mg \, kg^{-1}$ of arsenic, which is lower than many other Chinese and American coals (Table 3.22).

The high-arsenic coals of southwest Guizhou province contain arsenian pyrite, arsenopyrite, trace amounts of getchellite ($AsSbS_3$), and As(V)-bearing compounds including iron phosphate and sulfide oxidation products, including jarosite and scorodite (Dai, Zeng and Sun, 2006; Belkin, 1998).

Nevertheless, the arsenic in the most arsenic-rich ($>30\,000$ mg kg^{-1}) coals is largely organically bound and probably originated from ancient gold-bearing 165–300 °C hydrothermal fluids ((Belkin, 1998; Ding *et al.*, 2001); Section 3.6). The fluids deposited the arsenic in faults and surrounding coal beds (Zhang *et al.*, 2004), 59.

In contrast to many other coals, western US and Canadian subbituminous coals typically lack arsenian pyrite. Huggins, Goodarzi and Lafferty (1996) used *X-ray absorption fine structure spectroscopy* (XAFS) and found that arsenic in several samples of the subbituminous coals primarily occurs as As(III) coordinating with oxygen functional groups in organic materials. XANES analyses of a Wyodak (Wyoming, USA) subbituminous coal with about 1.58 mg kg^{-1} of arsenic suggest that about 50 % of it occurs in *carboxyl* groups, whereas the rest exists as As(V) in oxidation products (Kolker, 2000), 172–173. Organic associations for arsenic also occur in lignites from the Alexandriisk deposits in the Ukraine and the Shurab coals of Tajikistan. In these coals, arsenic associates with humic and fulvic acids (Yudovich and Ketris, 2005), 165.

3.24.3 Shales and oil shales

Shales are very fine-grained sedimentary rocks consisting of 67 % or more clay-sized (<4 μm) particles. Unlike *mudstones*, *siltstones*, and *claystones*, shales are very *laminar* and *fissile*; that is, they readily split into thin, closely spaced, and parallel layers (Boggs, 1995), 181. Along with clay minerals, shales may have abundant microscopic quartz, feldspars, pyrite, hematite, calcite, dolomite, and/or organic matter (Boggs, 1995), 178.

Shales typically have higher arsenic concentrations than sandstones and limestones (Table 3.23). Turekian and Wedepohl (1961) estimated the average arsenic concentration of shales as 13 mg kg^{-1}. This value is still widely cited in the literature. In most shales, arsenic mainly occurs in pyrite (Erickson and Barnes, 2005). Some shales even contain multiple generations of arsenian pyrite and other arsenic-bearing sulfide minerals, including sulfides that resulted from diagenesis and hydrothermal fluids (Chernoff and Barton, 2001). Shales may also have other arsenic-bearing and -sorbing materials that are often less common in sandstones and limestones, including iron (oxy)(hydr)oxides, organic matter, and clay minerals (Smedley and Kinniburgh, 2002), 532. When compared with relatively low-surface area pure quartz sand grains, organic matter and clay minerals have greater capacities to sorb arsenic (Mok and Wai, 1994), 110.

Shales that formed from marine muds usually have higher arsenic concentrations than nonmarine shales (Smedley and Kinniburgh, 2002), 532. Deep ocean water allows very fine-grained, high-surface area particles to sorb arsenic, settle, and accumulate. Marine shales also tend to contain more sulfur, which may convert into arsenian pyrite during diagenesis. As with other rocks, arsenic released by the weathering of arsenian pyrite or other sulfide minerals in shales may contaminate local surface waters and groundwaters (Section 3.19).

Like iridium, arsenic is enriched in Cretaceous-*Tertiary* boundary shales from New Zealand (Brooks *et al.*, 1984; Strong *et al.*, 1987). The iridium is believed to have originated from an asteroid impact that caused the massive extinction at the end of the Cretaceous period about 65 million years ago. In contrast, most of the arsenic in the boundary shales probably had a terrestrial origin (Strong *et al.*, 1987). The extinction of marine organisms, especially plankton, from the impact may have been responsible for increased anoxic conditions in the oceans, which led to the precipitation of arsenic in the marine deposits (Brooks *et al.*, 1984), 541.

Oil shales are fine-grained organic-rich sedimentary rocks. Many of them are actually siltstones, fine-grained limestones, or impure coals (Boggs, 1995), 281. Oil shales contain kerogen, which is a group of organic compounds that may be converted to petroleum through distillation involving heating at about 350 °C under low-O$_2$ conditions (Boggs, 1995), 280; (Fruchter *et al.*, 1980). Important oil shale

Table 3.23 Arsenic in sedimentary rocks. Coals are listed in Table 3.22

Sedimentary rock	Arsenic concentrations (mg kg^{-1})[a]	Reference(s)
Anhydrite and gypsum: Walton area, Nova Scotia, Canada: 27 samples	< 1–10	Boyle and Jonasson (1973)
Berry Siltstone: Port Kembla, New South Wales, Australia: 2 samples from each depth	5.1 ± 0.9 (0–5 cm depth); 7.7 ± 1.9 (5–20 cm depth)	Martley, Gulson and Pfeifer (2004)
Black shale: Chu-Bu, South Korea: 10 samples	0.5–6.2	Lee *et al.* (1998)
Black shale: Chung-Joo, South Korea: 7 samples	0.5–8.4	Lee *et al.* (1998)
Black shale: Duk-Pyung, South Korea: 9 samples	5.0–110.0	Lee *et al.* (1998)
Black shale: Lower Illinois River Basin, Illinois, USA: 4 samples	12–21	Warner (2001)
Black shale: Zarshuran, northwestern Iran	200–82 000	Mehrabi, Yardley and Cann (1999), 677
Budgong Sandstone: Port Kembla, New South Wales, Australia:12 samples from each depth	3.2 ± 1.4 (0–5 cm depth); 3.2 ± 1.1 (5–20 cm depth)	Martley, Gulson and Pfeifer (2004)
Carbonates: 40 samples	0.1–20.1	Boyle and Jonasson (1973)
Carbonates: typical value	1	Turekian and Wedepohl (1961)
Cascade Creek Shale: Erie, Pennsylvania, USA: 2 samples	5.9; 9.9	Murnock (2002)
Claystone (noncarbonaceous, nonmarine): north-central USA: 8 samples	3–10	Tourtelot (1964)
Conglomerate: Cobalt area, Ontario, Canada: 2 samples	<5	Boyle and Jonasson (1973)
Conglomerate: El. Morro Formation: Mexico	65; 5400 (mineralized)	Armienta *et al.* (2001)
Conglomerate: Kanker district: Central-east India	10 000	Pandey *et al.* (2006)
Edwards Limestone: East Texas, USA	1.0	Abraham (1998)
Grayson Marl: East Texas, USA	4.0	Abraham (1998)
Gypsum and anhydrite (evaporites): 5 samples	0.1–10	Boyle and Jonasson (1973)
Gypsum: Carpathian Foreland: Ukraine	0.07–3.36	Boyko, Kosmus and Gessler (2000)
Iron formation and iron-rich sediments: 45 samples	1–2900	Boyle and Jonasson (1973)
Iron ore (sedimentary): Kanker district: Central-east India	100	Pandey *et al.* (2006)
Ironstone: Eastern England, United Kingdom: 12 samples	37–400	Palumbo-Roe *et al.* (2005)
Limestone (Suwannee): Florida, USA: 306 analyses, 20 cores	<0.01–54.10	Price and Pichler (2006)

(continued overleaf)

Table 3.23 *(continued)*

Sedimentary rock	Arsenic concentrations (mg kg^{-1})[a]	Reference(s)
Limestone and dolostone (localized gold-bearing hydrothermal alteration): Roberts Mountain Formation: north-central Nevada, USA: 10 samples	3–121	Van Moort, Hotchkis and Pwa (1995)
Limestone and dolostone: Cobalt area, Ontario, Canada	< 1	Boyle and Jonasson (1973)
Limestone conglomerate: Macumber Formation: Walton area, Nova Scotia, Canada: 7 samples	1–8	Boyle and Jonasson (1973)
Limestone: Carpathian Foreland, Ukraine	0.27–25.7	Boyko, Kosmus and Gessler (2000)
Limestone: Keno-Hill-Galena Hill area, Yukon, Canada: 3 samples	5–10	Boyle and Jonasson (1973)
Limestone: Lake travertines: Tuscany, Italy: 7 samples	127–257	Di Benedetto *et al.* (2006)
Limestone: Lower Illinois River Basin, Illinois, USA: 2 samples	4; 9.3	Warner (2001)
Limestone: Lukhumi deposit, Central Caucasus, Georgia: 54 samples	27	Zhabin *et al.* (1990)
Limestone: Macumber Formation: Walton area, Nova Scotia, Canada: 5 samples	< 1–10	Boyle and Jonasson (1973)
Limestone: Some localized gold-bearing hydrothermal alteration: Popovich Formation: north-central Nevada, USA: 13 samples	7–413	Van Moort, Hotchkis and Pwa (1995)
Manganese ore (sedimentary): Kanker district: Central-east India	150	Pandey *et al.* (2006)
Mill Creek Shale: Erie, Pennsylvania, USA	8.8	Murnock (2002)
Mudstone (limy): Rodeo Creek Formation: Gold quarry: north-central Nevada, USA: 2 samples	23; 84	Van Moort, Hotchkis and Pwa (1995)
Mudstone: Japan: Niigata Plain	8–12	Kubota, Ishiyama and Yokota (2000)
Native sulfur: Carpathian Foreland: Ukraine	0.02–0.04	Boyko, Kosmus and Gessler (2000)
Oil 'shale': Green River Formation, USA	48.0 ± 0.7	Fruchter *et al.* (1980)
Oil 'shale': Green River Formation, Wyoming, USA (Duplicate analyses with two different analytical methods)	72.3 ± 0.68 (graphite furnace atomic absorption), 68 ± 1.2 (inductively coupled argon plasma spectrometry)	Jaganathan, Mohan and Zingaro (1986)
Paluxy Sandstone (> 90 % quartz): East Texas, USA	< 1.0	Abraham (1998)

PawPaw Formation: Calcareous marl: East Texas, USA	18	Abraham (1998)
Pepper Shale: East Texas, USA	6.0	Abraham (1998)
Pierre Shale: South Dakota, USA: 29 samples	16 ± 8	Callender and Robbins (1993)
Phosphatic pebbles and cements: Dover Sandstone, Pensacola Mountains, Antarctica	80	Altschuler (1980)
Phosphorite (typical values)	0.4–188	Mandal and Suzuki (2002)
Phosphorite: 41 samples	3.4–100	Boyle and Jonasson (1973)
Phosphorite: Belkinsk Altai Sayan, Siberia, former USSR	5.7	Altschuler (1980)
Phosphorite: Bone Valley Formation, Florida, USA	12	Altschuler (1980)
Phosphorite: Brazil: 3 samples	62.4–100.3	Mirlean et al. (2003)
Phosphorite: Duwi Formation: Egypt: 18 samples	0.1–2.1	Baioumy (2005)
Phosphorite: Karatau, Kazakhstan	15	Altschuler (1980)
Phosphorite: Qulad Abdoun Basin, Morrocco	15	Altschuler (1980)
Phosphorite: Phosphoria Formation: Montana, Idaho, Wyoming, and Utah, USA: 60 samples	< 10–200, 40 (average)	Gulbrandsen (1966)
Phosphorite: southeast Jordan: 3 beds	7 (average); 9 (average); 7 (average)	Al-Hwaiti, Matheis and Saffarini (2005)
Phosphorite: Tamalyk Krasnoyarsk, Siberia, former USSR	7.8	Altschuler (1980)
Sandstone (typical values)	0.6–9	Mandal and Suzuki (2002)
Sandstone and conglomerate: Wolfville Formation: 7 samples	2–27	Boyle and Jonasson (1973)
Sandstone, arkose, and conglomerates: 15 samples	0.6–120	Boyle and Jonasson (1973)
Sandstone: Cheverie Formation: Walton area, Nova Scotia, Canada: 26 samples	< 1–32	Boyle and Jonasson (1973)
Sandstone: Graywacke and arkose: Cobalt area, Ontario, Canada	3	Boyle and Jonasson (1973)
Sandstone: Japan: Niigata Plain	3–5	Kubota, Ishiyama and Yokota (2000)
Sandstone: Typical average	1	Turekian and Wedepohl (1961), Faure (1998), 514
Shale (calcareous) and limestones: Kaimichi Formation: East Texas, USA	6.0	Abraham (1998)
Shale (calcareous) and thin limestones: Lake Waco Formation: East Texas, USA	8.0	Abraham (1998)

(continued overleaf)

Table 3.23 *(continued)*

Sedimentary rock	Arsenic concentrations (mg kg^{-1})[a]	Reference(s)
Shale (carbonaceous) and argillite: Horton Bluff Formation: Walton area, Nova Scotia, Canada: 38 samples	1–45	Boyle and Jonasson (1973)
Shale (carbonaceous, nonmarine): north-central USA: 14 samples	3–12	Tourtelot (1964)
Shale (copper-rich): Germany: Lake Süßer See area	402	Becker et al. (2001)
Shale (sandy): Tennycape Formation: Walton area, Nova Scotia, Canada: 2 samples	< 1; 2	Boyle and Jonasson (1973)
Shale (unaltered carbonaceous): Central Caucasus, Georgia: 36 samples	30	Zhabin et al. (1990)
Shale and argillite: 116 samples	0.3–500	Boyle and Jonasson (1973)
Shale and argillite: Cheverie Formation: Walton area, Nova Scotia, Canada: 17 samples	< 1–50	Boyle and Jonasson (1973)
Shale and clay: Kanker district: Central-east India	12	Pandey et al. (2006)
Shale and claystone (near shore depositional environment): Trancas and Soyatal Formations: Mexico	10–360	Armienta et al. (2001)
Shale and claystone (near shore marine): north-central USA: 32 samples	4–25	Tourtelot (1964)
Shale and claystone (offshore depositional environment): Tamaulipas Formations: Mexico	15–125	Armienta et al. (2001)
Shale and claystone (offshore marine): north-central USA: 53 samples	3–490	Tourtelot (1964)
Shale: Lower Illinois River Basin, Illinois, USA: 3 samples	5–55	Warner (2001)
Shale: New Albany Shale: Illinois Basin, Illinois, USA	4–45	Frost, Zierath and Shimp (1985)
Shale: northern Labrador, Canada: 75 samples	< 3–500	Boyle and Jonasson (1973)
Shale: typical average	13	Turekian and Wedepohl (1961); Faure (1998), 514
Shale: Weches Formation: East Texas, USA: 8 samples	100	Ledger and Judy (2003)
Siltstone (marly): Lena Group, Cantabrian Mountains, Spain: 12 samples	89 ± 94	Crespo et al. (2000)
Siltstone: Lena Group, Cantabrian Mountains, Spain: 18 samples	142 ± 236	Crespo et al. (2000)
St. Peter Sandstone: eastern Wisconsin, USA	< 10–500	Gotkowitz et al. (2001)
Walnut Creek Shale: Erie, Pennsylvania, USA	15.5	Murnock (2002)

[a]Single values denote single analyses or average concentrations sometimes with 1 SD. Values separated by commas denote duplicate analyses of the same sample. Analyses involving different samples are separated by semicolons. Low and high values are separated by hyphens. Available information on sample depths, analytical techniques, and composition is listed.

deposits occur in the northern Rocky Mountains of the USA (Green River Formation), Australia (the Rundle Shales of Queensland), South Africa, China, Europe, and elsewhere (North, 1990), 97; (Jones, Chapman and Jung, 1990). The average arsenic content of the oil shales of the Green River Formation, Wyoming, USA, is relatively high (48 mg kg^{-1}) when compared with most other sedimentary rocks (Table 3.23). Jaganathan, Mohan and Zingaro (1986) found that the majority (approximately 65 %) of the arsenic in a sample of Green River oil shale was located in skutterudite ((Co, Fe, Ni)As) and safflorite ((Co, Fe)As$_2$) (Jaganathan, Mohan and Zingaro, 1986). As(V), MMA(V), DMA(V), phenylarsonic acid ($C_6H_5AsO(OH)_2$), and unspecified organoarsenicals have also been detected in methanol extracts of Green River oil shales (Cullen and Reimer, 1989), 758. Further discussions on the environmental impacts of oil shale utilization are in Chapter 5.

3.24.4 Other sedimentary rocks

Sandstones primarily consist of sand-sized (63 μm to 2 mm in diameter) sediment grains that are usually cemented with calcite, quartz, or hematite. When compared with sandstones, conglomerates are coarser-grained rocks that have abundant well-rounded *gravels* (>2 mm). Most arsenic-bearing minerals are simply too fine grained or weather away too quickly to accumulate in sands and gravels. Sandstones and conglomerates generally consist of minerals that resist weathering, such as quartz and some feldspars. Quartz and feldspars typically contain very little arsenic and, not surprisingly, the average arsenic concentration of a sandstone is usually very low (about 0.5–1.0 mg kg^{-1}) (Smedley and Kinniburgh, 2002), 530; (Matschullat, 2000), 299.

If substantial arsenic is present in a sandstone or conglomerate, it may occur in hematite or other iron (oxy)(hydr)oxide cements or coatings on mineral grains. Arsenic may also be enriched in sandstones, conglomerates, and other *clastic sedimentary rocks* if hydrothermal or other secondary sulfide minerals are present (e.g. St. Peter Sandstone, (Gotkowitz *et al.*, 2001); arsenian pyrite cement in the Marshall Sandstone of the Michigan Basin, USA, (Kolker *et al.*, 2000; Szramek, Walter and McCall, 2004)). Sections of the St. Peter Sandstone in eastern Wisconsin, USA, are enriched in sulfide minerals and contain about 500 mg kg^{-1} of arsenic. In contrast, unmineralized portions typically have <10 mg kg^{-1} (Gotkowitz *et al.*, 2001). If present in sedimentary rocks, hydrothermal sulfide deposits often tend to concentrate in veins and faults more than the rock matrices.

Carbonate sedimentary rocks largely consist of calcite, dolomite, and/or other carbonate minerals. They often form from sediments in tropical areas, where the debris of calcite- and *aragonite*-bearing organisms are plentiful and silicate-rich sediments are scarce (e.g. the modern Bahamas). Limestones and dolostones are the most common carbonate rocks. Calcite is the dominant mineral in limestones, whereas dolomite is prevalent in dolostones. Although As(III) may partially substitute for carbon in the carbonate of calcite (Di Benedetto *et al.*, 2006; Fernández-Martínez *et al.*, 2006), most limestones unaltered by hydrothermal or other fluids contain very little arsenic. Pure dolomite, like quartz, is also ineffective in sorbing arsenic when compared with iron (oxy)(hydr)oxides (Thornburg and Sahai, 2004). On the average, carbonate sedimentary rocks contain only about 1.0–1.5 mg kg^{-1} of arsenic ((Matschullat, 2000), 299; Table 3.23).

Some limestones and dolostones are relatively rich in arsenic because they contain significant hydrothermal or diagenetic sulfide minerals. In particular, the Suwannee Limestone of Florida, USA, contains up to 54 mg kg^{-1} of arsenic (Table 3.23). Almost all of the arsenic is associated with diagenetic pyrite. The pyrites typically contain 100–11 200 mg kg^{-1} of arsenic (an average of 2300 mg kg^{-1} for 25 samples) (Price and Pichler, 2006).

Salt deposits or *evaporites* precipitate from evaporating seawater that becomes trapped in semi-isolated marine basins. Salty desert lakes, such as Great Salt Lake, Utah, USA, or those in Death Valley, California, USA, are also sites of evaporite deposition. Common salt minerals include halite (NaCl), sylvite (KCl), anhydrite, and gypsum. Elemental sulfur (S^0) also occurs in some subsurface salt deposits and may result from the microbial reduction of anhydrite (Blatt, Middleton and Murray, 1980), 564. Although As(V) may partially substitute for sulfate in gypsum and anhydrite (Fernández-Martínez *et al.*, 2006), most salt deposits contain 10 mg kg^{-1} or less of arsenic (Table 3.23).

Phosphorites are sedimentary rocks that contain at least 15–20 wt % P_2O_5 (Boggs, 1995), 266. The phosphate in phosphorites primarily occurs as *apatite* ($Ca_5(PO_4)_3(F,Cl,OH)$). Typically, phosphorites chemically precipitate in deep, cold marine waters. Due to chemical similarities, arsenate may partially substitute for phosphate and the arsenic concentrations of phosphorites may exceed 100 mg kg^{-1} ((Matschullat, 2000), 299; Table 3.23). However, arsenic concentrations in some phosphorites (e.g. southeast Jordan) are relatively low (7–9 mg kg^{-1}) and the arsenic is mostly associated with clay and carbonate minerals rather than phosphates (Al-Hwaiti, Matheis and Saffarini, 2005).

Iron formations (also called *banded iron formations*) are thin-layered, iron-rich, and chemically precipitated sedimentary and metamorphosed sedimentary rocks. They are found in Minnesota (USA), Canada, Brazil, Russia, Australia, South Africa, and elsewhere. Iron formations mostly formed between 3.8 and 1.8 billion years ago (Blatt, Middleton and Murray, 1980), 604. More than 1.8 billion years ago, the Earth's atmosphere lacked substantial O_2. The atmosphere was probably rich in nitrogen (as N_2) and CO_2. Under low-O_2 conditions, water-soluble Fe(II) readily weathered from rocks and dissolved in high concentrations in nearby ocean basins. As the Fe(II) circulated through the basins, it slowly oxidized to form water-insoluble Fe(III). The Fe(III) chemically precipitated as extensive (oxy)(hydr)oxide layers in the ocean sediments. During diagenesis, the layers converted to more stable hematite- and magnetite-rich iron formations. Once simple plants and other photosynthetic organisms extensively evolved and released O_2 into the atmosphere about 1.8 billion years ago, iron formations largely ceased to form. Fe(II) could no longer developed in large quantities and dissolved in ocean water. Today, sparsely water-soluble Fe(III) may accumulate in shallow marine environments to produce ironstones (Boggs, 1995), 261. Like other rocks and sediments, arsenic frequently sorbs onto hematite and other iron (oxy)(hydr)oxides in iron formations and ironstones (Table 3.23).

3.25   Metamorphic rocks

Most metamorphic rocks form from the heating of subsurface sedimentary, igneous, or previously existing metamorphic rocks at temperatures that range from approximately 200 °C to the onset of melting of silicate rocks. Depending on water content, rock chemistry, and pressure, most silicate rocks within 35 km of the surface begin to melt and form magmas at about 650–1100 °C (Hyndman, 1985), 511; (Winkler, 1979), 5. Metamorphic rocks may form at deep locations within the lower crust (*regional metamorphism*) or from heat or hot fluids radiating from nearby magmas (contact metamorphism) (Winkler, 1979), 2–3. *Cataclastic metamorphism* refers to the grinding of rocks in active faults (Winkler, 1979), 2. Unlike other forms of metamorphism, cataclastic metamorphism may occur well below 200 °C.

In metamorphosed sedimentary rocks, arsenic tends to occur in oxide and sulfide minerals (Bebout *et al.*, 1999), 69–70. Many metamorphic rocks simply inherit their arsenic from their precursor rocks. That is, unless arsenic-rich metamorphic fluids are introduced, *quartzites* metamorphosed from low-arsenic quartz-rich sandstones and marbles metamorphosed from low-arsenic limestones should have relatively little arsenic. In contrast, shales often contain more arsenic than sandstones and limestones (Table 3.23). Therefore, slates and *phyllites* that form from the metamorphism of shales should inherit at least some of the arsenic (Table 3.24).

Metamorphic rocks also obtain arsenic from hydrothermal or other fluids that precipitate arsenopyrite, arsenian pyrite, and other sulfide minerals before, during, or after metamorphism (Section 3.6; (Smedley

Table 3.24 *Arsenic in metamorphic rocks*

Metamorphic rock	Arsenic concentrations (mg kg^{-1})[a]	Reference(s)
Amphibolite (epidote): California, USA: 2 samples	0.30; 0.40	Bebout *et al.* (1999)
Amphibolite: California, USA: 3 samples	0.30–1.70	Bebout *et al.* (1999)
Amphibolite: Fortymile River Watershed, Alaska, USA: 6 samples	2–17	Crock *et al.* (1999)
Blueschist (epidote): California, USA: 4 samples	0.20–1.30	Bebout *et al.* (1999)
Blueschist (lawsonite): California, USA: 4 samples	4.80–19.80	Bebout *et al.* (1999)
Gneiss: 7 samples	0.5–4.1	Boyle and Jonasson (1973)
Gneiss: Broken Ridge, southern Kootenay terrane, British Columbia, Canada: 2 samples	1.8; 1005.6	Lett *et al.* (1998)
Gneiss: Fortymile River Watershed, Alaska, USA: 18 samples	2–4	Crock *et al.* (1999)
Gneiss: Malcantone watershed, Switzerland	27	Pfeifer *et al.* (2004)
Greenstone and amphibolite: 45 samples	0.4–45	Boyle and Jonasson (1973)
Greenstone: Keno-Hill-Galena Hill area, Yukon, Canada: 27 samples	2–25	Boyle and Jonasson (1973)
Hornfel: 2 samples	0.7; 11	Boyle and Jonasson (1973)
Marble: Fortymile River Watershed, Alaska, USA: 2 samples	4; 8.6	Crock *et al.* (1999)
Metabasalt (metamorphosed basalts): Bowen Island, British Columbia, Canada	10–1246 (especially high near arsenopyrite-bearing veins	Boyle, Turner and Hall (1998)
Metagabbro (metamorphosed gabbro): Fortymile River Watershed, Alaska, USA: 1 sample	4	Crock *et al.* (1999)
Metagraywacke: Fortymile River Watershed, Alaska, USA: 1 sample	3	Crock *et al.* (1999)
Metasedimentary rock (lawsonite–albite): California, USA: 8 samples	1.6–62.0	Bebout *et al.* (1999)
Quartzite: 4 samples	2.2–7.6	Boyle and Jonasson (1973)
Quartzite: Fortymile River Watershed, Alaska, USA: 9 samples	3–230	Crock *et al.* (1999)
Quartzite: Keno-Hill-Galena Hill area, Yukon, Canada: 39 samples	2–35	Boyle and Jonasson (1973)
Schist (hydrothermally altered): Burin Peninsula, Newfoundland, Canada: 5 samples	4–23	O'Brien, Dubé and O'Driscoll (1999)
Schist (mica): Malcantone watershed, Switzerland	920	Pfeifer *et al.* (2004)
Schist (mica): Mt. Mottarone, Italy	2.59 ± 0.52	Salvioli-Mariani, Toscani and Venturelli (2001)

(*continued overleaf*)

Table 3.24   *(continued*

Metamorphic rock	Arsenic concentrations $(mg\ kg^{-1})^a$	Reference(s)
Schist: Fortymile River Watershed, Alaska, USA: 11 samples	<0.1−5.3	Crock *et al.* (1999)
Schist: Harper area, southern Kootenay terrane, British Columbia, Canada: 2 samples	2.2; 2.5	Lett *et al.* (1998)
Schist: Keno-Hill-Galena Hill area, Yukon, Canada: 29 samples	2−15	Boyle and Jonasson (1973)
Schist: Otago Schist group: New Zealand	2−6 (unmineralized); 2−20 000 (mineralized)	Craw, Falconer and Youngson (2003)
Schist: Otago Schist group: Wangaloa coal mine area, southeast Otago, New Zealand: 9 samples	2−11	Black and Craw (2001)
Schist: Pramerkogel Formation (metamorphosed volcanic deposits): Eastern Alps, Austria	6 - ~2,000; 4000 near veins	Bojar *et al.* (2001)
Serpentinite: Northwest Himalayas	6−275	Hattori *et al.* (2005)
Skarn: Keno-Hill-Galena Hill area, Yukon, Canada: 6 samples	7.5−20	Boyle and Jonasson (1973)
Slate (Madiyi Formation): Hunan, China	1.4	Yang and Blum (1999aa)
Slate and phyllite: 75 samples	0.5−143	Boyle and Jonasson (1973)
Slate: Geum-Kwan area, South Korea: 2 samples	2.0; 4.0	Lee *et al.* (1998)
Slate: I-Won area, South Korea: 2 samples	2.5; 5.0	Lee *et al.* (1998)

^{a}Single values denote single analyses or average concentrations sometimes with 1 SD. Analyses involving different samples are separated by semicolons. Low and high values are separated by hyphens.

and Kinniburgh, 2002), 530). The slates of the Madiyi Formation of Hunan, China, are examples of metamorphic rocks that contain hydrothermal arsenic deposits (Yang and Blum, 1999aa).

The arsenic concentrations of metamorphic rocks generally decrease with increased metamorphism (Ryan *et al.*, 1996), 265. As temperatures increase, more arsenic volatilizes out of the rocks. Specifically, metamorphic rocks that formed at <45 km in an ancient subduction zone in California lost about 80−85 % of their arsenic as metamorphic temperatures and pressures increased from about 275 °C and .5 kilobars (kb) to approximately 750 °C and 12 kb (Bebout *et al.*, 1999). As discussed in Section 3.6.2, the volatilization of arsenic during metamorphism may be important in transferring the element from subduction zones back into the crust (Figure 3.2).

References

Abdel-Moati, A.R. .(1990) Speciation and behavior of arsenic in the Nile delta lakes. *Water Air and Soil Pollution*, **51**(1-2), 117–32.

Abdullah, M.I., Shiyu, Z. and Mosgren, K. (1995) Arsenic and selenium species in the oxic and anoxic waters of the Oslofjord, Norway. *Marine Pollution Bulletin*, **31**(1-3), 116–26.

Abraham, J. (1998) Spatial distribution of major and trace elements in shallow reservoir sediments: an example from Lake Waco, Texas. *Environmental Geology*, **36**(3-4), 349–63.

Acero, P., Ayora, C., Torrentó, C. and Nieto, J.-M. (2006) The behavior of trace elements during schwertmannite precipitation and subsequent transformation into goethite and jarosite. *Geochimica et Cosmochimica Acta*, **70**(16), 4130–39.

Aggett, J. and Aspell, A.C. (1979) Release of arsenic from geothermal sources in the Waikato catchment. *Trace Substances in Environmental Health*, **13**, 84–91.

Aggett, J. and O'Brien, G.A. (1985) Detailed model for the mobility of arsenic in lacustrine sediments based on measurements in Lake Ohakuri. *Environmental Science and Technology*, **19**(3), 231–38.

Aichele, S. (2004) *Arsenic, nitrate, and chloride in groundwater, Oakland County, Michigan*, U.S. Geological Survey, Water Resources Division, Lansing, MI.

Aiuppa, A., Avino, R., Brusca, L. *et al.* (2006) Mineral control of arsenic content in thermal waters from volcano-hosted hydrothermal systems: Insights from island of Ischia and Phlegrean fields (Campanian volcanic province, Italy). *Chemical Geology*, **229**(4), 313–30.

Aiuppa, A., D'Alessandro, W., Federico, C. *et al.* (2003) The aquatic geochemistry of arsenic in volcanic groundwaters from southern Italy. *Applied Geochemistry*, **18**(9), 1283–96.

Akeredolu, F.A., Barrie, L.A., Olson, M.P. *et al.* (1994) The flux of anthropogenic trace metals into the Arctic from the mid-latitudes in 1979/80. *Atmospheric Environment*, **28**(8), 1557–72.

Akter, K. and Naidu, R. (2006) Arsenic speciation in the environment, in *Managing Arsenic in the Environment: From Soil to Human Health* (eds R. Naidu, E. Smith, G. Owens *et al.*), CSIRO Publishing, Collingwood, pp. 61–74.

Alaerts, G.J. and Khouri, N. (2004) Arsenic contamination and groundwater: mitigation strategies and policies. *Hydrogeology Journal*, **12**(1), 103–14.

Alam, M.B. and Sattar, M.A. (2000) Assessment of arsenic contamination in soils and waters in some areas of Bangladesh. *Water Science and Technology*, **42**(7-8), 185–92.

Al-Hwaiti, M., Matheis, G. and Saffarini, G. (2005) Mobilization, redistribution and bioavailability of potentially toxic elements in Shidiya phosphorites, southeast Jordan. *Environmental Geology*, **47**(3), 431–44.

Altschuler, Z.S. (1980) *The Geochemistry of Trace Elements in Marine Phosphorites, Part 1: Characteristic Abundances and Enrichment*, SEPM Special Publication 29, SEPM, pp. 19–30.

Amirbahman, A., Kent, D.B., Curtis, G.P. and Davis, J.A. (2006) Kinetics of sorption and abiotic oxidation of arsenic(III) by aquifer materials. *Geochimica et Cosmochimica Acta*, **70**(3), 533–47.

Anawar, H.M., Akai, J., Komaki, K. *et al.* (2003) Geochemical occurrence of arsenic in groundwater of Bangladesh: Sources and mobilization processes. *Journal of Geochemical Exploration*, **77**(2-3), 109–31.

Anawar, H.M., Akai, J. and Sakugawa, H. (2004) Mobilization of arsenic from subsurface sediments by effect of bicarbonate ions in groundwater. *Chemosphere*, **54**(6), 753–62.

Anders, E. and Ebihara, M. (1982) Solar-system abundances of the elements. *Geochimica et Cosmochimica Acta*, **46**(11), 2363–80.

Anders, E. and Grevesse, N. (1989) Abundances of the elements: meteoritic and solar. *Geochimica et Cosmochimica Acta*, **53**(1), 197–214.

Anderson, L.C.D. and Bruland, K.W. (1991) Biogeochemistry of arsenic in natural waters: The importance of methylated species. *Environmental Science and Technology*, **25**(3), 420–27.

Apambire, W.B. and Hess, J.W. (2000) The aqueous geochemistry of fluoride in the upper regions of Ghana. "Abstracts with Programs. *The Geological Society of America*." with "*Geological Society of America Abstracts with Programs*", **32**(7), 7.

Apodaca, L.E., Driver, N.E. and Bails, J.B. (2000) Occurrence, transport, and fate of trace elements, blue river basin, Summit County, Colorado: an integrated approach. *Environmental Geology*, **39**(8), 901–13.

Appelo, C.A.J., Van Der Weiden, M.J.J., Tournassat, C. and Charlet, L. (2002) Surface complexation of ferrous iron and carbonate on ferrihydrite and the mobilization of arsenic. *Environmental Science and Technology*, **36**(14), 3096–3103.

Appleton, J.D., Williams, T.M., Orbea, H. and Carrasco, M. (2001) Fluvial contamination associated with arsenical gold mining in the Ponce Enríquez, Portovelo -Zaruma and Nambija areas, Ecuador. *Water Air and Soil Pollution*, **131**(1-4), 19–39.

Appleyard, S.J., Angeloni, J. and Watkins, R. (2006) Arsenic-rich groundwater in an urban area experiencing drought and increasing population density, Perth, Australia. *Applied Geochemistry*, **21**(1), 83–97.

Arai, Y., Lanzirotti, A., Sutton, S.R. *et al.* (2006) Spatial and temporal variability of arsenic solid-state speciation in historically lead arsenate contaminated soils. *Environmental Science and Technology*, **40**(3), 673–79.

Arai, Y. and Sparks, D.L. (2002) Residence time effects on arsenate surface speciation at the aluminum oxide-water interface. *Soil Science*, **167**(5), 303–14.

Aras, N.K., Zoller, W.H., Gordon, G.E. and Lutz, G.J. (1973) Instrumental photon activation analysis of atmospheric particulate material. *Analytical Chemistry*, **45**(8), 1481–90.

Archer, J., Hudson-Edwards, K.A., Preston, D.A. *et al.* (2005) Aqueous exposure and uptake of arsenic by riverside communities affected by mining contamination in the Río Pilcomayo basin, Bolivia. *Mineralogical Magazine*, **69**(5), 719–36.

Arehart, G.B. (1996) Characteristics and origin of sediment-hosted disseminated gold deposits: a review. *Ore Geology Reviews*, **11**(6), 383–403.

Armienta, M.A., Villaseñor, G., Rodriguez, R. *et al.* (2001) The role of arsenic-bearing rocks in groundwater pollution at Zimapán valley, México. *Environmental Geology*, **40**(4-5), 571–81.

Arnett, D. (1996) *Supernovae and Nucleosynthesis*, Princeton University Press, Princeton, NJ p. 598.

Arnórsson, S. (2003) Arsenic in surface- and up to 90 °C ground waters in a basalt area, N-Iceland: processes controlling its mobility. *Applied Geochemistry*, **18**(9), 1297–312.

Arthur, J.D., Cowart, J.B. and Dabous, A.A. (2000) Arsenic and uranium mobilization during aquifer storage and recovery in the Floridan aquifer system. Abstracts with Programs. *The Geological Society of America*, **32**(7), 356.

Ashley, P.M., Craw, D., Graham, B.P. and Chappell, D.A. (2003) Environmental mobility of antimony around mesothermal stibnite deposits, New South Wales, Australia and southern New Zealand. *Journal of Geochemical Exploration*, **77**(1), 1–14.

Åström, M. and Corin, N. (2000) Abundance, sources and speciation of trace elements in humus-rich streams affected by acid sulphate soils. *Aquatic Geochemistry*, **6**(3), 367–83.

Ayora, C. and Casas, J.M. (1986) Strata-bound As-Au mineralisation in pre-caradocian rocks from the Vall de Ribes, eastern Pyrenees, Spain. *Mineralium Deposita*, **21**(4), 278–87.

Ayotte, J.D., Nolan, B.T., Nuckols, J.R. *et al.* (2006) Modeling the probability of arsenic in groundwater in New England as a tool for exposure assessment. *Environmental Science and Technology*, **40**(11), 3578–85.

Azcue, J.M., Mudroch, A., Rosa, F. *et al.* (1995) Trace elements in water, sediments, porewater, and biota polluted by tailings from an abandoned gold mine in British Columbia, Canada. *Journal of Geochemical Exploration*, **52**(1-2), 25–34.

Azcue, J.M. and Nriagu, J.O. (1994) Arsenic: historical perspectives, in *Arsenic in the Environment: Part I: Cycling and Characterization* (ed. J.O. Nriagu), John Wiley & Sons, Ltd., New York, pp. 1–15.

Azcue, J.M. and Nriagu, J.O. (1995) Impact of abandoned mine tailings on the arsenic concentrations in Moira Lake, Ontario. *Journal of Geochemical Exploration*, **52**(1-2), 81–89.

Baioumy, H.M. (2005) Preliminary data on cadmium and arsenic geochemistry for some phosphorites in Egypt. *Journal of African Earth Sciences*, **41**(3), 266–74.

Ballantyne, J.M. and Moore, J.N. (1988) Arsenic geochemistry in geothermal systems. *Geochimica et Cosmochimica Acta*, **52**, 475–83.

Barbaris, B. and Betterton, E.A. (1996) Initial snow chemistry survey of the Mogollon Rim in Arizona. *Atmospheric Environment*, **30**(17), 3093–103.

Baron, D. and Palmer, C.D. (1996) Solubility of jarosite at 4-35 °C. *Geochimica et Cosmochimica Acta*, **60**(2), 185–95.

Baroni, F., Boscagli, A., Di Lella, L.A. *et al.* (2004) Arsenic in soil and vegetation of contaminated areas in southern Tuscany (Italy). *Journal of Geochemical Exploration*, **81**(1-3), 1–14.

Barrett, J., Ewart, D.K., Hughes, M.N. and Poole, R.K. (1993) Chemical and biological pathways in the bacterial oxidation of arsenopyrite. *FEMS Microbiology Reviews*, **11**(1-3), 57–62.

Barringer, J.L., Szabo, Z. and Barringer, T.H. (1998) Arsenic and mercury in soil, aquifer sediments, and ground water in the coastal plain of New Jersey. Abstracts with Programs. *The Geological Society of America*, **30**(7), 58.

Bar-Yosef, B., Chang, A.C. and Page, A.L. (2005) Mass balance modeling of arsenic processes in cropland soils. *Environmental Geochemistry and Health*, **27**(2), 177–84.

Bashkin, V.N. and Wongyai, K. (2002) Environmental fluxes of arsenic from lignite mining and power generation in northern Thailand. *Environmental Geology*, **41**(8), 883–88.

Beaton, A.P., Goodarzi, F. and Potter, J. (1991) The petrography, mineralogy and geochemistry of a Paleocene lignite from southern Saskatchewan, Canada. *International Journal of Coal Geology*, **17L**, 117–48.

Beauchemin, S. and Kwong, Y.T.J. (2006) Impact of redox conditions on arsenic mobilization from tailings in a wetland with neutral drainage. *Environmental Science and Technology*, **40**(20), 6297–303.

Bebout, G.E., Ryan, J.G., Leeman, W.P. and Bebout, A.E. (1999) Fractionation of trace elements by subduction-zone metamorphism–effect of convergent-margin thermal evolution. *Earth and Planetary Science Letters*, **171**(1), 63–81.

Becker, A., Klök, W., Friese, K. *et al.* (2001) Lake Süßer see as a natural sink for heavy metals from copper mining. *Journal of Geochemical Exploration*, **74**(1-3), 205–17.

Bednar, A.J., Garbarino, J.R., Ranville, J.F. and Wildeman, T.R. (2005) Effects of iron on arsenic speciation and redox chemistry in acid mine water. *Journal of Geochemical Exploration*, **85**(2), 55–62.

Behr, R.S. and Beane, J.E. (2002) Arsenic Plumes Where the 'Source' Contains No Arsenic. Three Case Studies of Apparent Desorption of Naturally Occurring Arsenic. Arsenic in New England: A Multidisciplinary Scientific Conference, May 29-31, 2002, National Institute of Environmental Health Sciences, Superfund Basic Research Program. Manchester, NH.

Belkin, H.E. (1998) Mineralogy and speciation of arsenic in coals of the Upper Permian Longtan Formation, Guizhou Province, P. R. China. Abstracts with Programs. *The Geological Society of America*, **30**(7), 59.

Belzile, N. (1988) The fate of arsenic in sediments of the Laurentian Trough. *Geochimica et Cosmochimica Acta*, **52**(9), 2293–302.

Belzile, N., Lecomte, P. and Tessier, A. (1989) Testing readsorption of trace elements during partial chemical extractions of bottom sediments. *Environmental Science and Technology*, **23**(8), 1015–20.

Belzile, N. and Tessier, A. (1990) Interactions between arsenic and iron oxyhydroxides in lacustrine sediments. *Geochimica et Cosmochimica Acta*, **54**(1), 103–9.

Benn, D.I. and Evans, D.J.A. (1998) *Glaciers and Glaciation*, Arnold, London.

Bennett, B. and Dudas, M.J. (2003) Release of arsenic and molybdenum by reductive dissolution of iron oxides in a soil with enriched levels of native arsenic. *Journal of Environmental Engineering and Science*, **2**(4), 265–72.

Berg, M., Tran, H.C., Nguyen, T.C. *et al.* (2001) Arsenic contamination of groundwater and drinking water in Vietnam: a human health threat. *Environmental Science and Technology*, **35**(13), 2621–26.

Bexfield, L.M. and Plummer, L.N. (2003) Occurrence of arsenic in ground water of the Middle Rio Grande Basin, central New Mexico, in *Arsenic in Ground Water* (eds A.H. Welch and K.G. Stollenwerk), Kluwer Academic Publishers, Boston, MA, pp. 295–327.

Bezard, B., Drossart, P., Lellouch, E. *et al.* (1989) Detection of arsine in Saturn. *Astrophysical Journal*, **346**, 509–13.

Bhanarkar, A.D., Rao, P.S., Gajghate, D.G. and Nema, P. (2005) Inventory of SO_2, PM and toxic metals emissions from industrial sources in Greater Mumbai, India. *Atmospheric Environment*, **39**(21), 3851–64.

Bhattacharya, P., Ahmed, K.M., Hasan, M.A. (2006) Mobility of arsenic in groundwater in part of Brahmanbaria district, NE Bangladesh, in *Managing Arsenic in the Environment: From Soil to Human Health* (eds R. Naidu, E. Smith, G. Owens *et al.*), CSIRO Publishing, Collingwood, pp. 95–115.

Bhattacharya, S.K., Bhattacharya, P., Jacks, G. and Sracek, A. (2000) Mobilization of arsenic from the holocene sedimentary aquifers: evidences from the stable isotopic studies on the groundwater from bengal delta plains, in *Arsenic in Groundwater of Sedimentary Aquifers*, Pre-Congress Workshop, 31st International Geological Congress (eds P. Bhattacharya and A.H. Welch), Groundwater Arsenic Research Group, Division of Land and Water Resources, Royal Institute of Technology, Stockholm, pp. 22–26.

Bhattacharya, P. and Jacks, G. (2000) Arsenic contamination in groundwater of the sedimentary aquifers in the Bengal Delta Plains: a review, in *Arsenic in Groundwater of Sedimentary Aquifers*, Pre-Congress Workshop, 31st International Geological Congress (eds P. Bhattacharya and A.H. Welch), Groundwater Arsenic Research Group, Division of Land and Water Resources, Royal Institute of Technology, Stockholm, pp. 18–21.

Bhattacharyya, P., Ghosh, A.K., Chakraborty, A. *et al.* (2003) Arsenic uptake by rice and accumulation in soil amended with municipal solid waste compost. *Communications in Soil Science and Plant Analysis*, **34**(19-20), 2779–90.

Bhumbla, D.K. and Keefer, R.F. (1994) Arsenic mobilization and bioavailability in soils, in *Arsenic in the Environment: Part I: Cycling and Characterization* (ed. J.O. Nriagu), John Wiley & Sons, Ltd., New York, pp. 51–82.

Birkeland, P.W. (1984) *Soils and Geomorphology*, Oxford University Press, New York, p. 372.

Birkle, P., Cid Vázquez, A.L. and Fong Aguilar, J.L. (2005) Legal aspects and technical alternatives for the treatment of reservoir brines at the Activo Luna oilfield, Mexico. *Water Environment Research*, **77**(1), 68–77.

Birkle, P. and Merkel, B. (2000) Environmental impact by spill of geothermal fluids at the geothermal field of Los Azufres, Michoacan, Mexico. *Water Air and Soil Pollution*, **124**(3-4), 371–410.

Black, A. and Craw, D. (2001) Arsenic, copper and zinc occurrence at the Wangaloa coal mine, southeast Otago, New Zealand. *International Journal of Coal Geology*, **45**, 181–93.

Blatt, H., Middleton, G. and Murray, R. (1980) *Origin of Sedimentary Rocks*, 2nd edn, Prentice-Hall, Englewood Cliffs, NJ, p. 782.

Bodénan, F., Baranger, P., Piantone, P. *et al.* (2004) Arsenic behaviour in gold-ore mill tailings, Massif Central, France: hydrogeochemical study and investigation of *in situ* redox signatures. *Applied Geochemistry*, **19**(11), 1785–1800.

Boggs, S. Jr. (1995) *Principles of Sedimentology and Stratigraphy*, 2nd edn, Prentice Hall, Upper Saddle River, NJ.

Bojar, H.-P., Bojar, A.-V., Mogessie, A. *et al.* (2001) Evolution of veins and sub-economic ore at Strassegg, Paleozoic of Graz, eastern Alps, Austria: evidence for local fluid transport during metamorphism. *Chemical Geology*, **175**(3-4), 757–77.

Bone, S.E., Gonneea, M.E. and Charette, M.A. (2006) Geochemical cycling of arsenic in a coastal aquifer. *Environmental Science and Technology*, **40**(10), 3273–78.

Bonn, B.A. (1999) *Selected Elements and Organic Chemicals in Bed Sediment and Fish Tissue of the Tualatin River Basin, Oregon, 1992–96* Water-Resources Investigations Report 99-4107, U.S. Geological Survey, Portland, OR.

Borba, R.P., Figueiredo, B.R. and Matschullat, J. (2003) Geochemical distribution of arsenic in waters, sediments and weathered gold mineralized rocks from Iron Quadrangle, Brazil. *Environmental Geology*, **44**(1), 39–52.

Bostick, B.C., Chen, C. and Fendorf, S. (2004) Arsenite retention mechanisms within estuarine sediments of Pescadero, CA. *Environmental Science and Technology*, **38**(12), 3299–304.

Bostick, B.C. and Fendorf, S. (2003) Arsenite sorption on troilite (FeS) and pyrite (FeS2). *Geochimica et Cosmochimica Acta*, **67**(5), 909–21.

Bothe, J.V. Jr. and Brown, P.W. (1999) Arsenic immobilization by calcium arsenate formation. *Environmental Science and Technology*, **33**(21), 3806–811.

Boulet, M.P. and Larocque, A.C.L. (1996) A mineralogical and geochemical study of the Cleveland mine-tailings in New Mexico, U.S.A. Abstracts with Programs. *The Geological Society of America*, **28**(7), 518.

Bouška, V. and Pešek, J. (1999) Quality parameters of lignite of the North Bohemian Basin in the Czech Republic in comparison with the world average lignite. *International Journal of Coal Geology*, **40**(2-3), 211–35.

Bowell, R.J., Morley, N.H. and Din, V.K. (1994) Arsenic speciation in soil porewaters from the Ashanti mine, Ghana. *Applied Geochemistry*, **9**(1), 15–22.

Bowell, R.J., Warren, A., Minjera, H.A. and Kimaro, N. (1995) Environmental impact of former gold mining on the Orangi River, Serengeti N.P., Tanzania. *Biogeochemistry*, **28**(3), 131–60.

Boyko, T., Kosmus, V. and Gessler, V.V. (2000) Geochemistry of Se, Te and As in the Neogene sulfate-carbonate deposits in the Carpathian Foreland (in Russian). *Geologiya i Geokhimiya Goryuchikh Kopalin*, **3**, 52–61.

Boyle, R.W. and Jonasson, I.R. (1973) Geochemistry of arsenic and its use as indicator element in geochemical prospecting. *Journal of Geochemical Exploration*, **2**(3), 251–96.

Boyle, D.R., Turner, R.J.W. and Hall, G.E.M. (1998) Anomalous arsenic concentrations in groundwaters of an island community, Bowen Island, British Columbia. *Environmental Geochemistry and Health*, **20**(4), 199–212.

Brady, L.L. and Hatch, J.R. (1997) Chemical analyses of middle and upper pennsylvanian coals from southeastern Kansas. *Kansas Geological Survey Bulletin*, **240**(4), 43–57.

Brandvold, L. (2001) Arsenic in ground water in the Socorro Basin, New Mexico. *New Mexico Geology*, **23**(1), 2–8.

Brannon, J.M. and Patrick, W.H. Jr. (1987) Fixation, transformation, and mobilization of arsenic in sediments. *Environmental Science and Technology*, **21**(5), 450–59.

Breit, G.N., Foster, A.L., Sanzolone, R.F. *et al.* (2000) Arsenic cycling in eastern Bangladesh: the role of phyllosilicates. Abstracts with Programs. *The Geological Society of America*, **32**(7), 192.

Bright, D.A., Dodd, M. and Reimer, K.J. (1996) Arsenic in subarctic lakes influenced by gold mine effluent: the occurrence of organoarsenicals and 'hidden' arsenic. *Science of the Total Environment*, **180**(2), 165–82.

Brooks, R.R., Reeves, R.D., Yang, X.-H. *et al.* (1984) Elemental anomalies at the Cretaceous-Tertiary boundary, Woodside Creek, New Zealand. *Science*, **226**(4674), 539–42.

Brown, C.J. and Chute S.K. (2001) Arsenic in Bedrock Wells in Connecticut. *USGS Workshop on Arsenic in the Environment*, February 21–22, 2001, Denver, CO.

Buchanan, P.C., Zolensky, M.E. and Reid, A.M. (1997) Petrology of Allende dark inclusions. *Geochimica et Cosmochimica Acta*, **61**(8), 1733–43.

Buddhawong, S., Kuschk, P., Mattusch, J. *et al.* (2005) Removal of arsenic and zinc using different laboratory model wetland systems. *Engineering in Life Sciences*, **5**(3), 247–52.

Bundschuh, J., Bonorino, G., Viero, A.P. (2000) Arsenic and other trace elements in sedimentary aquifers in the Chaco-Pampean plain, in *Arsenic in Groundwater of Sedimentary Aquifers*, Pre-Congress Workshop, 31st International Geological Congress (eds P. Bhattacharya and A.H. Welch.), Groundwater Arsenic Research Group, Division of Land and Water Resources, Royal Institute of Technology, Stockholm, pp. 27–32.

Burbidge, E.M., Burbidge, G.R., Fowler. W.A. and Hoyle, F. (1957) Synthesis of the elements in stars. *Reviews of Modern Physics*, **29**(4), 547–650.

Burgess, W.G. and Pinto, L. (2005) Preliminary observations on the release of arsenic to groundwater in the presence of hydrocarbon contaminants in UK aquifers. *Mineralogical Magazine*, **69**(5), 887–96.

Burkel, R.S. and Stoll, R.C. (1999) Naturally occurring arsenic in sandstone aquifer water supply wells of north-eastern Wisconsin. *Ground Water Monitoring and Remediation*, **19**(2), 114–21.

Buschmann, J., Kappeler, A., Lindauer, U. *et al.* (2006) Arsenite and arsenate binding to dissolved humic acids: influence of pH, type of humic acid, and aluminum. *Environmental Science and Technology*, **40**(19), 6015–20.

Callender, E. and Robbins, J.A. (1993) Transport and accumulation of radionuclides and stable elements in a Missouri River reservoir. *Water Resources Research*, **29**(6), 1787–804.

Camm, G.S., Butcher, A.R., Pirrie, D. *et al.* (2003) Secondary mineral phases associated with a historic arsenic calciner identified using automated scanning electron microscopy: a pilot study from Cornwall, UK. *Minerals Engineering*, **16**(Suppl. 11), 1269–77.

Campbell, K.M., Malasarn, D., Saltikov, C.W. *et al.* (2006) Simultaneous microbial reduction of iron(III) and arsenic(V) in suspensions of hydrous ferric oxide. *Environmental Science and Technology*, **40**(19), 5950–55.

Campos, V. (2002) Arsenic in groundwater affected by phosphate fertilizers at São Paulo, Brazil. *Environmental Geology*, **42**(1), 83–87.

Cancès, B., Juillot, F., Morin, G. *et al.* (2005) XAS evidence of As(V) association with iron oxyhydroxides in a contaminated soil at a former arsenical pesticide processing plant. *Environmental Science and Technology*, **39**(24), 9398–405.

Canet, C., Prol-Ledesma, R.M., Rubio-Ramos, M.A. *et al.* (2003) Mn-Ba-Hg mineralization related to the shallow submarine hydrothermal vents of Bahia Concepcion, Baja California Sur, Mexico. Abstracts with Programs. *The Geological Society of America*, **35**(6), 579.

Cappuyns, V., Van Herreweghe, S., Swennen, R. *et al.* (2002) Arsenic pollution at the industrial site of Reppel-Bocholt (north Belgium). *Science of the Total Environment*, **295**(1-3), 217–40.

Capuano, R.M. and Moore, J.N. (1980) Hg and As soil geochemistry as a technique for mapping permeable structures over a hot-water geothermal system. Abstracts with Programs. *The Geological Society of America*, **12**(6), 269.

Carbonell-Barrachina, A.A., Rocamora, A., García-Gomis, C. *et al.* (2004) Arsenic and zinc biogeochemistry in pyrite mine waste from the Aznalcóllar environmental disaster. *Geoderma*, **122**(2-4 SPEC. IIS.), 195–203.

Carlson, L., Bigham, J.M., Schwertmann, U. *et al.* (2002) Scavenging of as from acid mine drainage by schwertmannite and ferrihydrite: a comparison with synthetic analogues. *Environmental Science and Technology*, **36**(8), 1712–19.

Carrillo-Chavez, A., Drever, J.I. and Martinez, M. (2000) Arsenic content and groundwater geochemistry of the San Antonio-El Triunfo, Carrizal and Los Planes aquifers in southernmost Baja California, Mexico. *Environmental Geology*, **39**(11), 1295–303.

Carrillo-Chavez, A., Gonzalez-Partida, E. and Morton-Bermea, O. (2002) Heavy metals distribution in rock, sediments, groundwater, mine tailings and leaching experiments in Mineral de Pozos historical mining site, central Mexico. Abstracts with Programs. *The Geological Society of America*, **34**(6), 415–16.

Casas, J.M., Rosas, H., Solé, M. and Lao, C. (2003) Heavy metals and metalloids in sediments from the Llobregat basin, Spain. *Environmental Geology*, **44**, 325–32.

Chagué-Goff, C., Rosen, M.R. and Eser, P. (1999) Sewage effluent discharge and geothermal input in a natural wetland, Tongariro delta, New Zealand. *Ecological Engineering*, **12**(1-2), 149–70.

Chaillou, G., Schäfer, J., Anschutz, P. *et al.* (2003) The behaviour of arsenic in muddy sediments of the Bay of Biscay (France). *Geochimica et Cosmochimica Acta*, **67**(16), 2993–3003.

Chakraborti, D., Rahman, M.M., Paul, K. *et al.* (2002) Arsenic calamity in the Indian subcontinent: What lessons have been learned? *Talanta*, **58**(1), 3–22.

Chakrapani, G.J. (2005) Major and trace element geochemistry in Upper Ganga River in the Himalayas, India. *Environmental Geology*, **48**(2), 189–201.

Chang, M.J., Walker, K., McDaniel, R.L. and Connell, C.T. (2005) Impaction collection and slurry sampling for the determination of arsenic, cadmium, and lead in sidestream cigarette smoke by inductively coupled plasma-mass spectrometry. *Journal of Environmental Monitoring*, **7**(12), 1349–54.

Chapman, C.R. (1999) Asteroids, in *The New Solar System* (eds J.K. Beatty, C.C. Petersen and A. Chaikin) 4th edn, Sky Publishing Corporation, Cambridge University Press, Cambridge, MA, pp. 337–50.

Chatterjee, A. and Mukherjee, A. (1999) Hydrogeological investigation of ground water arsenic contamination in south Calcutta. *Science of the Total Environment*, **225**(3), 249–62.

Chein, H., Hsu, Y.-D., Aggarwal, S.G. *et al.* (2006) Evaluation of arsenical emission from semiconductor and opto-electronics facilities in Hsinchu, Taiwan. *Atmospheric Environment*, **40**(10), 1901–7.

Chen, S.-L., Dzeng, S.R., Yang, M.-H. *et al.* (1994) Arsenic species in groundwaters of the Blackfoot disease area, Taiwan. *Environmental Science and Technology*, **28**(5), 877–81.

Chen, M., Ma, L.Q. and Harris, W.G. (1999) Baseline concentrations of 15 trace elements in Florida surface soils. *Journal of Environmental Quality*, **28**(4), 1173–81.

Chen, M., Ma, L.Q. and Harris, W.G. (2001) Distribution of Pb and As in soils at a shooting facility in central Florida. *Annual Proceedings Soil and Crop Science Society of Florida*, **60**, 15–20.

Chen, C.Y., Stemberger, R.S., Klaue, B. *et al.* (2000) Accumulation of heavy metals in food web components across a gradient of lakes. *Limnology and Oceanography*, **45**(7), 1525–36.

Chernoff, C.B. and Barton, M.D. (2001) Trace elements in black shale Fe-sulfides during diagenesis and metamorphism. Abstracts with Programs. *The Geological Society of America*, **33**(6), 129.

Chirenje, T., Ma, L.Q., Chen, M. and Zillioux, E.J. (2003) Comparison between background concentrations of arsenic in urban and non-urban areas of Florida. *Advances in Environmental Research*, **8**(1), 137–46.

Chirenje, T., Ma, L.Q., Clark, C. and Reeves, M. (2003) Cu, Cr and As distribution in soils adjacent to pressure-treated decks, fences and poles. *Environmental Pollution*, **124**(3), 407–17.

Chowdhury, T.R., Basu, G.K., Mandal, B.K. *et al.* (1999) Arsenic poisoning in the Ganges Delta. *Nature*, **401**(6753), 545–46, discussion 546.

Christensen, O.D., Capuano, R.M. and Moore, J.N. (1983) Trace–element distribution in an active hydrothermal system, Roosevelt Hot Springs Thermal Area, Utah. *Journal of Volcanology and Geothermal Research*, **16**(1-2), 99–129.

Christensen, V.G. and Juracek, K.E. (2001) Variability of metals in reservoir sediment from two adjacent basins in the central Great Plains. *Environmental Geology*, **40**(4-5), 470–81.

Clayton, C.A., Pellizzari, E.D., Whitmore, R.W. *et al.* (1999) National human exposure assessment survey (NHEXAS): distributions and associations of lead, arsenic and volatile organic compounds in EPA region 5. *Journal of Exposure Analysis and Environmental Epidemiology*, **9**(5), 381–92.

Cleverley, J.S. and Benning, L.G. (2003) From thermodata to models; a case study for arsenic in geothermal systems. Abstracts with Programs. *The Geological Society of America*, **35**(6), 47.

Clifford, D.A. and Ghurye, G.L. (2002) Metal-oxide adsorption, ion exchange, and coagulation-microfiltration for arsenic removal from water, in *Environmental Chemistry of Arsenic* (ed. W.T. Frankenberger Jr.), Marcel Dekker, New York, pp. 217–45.

Cline, J.S. (2001) Timing of gold and arsenic sulfide mineral deposition at the Getchell Carlin-type gold deposit, north-central Nevada. *Economic Geology*, **96**(1), 75–89.

Cobb, G.P., Abel, M.T., Rainwater, T.R. *et al.* (2006) Metal distributions in New Orleans following hurricanes Katrina and Rita: a continuation study. *Environmental Science and Technology*, **40**(15), 4571–77.

Coker, V.S., Gault, A.G., Pearce, C.I. *et al.* (2006) XAS and XMCD evidence for species-dependent partitioning of arsenic during microbial reduction of ferrihydrite to magnetite. *Environmental Science and Technology*, **40**(24), 7745–750.

Çolak, M., Gemici, U. and Tarcan, G. (2003) The effects of colemanite deposits on the arsenic concentrations of soil and ground water in Igdeköy-emet, Kütahya, turkey. *Water Air and Soil Pollution*, **149**(1-4), 127–43.

Coleman, S.L. and Bragg, L.J. (1990) *Distribution and Mode of Occurrence of Arsenic in Coal*. Geological Society of America Special Paper, p. 248.

Cooper, C.M. and Gillespie, W.B. Jr. (2001) Arsenic and mercury concentrations in major landscape components of an intensively cultivated watershed. *Environmental Pollution*, **111**, 67–74.

Courtin-Nomade, A., Bril, H., Neel, C. and Lenain, J.-F. (2003) Arsenic in iron cements developed within tailings of a former metalliferous mine — Enguialès, Aveyron, France. *Applied Geochemistry*, **18**(3), 395–408.

Craig, P.J., Eng, G. and Jenkins, R.O (2003) Occurrence and pathways of organometallic compounds in the environment — general considerations, in *Organometallic Compounds in the Environment*, 2nd edn (ed. P.J. Craig), John Wiley & Sons, Ltd, West Sussex, pp. 1–55.

Craig, J.R., Vaughan, D.J. and Skinner, B.J. (2001) *Resources of the Earth: Origin, Use, and Environmental Impact*, Prentice Hall, Upper Saddle River, NJ, p. 520.

Craw, D., Chappell, D., Nelson, M. and Walrond, M. (1999) Consolidation and incipient oxidation of alkaline arsenopyrite-bearing mine tailings, Macraes Mine, New Zealand. *Applied Geochemistry*, **14**(4), 485–98.

Craw, D., Falconer, D. and Youngson, J.H. (2003) Environmental arsenopyrite stability and dissolution: theory, experiment, and field observations. *Chemical Geology*, **199**(1-2), 71–82.

Craw, D., Koons, P.O. and Chappell, D.A. (2002) Arsenic distribution during formation and capping of an oxidised sulphidic minesoil, Macraes Mine, New Zealand. *Journal of Geochemical Exploration*, **76**(1), 13–29.

Craw, D., Koons, P.O., Horton, T. and Chamberlain, C.P. (2002) Tectonically driven fluid flow and gold mineralisation in active collisional orogenic belts: comparison between New Zealand and western Himalaya. *Tectonophysics*, **348**(1-3), 135–53.

Crespo, J.L., Moro, M.C., Fadón, O. *et al.* (2000) The Salamon gold deposit (Leon, Spain). *Journal of Geochemical Exploration*, **71**(2), 191–208.

Criaud, A. and Fouillac, C. (1989) The distribution of arsenic (III) and arsenic (V) in geothermal waters: examples from the massif central of France, the island of Dominica in the Leeward islands of the Caribbean, the valles Caldera of New Mexico, USA, and southwest Bulgaria. *Chemical Geology*, **76**(3-4), 259–69.

Crock, J.G., Gough, L.P., Wanty, R.B. *et al.* (1999) *Regional Geochemical Results from the Analyses of Rock, Water, Soil, Stream Sediment, and Vegetation Samples — Fortymile River Watershed, East-Central, Alaska, 1998 Sampling*, U.S. Geological Survey, Reston, Virginia, Open-File Report 00-511.

Crowley, S.S., Warwick, P.D., Ruppert, L.F. and Pontolillo, J. (1997) The origin and distribution of HAPs elements in relation to maceral composition of the A1 lignite bed (Paleocene, Calvert Bluff Formation, Wilcox Group), Calvert mine area, east-central Texas. *International Journal of Coal Geology*, **34**, 327–43.

Crowley, S.S., Warwick, P.D., SanFilipo, J.R. *et al.* (1998) Quality of lignites in three areas of the Gulf Coast province. Abstracts with Programs. *The Geological Society of America*, **30**(7), 175.

Crusius, J., Pieters, R., Leung, A. *et al.* (2003) Tale of two pit lakes: initial results of a three-year study of the main zone and waterline pit lakes near Houston, British Columbia, Canada. *Mining Engineering*, **55**(2), 43–48.

Cruz, R., Lázaro, I., González, I. and Monroy, M. (2005) Acid dissolution influences bacterial attachment and oxidation of arsenopyrite. *Minerals Engineering*, **18**, 1024–31.

Culbertson, C.W., Moll, D.M., Backer, L.C. *et al.* (2002) A Pilot Study of Arsenic Speciation in Domestic Well-Water Supplies in Maine. Arsenic in New England: A Multidisciplinary Scientific Conference, May 29–31, 2002, National Institute of Environmental Health Sciences, Superfund Basic Research Program, Manchester, NH.

Cullen, W.R. and Reimer, K.J. (1989) Arsenic speciation in the environment. *Chemical Reviews*, **89**(4), 713–64.

Cummings, D.E., March, A.W., Bostick, B. *et al* (2000) Evidence for microbial Fe(III) reduction in anoxic, mining-impacted lake sediments (Lake Coeur d'Alene, Idaho). *Applied and Environmental Microbiology*, **66**(1), 154–62.

Cuny, D., Davranche, L., Thomas, P. *et al.* (2004) Spatial and temporal variations of trace element contents in Xanthoria parietina thalli collected in a highly industrialized area in northern France as an element for a future epidemiological study. *Journal of Atmospheric Chemistry*, **49**(1-3), 391–401.

Cutter, G.A. and Cutter, L.S. (2006) Biogeochemistry of arsenic and antimony in the North Pacific Ocean. *Geochemistry, Geophysics, Geosystems*, **7**(5), 12.

Cutter, G.A., Cutter, L.S., Featherstone, A.M. and Lohrenz, S.E. (2001) Antimony and arsenic biogeochemistry in the western Atlantic Ocean. *Deep-Sea Research Part II-Topical Studies in Oceanography*, **48**(13), 2895–915.

Dai, S., Ren, D., Tang, Y. *et al.* (2005) Concentration and distribution of elements in Late Permian coals from western Guizhou Province, China. *International Journal of Coal Geology*, **61**(1-2), 119–37.

Dai, S., Zeng, R. and Sun, Y. (2006) Enrichment of arsenic, antimony, mercury, and thallium in a Late Permian anthracite from Xingren, Guizhou, southwest China. *International Journal of Coal Geology*, **66**(3), 217–26.

Dalrymple, G.B. (1991) *The Age of the Earth*, Stanford University Press, Stanford, CA, p. 474.

Daniele, L. (2004) Distribution of arsenic and other minor trace elements in the groundwater of Ischia Island (southern Italy). *Environmental Geology*, **46**(1), 96–103.

Davis, A., Bellehumeur, T., Hunter, P. *et al.* (2006) The nexus between groundwater modeling, pit lake chemogenesis and ecological risk from arsenic in the Getchell main pit, Nevada, U.S.A. *Chemical Geology*, **228**(1-3 Special Issue), 175–96.

Davis, A., De Curnou, P. and Eary, L.E. (1997) Discriminating between sources of arsenic in the sediments of a tidal waterway, Tacoma, Washington. *Environmental Science and Technology*, **31**(7), 1985–91.

Davis, A., Kempton, J.H., Nicholson, A. and Yare, B. (1994) Groundwater transport of arsenic and chromium at a historical tannery, Woburn, Massachusetts, USA. *Applied Geochemistry*, **9**(5), 569–82.

Deacon, J.R. and Driver, N.E. (1999) Distribution of trace elements in streambed sediment associated with mining activities in the Upper Colorado River Basin, Colorado, USA, 1995–96. *Archives of Environmental Contamination and Toxicology*, **37**(1), 7–18.

Deb, M.K., Thakur, M., Mishra, R.K. and Bodhankar, N. (2002) Assessment of atmospheric arsenic level in airborne dust particulates of an urban city of central India. *Water Air and Soil Pollution*, **140**(1-4), 57–71.

DeLemos, J.L., Bostick, B.C., Renshaw, C.E. *et al.* (2006) Landfill-stimulated iron reduction and arsenic release at the Coakley Superfund Site (NH). *Environmental Science and Technology*, **40**(1), 67–73.

Delgado-Morales, W., Zingaro, R.A. and Mohan, M.S. (1994) Analysis and removal of arsenic from natural-gas using potassium peroxydisulfate and polysulfide absorbents. *International Journal of Environmental Analytical Chemistry*, **54**(3), 203–20.

Dellwig, O., Böttcher, M.E., Lipinski, M. and Brumsack, H.-J. (2002) Trace metals in Holocene coastal peats and their relation to pyrite formation (NW Germany). *Chemical Geology*, **182**(2-4), 423–42.

Delsemme, A. (1998) *Our Cosmic Origins: from the Big Bang to the Emergence of Life and Intelligence*, Cambridge University Press, Cambridge, p. 322.

De Mora, S., Sheikholeslami, M.R., Wyse, E. *et al.* (2004) An assessment of metal contamination in coastal sediments of the Caspian Sea. *Marine Pollution Bulletin*, **48**(1-2), 61–77.

De Temmerman, L., Vanongeval, L., Boon, W. *et al.* (2003) Heavy metal content of arable soils in northern Belgium. *Water Air and Soil Pollution*, **148**(1-4), 61–76.

Di Benedetto, F., Costagliola, P., Benvenuti, M. *et al.* (2006) Arsenic incorporation in natural calcite lattice: evidence from electron spin echo spectroscopy. *Earth and Planetary Science Letters*, **246**(3-4), 458–65.

Diamond, M.L. (1995) Application of a mass balance model to assess in-place arsenic pollution. *Environmental Science and Technology*, **29**(1), 29–42.

Diawara, D.M., Litt, J.S., Unis, D. *et al.* (2006) Arsenic, cadmium, lead, and mercury in surface soils, Pueblo, Colorado: implications for population health risk. *Environmental Geochemistry and Health*, **28**(4), 297–315.

Diehl, S.F., Goldhaber, M.B. and Hatch, J.R. (2004) Modes of occurrence of mercury and other trace elements in coals from the warrior field, Black Warrior Basin, northwestern Alabama. *International Journal of Coal Geology*, **59**(3-4), 193–208.

Ding, Z., Zheng, B., Long, J. *et al.* (2001) Geological and geochemical characteristics of high arsenic coals from endemic arsenosis areas in southwestern Guizhou Province, China. *Applied Geochemistry*, **16**(11-12), 1353–60.

Dixit, S. and Hering, J.G. (2003) Comparison of arsenic(V) and arsenic(III) sorption onto iron oxide minerals: Implications for arsenic mobility. *Environmental Science and Technology*, **37**(18), 4182–89.

Dixit, S. and Hering, J.G. (2006) Sorption of Fe(II) and As(III) on goethite in single- and dual-sorbate systems. *Chemical Geology*, **228**(1-3 Special Issue), 6–15.

Doherty, K.A. and Hon, R. (2003) Migration pathways of arsenic in high arsenic zones in central Massachusetts. Abstracts with Programs. *The Geological Society of America*, **35**(3), 33.

Domènech, C., De Pablo, J. and Ayora, C. (2002) Oxidative dissolution of pyritic sludge from the Aznalcóllar mine (SW Spain). *Chemical Geology*, **190**(1-4), 339–53.

Donahue, R. and Hendry, M.J. (2003) Geochemistry of arsenic in uranium mine mill tailings, Saskatchewan, Canada. *Applied Geochemistry*, **18**(11), 1733–50.

Donkor, A.E., Bonzongo, J.-C.J., Nartey, V.K. and Adotey, D.K. (2005) Heavy metals in sediments of the gold mining impacted Pra River Basin, Ghana, West Africa. *Soil and Sediment Contamination*, **14**, 479–503.

Doušová, B., Koloušek, D., Kovanda, F. *et al.* (2005) Removal of As(V) species from extremely contaminated mining water. *Applied Clay Science*, **28**(1-4 Special Issue), 31–42.

Dove, P.M. and Rimstidt, J.D. (1985) The solubility and stability of scorodite, $FeAsO_4.2H_2O$. *American Mineralogist*, **70**(7-8), 838–44.

Doyle, M.O. and Otte, M.L. (1997) Organism-induced accumulation of iron, zinc and arsenic in wetland soils. *Environmental Pollution*, **96**(1), 1–11.

Drever, J.I. (1997) *The Geochemistry of Natural Waters: Surface and Groundwater Environments.*, Prentice Hall, Upper Saddle River, NJ, p. 436.

Driscoll, D.G. and Hayes T.S. (1995) *Arsenic Loads in Spearfish Creek, Western South Dakota, Water Years 1989–91* U.S. Geological Survey Water-Resources Investigations Report 95-4080, U.S. Geological Survey, Rapid City.

Duker, A.A., Carranza, E.J.M. and Hale, M. (2005) Spatial relationship between arsenic in drinking water and Mycobacterium ulcerans infection in the Amansie West district, Ghana. *Mineralogical Magazine*, **69**(5), 707–17.

Dumouchelle, D.H. (1998) *Selected Ground-water-quality Data of the Lockport Dolomite in Darke, Miami, Montgomery, and Preble Counties, Ohio.* U.S. Geological Survey Open File Report 98-655, U.S. Geological Survey, Reston, VA.

Dunbar, N.W., Chapin, C.E. and Chapin, D.J. (1995) Arsenic enrichment during potassium metasomatism and hydrothermal processes in the Socorro, NM, area; implications for tracing ground-water flow. *New Mexico Geology*, **17**(2), 26–27.

Duquesne, K., Lebrun, S., Casiot, C. *et al.* (2003) Immobilization of arsenite and ferric iron by Acidithiobacillus ferrooxidans and its relevance to acid mine drainage. *Applied and Environmental Microbiology*, **69**(10), 6165–73.

Durant, J.L., Ivushkina, T., MacLaughlin, K. *et al.* (2004) Elevated levels of arsenic in the sediments of an urban pond: sources, distribution and water quality impacts. *Water Research*, **38**(13), 2989–3000.

Earman, S. and Hershey, R.L. (2004) Water quality impacts from waste rock at a Carlin-type gold mine, Elko County, Nevada. *Environmental Geology*, **45**(8), 1043–53.

Eatough, D.J., Eatough, M., Lewis, L.J. *et al.* (1996) Apportionment of sulfur oxides at canyonlands during the winter of 1990 — II. fingerprints of emissions from point and regional sources impacting canyonlands. *Atmospheric Environment*, **30**(2), 283–94.

Eby, G.N. (2004) *Principles of Environmental Geochemistry*, Brooks/Cole-Thomson Learning, Pacific Grove, CA, p. 514.

Edward Aul and Associates, Incorporated and E. H. Pechan and Associates (1993) *Emission Factor Documentation for AP-42 Section 11 Waste Oil Combustion*, Office of Air Quality Planning and Standards, Office of Air and Radiation, U.S. Environmental Protection Agency, Research Triangle Park.

Elbaz-Poulichet, F., Nagy, A. and Cserny, T. (1997) The distribution of redox sensitive elements (U, As, Sb, V and Mo) along a river-wetland-lake system (Balaton region, Hungary). *Aquatic Geochemistry*, **3**(3), 267–82.

Entsch, B., Boto, K.G., Sim, R.G. and Wellington, J.T. (1983) Phosphorus and nitrogen in coral reef sediments. *Limnology and Oceanography*, **28**(3), 465–76.

Erickson, M.L. and Barnes, R.J. (2005) Glacial sediment causing regional-scale elevated arsenic in drinking water. *Ground Water*, **43**(6), 796–805.

Eshete, D.W. and Chyi, L.L. (2003) Source identification and natural attenuation of arsenic contaminated groundwater in northeastern Ohio. Abstracts with Programs. *The Geological Society of America*, **35**(6), 565.

Eskenazy, G.M. (1995) Geochemistry of arsenic and antimony in Bulgarian coals. *Chemical Geology*, **119**(1-4), 239–54.

Evangelou, V.P., Seta, A.K. and Holt, A. (1998) Potential role of bicarbonate during pyrite oxidation. *Environmental Science and Technology*, **32**(14), 2084–91.

Farag, A.M., Woodward, D.F., Goldstein, J.N. *et al.* (1998) Concentrations of metals associated with mining waste in sediments, biofilm, benthic macroinvertebrates, and fish from the Coeur d'Alene River Basin, Idaho. *Archives of Environmental Contamination and Toxicology*, **34**(2), 119–27.

Farinha, M.M., Šlejkovec, Z., van Elteren, J.T. *et al.* (2004) Arsenic speciation in lichens and in coarse and fine airborne particulate matter by HPLC-UV-HG-AFS. *Journal of Atmospheric Chemistry*, **49**(1-3), 343–53.

Faure, G. (1998) *Principles and Applications of Geochemistry*, 2nd edn, Prentice Hall, Upper Saddle River, NJ, 600 pp.

Faure, K. and Brathwaite, R.L. (2006) Mineralogical and stable isotope studies of gold-arsenic mineralisation in the Sams Creek peralkaline porphyritic granite, South Island, New Zealand. *Mineralium Deposita*, **40**(8), 802–27.

Feely, R.A., Trefey, J.H., Massoth, G.J. and Metz, S. (1991) A comparison of the scavenging of phosphorus and arsenic form seawater by hydrothermal iron oxyhydroxides in the Atlantic and Pacific Oceans. *Deep-Sea Research*, **38**(6), 617–23.

Feng, X., Hong, Y., Hong, B. and Ni, J. (2000) Mobility of some potentially toxic trace elements in the coal of Guizhou, China. *Environmental Geology*, **39**(3-4), 372–77.

Fernández-Martínez, A., Román-Ross, G., Cuello, G.J. *et al.* (2006) Arsenic uptake by gypsum and calcite: modelling and probing by neutron and X-ray scattering. *Physica B: Condensed Matter*, **385-386**, 935–37.

Fetter, C.W. (1993) *Contaminant Hydrology*, Prentice Hall, Upper Saddle River, NJ, p. 458.

Filippi, M., Goliáš, V. and Pertold, Z. (2004) Arsenic in contaminated soils and anthropogenic deposits at the Mokrsko, Roudný, and Kašperské Hory gold deposits, Bohemian Massif (CZ). *Environmental Geology*, **45**(5), 716–30.

Finkelman, R.B. and Gross, P.M.K. (1999) The types of data needed for assessing the environmental and human health impacts of coal. *International Journal of Coal Geology*, **40**(2-3), 91–101.

Finkelman, R.B., Orem, W., Castranova, V. *et al.* (2002) Health impacts of coal and coal use: possible solutions. *International Journal of Coal Geology*, **50**(1-4), 425–43.

Finkelstein, D.B., Munhall, A., Pratt, L.M. and Bauer, C.E. (2004) A baseline study of evaporative water chemistry and microbial mat diversity from alkaline lakes in Warner Valley, Oregon. Abstracts with Programs. *The Geological Society of America*, **36**(5), 87.

Fisher, R.S. (2002) *Groundwater Quality in Kentucky: Arsenic*, Kentucky Geological Survey Information Circular 5, Series XII, Lexington, KY.

Folkes, D.J., Kuehster, T.E. and Litle, R.A. (2001) Contributions of pesticide use to urban background concentrations of arsenic in Denver, Colorado, U.S.A. *Environmental Forensics*, **2**(2), 127–39.

Ford, R.G. (2002) Rates of hydrous ferric oxide crystallization and the influence on coprecipitated arsenate. *Environmental Science and Technology*, **36**(11), 2459–63.

Ford, R.G., Wilkin, R.T. and Hernandez, G. (2006) Arsenic cycling within the water column of a small lake receiving contaminated ground-water discharge. *Chemical Geology*, **228**(1-3 Special Issue), 137–55.

Foster, A.L. (2003) Spectroscopic investigations of arsenic species in solid phases, in *Arsenic in Ground Water* (eds A.H., Welch and K.G. Sollenwerk) Kluwer Academic Publishers, Boston, MA, pp. 27–65.

Foster, A.L., Brown, G.E., Tingle, T.N. and Parks, G.A. (1998) *Jr* Quantitative arsenic speciation in mine tailings using X-ray absorption spectroscopy. *American Mineralogist*, **83**(5-6), 553–68.

Foster, S., Maher, W., Taylor, A. *et al.* (2005) Distribution and speciation of arsenic in temperate marine saltmarsh ecosystems. *Environmental Chemistry*, **2**(3), 177–89.

Foust, R.D. Jr., Mohapatra, P., Compton-O'Brien, A.-M. and Reifel, J. (2004) Groundwater arsenic in the Verde Valley in central Arizona, USA. *Applied Geochemistry*, **19**(2), 251–55.

Fowles, J. and Dybing, E. (2003) Application of toxicological risk assessment principles to the chemical constituents of cigarette smoke. *Tobacco Control*, **12**(4), 424–30.

Fox, P.M. and Doner, H.E. (2003) Accumulation, release, and solubility of arsenic, molybdenum, and vanadium in wetland sediments. *Journal of Environmental Quality*, **32**(6), 2428–35.

Francesconi, K.A. and Edmonds, J.S. (1994) Biotransformation of arsenic in the marine environment, in *Arsenic in the Environment: Part I: Cycling and Characterization*, (ed. J.O. Nriagu), John Wiley & Sons, Ltd, New York, pp. 221–61.

Francesconi, K.A. and Kuehnelt, D. (2002) Arsenic compounds in the environment, in *Environmental Chemistry of Arsenic* (ed. W.T. Frankenberger Jr.), Marcel Dekker, New York, pp. 51–94.

Frankenberger, W.T. Jr. and Arshad, M. (2002) Volatilization of arsenic, in *Environmental Chemistry of Arsenic* (ed. W.T. Frankenberger Jr.), Marcel Dekker, New York, pp. 363–80.

Frau, F., Rossi, A., Ardau, C. *et al.* (2005) Determination of arsenic speciation in complex environmental samples by the combined use of TEM and XPS. *Microchimica Acta*, **151**(3-4), 189–201.

Freeze, R.A. and Cherry, J.A. (1979) *Groundwater*, Prentice-Hall, Englewood Cliffs, NJ, p. 604.

Freitas, M.C. and Pacheco, A.M.G. (2007) Elemental concentrations of aerosols near Portuguese power plants by INAA and PIXE. *Journal of Radioanalytical and Nuclear Chemistry*, **271**(1), 185–89.

Freitas, M.D., Pacheco, A.M.G. and Ventura, M.G. (2004) Anthropogenic sources of PM2.5's arsenic, lead, mercury and nickel in northern metropolitan Lisbon, Portugal. *Nuclear Instruments and Methods in Physics Research*, **B**(219-220), 153–56.

Frenzel, S.A. (2002) *Priority-Pollutant Trace Elements in Streambed Sediments of the Cook Inlet Basin, Alaska, 1998–2000* U.S. Geological Survey Water-Resources Investigations Report 02-4163, U.S. Geological Survey, Anchorage, AK.

Frey, H.C. and Zhao, Y. (2004) Quantification of variability and uncertainty for air toxic emission inventories with censored emission factor data. *Environmental Science and Technology*, **38**(22), 6094–100.

Frisbie, S.H., Ortega, R., Maynard, D.M. and Sarkar, B. (2002) The concentrations of arsenic and other toxic elements in Bangladesh's drinking water. *Environmental Health Perspectives*, **110**(11), 1147–53.

Frost, J.K., Zierath, D.L. and Shimp, N.F. (1985) *Chemical Composition and Geochemistry of the New Albany Shale Group (Devonian-Mississippian) in Illinois.*, Illinois State Geological Survey, Champaign, IL.

Fruchter, J.S., Wilkerson, C.L., Evans, J.C. and Sanders, R.W. (1980) Elemental partitioning in an aboveground oil shale retort pilot plant. *Environmental Science and Technology*, **14**(11), 1374–81.

Fuller, C.C. and Davis, J.A. (2003) Section II. Evaluation of the Processes Controlling Dissolved Arsenic in Whitewood Creek, South Dakota, U.S. Geological Survey Professional Paper1681, pp. 27–48.

Galasso, J.L., Siegel, F.R. and Kravitz, J.H. (2000) Heavy metals in eight 1965 cores from the Novaya Zemlya Trough, Kara Sea, Russian Arctic. *Marine Pollution Bulletin*, **40**(10), 839–52.

Gao, S., Fujii, R., Chalmers, A.T. and Tanji, K.K. (2004) Evaluation of adsorbed arsenic and potential contribution to shallow groundwater in Tulare Lake Bed area, Tulare Basin, California. *Soil Science Society of America Journal*, **68**(1), 89–95.

Gao, S., Goldberg, S., Herbel, M.J. *et al.* (2006) Sorption processes affecting arsenic solubility in oxidized surface sediments from Tulare Lake bed, California. *Chemical Geology*, **228**(1-3 Special Issue), 33–43.

Gao, S., Luo, T.-C., Zhang, B.-R. *et al.* (1998) Chemical composition of the continental crust as revealed by studies in east China. *Geochimica et Cosmochimica Acta*, **62**(11), 1959–75.

García, A., Armienta, M.A. and Cruz, O. (2001) *Sources, Distribution and Fate of Arsenic Along the Tolimán River, Zimapán, Mexico*, IAHS, IAHS-AISH Publication 266, pp. 57–64.

García-Sanchez, A., Alvarez-Ayuso, E. and Rodriguez-Martin, F. (2002) Sorption of As(V) by some oxyhydroxides and clay minerals. Application to its immobilization in two polluted mining soils. *Clay Minerals*, **37**(1), 187–94.

García-Sánchez, A., Moyano, A. and Mayorga, P. (2005) High arsenic contents in groundwater of central Spain. *Environmental Geology*, **47**(6), 847–54.

Gault, A.G., Islam, F.S., Polya, D.A. *et al.* (2005) Microcosm depth profiles of arsenic release in a shallow aquifer, West Bengal. *Mineralogical Magazine*, **69**(5), 855–63.

Gayer, R.A., Rose, M., Dehmer, J. and Shao, L.-Y. (1999) Impact of sulphur and trace element geochemistry on the utilization of a marine-influenced coal-case study from the South Wales Variscan Foreland Basin. *International Journal of Coal Geology*, **40**(2-3), 151–74.

Gelova, G.A. (1977) *Hydrogeochemistry of Ore Elements*, Nedra, Moscow.

Génin, J.-.M.R., Bourrié, G., Trolard, F. *et al.* (1998) Thermodynamic equilibria in aqueous suspensions of synthetic and natural Fe(II)-Fe(III) green rusts: occurrences of the mineral in hydromorphic soils. *Environmental Science and Technology*, **32**(8), 1058–68.

Ghimire, K.N., Inoue, K., Yamaguchi, H. *et al.* (2003) Adsorptive separation of arsenate and arsenite anions from aqueous medium by using orange waste. *Water Research*, **37**(20), 4945–53.

Gieré, R., Sidenko, N.V. and Lazareva, E.V. (2003) The role of secondary minerals in controlling the migration of arsenic and metals from high-sulfide wastes (Berikul gold mine, Siberia). *Applied Geochemistry*, **18**(9), 1347–59.

Gihring, T.M. (2001) Arsenic biogeochemistry. Master's Thesis, University of Wisconsin, Madison, WI.

Gihring, T.M., Edwards, K.J., Bond, P.L. *et al.* (1999) Microbial interactions with arsenopyrite during oxidative dissolution. Abstracts with Programs. *The Geological Society of America*, **31**(7), 448.

Gilbert, S.E., Cooke, D.R. and Hollings, P. (2003) The effects of hardpan layers on the water chemistry from the leaching of pyrrhotite-rich tailings material. *Environmental Geology*, **44**(6), 687–97.

Gill, T.E., Gillette, D.A., Niemeyer, T. and Winn, R.T. (2002) Elemental geochemistry of wind-erodible playa sediments, Owens Lake, California. *Nuclear Instruments and Methods in Physics Research, Section B: Beam Interactions with Materials and Atoms*, **189**(1-4), 209–13.

Giordano, S., Adamo, P., Sorbo, S. and Vingiani, S. (2005) Atmospheric trace metal pollution in the Naples urban area based on results from moss and lichen bags. *Environmental Pollution*, **136**(3), 431–42.

Glass, R.L. and Frenzel, S.A. (2001) Distribution of Arsenic in Water and Streambed Sediments, Cook Inlet Basin, Alaska, U.S. Geological Survey Fact Sheet FS-083-01.

Gleisner, M. and Herbert, R.B. Jr. (2002) Sulfide mineral oxidation in freshly processed tailings: batch experiments. *Journal of Geochemical Exploration*, **76**(3), 139–53.

Goh, K.-H. and Lim, T.-T. (2004) Geochemistry of inorganic arsenic and selenium in a tropical soil: effect of reaction time, pH, and competitive anions on arsenic and selenium adsorption. *Chemosphere*, **55**(6), 849–59.

Goldhaber, M.B., Hatch, J.R., Pashin, J.C. *et al.* (1997) Anomalous arsenic and fluorine concentrations in Carboniferous coal, Black Warrior Basin, Alabama; evidence for fluid expulsion during Alleghanian thrusting? Abstracts with Programs. *The Geological Society of America*, **29**(6), 51.

Goldhaber, M.B., Irwin, E., Atkins, B. *et al.* (2001) *Arsenic in Stream Sediments of Northern Alabama*, U.S. Geological Survey Miscellaneous Field Studies Map, Reston, VA.

Goldhaber, M.B., Lee, R.C., Hatch, J.R. (2003) Role of large scale fluid-flow in subsurface arsenic enrichment, in *Arsenic in Ground Water* (eds A.H. Welch and K.G. Stollenwerk), Kluwer Academic Publishers, Boston, MA, pp. 127–64.

Gong, Z., Lu, X., Watt, C. *et al.* (2006) Speciation analysis of arsenic in groundwater from Inner Mongolia with an emphasis on acid-leachable particulate arsenic. *Analytica Chimica Acta*, **555**(1), 181–87.

Gongalsky, K.B. (2003) Impact of pollution caused by uranium production on soil macrofauna. *Environmental Monitoring and Assessment*, **89**(2), 197–219.

González, A.Z.I., Krachler, M., Cheburkin, A.K. and Shotyk, W. (2006) Spatial distribution of natural enrichments of arsenic, selenium, and uranium in a minerotrophic peatland, Gola di Lago, Canton Ticino, Switzerland. *Environmental Science and Technology*, **40**(21), 6568–74.

Goodarzi, F. and Huggins, F.E. (2005) Speciation of arsenic in feed coals and their ash byproducts from Canadian power plants burning sub-bituminous and bituminous coals. *Energy and Fuels*, **19**(3), 905–15.

Goodarzi, F. and Swaine, D.J. (1993) Chalcophile elements in western Canadian coals. *International Journal of Coal Geology*, **24**(1-4), 281–92.

Gotkowitz, M.B., Simo, J.A., Schreiber, M.E. *et al.* (2001) Geologic and geochemical controls on arsenic release to groundwater in eastern Wisconsin. Abstracts with Programs. *The Geological Society of America*, **33**(6), 54.

Gough, L.P., Wanty, R.B., Wang, B. *et al.* (1999) Regional geochemical studies related to placer gold resources, Fortymile River area, East-Central Alaska. Abstracts with Programs. *The Geological Society of America*, **31**(6), 58.

Govil, P.K., Reddy, G.L.N. and Gnaneswara Rao, T. (1999) Environmental pollution in India: heavy metals and radiogenic elements in Nacharam Lake. *Journal of Environmental Health*, **61**(8), 23–28.

Grafe, M., Eick, M.J. and Grossl, P.R. (2001) Adsorption of arsenate (V) and arsenite (III) on goethite in the presence and absence of dissolved organic carbon. *Soil Science Society of America Journal*, **65**(6), 1680–87.

Gräfe, M. and Sparks, D.L. (2006) Solid phase speciation of arsenic, in *Managing Arsenic in the Environment: From Soil to Human Health* (eds R. Naidu, E. Smith, G. Owens *et al.*), CSIRO Publishing, Collingwood, pp. 75–91.

Graney, J.R. and Eriksen, T.M. (2004) Metals in pond sediments as archives of anthropogenic activities: a study in response to health concerns. *Applied Geochemistry*, **19**(7), 1177–88.

Gray, J.E. and Sanzolone R.F. (eds). (1996) *Environmental Studies of Mineral Deposits in Alaska*, U.S. Geological Survey Bulletin 2156, United States Government Printing Office, Washington, DC.

Grip, E. (1961) Geology of the nickel deposit at Lainijaur in northern Sweden and a summary of other nickel deposits in Sweden. *Sveriges Geologiska Undersoekning. Serie C, Avhandlingar och Uppsatser*, **1**(557), 79. pp.

Gulbrandsen, R.A. (1966) Chemical composition of phosphorites of the Phosphoria Formation. *Geochimica et Cosmochimica Acta*, **30**, 769–78.

Guo, H. and Wang, Y. (2005) Geochemical characteristics of shallow groundwater in Datong Basin, northwestern China. *Journal of Geochemical Exploration*, **87**(3), 109–20.

Hammarstrom, J.M., Seal, R.R. II, Meier, A.L. and Jackson, J.C. (2003) Weathering of sulfidic shale and copper mine waste: secondary minerals and metal cycling in Great Smoky Mountains National Park, Tennessee, and North Carolina, USA. *Environmental Geology*, **45**(1), 35–57.

Han, F.X., Kingery, W.L., Selim, H.M. *et al.* (2004) Arsenic solubility and distribution in poultry waste and long-term amended soil. *Science of the Total Environment*, **320**(1), 51–61.

Hanor, J.S. (1998) Pathways of shallow subsurface migration of saline oil field wastes at a commercial disposal site in south Louisiana. *Gulf Coast Association of Geological Societies Transactions*, **48**, 107–18.

Hansen, H., Larssen, T., Seip, H.M. and Vogt, R.D. (2001) Trace metals in forest soils at four sites in southern China. *Water Air and Soil Pollution*, **130**(1–4 Pt III), 1721–26.

Harrington, J.M., Fendorf, S.E. and Rosenzweig, R.F. (1998) Biotic generation of arsenic(III) in metal(loid)-contaminated freshwater lake sediments. *Environmental Science and Technology*, **32**(16), 2425–30.

Harrington, J.M., Laforce, M.J., Rember, W.C. *et al.* (1998) Phase associations and mobilization of iron and trace elements in Coeur d'Alene Lake, Idaho. *Environmental Science and Technology*, **32**(5), 650–56.

Harvey, M.C., Schreiber, M.E., Rimstidt, J.D. and Griffith, M.M. (2006) Scorodite dissolution kinetics: implications for arsenic release. *Environmental Science and Technology*, **40**(21), 6709–14.

Haskin, L. and Warren, P. (1991) Lunar chemistry, in *Lunar Sourcebook: A User's Guide to the Moon* (eds G. Heiken, D. Vaniman and B.M. French), Cambridge University Press, Cambridge, Chapter 8.

Hattori, K.H. and Guillot, S. (2003) Volcanic fronts form as a consequence of serpentinite dehydration in the forearc mantle wedge. *Geology*, **31**(6), 525–28.

Hattori, K., Takahashi, Y., Guillot, S. and Johanson, B. (2005) Occurrence of arsenic (V) in forearc mantle serpentinites based on X-ray absorption spectroscopy study. *Geochimica et Cosmochimica Acta*, **69**(23), 5585–96.

He, B., Liang, L. and Jiang, G. (2002) Distributions of arsenic and selenium in selected Chinese coal mines. *Science of the Total Environment*, **296**(1–3), 19–26.

Heal, M.R., Hibbs, L.R., Agius, R.M. and Beverland, I.J. (2005) Total and water-soluble trace metal content of urban background PM 10, PM2.5 and black smoke in Edinburgh, UK. *Atmospheric Environment*, **39**(8), 1417–30.

Hedberg, E., Gidhagen, L. and Johansson, C. (2005) Source contributions to PM10 and arsenic concentrations in Central Chile using positive matrix factorization. *Atmospheric Environment*, **39**(3), 549–61.

Hedenquist, J.W. and Henley, R.W. (1985) Hydrothermal eruptions in the Waiotapu geothermal system, New Zealand: their origin, associated breccias, and relation to precious metal mineralization. *Economic Geology*, **80**(6), 1640–68.

Heinrichs, G. and Udluft, P. (1999) Natural arsenic in Triassic rocks: a source of drinking-water contamination in Bavaria, Germany. *Hydrogeology Journal*, **7**(5), 468–76.

Helena, P., Franc, B. and Cvetka, R.L. (2004) Monitoring of short-term heavy metal deposition by accumulation in epiphytic lichens (Hypogymnia physodes (L.) Nyl.). *Journal of Atmospheric Chemistry*, **49**, 223–30.

Helle, S., Alfaro, G., Keim, U. and Tascón, J.M.D. (2000) Mineralogical and chemical characterisation of coals from southern Chile. *International Journal of Coal Geology*, **44**(1), 85–94.

Helmke, P.A., Koons, R.D., Schomberg, P.J. and Iskandar, I.K. (1977) Determination of trace element contamination of sediments by multielement analysis of clay-size fraction. *Environmental Science and Technology*, **11**(10), 984–89.

Helz, G.R., Tossell, J.A., Charnock, J.M. *et al.* (1995) Oligomerization in As(III) sulfide solutions: theoretical constraints and spectroscopic evidence. *Geochimica et Cosmochimica Acta*, **59**(22), 4591–604.

212 Arsenic

Herbel, M. and Fendorf, S. (2006) Biogeochemical processes controlling the speciation and transport of arsenic within iron coated sands. *Chemical Geology*, **228**(1-3 Special Issue), 16–32.

Hering, J.G. and Chiu, V.Q. (2000) Arsenic occurrence and speciation in municipal ground-water-based supply system. *Journal of Environmental Engineering*, **126**(5), 471–74.

Hering, J.G. and Kneebone, P.E. (2002) Biogeochemical controls on arsenic occurrence and mobility in water supplies, in *Environmental Chemistry of Arsenic* (ed. W.T. Frankenberger Jr.), pp. 155–81. Marcel Dekker, New York.

Hering, J.G., Wilkie, J.A., Kneebone, P.E. and Salmassi, T. (1999) Redox cycling of arsenic in gothermally-influenced surface waters. Abstracts with Programs. *The Geological Society of America*, **31**(7), 448.

Hernández-García, M.E. and Custodio, E. (2004) Natural baseline quality of Madrid Tertiary detrital aquifer groundwater (Spain): a basis for aquifer management. *Environmental Geology*, **46**(2), 173–88.

Hidy, G.M., Lachenmyer, C., Chow, J. and Watson, J. (2000) Urban outdoor-indoor PM2.5 concentrations and personal exposure in the Deep South Part II. Inorganic chemistry. *Aerosol Science and Technology*, **33**(4), 357–75.

Hinkle, S.R. and Polette D.J. (1999) *Arsenic in Ground Water of the Willamette Basin, Oregon* U.S. Geological Survey Water-Resources Investigations Report 98-4205, U.S. Geological Survey, Portland, OR.

Hirner, A.V., Feldmann, J., Krupp, E. *et al.* (1998) Metal(loid)organic compounds in geothermal gases and waters. *Organic Geochemistry*, **29**(5-7 pt 2), 1765–78.

Hollibaugh, J.T., Carini, S., Güleyuk, H. *et al.* (2005) Arsenic speciation in Mono Lake, California: response to seasonal stratification and anoxia. *Geochimica et Cosmochimica Acta*, **69**(8), 1925–37.

Hollings, P., Hendry, M.J. and Kerrich, R. (1999) Sequential filtration of surface and ground waters from the Rabbit Lake uranium mine, northern Saskatchewan, Canada. *Water Quality Research Journal of Canada*, **34**(2), 221–47.

Holm, T.R. and Curtiss, C.D. (1989) A comparison of oxidation-reduction potentials calculated from the As(V)/As(III) and Fe(III)/Fe(II) couples with measured platinum-electrode potentials in groundwater. *Journal of Contaminant Hydrology*, **5**(1), 67–81.

Holm, T.R., W.R. Kelly, S.D. Wilson, G.S. *et al.* (2004) *Arsenic Geochemistry and Distribution in the Mahomet Aquifer* Waste Management and Research Center Report RR-107, Illinois State Water Survey and Illinois Waste Management and Research Center, Chicago, IL.

Holmes, C.W. and Miller, R. (2002) Atmospheric transport and deposition of arsenic and potential link to Florida groundwater. Abstracts with Programs. *The Geological Society of America*, **34**(6), 295.

Hon, R., Mayo, M.J. and Brandon, W.C. (2003) Arsenic equilibria in ground water at landfills, central Massachusetts. Abstracts with Programs. *The Geological Society of America*, **35**(6), 48.

Hossain, M.F. (2006) Arsenic contamination in Bangladesh-an overview. *Agriculture, Ecosystems and Environment*, **113**, 1–16.

Howard, A.G., Apte, S.C., Comber, S.D.W. and Morris, R.J. (1988) Biogeochemical control of the summer distribution and speciation of arsenic in the Tamar estuary. *Estuarine Coastal and Shelf Science*, **27**(4), 427–43.

Hower, J.C., Calder, J.H., Eble, C.F. *et al.* (2000) Metalliferous coals of the Westphalian A Joggins Formation, Cumberland Basin, Nova Scotia, Canada: petrology, geochemistry, and palynology. *International Journal of Coal Geology*, **42**(2-3), 185–206.

Huang, W.W., Martin, J.M., Seyler, P. *et al.* (1988) Distribution and behaviour of arsenic in the Huang He (Yellow River) estuary and Bohai Sea. *Marine Chemistry*, **25**(1), 75–91.

Huang, J.-H. and Matzner, E. (2006) Dynamics of organic and inorganic arsenic in the solution phase of an acidic fen in Germany. *Geochimica et Cosmochimica Acta*, **70**(8), 2023–33.

Hudson-Edwards, K.A., Macklin, M.G., Miller, J.R. and Lechler, P.J. (2001) Sources, distribution and storage of heavy metals in the Río Pilcomayo, Bolivia. *Journal of Geochemical Exploration*, **72**(3), 229–50.

Huggins, F.E. (2002) Overview of analytical methods for inorganic constituents in coal. *International Journal of Coal Geology*, **50**(1-4), 169–214.

Huggins, F.E., Goodarzi, F. and Lafferty, C.J. (1996) Mode of occurrence of arsenic in subbituminous coals. *Energy and Fuels*, **10**(4), 1001–4.

Huggins, F.E. and Huffman, G.P. (1996) Modes of occurrence of trace elements in coal from XAFS spectroscopy. *International Journal of Coal Geology*, **32**(1-4), 31–53.

Huggins, F.E., Huffman, G.P., Kolker, A. *et al.* (2002) Combined application of XAFS spectroscopy and sequential leaching for determination of arsenic speciation in coal. *Energy and Fuels*, **16**(5), 1167–172.

Huggins, F.E., Shah, N., Zhao, J. *et al.* (1993) Nondestructive determination of trace element speciation in coal and coal ash by XAFS spectroscopy. *Energy and Fuels*, **7**(4), 482–89.

Huggins, F.E., Srikantapura, S., Parekh, B.K. *et al.* (1997) XANES spectroscopic characterization of selected elements in deep-cleaned fractions of Kentucky no. 9 coal. *Energy and Fuels*, **11**(3), 691–700.

Huisman, D.J., Vermeulen, F.J.H., Baker, J. *et al.* (1997) A geological interpretation of heavy metal concentrations in soils and sediments in the southern Netherlands. *Journal of Geochemical Exploration*, **59**, 163–74.

Hutton, C., Bryce, D.W., Russeau, W. *et al.* (2005) Aqueous and solid-phase speciation of arsenic in Cornish soils. *Mineralogical Magazine*, **69**(5), 577–89.

Hyndman, D.W. (1985) *Petrology of Igneous and Metamorphic Rocks*, 2nd edn, McGraw-Hill, New York, p. 786.

Ikem, A., Egiebor, N.O. and Nyavor, K. (2003) Trace elements in water, fish and sediment from Tuskegee Lake, southeastern USA. *Water Air and Soil Pollution*, **149**(1-4), 51–75.

Ingersoll, G.P. (2000) *Snowpack chemistry at selected sites in Colorado and New Mexico During Winter 1999–2000* U.S. Geological Survey Open-File Report 00-394, U.S. Geological Survey, Denver, CO.

Inskeep, W.P., Macur, R.E., Harrison, G. *et al.* (2004) Biomineralization of As(V)-hydrous ferric oxyhydroxide in microbial mats of an acid-sulfate-chloride geothermal spring, Yellowstone National Park. *Geochimica et Cosmochimica Acta*, **68**(15), 3141–55.

Inskeep, W.P., McDermott, T.R. and Fendorf, S. (2002) Arsenic (V)/(III) cycling in soils and natural waters: chemical and microbiological processes, in *Environmental Chemistry of Arsenic* (ed. W.T. Frankenberger Jr.), Marcel Dekker, New York, pp. 183–215.

Irgolic, K.J., Spall, D., Puri, B.K. *et al.* (1991) Determination of arsenic and arsenic compounds in natural-gas samples. *Applied Organometallic Chemistry*, **5**(2), 117–24.

Islam, F.S., Boothman, C., Gault, A.G. *et al.* (2005) Potential role of the Fe(III)-reducing bacteria Geobacter and Geothrix in controlling arsenic solubility in Bengal delta sediments. *Mineralogical Magazine*, **69**(5), 865–75.

Islam, F.S., Gault, A.G., Boothman, C. *et al.* (2004) Role of metal-reducing bacteria in arsenic release from Bengal Delta sediments. *Nature*, **430**(6995), 68–71.

Islam, Md.R., Lahermo, P., Salminen, R. *et al.* (2000) Lake and reservoir water quality affected by metals leaching from tropical soils, Bangladesh. *Environmental Geology*, **39**(10), 1083–89.

Ito, H., Masuda, H. and Kusakabe, M. (2003) Some factors controlling arsenic concentrations of groundwater in the northern part of Osaka Prefecture. *Journal of Groundwater Hydrology*, **45**(1), 3–18.

Jacks, G. and Bhattacharya, P. (2000) Redox conditions in the soil zone in the Bengal delta-facts and speculations, in *Arsenic in Groundwater of Sedimentary Aquifers*, Pre-Congress Workshop, 31st International Geological Congress (eds P. Bhattacharya and A.H. Welch), Groundwater Arsenic Research Group, Division of Land and Water Resources, Royal Institute of Technology, Stockholm, pp. 54–55.

Jackson, T.A. and Bistricki, T. (1995) Selective scavenging of copper, zinc, lead, and arsenic by iron and manganese oxyhydroxide coatings on plankton in lakes polluted with mine and smelter wastes: results of energy dispersive X-ray micro- analysis. *Journal of Geochemical Exploration*, **52**(1-2), 97–125.

Jaganathan, J., Mohan, M.S. and Zingaro, R.A. (1986) Identification of arsenic-bearing minerals in a sample of Green River oil shale. *Fuel*, **65**(2), 266–69.

Jain, P., Kim, H. and Townsend, T.G. (2005) Heavy metal content in soil reclaimed from a municipal solid waste landfill. *Waste Management*, **25**(1), 25–35.

Jakubick, A.T., Jenk, U. and Kahnt, R. (2002) Modelling of mine flooding and consequences in the mine hydrogeological environment: flooding of the Koenigstein mine, Germany. *Environmental Geology*, **42**(2-3), 222–34.

Jambor, J.L. and Dutrizac, J.E. (1998) Occurrence and constitution of natural and synthetic ferrihydrite, a widespread iron oxyhydroxide. *Chemical Reviews*, **98**(7), 2549–585.

Janney, D.E., Cowley, J.M. and Buseck, P.R. (2000) Structure of synthetic 2-line ferrihydrite by electron nanodiffraction. *American Mineralogist*, **85**(9), 1180–87.

Jay, J.A., Blute, N.K., Lin, K. *et al.* (2005) Controls on arsenic speciation and solid-phase partitioning in the sediments of a two-basin lake. *Environmental Science and Technology*, **39**(23), 9174–81.

Jdid, E.A., Blazy, P., Kamoun, S. *et al.* (1999) Environmental impact of mining activity on the pollution of the Medjerda River, north-west Tunisia. *Bulletin of Engineering Geology and the Environment*, **57**(3), 273–80.

Jing, C., Korfiatis, G.P. and Meng, X. (2003) Immobilization mechanisms of arsenate in iron hydroxide sludge stabilized with cement. *Environmental Science and Technology*, **37**(21), 5050–56.

Jo, I.S. and Koh, M.H. (2004) Chemical changes in agricultural soils of Korea: data review and suggested counter-measures. *Environmental Geochemistry and Health*, **26**(2), 105–17.

Johannesson, K.H., Lyons, W.B., Huey, S. *et al.* (1997) Oxyanion concentrations in eastern Sierra Nevada rivers — 2. Arsenic and phosphate. *Aquatic Geochemistry*, **3**(1), 61–97.

Johnson, D.M. (2001) Arsenic Occurrence in Wisconsin's Groundwater. American Water Resources Association-Wisconsin Section, Conference March 29-30, 2001.

Johnson, A. and Golding, S. (2002) Results and Recommendations from Monitoring Arsenic Levels in 303(d) Listed Rivers in Washington, Washington State Department of Ecology Report, Publication No. 02-03-045, Olympia.

Jonas, J.P. and Gammons, C.H. (2000) Iron cycling in the Berkeley Pit-lake, Butte, Montana. Abstracts with Programs. *The Geological Society of America*, **32**(5), 13.

Jones, D.R., Chapman, B.M. and Jung, R.F. (1990) Column leaching of unretorted and retorted oil shales and claystone form the Rundle deposit: water leaching. *Water Research*, **24**(2), 131–41.

Jones, C.A., Inskeep, W.P. and Neuman, D.R. (1997) Arsenic transport in contaminated mine tailings following liming. *Journal of Environmental Quality*, **26**(2), 433–39.

Jung, M.C., Thornton, I. and Chon, H.-T. (2002) Arsenic, Sb and Bi contamination of soils, plants, waters and sediments in the vicinity of the Dalsung Cu-W mine in Korea. *Science of the Total Environment*, **295**(1-3), 81–89.

Kampf, S.K., Salazar, M. and Tyler, S.W. (2002) Preliminary investigations of effluent drainage from mining heap leach facilities. *Vadose Zone Journal*, **1**, 186–96.

Kanivetsky, R. (2000) Arsenic in Minnesota Ground Water; Hydrochemical Modeling of the Quaternary Buried Artesian Aquifer and Cretaceous Aquifer Systems, Report of Investigations — Minnesota Geological Survey, St. Paul, MN.

Karayigit, A.I., Gayer, R.A., Engin Ortac, F. and Goldsmith, S. (2001) Trace elements in the lower Pliocene fossiliferous Kangal lignites, Sivas, Turkey. *International Journal of Coal Geology*, **47**(2), 73–89.

Karayigit, A.I., Spears, D.A. and Booth, C.A. (2000a) Antimony and arsenic anomalies in the coal seams from the Gokler coalfield, Gediz, Turkey. *International Journal of Coal Geology*, **44**(1), 1–17.

Karayigit, A.I., Spears, D.A. and Booth, C.A. (2000b) Distribution of environmental sensitive trace elements in the Eocene Sorgun coals, Turkey. *International Journal of Coal Geology*, **42**(4), 297–314.

Keimowitz, A.R., Zheng, Y., Chillrud, S.N. *et al.* (2005) Arsenic redistribution between sediments and water near a highly contaminated source. *Environmental Science and Technology*, **39**(22), 8606–13.

Kelley, J.A., Jaffe, D.A., Baklanov, A. and Mahura, A. (1995) Heavy metals on the Kola Peninsula: aerosol size distribution. *Science of the Total Environment*, **160-161**, 135–38.

Kelly, W.R., Holm, T.R., Wilson, S.D. and Roadcap, G.S. (2005) Arsenic in glacial aquifers: sources and geochemical controls. *Ground Water*, **43**(4), 500–10.

Kerr, L.C., Craw, D. and Youngson, J.H. (1999) Arsenopyrite compositional variation over variable temperatures of mineralization, Otago Schist, New Zealand. *Economic Geology*, **94**(1), 123–28.

Keskin, S.S. (1995) Time-Series Trends of Trace Elements in an Ice Core from Antarctica. Ph.D. Dissertation, Massachusetts Institute of Technology.

Kesler, S.E., Riciputi, L.C. and Ye, Z. (2005) Evidence for a magmatic origin for Carlin-type gold deposits: isotopic composition of sulfur in the Betze-Post-Screamer deposit, Nevada, USA. *Mineralium Deposita*, **40**(2), 127–36.

Kettles, I.M. and Shilts, W.W. (1994) Composition of glacial sediments in Canadian shield terrane, southeastern Ontario and southwestern Quebec: applications to acid rain research and mineral exploration. *Geological Survey of Canada Bulletin*, **463,** Report, 58 pp.

Kim, M.-J., Nriagu, J. and Haack, S. (2000) Carbonate ions and arsenic dissolution by groundwater. *Environmental Science and Technology*, **34**(15), 3094–310.

Kim, M.-J., Nriagu, J. and Haack, S. (2002) Arsenic species and chemistry in groundwater of southeast Michigan. *Environmental Pollution*, **120**(2), 379–90.

Kim, M.-J., Nriagu, J. and Haack, S. (2003) Arsenic behavior in newly drilled wells. *Chemosphere*, **52**(3), 623–33.

Kimball, B.A., Callender, E. and Axtmann, E.V. (1995) Effects of colloids on metal transport in a river receiving acid mine drainage, Upper Arkansas River, Colorado, U.S.A. *Applied Geochemistry*, **10**(3), 285–306.

Kirk, M.F., Holm, T.R., Park, J. *et al.* (2004) Bacterial sulfate reduction limits natural arsenic contamination in groundwater. *Geology*, **32**(11), 953–56.

Kizilshtein, L.Y. and Kholodkov, Y.I. (1999) Ecologically hazardous elements in coals of the Donets Basin. *International Journal of Coal Geology*, **40**(2-3), 189–97.

Klassen, R.A. (2004) Geological factors affecting the distribution of trace metals in glacial sediments of central Newfoundland. *Environmental Geology*, **33**(2-3), 154–69.

Klein, C. (2002) *The 22nd Edition of the Manual of Mineral Science (after James D. Dana)*, John Wiley & Sons, Ltd, New York, p. 641.

Klinchuch, L.A., Delfino, T.A., Jefferson, J.L. and Waldron, J.M. (1999) Does biodegradation of petroleum hydrocarbons affect the occurrence or mobility of dissolved arsenic in groundwater? *Environmental Geosciences*, **6**(1), 9–24.

Klumpp, A., Hintemann, T., Santana Lima, J. and Kandeler, E. (2003) Bioindication of air pollution effects near a copper smelter in Brazil using mango trees and soil microbiological properties. *Environmental Pollution*, **126**(3), 313–21.

Kneebone, P.E. and Hering, J.G. (2000) Behavior of arsenic and other redox-sensitive elements in Crowley Lake, CA: a reservoir in the Los Angeles aqueduct system. *Environmental Science and Technology*, **34**(20), 4307–12.

Kneebone, P.E., O'Day, P.A., Jones, N. and Hering, J.G. (2002) Deposition and fate of arsenic in iron- and arsenic-enriched reservoir sediments. *Environmental Science and Technology*, **36**(3), 381–86.

Ko, I., Kim, J.-Y. and Kim, K.-W. (2004) Arsenic speciation and sorption kinetics in the as-hematite-humic acid system. *Colloids and Surfaces A-Physicochemical and Engineering Aspects*, **234**(1-3), 43–50.

Koeberl, C., Bottomley, R., Glass, B.P. and Storzer, D. (1997) Geochemistry and age of ivory coast tektites and microtektites. *Geochimica et Cosmochimica Acta*, **61**(8), 1745–72.

Kolker, A. (2000) Arsenic content of pyrite in sedimentary aquifers: the Marshall Sandstone of southeastern Michigan, USA, in *Arsenic in Groundwater of Sedimentary Aquifers*, Pre-Congress Workshop, 31st International Geological Congress (eds P. Bhattacharya and A.H. Welch), Groundwater Arsenic Research Group, Division of Land and Water Resources, Royal Institute of Technology, Stockholm, pp. 65–68.

Kolker, A., Haack, S.K. and Cannon, W.F. (2003) Arsenic in southeastern Michigan, in *Arsenic in Ground Water* (eds A.H. Welch and K.G. Stollenwerk), Kluwer Academic Publishers, Boston, MA, pp. 281–94.

Kolker, A., Huggins, F.E., Palmer, C.A. *et al.* (2000) Mode of occurrence of arsenic in four US coals. *Fuel Processing Technology*, **63**(2), 167–78.

Komnitsas, K., Xenidis, A. and Adam, K. (1995) Oxidation of pyrite and arsenopyrite in sulphidic spoils in Lavrion. *Minerals Engineering*, **8**(12), 1443–54.

Korotev, R.L., Jolliff, B.L., Zeigler, R.A. *et al.* (2003) Feldspathic lunar meteorites and their implications for compositional remote sensing of the lunar surface and the composition of the lunar crust. *Geochimica et Cosmochimica Acta*, **67**(24), 4895–923.

Koski, A. and Wood, S. (2003) Aqueous geochemistry of thermal waters in the Alvord Basin, Oregon. Abstracts with Programs. *The Geological Society of America*, **35**(6), 407.

Krause, E. and Ettel, V.A. (1988) Solubility and stability of scorodite, FeAsO4.2H2O: new data and further discussion. *American Mineralogist*, **73**(7-8), 850–54.

Krauskopf, K.B. and Bird D.K. (1995) *Introduction to Geochemistry*, 3rd edn, McGraw-Hill, Boston, MA.

Krot, A.N., Keil, K., Goodrich, C.A. (2004) Classification of meteorites, in *Treatise on Geochemistry: Volume 1: Meteorites, Comets, and Planets* (eds A.M. Davis, H.D. Holland and K.K. Turekian Editors-in-Chief), Elsevier Science, pp. 83–128.

Kubota, Y., Ishiyama, Y. and Yokota, D. (2000) Arsenic distribution in the surface geology of the Niigata Plain, central Japan: source supply of arsenic in arsenic contaminated ground water problem: Part 1. *Earth Science*, **54**(6), 369–79.

Kubota, Y., Yokota, D. and Ishiyama, Y. (2001) Arsenic concentration in hot spring waters from the Niigata plain and Shinji lowland, Japan: source supply of arsenic in arsenic contaminated ground water problem: Part 2. *Earth Science*, **55**(1), 11–22.

Kuehnelt, D. and Goessler W. (2003) Organoarsenic compounds in the terrestrial environment, in *Organometallic Compounds in the Environment*, 2nd edn (ed. P.J. Craig), John Wiley & Sons, Ltd, West Sussex, pp. 223–75.

Kutle, A., Oreščanin, V., Obhodaš, J. and Valković, V. (2004) Trace element distribution in geochemical environment of the island Krk and its influence on the local population. *Journal of Radioanalytical and Nuclear Chemistry*, **259**(2), 271–76.

Kuwabara, J.S., Chang, C.C.Y. and Pasilis, S.P. (2003) Section 1. Effects of Benthic Flora on Arsenic Transport in Whitewood Creek, South Dakota, U.S. Geological Survey Professional Paper 1681, pp. 1–26.

Kwong, Y.T.J. (2003) Characteristics of impounded tailings at Mount Nansen, Yukon territory: implications for remediation, in *Proceedings-Assessment and Remediation of Contaminated Sites in Arctic and Cold Climates,* (eds M. Nahir, K. Biggar and G. Cotta), University of Alberta, Edmonton, pp. 57–62.

La Force, M.J., Hansel, C.M. and Fendorf, S. (2000) Arsenic speciation, seasonal transformations and co-distribution with iron in a mine waste-influenced palustrine emergent wetland. *Environmental Science and Technology*, **34**(18), 3937–43.

Lai, H.K., Kendall, M., Ferrier, H. *et al.* (2004) Personal exposures and microenvironment concentrations of PM2.5, VOC, NO2 and CO in Oxford, UK. *Atmospheric Environment*, **38**(37), 6399–410.

Landrum, K.E. (1994) Accumulation and trace-metal variability of estuarine sediments, Barataria Basin, Louisiana. *Transactions of the Gulf Coast Association of Geological Societies*, **XLIV**, 365–72.

Landsberger, S., Vermette, V.G., Stuenkel, D. *et al.* (1992) Elemental source signatures of aerosols from the Canadian high Arctic. *Environmental Pollution*, **75**(2), 181–87.

Langmuir, D. (1997) *Aqueous Environmental Geochemistry*, Prentice Hall, Upper Saddle River, NJ, p. 600.

Langmuir, D., Mahoney, J. and Rowson, J. (2006) Solubility products of amorphous ferric arsenate and crystalline scorodite ($FeAsO_4 \cdot 2H_2O$) and their application to arsenic behavior in buried mine tailings. *Geochimica et Cosmochimica Acta*, **70**(12), 2942–56.

Langner, H.W., Jackson, C.R., Mcdermott, T.R. and Inskeep, W.P. (2001) Rapid oxidation of arsenite in a hot spring ecosystem, Yellowstone National Park. *Environmental Science and Technology*, **35**(16), 3302–309.

Lawrence, L.J., Smith-Munro, V., Ramsden, A.R. *et al.* (1999) Geology and mineralogy of the Lorena gold mine, Cloncurry District, northwest Queensland. *Journal and Proceedings of the Royal Society of New South Wales*, **132**(1-2), 29–35.

Le, X.C. (2002) Arsenic speciation in the environment and humans, in *Environmental Chemistry of Arsenic* (ed. W.T. Frankenberger Jr.), Marcel Dekker, New York, pp. 95–116.

Le Guern, C., Baranger, P., Crouzet, C. *et al.* (2003) Arsenic trapping by iron oxyhydroxides and carbonates at hydrothermal spring outlets. *Applied Geochemistry*, **18**(9), 1313–23.

Ledger, E.B. and Judy, K. (2003) Elevated arsenic levels in the Weches Formation, Nacogdoches County, Texas. *Transactions of the Gulf Coast Association of Geological Societies*, **LIII**, 453–61.

Lee, P.K.H., Brook, J.R., Dabek-Zlotorzynska, E. and Mabury, S.A. (2003) Identification of the major sources contributing to PM2.5 observed in Toronto. *Environmental Science and Technology*, **37**(21), 4831–40.

Lee, J.-S., Chon, H.-T., Kim, J.-S. *et al.* (1998) Enrichment of potentially toxic elements in areas underlain by black shales and slates in Korea. *Environmental Geochemistry and Health*, **20**(3), 135–47.

Lee, L. and M.B. Goldhaber (2001) *The Distribution of MVT-related Metals in Ground Water of the Ozark Plateaus Region of the United States* U.S. Geological Survey Open File Report 01-171, U.S. Geological Survey, Denver, CO.

Lee, Y.C. and Hills, P.R. (2003) Cool season pollution episodes in Hong Kong, 1996–2002. *Atmospheric Environment*, **37**(21), 2927–39.

Lee, C.H., Lee, H.K. and Lee, J.C. (2001) Hydrogeochemistry of mine, surface and groundwaters from the Sanggok mine creek in the upper Chungju Lake, Republic of Korea. *Environmental Geology*, **40**(4-5), 482–94.

Lee, J.S. and Nriagu, J. (2003) Arsenic carbonate complexes in aqueous systems. *ACS Symposium Series*, **835**, 33–41.

Lengke, M.F. and Tempel, R.N. (2002) Reaction rates of natural orpiment oxidation at 25 to 40 °C and pH 6.8 to 8.2 and comparison with amorphous As_2S_3 oxidation. *Geochimica et Cosmochimica Acta*, **66**(18), 3281–91.

Lengke, M.F. and Tempel, R.N. (2005) Geochemical modeling of arsenic sulfide oxidation kinetics in a mining environment. *Geochimica et Cosmochimica Acta*, **69**(2), 341–56.

Leoni, L. and Sartori, F. (1997) Heavy metal and arsenic distributions in sediments of the Elba-Argentario basin, southern Tuscany, Italy. *Environmental Geology*, **32**(2), 83–92.

Lerouge, C., Bouchot, V. and Guerrot, C. (2000) Fluids and the W (+/-As,Au) ore deposits of the Enguialès-Leucamp District, La Châtaigneraie, French Massif Central. *Journal of Geochemical Exploration*, **69-70**, 343–47.

Lett, R.E., Bobrowsky, P., Cathro, M. and Yeow, A. (1998) Geochemical pathfinders for massive sulphide deposits in the southern Kootenay Terrane, *Geological Fieldwork 1997*, British Columbia Geological Division, Victoria, BC, pp. 15-1–15-9.

Lewis, M.A., Boustany, R.G., Dantin, D.D. *et al.* (2002) Effects of a coastal golf complex on water quality, periphyton, and seagrass. *Ecotoxicology and Environmental Safety*, **53**(1), 154–62.

Lide, D.R. (ed.) (2007) *CRC Handbook of Chemistry and Physics*, 88th edn, CRC Press, Boca Raton, FL.

Lima, A., Cicchella, D. and Di Francia, S. (2003) Natural contribution of harmful elements in thermal groundwaters of Ischia Island (southern Italy). *Environmental Geology*, **43**(8), 930–40.

Lin, H.-T., Wang, M.C. and Li, G.-C. (2002) Effect of water extract of compost on the adsorption of arsenate by two calcareous soils. *Water Air and Soil Pollution*, **138**(1-4), 359–74.

Ljung, K., Otabbong, E. and Selinus, O. (2006) Natural and anthropogenic metal inputs to soils in urban Uppsala, Sweden. *Environmental Geochemistry and Health*, **28**(4), 353–64.

Llobet, J.M., Schuhmacher, M. and Domingo, J.L (2002) Spatial distribution and temporal variation of metals in the vicinity of a municipal solid waste incinerator after a modernization of the flue gas cleaning systems of the facility. *Science of the Total Environment*, **284**(1-3), 205–14.

Loredo, J., Ordo'nez, A., A'lvarez, R. and Garcia-Iglesias, J. (2004) The potential for arsenic mobilisation in the Caudal River catchment, north-west Spain. *Transactions of the Institution of Mining and Metallurgy Section B: Applied Earth Science*, **113**(1), B65–B75.

Loredo, J., Ordo'nez, A., Baldo, C. and García-Iglesias, J. (2003) Arsenic mobilization from waste piles of the El Terronal mine, Asturias, Spain. *Geochemistry: Exploration, Environment, Analysis*, **3**(3), 229–37.

Loredo, J., Ordo'ñez, A., Gallego, J.R. *et al.* (1999) Geochemical characterisation of mercury mining spoil heaps in the area of Mieres (Asturias, northern Spain). *Journal of Geochemical Exploration*, **67**(1-3), 377–90.

Loring, D.H., Dahle, S., Naes, K. *et al.* (1998) Arsenic and other trace metals in sediments from the Kara Sea and the Ob and Yenisey estuaries, Russia. *Aquatic Geochemistry*, **4**(2), 233–52.

Loring, D.H., Naes, K., Dahle, S. *et al.* (1995) Arsenic, trace metals, and organic microcontaminants in sediments from the Pechora Sea, Russia. *Marine Geology*, **128**(3-4), 153–67.

Lu, X. and Zhang, X. (2005) Environmental geochemistry study of arsenic in western Hunan mining area, P.R. China. *Environmental Geochemistry and Health*, **27**(4), 313–20.

Lucaciu, A., Timofte, L., Culicov, O. *et al.* (2004) Atmospheric deposition of trace elements in Romania studied by the moss biomonitoring technique. *Journal of Atmospheric Chemistry*, **49**(1-3), 533–48.

Lund, U. and Fobian, A. (1991) Pollution of two soils by arsenic, chromium and copper, Denmark. *Geoderma*, **49**(1-2), 83–103.

Maher, W.A. (1985) Arsenic in coastal waters of South Australia. *Water Research*, **19**(7), 933–34.

Mains, D. and Craw, D. (2005) Composition and mineralogy of historic gold processing residues, east Otago, New Zealand. *New Zealand Journal of Geology and Geophysics*, **48**(4), 641–47.

Maki, T., Takeda, N., Hasegawa, H. and Ueda, K. (2006) Isolation of monomethylarsonic acid-mineralizing bacteria from arsenic contaminated soils of Ohkunoshima Island. *Applied Organometallic Chemistry*, **20**(9), 538–44.

Malm, W.C. and Sisler, J.F. (2000) Spatial patterns of major aerosol species and selected heavy metals in the United States. *Fuel Processing Technology*, **65**, 473–501.

Mandal, B.K. and Suzuki, K.T. (2002) Arsenic round the world: a review. *Talanta*, **58**(1), 201–35.

Mannio, J., Jarvinen, O., Tuominen, R. and Verta, M. (1995) Survey of trace elements in lake waters of Finnish Lapland using the ICP-MS technique. *Science of the Total Environment*, **160-161**, 433–39.

Mardon, S.M. and Hower, J.C. (2004) Impact of coal properties on coal combustion by-product quality: examples from a Kentucky power plant. *International Journal of Coal Geology*, **59**(3-4), 153–69.

Maret, T.R. and Skinner K.D. (2000) *Concentrations of Selected Trace Elements in Fish Tissue and Streambed Sediment in the Clark Fork-Pend Oreille and Spokane River Basins, Washington, Idaho, and Montana, 1998* U.S. Geological Survey Water-Resources Investigations Report 00-4159, U.S. Geological Survey, Boise, ID.

Marszałek, H. and Wsik, M. (2000) Influence of arsenic-bearing gold deposits on water quality in Zloty Stok mining area (SW Poland). *Environmental Geology*, **39**(8), 888–92.

Martin, J.M., Guan, D.M., Elbaz-Poulichet, F. *et al.* (1993) Preliminary assessment of the distributions of some trace elements (As, Cd, Cu, Fe, Ni, Pb, and Zn) in a pristine aquatic environment: the Lena River estuary (Russia). *Marine Chemistry*, **43**, 185–99.

Martin, D.B. and Hartman, W.A. (1984) Arsenic, cadmium, lead, mercury, and selenium in sediments of riverine and pothole wetlands of the north central United States. *Journal of the Association of Official Analytical Chemists*, **67**(6), 1141–46.

Martin, A.J. and Pedersen, T.F. (2002) Seasonal and interannual mobility of arsenic in a lake impacted by metal mining. *Environmental Science and Technology*, **36**(7), 1516–23.

Martley, E., Gulson, B.L. and Pfeifer, H.-R. (2004) Metal concentrations in soils around the copper smelter and surrounding industrial complex of Port Kembla, NSW, Australia. *Science of the Total Environment*, **325**(1-3), 113–27.

Mascaro, I., Benvenuti, B., Corsini, F. *et al.* (2001) Mine wastes at the polymetallic deposit of Fenice Capanne (southern Tuscany, Italy). Mineralogy, geochemistry, and environmental impact. *Environmental Geology*, **41**, 417–29.

Massacci, P., Serranti, S., Ghiani, M. *et al.* (2003) Reduction of the as content in colemanite concentrates. *Proceedings — International Mineral Processing Congress*, Vol. 22, International Mineral Processing Council, pp. 543–51.

Mastalerz, M., Hower, J.C., Drobniak, A. *et al.* (2004) From in-situ coal to fly ash: a study of coal mines and power plants from Indiana. *International Journal of Coal Geology*, **59**(3-4), 171–92.

Mastalerz, M. and Padgett, P.L. (1999) From *in situ* coal to the final coal product: a case study of the Danville coal member (Indiana). *International Journal of Coal Geology*, **41**(1-2), 107–23.

Masuda, H., Yamatani, Y. and Okai, M. (2005) Transformation of arsenic compounds in modern intertidal sediments of Iriomote Island, Japan. *Journal of Geochemical Exploration*, **87**(2), 73–81.

Matis, K.A., Zouboulis, A.I., Zamboulis, D. and Valtadorou, A.V. (1999) Sorption of As(V) by goethite particles and study of their flocculation. *Water Air and Soil Pollution*, **111**(1-4), 297–316.

Matschullat, J. (2000) Arsenic in the geosphere — a review. *Science of the Total Environment*, **249**(1-3), 297–312.

Matschullat, J., Perobelli Borba, R., Deschamps, E. *et al.* (2000) Human and environmental contamination in the Iron Quadrangle, Brazil. *Applied Geochemistry*, **15**(2), 181–90.

Matsumoto, T. and Hosoda, T. (2000) Arsenic contamination in groundwater and hydrogeological background in Samta village, western Bangladesh, in *Arsenic in Groundwater of Sedimentary Aquifers*, Pre-Congress Workshop, 31st International Geological Congress (eds P. Bhattacharya and A.H., Welch), Groundwater Arsenic Research Group, Division of Land and Water Resources, Royal Institute of Technology, Stockholm, pp. 69–70.

May, T.W., Wiedmeyer, R.H., Gober, J. and Larson, S. (2001) Influence of mining-related activities on concentrations of metals in water and sediment from streams of the Black Hills, South Dakota. *Archives of Environmental Contamination and Toxicology*, **40**(1), 1–9.

Mayo, M.J., Hon, R., Brandon, W.C. and Germansderfer, I. (2003) Elevated arsenic in groundwater at landfill sites in northern central Massachusetts. Abstracts with Programs. *The Geological Society of America*, **35**(3), 4.

Mazur, L., Milanes, C., Randles, K. and Salocks, C. (2004) *Used Oil in Bunker Fuel: A Review of Potential Human Health Implications*, Office of Environmental Health Hazard Assessment, California Environmental Protection Agency, Sacramento, CA.

McArthur, J.M., Ravenscroft, P., Hoque, B.A. and Safiullah, S. (2001) Pollution of groundwater by arsenic in sedimentary aquifers. Abstracts with Programs. *The Geological Society of America*, **33**(6), 115.

McBeth, I., Reddy, K.J. and Skinner, Q.D. (2003) Chemistry of trace elements in coalbed methane product water. *Water Research*, **37**(4), 884–90.

McBride, M.B., Richards, B.K., Steenhuis, T. and Spiers, G. (1999) Long-term leaching of trace elements in a heavily sludge-amended silly clay loam soil. *Soil Science*, **164**(9), 613–23.

McConnell, J.W. (1999) *Results of Geochemical Mapping Employing High-Density Lake-Sediment and Water Sampling in Central Labrador* Report 99-1, Current Research Newfoundland Department of Mines and Energy, Geological Survey, pp. 17–39.

McCreadie, H., Jambor, J.L., Blowes, D.W. (1998) Geochemical behavior of autoclave-produced ferric arsenates: jarosite in a gold-mine tailings impoundment, in *Waste Characterization and Treatment* (ed. W., Petruk), Society for Mining, Metallurgy, and Exploration, Littleton, Colorado, pp. 61–78.

McDonough, W.F. (2004) Composition model for the Earth's core, in *Treatise on Geochemistry: Volume 2: The Mantle and Core* (eds R.W. Carlson, H.D. Holland and K.K. Turekian Editors-in-Chief), Elsevier Science, pp. 547–68.

McDonough, W.F. and Sun, S.-S. (1995) The composition of the earth. *Chemical Geology*, **120**(3-4), 223–53.

McGregor, R.G. and Blowes, D.W. (2002) The physical, chemical and mineralogical properties of three cemented layers within sulfide-bearing mine tailings. *Journal of Geochemical Exploration*, **76**(3), 195–207.

McGregor, R.G., Blowes, D.W., Jambor, J.L. and Robertson, W.D. (1998a) The solid-phase controls on the mobility of heavy metals at the Copper Cliff tailings area, Sudbury, Ontario, Canada. *Journal of Contaminant Hydrology*, **33**(3-4), 247–71.

McGregor, R.G., Blowes, D.W., Jambor, J.L. and Robertson, W.D. (1998b) Mobilization and attenuation of heavy metals within a nickel mine tailings impoundment near Sudbury, Ontario, Canada. *Environmental Geology*, **36**(3-4), 305–19.

McLaren, R.G., Megharaj, M. and Naidu, R. (2005) Fate of arsenic in the soil environment, in *Managing Arsenic in the Environment: From Soil to Human Health* (eds R., Naidu, E., Smith, G., Owens *et al.*), CSIRO Publishing, Collingwood, pp. 157–82.

McTigue, D.F., Stein, C.L. and Brandon, W.C. (2001) Hydrologic controls on arsenic transport in a central Massachusetts overburden aquifer. Abstracts with Programs. *The Geological Society of America*, **33**(6), 53.

Mehrabi, B., Yardley, B.W.D. and Cann, J.R. (1999) Sediment-hosted disseminated gold mineralisation at Zarshuran, NW Iran. *Mineralium Deposita*, **34**(7), 673–96.

Meliker, J.R., Nriagu, J.O., Wahl, R. *et al.* (2002) Arsenic in Groundwater in Michigan: Standardized Mortality Ratio Analysis and Development of a Space-Time Information System. Arsenic in New England: A Multidisciplinary Scientific Conference, May 29-31, 2002, National Institute of Environmental Health Sciences, Superfund Basic Research Program, Manchester, NH.

Méndez, M. and Armienta, M.A. (2003) Arsenic phase distribution in Zimapán mine tailings, Mexico. *Geofisica Internacional*, **42**(1), 131–40.

Mendoza, O.T., Hernández, Ma.A.A., Abundis, J.G. and Mundo, N.F. (2006) Geochemistry of leachates from the El Fraile sulfide tailings piles in Taxco, Guerrero, southern Mexico. *Environmental Geochemistry and Health*, **28**(3), 243–55.

Merwin, I., Pruyne, L.P.T., Ebel, J.G. Jr. *et al.* (1994) Persistence, phytotoxicity, and management of arsenic, lead and mercury residues in old orchard soils of New York State. *Chemosphere*, **29**(6), 1361–67.

Metz, S. and Trefry, J.H. (1993) Field and laboratory studies of metal uptake and release by hydrothermal precipitates. *Journal of Geophysical Research*, **98**(B6), 9661–666.

Middelburg, J.J., Hoede, D., Van Der Sloot, H.A. *et al.* (1988) Arsenic, antimony and vanadium in the North Atlantic Ocean. *Geochimica et Cosmochimica Acta*, **52**(12), 2871–78.

Mielke, R.E., Southam, G. and Nordstrom, D.K. (2000) Arsenic resistant/oxidizing bacteria in acidic geothermal environments, Yellowstone National Park, Wyoming. Abstracts with Programs. *The Geological Society of America*, **32**(7), 190.

Migon, C. and Mori, C. (1999) Arsenic and antimony release from sediments in a Mediterranean estuary. *Hydrobiologia*, **392**(1), 81–88.

Mihaljevič, M., Sisr, L., Ettler, V. *et al.* (2004) Oxidation of As-bearing gold ore- a comparison of batch and column experiments. *Journal of Geochemical Exploration*, **81**, 59–70.

Millward, G.E., Kitts, H.J., Ebdon, L. *et al.* (1997) Arsenic in the thames plume, UK. *Marine Environmental Research*, **44**(1), 51–67.

Minami, H. and Kato, Y. (1997) Remobilization of arsenic in sub-oxic sediments from the seafloor of the continental margin. *Journal of Oceanography*, **53**(6), 553–62.

Minnesota Pollution Control Agency (1998) *Arsenic in Minnesota's Ground Water*, Ground Water Monitoring and Assessment Program, St. Paul, MN.

Mirlean, N., Andrus, V.E., Baisch, P. *et al.* (2003) Arsenic pollution in Patos Lagoon estuarine sediments, Brazil. *Marine Pollution Bulletin*, **46**(11), 1480–84.

Mitsunobu, S., Harada, T. and Takahashi, Y. (2006) Comparison of antimony behavior with that of arsenic under various soil redox conditions. *Environmental Science and Technology*, **40**(23), 7270–76, and supplements.

Mittlefehldt, D.W. (2004) Achondrites, in *Treatise on Geochemistry: Volume 1: Meteorites, Comets, and Planets* (eds A.M. Davis, H.D. Holland and K.K. Turekian Editors-in-Chief) Elsevier Science, pp. 291–324.

Mohammad, S., Ahmed, K.M., Saunders, J.A. and Lee, M.K. (2003) Occurrence and distribution of arsenic in groundwater of Singair Thana, Manikganj District, Bangladesh. Abstracts with Programs. *The Geological Society of America*, **35**(2), 6.

Mok, W.M., Riley, J.A. and Wai, C.M. (1988) Arsenic speciation and quality of groundwater in a lead-zinc mine, Idaho. *Water Research*, **22**(6), 769–74.

Mok, W.-M. and Wai, C.M. (1990) Distribution and mobilization of arsenic and antimony species in the Coeur d'Alene River, Idaho. *Environmental Science and Technology*, **24**(1), 102–108.

Mok, W.M. and Wai, C.M. (1994) Mobilization of arsenic in contaminated river waters, in *Arsenic in the Environment: Part I: Cycling and Characterization* (eds J.O. Nriagu), John Wiley & Sons, Ltd, New York, pp. 99–117.

Moldovan, B.J., Hendry, M.J. and Jiang, D.T. (2001) Geochemical controls on arsenic in uranium mine tailings. Abstracts with Programs. *The Geological Society of America*, **33**(6), 187.

Moncur, M.C., Ptacek, C.J., Blowes, D.W. *et al.* (2002) Impact of mine drainage on a lake from an abandoned tailings impoundment. Abstracts with Programs. *The Geological Society of America*, **34**(6), 51.

Moore, J.N. and Woessner, W.W. (2003) Arsenic contamination in the water supply of Milltown, Montana, in *Arsenic in Ground Water* (eds A.H. Welch and K.G. Stollenwerk), Kluwer Academic Publishers, Boston, MA, pp. 329–50.

Morace, J.L., Fuhrer, G.J., Rinella, J.F., McKenzie, S.W. *et al.* (1999) *Surface-Water-Quality Assessment of the Yakima River Basin, Washington: Overview of Major Findings, 1987–91* U.S. Geological Survey Water-Resources Investigations Report 98-4113, U.S. Geological Survey, Portland, OR.

Morin, G., Juillot, F., Casiot, C. *et al.* (2003) Bacterial formation of tooeleite and mixed arsenic(III) or arsenic(V)–iron(III) gels in the Carnoulès acid mine drainage, France. A XANES, XRD, and SEM study. *Environmental Science and Technology*, **37**(9), 1705–12.

Morrell, J.J. and Huffman, J. (2004) Copper, chromium, and arsenic levels in soils surrounding posts treated with chromated copper arsenate (CCA). *Wood and Fiber Science*, **36**(1), 119–28.

Morrell, J.J., Keefe, D. and Baileys, R.T. (2003) Copper, zinc, and arsenic in soil surrounding douglas-fir poles treated with ammoniacal copper zinc arsenate (ACZA). *Journal of Environmental Quality*, **32**(6), 2095–99.

Morrison, R.J., Gangaiya, P., Naqasima, M.R. and Naidu, R. (1997) Trace metal studies in the great astrolabe lagoon, Fiji, a pristine marine environment. *Marine Pollution Bulletin*, **34**(5), 353–56.

Morton, O., Carrillo, C.A. and Hernandez, E. (2000) Geochemical characterization of historical mine tailings and groundwater in the Guanajuato mining district, central Mexico: environmental considerations. Abstracts with Programs. *The Geological Society of America*, **32**(7), 488.

Mozgova, N.N., Borodaev, Yu.S., Gablina, I.F. *et al.* (2005) Mineral assemblages as indicators of the maturity of oceanic hydrothermal sulfide mounds. *Lithology and Mineral Resources*, **40**(4), 293–319.

Mroczek, E.K. (2005) Contributions of arsenic and chloride from the Kawerau geothermal field to the Tarawera River, New Zealand. *Geothermics*, **34**(2 Special Issue), 223–38.

Mucci, A., Richard, L.-F., Lucotte, M. and Guignard, C. (2000) The differential geochemical behavior of arsenic and phosphorus in the water column and sediments of the Saguenay Fjord estuary, Canada. *Aquatic Geochemistry*, **6**(3), 293–324.

Mueller, S.H., Goldfarb, R.J., Farmer, G.L. *et al.* (2001) A seasonal study of the arsenic and groundwater geochemistry in Fairbanks, Alaska, in *Mineral Deposits at the Beginning of the 21st Century* (eds A. Piestrzynski *et al.*), Society for Geology Applied to Mineral Deposits, pp. 1043–46.

Mueller, S., Goldfarb, R. and Verplanck, P. (2001) *Ground-water Studies in Fairbanks, Alaska- A Better Understanding of Some of the United States' Highest Natural Arsenic Concentrations*, U.S. Geological Survey Fact Sheet FS-111-01.

Mukherjee, S. and Srivastava, S.K. (2005) Trace elements in high-sulfur Assam coals from the Makum coalfield in the northeastern region of India. *Energy and Fuels*, **19**(3), 882–91.

Mukhopadhyay, P.K., Goodarzi, F., Crandlemire, A.L. *et al.* (1998) Comparison of coal composition and elemental distribution in selected seams of the Sydney and Stellarton Basins, Nova Scotia, eastern Canada. *International Journal of Coal Geology*, **37**(1-2), 113–41.

Mukhopadhyay, P.K., Lajeunesse, G., Crandlemire, A.L. and Finkelman, R.B. (1999) Mineralogy and geochemistry of selected coal seams and their combustion residues from the Sydney area, Nova Scotia, Canada. *International Journal of Coal Geology*, **40**(2-3), 253–54.

Muloin, T. and Dudas, M.J. (2005) Aqueous phase arsenic in weathered shale enriched in native arsenic. *Journal of Environmental Engineering and Science*, **4**(6), 461–68.

Murnock, J.M. (2002) *Vertical and Seasonal Distributions and Relationships of Arsenic in Presques Isle Sediments*. Master of Science Thesis, Gannon University, Erie, Pennsylvania, p. 81.

Neal, C., Williams, R.J., Neal, M. *et al.* (2000) The water quality of the river Thames at a rural site downstream of Oxford. *Science of the Total Environment*, **251-252**, 441–57.

Nédez, C., Boitiaux, J.-P., Cameron, C.J. and Didillon, B. (1996) Optimization of the textural characteristics of an alumina to capture contaminants in natural gas. *Langmuir*, **12**(16), 3927–31.

Néel, C., Bril, H., Courtin-Nomade, A. and Dutreuil, J.-P. (2003) Factors affecting natural development of soil on 35-year-old sulphide-rich mine tailings. *Geoderma*, **111**(1-2), 1–20.

Neff, J.M. (1997) Ecotoxicology of arsenic in the marine environment. *Environmental Toxicology and Chemistry*, **16**(5), 917–27.

Nesbitt, H.W. and Muir, I.J. (1998) Oxidation states and speciation of secondary products on pyrite and arsenopyrite reacted with mine waste waters and air. *Mineralogy and Petrology*, **62**, 123–44.

Nesbitt, H.W., Muir, I.J. and Pratt, A.R. (1995) Oxidation of arsenopyrite by air and air-saturated, distilled water, and implications for mechanism of oxidation. *Geochimica et Cosmochimica Acta*, **59**(9), 1773–86.

Nicholas, D.R., Ramamoorthy, S., Palace, V. *et al.* (2003) Biogeochemical transformations of arsenic in circumneutral freshwater sediments. *Biodegradation*, **14**(2), 123–37.

Nichols, E.M. (2003) Monthly variations in metal and arsenic concentrations before and after the draw-down of the Milltown reservoir. Abstracts with Programs. *The Geological Society of America*, **35**(5), 37.

Nicholson, R. (1999) Prediction of acidic drainage and metal leaching from sulfide mine waste: beyond yes or no answers. Abstracts with Programs. *The Geological Society of America*, **31**(7), 334.

Nicholson, A.D., Hanna, T., Mansanti, J. *et al.* (1996) Evolution of groundwater chemistry during dewatering: Lone Tree Mine, Nevada. Abstracts with Programs. *The Geological Society of America*, **28**(7), 467.

Nimick, D.A. (1998) Arsenic hydrogeochemistry in an irrigated river valley — a reevaluation. *Ground Water*, **36**(5), 743–53.

Nimick, D.A., Gammons, C.H., Cleasby, T.E. *et al.* (2003) Diel cycles in dissolved metal concentrations in streams: occurrence and possible causes. *Water Resources Research*, **39**(9), HWC21–HWC217.

Nishikawa, O., Okrugin, V., Belkova, N. *et al.* (2006) Crystal symmetry and chemical composition of yukonite: TEM study of specimens collected from Nalychevske hot springs, Kamchatka, Russia and from Venus mine, Yukon Territory, Canada. *Mineralogical Magazine*, **70**(1), 73–81.

Noble, D.C., Ressel, M.W. and Connors, K.A. (1998) Magmatic As, Sb, Cs and other volatile elements in glassy silicic rocks. Abstracts with Programs. *The Geological Society of America*, **30**(7), 377.

Noll, K.S. and Larson, H.P. (1991) The spectrum of Saturn from 1990–2230/cm — Abundances of AsH_3, CH_3D, CO, GeH_4, NH_3, and PH_3. *Icarus*, **89**, 168–89.

Noll, K.S., Larson, H.P. and Geballe, T.R. (1990) The abundance of AsH_3 in Jupiter. *Icarus*, **83**(2), 494–99.

Noll, P.D. Jr., Newsom, H.E., Leeman, W.P. and Ryan, J.G. (1996) The role of hydrothermal fluids in the production of subduction zone magmas: evidence from siderophile and chalcophile trace elements and boron. *Geochimica et Cosmochimica Acta*, **60**(4), 587–611.

Nordstrom, D.K. (2002) Worldwide occurrences of arsenic in ground water. *Science*, **296**(5576), 2143–45.

Nordstrom, D.K. and Alpers, C.N. (1999) Negative pH, efflorescent mineralogy, and consequences for environmental restoration at the Iron Mountain Superfund site, California. *Proceedings of the National Academy of Sciences of the United States of America*, **96**(7), 3455–62.

Nordstrom, D.K. and Archer, D.G. (2003) Arsenic thermodynamic data and environmental geochemistry, in *Arsenic in Ground Water* (eds A.H., Welch and K.G., Stollenwerk), Kluwer Academic Publishers, Boston, MA, pp. 1–25.

Nordstrom, D.K., Ball, J.W. and McCleskey, R.B. (2003) Orpiment solubility equilibrium and arsenic speciation for a hot spring at Yellowstone National Park using revised thermodynamic data. Abstracts with Programs. *The Geological Society of America*, **35**(6), 47.

Nordstrom, D.K. and Parks, G.A. (1987) Solubility and stability of scorodite, FeAsO4.2H2O: discussion. *American Mineralogist*, **72**(7-8), 849–51.

North, F.K. (1990) *Petroleum Geology*, Unwin Hyman, Boston, MA, p. 631.

Nriagu, J.O. (1989) A global assessment of natural sources of atmospheric trace metals. *Nature*, **338**(6210), 47–49.

Nriagu, J.O. (2002) Arsenic poisoning through the ages, in *Environmental Chemistry of Arsenic* (ed. W.T. Frankenberger Jr.), Marcel Dekker, New York, pp. 1–26.

Nriagu, J.O. and Pacyna, J.M. (1988) Quantitative assessment of worldwide contamination of air, water and soils by trace metals. *Nature*, **333**(6169), 134–39.

O'Brien, S.J., Dubé, B. and O'Driscoll C.F. (1999) High-Sulphidation, Epithermal-Style Hydrothermal Systems in Late Neoproterozoic Avalonian rocks on the Burin Peninsula, Newfoundland: Implications for Gold Exploration Report 99-1, Current Research Newfoundland Department of Mines and Energy Geological Survey, pp. 275–96.

O'Day, P.A., Vlassopoulos, D., Root, R. and Rivera, N. (2004) The influence of sulfur and iron on dissolved arsenic concentrations in the shallow subsurface under changing redox conditions. *Proceedings of the National Academy of Sciences of the United States of America*, **101**(38), 13703–708.

Ódor, L., Wanty, R.B., Horváth, I. and Fügedi, U. (1998) Mobilization and attenuation of metals downstream from a base-metal mining site in the Matra Mountains, northeastern Hungary. *Journal of Geochemical Exploration*, **65**(1 pt 2), 47–60.

Ohnuki, T., Sakamoto, F., Kozai, N. *et al.* (2004) Mechanisms of arsenic immobilization in a biomat from mine discharge water. *Chemical Geology*, **212**(3-4 SPEC.ISS.), 279–90.

Ölmez, E.I., Kut, D., Bilge, A.N. and Ölmez, I. (2004) Regional elemental signatures related to combustion of lignites. *Journal of Radioanalytical and Nuclear Chemistry*, **259**(2), 227–31.

Ona-Nguema, G., Abdelmoula, M., Jorand, F. *et al.* (2002) Iron(II,III) hydroxycarbonate green rust formation and stabilization from lepidocrocite bioreduction. *Environmental Science and Technology*, **36**(1), 16–20.

Ona-Nguema, G., Morin, G., Juillot, F. *et al.* (2005) *Jr.* EXAFS analysis of arsenite adsorption onto two-line ferrihydrite, hematite, goethite, and lepidocrocite. *Environmental Science and Technology*, **39**(23), 9147–55.

O'Reilly Wiese, S.B., Bubb, J.M. and Lester, J.N. (1995) The significance of sediment metal concentrations in two eroding Essex salt marshes. *Marine Pollution Bulletin*, **30**(3), 190–99.

Oremland, R.S. and Stolz, J.F. (2003) The ecology of arsenic. *Science*, **300**(5621), 939–44.

Oremland, R.S., Stolz, J.F. and Hollibaugh, J.T. (2004) The microbial arsenic cycle in Mono Lake, California. *FEMS Microbiology Ecology*, **48**(1), 15–27.

Ortega-Guerrero, A. (2003) Arsenic in groundwater at the southernmost end of the Cordilleran. Abstracts with Programs. *The Geological Society of America*, **35**(4), 24.

Oyarzun, R., Guevara, S., Oyarzún, J. *et al.* (2006) The As-contaminated Elqui river basin: a long lasting perspective (1975–1995) covering the initiation and development of Au-Cu-As mining in the high Andes of northern Chile. *Environmental Geochemistry and Health*, **28**(5), 431–43.

Oyarzun, R., Lillo, J., Higueras, P. *et al.* (2004) Strong arsenic enrichment in sediments from the Elqui watershed, northern Chile: industrial (gold mining at El Indio-Tambo district) vs. geologic processes. *Journal of Geochemical Exploration*, **84**(2), 53–64.

Pagel, B.E.J. (1997) *Nucleosynthesis and Chemical Evolution of Galaxies*, Cambridge University Press, Cambridge, p. 378.

Palme, H. and Jones, A. (2004) Solar System abundances of the elements, in *Treatise on Geochemistry: Volume 2: The Mantle and Core* (eds R.W. Carlson, H.D. Holland and K.K. Turekian Editors-in-Chief), Elsevier Science, pp. 41–61.

Palme, H. and O'Neill, H.St.C. (2004) Cosmochemical estimates of mantle composition, in *Treatise on Geochemistry: Volume 2: The Mantle and Core* (eds R.W. Carlson, H.D. Holland and K.K. Turekian Editors-in-Chief), Elsevier Science, pp. 1–38.

Palmer, C.A. (1990) Determination of twenty-nine elements in eight argonne premium coal samples by instrumental neutron activation analysis. *Energy and Fuels*, **4**, 436–39.

Palmer, N.E., Freudenthal, J.H. and von Wandruszka, R. (2006) Reduction of arsenates by humic materials. *Environmental Chemistry*, **3**(2), 131–36.

Palmer, C.A., Lyons, P.C., Brown, Z.A. and Moe, J.S. (1990) *The Use of Rare Earth and Trace Element Concentrations in Vitrinite Concentrates and Companion Whole Coals (hvA bituminous) to Determine Organic and Inorganic Associations*, Geological Society of America Special Paper 248, pp. 55–62.

Palumbo-Roe, B., Cave, M.R., Klinck, B.A. *et al.* (2005) Bioaccessibility of arsenic in soils developed over Jurassic ironstones in eastern England. *Environmental Geochemistry and Health*, **27**(2), 121–30.

Pandey, P.K., Sharma, R., Roy, M. *et al.* (2006) Arsenic contamination in the Kanker district of central-east India: geology and health effects. *Environmental Geochemistry and Health*, **28**(5), 409–20.

Panno, S.V., Sayre, E.V. and Harbottle, G. (1985) Alteration of shale of the Davis formation overlying a Pb-Zn deposit of the viburnum trend, S.E. Missouri. Abstracts with Programs. *The Geological Society of America*, **17**(7), 684.

Panov, B.S., Dudik, A.M., Shevchenko, O.A. and Matlak, E.S. (1999) On pollution of the biosphere in industrial areas: the example of the Donets coal Basin. *International Journal of Coal Geology*, **40**(2-3), 199–210.

Papanicolaou, C., Kotis, T., Foscolos, A. and Goodarzi, F. (2004) Coals of Greece: a review of properties, uses and future perspectives. *International Journal of Coal Geology*, **58**(3), 147–69.

Parker, J.T.C., Fossum, K.D. and Ingersoll, T.L. (2000) Chemical characteristics of urban stormwater sediments and implications for environmental management, Maricopa County, Arizona. *Environmental Management*, **26**(1), 99–115.

Parker, R., Herbert, B.E., Brandenberger, J. and Louchouarn, P. (2001) Ground water discharge from mid-Tertiary rhyolitic ash-rich sediments as the source of elevated arsenic in South Texas surface waters. Abstracts with Programs. *The Geological Society of America*, **33**(6), 53.

Pascua, C., Charnock, J., Polya, D.A. *et al.* (2005) Arsenic-bearing smectite from the geothermal environment. *Mineralogical Magazine*, **69**(5), 897–906.

Patel, K.S., Shrivas, K., Brandt, R. *et al.* (2005) Arsenic contamination in water, soil, sediment and rice of central India. *Environmental Geochemistry and Health*, **27**(2), 131–45.

Paulu, C.A., Moll, D.M., Backer, L.C. *et al.* (2002) Exposure to Arsenic Via Bathing and Other Contact in Households that Use Bottled Water or Point-of-Use Treatment Devices for Drinking Water. Arsenic in New England: A Multidisciplinary Scientific Conference, May 29-31, 2002, National Institute of Environmental Health Sciences, Superfund Basic Research Program, Manchester, MA.

Pedersen, H.D., Postma, D. and Jakobsen, R. (2006) Release of arsenic associated with the reduction and transformation of iron oxides. *Geochimica et Cosmochimica Acta*, **70**(16), 4116–29.

Peña-Vázquez, E., Villanueva-Alonso, J., Bermejo-Barrera, A. and Bermejo-Barrera, P. (2006) Arsenic and antimony distribution in the Ría de Arousa: before and after the Prestige oil tanker sinking. *Journal of Environmental Monitoring*, **8**(6), 641–48.

Pentecost, A., Jones, B. and Renaut, R.W. (2003) What is a hot spring? *Canadian Journal of Earth Sciences*, **40**(11), 1443–46.

Percival, T.J., Radtke, A.S., Bagby, W.C. *et al.* (1989) Bau, east Malaysia: arsenic-rich sedimentary-rock hosted gold deposits spatially and genetically associated with epizonal magmatism. Abstracts with Programs. *The Geological Society of America*, **21**(6), 294.

Perkins, D. (1998) *Mineralogy*, Prentice Hall, Upper Saddle River, NJ, p. 484.

Peryea, F.J. and Kammereck, R. (1997) Phosphate-enhanced movement of arsenic out of lead arsenate-contaminated topsoil and through uncontaminated subsoil. *Water Air and Soil Pollution*, **93**(1-4), 243–54.

Peters, S.C. and Blum, J.D. (2003) The source and transport of arsenic in a bedrock aquifer, New Hampshire, USA. *Applied Geochemistry*, **18**(11), 1773–87.

Peterson, M.L. and Carpenter, R. (1986) Arsenic distributions in porewaters and sediments of Puget Sound, Lake Washington, the Washington Coast and Saanich Inlet, BC. *Geochimica et Cosmochimica Acta*, **50**(3), 353–69.

Petrunic, B.M. and Al, T.A. (2005) Mineral/water interactions in tailings from a tungsten mine, Mount Pleasant, New Brunswick. *Geochimica et Cosmochimica Acta*, **69**(10), 2469–83.

Pfeifer, H.-R., Gueye-Girardet, A., Reymond, D. *et al.* (2004) Dispersion of natural arsenic in the Malcantone watershed, southern Switzerland: Field evidence for repeated sorption-desorption and oxidation-reduction processes. *Geoderma*, **122**(2-4 SPEC. IIS.), 205–34.

Pfeifer, H.-R. and Zobrist, J. (2002) Arsenic in deep groundwater of Switzerland and their environmental impact and health risk. *Geochimica et Cosmochimica Acta*, **66**(15A), 597.

Phoenix, V.R., Renaut, R.W., Jones, B. and Ferris, F.G. (2005) Bacterial S-layer preservation and rare arsenic-antimony-sulphide bioimmobilization in siliceous sediments from Champagne pool hot spring, Waiotapu, New Zealand. *Journal of the Geological Society*, **162**(2), 323–31.

Pichler, T., Hendry, M.J. and Hall, G.E.M. (2001) The mineralogy of arsenic in uranium mine tailings at the Rabbit Lake in-pit facility, northern Saskatchewan, Canada. *Environmental Geology*, **40**(4-5), 495–506.

Pichler, T. and Veizer, J. (1999) Precipitation of Fe(III) oxyhydroxide deposits from shallow-water hydrothermal fluids in Tutum Bay, Ambitle Island, Papua New Guinea. *Chemical Geology*, **162**(1), 15–31.

Pichler, T., Veizer, J. and Hall, G.E.M. (1999) Natural input of arsenic into a coral-reef ecosystem by hydrothermal fluids and its removal by Fe(III) oxyhydroxides. *Environmental Science and Technology*, **33**(9), 1373–78.

Pilgrim, W. and Schroeder, B. (1997) Multi-media concentrations of heavy metals and major ions from urban and rural sites in New Brunswick, Canada. *Environmental Monitoring and Assessment*, **47**(1), 89–108.

Pirrie, D., Power, M.R., Wheeler, P.D. *et al.* (2002) Geochemical signature of historical mining: Fowey estuary, Cornwall, UK. *Journal of Geochemical Exploration*, **76**(1), 31–43.

Piske, B. (1990) *Hydrogeologic evaluation of tailings deposits at the Coeur d'Alene River delta.* Master's Thesis, University of Idaho, Moscow.

Pizarro, I., Gomez, Ma.M., Cámara, C. and Palacios, Ma.A. (2003) Distribution of arsenic species in environmental samples collected in northern Chile. *International Journal of Environmental Analytical Chemistry*, **83**(10), 2879–90.

Planer-Friedrich, B., Armienta, M.A. and Merkel, B.J. (2001) Origin of arsenic in the groundwater of the Rioverde Basin, Mexico. *Environmental Geology*, **40**(10), 1290–98.

Planer-Friedrich, B., Lehr, C., Matschullat, J. *et al.* (2006) Speciation of volatile arsenic at geothermal features in Yellowstone National Park. *Geochimica et Cosmochimica Acta*, **70**(10), 2480–91.

Planer-Friedrich, B., London, J., McCleskey, R.B. *et al.* (2007) Thioarsenates in geothermal waters of Yellowstone National Park: determination, preservation, and geochemical importance. *Environmental Science and Technology*, **41**(15), 5245–51.

Pokrovski, G., Gout, R., Schott, J. *et al.* (1996) Thermodynamic properties and stoichiometry of As(III) hydroxide complexes at hydrothermal conditions. *Geochimica et Cosmochimica Acta*, **60**(5), 737–49.

Pokrovski, G.S., Kara, S. and Roux, J. (2002) Stability and solubility of arsenopyrite, FeAsS, in crustal fluids. *Geochimica et Cosmochimica Acta*, **66**(13), 2361–78.

Polissar, A.V., Hopke, P.K. and Poirot, R.L. (2001) Atmospheric aerosol over Vermont: chemical composition and sources. *Environmental Science and Technology*, **35**(23), 4604–21.

Polya, D.A., Gault, A.G., Diebe, N. *et al.* (2005) Arsenic hazard in shallow Cambodian groundwaters. *Mineralogical Magazine*, **69**(5), 807–23.

Pope, L.M., Rosner, S.M., Hoffman, D.C. and Ziegler, A.C. (2004) Summary of Available State Ambient Stream-water-quality Data, 1990–98, and Limitations for National Assessment U.S. Geological Survey Water-Resources Investigations Report 03-4316, U.S. Geological Survey.

Pouschat, P. and Zagury, G.J. (2006) In vitro gastrointestinal bioavailability of arsenic in soils collected near CCA-treated utility poles. *Environmental Science and Technology*, **40**(13), 4317–23.

Power, M.R., Pirrie, D., Jedwab, J. and Stanley, C.J. (2004) Platinum-group element mineralization in an As-rich magmatic sulphide system, Talnotry, southwest Scotland. *Mineralogical Magazine*, **68**(2), 395–411.

Pratt, G.C., Palmer, K., Wu, C.Y. *et al.* (2000) An assessment of air toxics in Minnesota. *Environmental Health Perspectives*, **108**(9), 815–25.

Presley, S.M., Rainwater, T.R., Austin, G.P. *et al.* (2006) Assessment of pathogens and toxicants in New Orleans, LA following Hurricane Katrina. *Environmental Science and Technology*, **40**(2), 468–74.

Press, F. and Siever, R. (2001) *Understanding Earth*, 3rd edn, W.H. Freeman & Company, New York.

Price, R.E. and Pichler, T. (2005) Distribution, speciation and bioavailability of arsenic in a shallow-water submarine hydrothermal system, Tutum Bay, Ambitle Island, PNG. *Chemical Geology*, **224**(1-3), 122–35.

Price, R.E. and Pichler, T. (2006) Abundance and mineralogical association of arsenic in the Suwannee Limestone (Florida): implications for arsenic release during water-rock interaction. *Chemical Geology*, **228**(1-3 Special Issue), 44–56.

Priyadarshi, N. (2004) Distribution of arsenic in Permian coals of north Karanpura coalfield, Jharkhand. *Journal of the Geological Society of India*, **63**(5), 533–36.

Querol, X., Whateley, M.K.G., Fernández-Turiel, J.L. and Tuncali, E. (1997) Geological controls on the mineralogy and geochemistry of the Beypazari Lignite, central Anatolia, Turkey. *International Journal of Coal Geology*, **33**(3), 255–71.

Quiseft, J.P., Toutain, J.P., Bergametti, G. *et al.* (1989) Evolution versus cooling of gaseous volcanic emissions from Momotombo Volcano, Nicaragua: thermochemical model and observations. *Geochimica et Cosmochimica Acta*, **53**, 2591–608.

Rabbi, F. (1994) *Trace Element Geochemistry of Bottom Sediments and Waters from the Lateral Lakes of Coeur d'Alene River, Kootenai County, North Idaho*. Dissertation, University of Idaho, Moscow.

Rahman, M.M., Sengupta, M.K. and Chowdhury, U.K (2006) Arsenic contamination incidents around the world, in *Managing Arsenic in the Environment: From Soil to Human Health* (eds R. Naidu, E. Smith, G. Owens *et al.*), CSIRO Publishing, Collingwood, pp. 3–30.

Rancourt, D.G., Fortin, D., Pichler, T. *et al.* (2001) Mineralogy of a natural As-rich hydrous ferric oxide coprecipitate formed by mixing of hydrothermal fluid and seawater: implications regarding surface complexation and color banding in ferrihydrite deposits. *American Mineralogist*, **86**(7-8), 834–51.

Randall, S.R., Sherman, D.M. and Ragnarsdottir, K.V. (2001) Sorption of As(V) on green rust ($Fe_4(II)Fe_2(III)$ $(OH)_{12}SO_4.3H_2O$) and lepidocrocite (γ-FeOOH): surface complexes from EXAFS spectroscopy. *Geochimica et Cosmochimica Acta*, **65**(7), 1015–23

Rasilainen, K., Nurmi, P.A. and Bornhorst, T.J. (1993) Rock geochemical implications for gold exploration in the late Archean Hattu Schist belt, Ilomantsi, eastern Finland. *Special Paper–Geological Survey of Finland*, **17**, 353–62.

Rawlins, B.G., Lister, T.R. and Mackenzie, A.C. (2002) Trace-metal pollution of soils in northern England. *Environmental Geology*, **42**(6), 612–20.

Razo, I., Carrizales, L., Castro, J. *et al.* (2004) Arsenic and heavy metal pollution of soil, water and sediments in a semi-arid climate mining area in Mexico. *Water Air and Soil Pollution*, **152**(1-4), 129–52.

Redman, A.D., Macalady, D.L. and Ahmann, D. (2002) Natural organic matter affects arsenic speciation and sorption onto hematite. *Environmental Science and Technology*, **36**(13), 2889–96.

Reeve, A.S., Horesh, M., Warner, B. and Yates, M. (2001) Geochemical evaluation of sources for arsenic in ground water. Abstracts with Programs. *The Geological Society of America*, **33**(1), 61.

Refait, P.H., Abdelmoula, M. and Génin, J.-M.R. (1998) Mechanisms of formation and structure of green rust one in aqueous corrosion of iron in the presence of chloride ions. *Corrosion Science*, **40**(9), 1547–60.

Reich, M. and Becker, U. (2006) First-principles calculations of the thermodynamic mixing properties of arsenic incorporation into pyrite and marcasite. *Chemical Geology*, **225**(3-4), 278–90.

Reich, M., Kesler, S.E., Utsunomiya, S. *et al.* (2005) Solubility of gold in arsenian pyrite. *Geochimica et Cosmochimica Acta*, **69**(11), 2781–96.

Reid, K.D., Goff, F. and Counce, D.A. (2003) Arsenic concentration and mass flow rate in natural waters of the Valles Caldera and Jemez mountains region, New Mexico. *New Mexico Geology*, **25**(3), 75–81.

Reis, A.P., Sousa, A.J., Da Silva, E.F. and Fonseca, E.C. (2005) Application of geostatistical methods to arsenic data from soil samples of the Cova dos Mouros mine (Vila Verde-Portugal). *Environmental Geochemistry and Health*, **27**(3), 259–70.

Reisinger, H.J., Burris, D.R. and Hering, J.G. (2005) Remediating subsurface arsenic contamination with monitored natural attenuation. *Environmental Science and Technology*, **39**(22), 458A–464A.

Ren, D., Xu, D. and Zhao, F. (2004) A preliminary study on the enrichment mechanism and occurrence of hazardous trace elements in the Tertiary lignite from the Shenbei coalfield, China. *International Journal of Coal Geology*, **57**(3-4), 187–96.

Ren, D., Zhao, F., Wang, Y. and Yang, S. (1999) Distributions of minor and trace elements in Chinese coals. *International Journal of Coal Geology*, **40**(2-3), 109–18.

Rencz, A.N., O'Driscoll, N.J., Hall, G.E.M. *et al.* (2003) Spatial variation and correlations of mercury levels in the terrestrial and aquatic components of a wetland dominated ecosystem: Kejimkujik Park, Nova Scotia, Canada. *Water Air and Soil Pollution*, **143**(1-4), 271–85.

Revazyan, R.H. (1999) Technological ecosystems and creation of anti-filter barriers preventing pollution of groundwater, in *Sudbury '99: Mining and the Environment II*, Conference Proceedings (eds D.E. Goldsack, N. Belzile, P. Yearwood and G.J. Hall), Vol. **3**, pp. 1225–29.

Reynolds, J.G., Naylor, D.V. and Fendorf, S.E. (1999) Arsenic sorption in phosphate-amended soils during flooding and subsequent aeration. *Soil Science Society of America Journal*, **63**(5), 1149–156.

Rice, K.C. (1999) Trace-element concentrations in streambed sediment across the conterminous United States. *Environmental Science and Technology*, **33**(15), 2499–504.

Riedel, G.F., Sanders, J.G. and Osman, R.W. (1998) The effect of biological and physical disturbances on the transport of arsenic from contaminated estuarine sediments. *Estuarine Coastal and Shelf Science*, **25**, 693–706.

Rimstidt, J.D. and Vaughan, D.J. (2003) Pyrite oxidation: A state-of-the-art assessment of the reaction mechanism. *Geochimica et Cosmochimica Acta*, **67**(5), 873–80.

Roberts, M.D., Herbert, B.E. and Louchouarn, P. (2001) Organoarsenicals: the missing arsenic sink. Abstracts with Programs. *The Geological Society of America*, **33**(6), 53.

Robins, R.G. (1987) Solubility and stability of scorodite, $FeAsO_4.2H_2O$: discussion. *American Mineralogist*, **72**(7-8), 842–44.

Robins, R.G. and Tozawa, K. (1982) Arsenic removal from gold processing waste waters: the potential ineffectiveness of lime. *CIM Bulletin*, **75**(840), 171–74.

Rognerud, S. and Fjeld, E. (2001) Trace element contamination of Norwegian lake sediments. *Ambio*, **30**(1), 11–19.

Romero, L., Alonso, H., Campano, P. *et al.* (2003) Arsenic enrichment in waters and sediments of the Rio Loa (second region, Chile). *Applied Geochemistry*, **18**(9), 1399–1416.

Rösner, U. (1998) Effects of historical mining activities on surface water and groundwater — an example from northwest Arizona. *Environmental Geology*, **33**(4), 224–30.

Ross, C.S., Bain, J.G. and Blowes, D.W. (1999) Transport and attenuation of arsenic from a gold mine tailings impoundment, in *Sudbury '99: Mining and the Environment II*, Conference Proceedings (eds. D.E. Goldsack, N. Belzile, P. Yearwood and G.J. Hall, Vol. **2**, pp. 745–54.

Roychowdhury, T., Uchino, T., Tokunaga, H. and Ando, M. (2002) Arsenic and other heavy metals in soils from an arsenic-affected area of West Bengal, India. *Chemosphere*, **49**(6), 605–18.

Ruppert, L.F., Hower, J.C. and Eble, C.F. (2005) Arsenic-bearing pyrite and marcasite in the fire clay coal bed, middle Pennsylvanian Breathitt Formation, eastern Kentucky. *International Journal of Coal Geology*, **63**(Special Issue 1-2), 27–35.

Ruppert, L.F., Minkin, J.A., McGee, J.J. and Cecil, C.B. (1992) An unusual occurrence of arsenic-bearing pyrite in the Upper Freeport coal bed, west-central Pennsylvania. *Energy and Fuels*, **6**(2), 120–25.

Ryan, J., Mooris, J., Behout, G. and Leeman, B. (1996) Describing chemical fluxes in subduction zones: Insights from 'depth-profiling' studies of arc and forearc rocks, in *Subduction: Top to Bottom* (eds G.E. Bebout, D.W. Scholl, S.H. Kirby and J.P. Platt), American Geophysical Union, Geophysical Monograph, Vol. **96**, pp. 263–68.

Ryu, J.-H., Gao, S., Dahlgren, R.A. and Zierenberg, R.A. (2002) Arsenic distribution, speciation and solubility in shallow groundwater of Owens Dry Lake, California. *Geochimica et Cosmochimica Acta*, **66**(17), 2981–994.

Sadiq, M. (1997) Arsenic chemistry in soils: an overview of thermodynamic predictions and field observations. *Water Air and Soil Pollution*, **93**(1-4), 117–36.

Sadiq, M., Locke, A., Spiers, G. and Pearson, D.A.B. (2002) Geochemical behavior of arsenic in Kelly Lake, Ontario. *Water Air and Soil Pollution*, **141**(1-4), 299–312.

Salters, V. and Stracke, A. (2004) Composition of the depleted mantle. *Geochemistry, Geophysics, Geosystems*, **5**(5), 1525–2027.

Salvioli-Mariani, E., Toscani, L. and Venturelli, G. (2001) Weathering of granodiorite and micaschists, and soil pollution at Mt. Mottarone (northern Italy). *Mineralogical Magazine*, **65**(3), 415–25.

Salzsauler, K.A., Sidenko, N.V. and Sherriff, B.L. (2005) Arsenic mobility in alteration products of sulfide-rich, arsenopyrite-bearing mine wastes, Snow Lake, Manitoba, Canada. *Applied Geochemistry*, **20**(12), 2303–14.

Samanta, G., Chowdhury, T.R., Mandal, B.K. *et al.* (1999) Flow injection hydride generation atomic absorption spectrometry for determination of arsenic in water and biological samples from arsenic-affected districts of West Bengal, India, and Bangladesh. *Microchemical Journal*, **62**(1), 174–91.

Sanders, J.G., Riedel, G.F. and Osman, R.W. (1994) Arsenic cycling and its impact in estuarine and coastal marine ecosystems, in *Arsenic in the Environment: Part I: Cycling and Characterization* (ed. J.O., Nriagu), John Wiley & Sons, Ltd, New York, pp. 289–308.

Santosa, S.J., Mokudai, H., Takahashi, M. and Tanaka, S. (1996) The distribution of arsenic compounds in the ocean: biological activity in the surface zone and removal processes in the deep zone. *Applied Organometallic Chemistry*, **10**(9), 697–705.

Santosa, S.J., Wada, S., Mokudai, H. and Tanaka, S. (1997) The contrasting behaviour of arsenic and germanium species in seawater. *Applied Organometallic Chemistry*, **11**(5), 403–14.

Santosa, S.J., Wada, S. and Tanaka, S. (1994) Distribution and cycle of arsenic compounds in the ocean. *Applied Organometallic Chemistry*, **8**, 273–83.

Savage, K.S., Bird, D.K. and Ashley, R.P. (2000) Legacy of the California gold rush: environmental geochemistry of arsenic in the southern mother lode gold district. *International Geology Review*, **42**(5), 385–415.

Savage, K.S., Bird, D.K. and O'Day, P.A. (2005) Arsenic speciation in synthetic jarosite. *Chemical Geology*, **215**(1-4 SPEC. ISS.), 473–98.

Savage, K.S., Tingle, T.N., O'Day, P.A. *et al.* (2000) Arsenic speciation in pyrite and secondary weathering phases, Mother Lode gold district, Tuolumne County, California. *Applied Geochemistry*, **15**(8), 1219–44.

Scazzola, R., Matteucci, G., Guerzoni, S. *et al.* (2004) Evaluation of trace metal fluxes to soils in hinterland of Porto Marghera industrial zone: comparisons with direct measurements in the lagoon of Venice. *Water Air and Soil Pollution*, **153**(1-4), 195–203.

Schaufelberger, F.A. (1994) Arsenic minerals formed at low temperatures, in *Arsenic in the Environment: Part I: Cycling and Characterization* (ed. J.O. Nriagu), John Wiley & Sons, Ltd, New York, pp. 403–15.

Schieweck, A., Lohrengel, B., Siwinski, N. *et al.* (2005) Organic and inorganic pollutants in storage rooms of the lower Saxony state museum Hanover, Germany. *Atmospheric Environment*, **39**(33), 6098–108.

Schiff, K.C. and Weisberg, S.B. (1999) Iron as a reference element for determining trace metal enrichment in southern California coastal shelf sediments. *Marine Environmental Research*, **48**(2), 161–76.

Schobert, H.H. (1995) *Coal Science and Technology 23: Lignites of North America*, Elsevier Science, Amsterdam.

Schreiber, M.E., Gotkowitz, M.B., Simo, J.A. and Freiberg, P.G. (2003) Mechanisms of arsenic release to water from naturally occurring sources, eastern Wisconsin, in *Arsenic in Ground Water* (eds A.H. Welch and K.G. Stollenwerk), Kluwer Academic Publishers, Boston, pp. 259–80.

Schreiber, M.E., Simo, J.A. and Freiberg, P.G. (2000) Stratigraphic and geochemical controls on naturally occurring arsenic in groundwater, eastern Wisconsin, USA. *Hydrogeology Journal*, **8**(2), 161–76.

Schroth, A.W., Parnell, R.A. Jr. and Ketterer, M.E. (2000) The influence of wasterock removal on geochemical pathways and processes in an acid mine drainage system, Mount Alta Mine. Abstracts with Programs. *The Geological Society of America*, **32**(7), 124.

Schwertmann, U., Friedl, J. and Stanjek, H. (1999) From Fe(III) ions to ferrihydrite and then to hematite. *Journal of Colloid and Interface Science*, **209**(1), 215–23.

Scott, K.M. (1986) Sulphide geochemistry and wall rock alteration as a guide to mineralization, Mammoth Area, NW Queensland, Australia. *Journal of Geochemical Exploration*, **25**, 283–308.

Seal R.R. II, Hammarstrom, J.M., Foley, N.K. and Alpers, C.N. (2002) Geoenvironmental models for seafloor massive sulfide deposits, in *Progress on Geoenvironmental Models for Selected Mineral Deposit Types* (eds. R.R. Seal II and N.K. Foley), Chapter L, U.S. Geological Survey Open-File Report 02-195, US Geological Survey, Reston, VA, USA 20192.

Seames, W.S. and Wendt, J.O.L. (2000) The partitioning of arsenic during pulverized coal combustion. *Symposium International on Combustion*, **28**(2), 2305–312.

Senn, D.B. and Hemond, H.F. (2002) Nitrate controls on iron and arsenic in an urban lake. *Science*, **296**(5577), 2373–76.

Serfor-Armah, Y., Nyarko, B.J.B., Adotey, D.K. *et al.* (2006) Levels of arsenic and antimony in water and sediment from Prestea, a gold mining town in Ghana and its environs. *Water Air and Soil Pollution*, **175**(1-4), 181–92.

Seyler, P. and Martin, J.-M. (1989) Biogeochemical processes affecting arsenic species distribution in a permanently stratified lake. *Environmental Science and Technology*, **23**(10), 1258–63.

Seyler, P. and Martin, J.-M. (1991) Arsenic and selenium in a pristine river-estuarine system: the Krka (Yugoslavia). *Marine Chemistry*, **34**(1-2), 137–51.

Shahnewaz, M., Saunders, J.A., Lee, M.K. *et al.* (2003) Naturally occurring arsenic in Holocene alluvial aquifer and is implications for biogeochemical linkage among arsenic, iron and sulfur form source to sink. Abstracts with Programs. *The Geological Society of America*, **35**(6), 247.

Shanks, W.C. III and Lichte, F.E. (1996) Trace element and stable isotope geochemistry of environmental minerals and acid rock drainage: Patagonia Mountains, Arizona. *Abstract with Programs — Geological Society of America*, **28**(7), 529.

Sharma, M., Tobschall, H.J. and Singh, I.B. (2003) Environmental impact assessment in the Moradabad industrial area (rivers Ramganga-Ganga interfluve), Ganga Plain, India. *Environmental Geology*, **43**(8), 957–67.

Shaw, G.E. (1991) Aerosol chemical components in Alaska air masses 1. Aged pollution. *Journal of Geophysical Research*, **96**(D12), 22-357–22-368.

Shih, M.-C. (2005) An overview of arsenic removal by pressure-driven membrane processes. *Desalination*, **172**(1), 85–97.

Shih, C.-J. and Lin, C.-F. (2003) Arsenic contaminated site at an abandoned copper smelter plant: waste characterization and solidification/stabilization treatment. *Chemosphere*, **53**(7), 691–703.

Shiikawa, M. (1983) The role of mercury, arsenic and boron as pathfinder elements in geochemical exploration for geothermal energy. *Journal of Geochemical Exploration*, **19**, 337–38.

Shinn, E.A. (2001) Transatlantic soil dust: a case history of science and public education, *Earth System Processes: Programmes with Abstracts*, Geological Society of America and Geological Society of London.

Shotyk, W., Goodsite, M.E., Roos-Barraclough, F. *et al.* (2003) Anthropogenic contributions to atmospheric Hg, Pb and As accumulation recorded by peat cores from southern Greenland and Denmark dated using the 14C 'bomb pulse curve'. *Geochimica et Cosmochimica Acta*, **67**(21), 3991–4011.

Shraim, A., Cui, X., Li, S. *et al.* (2003) Arsenic speciation in the urine and hair of individuals exposed to airborne arsenic through coal-burning in Guizhou, PR china. *Toxicology Letters*, **137**(1-2), 35–48.

Shumilin, E., Páez-Osuna, F., Green-Ruiz, C. *et al.* (2001) Arsenic, antimony, selenium and other trace elements in sediments of the La Paz lagoon, peninsula of Baja California, Mexico. *Marine Pollution Bulletin*, **42**(3), 174–78.

Sidle, W.C. (2002) $^{18}O_{SO4}$ and $^{18}O_{H2O}$ as prospective indicators of elevated arsenic in the Goose River ground-watershed, Maine. *Environmental Geology*, **42**(4), 350–59.

Sidle, W.C. and Fischer, R.A. (2003) Detection of 3H and 85Kr in groundwater from arsenic-bearing crystalline bedrock of the Goose River Basin, Maine. *Environmental Geology*, **44**(7), 781–89.

Sidle, W.C., Wotten, B. and Murphy, E. (2001) Provenance of geogenic arsenic in the Goose River Basin, Maine, USA. *Environmental Geology*, **41**(1-2), 62–73.

Signorelli, S. (1997) Arsenic in volcanic gases. *Environmental Geology*, **32**(4), 239–44.

Simón, M., Dorronsoro, C., Ortiz, I. *et al.* (2002) Pollution of carbonate soils in a Mediterranean climate due to a tailings spill. *European Journal of Soil Science*, **53**(2), 321–30.

Simon, G., Huang, H., Penner-Hahn, J.E. *et al.* (1999) Oxidation state of gold and arsenic in gold-bearing arsenian pyrite. *American Mineralogist*, **84**, 1071–79.

Sims, K.W.W., Newsom, H.E. and Gladney, E.S. (1990) Chemical fractionation during formation of the Earth's core and continental crust: clues from As, Sb, W, and Mo, in *Origin of the Earth* (eds H.E. Newson and J.H. Jones), Oxford University Press, New York; Lunar and Planetary Institute, Houston, pp. 291–317.

Sjöblom, A., Håkansson, K. and Allard, B. (2004) River water metal speciation in a mining region — the influence of wetlands, liming, tributaries, and groundwater. *Water Air and Soil Pollution*, **152**(1-4), 173–94.

Skanes, M., Kerr, A. and Sylvester, P.J. (2004) The VBE-2 Gold Prospect, Northern Labrador: Geology, Petrology and Mineral Geochemistry Report 04-1, Current Research Newfoundland Department of Mines and Energy, Geological Survey, pp. 43–61.

Skjelkvåle, B.L., Andersen, T., Fjeld, E. *et al.* (2001) Heavy metal surveys in Nordic lakes; concentrations, geographic patterns and relation to critical limits. *Ambio*, **30**(1), 2–10.

Sky and Telescope (1989) Arsenic in the gas giants, **78**(2), 133.

Šlejkovec, Z. and Kanduč, T. (2005) Unexpected arsenic compounds in low-rank coals. *Environmental Science and Technology*, **39**(10), 3450–54.

Smedley, P.L. (1996) Arsenic in rural groundwater in Ghana. *Journal of African Earth Sciences*, **22**(4), 459–70.

Smedley, P.L. and Kinniburgh, D.G. (2002) A review of the source, behaviour and distribution of arsenic in natural waters. *Applied Geochemistry*, **17**(5), 517–68.

Smedley, P.L., Kinniburgh, D.G., Macdonald, D.M.J. *et al.* (2005) Arsenic associations in sediments from the loess aquifer of La Pampa, Argentina. *Applied Geochemistry*, **20**(5), 989–1016.

Smedley, P.L., Nicolli, H.B., Macdonald, D.M.J. *et al.* (2002) Hydrogeochemistry of arsenic and other inorganic constituents in groundwaters from La Pampa, Argentina. *Applied Geochemistry*, **17**(3), 259–84.

Smedley, P.L., Zhang, M., Zhang, G. and Luo, Z. (2003) Mobilisation of arsenic and other trace elements in fluvio-lacustrine aquifers of the Huhhot Basin, Inner Mongolia. *Applied Geochemistry*, **18**(9), 1453–77.

Smith. S.J. (2005) *Naturally Occurring Arsenic in Ground Water, Norman, Oklahoma, 2004, and Remediation Options for Produced Water*, U.S. Geological Survey Fact Sheet 2005–3111.

Smith, A.M.L., Dubbin, W.E., Wright, K. and Hudson-Edwards, K.A. (2006) Dissolution of lead- and lead-arsenic-jarosites at pH 2 and 8 and 20 °C: Insights from batch experiments. *Chemical Geology*, **229**(4), 344–61.

Sohrin, Y., Matsui, M., Kawashima, M. *et al.* (1997) Arsenic biogeochemistry affected by eutrophication in Lake Biwa, Japan. *Environmental Science and Technology*, **31**(10), 2712–720.

Sonderegger, J.L. and Ohguchi, T. (1988) Irrigation related arsenic contamination of a thin, alluvial aquifer, Madison River valley, Montana, U.S.A. *Environmental Geology and Water Sciences*, **11**(2), 153–61.

Sonderegger, J.L. and Sholes, B.R. (1989) Arsenic contamination of aquifers caused by irrigation with diluted geothermal water. Abstracts with Programs. *The Geological Society of America*, **21**(5), 147.

Southam, G. and Saunders, J.A. (2005) The geomicrobiology of ore deposits. *Economic Geology*, **100**(6), 1067–84.

Spadoni, M., Voltaggio, M. and Cavarretta, G. (2005) Recognition of areas of anomalous concentration of potentially hazardous elements by means of a subcatchment-based discriminant analysis of stream sediments. *Journal of Geochemical Exploration*, **87**(3), 83–91.

Spergel, D.N., Verde, L., Peiris, H.V. *et al.* (2003) First-year wilkinson microwave anisotropy probe (WMAP) observations: determination of cosmological parameters. *Astrophysical Journal Supplement Series*, **148**, 175–94.

Spliethoff, H.M., Mason, R.P. and Hemond, H.F. (1995) Interannual variability in the speciation and mobility of arsenic in a dimictic lake. *Environmental Science and Technology*, **29**(8), 2157–161.

Spongberg, A.L. and Becks, P.M. (2000) Inorganic soil contamination from cemetery leachate. *Water Air and Soil Pollution*, **117**(1-4), 313–27.

Spotts, E., Schafer, W.M., Luckay, C.F. and Mitchell, T.S. (1997) Determination of runoff metal loading from reclaimed and unreclaimed tailings, in *Tailings and Mine Waste '97: Proceedings of the International Conference on Tailings and Mine Waste* (ed. J.D. Nelson) Vol. **4**, A.A. Balkema, Rotterdam, 583–92.

Sracek, O., Bhattacharya, P., Jacks, G. *et al.* (2004) Behavior of arsenic and geochemical modeling of arsenic enrichment in aqueous environments. *Applied Geochemistry*, **19**(2), 169–80.

Sriwana, T., Van Bergen, M.J., Sumarti, S. *et al.* (1998) Volcanogenic pollution by acid water discharges along Ciwidey River, West Java (Indonesia). *Journal of Geochemical Exploration*, **62**(1-3), 161–82.

Stanton, M.R., Sanzolone, R.E., Sutley, S.J. *et al.* (2001) Abundance, residence, and mobility of arsenic in Santa Fe Group sediments, Albuquerque Basin, New Mexico. Abstracts with Programs. *The Geological Society of America*, **33**(5), 2.

Stauder, S., Raue, B. and Sacher, F. (2005) Thioarsenates in sulfidic waters. *Environmental Science and Technology*, **39**(16), 5933–39.

Stauffer, R.E. and Thompson, J.M. (1984) Arsenic and antimony in geothermal waters of Yellowstone National Park, Wyoming, U.S.A. *Geochimica et Cosmochimica Acta*, **48**(12), 2547–61.

Sterckeman, T., Douay, F., Proix, N. *et al.* (2002) Assessment of the contamination of cultivated soils by eighteen trace elements around smelters in the north of France. *Water Air and Soil Pollution*, **135**(1-4), 173–94.

Stipp, S.L.S., Hansen, M., Kristensen, R. *et al.* (2002) Behaviour of Fe-oxides relevant to contaminant uptake in the environment. *Chemical Geology*, **190**(1-4), 321–37.

Stoeppler, M., Burow, M. and Backhaus, F. (1986) Arsenic in seawater and brown algae of the Baltic and the North Sea. *Marine Chemistry*, **18**(2-4), 321–34.

Stollenwerk, K.G. (2003) Geochemical processes controlling transport of arsenic in groundwater: a review of adsorption, in *Arsenic in Ground Water* (eds A.H., Welch and K.G. Stollenwerk), Kluwer Academic Publishers, Boston, MA, pp. 67–100.

Stollenwerk, K.G. and Colman, J.A. (2003) Natural remediation potential of arsenic-contaminated ground water, in *Arsenic in Ground Water* (eds A.H. Welch and K.G. Stollenwerk), Kluwer Academic Publishers, Boston, MA, pp. 351–80.

Strong, C.P., Brooks, R.R., Wilson, S.M. *et al.* (1987) A new Cretaceous-Tertiary boundary site at Flaxbourne River, New Zealand: biostratigraphy and geochemistry. *Geochimica et Cosmochimica Acta*, **51**, 2769–77.

Su, C. and Puls, R.W. (2001) Arsenate and arsenite removal by zerovalent iron: kinetics, redox transformation, and implications for *in situ* groundwater remediation. *Environmental Science and Technology*, **35**(7), 1487–92.

Sullivan, K.A. and Aller, R.C. (1996) Diagenetic cycling of arsenic in Amazon shelf sediments. *Geochimica et Cosmochimica Acta*, **60**(9), 1465–77.

Swartz, R.J. and Thyne, G.D. (1995) The relative contributions of human activities and nature to elevated dissolved arsenic in ground water in the Kern Water Bank, southern San Joaquin Valley, CA. *AAPG Bulletin*, **79**(4), 599.

Sweet, C.W., Weiss, A. and Vermette, S.J. (1998) Atmospheric deposition of trace metals at three sites near the Great Lakes. *Water Air and Soil Pollution*, **103**(1-4), 423–39.

Szramek, K., Walter, L.M. and McCall, P. (2004) Arsenic mobility in groundwater/surface water systems in carbonate-rich Pleistocene glacial drift aquifers (Michigan). *Applied Geochemistry*, **19**(7), 1137–55.

Tadanier, C.J., Schreiber, M.E. and Roller, J.W. (2005) Arsenic mobilization through microbially mediated deflocculation of ferrihydrite. *Environmental Science and Technology*, **39**(9), 3061–68.

Takeuchi, M., Terada, A., Nanba, K. *et al.* (2005) Distribution and fate of biologically formed organoarsenicals in coastal marine sediment. *Applied Organometallic Chemistry*, **19**(8), 945–51.

Tariq, J., Ashraf, M., Jaffar, M. and Afzal, M. (1996) Pollution status of the Indus River, Pakistan, through heavy metal and macronutrient contents of fish, sediment and water. *Water Research*, **30**(6), 1337–44.

Tarvainen, T., Lahermo, P. and Mannio, J. (1997) Sources of trace metals in streams and headwater lakes in Finland. *Water Air and Soil Pollution*, **94**(1-2), 1–32.

Tayler, R.J. (1988) Nucleosynthesis and the origin of the elements. *Philosophical Transactions of the Royal Society of London*, **A 325**, 391–403.

Taylor, H.E., Antweiler, R.C., Roth, D.A. *et al.* (2001) The occurrence and distribution of selected trace elements in the Upper Rio Grande and tributaries in Colorado and northern New Mexico. *Archives of Environmental Contamination and Toxicology*, **41**, 410–26.

Taylor, H.E. and Shiller, A.M. (1995) Mississippi River methods comparison study: implications for water quality monitoring of dissolved trace elements. *Environmental Science and Technology*, **29**(5), 1313–17.

Tazaki, K., Okrugin, V., Okuno, M. *et al.* (2003) *Heavy Metallic Concentration in Microbial Mats Found at Hydrothermal Systems, Kamchatka, Russia*. Science Reports of the Kanazawa University, Vol. 47, Issue 1-2, 1–48.

Temerev, S.V. and Yu Kondakova, I. (2006) Determination of arsenic in surface waters of the Ob River basin. *Journal of Analytical Chemistry*, **61**(2), 186–89.

Temgoua, E. and Pfeifer, H.-R. (2002) Arsenic in spring waters and soils in southern Switzerland: evidence of complex weathering and redeposition processes. *Geochimica et Cosmochimica Acta*, **66**(15A), 768.

Thielen, F., Zimmermann, S., Baska, F. *et al.* (2004) The intestinal parasite Pomphorhynchus laevis (acanthocephala) from barbel as a bioindicator for metal pollution in the Danube River near Budapest, Hungary. *Environmental Pollution*, **129**(3), 421–29.

Thiros, S.A. (2003) Quality and Sources of Shallow Ground Water in Areas of Recent Residential Development in Salt Lake Valley U.S. Geological Survey Water-Resources Investigations Report 03-4028, U.S. Geological Survey, Salt Lake City, UT.

Thomas, M.A. (2003) Arsenic in Midwestern Glacial Deposits- Occurrence and Relation to Selected Hydrogeologic and Geochemical Factors U.S. Geological Survey Water-Resources Investigations Report 03-4228, Columbus, OH.

Thompson, T.S., Le, M.D., Kasick, A.R. and Macaulay, T.J. (1999) Arsenic in well water supplies in Saskatchewan. *Bulletin of Environmental Contamination and Toxicology*, **63**(4), 478–83.

Thornburg, K. and Sahai, N. (2004) Arsenic occurrence, mobility, and retardation in sandstone and dolomite formations of the Fox River valley, eastern Wisconsin. *Environmental Science and Technology*, **38**(19), 5087–94.

Tingle, T.N., Waychunas, G.A., Bird, D.K. and O'Day, P. (1996) X-ray absorption spectroscopy (EXAFS) of arsenic solid solution in pyrite, Clio Mine, Mother Lode gold district, Tuolumne County. Abstracts with Programs. *The Geological Society of America*, **28**(7), 518.

Togashi, S., Imai, N., Okuyama-Kusunose, Y. *et al.* (2000) Young upper crustal chemical composition of the orogenic Japan arc. *Geochemistry, Geophysics, Geosystems*, **G3**, 1.

Tongesayi, T. and Smart, R.B. (2006) Arsenic speciation: reduction of arsenic(V) to arsenic(III) by fulvic acid. *Environmental Chemistry*, **3**(2), 137–41.

Tourtelot, H.A. (1964) Minor-element composition and organic carbon content of marine and nonmarine shales of Late Cretaceous age in the western interior of the United States. *Geochimica et Cosmochimica Acta*, **28**, 1579–1604.

Townsend, T., Solo-Gabriele, H., Tolaymat, T. *et al.* (2003) Chromium, copper, and arsenic concentrations in soil underneath CCA-treated wood structures. *Soil and Sediment Contamination*, **12**(6), 779–98.

Traversy, W.J., Goulden, P.D., Sheikh, Y.M. and Leacock, J.R. (1975) *Levels of Arsenic and Selenium in the Great Lakes Region*, Scientific Series, Vol. 58, Environment Canada, Inland Waters Directorate, Ontario Region, Burlington, ON, p. 18.

Trefry, J.H., Rember, R.D., Trocine, R.P. and Brown, J.S. (2003) Trace metals in sediments near offshore oil exploration and production sites in the Alaskan Arctic. *Environmental Geology*, **45**(2), 149–60.

Trolard, F., Génin, J.-M.R., Abdelmoula, M. *et al.* (1997) Identification of a green rust mineral in a reductomorphic soil by Mössbauer and Raman spectroscopies. *Geochimica et Cosmochimica Acta*, **61**(5), 1107–11.

Turekian, K.K. and Wedepohl, K.H. (1961) Distribution of the elements in some major units of the Earth's crust. *Geological Society of America Bulletin*, **72**, 175–92.

Turner, S.J., Flindell, P.A., Hendri, D. *et al.* (1994) Sediment-hosted gold mineralisation in the Ratatotok district, North Sulawesi, Indonesia. *Journal of Geochemical Exploration*, **50**(1-3), 317–36.

Tuttle, M.L.W., Goldhaber, M.B., Ruppert, L.F. and Hower J.C. (2002) Arsenic in Rocks and Stream Sediments of the Central Appalachian Basin, Kentucky U.S. Geological Survey Open-File Report 02-28, U.S. Geological Survey.

U.S. Geological Survey Coal Quality Database (2006) U.S. Geological Survey, http://energy.er.usgs.gov/products/databases/CoalQual/intro.htm. Accessed on February 2, 2006.

US National Committee for Geochemistry. Panel on the Trace Element Geochemistry of Coal Resource Development Related to Health (1980) *Trace Element Geochemistry of Coal Resource Development Related to Environmental Quality and Health,* National Academy of Sciences, National Academy Press, Washington, DC, p. 153.

Valette-Silver, N., Hameedi, M.J., Efurd, D.W. and Robertson, A. (1999) Status of the contamination in sediments and biota from the western Beaufort Sea (Alaska). *Marine Pollution Bulletin*, **38**(8), 702–22.

Van Herreweghe, S., Swennen, R., Vandecasteele, C. and Cappuyns, V. (2003) Solid phase speciation of arsenic by sequential extraction in standard reference materials and industrially contaminated soil samples. *Environmental Pollution*, **122**(3), 323–42.

Vaniman, D., Dietrich, J., Taylor, G.J. and Heiken, G. (1991) Exploration, samples, and recent concepts of the Moon, in *Lunar Sourcebook: A User's Guide to the Moon* (eds G. Heiken, D. Vaniman and B.M. French), Cambridge University Press, Cambridge, Chapter 2.

Van Moort, J.C., Hotchkis, M.A.C. and Pwa, A. (1995) EPR spectra and lithogeochemistry of jasperoids at Carlin, Nevada: distinction between auriferous and barren rocks. *Journal of Geochemical Exploration*, **55**(1-3), 283–99.

Van der Weijden, C.H., Hoede, D., Middelburg, J.J. *et al.* (1988) Arsenic, antimony and vanadium in the North Atlantic. *Chemical Geology*, **70**(1-2), 19.

Varnavas, S.P. and Cronan, D.S. (1988) Arsenic, antimony and bismuth in sediments and waters from the Santorini hydrothermal field, Greece. *Chemical Geology*, **67**(3-4), 295–305.

Veal, D.J. (1966) Nondestructive activation analysis of crude oils for arsenic to one part per billion, and simultaneous determination of five other trace elements. *Analytical Chemistry*, **38**(8), 1080–83.

Vedal, S. (1995) *Health Effects of Inhalable Particles: Implications for British Columbia*, Prepared for the Air Resources Branch, British Columbia Ministry of Environment, Lands and Parks.

Verplanck, P.L., S.H. Mueller, E.K. Youcha, R.J. *et al.* (2003) Chemical Analyses of Ground and Surface Waters, Ester Dome, Central Alaska, 2000–2001 U.S. Geological Survey Open-File Report 03-244, U.S. Geological Survey, Boulder, CO.

Viollier, E., Jezequel, D., Michard, G. *et al.* (1995) Geochemical study of a Crater lake (Pavin Lake, France): trace — element behaviour in the monimolimnion. *Chemical Geology*, **125**(1-2), 61–72.

Wagner, N.J. and Hlatshwayo, B. (2005) The occurrence of potentially hazardous trace elements in five Highveld coals, South Africa. *International Journal of Coal Geology*, **63**(3-4), 228–46.

Walker, F.P., Schreiber, M.E. and Rimstidt, J.D. (2006) Kinetics of arsenopyrite oxidative dissolution by oxygen. *Geochimica et Cosmochimica Acta*, **70**(7), 1668–76.

Wallerstein, G., Iben, I. Jr., Parker, P. *et al.* (1997) Synthesis of the elements in stars: forty years of progress. *Reviews of Modern Physics*, **69**(4), 995–1084.

Walraven, N., Van Os, B.J.H., Klaver, G.Th. *et al.* (1997) Trace element concentrations and stable lead isotopes in soils as tracers of lead pollution in Graft-de Rijp, the Netherlands. *Journal of Geochemical Exploration*, **59**(1), 47–58.

Wang, C.-F., Chang, C.-Y., Tsai, S.-F. and Chiang, H.-L. (2005) Characteristics of road dust from different sampling sites in northern Taiwan. *Journal of the Air and Waste Management Association*, **55**(8), 1236–44.

Wang, S. and Mulligan, C.N. (2006) Effect of natural organic matter on arsenic release from soils and sediments into groundwater. *Environmental Geochemistry and Health*, **28**(3), 197–214.

Wang, J., Zhao, L. and Wu, Y. (1998) Environmental geochemical study on arsenic in arseniasis areas in Shanyin and Yingxian, Shanxi China. *Geoscience*, **12**(2), 243–48.

Ward, C.R., Spears, D.A., Booth, C.A. *et al.* (1999) Mineral matter and trace elements in coals of the Gunnedah Basin, New South Wales, Australia. *International Journal of Coal Geology*, **40**(4), 281–308.

Warner, K.L. (2001) Arsenic in glacial drift aquifers and the implication for drinking water — Lower Illinois River Basin. *Ground Water*, **39**(3), 433–42.

Warren, C., Burgess, W.G. and Garcia, M.G. (2005) Hydrochemical associations and depth profiles of arsenic and fluoride in Quaternary loess aquifers of northern Argentina. *Mineralogical Magazine*, **69**(5), 877–86.

Warren, P.H., Kallemeyn, G.W. and Kyte, F.T. (1999) Origin of planetary cores: evidence from highly siderophile elements in Martian meteorites. *Geochimica et Cosmochimica Acta*, **63**(13-14), 2105–22.

Warwick, P.D., Crowley, S.S., Ruppert, L.F. and Pontolillo, J. (1997) Petrography and geochemistry of selected lignite beds in the Gibbons Creek mine (Manning Formation, Jackson Group, Paleocene) of east-central Texas. *International Journal of Coal Geology*, **34**(3-4), 307–26.

Warwick, P., Inam, E. and Evans, N. (2005) Arsenic's interaction with humic acid. *Environmental Chemistry*, **2**(2), 119–24.

Waslenchuk, D.G. (1979) The geochemical controls on arsenic concentrations in southeastern United States rivers. *Chemical Geology*, **24**, 315–25.

Wasson, J.T. (1999) Trapped melt in IIIAB irons; solid/liquid elemental partitioning during the fractionation of the IIIAB magma. *Geochimica et Cosmochimica Acta*, **63**(18), 2875–89.

Wasson, J.T., Choi, B.-G., Jerde, E.A. and Ulff-Møller, F. (1998) Chemical classification of iron meteorites: XII. New members of the magmatic groups. *Geochimica et Cosmochimica Acta*, **62**(4), 715–24.

Wasson, J.T. and Kallemeyn, G.W. (1988) Compositions of chondrites. *Transactions of the Royal Society of London, Series A*, **325**, 535–44.

Wasson, S.J., Linak, W.P., Gullett, B.K. *et al.* (2005) Emissions of chromium, copper, arsenic, and PCDDs/Fs from open burning of CCA-treated wood. *Environmental Science and Technology*, **39**(22), 8865–76.

Wasson, J.T. and Richardson, J.W. (2001) Fractionation trends among IVA iron meteorites: contrast with IIIAB trends. *Geochimica et Cosmochimica Acta*, **65**(6), 951–70.

Waychunas, G.A., Kim, C.S. and Banfield, J.F. (2005) Nanoparticulate iron oxide minerals in soils and sediments: unique properties and contaminant scavenging mechanisms. *Journal of Nanoparticle Research*, **7**(4-5), 409–33.

Webster, J.G. and Nordstrom, D.K. (2003) Geothermal arsenic, in *Arsenic in Ground Water* (eds A.H., Welch and K.G. Stollenwerk) Kluwer Academic Publishers, Boston, pp. 101–25.

Wedepohl, K.H. (1995) The composition of the continental crust. *Geochimica et Cosmochimica Acta*, **59**(7), 1217–32.

Weiss, H.V. and Bertine, K.K. (1973) Simultaneous determination of manganese, copper, arsenic, cadmium, antimony and mercury in glacial ice by radioactivation. *Analytica Chimica Acta*, **65**(2), 253–59.

Welch, A.H., Helsel, D.R., Focazio, M.J. and Watkins, S.A. (1998) Arsenic in ground water supplies of the United States, in *Arsenic Exposure and Health Effects* (eds W.R. Campbell, C.O. Abernathy and R.L. Calderon), Elsevier Science, New York, pp. 9–17.

Welch, A.H., Westjohn, D.B., Helsel, D.R. and Wanty, R.B. (2000) Arsenic in ground water of the United States: occurrence and geochemistry. *Ground Water*, **38**(4), 589–604.

Wellman, D.E., Reid, D.A. and Ulery, A.L. (1999) Elevated soil arsenic levels at a former crude oil storage facility-assessment, remediation, and possible sources. *Soil and Sediment Contamination*, **8**(3), 329–41.

Wennrich, R., Mattusch, J., Morgenstern, P. *et al.* (2004) Characterization of sediments in an abandoned mining area; a case study of Mansfeld region, Germany. *Environmental Geology*, **45**(6), 818–33.

Widerlund, A. and Ingri, J. (1995) Early diagenesis of arsenic in sediments of the Kalix River estuary, northern Sweden. *Chemical Geology*, **125**(3-4), 185–96.

Wiertz, J.V., Mateo, M. and Escobar, B. (2006) Mechanism of pyrite catalysis of As(III) oxidation in bioleaching solutions at 30 °C and 70 °C. *Hydrometallurgy*, **83**(1-4), 35–39.

Wilcox, R. (1997) Concentrations of Selected Trace Elements and other Constituents in the Rio Grande and in Fish Tissue in the Vicinity of Albuquerque, New Mexico 1994–1996 U.S. Geological Survey Open-File Report 97–0667, U.S. Geological Survey.

Wilkin, R.T. and Ford, R.G. (2006) Arsenic solid-phase partitioning in reducing sediments of a contaminated wetland. *Chemical Geology*, **228**(1-3 Special Issue), 156–74.

Williams, M. (2001) Arsenic in mine waters: international study. *Environmental Geology*, **40**(3), 267–78.

Williams, T.M., Rawlins, B.G., Smith, B. and Breward, N. (1998) In-vitro determination of arsenic bioavailability in contaminated soil and mineral beneficiation waste from Ron Phibun, southern Thailand: a basis for improved human risk assessment. *Environmental Geochemistry and Health*, **20**(4), 169–77.

Winkler, H.J.F. (1979) *Petrogenesis of Metamorphic Rocks*, 5th edn Springer-Verlag, New York, p. 348.

Wirth, E.F., Scott, G.I., Fulton, M.H. *et al.* (1996) *In situ* monitoring of dredged material spoil sites using the oyster Crassostrea virginica. *Archives of Environmental Contamination and Toxicology*, **30**(3), 340–48.

Wong, H.K.T., Gauthier, A., Beauchamp, S. and Tordon, R. (2002) Impact of toxic metals and metalloids from the Caribou gold-mining areas in Nova Scotia, Canada. *Geochemistry: Exploration, Environment, Analysis*, **2**(3), 235–41.

Woo, N.C. and Choi, M.C. (2001) Arsenic and metal contamination of water resources from mining wastes in Korea. *Environmental Geology*, **40**(3), 305–11.

Wretblad, P.E. (1941) Minerals of the Varutraesk pegmatite. *Geologiska Foereningen i Stockholm Foerhandlingar*, **63**, 424.

Xiankun, L., Jing, L. and Xinian, M. (1994) Arsenic in several Chinese estuaries and coastal seas, in *Arsenic in the Environment: Part I: Cycling and Characterization* (ed. J.O. Nriagu), John Wiley & Sons, Ltd, New York, pp. 309–36.

Xiao, T., Guha, J., Boyle, D. *et al.* (2004) Environmental concerns related to high thallium levels in soils and thallium uptake by plants in southwest Guizhou, China. *Science of the Total Environment*, **318**(1-3), 223–44.

Xiaoying, Y. (2001) Humic acids from endemic arsenicosis areas in Inner Mongolia and from the blackfoot-disease areas in Taiwan: a comparative study. *Environmental Geochemistry and Health*, **23**(1), 27–42.

Yamasaki, Y. and Hata, Y. (2000) Changes and their factors of concentrations of arsenic and boron in the process of groundwater recharge in the lower Lluta River Basin, Chile. *Journal of Groundwater Hydrology*, **42**(4), 341–53.

Yamazaki, C., Ishiga, H., Ahmed, F. *et al.* (2003) Vertical distribution of arsenic in Ganges delta sediments in Deuli Village, Bangladesh. *Soil Science and Plant Nutrition*, **49**(4), 567–74.

Yan, X.-P., Kerrich, R. and Hendry, M.J. (2000) Distribution of arsenic(III), arsenic(V) and total inorganic arsenic in porewaters from a thick till and clay-rich aquitard sequence, Saskatchewan, Canada. *Geochimica et Cosmochimica Acta*, **64**(15), 2637–48.

Yan-Chu, H. (1994) Arsenic distribution in soils, in *Arsenic in the Environment: Part I: Cycling and Characterization* (ed. J.O. Nriagu), John Wiley & Sons, Ltd., New York, pp. 17–49.

Yang, S.X. and Blum, N. (1999a) A fossil hydrothermal system or a source -bed in the Madiyi formation near the Xiangxi Au-Sb-W deposit, NW Hunan, PR China? *Chemical Geology*, **155**(1-2), 151–69.

Yang, S.X. and Blum, N. (1999b) Arsenic as an indicator element for gold exploration in the region of the Xiangxi Au-Sb-W deposit, NW Hunan, PR China. *Journal of Geochemical Exploration*, **66**(3), 441–56.

Yin, X., Liu, X., Sun, L. *et al.* (2006) A 1500-year record of lead, copper, arsenic, cadmium, zinc level in Antarctic seal hairs and sediments. *Science of the Total Environment*, **371**(1-3), 252–57.

Yokel, J. and Delistraty, D.A. (2003) Arsenic, lead, and other trace elements in soils contaminated with pesticide residues at the Hanford site USA. *Environmental Toxicology*, **18**(2), 104–14.

Yokoyama, T., Takahashi, Y. and Tarutani, T. (1993) Simultaneous determination of arsenic and arsenious acids in geothermal water. *Chemical Geology*, **103**(1-4), 103–11.

Yu, Y., Zhu, Y. and Gao, Z. (2004) Stability of arsenopyrite and As(III) in low-temperature acidic solutions. *Science in China, Series D: Earth Sciences*, **47**(5), 427–36.

Yudovich, Ya.E. and Ketris, M.P. (2005) Arsenic in coal: a review. *International Journal of Coal Geology*, **61**(3-4), 141–96.

Zachariás, J., Frýda, J., Paterova, B. and Mihaljevič, M. (2004) Arsenopyrite and As-bearing pyrite from the Roudný Deposit, Bohemian Massif. *Mineralogical Magazine*, **68**(1), 31–46.

Zagury, G.J., Samson, R. and Deshênes, L. (2003) Occurrence of metals in soil and ground water near chromated copper arsenate-treated utility poles. *Journal of Environmental Quality*, **32**(2), 507–14.

Zänker, H., Moll, H., Richter, W. *et al.* (2002) The colloid chemistry of acid rock drainage solution from an abandoned Zn-Pb-Ag mine. *Applied Geochemistry*, **17**(5), 633–48.

Zarcinas, B.A., Ishak, C.F., McLaughlin, M.J. and Cozens, G. (2004) Heavy metals in soils and crops in Southeast Asia. 1. Peninsular Malaysia. *Environmental geochemistry and health*, **26**(4), 343–57.

Zarcinas, B.A., Pongsakul, P., McLaughlin, M.J. and Cozens, G. (2004) Heavy metals in soils and crops in Southeast Asia. 2. Thailand. *Environmental geochemistry and health*, **26**(4), 359–71.

Zereini, F., Alt, F., Messerschmidt, J. *et al.* (2005) Concentration and distribution of heavy metals in urban airborne particulate matter in Frankfurt am Main, Germany. *Environmental Science and Technology*, **39**(9), 2983–89.

Zhabin, A.G., Samsonova, N.S., Chuchua, I.B. *et al.* (1990) Ore-bearing metasomatically altered limestones of the black shale association. *International Geology Review*, **32**(11), 1145–55.

Zhang, H. (2004) Heavy-metal pollution and arseniasis in Hetao region, China. *Ambio*, **33**(3), 138–40.

Zhang, H., Davison, W., Mortimer, R.J.G. *et al.* (2002) Localised remobilization of metals in a marine sediment. *The Science of the Total Environment*, **296**, 175–87.

Zhang, J., Ren, D., Zhu, Y. *et al.* (2004) Mineral matter and potentially hazardous trace elements in coals from Qianxi Fault Depression Area in southwestern Guizhou, China. *International Journal of Coal Geology*, **57**, 49–61.

Zhao, Y. and Frey, H.C. (2004) Development of probabilistic emission inventories of air toxics for Jacksonville, Florida. *Journal of the Air and Waste Management Association*, **54**(11), 1405–21.

Zheng, B., Ding, Z., Huang, R. *et al.* (1999) Issues of health and disease relating to coal use in southwestern China. *International Journal of Coal Geology*, **40**(2-3), 119–32.

Zhuang, X., Querol, X., Zeng, R. *et al.* (2000) Mineralogy and geochemistry of coal from the Liupanshui mining district, Guizhou, south China. *International Journal of Coal Geology*, **45**(1), 21–37.

Ziegler, A.C., Ross, H.C., Trombley, T.J. and Christensen, V.G. (2001) *Effects of Artificial Recharge on Water Quality in the Equus Beds Aquifer, South-Central Kansas, 1995–2000*, U.S. Geological Survey Fact Sheet 096–01.

Zobrist, J., Dowdle, P.R., Davis, J.A. and Oremland, R.S. (2000) Mobilization of arsenite by dissimilatory reduction of adsorbed arsenate. *Environmental Science and Technology*, **34**(22), 4747–53.

Zoller, W.H., Parrington, J.R. and Phelan Kotra, J.M. (1983) Iridium enrichment in airborne particles from Kilauea volcano: January 1983. *Science*, **222**(4628), 1118–21.

Further reading

Azcue, J.M., Nriagu, J.O. and Schiff, S. (1994) Role of sediment porewater in the cycling of arsenic in a mine-polluted lake. *Environment International*, **20**(4), 517–27.

Kubota, H., Urabe, T., Yamada, R. and Tanimura. S. (2004) Exploration indices and mineral potential map of the Kuroko deposits in northeast Japan. *Resource Geology*, **54**(4), 387–97.

Yan, R., Gauthier, D. and Flamant, G. (2000) Possible interactions between As, Se, and Hg during coal combustion. *Combustion and Flame*, **120**(1-2), 49–60.

4

Toxicology and Epidemiology of Arsenic and its Compounds

MICHAEL F. HUGHES, DAVID J. THOMAS, and ELAINA M. KENYON

US Environmental Protection Agency
Office of Research and Development
National Health and Environmental Effects Research Laboratory

Disclaimer: This article has been reviewed in accordance with the policy of the National Health and Environmental Effects Research Laboratory, US Environmental Protection Agency, and approved for publication. Approval does not signify that the contents necessarily reflect the views and policies of the Agency, nor does mention of trade names or commercial products constitute endorsement or recommendation for use.

4.1 Introduction

Arsenic is an element that is found in various chemical forms and oxidation states in the environment. Its origin is geological, being the twentieth most abundant element in the Earth's crust. Arsenic is a common environmental contaminant found naturally in water and food. While noted for its use in intentional poisonings of and by royalty (the 'King of Poisons'), arsenic has uses as a medicinal agent, as a pesticide, a growth promoter, in semiconductors, in the manufacture of glass, and other products (Chapter 5). From its use and disposal in the United States, arsenic is the top ranked hazardous substance on the *Superfund National Priorities List* by the Agency for Toxic Substances and Disease Registry (ATSDR) in 2007. Arsenic of geological origin has caused a major environmental crisis in developing countries, such as Bangladesh and India (Chakraborti *et al.*, 2004). Millions of people worldwide are at risk for the development of cardiovascular disease, diabetes, cancer, and other adverse health effects from drinking arsenic-contaminated groundwater. Arsenic is an important element that has impacts on public health, commercial interests, the environment, and geopolitics.

Arsenic Edited by Kevin R. Henke
© 2009 John Wiley & Sons, Ltd

Table 4.1   *Examples of arsenic compounds of environmental and biological interest.*

Inorganic arsenicals	Organic arsenicals
Arsenic trioxide — As_2O_3 (white arsenic)	Monomethylarsonic acid — $CH_3As(O)(OH)_2$ (mono-
Arsenious acid — H_3AsO_3	and disodium salts; e.g. MSMA
Arsenic acid — H_3AsO_4	(monosodiummethanearsonate) and DSMA
	(disodiummethylarsonate))
Inorganic Arsenite — $H_{3-x}AsO_3{}^{x-}$, where $x = 0, 1,$ 2, or 3.	Monomethylarsonous acid — $CH_3As(OH)_2$
Inorganic Arsenate — $H_{3-x}AsO_4{}^{x-}$, where $x = 0, 1,$ 2, or 3.	Dimethylarsinic acid — $(CH_3)_2As(O)OH$ (cacodylic acid)
Lead arsenate — $PbHAsO_4$	Trimethylarsine oxide — $(CH_3)_3AsO$
Arsenopyrite — FeAsS	Arsenobetaine — $(CH_3)_3As^+CH_2COO^-$
Arsine — AsH_3	Copper acetoarsenite — $Cu_4C_4H_6As_6O_{16}$ (Paris green)

4.2   Physical and chemical properties of arsenic

Arsenic is a *metalloid*, having several, but not all of the physical and chemical properties of metals (Chapter 2). It can exist in four *valence states*:-3 (*arsenides*), 0 (elemental), $+3$ (trivalent, *arsenites*) and $+5$ (pentavalent, *arsenates*). The arsenides are usually bound with other metals such as gallium. Gallium arsenide, an alloylike or intermetallic compound, is used in the semiconductor industry (Carter, Aposhian and Gandolfi, 2003; Chapter 5). Another arsenic compound with a -3 valence state is arsine. Arsine is a colorless gas that has several industrial uses and is an acutely potent *hemolytic agent* (Carter, Aposhian and Gandolfi, 2003; Klimecki and Carter, 1995). The elemental form of arsenic is rarely encountered. In most cases, and for toxicological purposes, the most relevant oxidation states of arsenic are the $+3$ and $+5$ forms. These can also be identified as As(III) and As(V), respectively. The toxicological importance of these two oxidation states is that trivalent forms of arsenic, both inorganic and organic, are generally more toxic than the pentavalent forms. Arsenic binds *covalently* to oxygen, sulfur, hydrogen, and itself to form inorganic *arsenicals* and to carbon to form *organoarsenicals*. Examples of arsenic compounds of interest are listed in Table 4.1.

4.3   Exposure to arsenic

Humans encounter arsenic from natural and anthropogenic sources. Because arsenic is a natural component of the Earth's crust, it is detected in *rocks*, *sediments*, and *soils* with the amount depending on the geological history of the area (National Research Council (NRC), 1997; Chapter 3). In the United States, ambient arsenic air levels in rural and urban areas range from $<1–3$ ng m^{-3} and $20–30$ ng m^{-3}, respectively (ATSDR, 2007). The levels of arsenic in freshwater (surface and *groundwater*), typically inorganic in form, range from 1 to $10\,\mu$g L^{-1}) (World Health Organization (WHO), 2001). However, there are areas in the world where the arsenic levels in drinking water (primarily groundwater) are excessively high ($>100\,\mu$g L^{-1}) (Nordstrom, 2002). These areas include Bangladesh, Mexico, Vietnam, parts of the United States and several other countries (Appendix D). Arsenic is found naturally in food and the amount and its form (inorganic or organic) depends on its source. For most people, excluding sources of arsenic pollution and drinking water contamination, the diet is the major source of arsenic ($\sim$50 µg/day) (Abernathy, Thomas and Calderon, 2003). Foods of marine origin, such as fish and shellfish, tend to have the highest levels of arsenic, primarily organic in form (e.g. arsenobetaine (($CH_3)_3As^+CH_2CH_2OH$)

and arsenocholine ($(CH_3)_3As^+CH_2CH_2O^-$)). Tuna and shrimp have levels of arsenic ranging from 0.6 to 1.5 $\mu g\,g^{-1}$ and 0.3–2.7 $\mu g\,g^{-1}$, respectively (Tao and Bolger, 1999). Arsenosugars are found in marine algae, mussels, oysters, and clams (Le, Lu and Li, 2004). Compared to inorganic arsenic, the organic forms of arsenic in seafood are considered relatively nontoxic. Inorganic arsenic is the predominant form of arsenic in foods such as rice (74 $ng\,g^{-1}$), flour (11 $ng\,g^{-1}$), juice (9 $ng\,g^{-1}$) and spinach (6 $ng\,g^{-1}$) (Schoof, Yost and Eickhoff, 1999).

Current and past uses of arsenic include pesticides, wood preservatives, munitions, semiconductors, antimicrobials for growth promotion in animals, and anticancer agents (Table 4.2; Chapter 5). Although production of arsenic ceased in the United States in 1985, it was the world's largest consumer of arsenic in 2003 (ATSDR, 2007). People are exposed to arsenic from its use today as well as from its use years ago.

Arsenicals of recent interest include the wood preservatives chromated copper arsenate (CCA) (Chapter 5). Because of the concern over the potential toxic effects of arsenic in the preservative, in an agreement with the US Environmental Protection Agency (EPA), the wood preservative industry voluntary phased out the use of CCA in wood for residential use in 2003 (Katz and Salem, 2005). However, CCA-treated wood can still be used in industrial applications. A problem in the future will be how to safely dispose of CCA-treated wood (Chapter 7).

A unique use of arsenic, in the form of arsenic trioxide (As_2O_3), is for the treatment of cancer. Relapsed or refractory cases of *acute promyelocytic leukemia* have been successfully treated with this arsenical. Its use in the United States was approved by the US Food and Drug Administration (FDA) in 2000 (Antman, 2001).

Arsenic is a by-product of the smelting of nonferrous ores and *coal* combustion. Areas near smelters may be contaminated with arsenic in the air, soil, house dust, and water (Polissar *et al.*, 1990; Hwang *et al.*, 1997). People living in Guizhou, China, use coal in their homes for heating, daily cooking, and crop-drying. Coal from this region of China contains high levels of arsenic (100–9000 $\mu g\,g^{-1}$) (Liu *et al.*, 2002; Chapter 3). These people are exposed to arsenic in the air and their food (from arsenic in the air falling on and coating it). Although the levels of arsenic in drinking water are considered in the normal range, there are individuals in this population that are showing the signs of *chronic* arsenic poisoning.

Lead arsenate was once used as an insecticide in apple orchards. Highly arsenic-contaminated soils of apple orchards have been reported in the US state of Washington (660 $\mu g\,g^{-1}$) (Peryea and Creger, 1994). Robinson *et al.* (2007) report that 50 % of the orchard sites examined in the Great Valley region of Virginia and West Virginia, United States, had soil levels of arsenic that exceeded the preliminary remediation goal (PRG) screening guideline of 22 $mg\,kg^{-1}$ (dry mass) for arsenic in residential soil.

Table 4.2 *Examples of present and past uses of arsenicals (also see Chapters 2 and 5).*

Application	Arsenical
Herbicide	Dimethylarsinic acid
Insecticide	Lead arsenate
War gas	Lewisite
Pigment, pesticide	Copper acetoarsenite
Antimicrobial feed additive	Roxarsone
Antiprotozoal drug	Melarsoprol
Antimicrobial drug	Salvarsan
Semiconductor synthesis	Arsine
Semiconductor	Gallium arsenide
Tonic	Fowler's solution (1 % potassium arsenite)

(Until recently, this PRG was used by Region 9 of the US EPA for Superfund sites in Arizona, California, Hawaii, Nevada, US Pacific Islands, and Tribal Nations, Appendix E). As farmlands and orchards once treated with arsenical pesticides are being developed for residential use, occupants of these homes could be exposed to arsenic (Belluck *et al.*, 2003).

4.4   Arsenic disposition and biotransformation in mammals

4.4.1   Introduction

The *metabolism* and disposition of inorganic arsenic is largely dependent on its valence state. The two most common valence states of inorganic arsenic to which humans might be environmentally exposed are As(III) and As(V). Since these two forms are readily interconverted, studies cited in this section were evaluated with particular attention to whether the used methods were appropriate to insure that inorganic arsenic was maintained in the intended valence state. Arsenic metabolism is also characterized by relatively large qualitative and quantitative interspecies differences compared to other metalloids and metals. Given the relatively large interspecies differences in arsenic metabolism and that there is considerable information on human metabolism of arsenicals, discussion of animal studies focuses on areas where human data are inadequate or where animal data can serve to aid in the interpretation of toxic effects caused by inorganic arsenic.

4.4.2   Respiratory deposition and absorption

Human inhalation exposure to inorganic arsenic can occur as a consequence of industrial activity, during cigarette smoking, and energy production. Arsenic in air exists on particulate matter and thus respiratory absorption of arsenic is a two part process: (1) deposition of the particles onto airway and lung surfaces and (2) absorption of arsenic from the deposited particulates. The extent of deposition of inhaled arsenic will depend largely on the size of the inhaled particulates and absorption of the deposited arsenic is highly dependent on the solubility of the chemical form of the arsenic.

Both human and animal data are insufficient to quantitatively estimate regional arsenic deposition in the respiratory tract. Occupational studies, in which both the concentration of inorganic arsenic in the breathing zone and urinary excretion of inorganic arsenic and its *metabolites*, provide information on arsenic absorption. These studies (e.g. Vahter, Friberg and Rahnster, 1986; Yamauchi *et al.*, 1989; Offergelt, Roels and Buchet, 1992; Hakala and Pyy, 1995; Yager, Hicks and Fabianova, 1997) demonstrate that excretion of inorganic arsenic and sometimes total arsenicals and methylated metabolites are significantly increased in workers exposed to higher levels of inorganic arsenic in their breathing zone compared to unexposed workers. This indicates that arsenic is absorbed from the respiratory tract, but does not provide sufficient information to quantitatively estimate arsenic absorption.

A comparison of studies that relate occupational arsenic exposure in different industrial environments to urinary arsenic excretion suggests differences in respiratory absorption depending on the form of arsenic. Using equations relating urinary arsenic excretion to air concentrations, Yager, Hicks and Fabianova (1997) reported that the predicted urinary arsenic output for workers exposed to $10\,\mu\mathrm{gm}^{-3}$ of arsenic was more than one-third lower in workers performing boiler maintenance in a coal-fired power plant compared to copper smelter workers in several studies. This finding was attributed to the fact that the arsenic in the coal *flyash* of their study was predominantly in the form of calcium arsenate, whereas the form of arsenic present in the copper smelter work environment was arsenic trioxide. Such an interpretation is consistent with the much greater retention of calcium arsenate in hamster lung compared to arsenic trioxide (Pershagen, Lind and Bjorklund, 1982).

Intratracheal instillation studies in laboratory animals provide more direct information on the extent of absorption of various chemical forms of inorganic arsenic. In general, solubility appears to be the most important physicochemical property determining the extent of *lung clearance*, although *wetting capacity*, and *pulmonary toxicity* may also have an important influence. Pershagen, Lind and Bjorklund (1982) found that lung concentrations of arsenic in hamsters given weekly intratracheal instillations of arsenic trioxide, arsenic trisulfide, and calcium arsenate each differed by a factor of approximately 10-fold after four weeks. The much more rapid clearance of arsenic trioxide was attributed to its much greater *in vivo* solubility compared to the other two arsenicals. The authors speculated that the much slower clearance of calcium arsenate relative to arsenic trisulfide was a consequence of its higher wetting capacity; this would result in more calcium arsenate being transported to the alveolar regions of the lung where clearance is slower. The authors also indicated that the pulmonary toxicity of calcium arsenate may have impaired normal clearance mechanisms, which would have prolonged lung retention. Marafante and Vahter (1987) reported that the extent of absorption of inorganic arsenicals from the lungs of hamsters after intratracheal instillation was directly correlated with their *in vivo* solubility as determined by the amount of radiolabeled arsenical retained at an intramuscular injection site. The lung retention of arsenic ($2\,mg\,As\,kg^{-1}$) three days after an intratracheal instillation of sodium arsenite, sodium arsenate, arsenic trisulfide, and lead arsenate was 0.06, 0.02, 1.3, and 45.5 % of the dose, respectively.

Minimal data are available from typical inhalation studies in laboratory animals to allow evaluation of extent or dose-dependency in inhaled arsenic absorption. Beck, Slayton and Farr (2002) reported a study in which rabbits were exposed to 0.05, 0.1, 0.22, or $1.1\,mg\,m^{-3}$ of arsenic trioxide 8 hours/day, seven days/week for eight weeks. The particle size (*mass median aerodynamic diameter*, MMAD) ranged from 3.2 to 4.1 µm. On the basis of minimal elevation of inorganic arsenic in plasma until exposure levels were at or above $0.22\,mg\,m^{-3}$, the authors concluded that systemic uptake of arsenic trioxide following inhalation exposure was low and did not contribute significantly to body burden until relatively high levels of exposure were achieved.

4.4.3 Gastrointestinal absorption

Arsenic can be absorbed from the gastrointestinal tract following ingestion of arsenic-containing food, water, beverages, or medicines, or as a result of inhalation and subsequent mucociliary clearance and swallowing. The bioavailability of ingested inorganic arsenic will vary depending on the matrix in which it is ingested (e.g. food, water, beverages, or soil), the solubility of the arsenical compound, and the presence of other food constituents and nutrients in the gastrointestinal tract.

Controlled ingestion studies in humans indicate that both As(III) and As(V) are well absorbed from the gastrointestinal tract. For example, Pomroy *et al.* (1980) reported that healthy male human volunteers excreted 62 % of a 0.06 ng dose of [74]As-arsenic acid (inorganic As(V), protonated arsenate, H_3AsO_4) in urine over a period of seven days, whereas only 6 % of the dose was excreted in the feces. Few other controlled human ingestion studies have actually reported data on both urine and fecal elimination of arsenic. However, between 45 and 75 % of the dose of various As(III) and As(V) forms of arsenic are excreted in the urine within a few days, which suggests that gastrointestinal absorption is both relatively rapid and extensive (Tam, Charbonneau and Bryce, 1979; Yamauchi and Yamamura, 1979; Buchet, Lauwerys and Roels, 1981a; Buchet, Lauwerys and Roels, 1981b; Lee, 1999).

Both As(V) and As(III) are rapidly and extensively absorbed from the gastrointestinal tract of common laboratory animals following a single oral dose. Based on the mouse data of Vahter and Norin (1980), inorganic As(III) is more extensively absorbed from the gastrointestinal tract compared to inorganic As(V) at lower doses (e.g. $0.4\,mg\,As\,kg^{-1}$), whereas the reverse is true at higher doses (e.g. $4.0\,mg\,As\,kg^{-1}$);

the latter is consistent with the more rapid whole body clearance observed with inorganic As(III) compared to inorganic As(V) at lower (0.5 mg As kg^{-1}), but not higher (5.0 mg As kg^{-1}) doses reported by Hughes *et al.* (1999). Based on comparison of whole body clearance of inorganic As(V) (0.5 or 5.0 mg As kg^{-1}) following a single oral or *intraperitoneal dose*, these authors suggested that there might be dose-dependent differences in the rate of gastrointestinal absorption for inorganic As(V) (Hughes *et al.*, 1999). There are some *in vitro* studies that may partially explain dose-dependent differences in rates of gastrointestinal absorption, as well as differences between inorganic As(III) and inorganic As(V). Using isolated perfused rat small intestine, Gonzalez, Aguilar and Para (1995) demonstrated that uptake of As(V) results from a saturable transport process and that the addition of phosphate markedly decreased arsenic absorption, most likely because arsenate (inorganic As(V)) and phosphate can share the same transport mechanism.

Studies with mice conducted by Odanaka, Matano and Goto (1980) suggest that much less As(V) is absorbed from the gastrointestinal tract following oral administration compared to the results of Vahter and Norin (1980), or 49 % of the dose (5 μg g^{-1}) in urine compared to 89 % of the dose (4 mg As kg^{-1}) excreted in urine, respectively. This difference may be attributable to the fact that the mice in the study of Vahter and Norin were not fed for at least two hours before and 48 hours after dosing, whereas the mice in the Odanaka *et al.* studies were not food restricted. Kenyon, Hughes and Levander (1997) found that feeding a diet lower in fiber or 'bulk' to female mice increased absorption of inorganic As(V) by ~10 % compared to a standard rodent chow diet.

The bioavailability of arsenic from soils has been assessed using various animal models because this can be a significant issue in *risk assessment* for contaminated industrial or agricultural sites, where there is potential for arsenic exposure via soil ingestion. In general, these studies indicate that oral bioavailability of arsenic in a soil or dust vehicle is considerably lower compared to the pure soluble salts typically used in toxicity studies (Groen *et al.*, 1994; Freeman *et al.*, 1993; Freeman *et al.*, 1995; Roberts *et al.*, 2002; (US EPA, 1996). Davis, Ruby and Bergstrom (1992) have pointed out that this is due mainly to mineralogical factors, which control solubility in the gastrointestinal tract, such as the solubility of the arsenic-bearing mineral itself and encapsulation within insoluble matrices (e.g. silica). Yang *et al.*, (2002, 2005) compared inorganic As(III) and inorganic As(V) with respect to their relative adsorption, oxidation, and bioaccessibility (as a surrogate for bioavailability) in soils. In general, inorganic As(III) is more readily bioaccessible than inorganic As(V) initially; however, aging the soils by holding them in containers and maintaining a moisture content of 30 % for six months (which results in oxidation of inorganic As(III) to inorganic As(V)) increases bioaccessibility considerably Yang *et al.* (2005). They also report that lower soil pH and higher iron content decrease the bioaccessibility of arsenic from soil (Yang *et al.*, 2002).

4.4.4   Dermal absorption

Dermal absorption of inorganic arsenic is generally very low compared to oral absorption and as with oral absorption, the matrix (water, soil, dust, or CCA-treated wood) can have an effect (Lowney *et al.*, 2005). Wester *et al.* (1993) studied the percutaneous absorption of arsenic acid (H$_3$AsO$_4$) from water and soil both *in vivo* using rhesus monkeys and *in vitro* with human skin. *In vivo* absorption of arsenic acid from water (loading 5 μl cm^{-2} skin area) was 6 % at the low dose (0.000 024 μg cm^{-2}) and 2 % at the high dose (2.1 μg cm^{-2}). Absorption from soil (loading 0.04 g soil/cm^2 skin area) *in vivo* was 5 % at the low dose (0.000 04 μg cm^{-2}) and 3 % at the high dose (0.6 μg cm^{-2}). Thus, *in vivo* using the rhesus monkey, percutaneous absorption of arsenic acid is low from either soil or water vehicles and does not differ appreciably at doses more than 10,000-fold apart. Wester *et al.* (1993) also reported that using human skin, 2 % was absorbed at the low dose from water and 1 % from soil (absorption based on combined arsenic in receptor fluid plus skin) over a 24-hour period. Utilizing comparable methods, Wester *et al.* (2004) concluded that absorption of inorganic arsenic from residue of CCA-treated wood is also low (0.04 %). Recent studies utilizing soils and the rhesus monkey as the animal model and similar methods indicate that absorption is in the range of 0.5–1 % for most soils compared to 5 % for soluble arsenic in aqueous solution (Lowney *et al.*, 2007).

4.5   Systemic clearance of arsenic and binding to blood components

Once in systemic circulation, arsenic is rapidly cleared from blood and distributed to all major organs in the body in most common laboratory animals, including mice, rabbits, and hamsters (Vahter and Norin, 1980; Marafante, Bertolero and Edel, 1982; Marafante, Vahter and Envall, 1985; Yamauchi and Yamamura, 1985; Kenyon, Del Razo and Hughes, 2005a; Kenyon, Del Razo and Hughes, 2005b). In humans, inorganic arsenic is also rapidly cleared from blood with a half-life of about two hours. It is for this reason that blood arsenic is considered to be a useful bioindicator only for very recent or relatively high level exposures (Ellenhorn, 1997). Data on systemic clearance of arsenic are naturally rather sparse in humans; however, Pomroy *et al.* (1980) studied the whole body retention of (^{74}As) (6.4 µCi, 0.06 ng arsenic) administered once orally as arsenic acid in healthy male volunteers (age 28–60 years) using whole body counting for periods of up to 103 days. While the averaged whole body clearance data for the six subjects in the study were best described by a *triexponential model*, it is noteworthy that interindividual variation was quite high. It was reported that 65.9 % of the dose was cleared with a half-life of 2.1 days, 30.4 % with a half-life of 9.5 days and 3.7 % with a half-life of 38.4 days.

The systemic clearance of arsenic has been evaluated in mice in several studies and both dose-dependent and valence state-specific differences have been observed (Vahter and Norin, 1980; Hughes *et al.*, 1999; Lindgren, Vahter and Dencker, 1982). Hughes *et al.* (1999) did not observe major differences in whole body clearance of (^{73}As)-inorganic As(V) compared to (^{73}As)-inorganic As(III) when administered at the same single oral dose (0.5 or 5 mg As kg^{-}) to three different strains of female mice over a 24-hour period. However, with increased oral dose, clearance of (^{73}As) was slower and tissue retention of (^{73}As) was higher. Lindgren, Vahter and Dencker (1982) compared the whole body retention of inorganic As(V) and inorganic As(III) administered intravenously as sodium salts to male mice at a dose of 0.4 mg As kg^{-1}. Retention was higher in inorganic As(III)-treated mice compared to inorganic As(V)-treated mice at all times measured, that is, 44 % versus 20 % at 6 hours postdosing, 14 % versus 3 % at 24 hours, and 6 % versus 2 % at 72 hours postdosing. Vahter and Norin (1980) earlier reported that in male mice dosed orally with 0.4 mg As kg^{-1} inorganic As(V) or inorganic As(III) that whole body retention was similar over the 35-day course of the experiment, but that retention was consistently slightly higher in inorganic As(III)-dosed mice. In contrast and similar to what was observed with intravenously dosed mice in the Lindgren, Vahter and Dencker (1982) study, whole body retention in inorganic As(III)-dosed mice was clearly consistently higher compared to inorganic As(V)-dosed mice when dosed orally with 4 mg As kg^{-1}. Thirty-five days after administration, the high/low dose retention ratios were 11 and 6 for inorganic As(III)- and inorganic As(V)-dosed mice, respectively.

Although clearance of both inorganic As(V) and inorganic As(III) from blood is rapid, both valance state and dose-dependent differences have been observed (Vahter and Norin, 1980; Kenyon, Del Razo and Hughes, 2005a; Kenyon, Del Razo and Hughes, 2005b). Vahter and Norin (1980) reported that at a high oral dose of arsenic (4 mg As kg^{-1}), inorganic As(III)-dosed mice had a higher *erythrocyte* to plasma ratio of approximately 2–3, whereas in inorganic As(V)-dosed mice the ratio was much closer to one. No such difference was observed at a lower oral dose (0.4 mg As kg^{-1}) of inorganic As(V) or inorganic As(III). Delnomdedieu *et al.* (1994b) investigated the uptake of inorganic As(III) compared to inorganic As(V) in intact rabbit erythrocytes. They reported that ~76 % of inorganic As(III) compared to ~25 % of inorganic As(V) was taken up within 0.5 hour, and that inorganic As(III) subsequently bound with intracellular *glutathione* (GSH), whereas inorganic As(V) entered the phosphate pathway, depleting adenosine-5'-triphosphate ($C_{10}H_{16}N_5O_{13}P_3$, ATP) and increasing inorganic phosphate levels.

Rats are notable exceptions to rapid blood and systemic clearance among species and the presence of arsenic in blood is prolonged due to specific binding of *dimethylarsinous acid* (DMA(III)) to the alpha (α) chain of rat hemoglobin (Hb). Rat Hb also has a 16-fold higher binding affinity for DMA(III) compared to human Hb. In addition, under subchronic exposure conditions, DMA(III) accumulates in rat erythrocytes bound to *cysteine*-13 α (Cys-13α) of Hb whether the rats are fed inorganic As(V), *monomethylarsonic acid* (MMA(V)), or *dimethylarsinic acid* (DMA(V)) in the diet (Lu *et al.*, 2004; Lu *et al.*, 2007). Binding to rat Hb greatly increases the biological half-life of inorganic arsenic and DMA(V) in rats (weeks) compared to mice (days) (Vahter, 1981; Vahter, Marafante and Dencker, 1984). Interestingly, differences among animal species in patterns of erythrocyte uptake have also been reported for DMA(V). Shiobara, Ogra and Suzuki (2001) reported that DMA(V) was essentially not taken up at all when incubated with erythrocytes of mice, rats, hamsters, or humans. DMA(III) was readily taken up by erythrocytes, but the pattern differed among species with the rat being most efficient, followed by hamster, human, and mouse erythrocytes.

4.6 Tissue distribution

Arsenic and its metabolites are widely distributed in tissues of both humans and experimental animals, although both dose-dependent and metabolite-specific patterns of accumulation are observed in different tissues (World Health Organization (WHO), 2001; Kenyon, Del Razo and Hughes, 2005a; Kenyon *et al.*, 2005b; Hughes *et al.*, 2003). Numerous mechanistic studies have documented basic biochemical differences in the interaction of As(V) compared to inorganic As(III) with cellular components and this can be an important determinant in observed differences in tissue distribution. As(V) can act as a phosphate analog. At the molecular level this means that inorganic As(V) can compete with phosphate for active transport processes. This is why the addition of phosphate can decrease intestinal uptake (Gonzalez, Aguilar and Para, 1995) and renal tubular reabsorption (Ginsburg and Lotspeich, 1963) of inorganic As(V). As(V) can also substitute for phosphate in the hydroxy*apatite* crystal of bone, which accounts for the higher concentrations of arsenic-derived radioactivity in bone after administration of (^{74}As)-inorganic As(V) compared to (^{74}As)-inorganic As(III) (Lindgren, Vahter and Dencker, 1982). As(III) reacts readily with the *sulfhydryl* groups in a variety of essential *enzymes* and proteins. It is the affinity of inorganic As(III) for sulfhydryl groups, and *vicinal dithols* in particular, that accounts for its accumulation in keratin-rich tissues, such as skin, hair, and nails. As(III) also interacts with the ubiquitous sulfhydryl-containing cellular tripeptide GSH at many different levels in the *methylation* process. These include, but may not be limited to, reduction of As(V) to As(III) following the addition of a methyl group and formation of complexes with As(III) arsenicals, which may be substrates for methylation and the form in which arsenic is excreted into bile (Styblo, Delnomdedieu and Thomas, 1996; Hayakawa *et al.*, 2005; Kala *et al.*, 2000).

Analysis of postmortem human tissues reveals that arsenic is widely distributed in the body following either long-term relatively low-level exposure or poisoning (Raie, 1996; Gerhardsson *et al.*, 1988; Dang, Jaiswal and Somasundaram, 1983). Dang, Jaiswal and Somasundaram (1983) used *neutron activation analysis* (NAA) to measure total arsenic in various tissues of individuals (age and sex not specified) dying in accidents in the Bombay (Mumbai) area of India. Notable results from this study are that arsenic concentrations are quite low in both blood and brain relative to other tissues and that the arsenic concentration in any given tissue was quite variable.

Yamauchi and Yamamura (1983) reported levels of total arsenic and major arsenic metabolites in a variety of human tissues obtained from adult patients (age 36–79) dying of cerebral hemorrhage, pneumonia, or cancer in Kawasaki, Japan. No sex-dependent differences in arsenical tissue levels were observed and inorganic arsenic was the predominant form of arsenic in tissues followed by DMA (note: abbreviations without valence state indicate total arsenical; in this case, DMA(III) + DMA(V)). MMA levels were uniformly low and detected only in the liver and kidneys. It is interesting to note that total arsenic levels were higher than those reported in the Indian study of Dang, Jaiswal and Somasundaram (1983) and levels in the brain tended to be more comparable to arsenic levels in other tissues. Interindividual variation in total tissue arsenic was also quite high as observed in the Dang *et al.* study.

Raie (1996) compared tissue arsenic levels in infants (one day to five months) and adults from the Glasgow, Scotland area using NAA. Mean levels of arsenic (micrograms per gram dry weight $\mu g\, g^{-1}$ dry) in liver, lung, and spleen in infants versus adults were 0.0099 versus 0.048, 0.007 versus 0.044, and 0.0049 versus 0.015, respectively. These data suggest that arsenic accumulates in tissues with age, which is consistent with observations in laboratory animals. Studies have been conducted in humans with the goal of determining whether there are differences in tissue arsenic accumulation (and other metals) in differing disease states. Warren, Horksy and Gould (1983) compared trace element levels in brain and other tissues of multiple sclerosis and non–multiple sclerosis patients and found no significant difference in any tissue arsenic levels. Narang and Datta (1983) have reported that concentrations of arsenic in both the liver and brain of patients who died of fulminant hepatitis are high compared to patients who died of non-hepatic-related causes. Collecchi, Esposito and Brera (1986) compared the distribution of arsenic and cobalt in cancerous and noncancerous laryngeal tissue and plasma of patients with and without laryngeal cancer. Malignant tissue had significantly higher levels of arsenic compared to normal tissue and plasma arsenic levels were also significantly higher in cancer patients compared to controls.

Studies in rabbits, rats, mice, hamsters, and monkeys demonstrate that arsenic, administered orally or parenterally, as either As(III) or As(V), is rapidly distributed throughout the body. Many of these studies have used radiolabeled arsenic and it is noteworthy that arsenic-derived radioactivity is generally present in all examined tissues (Marafante, Bertolero and Edel, 1982; Kenyon, Del Razo and Hughes, 2005a; Kenyon, Del Razo and Hughes, 2005b; Lindgren, Vahter and Dencker, 1982; Vahter *et al.*, 1982; Vahter and Marafante, 1985).

Numerous studies also reveal that skin, hair, and tissues high in *squamous epithelium* (e.g. mucosa of the oral cavity, esophagus, stomach, and small intestine) have a strong tendency to accumulate and maintain higher levels of arsenic (e.g. Yamauchi and Yamamura, 1985; Lindgren, Vahter and Dencker, 1982). This is apparently a function of the binding of arsenic to keratin in these tissues (Lindgren, Vahter and Dencker, 1982). Autoradiographic studies have also revealed a tendency for arsenic to accumulate in the epididymis, thyroid, and lens of the eye of mice (Lindgren, Vahter and Dencker, 1982).

Arsenic is capable of crossing the blood-brain barrier since it is found in brain tissue after oral or parenteral administration of inorganic As(V) or inorganic As(III) in all studied species. However, the levels are uniformly low both across time and relative to other tissues, which indicate that arsenic does not readily cross the blood-brain barrier or accumulate in brain tissue following *acute* dosing (Marafante, Bertolero and Edel, 1982; Yamauchi and Yamamura, 1985; Lindgren, Vahter and Dencker, 1982; Dang, Jaiswal and Somasundaram, 1983; Yamauchi and Yamamura, 1983; Vahter *et al.*, 1982; Vahter and Marafante, 1985).

The discussion and studies cited previously generally reflect overall tissue distribution of total arsenic after acute exposure in the case of laboratory animals or unknown exposures in the case of humans. Advances in analytical technology in the last decade have facilitated the identification of tissue-specific patterns of metabolite distribution and accumulation in laboratory animals. Kenyon, Del Razo and Hughes (2005a) found that inorganic arsenic was the predominant form of arsenic in the liver and kidney up to two hours post administration of 10 or 100 μ mol As kg^{-1} as inorganic As(V) to female mice, whereas

DMA was the predominant metabolite at later times. Three- to fourfold higher levels of MMA were achieved in the kidneys by one hour postdosing compared to other tissues. DMA was the predominant metabolite in the lungs from two hours postdosing onward. Similar patterns of tissue distribution were observed in the livers, lungs, and kidneys in female mice given a single oral dose of $100\,\mu mol\ As\ kg^{-1}$ as inorganic As(III) (Kenyon, Del Razo and Hughes, 2005b).

Studies in laboratory animals also demonstrate tissue- and metabolite-specific patterns of accumulation following longer term exposure. Hughes *et al.* (2003) reported that accumulation of arsenic-derived radioactivity was highest in bladder, kidney, and skin following nine repeated daily gavage doses of $0.5\,mg\ As\ kg^{-1}$ as (^{73}As)-inorganic As(V). Based on measurements at eight days after the end of exposure, loss of radioactivity was most rapid in the lungs and slowest in the skin. Speciated arsenical analysis after nine days of exposure revealed that the distribution of metabolites varied by tissue. For inorganic arsenic, MMA, and DMA, the distribution in the bladder was 11.6, 0, and 88.4%, respectively. For the kidney, it was 56.6, 12.5, and 31.0%, respectively. In the liver, the distribution was 35.5, 23.4, and 34.1%, respectively, with a trimethylated metabolite comprising ~7% detected in the liver, but not in other tissues. In the lung, the metabolite distribution was 30.4, 3.3, and 66.3%, respectively, for inorganic arsenic, MMA, and DMA. Kenyon *et al.* (2008) also observed tissue- and metabolite-specific patterns of accumulation in female mice exposed to 0.5, 2, 10, or $50\,\mu g\ g^{-1}$ inorganic As(V) in drinking water for 90 days. The highest total arsenic accumulation occurred in the kidney, bladder, and lung. MMA was preferentially sequestered in the kidney at the two highest exposure levels, whereas DMA preferentially accumulated in the lung. The bladder had relatively high levels of both inorganic arsenic and DMA. Interestingly, both bladder and lungs are target organs for cancer in humans.

4.7 Placental transfer and distribution in the fetus

Case reports of arsenic poisoning in pregnant women resulting in the death of the fetus accompanied by toxic levels of arsenic in fetal tissues demonstrate that As(III) as arsenic trioxide (As_2O_3) readily passes through the placenta (Lugo, Cassady and Palmisano, 1969; Bollinger, van Zijl and Louw, 1992). Concha *et al.* (1998a) found that arsenical concentrations were similar in cord blood and maternal blood (~$9\,\mu gl^{-1}$) of maternal-infant pairs exposed to high arsenic-containing drinking water (~$200\,\mu gl^{-1}$). Hall *et al.* (2007) also reported maternal and cord blood total arsenic and metabolite distribution were similar in a Bangladesh population. Hopenhayn *et al.* (2003) concluded that total urinary arsenic increased with increasing weeks of gestation in a population of Chilean women exposed to arsenic at $40\,\mu g\,L^{-1}$ in drinking water and that most of the increase in urinary arsenic was accounted for by DMA. Concha *et al.* (1998a) also found that DMA accounted for most of the arsenic in maternal and fetal plasma.

Both older and more recent studies have documented the ability of inorganic As(III) and inorganic As(V) to cross the placenta in laboratory animals (Lindgren *et al.*, 1984; Hood, Vedel-Macrander and Zaworotko, 1987; Hood *et al.*, 1988; Jin *et al.*, 2006; Devesa *et al.*, 2006). Lindgren *et al.* (1984) reported that in pregnant mice given a single intravenous injection ($4\,mg\ As\ kg^{-1}$) of inorganic As(III) or inorganic As(V), both forms passed through the placenta easily and to approximately the same extent. Hood, Vedel-Macrander and Zaworotko (1987) compared the fetal uptake of inorganic As(V) following oral ($40\,\mu g\ g^{-1}$) or intraperitoneal ($20\,\mu g\ g^{-1}$) administration to pregnant mice on day 18 of gestation. Arsenic levels peaked later and were over fivefold lower in the fetuses of mice dosed orally, most likely reflecting both slower uptake from the gastrointestinal tract and a greater opportunity for methylation in the liver prior to reaching the systemic circulation. The quantity of dimethylated metabolite present in the fetuses rose over time (to ~80% of total metabolites present for both routes of administration) and remained relatively constant from ~10 hours postdosing until the termination of the study at 24 hours postdosing. Hood *et al.* (1988) also compared the fetal uptake of inorganic As(III) following oral ($25 \, \mu g \, g^{-1}$) or intraperitoneal ($8 \, \mu g \, g^{-1}$) administration to 18 days old pregnant mice. As was the case with inorganic As(V), injected mice achieved both higher fetal and placental levels of arsenic more quickly compared to mice dosed orally. Both valence forms followed similar time-course trends after oral administration. However, levels of arsenic in fetuses of dams injected with inorganic As(III) plateaued from 12 to 24 hours after dosing, whereas levels of arsenic in fetuses of dams injected with inorganic As(V) peaked at 2–4 hours postdosing and then declined quickly. The proportion of arsenic present in fetuses as methylated metabolite increased over time to 88 % and 79 % following oral and intraperitoneal administration, respectively. A higher fraction of monomethylated arsenic was present in fetuses of dams dosed with inorganic As(III) compared to inorganic As(V). The authors concluded that much of the arsenic reaching the fetus has already been transformed to methylated metabolites.

Placental transfer and distribution of arsenicals in the fetus have also been studied in mice exposed via drinking water throughout gestation (Jin *et al.*, 2006) or during the major period of organogenesis (Devesa *et al.*, 2006). Jin *et al.* (2006) exposed albino mice to either 10 or $30 \, mg \, L^{-1}$ inorganic As(III) or inorganic As(V) throughout gestation and compared the distribution of metabolites in the liver and brain of dams and pups. Overall, arsenical levels in the liver and brain of both dams and pups were significantly higher in the $10 \, mg \, L^{-1}$ inorganic As(III)- compared to the inorganic As(V)-exposed group, whereas this was not the case with the $30 \, mg \, L^{-1}$ inorganic As(III)- and inorganic As(V)-exposed groups. Among the metabolites, DMA was the predominant one in both neonatal liver and brain. Devesa *et al.* (2006), using somewhat higher exposure levels of inorganic As(III) in drinking water (42.5 and $85 \, mg \, L^{-1}$) from gestation days 8–18, reported that arsenical concentrations did not differ significantly between exposure levels for maternal blood or for fetal lung, liver, blood, and placenta. DMA was generally the predominant metabolite in fetal tissues and there were no dose-dependent differences in the percentage distribution of metabolites within tissues. The findings of Devesa *et al.* (2006) for inorganic As(III) are consistent with those of Jin *et al.* (2006).

4.8   Arsenic biotransformation

4.8.1   Introduction

The study of the biological methylation of inorganic arsenic has its roots in the widespread use of inorganic arsenicals as pigments in the nineteenth century (Chapter 5). The mildewing of wallpapers colored with arsenic compounds was associated with a distinctive odor and adverse health effects were attributed to exposure to these arsenic-containing fumes (Chasteen, Wiggli and Bentley, 2002; Meharg, 2005; Chapter 5). By the 1890s, Bartoleomo Gosio had shown that the garlic-like odor associated with the mildewing of arsenic-laden wallpaper was the product of a biological reaction in which a fungus converted inorganic arsenic into a volatile arsenical (Gosio, 1892). So-called Gosio gas was thought to be diethylarsine; however, Frederick Challenger and associates later identified it as *trimethylarsine* (Challenger, Higginbottom and Ellis, 1933). In subsequent work, Challenger identified many of the intermediates in the pathway from inorganic arsenic to trimethylarsine and postulated a Scheme 4.1 for the production of methylated arsenicals from inorganic arsenic, as summarized below (see also Chapter 2):

$$\text{Inorganic As(III)} + CH_3^+ \rightarrow CH_3As(V) + 2e^- \rightarrow CH_3As(III) + CH_3^+$$

$$\rightarrow (CH_3)_2As(V) + 2e^- \rightarrow (CH_3)_2As(III) + CH_3^+ \rightarrow (CH_3)_3As(V) + 2e^- \rightarrow (CH_3)_3As(III) \quad (4.1)$$

In the *Challenger mechanism* or scheme, oxidative methylation of arsenicals containing As(III) produces a methylated product that contains As(V). Because methylation is oxidative, As(V) must be reduced to trivalency before it can be methylated. Hence, the pathway for the formation of mono-, di-, and trimethylated arsenic species consists of alternating oxidation and reduction reactions (Chapter 2).

4.8.2 Arsenic methylation in humans and other mammals

The universality of the methylation of arsenic in higher organisms has been demonstrated in the past three decades. Using a newly available analytical technique, Braman and Foreback (1973) found that human urine contained inorganic, methylated, and dimethylated arsenicals. However, it was uncertain from this work whether the presence of methylated arsenicals in human urine reflected exposure to these compounds or metabolic production from inorganic arsenic. Work by Eric Crecelius and associates demonstrated that humans rapidly convert ingested or inhaled inorganic arsenic to methylated species (Crecelius, 1977; Smith, Crecelius and Reading, 1977). Hence, shortly after exposure to inorganic arsenic, urine from exposed individuals will contain inorganic, monomethyl, and dimethyl arsenic.

4.8.3 Significance of arsenic methylation

Originally, it was commonly held that the methylation was a means for the detoxification of inorganic arsenic. This evaluation was based largely on comparisons of the acute toxicities of As(III), MMA(V), and DMA(V) (Yamauchi and Fowler, 1994). However, because the Challenger scheme (Section 4.8 and Chapter 2) for the metabolism of inorganic arsenic proposed that methylated species containing As(III) were obligatory intermediates in the pathway, researchers examined the toxic effects of *monomethylarsonous acid* (MMA(III)) and DMA(III). These latter compounds were found to be more potent *cytotoxins* and *genotoxins* and inhibitors of enzyme activities than inorganic As(III) (Styblo *et al.*, 2000; Thomas, Styblo and Lin, 2001). Thus, it was likely that the formation of reactive methylated species containing As(III) should be considered as an activation process, which could increase the toxic potential of arsenic ingested as inorganic arsenic. Evidence that humans exposed to inorganic arsenic excrete MMA(III) and DMA(III) in urine (Le *et al.*, 2000; Del Razo *et al.*, 2001b) was consistent with the prediction of the Challenger scheme for methylation. Differences among individuals in the formation of reactive intermediates could be a determinant of the risk associated with exposure to inorganic arsenic. For example, in a population chronically exposed to inorganic arsenic in drinking water, individuals with skin lesions characteristic of chronic arsenic toxicity have been reported to have significantly higher percentages of inorganic arsenic and MMA and lower percentages of DMA in urine than do individuals lacking these skin lesions (Valenzuela *et al.*, 2005). Taken together, these factors have provided the impetus for elucidating the molecular basis for arsenic methylation in humans and other mammals.

4.8.4 Molecular basis of the metabolism of inorganic arsenic

4.8.4.1 Introduction

Subsequent work on the molecular basis of arsenic methylation can be divided into two areas based on the Challenger scheme, which posits the existence of two distinct processes in arsenic methylation. These are the reduction of As(V) to As(III) and the oxidative methylation of arsenicals containing As(III) to As(V).

4.8.4.2 Reduction of As(V)

Cullen and associates examined the reduction of As(V) by *thiols*, especially GSH (Cullen, McBride and Reglinski, 1984a; Cullen, McBride and Reglinski, 1984b). They suggested that this monothiol could play an important role in the reduction of As(V). Later work examined the reduction of As(V) by GSH in chemically defined systems and in intact rabbit erythrocytes (Delnomdedieu *et al.*, 1994b; Scott *et al.*, 1993; Delnomdedieu *et al.*, 1993; Delnomdedieu *et al.*, 1994a). These studies showed that inorganic

As(V) was sequentially reduced and complexed by GSH in the following reaction scheme:

$$\text{Inorganic As(V)} + 2\ \text{GSH} \rightarrow H_3AsO_3 + 3\ \text{GSH} \rightarrow \text{As(III)(GS)}_3 \tag{4.2}$$

Hence, reduction of inorganic As(V) and complexation of inorganic As(III) by GSH was a plausible mechanism in cellular environments. However, some data suggest that the rate for complexation of inorganic As(III) by GSH may be too slow to account for the rates of formation that are observed in cells (Spuches, 2005). This suggests that this reaction could be enzymatically catalyzed, possibly by members of the glutathione-*S*-transferase family (Thomas, 2007).

Three *phosphorylytic* enzymes –purine nucleoside phosphorylase (PNP), glyceraldehyde-3-phosphate dehydrogenase (GAPDH), and glycogen phosphorylase (GP) – have been shown to catalyze the thiol-dependent reduction of inorganic As(V) to inorganic As(III) (Gregus and Németi, 2002; Gregus and Németi, 2005; Németi, Csanaky and Gregus, 2003; Németi and Gregus, 2005; Németi and Gregus, 2007). For PNP, the *reductant* is dihydrolipoic acid (DHLA); for GAPDH and GP, the reductant is GSH. These enzymes may catalyze the formation of unstable arsenoesters, which decompose to yield inorganic As(III). Although these reactions are chemically possible, there is considerable uncertainty whether or not these enzymes contribute significantly to the reduction of inorganic As(V) in cells. A GSH-dependent *reductase* converting MMA(V) to MMA(III) was partially purified from rabbit and human livers (Zakharyan and Aposhian, 1999). This enzyme had considerable sequence *homology* to glutathione-*S*-transferase omega (GSTO) (Zakharyan *et al.*, 2001), suggesting that reduction of MMA(V) was a function of GSTO. However, studies in GSTO knockout mice found no effect of the *null genotype* on the disposition and metabolism of arsenicals in tissues after treatment with inorganic As(V) (Chowdhury *et al.*, 2006).

4.8.4.3 *Arsenic methyltransferases*

Early work on the molecular basis of arsenic methylation used *in vitro* assays containing subcellular fractions to define the conditions under which methylation of inorganic arsenic occurred (Buchet and Lauwerys, 1985; Buchet and Lauwerys, 1988; Hirata *et al.*, 1989; Smith, Jones and Shirachi, 1992). These studies indicated that the presence of GSH and *S*-adenosylmethionine (AdoMet) in reaction mixtures increased the rates of formation of methylated arsenicals. An AdoMet-dependent *methyltransferase* purified from rabbit liver *cytosol* catalyzed reactions that methylated either inorganic As(III) or MMA(III) (Zakharyan *et al.*, 1995, 1999). The activity of this enzyme could be stimulated by addition of GSH or D,L-dithiothreitol (DTT), a nonphysiological dithiol. These two activities shared by this ~60-kDa protein were designated as arsenite methyltransferase (E.C. 2.1.1.137) and methylarsonate methyltransferase (E.C. 2.1.1.138). Subsequent work described the purification of this enzyme from other tissue sources (Wildfang and Zakharyan, 1998); however, it has not been fully sequenced and its gene has not been cloned.

Another protein (~42 kDa) with arsenic methyltransferase activity has been purified from the liver cytosol of adult male Fischer 344 rats (Lin *et al.*, 2002). The gene for the rat protein has been cloned and the recombinant protein *expressed* (Walton *et al.*, 2003). This protein was initially designated as cyt19 but is now assigned the systematic name, arsenic (+3 oxidation state) methyltransferase (AS3MT) (Thomas *et al.*, 2007). Subsequent work has identified genes encoding AS3MT-like proteins in rat, mouse, and human genomes. Across these species, the primary sequence of the protein is highly conserved (Figure 4.1). Especially notable is the complete conservation of five critical *cysteinyl* residues in the rat, mouse, and human proteins. In rat and mouse AS3MT, site-directed *mutagenesis* replacing of the cysteinyl residue in the 156/157 position of the protein with *serine* leads to a complete loss of catalytic activity (Li *et al.*, 2005; Fomenko *et al.*, 2007). Studies of altered expression of AS3MT suggest that this protein normally functions as an arsenic methyltransferase. *Heterologous expression* of AS3MT in a human

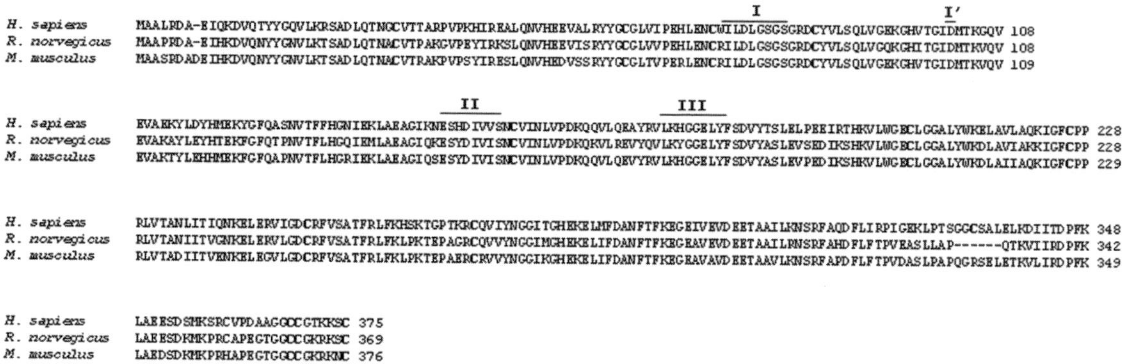

	I	I′
H. sapiens	MAALRDA-EIQKDVQTYYGQVLKRSADLQTMGCVTTARPVPKGHIREALQMVHEEVALRYYGCGLVIPEHLENCMIIDLGSGSGRDCYVLSQLVGKKGHWTGIDMTKGQV	108
R. norvegicus	MAAPRDA-EIHKDVQNYYGNVIKTSADLQTNACVTPARGVPKEYIRKSLQMVHEEVISRYYGCGLVVPEHLENCRIIDLGSGSGRDCYVLSQLVGKHGHITGIDMTKGQV	108
M. musculus	MAASRDADEIHKDVQNYYGNVIKTSADLQTNACVTRAKPVPSYIRESLQMVHEDVSSRYYGCGLIVPERLENCRIIDLGSGSGRDCYVLSQLVGKHGHWTGIDMTKGQV	109

	II	III
H. sapiens	EVAEKYLDYHMEKYGFQASNVTFFHGNIEKLAEAGIKNESHDIVVSMCVINLVPDKQLQRAYRVLKHGGELYFSDVYTSLELPEKIRTHKVLWGECLGGALVKKELAVLAQKIGFCPP	228
R. norvegicus	EVAKAYLEYHMEKFGFQIPNVTFLHGQIEHLAEAGIQKHSTDIVISMCVINLVPDKQRKVLREVYQVLKYGGELYFSDVYASLEVSDIDSHKVLWGECLGGALVMRDLAVLARKIGFCPP	228
M. musculus	EVAKTYLEHHMEKFGFQAPNVTFLHGRIEKLAEAGIQSHSTDIVISMCVINLVPDKQQVLQRVYRVLKHGGELYFSDVYASLEVPEDIDSHKVLWGECLGGALVMRDLAIILAQKIGFCPP	229

H. sapiens	RLVTANLITIQKRELERVLGDCRFVSATFRLFKESKTGPTGSCGVITNGGITGHEGELMFDANFTFKEGIVEVDEKTAALLNSRPAQDFLIRPIGEKLPTSGGCSALELKDIITDPFK	348
R. norvegicus	RLVTANIITVGKRGLERVLGDCRFVSATFRLFKLPKTGPAGRCGVVTNGGIMGHEGELIFDANFTFKEGEAVEVDEKTAALLNSRPAHDFLPTFVEASLLAP------QIKVIIPDPFK	342
M. musculus	RLVTADIITVENKRLEGVLGDCRFVSATFRLFKLPKTEPAHRCKVVTNGGIKGHEGELIFDANFTFKEGEAVAVDEKTAAVIENSRPAPDFLPTFVDASLPAPQGRSELETKVLIRDPFK	349

H. sapiens	LAHEDSHMHSRCVPDAAGCGCGTKRSC	375
R. norvegicus	LAHESDHMKPRCAPEGTGCGCGKRKSC	369
M. musculus	LAHESDHMKPHHAPEGTGCGCGKRKMC	376

Figure 4.1 Sequence alignments for *Homo sapiens* (human), *Rattus norvegicus* (rat), and *Mus musculus* (mouse) arsenic (+3 oxidation state) methyltransferase (AS3MT). Positions of conserved nonnucleic acid sequence motifs (I, I′, II, and III) indicated. Cysteinyl residues conserved in all three sequences marked with C. Accession numbers for sequences from NCBI GenBank: *H. sapiens* (Q9HBK9); *R. norvegicus* (NP_543166); *M. musculus* (AAH13468).

cell line that does not methylate inorganic As(III) confers an arsenic methylating *phenotype* (Drobná *et al.*, 2005). Suppression of *AS3MT* gene expression in cultured human *hepatoma cells* by ribonucleic acid (RNA) interference diminishes the capacity of these cells to methylate inorganic As(III) (Drobná *et al.*, 2006).

In *in vitro* assay systems, the methylating activity of *AS3MT* is dependent on the presence of a reductant. Although nonphysiological reductants, such as DTT, can support the catalytic activity of *AS3MT*, it is of more biological interest to identify physiological reductants that support catalysis by this protein. Thioredoxin (Tx), glutaredoxin (Gx), and DHLA have been identified as physiological reductants that can support methylation of inorganic As(III) by *AS3MT* (Waters *et al.*, (2004a)). Each of these reductants contains a dithiol pair that is oxidized to a disulfide during reduction of a target molecule. The dithiol pair of each reductant is regenerated in an enzymatic reaction that uses *nicotinamide adenine dinucleotide phosphate* (NADPH) as the source of reducing electrons. For Tx or DHLA, this reaction requires thioredoxin reductase and NADPH; for Gx, the reaction requires GSH, GSH reductase, and NADPH.

Omission of these dithiol reductants from reaction mixtures results in loss of catalytic activity of *AS3MT*. Replacement of each of these reductants with GSH as a reductant yields negligible rates for methylation of inorganic As(III). However, coaddition of each dithiol reductant with GSH leads to an overall increase in the rate of arsenic methylation. Notably, the presence of GSH markedly affects the ability of recombinant rat *AS3MT* to catalyze the formation of *trimethylarsine oxide* (TMAO) (Waters *et al.*, 2004b). Thus, although GSH may not be the primary reductant needed to support the catalytic activity of *AS3MT*, in cellular environments it may be a critical modulator of the rate of methylation and of the pattern of metabolites produced from inorganic As(III).

4.8.5 Reconciling experimental data and the Challenger scheme

As described above, the Challenger Scheme 4.1 posits that the pathway for the production of methylated arsenicals involves alternating steps of oxidative methylation using As(III) as the substrate and reduction of As(V) back to As(III) in preparation for the next round of oxidative methylation. Experimental data show that methylated arsenicals containing As(III) can be detected in the urine of individuals exposed to inorganic arsenic suggesting that the Challenger scheme is compatible with the biological process involved in arsenic methylation. However, to date there is no conclusive evidence on the molecular process involved in the reduction of As(V). As noted above, the enzymatic basis for arsenic reduction in higher organisms is not clear. The contribution of phosphorylytic enzymes that reduce inorganic As(V) in *in vitro* systems to inorganic As(III) in intact organisms has not been satisfactorily evaluated. The identity of GSTO as a MMA(V) reductase has not been corroborated in experiments with GSTO knockout mice. Reduction of As(V) by GSH has been demonstrated in a variety of systems, but it is still unclear whether the rates of reduction and complexation of arsenicals with GSH are sufficiently high to meet the demands for intracellular reduction. Complexation of As(III) with GSH could be catalyzed by glutathione-*S*-transferases. However, the experimental data on this issue remain unclear.

The capacity of *AS3MT* to catalyze each step in the reaction scheme that leads from inorganic As(III) to TMAO suggests that this enzyme catalyzes both oxidative methylation and reduction of As(V). However, it has not yet been possible to resolve these functions of the protein by kinetic analysis or by the generation of mutant proteins in which one or the other function (oxidative methylation or reduction) are disrupted. It has been suggested that the reactions catalyzed by *AS3MT* do not conform to the Challenger scheme. Hirano and associates (Hayakawa *et al.*, 2005) have suggested that the substrate for methylation reactions catalyzed by *AS3MT* is arsenic–GSH complexes. Here, the reaction Scheme 4.3

involves repeated methylation reactions involving As–GSH complexes:

$$As(III)(GS)_3 + CH_3^+ \rightarrow CH_3As(III)(GS)_2 + CH_3^+ \rightarrow (CH_3)_2As(III)(GS) + CH_3^+ \rightarrow (CH_3)_3As(III)$$

$$(4.3)$$

This scheme does not link the oxidation of As(III) with methylation and posits that the obligatory substrate for methylation is an As(III)–GSH complex. In a related analysis of *AS3MT*-catalyzed methylation, it has been suggested that As(III) bound to cellular proteins are the substrates for methylation (Naranmandura, Suzuki and Suzuki, 2006). Because *AS3MT* can catalyze the methylation of inorganic As(III) in reaction mixtures that do not contain GSH, but do contain dithiol reductants, it is unclear how the presence of GSH can be considered a requirement for *AS3MT*-catalyzed methylation. In cellular environments that contain the dithiol reductants, GSH, and a plethora of proteins that could bind As(III), it is quite possible that multiple reaction pathways may be involved in the production of methylated arsenicals. Further experimental work will be required to identify each of the molecular components of these pathways.

4.9   Arsenic excretion

Inorganic arsenic is eliminated primarily via the kidneys in humans as well as laboratory animals. Studies in adult human males voluntarily ingesting a known amount of either As(III) or As(V) indicate that 45–75 % of the dose is excreted in the urine within a few days to a week (Pomroy *et al.*, 1980; Tam, Charbonneau and Bryce, 1979; Buchet, Lauwerys and Roels, 1981a; Buchet, Lauwerys and Roels, 1981b; Lee, 1999). Relatively few studies in volunteers have included the measurement of arsenic in both feces and urine. However, Pomroy *et al.* (1980) reported that 6 % of a single oral dose of arsenic acid was excreted in the feces over a period of seven days compared to 62 % of the dose excreted in urine. No quantitative data are available that directly assess the significance of biliary excretion of As(III) or As(V) in humans.

Arsenic is excreted by routes other than urine and feces; however, these routes are quantitatively minor in general. Studies summarized in the older literature indicate that arsenic is excreted in sweat to some degree (World Health Organization (WHO), 1981). Given its capacity to accumulate in keratin-containing tissues, skin, hair, and nails can also be considered potential excretory routes for arsenic, although they would in general be quantitatively minor. However, the accessibility for collection of toe nail and hair, plus the much slower rate of accumulation of arsenic, make these tissues useful bioindicators to assess longer term arsenic exposure.

Arsenic is also excreted in human milk although levels are very low (Dang, Jaiswal and Somasundaram, 1983; World Health Organization (WHO), 1981; Concha, *et al.*, 1998b). Dang, Jaoswal and Somasundaram (1983) reported arsenic levels in breast milk of nursing mothers in the Bombay (Mumbai) area of India that were one to three months postpartum as ranging from 0.2 to 1.1 ng g^{-1} (ppb). In a study of three weeks to five months postpartum Andean women in Argentina consuming high arsenic drinking water ($\sim$200 μg L^{-1}), Concha *et al.* (1998b) found that the average concentration of arsenic in breast milk was quite low (3.1 ng g^{-1}) even when urinary arsenic excretion was high (230–300 μg L^{-1}). Significantly, low arsenic excretion in breast milk of nursing mothers led to a decrease in urinary arsenic concentration of their infants during the nursing period.

Urine is the primary route of elimination for both inorganic As(III) and inorganic As(V) in most common laboratory animals. With the exception of the rat, which exhibits slower overall elimination of arsenic, 50 % or more of a single oral dose of arsenic is usually eliminated in urine within 48 hours. Urine is also the primary route of elimination in species, such as the marmoset, which do not methylate arsenic (Vahter *et al.*, 1982). Comparison of urinary and fecal elimination in mice that have been given the same dose of arsenic by oral and parenteral routes (e.g. Vahter and Norin, 1980) reveals that only ~4–8 % of the dose is eliminated in feces irrespective of route of administration. This suggests that for both inorganic As(III) and inorganic As(V) that biliary elimination in mice is quite low (<3 % over 48 hours) and that most arsenic appearing in the feces following oral dosing was unabsorbed by the gastrointestinal tract.

Urinary elimination of inorganic As(V) in laboratory animals – at least for mice – does not appear to be capacity-limited or dose-dependent. Hughes, Menache and Thompson (1994) reported that 65–71 % of a single oral dose of inorganic As(V) was eliminated in the urine in 48 hours over a 10,000-fold dose range. Vahter and Norin (1980) did report a significant decrease in both urinary and total excretion of arsenic in mice when administered as inorganic As(III), which apparently is a function of the greater binding of inorganic As(III) in tissues with increasing dose. Csanaky and Gregus (2002) evaluated species differences in urinary and biliary excretion over a two-hour period following intravenous injection of 50 μmol kg^{-1} of either inorganic As(V) or inorganic As(III) in mice, rats, hamsters, guinea pigs, and rabbits. For all species examined, arsenic was preferentially excreted in urine when administered as inorganic As(V) and preferentially excreted into bile when administered as inorganic As(III); the only exception was the rabbit, which excreted more arsenic in urine when administered either inorganic As(V) or inorganic As(III). Interestingly, DMA(V) was preferentially excreted in urine of all species and not detectable in bile, whereas MMA(III) was excreted exclusively in bile and undetectable in urine whether inorganic As(III) or inorganic As(V) was administered.

Both species-specific and valence state-dependent differences have been demonstrated in the biliary excretion of arsenic. Studies in the older literature indicate that excretion of As(III) into the bile of rats is much more extensive compared to rabbits or dogs (Klaassen, 1974). The studies of Lindgren, Vahter and Dencker (1982) suggest that inorganic As(III) is excreted to a greater extent in the bile of mice compared to inorganic As(V); the authors also attribute the higher concentrations of arsenite-derived radioactivity in the duodenum of mice to greater biliary excretion of (^{74}As)-inorganic As(III). Excretion of inorganic As(III) into the bile of rats is also more rapid and efficient compared to inorganic As(V) — 19 versus 6 % of the dose in 2 hours (Gyurasics, Varga and Gregus, 1991). Mechanistic studies indicate that transport of either inorganic As(III) or inorganic As(V) into the bile of rats is dependent upon GSH. Agents that decrease hepatobiliary transport of GSH (e.g. diethyl maleate) or deplete tissue GSH (e.g. buthionine sulfoxime) also decrease hepatobiliary transport of arsenic (Gyurasics, Varga and Gregus, 1991; Alexander and Aaseth, 1985; Cui *et al.*, 2004; Csanaky and Gregus, 2005). Hepatobiliary transport of arsenic is dependent upon GSH because inorganic As(III) and MMA(III) both form *labile* GSH conjugates (inorganic As(III)-(GS)$_3$ and MMA(III)-(GS)$_2$, respectively) in the liver, which are actively transported into the bile by multidrug resistance protein-2 (mrp-2) (Kala *et al.*, 2000).

4.10   Effects of arsenic exposure

4.10.1   Acute exposure

Acute arsenic poisoning is an event that occurs primarily unintentionally, but cases of intentional exposure are known. In 2006, the American Association of Poison Control Centers reported 898 cases of nonpesticidal poisoning from arsenic (21 intentional) with one reported death and 338 reported cases of arsenic pesticide poisonings (two intentional) with no deaths (Bronstein *et al.*, 2007). The lethal dose of ingested inorganic arsenic in adult humans is estimated at 1–3 mg As kg^{-1} (Ellenhorn, 1997). The clinical symptoms of acute arsenic poisoning are dose-related and progress over a span of a few days (Ratnaike, 2003; Ibrahim *et al.*, 2006). Initially, the gastrointestinal system is involved with nausea, vomiting, and diarrhea. The latter can progress to bloody 'rice water' diarrhea. Following these symptoms, the patient may progress to shock from endothelial damage and loss of fluids. Other abnormalities that may arise are hematological and include bone marrow depression, pancytopenia, anemia, and basophilic stippling. Even though the gastrointestinal symptoms may improve, cardiac abnormalities may arise including a prolonged *QT interval* and ventricular arrhythmias. The peripheral and central nervous system may also become involved. A patient that survives acute arsenic poisoning may show neurological signs of light-headedness, weakness, delirium, and encephalopathy. Peripheral neuropathy may show in a few days or be delayed for several weeks. The symptoms are distal and are presented as burning and numbness in the hands and feet. The neuropathy involves both sensory and motor fibers. Nerve velocity conduction is diminished and fibers may eventually show signs of axonal degeneration. At higher exposures, an ascending weakness in the limbs may rapidly occur.

The most reliable diagnosis for acute arsenic poisoning is the measurement of 24-hour urinary arsenic (Hughes, 2006). Absorbed arsenic is rapidly excreted in urine with normal levels less than $50\,\mu g\,L^{-1}$. It is recommended that urinary arsenic be speciated. Measurement of total arsenic may overestimate arsenic exposure. This is because arsenic in seafood, which is primarily organic (e.g. arsenobetaine), is rapidly excreted in urine after it is absorbed. An overestimation of inorganic arsenic exposure may occur if urine is measured for total arsenic and the individual consumed arsenic-containing seafood a few days before the collection and analysis of the urine. Blood arsenic can be determined, but elevated levels of arsenic are only reliable for about 24 hours post exposure. This is because of the rapid clearance of arsenic from blood. Other laboratory tests that can be done include a blood count and a *comprehensive metabolic panel* (Ibrahim *et al.*, 2006). An electrocardiogram and nerve conduction testing can be performed to assess the cardiovascular and peripheral nervous systems, respectively.

Treatment of acute arsenic poisoning includes removal from the exposure source, supportive measures for loss of fluids, and chelation therapy (Ibrahim *et al.*, 2006). Chelators that can be used include dimercaprol or 2,3-dimercaptosuccinic acid. In cases of renal failure, hemodialysis should be considered.

4.10.2   Chronic exposure

Many organ systems in the human body can be affected by chronic exposure to arsenic (Agency for Toxic Substances and Disease Registry (ATSDR), 2007; World Health Organization (WHO), 2001; National Research Council (NRC), 1999, 2001). These include the skin, developing fetus, liver and the cardiovascular, pulmonary, nervous, and endocrine systems. These effects are dose-related and primarily arise from oral exposure to arsenic, although inhalation of arsenic may also result in adverse health effects. The chronic effects from dermal exposure to arsenic are not known.

4.11   Cardiovascular

4.11.1   Introduction

The chronic exposure to arsenic in drinking water is associated with the development of cardiovascular diseases (Navas-Acien *et al.*, 2005; Wang *et al.*, 2007). Cardiovascular disease, regardless of its etiology as a whole, is a major cause of death in the world. Many *epidemiology* studies have shown there is a dose–response relationship with arsenic exposure and the development of diseases that affect the cardiovascular system. *Subclinical disorders* that have been associated with long-term drinking water exposure to arsenic are carotid atherosclerosis and electrocardiogram abnormalities (Wang *et al.*, 2007). The latter include prolongation of the heart rate-corrected QT interval and increased QT dispersion. Clinical outcomes of long-term arsenic exposure include peripheral vascular disease (PVD), ischemic heart disease (IHD), and cerebrovascular disease (CVD).

4.11.2   Peripheral vascular disease

Tsai, Wang and Ko (1999) conducted an ecological study that examined the relationship between PVD mortality and exposure to arsenic in drinking water in a population of southwest Taiwan. The *standardized mortality ratio* (SMR) for PVD in males (3.56) and females (2.3), after adjustment for age, was significantly higher in the arsenic-exposed than in the control population. The PVD mortality decreased in the arsenic-exposed Taiwanese population when the water source was changed to one with lower levels of arsenic (Yang, 2006). Other countries, such as Mexico and Chile, have populations that are exposed to high levels of arsenic in drinking water. There is also an increased prevalence of PVD among these exposed populations (National Research Council (NRC), 1999).

Blackfoot disease (BFD) affects the peripheral vascular system and at one time was endemic along the southwest coast of Taiwan (Tseng, 2005; Tseng *et al.*, 2005). It is a result of progressive arterial occlusion predominantly in the lower extremities. Some of the initial symptoms include numbness and coldness in the extremities and intermittent claudication or absence of peripheral pulsation. These symptoms could develop to ulceration, gangrene, and spontaneous amputation of the affected extremity. For unknown reasons, BFD has principally been localized in Taiwan. BFD has been attributed to the population drinking water from deep artesian wells contaminated with arsenic. The median arsenic levels in these wells where BFD has been observed ranged from 0.7 to $0.93\,mg\,L^{-1}$. Humic substances in the artesian well water do not appear to have a role in BFD (Yu, Lee and Chen, 2002). The incidence level of the disease peaked in the mid-twentieth century with prevalence rates ranging from 6.5 to 18.9 per 1000 individuals.

4.11.3   Ischemic heart disease

IHD can result when there is diminished oxygen supply to the myocardium. Chen *et al.* (1996), in an ecological study of residents of several villages in the arsenic-endemic area of southwest Taiwan, reported an increased cumulative mortality from IHD with increasing arsenic exposure. Additional risk factors, such as age and cigarette smoking, were adjusted for the analysis. With $<0.1\,mg\,L^{-1}$ arsenic exposure, the mortality was 3.5 % and nearly doubled to 6.6 % with an arsenic exposure of $\geq 0.6\,mg\,L^{-1}$.

In a cohort study by Chen *et al.* (1996), a population of BFD patients from southwest Taiwan were compared to residents that were unaffected by this disease, but lived in the same area. The relative risk of IHD mortality increased with increasing arsenic-year exposure. These risks were adjusted for age, cigarette smoking, and other risk factors for IHD.

4.11.4   Cerebrovascular disease

The relative risk for CVD in areas of Taiwan with elevated arsenic drinking water levels ranged from 1.2 to 2.7 (per 1000 individuals exposed versus 1000 individuals nonexposed) (Navas-Acien *et al.*, 2006). Tsai, Wang and Ko (1999) conducted an ecological study of a population in southwest Taiwan in the arsenic-endemic area and a referrant population not residing in this specific area. The relative risk for CVD mortality in the arsenic-exposed population was 1.14 for males and 1.24 for females.

4.11.5   Atherosclerosis

Atherosclerosis is a pathogenic response of the intima of the arterial vessel walls to noxious stimuli. It is characterized by lipids depositing in the vessel walls, which leads to wall narrowing. This can progress to IHD. Exposure to arsenic in drinking water is associated with an increased prevalence of carotid atherosclerosis in a dose–response relationship. In a cross-sectional study, Wang *et al.* (2002) assessed a population in southwest Taiwan living in the arsenic-endemic area, for carotid atherosclerosis, using duplex ultrasonography. The risk factor for atherosclerosis from arsenic exposure, after adjustment for other risk factors (smoking, alcohol consumption, serum cholesterol, and others) was 1.8 and 3.1 for those with cumulative arsenic exposures of $0.2–19.9\,mg\,L^{-1}\,year^{-1}$ and $\geq 20\,mg\,L^{-1}\,year^{-1}$, respectively.

4.11.6 Hypertension

Long-term arsenic exposure is also associated with hypertension (Chen *et al.*, 2007). In a cross-sectional study, Chen *et al.* (1995) reported the prevalence of hypertension increased from 5 % in the control population to 29 % in the population with the highest cumulative arsenic exposure ($18.5\,mg\,L^{-1}\,year^{-1}$). Occupational studies or those from the general population, which are not highly exposed to arsenic, are less conclusive (Chen *et al.*, 2007). However, Kwok *et al.* (2007) conducted a cross-sectional study of a population from China. This was a large study (8790 pregnant woman) with participants that had been exposed to arsenic drinking water levels not as excessive as those in the arsenic-endemic areas of southwest Taiwan. The mean systolic blood pressure, when controlled for age and body weight, rose 1.9, 3.9, and 6.8-mm mercury for arsenic levels of 12–50, 51–100, and $>100\,\mu g\,L^{-1}$, respectively. This increase in systolic blood pressure was statistically significant with arsenic exposure. The mean diastolic blood pressure also increased significantly with arsenic exposure, but to a lesser extent.

4.12 Endocrine

The chronic exposure to arsenic, primarily from drinking water, has been associated with the development of type 2 diabetes (non–insulin dependent) in adults (Chen *et al.*, 2007). Epidemiological studies of this disease have been reported by Lai *et al.* (1994) and Tseng *et al.* (2000) in Taiwan and Rahman *et al.* (1998) in Bangladesh. Lai *et al.* (1994) found that the prevalence of diabetes mellitus was two times higher in villages consuming water contaminated with inorganic arsenic than the general population of Taiwan. The *multivariate-adjusted odds ratio*, which took into account risk factors for diabetes mellitus (e.g. *body mass index* and physical activity) was 6.6 and 10 for those with a cumulative arsenic exposure of $0.1–15$ and $\geq 15\,mg\,L^{-1}\,year^{-1}$, respectively, compared to the control population. In a cohort study by Tseng *et al.* (2000), they reported that the incidence of diabetes was correlated with age, body mass index, and cumulative arsenic exposure. After adjusting for age and body mass index, the relative risk was 2.1 for a cumulative arsenic exposure $> 17\,mg\,L^{-1}\,year^{-1}$ compared to an exposure of $< 17\,mg\,L^{-1}\,year^{-1}$. Rahman *et al.* (1998), in a cross-sectional study, reported the corresponding prevalence ratios of diabetes were 2.6, 3.9, and 8.8 with time-weighed average arsenic levels of <0.5, $0.5–1$, and $>1\,mg\,L^{-1}$ respectively, compared to the control population.

In a systematic review of the experimental and epidemiological data, Navas-Acien *et al.* (2006) argue that the evidence to date is inconclusive for a role of arsenic in the development of diabetes mellitus. Pooling the results from the Taiwan and Bangladesh studies, which were from extreme arsenic exposures, there was a risk of 2.5 for the development of diabetes. However, they point out, because of the types of studies conducted, particularly with characterization of the exposures, it was hard to interpret the association. In addition, the occupational studies that they examined showed increased and decreased morbidity from diabetes due to arsenic exposure. The general population studies, which typically had much lower arsenic exposures, were inconclusive.

4.13 Hepatic

The effects of arsenic on the liver have been known for over 100 years (Hutchinson, 1895). Therapeutic use of arsenic resulted in the development of ascites in patients. Noncirrhotic portal hypertension has been reported following the chronic use of Fowler's solution, a tonic containing 1 % potassium arsenite (Huet, Guillamue and Cote, 1975). Guha Mazumder (2005) observed hepatomegaly in 77 % of patients with arsenicosis from an area in West Bengal, India, with arsenic-contaminated drinking water. Liver function tests of several of these patients with hepatomegaly showed elevated serum enzymes. Further study of some of the patients revealed that noncirrhotic portal fibrosis was the predominant hepatic lesion. Analysis of the hepatic arsenic levels were significantly elevated in patients with the fibrosis. However, there was no correlation between hepatic arsenic levels and the severity of fibrosis or with the level of arsenic consumed in drinking water. An epidemiology study by Guha Mazumder (2005) revealed that a significant portion (10.2 %) of the study population that had consumed water contaminated with arsenic greater than $50 \mu g \, L^{-1}$ had hepatomegaly compared to the control population (As $<50 \mu g \, L^{-1}$). The study analysis was controlled for alcohol intake and exposure to viruses and parasites. There was a linear relationship between the incidence of hepatomegaly in males and females and increasing exposure to arsenic in drinking water. This trend was also greater in males than in females.

4.14 Neurological

Chronic arsenic exposure may result in peripheral neuritis and neuropathy, and central effects, such as impaired intellectual function. Nervous system effects from chronic exposure to arsenic have been observed in arsenic smelter workers (Feldman, Niles and Kelly-Hayes, 1979), beer drinkers in England (de Wolff and Edelbroek, 1994), and in children in Bangladesh exposed to arsenic in drinking water (Wasserman *et al.*, 2004). The peripheral neuropathy involves sensory and motor fibers and is distinguished by the dying back of axons with some evidence of segmental demyelination (de Wolff and Edelbroek, 1994). A classic case of peripheral neuritis caused by arsenic exposure occurred in northern England in the nineteenth century. These individuals drank beer that was contaminated with arsenic. The source of arsenic was from sulfuric acid, originally prepared from arsenical pyrites. The acid was used to prepare the sugar that was used in the brewing process. Many of these individuals complained of pain and weakness in the extremities, 'pins and needles' in the fingers and toes, difficulty in walking, and other effects (de Wolff and Edelbroek, 1994). Wasserman *et al.* (2004) have reported that Bangladeshi children exposed to drinking water with arsenic greater than $50 \mu g \, L^{-1}$ have decreased intelligence testing scores when compared with children exposed to lower levels of arsenic in drinking water.

4.15 Skin

The earliest known adverse effects associated with chronic arsenic exposure were skin lesions (Yoshida, Yamauchi and Fan Sun, 2004). The effect of arsenic on skin was first noted by several physicians in the late 1800s (Schwartz, 1997). They observed the effects in patients being treated with inorganic arsenic for various aliments. The effect is more widespread in populations that consume arsenic-contaminated drinking water and is related to dose (Yoshida, Yamauchi and Fan Sun, 2004). In southwest Taiwan, the prevalence rate of hyperpigmentation and keratoses was 18 % and 7 %, respectively (Tseng *et al.*, 1968). The most common change in skin is hyperpigmentation, which is thought to be due to increased melanin in the melanocytes (Maloney, 1996). Hyperpigmentation can occur in any body site, but may occur in areas that are more pigmented, such as the areola or groin. Some individuals have areas of hypopigmentation, giving the appearance of 'raindrops'. Hyperkeratotic papules are characteristically observed on the palms and soles of individuals that consume arsenic-contaminated water. These lesions may develop anywhere from 3 to 30 years after inorganic arsenic exposure (Maloney, 1996; Shannon and Strayer, 1989). Mees lines, which are white horizontal bands in the nails, may also appear.

4.16 Developmental

Arsenic can transfer from maternal blood to the placenta and developing fetus. Whether inorganic arsenic can cause developmental effects in humans has been debated (Desesso, 2001; Desesso *et al.*, 1998; Wang *et al.*, 2006; Vahter, 2008). It is known that parenteral administration of inorganic arsenic to rodents at maternally toxic doses causes malformations, principally neural tube defects, in offspring. However, these malformations are not observed in rodents exposed to arsenic via oral gavage, inhalation, or diet, and rabbits exposed by oral gavage (Desesso *et al.*, 1998). More recent studies suggests that the exposure of pregnant rats to nontoxic doses of inorganic As(III) in drinking water results in defects in fetal brain development and behavior in the offspring (Chattopadhyay *et al.*, 2002; Rogdríguez *et al.*, 2002).

In humans, following exposure to arsenic in the drinking water, adverse pregnancy outcomes, such as spontaneous abortion, still births, preterm birth rates, low birth weight, and increased infant mortality have been reported (Ahmad *et al.*, 2001; Hopenhayn-Rich, 2000). Ahmad *et al.* (2001) conducted a cross-sectional study of woman of reproductive age in Bangladesh. In the exposed population that had been drinking water containing levels of arsenic $> 0.05 \, \text{mg L}^{-1}$ for at least five years, there was a significantly greater adverse pregnancy outcome than the nonexposed population (As $< 0.02 \, \text{mg L}^{-1}$). In a retrospective study of infant mortality in two cities of Chile with a high (Antofagasta) and low (Valparaiso) exposure to arsenic in drinking water, Hopenhayn-Rich *et al.* (2000) reported a significant association between arsenic exposure and late fetal mortality (*rate ratio* (RR) = 1.7), neonatal mortality (RR = 1.53), and postnatal morality (RR = 1.26) after adjustment for location and calendar time. Overall, additional studies need to be conducted to investigate this potential relationship between arsenic *in utero* exposure and development effects in humans. Better control of confounding factors (such as exposure to other metals, maternal nutrition, and smoking) will result in a more definitive answer about whether this is an actual relationship.

4.17 Other organ systems

Inorganic arsenic has been reported to affect the lungs of individuals consuming high levels of arsenic in their drinking water (Mazumder *et al.*, 2005). The effect that has been observed is bronchiectasis, which is chronic dilation of bronchi or bronchioles due to inflammation or obstruction. Arsenic may also affect the hematological, reproductive, and immunological systems (Agency for Toxic Substances and Disease Registry (ATSDR), 2007; World Health Organization (WHO), 2001; National Research Council (NRC), 1999, 2001).

4.18   Cancer

4.18.1   Introduction

Inorganic arsenic is recognized as a human carcinogen by regulatory and public health organizations throughout the world. This recognition or classification is essentially based on epidemiology data. Most of the studies have focused on populations in Taiwan who were exposed to high arsenic levels in their drinking water, or from occupational studies of smelter workers. Only more recently have studies been conducted in areas, such as India, Bangladesh, Mexico, and Chile that have or have had sources of drinking water naturally contaminated with high levels of inorganic arsenic. The most common tumor types in humans associated with exposure to arsenic are skin, lung, and bladder. Other organs that may also be affected include the liver, kidney, and prostate (Agency for Toxic Substances and Disease Registry (ATSDR), 2007; World Health Organization (WHO), 2001; National Research Council (NRC), 1999, 2001).

4.18.2   Skin

The tumor types in skin most commonly associated with exposure to arsenic are Bowen's disease and squamous and basal cell carcinoma (Maloney, 1996; Shannon and Strayer, 1989). These tumors may develop 6–20 years following exposure to arsenic, although they generally develop following several years of exposure (mean latency of 14 years) (Shannon and Strayer, 1989). Bowen's disease, or carcinoma in situ, is a preinvasive form of squamous cell carcinoma. It is the most common form of skin cancer induced by arsenic (Maloney, 1996). Bowen's disease tumors can be observed on any part of the body, but are commonly found on sun-protected areas such as the trunk. The lesions may be solitary, but are usually multifocal and randomly distributed. The lesions may be sharply demarcated round or irregular plaques of various sizes (1 mm to >10 cm) (Shannon and Strayer, 1989). Squamous cell carcinomas that arise from arsenic exposure can develop *de novo* or from the arsenical keratoses or Bowen's disease. The cells originating from the latter are more aggressive than those from hyperkeratotic cells. Although squamous cell carcinomas can develop at any site, they are generally found on the extremities (Shannon and Strayer, 1989). Arsenic-induced basal cell carcinoma histologically resembles basal cell carcinoma not associated with arsenic exposure. The latter tumors are typically due to over exposure to the sun and are frequently found on the neck and head. The arsenic-related basal cell carcinomas are generally multiple and found on sun-protected areas such as the trunk.

In a southwest Taiwan population exposed to arsenic in drinking water, Tseng *et al.* (1968) conducted a cross-sectional study for development of skin cancer. The prevalence rate for skin cancer was 10.6 per 1000 individuals (population: 40 421) across all ages (20–70+ years). Males had a higher prevalence than females (16.1 versus 5.6). With increasing age, there was a greater prevalence of skin cancer. There was also a dose–response relationship for skin cancer when the study participants were grouped by village and exposure ($<0.3 \, \text{mg L}^{-1}$, $0.3–0.6 \, \text{mg L}^{-1}$, and $>0.6 \, \text{mg L}^{-1}$). For both men and women, there was a dose–response relationship between increasing arsenic exposure and prevalence of skin cancer.

In another Taiwan population, Hsueh *et al.* (1995) conducted a cross-sectional study and found a significant trend between skin cancer and chronic arsenic exposure with respect to duration of living in the endemic area, duration of drinking artesian well water, average arsenic exposure, and cumulative arsenic exposure (micrograms per gram-year, $\mu \text{g g}^{-1} \, \text{year}^{-1}$). In addition, risk factors for chronic liver disease and malnutrition in this population were statistically significant in the conducted *multiple logistic regression analysis*. Liver function and nutritional status may have a role in development of skin cancer from exposure to arsenic.

4.18.3 Lung

There is convincing epidemiologic evidence that inhalation of inorganic arsenic increases the risk for lung cancer (Agency for Toxic Substances and Disease Registry (ATSDR), 2007; World Health Organization (WHO), 2001; National Research Council (NRC), 1999). The majority of these studies have investigated occupational exposure to arsenic trioxide at copper smelters. During the smelting process of the ores, arsenic trioxide is released into the air. Other studies have examined arsenic pesticide workers and the development of lung cancer.

Lee-Feldstein (1986) studied a cohort of workers employed at a copper smelter. In the cohort employed between 1925 and 1947, the SMR for death from respiratory cancer was 203 for an estimated average exposure of $0.5\,mg\,m^{-3}$ arsenic, 292 for exposures of $7\,mg\,m^{-3}$ arsenic, and 444 for exposures of $62\,mg\,m^{-3}$ arsenic. At a different smelter site, Enterline *et al.* (1987) reported that the SMR from respiratory cancer increased from 144 at a time-weighed exposure level of $400\,\mu g\,As\,m^{-3}\,year^{-1}$ to 477 at an exposure level of $59\,000\,\mu g\,As\,m^{-3}\,years^{-1}$.

There does not appear to be one specific cell type of lung cancer in the exposed-workers (Agency for Toxic Substances and Disease Registry (ATSDR), 2007). Several tumor types (such as epidermoid carcinomas, small carcinomas, and adenocarcinomas) have all been found to increase following exposure to arsenic by inhalation.

More recently, an association between environmental exposure to arsenic and development of lung cancer has been proposed. Ferreccio *et al.* (2000) conducted a case-control study of a Chilean population exposed to arsenic in their drinking water. The odds ratio for lung cancer, which was adjusted for several confounders (e.g. smoking and copper smelter workers) ranged from 1 for exposures $<10\,\mu gl^{-1}$ to 9 for concentrations of $200–400\,\mu gl^{-1}$ (65-year average exposure).

4.18.4 Bladder

Medical treatment with arsenic for skin conditions as well as environmental exposure to arsenic in drinking water have been associated with the development of bladder cancer. Cuzick *et al.* (1992), in a cohort study, followed the health status of a group of individuals treated (2 weeks–12 years) with Fowler's solution (potassium arsenite) in England. There was a significant excess of bladder cancer mortality in this group (5 observed per 1.6 expected, SMR of 3.1). The total amount of arsenic ingested by these individuals ranged from 224 to 3325 mg. The latency period from first exposure to death ranged from 10 to 20 years. Also, a subcohort of this patient population was examined for arsenical keratoses and those that died from all types of cancer. All patients that died from cancer had arsenical keratoses. This suggests that the signs of arsenicism (keratoses) may be a biomarker for susceptibility for internal cancer following chronic exposure to arsenic.

In an ecological study of a population in southwest Taiwan, Wu *et al.* (1989) reported a significant dose–response relationship in age-adjusted mortality from bladder cancer. In males, the mortality rates (per 100,000) were 23, 61, and 93 and for females the rates were 26, 57, and 111 for mean arsenic levels of <0.3, $0.3–0.59$, and $\geqslant 0.6\,mg\,L^{-1}$ in drinking water.

4.19   Animal models for arsenic-induced cancer

Just within the past decade have definitive animal studies for arsenic-induced carcinogenesis emerged with positive results (Kitchin, 2001; Hughes, 2002; Rossman, 2003; Wanibuchi *et al.*, 2004; Cohen *et al.*, 2006; Waalkes, Liu and Diwan, 2007). Up until this point, standard lifetime cancer bioassays with inorganic arsenic had been conducted in several species of laboratory animals, but there were very few studies showing that arsenic was carcinogenic (Hughes, 2002; Rossman, 2003). Scientists began to study the metabolites of inorganic arsenic, principally DMA(V), for cocarcinogenic, promoting, and carcinogenic activities. Also, animal models that either were susceptible to spontaneous tumor development or genetically modified (e.g. Tg.AC and K6/ODC mice) were used in studies with the inorganic and organic arsenicals. DMA(V) has been shown to be a multiorgan tumor promoter in rats in a two-stage carcinogenicity bioassay, a developer of lymphomas and lung tumors in p53 heterozygous knockout and Mmh/OGG1 mice, respectively, and a promoter for skin carcinogenesis in K6/ODC mice (Wanibuchi *et al.*, 2004). In addition, DMA(V) is a complete carcinogen in rat urinary bladders following exposure from drinking water with $50–100\,mg\,L^{-1}$ (Wei *et al.*, 2002) or a diet of $100\,\mu g\,g^{-1}$ (Arnold *et al.*, 2006). However, mice did not develop tumors following dietary exposure to MMA(V) (up to $500\,\mu g\,g^{-1}$) (Arnold *et al.*, 2003). Liver tumors develop in rats administered TMAO ($200\,mg\,L^{-1}$) in drinking water (Shen *et al.*, 2003). The transgenic Tg.AC mouse carries the v-Ha-*ras* oncogene and is a genetically initiated model for skin carcinogenesis. Skin papillomas develop in Tg.AC mice that were exposed to inorganic As(III) ($200\,mg\,L^{-1}$ As) in drinking water and had 12-*O*-tetradecanoyl phorbol-13-acetate (TPA) applied to their skin (Germolec *et al.*, 1997). Papillomas did not develop in mice administered control water or those not treated topically with TPA. When coexposed with ultraviolet radiation, As(III) ($0.5–10\,mg\,L^{-1}$) administered in drinking water to mice for 26 weeks resulted in carcinogenic skin tumors (Burns *et al.*, 2004; Rossman *et al.*, 2001). More recently, a transplacental carcinogenesis mouse model has been developed by Waalkes *et al.* (2007). They chose to use this model because of the insensitivity of rodents to arsenic carcinogenesis and the prenatal period is a susceptible life stage. Pregnant C3H mice were administered inorganic As(III) (up to $85\,mg\,L^{-1}$) in drinking water from gestation day 8 to day 18. The adult male offspring exposed in utero to arsenic developed liver carcinomas and adrenal cortical adenomas. Similarly exposed female offspring developed ovarian tumors, lung carcinomas, and proliferative lesions of the uterus and oviduct. CD1 mice that are similarly treated with As(III) in utero, followed by postnatal exposure to diethylstilbestrol or tamoxifen, develop bladder and liver tumors (Waalkes *et al.*, 2007). The rat urinary bladder model and mouse transplacental model for arsenic-induced carcinogenesis offer opportunities to gain insight into the mechanism of action for arsenic because the tumors that develop (rat — urinary bladder; mouse — bladder, liver, and lung) in these animals correspond to those that develop in humans following exposure to inorganic arsenic.

4.20   Mechanism of action

4.20.1   Introduction

There are many proposed mechanisms of action for the noncancerous and cancerous effects of arsenic (Kitchin, 2001; Hughes, 2002; Rossman, 2003; Aposhian, 1989; Huang *et al.*, 2004). There may actually be more than one mechanism of action for arsenic and some may work together. As(V) and As(III) can interconvert and are also metabolized, so some of the effects following exposure to one may be due to the other or a metabolite of it. It should be recognized that the As(III) forms are more acutely potent than the As(V) forms. It has also been thought that the methylation of arsenic is a detoxification pathway. However, the trivalent organic arsenicals that are metabolically formed, MMA(III) and DMA(III), are potent *cytotoxicants* (Styblo *et al.*, 2000). They are more potent than inorganic As(V) and their pentavalent organic arsenic counterparts. In addition, the LD_{50} of MMA(III) ($2\,mg\,As\,kg^{-1}$) administered intraperitoneally is greater than that of inorganic As(III) ($8\,mg\,As\,kg^{-1}$) in hamsters (Petrick *et al.*, 2001).

4.20.2 Replacement of phosphate

Inorganic As(V) and phosphate have many similar physical and chemical properties (Dixon, 1997; Chapter 2). *In vitro* studies have shown that inorganic As(V) can replace phosphate in several biochemical activities (Hughes, 2002). An example is the replacement of inorganic As(V) for phosphate in the sodium pump and anion exchange transport system in human red blood cells (Kenney and Kaplan, 1988). *In vitro* formation of ATP can be uncoupled by inorganic As(V) by a mechanism termed *arsenolysis*. This has been observed at the subcellular, organelle, and cellular level (Hughes, 2002). Cells could become energy deprived in the presence of inorganic As(V) from arsenolysis because of the lower levels of ATP. An adverse cellular effect may then begin to develop.

4.20.3 Enzyme inhibition

Inorganic As(III) and other trivalent arsenicals are reactive toward sulfhydryl groups. As(III) reacts readily *in vitro* with GSH and cysteine forming stable complexes (Scott *et al.*, 1993; Delnomdedieu *et al.*, 1994a). However, the most stable complexes are those with *vicinal dithiols* (Delnomdedieu *et al.*, 1993). Many enzymes have thiol groups that are critical for catalytic activity. Binding of As(III) to these catalytic sites may inactivate the enzyme or inhibit its activity. Thioredoxin reductase is an NADPH-dependent flavoenzyme, which catalyzes the reduction of the disulfide thioredoxin and other substrates, such as dehydroascorbate, lipid peroxides, and proteins, such as disulfide isomerase. Thioredoxin reductase has redox-active cysteine residues, which are involved with its catalytic activity. This enzyme is one of several that regulate the cellular response to oxidative stress. As(III) and more potently MMA(III) inhibit thioredoxin reductase *in vitro* (Lin *et al.*, 2001). Inhibition of this enzyme by trivalent arsenicals could present *dysregulation* of cellular oxidant status and result in a potential adverse cellular effect.

4.20.4 Oxidative stress

Arsenic has been shown to induce oxidative stress (Shi, Shi and Liu, 2004; Hughes and Kitchin, 2006). Oxidative stress is a result of an imbalance between reactive oxygen species and the ability of a cell's antioxidant defense apparatus to respond. Oxidative stress can result in the damage of proteins, lipids, RNA, and deoxyribonucleic acid (DNA). In addition, since oxidant species have a role in cell signaling, a state of oxidative stress could potentially alter signaling within and between cells.

Heat shock or stress proteins are induced by arsenic *in vitro* and *in vivo* (Del Razo *et al.*, 2001a). With the use of *electron spin traps* and *fluorescent probes*, oxygenated radicals have been detected *in vitro* in cells exposed to inorganic and organic arsenicals. Formation of the oxidative species was decreased by the antioxidant butylated hydroxytoluene, the free radical scavenger dimethyl sulfoxide, and superoxide dismutase or nitric oxide synthase inhibitors. It appears that several oxidants, including superoxide and reactive nitrogen species, can be formed within cells in the presence of arsenic (Shi, Shi and Liu, 2004; Hughes and Kitchin, 2006).

DNA isolated from cells (Kessel *et al.*, 2002) and tissues and urine of animals (Vijayaraghavan *et al.*, 2001; Yamanaka *et al.*, 2001) treated with arsenic show lesions induced by oxidative stress. These lesions include 8-oxo-2'-deoxyguanosine and 8-hydroxy-2'-deoxyguanosine. These DNA lesions may lead to base-pair substitutions (guanine to thymidine and adenine to cytosine) during DNA synthesis, which could lead to altered gene products.

The mechanism for arsenic-induced oxidative stress injury is not known with certainty. It may occur as a consequence of free radical production due to redox cycling of As(III) and As(V), release of iron from ferritin (generating activated oxygen by the *Haber–Weiss reaction*), stimulation of NADPH oxidase, or inhibition of redox enzymes, such as GSH and thioredoxin reductases (Hughes and Kitchin, 2006).

4.20.5   Genotoxicity

Arsenic is not a point mutagen when tested in standard bacterial and mammalian cell mutation assays (Basu *et al.*, 2001). However, it is comutagenic *in vitro* with ultraviolet radiation (Li and Rossman, 1991). Arsenic does cause multilocus deletions in cells, resulting in cellular death (Hei, Liu and Waldren, 1998). Arsenic can damage DNA, causing effects such as nicks and single- and double-strand breaks. The strand breaks are observed *in vitro* in cultured human alveolar cells (Kato *et al.*, 1994) and *in vivo* in mouse lung (Yamanaka *et al.*, 1989) following exposure to DMA(V). This effect is thought to be due to oxidative stress or by inhibition of DNA repair (Hughes, 2002). Arsenicals are positive in cytogenetic assays with mammalian cells (Basu *et al.*, 2001). Genotoxic effects observed include chromosomal aberrations, sister chromatid exchanges, and micronuclei. These effects are observed in cells incubated with arsenic or in bone marrow cells isolated from animals administered arsenic trioxide.

4.20.6   Alteration of DNA repair

Inhibition of DNA repair may result in an adverse event within the cell. As(III) has been shown to inhibit DNA ligase *in vitro*, but it does not appear to occur by the direct interaction of As(III) with this repair enzyme (Li and Rossman, 1989). It has been suggested that As(III) alters redox levels or affects signal transduction pathways involved with DNA ligase (Hu, Su and Snow, 1998). The nucleotide excision repair system is also impaired by arsenic (Okui and Fujiwara, 1986; Hartwig *et al.*, 1997).

4.20.7   Signal transduction

Extracellular information can be sent intracellularly via a pathway consisting of a sequence of signaling molecules, which evokes a cellular response. An example of one of these pathways is the *mitogen activated protein kinase* (MAPK) pathway. The MAPK pathway consists of three pathways including extracellular-regulated protein kinase (Erk), c-Jun N-terminal kinase (JNK), and p38 kinase. There is differential activation of JNK and Erk by arsenic (Huang *et al.*, 2001). Low concentrations of inorganic As(III) induce Erk and eventually the *in vitro* transformation of mouse epidermal cells (Huang *et al.*, (1999a)). JNK is activated by higher concentrations of inorganic As(III), which cause mouse epidermal cells to undergo *apoptosis* (Huang *et al.*, 1999b).

4.20.8   Gene transcription

The transcription of genes is initiated by a stimulus transmitted by the signal transduction pathway (a downstream effect). Transcription factors are regulatory proteins that receive a signal (by binding to a molecule or protein), are activated, and bind to a specific DNA site. Transcription of genes proceeds and eventually, the cell responds, which may include proliferation or death. In UROtsa cells, a human urinary bladder epithelial cell line, the binding of the transcription factor activating protein (AP)-1 to DNA is increased by inorganic As(III) (Simeonova *et al.*, 2000). These cells also proliferate in the presence of inorganic As(III). The transcription factor nuclear factor κB (NFκB) is also activated by inorganic arsenic in mouse epidermal cells (Huang *et al.*, 2001). Activation of NFκB may have a role in development of cancer.

4.20.9   DNA methylation

Gene transcription can be regulated by the methylation of DNA. Inorganic arsenic has been shown to affect the methylation of DNA. Mass and Wang (1997) demonstrated that the tumor suppressor gene *p53* in the human adenomacarcinoma cell line A549 is hypermethylated in the presence of inorganic As(III). Cytosines within the sequence CpG were specifically hypermethylated. Whether or not the expression of this gene was affected is not known. As(III) transforms rat liver TRL1215 cells (Zhao *et al.*, 1997). The DNA of the inorganic As(III)-transformed cells was globally hypomethylated. The DNA of the cells incubated with a concentration of inorganic As(III) that did not transform the cells was not hypomethylated. DNA hypomethylation may result in aberrant gene expression.

4.20.10   Growth factors

Incubation of primary human keratinocytes with inorganic As(III) results in a proliferative response (Germolec *et al.*, 1997, 1996). In these cells, messenger ribonucleic acid (mRNA) transcripts and secretion of the *granulocyte macrophage-colony stimulating factor* (GM-CSF), *transforming growth factor-α* (TGF-α), and *tumor necrosis factor-α* were increased. Incubation of human keratinocytes with MMA(III) and DMA(III) causes similar results (Vega *et al.*, 2001). In epidermal tissue removed from Tg.AC mice treated with inorganic As(III) in drinking water, mRNA transcripts of GM-CSF and TGF-α were increased, which is consistent with the *in vitro* human keratinocyte studies described above (Germolec *et al.*, 1997).

4.21   Regulation of arsenic

The levels of arsenic are regulated in the workplace, air, water, food, and other media (Table 4.3; Appendix E). In the United States, arsenic is regulated by various federal governmental organizations, such as the US EPA, US FDA, US Occupational Safety and Health Administration (OSHA), and the US Department of Agriculture. Nongovernmental organizations, such as the American Conference of Governmental Industrial

Table 4.3   *Regulations and guidelines of arsenic levels in occupational, environmental, and biological media in the United States (also see Appendix E). Organizational acronyms are identified in the text and many of the terms are defined in Appendix B.*

Medium	Organization	Regulation/Guideline	Level of arsenic
Air	US OSHA	Permissible exposure limit (PEL) — time weighted average (TWA) of 8 h	$10 \, \mu g \, m^{-3}$
Air	ACGIH	Threshold limit value (TLV) — TWA	$10 \, \mu g \, m^{-3}$
Air	US EPA	Release threshold quantity	$15\,000$ lb ($AsCl_3$) 1000 lb (AsH_3)
Water	US EPA	Maximum contaminant level (MCL) — drinking water	$10 \, \mu g \, L^{-1}$
Water	US EPA	Water quality criteria	$0.018 \, \mu g \, L^{-1}$
Water	US EPA	Biosolids rule	$75 \, mg \, kg^{-1}$
Water	US FDA	Bottled water	$10 \, \mu g \, L^{-1}$
Food	US FDA	Tolerance levels in swine and chicken meat	$0.5–2 \, mg \, kg^{-1}$
Urine	ACGIH	Biological exposure index	$35 \, \mu g \, L^{-1}$

Hygienists (ACGIH), International Agency for Research on Cancer (IARC), and World Health Organization (WHO) have established guidelines on exposure and classifications of the carcinogenicity of arsenic.

In 1975, the US EPA set an interim primary drinking water regulation for arsenic at $50 \mu gl^{-1}$. This level was for public drinking water supplies and was based on a US Public Health Standard developed in 1943 (Abernathy *et al.*, 2000). After much debate and pressure from the US Congress, the US EPA in 2001 announced that the arsenic drinking water standard, termed the *Maximum Contaminant Level* (MCL) would be lowered to $10 \mu g$ As l^{-1} (Smith *et al.*, 2002). This is the same level as recommended by the WHO. The MCL is an enforceable standard and takes into account public health goals and the available technology and cost to attain this level. The nonenforceable *Maximum Contaminant Level Goal* (MCLG) for arsenic was set at $0 \mu gl^{-1}$ (Appendix E). The MCLG is derived solely from health effects data. The MCL for arsenic took effect in 2006. The US FDA regulates arsenic in bottled water at $10 \mu gl^{-1}$ (US FDA, 2005); Appendix E).

Arsenic is classified as a human carcinogen by several public health organizations. These include the National Toxicology Program (as a known human carcinogen), EPA (Group A- a known human carcinogen) and IARC (group 1 — carcinogenic to humans). These classifications are primarily determined from epidemiologic data, because of the few positive animal studies available at the time these classifications were made.

References

Abernathy, C.O., Dooley, I.S., Taft, J. and Orme-Zavaleta, J. (2000) Arsenic: moving towards a regulation, in *Toxicology and Risk Assessment* (eds H. Salem and E.J. Olagos), Taylor & Francis, Philadelphia, PA, pp. 211–22.

Abernathy, C.O., Thomas, D.J. and Calderon, R.L. (2003) Health effects and risk assessment of arsenic. *Journal of Nutrition*, **133**(5, Suppl 2), 1536S–38S.

Agency for Toxic Substances and Disease Registry (ATSDR) (2007) *Toxicological Profile for Arsenic*, U.S. Department of Health and Human Services.

Ahmad, S.A., Salim Ullah Sayed, M.H., Barua, S. *et al.* (2001) Arsenic in drinking water and pregnancy outcomes. *Environmental Health Perspectives*, **109**(6), 629–31.

Alexander, J. and Aaseth, J. (1985) Excretion of arsenic in rat bile: A role of complexing ligands containing sulphur and selenium. *Nutrition Research*, **5**(Suppl 1), S-515–S-519.

Antman, K.H. (2001) Introduction: the history of arsenic trioxide in cancer therapy. *Oncologist*, **6**(Suppl 2), 1–2.

Aposhian, H.V. (1989) Biochemical toxicology of arsenic, in *Reviews of Biochemical Toxicology* (eds E. Hodgson, J.R. Bend and R.M. Philpot), Elsevier, Amsterdam, pp. 265–99.

Arnold, L.L., Eldan, M., Nyska, A. *et al.* (2006) Dimethylarsinic acid: results of chronic toxicity/oncogenicity studies in F344 rats and in B6C3F1 mice. *Toxicology*, **223**(1–2), 82–100.

Arnold, L.L., Eldan, M., Van Gemert, M. *et al.* (2003) Chronic studies evaluating the carcinogenicity of monomethylarsonic acid in rats and mice. *Toxicology*, **190**(3), 197–219.

Basu, A., Mahata, J., Gupta, S. and Giri. A.K. (2001) Genetic toxicology of a paradoxical human carcinogen, arsenic: a review. *Mutation Research-Reviews in Mutation Research*, **488**(2), 171–94.

Beck, B.D., Slayton, T.M., Farr, C.H. *et al.* (2002) Systemic uptake of inhaled arsenic in rabbits. *Human and Experimental Toxicology*, **21**(4), 205–15.

Belluck, D.A., Benjamin, S.L., Baveye, P. *et al.* (2003) Widespread arsenic contamination of soils in residential areas and public spaces: an emerging regulatory or medical crisis?. *International Journal of Toxicology*, **22**(2), 109–28.

Bollinger, C.T., van Zijl, P. and Louw, J.A. (1992) Multiple organ failure with the adult respiratory distress syndrome in homicidal arsenic poisoning. *Respiration*, **59**, 57–61.

Braman, R.S. and Foreback, C.C. (1973) Methylated forms of arsenic in the environment. *Science*, **182**(4118), 1247–49.

Bronstein, A.C., Spyker, D.A., Cantilena, L.R., Jr. *et al.* (2007) Jr. 2006 annual report of the American Association of Poison Control Centers' National Poison Data System (NPDS). *Clinical Toxicology*, **45**(8), 815–917.

Buchet, J.P. and Lauwerys, R. (1985) Study of inorganic arsenic methylation by rat liver *in vitro*: Relevance for the interpretation of observations in man. *Archives of Toxicology*, **57**(2), 125–29.

Buchet, J.P. and Lauwerys, R. (1988) Role of thiols in the in-vitro methylation of inorganic arsenic by rat liver cytosol. *Biochemical Pharmacology*, **37**(16), 3149–53.

Buchet, J.P., Lauwerys, R. and Roels, H. (1981a) Comparison of the urinary excretion of arsenic metabolites after a single oral dose of sodium arsenite, monomethylarsonate, or dimethylarsinate in man. *International Archives of Occupational and Environmental Health*, **48**(1), 71–79.

Buchet, J.P., Lauwerys, R. and Roels, H. (1981b) Urinary excretion of inorganic arsenic and its metabolites after repeated ingestion of sodium metaarsenite by volunteers. *International Archives of Occupational and Environmental Health*, **48**(2), 111–18.

Burns, F.J., Uddin, A.N., Wu, F. *et al.* (2004) Arsenic-induced enhancement of ultraviolet radiation carcinogenesis in mouse skin: a dose-response study. *Environmental Health Perspectives*, **112**(5), 599–603.

Carter, D.E., Aposhian, H.V. and Gandolfi, A.J. (2003) The metabolism of inorganic arsenic oxides, gallium arsenide, and arsine: a toxicochemical review. *Toxicology and Applied Pharmacology*, **193**(3), 309–34.

Chakraborti, D., Sengupta, M.K., Rahman, M.M. *et al.* (2004) Groundwater arsenic contamination and its health effects in the Ganga-Meghna-Brahmaputra plain. *Journal of Environmental Monitoring*, **6**(6), 74N–83N.

Challenger, F. (1945) Biological methylation. *Chemical Reviews*, **36**(3), 315–61.

Challenger, F. (1951) Biological methylation. *Advances in Enzymology and Related Subjects of Biochemistry*, **12**, 429–91.

Challenger, F., Higginbottom, C. and Ellis, L. (1933) The formation of organo-metalloidal compounds by microorganisms. Part I. Trimethylarsine and dimethylarsine. *Journal of the Chemical Society*, 95–101.

Chasteen, T.G., Wiggli, M. and Bentley, R. (2002) Of garlic, mice and Gmelin: The odor of trimethylarsine. *Applied Organometallic Chemistry*, **16**(6), 281–86.

Chattopadhyay, S., Bhaumik, S., Nag Chaudhury, A. and Das Gupta, S. (2002) Arsenic induced changes in growth development and apoptosis in neonatal and adult brain cells *in vivo* and in tissue culture. *Toxicology Letters*, **128**(1-3), 73–84.

Chen, C.-J., Chiou, H.-Y., Chiang, M.-H. *et al.* (1996) Dose-response relationship between ischemic heart disease mortality and long-term arsenic exposure. *Arteriosclerosis, Thrombosis, and Vascular Biology*, **16**(4), 504–10.

Chen, C.-J., Hsueh, Y.-M., Lai, M.-S. *et al.* (1995) Increased prevalence of hypertension and long-term arsenic exposure. *Hypertension*, **25**(1), 53–60.

Chen, C.-J., Wang, S.-L., Chiou, J.-M. *et al.* (2007) Arsenic and diabetes and hypertension in human populations: A review. *Toxicology and Applied Pharmacology*, **222**(3), 298–304.

Chowdhury, U.K., Zakharyan, R.A., Hernandez, A. *et al.* (2006) Glutathione-S-transferase-omega (MMA(V) reductase) knockout mice: Enzyme and arsenic species concentrations in tissues after arsenate administration. *Toxicology and Applied Pharmacology*, **216**(3), 446–57.

Cohen, S.M., Arnold, L.L., Eldan, M. *et al.* (2006) Methylated arsenicals: the implications of metabolism and carcinogenicity studies in rodents to human risk assessment. *Critical Reviews in Toxicology*, **36**(2), 99–133.

Collecchi, P., Esposito, M. and Brera, S. (1986) The distribution of arsenic and cobalt in patients with laryngeal carcinoma. *Journal of Applied Toxicology*, **6**(4), 287–89.

Concha, G., Vogler, G., Lezcano, D. *et al.* (1998a) Exposure to inorganic arsenic metabolites during early human development. *Toxicological Sciences*, **44**(2), 185–90.

Concha, G., Vogler, G., Nermell, B. and Vahter, M. (1998b) Low-level arsenic excretion in breast milk of native Andean women exposed to high levels of arsenic in the drinking water. *International Archives of Occupational and Environmental Health*, **71**(1), 42–46.

Crecelius, E.A. (1977) Changes in the chemical speciation of arsenic following ingestion by man. *Environmental Health Perspectives*, **19**, 147–50.

Csanaky, I. and Gregus, Z. (2002) Species variations in the biliary and urinary excretion of arsenate, arsenite and their metabolites. *Comparative Biochemistry and Physiology C-Toxicology and Pharmacology*, **131**(3), 355–65.

Csanaky, I. and Gregus, Z. (2005) Role of glutathione in reduction of arsenate and of γ - glutamyltranspeptidase in disposition of arsenite in rats. *Toxicology*, **207**(1), 91–104.

Cui, X., Kobayashi, Y., Hayakawa, T. and Hirano, S. (2004) Arsenic speciation in bile and urine following oral and intravenous exposure to inorganic and organic arsenics in rats. *Toxicological Sciences*, **82**(2), 478–87.

Cullen, W.R., McBride, B.C. and Reglinski, J. (1984a) The reduction of trimethylarsine oxide to trimethylarsine by thiols: a mechanistic model for the biological reduction of arsenicals. *Journal of Inorganic Biochemistry*, **21**(1), 45–60.

Cullen, W.R., McBride, B.C. and Reglinski, J. (1984b) The reaction of methylarsenicals with thiols: Some biological implications. *Journal of Inorganic Biochemistry*, **21**(3), 179–93.

Cuzick, J., Sasieni, P. and Evans, S. (1992) Ingested arsenic, keratoses, and bladder cancer. *American Journal of Epidemiology*, **136**(4), 417–21.

Dang, H.S., Jaiswal, D.D. and Somasundaram, S. (1983) Distribution of arsenic in human tissues and milk. *Science of the Total Environment*, **29**(1–2), 171–75.

Davis, A., Ruby, M.V. and Bergstrom, P.D. (1992) Bioavailability of arsenic and lead in soils from the Butte, Montana, mining district. *Environmental Science and Technology*, **26**(3), 461–68.

de Wolff, F.A. and Edelbroek, P.M. (1994) Neurotoxicity of arsenic and its compounds, in *Handbook of Clinical Neurology: Intoxications of the Nervous System, Part I* (ed. F.A. de Wolff), Elsevier, Amsterdam, pp. 283–91.

Del Razo, L.M., Quintanilla-Vega, B., Brambila-Colombres, E. *et al.* (2001a) Stress proteins induced by arsenic. *Toxicology and Applied Pharmacology*, **177**(2), 132–48.

Del Razo, L.M., Styblo, M., Cullen, W.R. and Thomas, D.J. (2001b) Determination of trivalent methylated arsenicals in biological matrices. *Toxicology and Applied Pharmacology*, **174**(3), 282–93.

Delnomdedieu, M., Basti, M.M., Otvos, J.D. and Thomas, D.J. (1993) Transfer of arsenite from glutathione to dithiols: a model of interaction. *Chemical Research in Toxicology*, **6**(5), 598–602.

Delnomdedieu, M., Basti, M.M., Otvos, J.D. and Thomas, D.J. (1994a) Reduction and binding of arsenate and dimethylarsinate by glutathione: a magnetic resonance study. *Chemico-Biological Interactions*, **90**(2), 139–55.

Delnomdedieu, M., Basti, M.M., Styblo, M. *et al.* (1994b) Complexation of arsenic species in rabbit erythrocytes. *Chemical Research in Toxicology*, **7**(5), 621–27.

Desesso, J.M. (2001) Teratogen update: inorganic arsenic. *Teratology*, **64**(3), 170–73.

Desesso, J.M., Jacobson, C.F., Scialli, A.R. *et al.* (1998) An assessment of the developmental toxicity of inorganic arsenic. *Reproductive Toxicology*, **12**(4), 385–433.

Devesa, V., Adair, B.M., Liu, J. *et al.* (2006) Arsenicals in maternal and fetal mouse tissues after gestational exposure to arsenite. *Toxicology*, **224**(1–2), 147–55.

Dixon, H.B.F. (1997) The biochemical action of arsonic acids especially as phosphate analogues. *Advances in Inorganic Chemistry*, **44**, 191–227.

Drobná, Z., Waters, S.B., Devesa, V. *et al.* (2005) Metabolism and toxicity of arsenic in human urothelial cells expressing rat arsenic (+3 oxidation state)-methyltransferase. *Toxicology and Applied Pharmacology*, **207**(2), 147–59.

Drobná, Z., Xing, W., Thomas, D.J. and Stýblo, M. (2006) shRNA silencing of AS3MT expression minimizes arsenic methylation capacity of HepG2 cells. *Chemical Research in Toxicology*, **19**(7), 894–98.

Ellenhorn, M.J. (1997) Arsenic, *Ellenhorn's Medical Toxicology: Diagnosis and Treatment of Human Poisoning*, 2nd edn, Williams and Wilkins, Baltimore, MD, pp. 1538–42.

Enterline, P.E., Henderson, V.L. and Marsh, G.M. (1987) Exposure to arsenic and respiratory cancer: a reanalysis. *American Journal of Epidemiology*, **125**(6), 929–38.

Feldman, R.G., Niles, C.A. and Kelly-Hayes, M. (1979) Peripheral neuropathy in arsenic smelter workers. *Neurology*, **29**(7), 939–44.

Ferreccio, C., Gonzáles, C., Milosavjlevic, V. *et al.* (2000) Lung cancer and arsenic concentrations in drinking water in Chile. *Epidemiology*, **11**(6), 673–79.

Fomenko, D.E., Xing, W., Thomas, D.J. and Gladyshev, V.N. (2007) Large-scale identification of catalytic redox-active cysteines by detecting sporadic cysteine/selenocysteine pairs in homologous sequences. *Science*, **315**, 387–89.

Freeman, G.B., Johnson, J.D., Killinger, J.M. *et al.* (1993) Bioavailability of arsenic in soil impacted by smelter activities following oral administration in rabbits. *Fundamental and Applied Toxicology*, **21**(1), 83–88.

Freeman, G.B., Schoof, R.A., Ruby, M.V. *et al.* (1995) Bioavailability of arsenic in soil and house dust impacted by smelter activities following oral administration in cynomolgus monkeys. *Fundamental and Applied Toxicology*, **28**(2), 215–22.

Gerhardsson, L., Dahlgren, E., Eriksson, A. *et al.* (1988) Fatal arsenic poisoning - A case report. *Scandinavian Journal of Work, Environment and Health*, **14**(2), 130–33.

Germolec, D.R., Spalding, J., Boorman, G.A. *et al.* (1997) Arsenic can mediate skin neoplasia by chronic stimulation of keratinocyte-derived growth factors. *Mutation Research-Reviews in Mutation Research*, **386**(3), 209–18.

Germolec, D.R., Yoshida, T., Gaido, K. *et al.* (1996) Arsenic induces overexpression of growth factors in human keratinocytes. *Toxicology and Applied Pharmacology*, **141**(1), 308–18.

Ginsburg, J.M. and Lotspeich, W.D. (1963) Interrelations of arsenate and phosphate in the dog kidney. *American Journal of Physiology*, **205**(4), 707–14.

Gonzalez, M.J., Aguilar, M.V. and Para, M.C.M. (1995) Gastrointestinal absorption of inorganic arsenic (V): the effect of concentration and interactions with phosphate and dichromate. *Veterinary and Human Toxicology*, **37**(2), 131–36.

Gosio, B. (1892) Action of microphytes on solid compounds of arsenic: a recapitulation. *Science*, **19**, 104–6.

Gregus, Z. and Nemeti, B. (2002) Purine nucleoside phosphorylase as a cytosolic arsenate reductase. *Toxicological Sciences*, **70**(1), 13–19.

Gregus, Z. and Nemeti, B. (2005) The glycolytic enzyme glyceraldehyde-3-phosphate dehydrogenase works as an arsenate reductase in human red blood cells and rat liver cytosol. *Toxicological Sciences*, **85**(2), 859–69.

Groen, K., Vaessen, H.A., Kliest, J.J. *et al.* (1994) Bioavailability of inorganic arsenic from bog ore-containing soil in the dog. *Environmental Health Perspectives*, **102**(2), 182–84.

Guha Mazumder, D.N. (2005) Effect of chronic intake of arsenic-contaminated water on liver. *Toxicology and Applied Pharmacology*, **206**(2), 169–75.

Gyurasics, A., Varga, F. and Gregus, Z. (1991) Glutathione-dependent biliary excretion of arsenic. *Biochemical Pharmacology*, **42**(3), 465–68.

Hakala, E. and Pyy, L. (1995) Assessment of exposure to inorganic arsenic by determining the arsenic species excreted in urine. *Toxicology Letters*, **77**(1-3), 249–58.

Hall, M., Gamble, M., Slavkovich, V. *et al.* (2007) Determinants of arsenic metabolism: Blood arsenic metabolites, plasma folate, cobalamin, and homocysteine concentrations in maternal-newborn pairs. *Environmental Health Perspectives*, **115** (10), 1503–9.

Hartwig, A., Groblinghoff, U.D., Beyersmann, D. *et al.* (1997) Interaction of arsenic(III) with nucleotide excision repair in UV-irradiated human fibroblasts. *Carcinogenesis*, **18**(2), 399–405.

Hayakawa, T., Kobayashi, Y., Cui, X. and Hirano, S. (2005) A new metabolic pathway of arsenite: arsenic-glutathione complexes are substrates for human arsenic methyltransferase Cyt19. *Archives of Toxicology*, **79**(4), 183–91.

Hei, T.K., Liu, S.U.X. and Waldren, C. (1998) Mutagenicity of arsenic in mammalian cells: role of reactive oxygen species. *Proceedings of the National Academy of Sciences of the United States of America*, **95**(14), 8103–7.

Hirata, M., Mohri, T., Hisanaga, A. and Ishinishi, N. (1989) Conversion of arsenite and arsenate to methylarsenic and dimethylarsenic compounds by homogenates prepared from livers and kidneys of rats and mice. *Applied Organometallic Chemistry*, **3**(4), 335–41.

Hood, R.D., Vedel-Macrander, G.C. and Zaworotko, M.J. (1987) Distribution, metabolism, and fetal uptake of pentavalent arsenic in pregnant mice following oral or intraperitoneal administration. *Teratology*, **35**(1), 19–25.

Hood, R.D., Vedel, G.C., Zaworotko, M.J. *et al.* (1988) Uptake, distribution, and metabolism of trivalent arsenic in the pregnant mouse. *Journal of Toxicology and Environmental Health*, **25**(4), 423–34.

Hopenhayn-Rich, C., Browning, S.R., Hertz-Picciotto, I. *et al.* (2000) Chronic arsenic exposure and risk of infant mortality in two areas of Chile. *Environmental Health Perspectives*, **108**(7), 667–73.

Hopenhayn, C., Huang, B., Christian, J. *et al.* (2003) Profile of urinary arsenic metabolites during pregnancy. *Environmental Health Perspectives*, **111**(16), 1888–91.

Hsueh, Y.-M., Cheng, G.-S., Wu, M.-M. *et al.* (1995) Multiple risk factors associated with arsenic-induced skin cancer: effects of chronic liver disease and malnutritional status. *British Journal of Cancer*, **71**(1), 109–14.

Hu, Y., Su, L. and Snow, E.T. (1998) Arsenic toxicity is enzyme specific and its affects on ligation are not caused by the direct inhibition of DNA repair enzymes. *Mutation Research-DNA Repair*, **408**(3), 203–18.

Huang, C., Ke, Q., Costa, M. and Shi, X. (2004) Molecular mechanisms of arsenic carcinogenesis. *Molecular and Cellular Biochemistry*, **255**(1–2), 57–66.

Huang, C., Li, J., Ding, M. *et al.* (2001) Arsenic-induced NFB transactivation through Erks- and JNKs-dependent pathways in mouse epidermal JB6 cells. *Molecular and Cellular Biochemistry*, **222**(1-2), 29–34.

Huang, C., Ma, W.-Y., Li, J. *et al.* (1999a) Requirement of Erk, but not JNK, for arsenite-induced cell transformation. *Journal of Biological Chemistry*, **274**(21), 14595–601.

Huang, C., Ma, W.-Y., Li, L. and Dong, Z. (1999b) Arsenic induces apoptosis through a c-Jun NH2-terminal kinase-dependent, p53-independent pathway. *Cancer Research*, **59**(13), 3053–58.

Huet, P.M., Guillaume, E. and Cote, J. (1975) Noncirrhotic presinusoidal portal hypertension associated with chronic arsenical intoxication. *Gastroenterology*, **68**(5), 1270–77.

Hughes, M.F. (2002) Arsenic toxicity and potential mechanisms of action. *Toxicology Letters*, **133**(1), 1–16.

Hughes, M.F. (2006) Biomarkers of exposure: A case study with inorganic arsenic. *Environmental Health Perspectives*, **114**(11), 1790–96.

Hughes, M.F. and Kitchin, K.T. (2006) Arsenic, oxidative stress and carcinogenesis, in *Oxidative Stress, Disease and Cancer* (ed. K.K. Singh), Imperial College Press, London, pp. 825–50.

Hughes, M.F., Kenyon, E.M., Edwards, B.C. *et al.* (1999) Strain-dependent disposition of inorganic arsenic in the mouse. *Toxicology*, **137**(2), 95–108.

Hughes, M.F., Kenyon, E.M., Edwards, B.C. *et al.* (2003) Accumulation and metabolism of arsenic in mice after repeated oral administration of arsenate. *Toxicology and Applied Pharmacology*, **191**(3), 202–10.

Hughes, M.F., Menache, M. and Thompson, D.J. (1994) Dose-dependent disposition of sodium arsenate in mice following acute oral exposure. *Fundamental and Applied Toxicology*, **22**(1), 80–89.

Hutchinson, J. (1895) Diet and therapeutics. *Archives of Surgery*, **6**, 389–95.

Hwang, Y.-H., Bornschein, R.L., Grote, J. *et al.* (1997) Environmental arsenic exposure of children around a former copper smelter site. *Environmental Research*, **72**(1), 72–81.

Ibrahim, D., Froberg, B., Wolf, A. and Rusyniak, D.E. (2006) Heavy metal poisoning: Clinical presentations and pathophysiology. *Clinics in Laboratory Medicine*, **26**(1), 67–97.

Jin, Y., Xi, S., Li, X. *et al.* (2006) Arsenic speciation transported through the placenta from mother mice to their newborn pups. *Environmental Research*, **101**(3), 349–55.

Kala, S.V., Neely, M.W., Kala, G. *et al.* (2000) The MRP2/cMOAT transporter and arsenic-glutathione complex formation are required for biliary excretion of arsenic. *Journal of Biological Chemistry*, **275**(43), 33404–8.

Kato, K., Hayashi, H., Hasegawa, A. *et al.* (1994) DNA damage induced in cultured human alveolar (L-132) cells by exposure to dimethylarsinic acid. *Environmental Health Perspectives*, **102**(Suppl 3), 285–88.

Katz, S.A. and Salem, H. (2005) Chemistry and toxicology of building timbers pressure-treated with chromated copper arsenate: a review. *Journal of Applied Toxicology*, **25**(1), 1–7.

Kenney, L.J. and Kaplan, J.H. (1988) Arsenate substitutes for phosphate in the human red cell sodium pump and anion exchanger. *Journal of Biological Chemistry*, **263**(17), 7954–60.

Kenyon, E.M., Del Razo, L.M. and Hughes, M.F. (2005a) Tissue distribution and urinary excretion of inorganic arsenic and its methylated metabolites in mice following acute oral administration of arsenate. *Toxicological Sciences*, **85**(1), 468–75.

Kenyon, E.M., Del Razo, L.M., Hughes, M.F. and Kitchin, K.T. (2005b) An integrated pharmacokinetic and pharmacodynamic study of arsenite action: 2. Heme oxygenase induction in mice. *Toxicology*, **206**(3), 389–401.

Kenyon, E.M., Hughes, M.F., Adair, B. *et al.* (2008) Tissue distribution and urinary excretion of inorganic arsenic and its methylated metabolites in C57Bl/6 mice following subchronic exposure to arsenate (AsV) in drinking water. *Toxicology and Applied Pharmacology*, **232**(3), 448–55.

Kenyon, E.M., Hughes, M.F. and Levander, O.A. (1997) Influence of dietary selenium on the disposition of arsenate in the female B6C3F1 mouse. *Journal of Toxicology and Environmental Health-Part A*, **51**(3), 279–99.

Kessel, M., Liu, S.X., Xu, A. *et al.* (2002) Arsenic induces oxidative DNA damage in mammalian cells. *Molecular and Cellular Biochemistry*, **234–235**, 301–8.

Kitchin, K.T. (2001) Recent advances in arsenic carcinogenesis: modes of action, animal model systems, and methylated arsenic metabolites. *Toxicology and Applied Pharmacology*, **172**(3), 249–61.

Klaassen, C.D. (1974) Biliary excretion of arsenic in rats, rabbits, and dogs. *Toxicology and Applied Pharmacology*, **29**(3), 447–57.

Klimecki, W.T. and Carter, D.E. (1995) Arsine toxicity: chemical and mechanistic implications. *Journal of Toxicology and Environmental Health*, **46**(4), 399–409.

Kwok, R.K., Mendola, P., Liu, Z.Y. *et al.* (2007) Drinking water arsenic exposure and blood pressure in healthy women of reproductive age in Inner Mongolia, China. *Toxicology and Applied Pharmacology*, **222**(3), 337–43.

Lai, M.-S., Hsueh, Y.-M., Chen, C.-J. *et al.* (1994) Ingested inorganic arsenic and prevalence of diabetes mellitus. *American Journal of Epidemiology*, **139**(5), 484–92.

Le, X.C., Lu, X. and Li, X.-F. (2004) Arsenic speciation. *Analytical Chemistry*, **76**(1), 26A–33A.

Le, X.C., Ma, M., Lu, X. *et al.* (2000) Determination of monomethylarsonous acid, a key arsenic methylation intermediate, in human urine. *Environmental Health Perspectives*, **108**(11), 1015–18.

Lee, E. (1999) *A physiologically based pharmacokinetic model for the ingestion of arsenic in humans*. Dissertation in Environmental Toxicology, University of California at Irvine.

Lee-Feldstein, A. (1986) Cumulative exposure to arsenic and its relationship to respiratory cancer among copper smelter employees. *Journal of Occupational Medicine*, **28**(4), 296–302.

Li, J.-H. and Rossman, T.G. (1989) Inhibition of DNA ligase activity by arsenite: a possible mechanism of its comutagenesis. *Molecular Toxicology*, **2**(1), 1–9.

Li, J.-H. and Rossman, T.G. (1991) Comutagenesis of sodium arsenite with ultraviolet radiation in Chinese hamster V79 cells. *Biology of Metals*, **4**(4), 197–200.

Li, J., Waters, S.B., Drobna, Z. *et al.* (2005) Arsenic (+3 oxidation state) methyltransferase and the inorganic arsenic methylation phenotype. *Toxicology and Applied Pharmacology*, **204**(2), 164–69.

Lin, S., Del Razo, L.M., Styblo, M. *et al.* (2001) Arsenicals inhibit thioredoxin reductase in cultured rat hepatocytes. *Chemical Research in Toxicology*, **14**(3), 305–11.

Lin, S., Shi, Q., Brent Nix, F. *et al.* (2002) A novel S-adenosyl-L-methionine:arsenic(III) methyltransferase from rat liver cytosol. *Journal of Biological Chemistry*, **277**(13), 10795–803.

Lindgren, A., Danielsson, B.R.G., Dencker, L. and Vahter, M. (1984) Embryotoxicity of arsenite and arsenate: Distribution in pregnant mice and monkeys and effects on embryonic cells *in vitro*. *Acta Pharmacologica et Toxicologica*, **54**(4), 311–20.

Lindgren, A., Vahter, M. and Dencker, L. (1982) Autoradiographic studies on the distribution of arsenic in mice and hamsters administered [74]As-arsenite or -arsenate. *Acta Pharmacologica et Toxicologica*, **51**(3), 253–65.

Liu, J., Zheng, B., Aposhian, H.V. *et al.* (2002) Chronic arsenic poisoning from burning high-arsenic-containing coal in Guizhou, China. *Environmental Health Perspectives*, **110**(2), 119–22.

Lowney, Y.W., Ruby, M.V., Wester, R.C. *et al.* (2005) Percutaneous absorption of arsenic from environmental media. *Toxicology and Industrial Health*, **21**(1–2), 1–14.

Lowney, Y.W., Wester, R.C., Schoof, R.A. *et al.* (2007) Dermal absorption of arsenic from soils as measured in the rhesus monkey. *Toxicological Sciences*, **100**(2), 381–92.

Lu, M., Wang, H., Li, X.-F. *et al.* (2004) Evidence of hemoglobin binding to arsenic as a basis for the accumulation of arsenic in rat blood. *Chemical Research in Toxicology*, **17**(12), 1733–42.

Lu, M., Wang, H., Li, X.-F. *et al.* (2007) Binding of dimethylarsinous acid to Cys-13α of rat hemoglobin is responsible for the retention of arsenic in rat blood. *Chemical Research in Toxicology*, **20**(1), 27–37.

Lugo, G., Cassady, G. and Palmisano, P. (1969) Acute maternal arsenic intoxication with neonatal death. *American Journal of Diseases of Children*, **117**(3), 328–30.

Maloney, M.E. (1996) Arsenic in dermatology. *Dermatologic Surgery*, **22**(3), 301–4.

Marafante, E., Bertolero, F. and Edel, J. (1982) Intracellular interaction and biotransformation of arsenite in rats and rabbits. *Science of the Total Environment*, **24**(1), 27–39.

Marafante, E. and Vahter, M. (1987) Solubility, retention, and metabolism of intratracheally and orally administered inorganic arsenic compounds in the hamster. *Environmental Research*, **42**(1), 72–82.

Marafante, E., Vahter, M. and Envall, J. (1985) The role of the methylation in the detoxication of arsenate in the rabbit. *Chemico-Biological Interactions*, **56**(2–3), 225–38.

Mass, M.J. and Wang, L. (1997) Arsenic alters cytosine methylation patterns of the promoter of the tumor suppressor gene p53 in human lung cells: a model for a mechanism of carcinogenesis. *Mutation Research-Reviews in Mutation Research*, **386**(3), 263–77.

Mazumder, D.N.G., Steinmaus, C., Bhattacharya, P. *et al.* (2005) Bronchiectasis in persons with skin lesions resulting from arsenic in drinking water. *Epidemiology*, **16**(6), 760–65.

Meharg, A.A. (2005) *Venomous Earth: How Arsenic Caused the World's Worst Mass Poisoning*, Macmillan, New York, 192.

Narang, A.P. and Datta, D.V. (1983) Brain arsenic concentrations in fulminant hepatitis. *The Journal of the Association of Physicians of India*, **31**(8), 518–19.

Naranmandura, H., Suzuki, N. and Suzuki, K.T. (2006) Trivalent arsenicals are bound to proteins during reductive methylation. *Chemical Research in Toxicology*, **19**(8), 1010–18.

National Research Council (NRC) (1977) *Arsenic*, National Research Council, National Academy of Sciences, Washington, DC.

National Research Council (NRC) (1999) *Arsenic in Drinking Water*, National Research Council, National Academy of Sciences, Washington, DC.

National Research Council (NRC) (2001) *Arsenic in Drinking Water, 2001 Update*, National Research Council, National Academy of Sciences, Washington, DC.

Navas-Acien, A., Sharrett, A.R., Silbergeld, E.K. *et al.* (2005) Arsenic exposure and cardiovascular disease: A systematic review of the epidemiologic evidence. *American Journal of Epidemiology*, **162**(11), 1037–49.

Navas-Acien, A., Silbergeld, E.K., Streeter, R.A. *et al.* (2006) Arsenic exposure and type 2 diabetes: a systematic review of the experimental and epidemiologic evidence. *Environmental Health Perspectives*, **114**(5), 641–48.

Nemeti, B. and Gregus, Z. (2005) Reduction of arsenate to arsenite by human erythrocyte lysate and rat liver cytosol –Characterization of a glutathione-and NAD-dependent arsenate reduction linked to glycolysis. *Toxicological Sciences*, **85**(2), 847–58.

Nemeti, B. and Gregus, Z. (2007) Glutathione-dependent reduction of arsenate by glycogen phosphorylase — A reaction coupled to glycogenolysis. *Toxicological Sciences*, **100**(1), 36–43.

Nemeti, B., Csanaky, I. and Gregus, Z. (2003) Arsenate reduction in human erythrocytes and rats — Testing the role of purine nucleoside phosphorylase. *Toxicological Sciences*, **74**(1), 22–31.

Nordstrom, D.K. (2002) Worldwide occurrences of arsenic in ground water. *Science*, **296**(5576), 2143–44.

Odanaka, Y., Matano, O. and Goto, S. (1980) Biomethylation of inorganic arsenic by the rat and some laboratory animals. *Bulletin of Environmental Contamination and Toxicology*, **24**(3), 452–59.

Offergelt, J.A., Roels, H., Buchet, J.P. *et al.* (1992) Relation between airborne arsenic trioxide and urinary excretion of inorganic arsenic and its methylated metabolites. *British Journal of Industrial Medicine*, **49**(6), 387–93.

Okui, T. and Fujiwara, Y. (1986) Inhibition of human excision DNA repair by inorganic arsenic and the co-mutagenic effect in V79 Chinese hamster cells. *Mutation Research*, **172**(1), 69–76.

Pershagen, G., Lind, B. and Bjorklund, N.E. (1982) Lung retention and toxicity of some inorganic arsenic compounds. *Environmental Research*, **29**(2), 425–34.

Peryea, F.J. and Creger, T.L. (1994) Vertical distribution of lead and arsenic in soils contaminated with lead arsenate pesticide residues. *Water, Air, and Soil Pollution*, **78**(3–4), 297–306.

Petrick, J.S., Jagadish, B., Mash, E.A. and Aposhian, H.V. (2001) Monomethylarsonous acid (MMA(III)) and arsenite: LD50 in hamsters and *in vitro* inhibition of pyruvate dehydrogenase. *Chemical Research in Toxicology*, **14**(6), 651–56.

Polissar, L., Lowry-Coble, K., Kalman, D.A. *et al.* (1990) Pathways of human exposure to arsenic in a community surrounding a copper smelter. *Environmental Research*, **53**(1), 29–47.

Pomroy, C., Charbonneau, S.M., McCullough, R.S. and Tam, G.K.H. (1980) Human retention studies with 74As. *Toxicology and Applied Pharmacology*, **53**(3), 550–56.

Rahman, M., Tondel, M., Ahmad, S.A. and Axelson, O. (1998) Diabetes mellitus associated with arsenic exposure in Bangladesh. *American Journal of Epidemiology*, **148**(2), 198–203.

Raie, R.M. (1996) Regional variation in As, Cu, Hg, and Se and interaction between them. *Ecotoxicology and Environmental Safety*, **35**(3), 248–52.

Ratnaike, R.N. (2003) Acute and chronic arsenic toxicity. *Postgraduate Medical Journal*, **79**(933), 391–96.

Roberts, S.M., Weimar, W.R., Vinson, J.R.T. *et al.* (2002) Measurement of arsenic bioavailability in soil using a primate model. *Toxicological Sciences*, **67**(2), 303–10.

Robinson, G.R., Jr., Larkins, P., Boughton, C.J., Reed, B.W. *et al.* (2007) Assessment of contamination from arsenical pesticide use on orchards in the Great Valley region, Virginia and West Virginia, USA. *Journal of Environmental Quality*, **36**(3), 654–63.

Rodríguez, V.M., Carrizales, L., Mendoza, M.S. *et al.* (2002) Effects of sodium arsenite exposure on development and behavior in the rat. *Neurotoxicology and Teratology*, **24**(6), 743–50.

Rossman, T.G. (2003) Mechanism of arsenic carcinogenesis: an integrated approach. *Mutation Research-Fundamental and Molecular Mechanisms of Mutagenesis*, **533**(1-2), 37–65.

Rossman, T.G., Uddin, A.N., Burns, F.J. and Bosland, M.C. (2001) Arsenite is a cocarcinogen with solar ultraviolet radiation for mouse skin: an animal model for arsenic carcinogenesis. *Toxicology and Applied Pharmacology*, **176**(1), 64–71.

Schoof, R.A., Yost, L.J., Eickhoff, J. *et al.* (1999) A market basket survey of inorganic arsenic in food. *Food and Chemical Toxicology*, **37**(8), 839–46.

Schwartz, R.A. (1997) Arsenic and the skin. *International Journal of Dermatology*, **36**(4), 241–50.

Scott, N., Hatlelid, K.M., MacKenzie, N.E. and Carter, D.E. (1993) Reactions of arsenic(III) and arsenic(V) species with glutathione. *Chemical Research in Toxicology*, **6**(1), 102–6.

Shannon, R.L. and Strayer, D.S. (1989) Arsenic-induced skin toxicity. *Human Toxicology*, **8**(2), 99–104.

Shen, J., Wanibuchi, H., Salim, E.I. *et al.* (2003) Liver tumorigenicity of trimethylarsine oxide in male Fischer 344 rats — association with oxidative DNA damage and enhanced cell proliferation. *Carcinogenesis*, **24**(11), 1827–35.

Shi, H., Shi, X. and Liu, K.J. (2004) Oxidative mechanism of arsenic toxicity and carcinogenesis. *Molecular and Cellular Biochemistry*, **255**(1-2), 67–78.

Shiobara, Y., Ogra, Y. and Suzuki, K.T. (2001) Animal species difference in the uptake of dimethylarsinous acid (DMA(III)) by red blood cells. *Chemical Research in Toxicology*, **14**(10), 1446–52.

Simeonova, P.P., Wang, S., Toriuma, W. *et al.* (2000) Arsenic mediates cell proliferation and gene expression in the bladder epithelium association with activating protein-1 transactivation. *Cancer Research*, **60**(13), 3445–53.

Smith, T.J., Crecelius, E.A. and Reading, J.C. (1977) Airborne arsenic exposure and excretion of methylated arsenic compounds. *Environmental Health Perspectives*, **19**, 89–93.

Smith, M.S., Jones, P.R. and Shirachi, D.Y. (1992) The biomethylation of sodium arsenate by rat liver cytosol determined by mass spectrometry. *Proceedings of the Western Pharmacology Society*, **35**, 53–55.

Smith, A.H., Lopipero, P.A., Bates, M.N. and Steinmaus, C.M. (2002) Arsenic epidemiology and drinking water standards. *Science*, **296**(5576), 2145–46.

Spuches, A.M., Kruszyna, H.G., Rich, A.M. and Wilcox, D.E. (2005) Thermodynamics of the As(III)-thiol interaction: arsenite and monomethylarsenite complexes with glutathione, dihydrolipoic acid, and other thiol ligands. *Inorganic Chemistry*, **44**(8), 2964–72.

Styblo, M., Del Razo, L.M., Vega, L. *et al.* (2000) Comparative toxicity of trivalent and pentavalent inorganic and methylated arsenicals in rat and human cells. *Archives of Toxicology*, **74**(6), 289–99.

Styblo, M., Delnomdedieu, M. and Thomas, D.J. (1996) Mono- and dimethylation of arsenic in rat liver cytosol *in vitro*. *Chemico-Biological Interactions*, **99**(1-3), 147–64.

Tam, G.K.H., Charbonneau, S.M. and Bryce, F. (1979) Metabolism of inorganic arsenic (74As) in humans following oral ingestion. *Toxicology and Applied Pharmacology*, **50**(2), 319–22.

Tao, S.S.-H. and Bolger, P.M. (1999) Dietary arsenic intakes in the United States: FDA Total Diet Study, September 1991-December 1996. *Food Additives and Contaminants*, **16**(8), 465–72.

Thomas, D.J. (2007) Molecular processes in cellular arsenic metabolism. *Toxicology and Applied Pharmacology*, **222**(3), 365–73.

Thomas, D.J., Li, J., Waters, S.B. *et al.* (2007) Arsenic (+3 oxidation state) methyltransferase and the methylation of arsenicals. *Experimental Biology and Medicine*, **232**(1), 3–13.

Thomas, D.J., Styblo, M. and Lin, S. (2001) The cellular metabolism and systemic toxicity of arsenic. *Toxicology and Applied Pharmacology*, **176**(2), 127–44.

Tsai, S.-M., Wang, T.-N. and Ko, Y.-C. (1999) Mortality for certain diseases in areas with high levels of arsenic in drinking water. *Archives of Environmental Health*, **54**(3), 186–93.

Tseng, C.-H. (2005) Blackfoot disease and arsenic: A never-ending story. *Journal of Environmental Science and Health-Part C Environmental Carcinogenesis and Ecotoxicology Reviews*, **23**(1), 55–74.

Tseng, W.P., Chu, H.M., How, S.W. *et al.* (1968) Prevalence of skin cancer in an endemic area of chronic arsenicism in Taiwan. *Journal of the National Cancer Institute*, **40**(3), 453–63.

Tseng, C.-H., Huang, Y.-K., Huang, Y.-L. *et al.* (2005) Arsenic exposure, urinary arsenic speciation, and peripheral vascular disease in blackfoot disease-hyperendemic villages in Taiwan. *Toxicology and Applied Pharmacology*, **206**(3), 299–308.

Tseng, C.-H., Tai, T.-Y., Chong, C.-K. *et al.* (2000) Long-term arsenic exposure and incidence of non-insulin-dependent diabetes mellitus: A cohort study in arseniasis-hyperendemic villages in Taiwan. *Environmental Health Perspectives*, **108**(9), 847–51.

U.S. Environmental Protection Agency (US EPA) (1996) Bioavailability of Arsenic and Lead in Environmental Substrates. 1. Results of an Oral Dosing Study of Immature Swine. EPA 910/R-96-002, Superfund/Office of Environmental Assessment.

U.S. Food and Drug Administration (US FDA) (2005) Beverages: bottled water. Final rule. *Federal Register*, **66**, 6976–7066.

Vahter, M. (1981) Biotransformation of trivalent and pentavalent inorganic arsenic in mice and rats. *Environmental Research*, **25**(2), 286–93.

Vahter, M. (2008) Health effects of early life exposure to arsenic. *Basic and Clinical Pharmacology and Toxicology*, **102**(2), 204–11.

Vahter, M., Friberg, L. and Rahnster, B. (1986) Airborne arsenic and urinary excretion of metabolites of inorganic arsenic among smelter workers. *International Archives of Occupational and Environmental Health*, **57**(2), 79–91.

Vahter, M. and Marafante, E. (1985) Reduction and binding of arsenate in marmoset monkeys. *Archives of Toxicology*, **57**(2), 119–24.

Vahter, M., Marafante, E. and Dencker, L. (1984) Tissue distribution and retention of ^{74}As-dimethylarsinic acid in mice and rats. *Archives of Environmental Contamination and Toxicology*, **13**(3), 259–64.

Vahter, M., Marafante, E., Lindgren, A. and Dencker, L. (1982) Tissue distribution and subcellular binding of arsenic in marmoset monkeys after injection of ^{74}As-arsenite. *Archives of Toxicology*, **51**(1), 65–77.

Vahter, M. and Norin, H. (1980) Metabolism of ^{74}As-labeled trivalent and pentavalent inorganic arsenic in mice. *Environmental Research*, **21**(2), 446–57.

Valenzuela, O.L., Borja-Aburto, V.H., Garcia-Vargas, G.G. *et al.* (2005) Urinary trivalent methylated arsenic species in a population chronically exposed to inorganic arsenic. *Environmental Health Perspectives*, **113**(3), 250–54.

Vega, L., Styblo, M., Patterson, R. *et al.* (2001) Differential effects of trivalent and pentavalent arsenicals on cell proliferation and cytokine secretion in normal human epidermal keratinocytes. *Toxicology and Applied Pharmacology*, **172**(3), 225–32.

Vijayaraghavan, M., Wanibuchi, H., Karim, R. *et al.* (2001) Dimethylarsinic acid induces 8-hydroxy-2'-deoxyguanosine formation in the kidney of NCI-Black-Reiter rats. *Cancer Letters*, **165**(1), 11–17.

Waalkes, M.P., Liu, J. and Diwan, B.A. (2007) Transplacental arsenic carcinogenesis in mice. *Toxicology and Applied Pharmacology*, **222**(3), 271–80.

Walton, F.S., Waters, S.B., Jolley, S.L. *et al.* (2003) Selenium compounds modulate the activity of recombinant rat AsIII-methyltransferase and the methylation of arsenite by rat and human hepatocytes. *Chemical Research in Toxicology*, **16**(3), 261–65.

Wang, A., Holladay, S.D., Wolf, D.C. *et al.* (2006) Reproductive and developmental toxicity of arsenic in rodents: a review. *International Journal of Toxicology*, **25**(5), 319–31.

Wang, C.-H., Hsiao, C.K., Chen, C.-L. *et al.* (2007) A review of the epidemiologic literature on the role of environmental arsenic exposure and cardiovascular diseases. *Toxicology and Applied Pharmacology*, **222**(3), 315–26.

Wang, C.-H., Jeng, J.-S., Yip, P.-K. *et al.* (2002) Biological gradient between long-term arsenic exposure and carotid atherosclerosis. *Circulation*, **105**(15), 1804–9.

Wanibuchi, H., Salim, E.I., Kinoshita, A. *et al.* (2004) Understanding arsenic carcinogenicity by the use of animal models. *Toxicology and Applied Pharmacology*, **198**(3), 366–76.

Warren, H.V., Horksy, S.J. and Gould, C.E. (1983) Quantitative analysis of zinc, copper, lead, molybdenum, bismuth, mercury and arsenic in brain and other tissues from multiple sclerosis and non-multiple sclerosis cases. *Science of the Total Environment*, **29**(1-2), 163–69.

Wasserman, G.A., Liu, X., Parvez, F. *et al.* (2004) Water arsenic exposure and children's intellectual function in Araihazar, Bangladesh. *Environmental Health Perspectives*, **112**(13), 1329–33.

Wasserman, G.A., Liu, X., Parvez, F. *et al.* (2007) Water arsenic exposure and intellectual function in 6-year-old children in Araihazar, Bangladesh. *Environmental Health Perspectives*, **115**(2), 285–89.

Waters, S.B., Devesa, V., Del Razo, L.M. *et al.* (2004a) Endogenous reductants support the catalytic function of recombinant rat Cyt19, an arsenic methyltransferase. *Chemical Research in Toxicology*, **17**(3), 404–9.

Waters, S.B., Devesa, V., Fricke, M.W. *et al.* (2004b) Glutathione modulates recombinant rat arsenic (+3 oxidation state) methyltransferase-catalyzed formation of trimethylarsine oxide and trimethylarsine. *Chemical Research in Toxicology*, **17**(12), 1621–29.

Wei, M., Wanibuchi, H., Morimura, K. *et al.* (2002) Carcinogenicity of dimethylarsinic acid in male F344 rats and genetic alterations in induced urinary bladder tumors. *Carcinogenesis*, **23**(8), 1387–97.

Wester, R.C., Hui, X., Barbadillo, S. *et al.* (2004) *In vivo* percutaneous absorption of arsenic from water and CCA-treated wood residue. *Toxicological Sciences*, **79**(2), 287–95.

Wester, R.C., Maibach, H.I., Sedik, L. *et al.* (1993) *In vivo* and *in vitro* percutaneous absorption and skin decontamination of arsenic from water and soil. *Fundamental and Applied Toxicology*, **20**(3), 336–40.

Wildfang, E., Zakharyan, R.A. and Aposhian, H.V. (1998) Enzymatic methylation of arsenic compounds: VI. Characterization of hamster liver arsenite and methylarsonic acid methyltransferase activities *in vitro*. *Toxicology and Applied Pharmacology*, **152**(2), 366–75.

World Health Organization (WHO) (1981) *Arsenic*. Environmental Health Criteria, Vol. **18**, World Health Organization, International Programme on Chemical Safety, Geneva.

World Health Organization (WHO) (2001) *Arsenic and Arsenic Compounds*, Environmental Health Criteria, Vol. **224**, 2nd edn, World Health Organization, International Programme on Chemical Safety, Geneva.

Wu, M.-M., Kuo, T.-L., Hwang, Y.-H. and Chen, C.-J. (1989) Dose-response relation between arsenic concentration in well water and mortality from cancers and vascular diseases. *American Journal of Epidemiology*, **130**(6), 1123–32.

Yager, J.W., Hicks, J.B. and Fabianova, E. (1997) Airborne arsenic and urinary excretion of arsenic metabolites during boiler cleaning operations in a Slovak coal-fired power plant. *Environmental Health Perspectives*, **105**(8), 836–42.

Yamanaka, K., Hasegawa, A., Sawamura, R. and Okada, S. (1989) Dimethylated arsenics induce DNA strand breaks in lung via the production of active oxygen in mice. *Biochemical and Biophysical Research Communications*, **165**(1), 43–50.

Yamanaka, K., Takabayashi, F., Mizoi, M. *et al.* (2001) Oral exposure of dimethylarsinic acid, a main metabolite of inorganic arsenics, in mice leads to an increase in 8-oxo-2'-deoxyguanosine level, specifically in the target organs for arsenic carcinogenesis. *Biochemical and Biophysical Research Communications*, **287**(1), 66–70.

Yamauchi, H. and Fowler, B.A. (1994) Toxicity and metabolism of inorganic and methylated arsenicals, in *Arsenic in the Environment: Part II: Human Health and Ecosystem Effects* (ed. J.O. Nriagu), John Wiley & Sons, Inc., New York, pp. 35–43.

Yamauchi, H., Takahashi, K., Mashiko, M. and Yamamura, Y. (1989) Biological monitoring of arsenic exposure of gallium arsenide- and inorganic arsenic-exposed workers by determination of inorganic arsenic and its metabolites in urine and hair. *American Industrial Hygiene Association Journal*, **50**(11), 606–12.

Yamauchi, H. and Yamamura, Y. (1979) Dynamic change of inorganic arsenic and methylarsenic compounds in human urine after oral intake as arsenic trioxide. *Industrial Health*, **17**(2), 79–83.

Yamauchi, H. and Yamamura, Y. (1983) Concentration and chemical species of arsenic in human tissue. *Bulletin of Environmental Contamination and Toxicology*, **31**(3), 267–77.

Yamauchi, H. and Yamamura, Y. (1985) Metabolism and excretion of orally administered arsenic trioxide in the hamster. *Toxicology*, **34**(2), 113–21.

Yang, C.-Y. (2006) Does arsenic exposure increase the risk of development of peripheral vascular diseases in humans? *Journal of Toxicology and Environmental Health-Part A: Current Issues*, **69**(19), 1797–804.

Yang, J.-K., Barnett, M.O., Jardine, P.M. *et al.* (2002) Adsorption, sequestration, and bioaccessibility of As(V) in soils. *Environmental Science and Technology*, **36**(21), 4562–69.

Yang, J.-K., Barnett, M.O., Zhuang, J. *et al.* (2005) Adsorption, oxidation, and bioaccessibility of As(III) in soils. *Environmental Science and Technology*, **39**(18), 7102–10.

Yoshida, T., Yamauchi, H. and Fan Sun, G. (2004) Chronic health effects in people exposed to arsenic via the drinking water: Dose-response relationships in review. *Toxicology and Applied Pharmacology*, **198**(3), 243–52.

Yu, H.-S., Lee, C.-H. and Chen, G.-S. (2002) Peripheral vascular diseases resulting from chronic arsenical poisoning. *Journal of Dermatology*, **29**(3), 123–30.

Zakharyan, R.A. and Aposhian, H.V. (1999) Enzymatic reduction of arsenic compounds in mammalian systems: the rate-limiting enzyme of rabbit liver arsenic biotransformation is MMA(V) reductase. *Chemical Research in Toxicology*, **12**(12), 1278–83.

Zakharyan, R.A., Ayala-Fierro, F., Cullen, W.R. *et al.* (1999) Enzymatic methylation of arsenic compounds. VII. Monomethylarsonous acid (MMA(III)) is the substrate for MMA methyltransferase of rabbit liver and human hepatocytes. *Toxicology and Applied Pharmacology*, **158**(1), 9–15.

Zakharyan, R.A., Sampayo-Reyes, A., Healy, S.M. *et al.* (2001) Human monomethylarsonic acid (MMA (V)) reductase is a member of the glutathione-S-transferase superfamily. *Chemical Research in Toxicology*, **14**(8), 1051–57.

Zakharyan, R., Wu, Y., Bogdan, G.M. and Aposhian, H.V. (1995) Enzymatic methylation of arsenic compounds: Assay, partial purification, and properties of arsenite methyltransferase and monomethylarsonic acid methyltransferase of rabbit liver. *Chemical Research in Toxicology*, **8**(8), 1029–38.

Zhao, C.Q., Young, M.R., Diwan, B.A. *et al.* (1997) Association of arsenic-induced malignant transformation with DNA hypomethylation and aberrant gene expression. *Proceedings of the National Academy of Sciences of the United States of America*, **94**(20), 10907–912.

5

Arsenic in Human History and Modern Societies

KEVIN R. HENKE[1] and DAVID A. ATWOOD[2]

[1] University of Kentucky Center for Applied Energy Research
[2] Department of Chemistry, University of Kentucky

5.1 Introduction

Over the years, arsenic compounds have been widely used in a variety of pigments, medicines, alloys, pesticides, herbicides, glassware, embalming fluids, chemical warfare agents, and as a depilatory in leather manufacturing. Additionally, some of the compounds, such as arsenolite (arsenic trioxide, As_2O_3), are traditional poisons in murder and suicide. In the twentieth century, new applications for arsenic were developed, including: livestock dips and feed supplements, drugs for treating leukemia, semiconductors, and wood preservatives. Enormous amounts of arsenic were also released into the environment by mining, coal utilization, and ore smelting (Chapter 3). Beginning in the nineteenth and continuing into the twentieth century, medical personnel, environmentalists, and other groups became increasingly concerned about the potential impacts of arsenic on human health and the environment. As a result, government regulators in many countries now recognize the toxic properties of arsenic and its compounds (Chapter 4) and have restricted their use. Although its commercial applications have substantially declined in developed countries in recent years, arsenic contamination is still widespread. In 1999, the *National Priority List* (*Superfund* Program) of the U.S. Environmental Protection Agency (US EPA) identified 1209 sites in the United States that had serious environmental and human health risks (US EPA, 2002a), 2; Appendix E). After lead, arsenic was the most common inorganic contaminant (568 sites or 47% of the total (US EPA, 2002b), 2). Rather than always resulting from the improper disposal of arsenic-bearing wastes, much of the contamination at the Superfund sites occurred from the mobilization

Arsenic Edited by Kevin R. Henke

of naturally occurring arsenic in *rocks*, *sediments*, and *soils*. In particular, arsenic may be mobilized into the environment by the *oxidation* of arsenic-bearing sulfide *minerals* or from the desorption of arsenic from *(oxy)(hydr)oxide* minerals due to *reductive dissolution* or interactions with carbonate-bearing solutions produced by microbial decomposition of buried organic wastes (Chapter 3).

5.2   Early recognition and uses of arsenic by humans

Orpiment (As_2S_3) and realgar (As_4S_4), which have bright yellow and red colors, respectively, have long been recognized for their beauty and toxicity in pigments. Persia (Iran), China, and Asia Minor (modern Turkey) were especially important ancient mining sites for these and other arsenic compounds (Azcue and Nriagu, 1994), 5; (Nriagu, 2002), 5. The Iranian sources included deposits at Talmessi-Meskani in the central part of the country and the *hydrothermal* ores at Zarshuran in the west on the border with modern Turkey (Azcue and Nriagu, 1994), 5; (Mehrabi, Yardley and Cann, 1999). The ores at Zarshuran include up to arsenic sulfides (orpiment with some realgar and various trace sulfoarsenides) (Mehrabi, Yardley and Cann, 1999).

The term '*arsenic*' probably originated from the Persian word *az-zarnikh* or other modifications of its root word, *zar*, which referred to yellow or gold orpiment (Azcue and Nriagu, 1994), 3; (Meharg, 2005), 39. The ancient Greeks believed that metals and other substances had masculine or feminine properties. They referred to yellow orpiment pigments as $\alpha\rho\rho\varepsilon V I K O U$ (arrenikos or arsenikos), which means 'potent' or 'masculine' (Azcue and Nriagu, 1994), 3; (Meharg, 2005), 38-39. In Latin, the Greek name became *arsenicum*. Eventually, the French derived the term *arsenic*, which is also still used in English-speaking countries. Although the noun 'arsenic' often referred to orpiment or perhaps other arsenic compounds before the twentieth century, today the definition of the noun is technically restricted to element number 33 or samples of the element (Chapter 2).

Copper ores usually contain significant amounts of arsenic. *Native* copper (i.e. natural deposits of copper metal) occurs in some areas. As early as 15,000 BC, humans were commonly hammering native copper into various artifacts (Craig, Vaughan and Skinner, 2001), 51, such as tools, weapons, and jewelry. By 4000 BC, individuals were widely smelting copper sulfides, oxides, and carbonates into copper metal (Craig, Vaughan and Skinner, 2001), 57. The discovery of the smelting process was probably accidental as humans noticed metallic copper forming from certain rocks in ovens, kilns, and campfires.

Although early coppersmiths did not fully recognize the presence of arsenic, they realized that certain copper ores (those with higher concentrations of arsenic) tended to produce harder copper, which was better suited for tools and weapons. Coppersmiths were routinely exposed to garlic-smelling As_4O_6 ('As_2O_3') fumes, which produced undesirable health effects, including: polyneuritis and lameness through muscular atrophy. Of all of the Greek gods, only Hephaestus the metalworker (Vulcan in Latin) was physically imperfect. The deformities of Hephaestus probably originated from observations of the frequent symptoms associated with ancient metal smiths that had long-term exposure to poisonous arsenic and lead fumes (Azcue and Nriagu, 1994), 8; (Nriagu, 2002), 2.

The production of bronze, a copper–tin alloy, began in Iran between 3900 and 2900 BC, and quickly spread into India, the Middle East, and Europe (Craig, Vaughan and Skinner, 2001), 57. Arsenic in molten bronze tends to move toward the surface and produce a silvery polishable sheen. The ancient Egyptians and Chinese utilized this property to create bronze mirrors (Nriagu, 2002), 3. Furthermore, when copper oxide is mixed with arsenolite and heated, the resulting alloy has the appearance of silver. This observation led some alchemists to believe that arsenolite could transform copper into silver (Azcue and Nriagu, 1994), 3–4. Even in recent times, arsenic has often been added to copper and copper alloys to increase strength and resistance to corrosion (Azcue and Nriagu, 1994), 13.

5.3   Alchemy, development of methods to recover elemental arsenic, and the synthesis of arsenic compounds

Many discoveries involving arsenic chemistry were probably never recorded in ancient cultures or the manuscripts have been lost overtime. Geber (Jabir ibn-Haiyan), a supposed eight century AD Arabian alchemist, is credited with recovering elemental arsenic (As(0)) and discovering 'white arsenic' (actually, arsenolite or As_2O_3) by heating orpiment (Meharg, 2005), 46; (Nriagu, 2002), 6; (Azcue and Nriagu, 1994), 4. However, like many figures from ancient history, the existence of Geber has been questioned. Confirming Geber's existence and possible accomplishments are very difficult. His burial site is unknown. None of his original alchemy manuscripts, if they ever existed, have survived (Meharg, 2005), 45–46. Furthermore, many ancient writers tended to credit discoveries to long-dead celebrities that may have been more legend than reality.

Around AD 1250, Albertus Magnus (1193–1280), a German alchemist and Dominican, was the first individual to provide suitably documented evidence of the recovery of As(0) (Nriagu, 2002), 6; (Mandal and Suzuki, 2002), 201; (Meharg, 2005), 46–48. However, for at least the Europeans, it was not until 1540 that Vannocio Biringuccio (1480–1539), an Italian metallurgist, recognized the fundamental differences between As(0) and arsenic compounds. By 1641, Johann Schroeder (1600–1664) was able to readily convert arsenolite into As(0) with charcoal (Nriagu, 2002), 6; (Azcue and Nriagu, 1994), 4.

In the eighteenth and nineteenth centuries, European chemists began to synthesize a number of different arsenic compounds and widely report their results. Arsine (AsH_3) gas was discovered by Carl Wilhelm Scheele (1742–1786) in 1775. Small quantities of the gas killed the famous Bavarian chemist Arnold F. Gehlen (1775–1815) (Nriagu, 2002), 6. In 1757 while attempting to produce As(0) from powdered As_2O_3 and dry potassium acetate, Louis-Claude Cadet de Gassicourt ('Cadet', 1731–1799) produced a smelly and combustible brown fuming liquid ((Meharg, 2005), 61; also see Chapter 2). During 1837–1843, Robert Wilhelm Bunsen (1811–1899) extensively studied the components of Cadet's fuming liquid, which is a mixture of cacodyl oxide ($(CH_3)_2AsOAs(CH_3)_2$), $(CH_3)_4As_2$, and other arsenic compounds (Seyferth, 2001), 1494, 1496. In the process, Bunsen partially lost sight in one eye and almost died from arsenic exposure after an explosion occurred during the preparation of cacodyl cyanide ($(CH_3)_2AsCN$) (Seyferth, 2001), 1491. As discussed in Seyferth (2001), the synthesis and analysis of Cadet's fuming liquid led to important developments in organometallic chemistry.

5.4   Applications with arsenic

5.4.1   Medicinal applications: dangerous quackery and some important drugs

While the toxic nature of arsenic was understood from early times, it was also recognized that the toxicity might have some beneficial effects in curing diseases. Alchemist and physician Paracelsus (Phillip von Hohenheim, 1493–1541) noted that 'all substances are poisons; there is none which is not a poison. The right dose differentiates a poison and remedy.' Hippocrates (c. 460–c. 370 BC) mentioned that orpiment and realgar were used as escharotics (Nriagu, 2002), 4 and in treating ulcers (Gorby, 1994), 5. The Roman author and naturalist Pliny the Elder (AD 23–79) described the use of realgar in eye-washes and to treat sore throats, asthma, and coughs (Azcue and Nriagu, 1994), 9. Pedanius Dioscorides (AD c. 40–c. 90) claimed a number of different medical applications for orpiment, including killing head lice (Nriagu, 2002), 4. These methods were adopted and modified by Arab physicians. Persian alchemist Zakariyya al-Razi (Rhazes) (c. 850–925) recommended arsenic for treating asthma, respiration problems, ulcerations, and, when mixed with unslaked lime and opium, a treatment for dysentery

(Nriagu, 2002), 9; (Azcue and Nriagu, 1994), 3. Avicena of Persia (980–1037) advocated the use of arsenic for skin and lung diseases. Nikolaos Myrepsos of Arabia utilized arsenic to treat plague victims (Nriagu, 2002), 9. In sixteenth-century Europe, Paracelsus even described a recipe containing arsenic compounds that was placed in a necklace to supposedly protect the wearer against the plague (Azcue and Nriagu, 1994), 10.

Orpiment and realgar also had 'medicinal applications' in ancient India and China (Nriagu, 2002), 4-5. Many traditional Chinese medicines, which are still used today, contain unacceptably high concentrations of arsenic (Nriagu, 2002), 5. Indian texts written during the time of Buddha (563–483 BC) promoted the beneficial use of both orpiment and realgar for internal and external medical treatments (Azcue and Nriagu, 1994), 3. Some pregnant Indian women believed that arsenic potions increased their chances of having a son (Nriagu, 2002), 18.

Arsenic continued to be used in various 'pharmaceuticals' in the nineteenth century, including 'treatments' for almost every major disease, including: rheumatism, malaria, diabetes, scarlet fever, diphtheria, influenza, asthma, syphilis, hay fever, bronchitis, pneumonia, morning sickness, chorea, skin disorders, leukemia and other cancers, snake bites, rabies, high blood pressure, tuberculosis, and heart diseases (Nriagu, 2002), 10-11; (Gorby, 1994), 5-6. In particular, popular arsenic 'pharmaceuticals' included Fowler's (1 % potassium arsenite), Donovan's (arsenic iodide), and Valagin's solutions (arsenic trichloride) (Azcue and Nriagu, 1994), 10; (Nriagu, 2002), 11. Fowler's solution continued to be sold in the United States until the 1950s (Gorby, 1994), 6. As expected, the effectiveness of these arsenic drugs was highly variable. In many cases, the arsenic 'medicines' were more deadly and greater health threats than the diseases that they were supposedly curing (Nriagu, 2002), 11-12.

Since the 1850s, there have been reports from the Styria region of the Austrian Alps of individuals deliberately consuming arsenic. These individuals believed that arsenic would maintain health, improve their complexion, and increase their stamina in mountain climbing (Meharg, 2005), 101-102. Arsenic eaters would begin by consuming small amounts of the poison and increasing the dosages over the years. Despite consuming up to 400 mg day^{-1} of As_2O_3, fatal poisonings were rare (Meharg, 2005), 102. Individuals that stopped eating arsenic would soon suffer from withdraw symptoms that were similar to slight arsenic poisoning, including: loss of appetite, indigestion, painful bowels, and difficulty in breathing ((Meharg, 2005), 103; Chapter 4). The habits of the arsenic eaters caused many nineteenth-century European government officials, medical professionals, and others to seriously underestimate the dangers of arsenic in commercial products, drugs, food, and the environment.

At the end of the nineteenth and into the early twentieth centuries, a number of *organoarsenicals* were introduced as medicines. Arsphenamine (under the trade name Salvarsan 606) was the principal treatment for syphilis from 1909 until its replacement with penicillin in the 1940s (Gorby, 1994), 6; (Nriagu, 2002), 11; (Azcue and Nriagu, 1994), 10; (Meharg, 2005), 104–105. Today, arsenic medicines have been largely replaced by less toxic compounds (Gorby, 1994), 6. However, arsenic compounds are still recognized as having beneficial uses in treating certain diseases, including some types of leukemia ((Francesconi and Kuehnelt, 2004), 374; (Oremland and Stolz, 2003), 939; (Ben Zirar *et al.*, 2007); Chapter 4). For example, As_2O_3 is utilized to treat patients with acute promyelocytic leukemia (Zhu *et al.*, 1999; Shen *et al.*, 1997).

5.4.2   Pesticides and agricultural applications

5.4.2.1   *Pesticides and herbicides*

For centuries, farmers have noticed that certain arsenic compounds are effective in killing pests and unwanted plants. In 1600s, the *Chinese Encyclopedia of Medicine* noted that unspecified arsenic compounds could kill insects infesting rice crops (Azcue and Nriagu, 1994), 9. The widespread use of arsenic pesticides began in the nineteenth century supposedly when a frustrated farmer threw Paris Green (copper *arsenate*, $Cu(AsO_2)_2Cu(C_2H_3O_2)$) paint on his beetle-infested potato plants. Arsenic compounds were found to be effective in killing a large variety of insects, spiders, leeches, worms, and other pests. By 1872, 'London purple' (a mixture of calcium arsenate, *arsenite* (inorganic As(III)), and organic matter) largely replaced Paris green as a pesticide. Paris green and London purple were also fairly toxic to plants. The toxicity was reduced through the development of lead arsenate in 1892. Although there were health concerns over the use of lead and arsenate (As(V)) pesticides, proponents argued that they did more good than harm and that no suitable safer alternatives were available. In 1906, calcium arsenate became the dominant pesticide for protecting cotton crops until the development of organic insecticides in the 1940s (Nriagu, 2002), 12–13.

Arsenic-bearing pesticides and herbicides may accumulate in agricultural soils from long-term use (Embrick *et al.*, 2005; Merwin *et al.*, 1994; Folkes, Kuehster and Litle, 2001). The movement of arsenic into the natural environment was revealed by an increase in its average concentration in American tobacco from about $10 \, mg \, kg^{-1}$ in 1917 to more than $50 \, mg \, kg^{-1}$ by 1952 (Nriagu, 2002), 10. Although organoarsenicals often have low toxicity to humans, they can degrade into more toxic inorganic forms in the environment (Nriagu, 2002), 20. Arsenic contamination may also occur at manufacturing sites from spills and releases ((Cancès *et al.*, 2005; Arai *et al.*, 2006; Keimowitz *et al.*, 2005; Davis *et al.*, 1994); Chapter 3). Furthermore, serious human health and environmental problems can arise from the improper disposal of arsenic-bearing materials. In one serious incident, *groundwater* contaminated from an arsenic insecticide dump poisoned 11 individuals in western Minnesota, United States (Gorby, 1994), 3. Poisoning incidences and environmental damage resulted in the banning of inorganic arsenic in US pesticides during the 1980s and 1990s (Welch *et al.*, 2000), 590.

5.4.2.2 *Fertilizers*

Fertilizers may inherit arsenic from their source materials, including phosphate minerals and *mine tailings*. As discussed in Chapters 2 and 3, inorganic As(V) and phosphorus have similar chemical properties and trace amounts of As(V) often occur in phosphate minerals that are used in the manufacturing of fertilizers. Specifically, Campos (2002) listed the arsenic contents of four commercial synthetic fertilizers, which ranged from 0.86 ± 0.01 to $13.42 \pm 0.06 \, mg \, kg^{-1}$. Although arsenic concentrations in phosphate fertilizers may not be extremely high, the phosphate from the fertilizers could desorb and mobilize arsenic in soils, where the arsenic largely originated from earlier extensive applications of pesticides ((Peryea and Kammereck, 1997); Chapter 3).

Ironite, which is produced from mine tailings, is an iron and nitrogen supplement for plants in croplands, lawns, and gardens (Williams *et al.*, 2006). After studying three batches of ironite, Williams *et al.* (2006) concluded that arsenic was primarily associated with iron (oxy)(hydr)oxides and relict arsenopyrite (FeAsS). Arsenopyrite is unstable in aerated water and iron (oxy)(hydr)oxides may decompose and release arsenic under slightly reducing conditions ((Williams *et al.*, 2006), 4877–4878; Chapter 3). Under US EPA regulations, the *toxicity characteristic leaching procedure* (TCLP) is typically used to determine the toxic *hazardousness* of wastes and other solid materials (Appendix E). Although ironite is exempt from hazardous waste regulations (Dubey and Townsend, 2004), 5400; (US EPA, 2002a), the product often fails or nearly fails the TCLP for arsenic (Williams *et al.*, 2006; Dubey and Townsend, 2004). Therefore, ironite could release toxic concentrations of arsenic into local surface waters and groundwaters ((Williams *et al.*, 2006); Chapter 3).

5.4.2.3   *Livestock applications*

5.4.2.3.1   Roxarsone   Roxarsone ($C_6H_6AsNO_6$ or 3-nitro-4-hydroxyphenylarsonic acid) is often fed to poultry and other livestock to control intestinal parasites, improve feed efficiency, and accelerate growth and weight gain (Nriagu, 2002; Garbarino *et al.*, 2003), 20; (Gorby, 1994), 3. In 2000, 70 % of the chickens in the United States or about 5.8 billion animals were fed roxarsone, which typically results in about 10–50 mg kg^{-1} of arsenic (mostly as unaltered roxarsone) in chicken litter (Garbarino *et al.*, 2003; Brown, Slaughter and Schreiber, 2005).

Poultry litter is often applied to agricultural fields as fertilizer. Up to 70–90 % of roxarsone-originating arsenic in the litter is water-soluble (Garbarino *et al.*, 2003; Han *et al.*, 2004). Therefore, arsenic in poultry litter fertilizer could migrate into water and organisms surrounding application sites (Rutherford *et al.*, 2003). Microorganisms will also eventually degrade roxarsone into more toxic inorganic As(V) after its applied to fields under *aerobic* conditions (Garbarino *et al.*, 2003). In one study by Garbarino *et al.* (2003), laboratory water extracts of chicken litter contained 91 % roxarsone, 1.5 % *dimethylarsinate* (DMA, $(CH_3)_2AsO(OH)$), 1.1 % inorganic As(V), 0.8 % inorganic As(III), and 5.6 % uncharacterized arsenic compounds. Adding 50 wt % water to litter and composing it at 40 °C largely converts the roxarsone to inorganic As(V) in about 30 days (Garbarino *et al.*, 2003). Under *anaerobic* conditions, roxarsone eventually biodegrades to highly toxic inorganic As(III) and, to a lesser extent, inorganic As(V) (Cortinas *et al.*, 2006).

5.4.2.3.2   Livestock dips   Through much of the twentieth century, arsenic trioxide (As_2O_3) was used in dips to control parasites on cattle, sheep, and horses (Nriagu, 2002), 16; (Reisinger, Burris and Hering, 2005). Groundwaters around the locations of old dipping facilities are often contaminated with arsenic. In particular, groundwater with dissolved arsenic concentrations up to 1.1 mg L^{-1} occurs around a former cattle dipping facility at Eglin Air Force Base, Florida. The facility dates back to the 1930s (Reisinger, Burris and Hering, 2005).

5.4.3   Chemical weapons

Arsenic has been utilized in chemical weapons for centuries. The ancient Chinese added arsenic sulfides to 'smoke bombs'. Marcus Gracchus mixed arsenic with incendiaries to burn the Roman fleet (Nriagu, 2002), 5; (Azcue and Nriagu, 1994), 11. In the 1840s, British settlers in Australia poisoned food with arsenic to commit genocide against Aboriginal people (Nriagu, 2002), 18. Also in the nineteenth century, arsenic was routinely added to lead shot to increase its sphericity, which further increased its toxicity (Nriagu, 2002), 18-19. During World War I, lewisite, an arsine gas (2-chlorovinyldichloroarsine, $C_2H_2AsCl_3$), produced skin lesions on its victims ((Nriagu, 2002), 18; (Corwin, David and Goldberg, 1999), 36; Chapter 2). Less toxic arsenic gases are still used to control riots. Cacodylic acid (Silvicide or Orange Blue) was a forest defoliant in the United States–Vietnam War (Nriagu, 2002), 18. Soils also have been contaminated with arsenic from the development, manufacturing, storage, and disposal of chemical weapons (Corwin, David and Goldberg, 1999; Maki *et al.*, 2006).

5.4.4   Embalming fluids

Arsenic was used in embalming fluids in Europe from about AD 650 until the early twentieth century (Spongberg and Becks, 2000), 314. During the US Civil War (1861–1865), arsenic became a popular embalming chemical. Bodies needed to be preserved for medical schools or until they could be delivered to relatives and buried. In some cases, as much as 1.4 kg (three US pounds) of arsenic were added to every corpse (Fetter, 1993), 275. By 1910, the US federal government banned arsenic in embalming fluids because it could interfere with forensic examinations in suspected poisoning cases. The past use of arsenic embalming fluids could result in significant contamination of groundwaters, soils, and sediments in some cemeteries ((Fetter, 1993), 275; (Spongberg and Becks, 2000); Chapter 3).

5.4.5 Paints and dyes

Orpiment and realgar have been used as pigments and cosmetics for millennia (Nriagu, 2002), 4. The ancient Egyptians decorated walls with orpiment (Nriagu, 2002), 3. Orpiment was even found in a linen bag in King Tutankhamen's tomb, which dates to 1324 BC. The mineral does not naturally occur in Egypt, but must have been imported from Persia, Asia Minor, or elsewhere in Asia (Azcue and Nriagu, 1994), 9. The ancient Greeks used realgar in red pigments and orpiment in yellow (Nriagu, 2002), 4. In the fourth century BC, Aristotle referred to the red color of realgar by naming the mineral sandarach, where the root word 'sand' means red (Azcue and Nriagu, 1994), 3. At the same time, Aristotle recognized that realgar is poisonous (Azcue and Nriagu, 1994), 9. The Greek geographer Strabo (c. 63 BC–AD c. 24) also remarked that the mining of arsenic compounds at Pompeiopolis in Asia Minor was so toxic that only slaves were used to perform the task (Azcue and Nriagu, 1994), 9.

During the Middle Ages, orpiment was made into gold paint to decorate European manuscripts (Nriagu, 2002), 7. However, the orpiment paint tended to turn black from air exposure when applied to walls (Azcue and Nriagu, 1994), 8-9. Yellow orpiment was also mixed with blue pigments to form various shades of green. The frequent use of arsenic compounds in pigments probably contributed to widespread poisoning among medieval painters (Nriagu, 2002), 7.

Beginning in the 1800s, several additional arsenic compounds were extensively utilized as coloring agents. In addition to the natural orpiment and realgar, synthetic compounds included: Mineral Blue ($CuKAsO_4$), Scheele's Green ($CuHAsO_3$), and Paris Green (Azcue and Nriagu, 1994), 11; (Nriagu, 2002), 16. In the nineteenth and early twentieth centuries, arsenic pigments were found in a wide variety of products, including: Christmas ornaments, books, magazines, candles, lampshades, drinking glasses, fabrics, wrapping paper, toys, wallpaper, and paint. Arsenic pigments and coloring agents were responsible for a large number of unintentional poisonings in the nineteenth and twentieth centuries. In Paris alone from 1828 to 1829, about 4000 people acquired neuritis, digestive disorders, and skin diseases from exposure to arsenic. In 1842, 14 children in Dublin, Ireland were reportedly poisoned from ornaments colored with copper arsenite (Nriagu, 2002), 17. There were also accounts of individuals being poisoned while reading in bed from the light of arsenic-bearing candles (Meharg, 2005), 85. Numerous individuals were also poisoned from arsenic dyes in clothing and food (Meharg, 2005), 85–89. Poisoning from arsenic food coloring became so common in nineteenth century Scotland that widespread aversions to green-colored foods among the Scottish population continued well into the twentieth century (Meharg, 2005), 86–87. Like most other arsenic products, arsenic pigments and dyes were replaced by less toxic compounds in the late nineteenth and twentieth centuries.

Poisoning events involving arsenic pigments in wallpaper were common in nineteenth century Europe (Cullen and Reimer, 1989), 714; (Frankenberger and Arshad, 2002), 370. Mold growing on the wallpaper converted arsenic in green and other wallpaper pigments into 'Gosio gas', which consists of poisonous and volatile *trimethylarsine* ($(CH_3)_3As$) (Cullen and Reimer, 1989), 714; (Frankenberger and Arshad, 2002), 370; (Craig, Eng and Jenkins, 2003), 38. Individuals became ill or even died from inhaling the poisonous vapors.

The attractiveness of the bright arsenic colors and the development of cheap arsenic pigments and printing methods led to the widespread use of wallpaper in British and other European homes in the nineteenth century. A square meter of wallpaper would often contain 25–35 g of arsenic and by 1858 some 260 million km^2 of arsenic-bearing wallpaper hung in British homes (Meharg, 2005), 67–68. Wallpaper use further increased in 1861 after prices declined due to the British government abolishing duties on paper imports (Meharg, 2005), 67. As early as 1815, arsenic wallpaper was identified as a serious health hazard in Germany (Meharg, 2005), 69. Poisoning events, especially among children, became quite common in British homes and wallpaper factories, and continued to as late as 1931 when two English children died from wallpaper that had been hung in the nineteenth century (Meharg, 2005), 69. Despite numerous poisoning reports and warnings in the nineteenth-century popular press and medical journals, the British government took no action to protect the public from wallpaper and other arsenic products until 1895 (Meharg, 2005), 69, 89. Nevertheless, news about the dangers of arsenic-bearing products eventually spread throughout British society, which led to a decline in the sales of arsenic-bearing wallpapers and other products in the late nineteenth century. In some cases, businesses responded to the sale declines by offering nonarsenic alternatives (Meharg, 2005), 80–81. The elimination of arsenic in British wallpapers in the 1890s was due more to pressures from the market place than government regulations (Meharg, 2005), 69.

Englishman William Morris (1834–1896) was one of the leading wallpaper manufacturers of the nineteenth century. He and his family also made a fortune from the mining and smelting of copper and the recovery and selling of arsenic from their smelters (Meharg, 2005), 137-145. Although he and his family made a lot of money from arsenic and many individuals died from their businesses, Morris became a Marxist ecologist and an arch-critic of capitalism in the 1880s (Meharg, 2005), 144–146. Morris was a prolific writer. However, perhaps deliberately, he left few written records about any knowledge that he may have had about the fatal effects of his wallpaper products or how his wallpaper factories compared with the terrible pollution and deplorable working conditions that were commonly associated with the wallpaper industry (Meharg, 2005), 79, 137. His personal life also gave few clues about his attitudes toward the safety of his products. Although he used his wallpapers in some of his rooms, Morris preferred wall hangings, murals, or even whitewash on the walls of his homes (Meharg, 2005), 75–76. Like other wallpaper manufacturers, public concerns and demands caused Morris' company to begin to offer no arsenic alternatives in the 1870s (Meharg, 2005), 80–81. Some surviving letters from 1885 suggest that Morris was skeptical of wallpaper poisoning. However, contemporary newspapers were too full of well-documented poisoning accounts to honestly dismiss the issue (Meharg, 2005), 81–83.

5.4.6 Wood treatment

In the twentieth century, a variety of arsenate-based preservatives were developed and commercialized to protect wood from microorganisms, fungi, wood-feeding insects, and marine borers. The preservatives included chromated copper arsenate (CCA), ammoniacal copper arsenate, and ammoniacal copper zinc arsenate (Pedersen *et al.*, 2005), 332; (Cox, 1991), 2; (Hingston *et al.*, 2001), 54; (Morrell, Keefe and Baileys, 2003). Since their introduction in India in the 1930s, CCA preservatives became very popular in the United States, Europe, and elsewhere (Morrell and Huffman, 2004), 119; (Hingston *et al.*, 2001). In 1993, about 80 % of US lumber were treated with CCA (Sarahney, Wang and Alshawabkeh, 2005), 642. CCA is not a single compound, but various mixtures of arsenic, chromium, and copper oxides. Beginning in the 1960s, the 'C variety' was most commonly used in the United States. This variety consisted of a mixture of about 30–38 wt % As(V) pentoxide (As_2O_5), 44.5–50.5 wt % Cr(VI) oxide (CrO_3), and 17–21 wt % Cu(II) oxide (CuO) (Townsend *et al.*, 2003), 790. The compounds were applied to the wood by pressure treatment (Cox, 1991), 2. Typically, $1 m^3$ of CCA-treated wood contains about 1.41 kg of As(V) (Morrell and Huffman, 2004), 120. After treatment, the As(V) probably exists as $CrAsO_4$ (Nico *et al.*, 2004), 5253. However, once the wood is exposed to the elements or burned, the arsenic could partially reduce to more toxic and mobile As(III) (Helsen and Van den Bulck, 2004), 286; (Khan *et al.*, 2004).

Arsenic may be released from the *weathering* of CCA-treated wood during normal use overtime or from the landfilling, incineration, or recycling of wood debris (Chapters 3 and 7; (Khan *et al.*, 2006a; Khan *et al.*, 2006b)). Although there are reports of children experiencing arsenic poisoning from playing in soils near CCA-treated wood (Nriagu, 2002), 20, other studies indicate that arsenic exposure from the wood and associated soils is negligible (Pouschat and Zagury, 2006; Nico *et al.*, 2006). If spills occur at CCA manufacturing sites, arsenic would tend to collect on organic matter in the *A horizons* of soils and on iron and other (oxy)(hydr)oxides in the *B and C horizons* (Chapter 3; (Lund and Fobian, 1991)). Toxic arsenic fumes may be released from the burning of CCA-treated wood ((Wasson *et al.*, 2005); Chapter 7). In one case, arsenic from the burning of treated plywood in a poorly ventilated cabin poisoned a Wisconsin, US family (Gorby, 1994), 3.

Although TCLP leachates of the wood often exceed the $5 \, \mathrm{mg \, L^{-1}}$ regulatory limit for arsenic ((Townsend, Dubey and Solo-Gabriele, 2004), 171-172; (Stook *et al.*, 2005); Appendix E), spent CCA-treated wood is normally exempt under current US federal regulations dealing with the disposal of hazardous wastes (40 *US Code of Federal Regulations* (CFR) 261.4(b)(9)). Nevertheless, some US states (such as Minnesota) have adopted stricter disposal regulations. Furthermore, individual landfill operators may not accept the material (Oskoui, 2004), 241. If allowed, CCA-treated wood is often landfilled with construction and demolition wastes (Townsend, Dubey and Solo-Gabriele, 2004; Oskoui, 2004), 241.

Under a voluntary agreement between the US EPA and manufacturers, CCA was discontinued in virtually all consumer products in the United States after 2003 (Wasson *et al.*, 2005), 8865; (Townsend *et al.*, 2003), 781. Additionally, CCA wood preservatives have been banned in Norway and Denmark (Ottosen, Pedersen and Christensen, 2004), 296. Bans are also expected in other countries. Even with the implementation of voluntary and regulatory bans, the preservatives still commonly occur in fence posts, playground equipment, decking, and other outdoor wood (Cox, 1991), 2. The life expectancy of CCA-treated wood is at least 30 years in terrestrial environments and about 15 years in salt water (Christensen *et al.*, 2004), 228; (Hingston *et al.*, 2001), 54. Thus, the issues dealing with the disposal of CCA-treated wood will persist for decades. The challenge is then to find suitable alternative preservatives and develop effective and economical technologies that will properly dispose of arsenate-treated wood wastes (Illman and Yang, 2004), 259; (Helsen and Van Den Bulck, 2005). The amount of wood wastes is often quite high, especially at sites affected by natural disasters. Hurricane Katrina, which struck the New Orleans, US, region in 2005, produced about 12 million m^3 of wood debris, including 1740 metric tons of arsenic in the treated lumber (Dubey, Solo-Gabriele and Townsend, 2007).

Even if landfilling is still a widely legal option, the volume of wood wastes will probably begin to curtail landfilling and promote the development of alternative CCA-waste management technologies that involve incineration, biotreatment, and/or extraction (Chapter 7). Helsen and Van Den Bulck (2005) provide a concise review of landfilling alternatives for CCA-treated wood, especially thermal methods. Most of the alternatives are still undergoing commercialization and their future marketability will greatly depend on whether any local landfills are still willing to accept CCA-treat materials and, if so, the landfilling costs.

The toxicity of As(V) wood preservatives has led researchers to seek less hazardous alternatives, which may contain octaborate tetrahydrate or silver compounds (Townsend, Dubey and Solo-Gabriele, 2004) 175. When considering any alternative, the entire life cycle of the wood product must be considered including manufacturing, use, disposal, potential environmental impacts, and costs. Although alternative wood preservatives are expected to be less hazardous, *leaching* studies, and toxicity assays in Stook *et al.* (2005) show that alternative copper preservatives (alkaline copper quaternary, copper boron azole, copper citrate, and copper dimethyldithiocarbamate) could have greater negative impacts on aquatic environments than CCA. Furthermore, new preservatives are still likely to raise safety concerns related to how they are applied to wood, any undesirable odors or other negative impacts on wood quality, their long-term effectiveness, 'cradle to grave' costs, and regulatory requirements related to their production, use, and disposal (Townsend, Dubey and Solo-Gabriele, 2004).

5.4.7   Semiconductors

Highly pure elemental arsenic ($\geq$99.999 %) is used in the synthesis of gallium(III) and indium(III) arsenide (GaAs and InAs) semiconductors. InAs has applications in infrared devices and lasers, including those found in compact disk players. GaAs is utilized in solar cells, light-emitting diodes (LEDs), and other electronic equipment (Azcue and Nriagu, 1994; Brooks, 2008), 13. Although silicon is cheaper and less toxic, GaAs components are faster and have more applications (Azcue and Nriagu, 1994), 13.

The production of arsenide semiconductors uses highly toxic arsine gas (AsH_3) (Chein *et al.*, 2006). Small amounts of the gas are released into the atmosphere at the manufacturing facilities. Once in the atmosphere, arsine oxidizes to As_4O_6. The As_4O_6 subsequently deposits on atmospheric particles that could be inhaled (Chein *et al.*, 2006), 1901–1902.

5.5   Increasing health, safety, and environmental concerns

Over the centuries, numerous cases of arsenic poisoning from commercial products, mining, and industrial facilities have been recorded. In particular, arsenic emissions from coal combustion facilities and smelters have long been identified as serious environmental and health threats. In 1661, John Evelyn recognized that the burning of New Castle (English) coals caused arsenic poisoning among some of the local inhabitants (Nriagu, 2002), 15. Lead, copper, tin, and gold ores often contain significant amounts of arsenic, which are released during smelting ((Nriagu, 2002), 3; Chapter 3). In 1812, British physician John Paris noticed cancers in tin and copper workers and farm animals near copper smelters. He speculated that the cancers were due to arsenic associated with the metals (Nriagu, 2002), 8. Although nineteenth-century British courts and government officials usually did nothing to stop the deleterious impacts of arsenic on the environment and society, Thomas Garland was prosecuted in 1851 for killing livestock and damaging crops around his arsenic plant in Perranarworthal, England (Meharg, 2005), 133. Around 1902, ranchers began to complain that livestock within a 13 km radius of the Anaconda, Montana, US copper smelter were being poisoned. As_4O_6 emissions from the smelter were probably responsible (Nriagu, 2002), 15. The Anaconda Mining Company paid $US 330 000 in compensation to the ranchers and 'solved' the problem by raising a higher (180 m) smelter stack, which better dispersed the arsenic fumes into the environment (Meharg, 2005), 151.

As mentioned earlier, unintentional poisonings from food, water, and commercial products were fairly common in industrializing countries in the nineteenth and twentieth centuries. In 1858, arsenic was accidentally added to a batch of candy in Bradford, United Kingdom, which poisoned 200 people and killed 17 of them (Nriagu, 2002), 17. Arsenic-contaminated glucose was inadvertently put into some Manchester, United Kingdom, beer in 1900, which affected 6000 people and killed 80 (Nriagu, 2002), 17; (Gorby, 1994), 3. The pyrite, which was used to produce sulfuric acid as part of the production of glucose, contained natural arsenic (Gorby, 1994), 3. In 1857, 340 children were poisoned after drinking milk. The milk had been diluted with water that came from a boiler that had been 'cleaned' with alkaline arsenite (Nriagu, 2002), 17. Nearly a century later in 1955, dry milk with arsenic-contaminated sodium phosphate poisoned 13 419 Japanese children and 839 of them died (Nriagu, 2002), 20; (Gorby, 1994), 3. Also in Japan, soy sauce accidentally poisoned 400 individuals with arsenic in 1956 (Nriagu, 2002), 21. In 1898, arsenic from a gold mine in western Poland contaminated a town's water supply

(Nriagu, 2002), 15. The water contained up to $26 \, mg \, kg^{-1}$ of arsenic and about 60 people were poisoned. Symptoms included skin cancer (Reichensteiner's disease) (Nriagu, 2002), 15. As early as 1880, the Medical Society of London became so concerned about human exposure to toxic arsenic pigments in commercial products that they issued a warning and listed items that were likely to contain the poisons (Azcue and Nriagu, 1994), 11-12. Despite the eventual implementation of strict environmental and health regulations in many countries (Appendix E), poisoning catastrophes continued well into the late twentieth century. In 1984 alone, over 1200 cases of accidental arsenic poisoning were reported in the United States (Nriagu, 2002), 21.

One of the more well-known unintentional poisoning incidents of the twentieth century involved Clare Boothe Luce (1903–1987). She served as US Ambassador for Italy from 1954 to 1956. During her ambassadorship, she became ill (Shadegg, 1970), 262. An analysis of her urine detected high concentrations of arsenic. At first, the US government suspected that someone was trying to assassinate her. However, an investigation of her bedroom-office found that vibrations from walking into the room and from the embassy washing machine were causing dust to fall from the ceiling. The ceiling of the seventieth century embassy had lead arsenate in flaking green paint (Shadegg, 1970), 262–266; (Azcue and Nriagu, 1994), 11; (Cullen and Reimer, 1989), 714). She survived the poisoning and eventually died of a brain tumor in 1987.

5.6   Arsenic in crime

For centuries, arsenic has been widely used as a murder weapon (Goessler and Kuehnelt, 2002), 27. One of the earliest documented accounts of murder with arsenic involved Nero, who assassinated Britannicus to secure the throne of the Roman Empire in AD 55 (Nriagu, 2002), 5-6; (Gorby, 1994), 4; (Meharg, 2005), 41. After its synthesis and commercialization in Europe during the Middle Ages, arsenolite became a popular poison for suicide and murder. Arsenolite was cheap, tasteless, odorless, and small amounts would mimic death from natural causes (Nriagu, 2002), 8. Typically, the murder victim was given the poison over several months to create the appearance of a natural disease, such as cholera or dysentery (Meharg, 2005), 41. Once the victim was weakened, the murderer would administer the final fatal dose (Nriagu, 2002), 8. One of the first well-documented cases of deliberate arsenolite poisoning was when 'Charles the Bad' unsuccessfully tried to assassinate Charles VI, king of France in 1314 (Nriagu, 2002), 8. By 1383, arsenolite was widely available in France (Nriagu, 2002), 8. Eventually, the compound became so popular among the court of France that it was known as '*inheritance powder*' (*poudre de succession*); that is, individuals would kill off relatives so that they would not have to wait to obtain their inheritances (Nriagu, 2002), 9; (Gorby, 1994), 4. In the late seventeenth and early eighteenth centuries, Signora Toffana of Palermo and Naples offered her poisonous arsenic oil 'Manna of St. Nicholas of Bari' to wives who wished to eliminate their husbands (Nriagu, 2002), 8. She was eventually caught and executed by strangulation in 1709 (Meharg, 2005), 113. Over 600 individuals were murdered by various secret poisons (often called *Aqua della Toffana*) that were sold throughout Europe from 1630 to 1730 (Nriagu, 2002), 8. Sometimes arsenic was added to candles and lamps. Arsenolite begins to sublime at $135 \, °C$ and so if it is mixed in candle wax, for example, it will readily produce an odorless vapor that, with enough inhalation, would prove fatal. Leopold I of Austria might have been poisoned with such candles in 1670 (Azcue and Nriagu, 1994), 11.

In 1836, James Marsh developed an analytical method for detecting low concentrations of arsenic in various materials (Nriagu, 2002), 9. The technique allowed investigators to identify arsenic in autopsies and increase the number of murder convictions. Nevertheless, the Marsh and subsequent analytical techniques have often not been used unless physicians or other investigators first suspect arsenic poisoning. Even after the development of analytical techniques for arsenic and increased knowledge of the symptoms of arsenic poisoning, physicians and forensic scientists continued to misdiagnose arsenic poisoning as disease or death from natural causes (Young, 2000). In one mysterious case, famed American explorer Charles Francis Hall (1821–1871) apparently died from a severe illness during a voyage to the Arctic on the *USS Polaris*. In 1968, biographer Chauncey Loomis exhumed Hall's body from the Arctic permafrost and had Hall's finger nails and hair tested for arsenic (Petrides, 1999). The body contained high levels of the poison. Although Hall may have been accidentally poisoned with quack drugs, it is more likely that he was murdered by one or more unhappy crew members. Hall was a stern ship captain and had alienated a number of individuals during the voyage, including Sidney Budington (his second-in-command), Emil Bessels (leader of the scientific team), and Sergeant Frederick Meyer (the second science officer) (Petrides, 1999). Hall had ridiculed Budington during the voyage and Budington had no respect for him. Civilian Bessels thought that his scientific goals should determine the course of the voyage. Hall saw Bessels as a threat to his leadership and asserted his authority over him as ship captain. Meyer threatened to desert the ship after Hall ordered him to maintain the Captain's log and concentrate on ship navigation (Petrides, 1999), 25. In this case, Hall backed down after Bessels sided with Meyer. Although most of the crew were distraught by Hall's death, Meyer, Bessels, and Budington were not. Meyer and Budington showed open relief. Crewman Noah Hayes even claimed that Bessels laughed with joy over Hall's demise (Petrides, 1999), 26. Although it is likely that Hall was murdered with arsenic, there is inconclusive evidence to identify the killer(s).

In 1938–1939, advances in the forensic detection of arsenic allowed officials to uncover an extensive murder ring in and around Philadelphia, Pennsylvania, United States (Young, 2000). Four individuals were convicted of being involved in the arsenic poisoning of approximately 50–100 people for insurance money. Two of the convicts, Morris Bolber and Carina Favato, cooperated with authorities and received life in prison. The other two, cousins Paul and Herman Petrillo, were executed in 1940 and 1941, respectively (Young, 2000).

The common use of arsenic for murder inspired John Kesselring's 1939 play *Arsenic and Old Lace*, where some older women spike elderberry wine with arsenic and other poisons to kill off suitors (Nriagu, 2002), 8–9. In the late twentieth and early twenty-first centuries, arsenic continues to be a popular poison among murderers, including serial killers. For example, 718 people were poisoned in Argentina in 1987 when vandals added arsenic to meat in a butcher shop (Gorby, 1994), 3.

5.7 Poisoning controversies: Napoleon Bonaparte

Over the years, arsenic has been blamed for the illnesses and deaths of several politicians and other celebrities. In the case of Charles Francis Hall, suspicions of arsenic poisoning by accident or murder proved true. The arsenic poisoning of US Ambassador Clare Boothe Luce was unintentional and not an attempted assassination as first suspected. In contrast, the exhumation and autopsy of US President Zachary Taylor found no evidence of arsenic poisoning.

Perhaps, the most controversial death possibly involving arsenic is that of Napoleon Bonaparte (1769–1821), the French Emperor. Despite extensive studies over the past several decades, controversies still continue over Napoleon's death and whether accidental poisoning or assassination with arsenic was involved. Most historians have traditionally argued that Napoleon Bonaparte died of gastric ('stomach') cancer in exile on the island of St. Helena in 1821. The autopsy report of Napoleon's death was not questioned for many years because Napoleon's father probably died of the disease (Lugli *et al.*, 2007), 52. Since 1961, however, analyses of Napoleon's hair at the time of his death found high concentrations of arsenic in some, but not all, of the samples. Some experts suspected that Napoleon died of arsenic poisoning, which could have resulted from a deliberate assassination, inappropriate medical treatments, or accidental poisoning. In particular, Mari *et al.* (2004) argue that Napoleon died from *torsades de pointes*, which results from a critical lack of potassium in the body due to the improper administering of medicines containing arsenic and other toxic substances.

Other researchers have speculated that Napoleon unintentionally died from exposure to arsenic in his food, drink, or surroundings, in particular, green pigments in the wallpaper of the bedroom of his exile residence (Azcue and Nriagu, 1994), 11. As mentioned earlier, mold growing on wallpaper containing arsenic pigments may release potentially deadly concentrations of 'Gosio gas'. Although analyses of the green wallpaper from Napoleon's residence suggest that the As(III) concentration was high enough to possibly cause illness, the concentration was probably not sufficient to kill individuals (Jones and Ledingham, 1982).

Weider and Fournier (1999) are strongly convinced that Napoleon was murdered with arsenic and mercury. They argue that Napoleon was assassinated in two stages. At first, he was weakened with arsenic. Finally, he was killed with a mixture of calomel (HgCl) and a drink containing bitter almonds, which produced mercuric cyanide in his stomach. Weider and Fournier (1999) believe that Comte de Montholon was the leading suspect in the alleged assassination. In a letter to his wife which refers to one of Napoleon's outdoor pastimes, Montholon supposedly wrote that calomel would soon end Napoleon's 'gardening efforts' (Weider and Fournier, 1999).

Besides arsenic in hair samples, Weider and Fournier (1999) also list numerous symptoms suffered by Napoleon that they claim are consistent with arsenic poisoning. Nevertheless, other researchers remain unconvinced that arsenic had any role in Napoleon's death. Some individuals (e.g. (Lugli *et al.*, 2007; Lugli, Lugli and Horcic, 2005)), still argue that gastric cancer is the most likely explanation. Critics of the arsenic poisoning *hypotheses* note that the hair shampoos and cosmetics of Napoleon's time often contained arsenic and that the presence of arsenic in hair does not necessarily indicate that the individual consumed or inhaled high concentrations of the element before death. Although laboratory cleaning procedures are supposed to remove arsenic contamination from the surfaces of hair, studies indicate that the cleaning methods may be ineffective (Lin, Alber and Henkelmann, 2004). Furthermore, arsenic contamination may diffuse into hair cores and mix with any core arsenic that originated from inhalation and the consumption of food and water (Lin and Henkelmann, 2003) 619. Elevated concentrations of arsenic were also found in samples of Napoleon's hair from 1814, which was before he went into exile and became noticeably ill (Lin, Alber and Henkelmann, 2004). Although Napoleon had symptoms that might result from chronic arsenic poisoning, several key symptoms were absent, including: hyperkeratotic skin lessons, Mee's lines in the finger and toe nails, and cancer tumors in the lungs and bladder ((Lugli *et al.*, 2007), 54; Chapter 4). Measurements of Napoleon's pants sizes further indicate that he lost at least 10 kg in the last year of his life, which is consistent with gastric cancer (Lugli, Lugli and Horcic, 2005). Whether or not arsenic hastened Napoleon's death, autopsy descriptions of the interior of his stomach indicated the presence of cancer (Lugli *et al.*, 2007).

5.8 Arsenic in prospecting, mining, and markets

5.8.1 Arsenic as a pathfinder element in prospecting

As discussed in Chapter 3, arsenic commonly occurs in sulfide minerals within hydrothermal and other ore deposits. The oxidation of sulfides in ore deposits and at mines may produce arsenic-rich drainage that results in environmental damage. At the same time, the weathering of arsenic-bearing sulfides in ore deposits may have important applications in prospecting for valuable metals. Due to its close geochemical associations with many metals and nonmetals, arsenic may be used as a *pathfinder element* in prospecting for gold (Craw *et al.*, 2002), 146; (Lett *et al.*, 1998; McConnell, 1999) and possibly other valuable metals and nonmetals, such as: copper, silver, zinc, antimony, bismuth, and platinum (Boyle and Jonasson, 1973; Zhang *et al.*, 2000). Under oxidizing conditions, arsenic tends to be more soluble in water than gold and many other metals. Therefore, arsenic may mobilize and accumulate in soils, sediments, water, and vegetation surrounding a buried ore deposit. Hypothetically, arsenic released by a gold deposit in the shallow subsurface of a hill could accumulate in surface soils and plants at the base of the hill. *Glaciers* can also partially excavate arsenic-bearing ores and distribute water-soluble arsenic over larger areas in *tills* and other glacial sediments. By measuring the arsenic concentrations in tills, plants, and other samples and plotting the results on maps, buried ore deposits might be located.

5.8.2   Arsenic mining, production, and market trends

In recent years, arsenic in ore deposits has become more of an environmental nuisance rather than a valuable resource. The worldwide production of arsenic increased until the 1930s and then began to fluctuate and generally decline as less toxic substitutes were found for pesticides and other commercial products (Han *et al.*, 2003), 396. The average price of $\geq 99\%$ As(0) in 2006 was $US 1.21 kg^{-1} (Hetherington *et al.*, 2008), 147. In 2007, the leading producers of As_2O_3, the primary arsenic compound, were China, Chile, Morocco, Peru, Mexico, Kazakhstan, Russia, Belgium, and France (Brooks, 2008). As of May 2008, government Internet sites for Australia, Canada, and New Zealand provided no current information on their arsenic production, imports, and exports. Several European Union nations continued to produce, export, and import As(0) and As_2O_3 as of 2006, although records were often incomplete (Hetherington *et al.*, 2008). The largest European producers of As_2O_3 in 2006 were Belgium and France. Each produced about 1000 metric tons (one metric ton = 1000 kg) (Hetherington *et al.*, 2008), 147. By far, the largest known European exporters of As(0) were Belgium and Luxembourg with a total of 2542 metric tons (Hetherington *et al.*, 2008), 147. The United Kingdom was the largest known importer of As(0) in Europe in 2004 with about 165 metric tons. However, As(0) imports to the United Kingdom dramatically declined in 2005 to only three metric tons and then increased to 49 metric tons in 2006 (Hetherington *et al.*, 2008), 148. The largest known European importer of As(0) in 2006 was Spain with 143 metric tons (Hetherington *et al.*, 2008), 148.

The United States has not intentionally mined and produced arsenic or its compounds since 1985 (Brooks, 2008). Currently, American demands for arsenic and its compounds are satisfied through imports. For 2002–2005, the United States primarily imported As(0)^0 from China and Japan, and As_2O_3 mainly from China and Morocco. The United States consumed about $9 million worth of arsenic and its compounds in 2007 (Brooks, 2008).

Before 1900, New Hampshire was the leading US source of arsenic ores (Peters *et al.*, 2006), 73. The first manufacturing of arsenolite in the United States occurred in 1901 in Everett, Washington, and was a byproduct of gold and silver smelting (Loebenstein, 1994). During the first half of the twentieth century, most American production of arsenic compounds was at ore smelters in Montana and Utah. Peak arsenic production in the United States occurred in 1944 (Loebenstein, 1994). Loebenstein (1994) reviews further details of the history of the production, regulation, and use of arsenic and its compounds in the United States.

Inorganic arsenic was banned in US pesticides in the 1980s and 1990s (Welch *et al.*, 2000), 590. In 1998, the production of CCA wood preservatives accounted for more than 90% of the As_2O_3 used in the United States (Leist, Casey and Caridi, 2000), 126. However, as previously discussed in this chapter, the voluntary agreement between the US EPA and wood preservative manufacturers discontinued the use of CCA in virtually all American consumer products after 2003 (Wasson *et al.*, 2005), 8865; (Townsend *et al.*, 2003), 781. Except for the use of highly pure arsenic in the production of semiconductors, the utilization of arsenic in American commercial products declined during the second half of the twentieth century (Leist, Casey and Caridi, 2000), 126.

5.9 Arsenic in coal and oil shale utilization and their by-products

5.9.1 Coal cleaning and combustion

Coal is often combusted to produce electricity and generate steam for heating and other applications. Prior to combustion, coals are sometimes washed with aqueous solutions to remove pyrite (FeS_2) and other sulfide minerals that, if burned, would produce toxic gases, such as sulfur dioxide (SO_2) and volatile arsenic species (Chapters 3 and 7). While coal cleaning is effective with massive sulfide fillings and veins, the process often misses widely dispersed, fine-grained sulfides in coal matrices (Yudovich and Ketris, 2005), 185. Recovered pyrite and other sulfide minerals from coal cleaning processes require careful disposal. Sulfide minerals stored in piles on the open ground often oxidize and generate toxic drainage that may contaminate surface waters and groundwater ((He, Liang and Jiang, 2002), 24; Chapter 3). In some American power plants, sulfide minerals from coal cleaning processes are mixed with *bottom ash* prior to disposal in ponds or landfills (Hower *et al.*, 1996), 409.

Pulverized coal combustion systems are most commonly used in power plants. In pulverized coal combustion, temperatures typically reach around 1480 °C at atmospheric pressure. In the past couple of decades, fluidized bed combustion (FBC) technologies have been commercialized. These combustors often use *limestone* bed materials to capture sulfur gases. They operate at about 880 °C and usually at atmospheric pressure (Smoot and Smith, 1985), 38.

5.9.2 Arsenic behavior during combustion

Arsenic is semivolatile in combustion systems (Mukherjee and Kikuchi, 1999), 64. The volatilization of arsenic during the combustion of coal has been studied with thermodynamic modeling, laboratory experiments, pilot-scale combustion, and in some cases direct measurements of *flue gases* at commercial utilities (Yudovich and Ketris, 2005), 172. Depending on combustion conditions and the chemistry of the coal, some arsenic may accumulate in bottom ashes ((Yudovich and Ketris, 2005), 173; Table 5.1). However, because of its semivolatility, arsenic mostly concentrates in flue gases and then largely condenses onto high-*surface-area*, fine-grained *flyash* particles especially during post-combustion cooling ((Mukherjee and Kikuchi, 1999; Hower *et al.*, 1999), 64; (Guo, Yang and Liu, 2004; Llorens, Fernández-Turiel and Querol, 2001; Galbreath and Zygarlicke, 2004); Table 5.1).

The presence of limestone in FBC bed materials or the injection of hydrated lime ($Ca(OH)_2$, portlandite), lime (CaO), or calcite ($CaCO_3$) into flue gases may be effective in removing volatile arsenic (also see Section 5.9.3 Post-Combustion Flue Gas Treatment). In bench-scale experiments with lime and calcium silicates, (Sterling and Helble (2003) found that the calcium compounds effectively remove As_4O_6 ('As_2O_3') vapors in either nitrogen gas (N_2) or air at 600–1000 °C. In other studies, Mahuli *et al.* (1997) concluded that hydrated lime is most effective in removing volatile arsenic in flue gases at 600–1000 °C and Jadhav and Fan (2001) state that lime removed arsenic at 300–1000 °C (Chapter 7). In results that are similar to Sterling and Helble (2003) and Mahuli *et al.* (1997), Li *et al.* (2007) concluded that lime can sorb As_4O_6 vapors and form calcium arsenates as $Ca_3(AsO_4)_2$ at 600–1000 °C, as shown in the following reaction:

$$6CaO + As_4O_6 \,(gas) + 2O_2 \,(gas) \rightarrow 2Ca_3(AsO_4)_2 \qquad (5.1)$$

Table 5.1 *Arsenic in coal utilization byproducts. Compare with coal analyses in Table 3.22.*

Location	Material	Arsenic concentration (milligrams per kilogram, mg kg^{-1})[a]	References
Australia: New South Wales, Queensland, and Western Australia	Flyashes: four samples from four power stations	6.58–22.30	Jankowski *et al.* (2006)
Australia: New South Wales, Queensland, and Western Australia	Flyashes and bottom ashes from four power stations	7.7 (a primary flyash); 19 (flyashes); 0.4–11 (bottom ashes)	Shah *et al.* (2007)
Canada	Electrostatic precipitator flyashes: Subbituminous coal: four power plants	17.5–52.0	Goodarzi (2006)
Canada	Flyashes from high-sulfur bituminous coal: two power plants	600 (electrostatic precipitator flyash); 551 (baghouse)	Goodarzi (2006)
Canada	Flyashes from low-sulfur bituminous coal: two power plants	7.14 (electrostatic precipitator flyash); 27.80 (baghouse ash)	Goodarzi (2006)
Canada: Nova Scotia: Lingan Power Plant	Coal, coal cleaning wastes, and combustion byproducts	468 (coal, Sydney field, Nova Scotia); 1126 (precombustion mill rejects, mostly pyrite); 39 (bottom ash); 766 (flyash)	Mukhopadhyay, Lajeunesse and Crandlemire (1996)
Czech Republic: Počerady power plant	Flyash	44.3	Kapička *et al.* (1999)
Denmark: Stigsnæsvarket power plant	Flyash (alkaline)	23.0	Van der Hoek, Bonouvrie and Comans (1994)
Greece: Drama Basin	Coal ashes	102–606	Papanicolaou *et al.* (2004)
Greece: Megalopolis	Coal ashes	20–30	Papanicolaou *et al.* (2004)
Greece: Serres Basin	Coal ashes	856–1131	Papanicolaou *et al.* (2004)

Location	Sample	As	Reference
Netherlands: Amercentrale power plant	Flyash (acidic)	9.69	Van der Hoek, Bonouvrie and Comans (1994)
Thailand: Mae Moh power plant	Mae Moh lignite flyash	198–312	Bashkin and Wongyai (2002)
Turkey: Fluidized bed reactor	Lignite bottom ashes and flyashes: five samples of each	42 ± 15 to 327 ± 29 (bottom ashes); 208 ± 19 to 1360 ± 115 (flyashes)	Ölmez et al. (2004)
USA: American Electric power plant, Beverly, Ohio and First Energy Corp. Edgewater plant, Lorain, Ohio	Fluegas desulfurization; Duct injection: seven samples	8.0–378	Kost et al. (2005)
USA: First Energy Corp. Edgewater plant, Lorain, Ohio	Fluegas desulfurization : Lime injection multistage burner: 12 samples	55.1–386	Kost et al. (2005)
USA: Illinois, Kentucky, Michigan, Ohio, Tennessee, and West Virginia	FBC: cyclone material: 10 samples	5.2–147	Kost et al. (2005)
USA: Indiana	Baghouse flyash from combustion of low-sulfur Danville Coal Member of Dugger Formation: eight samples	9.3–21.6	Mastalerz et al. (2004)
USA: Indiana	Flyash from combustion of high-sulfur Springfield Member of Petersburg Formation: three samples	52.7; 66.3; 82.8	Mastalerz et al. (2004)

(continued overleaf)

Table 5.1 (Continued)

Location	Material	Arsenic concentration (milligrams per kilogram, mg kg^{-1})[a]	References
USA: Kentucky	Bituminous Dean (Fire Clay) seam: flyashes and bottom ash	115 (economizer flyash); 78.7–102 (mechanical collector flyashes); 540–1650 (electrostatic precipitator flyashes); 18.8 (bottom ash)	Mardon and Hower (2004)
USA: Kentucky, Michigan, Ohio, and West Virginia	FBC: bed material: six samples	5.4–213	Kost et al. (2005)
USA: North Dakota: Beulah	Lignite: Gasification ash	31	McCarthy, Hassett and Manz (1986)
USA: North Dakota: Beulah: Antelope Valley Station power plant	Bottom ash	12	McCarthy, Hassett and Manz (1986)
USA: North Dakota: Beulah: Antelope Valley Station power plant	Scrubber ash (flyash, lime, and fluegas desulfurization sludge)	36	McCarthy, Hassett and Manz (1986)
USA: Ohio: McCracken power plant	Hydrated lime spray dryer ash (FGD sludge and flyash) from an eastern US bituminous coal: four samples	31.9–44.9	Taerakul et al. (2006)
USA: Ohio and New York	FGD: spray dryer product: 13 samples	6.9–165	Kost et al. (2005)

[a]Single values denote single analyses or average concentrations sometimes with one standard deviation. Analyses involving different samples are separated by semicolons. Low and high values are separated by hyphens.

The ability of lime to sorb As_4O_6 increases as temperatures increase between 400 and 1000 °C (Li *et al.*, 2007). The presence of either sulfur dioxide (SO_2) or carbon dioxide (CO_2) gases did not substantially interfere with the ability of lime to capture As_4O_6. $CaSO_4$ is also capable of sorbing at least some As_4O_6 (Li *et al.*, 2007).

5.9.3 Postcombustion flue gas treatment

Baghouse and *electrostatic precipitator* treatment systems are designed to remove flyash particles from flue gas before the gas exits into the atmosphere. By capturing arsenic-bearing flyashes, baghouses, and electrostatic precipitators may significantly reduce arsenic emissions to the air from coal combustion facilities (Table 5.1). Nevertheless, some very fine-grained, arsenic-bearing flyash particles will readily pass through flue gas treatment systems and enter the atmosphere, where they may be inhaled (Yudovich and Ketris, 2005), 180-181.

Gaseous arsenic species and SO_2 may be removed from post-combustion flue gases with spray dryer *flue gas desulfurization* (FGD) systems (Table 5.1). Typically, hydrated lime is sprayed into the flue gas. The process produces a sludge that is rich in calcium sulfites and/or calcium sulfates (Taerakul *et al.*, 2006). In the process of removing sulfur from flue gases, hydrated lime or other injected absorbents (such as, calcite or lime) frequently capture arsenic through the formation of calcium arsenates (Yudovich and Ketris, 2005; Mahuli *et al.*, 1997), 186; (Taerakul *et al.*, 2006; Sterling and Helble, 2003; Li *et al.*, 2007); Chapter 7). Provided that arsenic and other potentially toxic *trace elements* are immobile, FGD sludges have applications in soil amending and commercial products, such as wallboard.

5.9.4 Arsenic chemistry in coal combustion byproducts

In low-calcium flyashes, arsenic is typically associated with iron oxides and iron-rich silicate glasses (Goodarzi, 2006; Mukhopadhyay, Lajeunesse and Crandlemire, 1996; Taerakul *et al.*, 2006; Sterling and Helble, 2003; Mukhopadhyay *et al.*, 1999; Seames and Wendt, 2000). Calcium in flyashes may react with arsenic in flue gas to form calcium arsenates on the surfaces of flyash particles (Goodarzi, 2006), 1423. However, calcium originating from coals may occur as silicates in flyashes and calcium silicates are not as effective in removing arsenic vapors as lime or hydrated lime (Sterling and Helble, 2003).

Arsenic in bottom ashes, flyashes, and other coal combustion byproducts is usually As(V) (Goodarzi and Huggins, 2005; Huggins *et al.*, 1993), 487–488; (Shah *et al.*, 2007). However, detectable As(III) occurs in some flyashes and As_4O_6 ('As_2O_3') vapor is often present in combustion flue gas (Sterling and Helble, 2003; Huggins *et al.*, 2004), 1112; (Mahuli *et al.*, 1997), 3226. As_2O_3 has a boiling point of about 460 °C at atmospheric pressure and is much more volatile than As(0) (Mukherjee and Kikuchi, 1999), 64; (Mahuli *et al.*, 1997), 3226. Any As(III) in flyash probably results from the condensation of As_4O_6 vapors (Sterling and Helble, 2003; Hirsch *et al.*, 2000), 1112.

Arsenic-bearing flyash particles and other particulate matter of less than 2.5 μm in diameter ($PM_{2.5}$) are capable of deeply penetrating into human lungs (Chapters 3 and 4). Huggins *et al.* (2004) used *X-ray absorption near edge structure* (XANES) *spectra* to identify As(III) species in $PM_{2.5}$ particulate matter from coal combustion plants. The identification of arsenic in flyash may provide important information on respiratory health concerns and how arsenic species form during coal combustion. Specifically, XANES and associated analyses indicated that the $PM_{2.5}$ fraction from the combustion of an eastern United States (Pittsburgh) bituminous coal contained 310 mg kg^{-1} of arsenic (97 % As(V) and 3 % As(III), ±3 %), whereas the particles coarser than 2.5 μm contained only 100 mg kg^{-1} of arsenic, but more As(III) (94 % As(V) and 6 % As(III), ±3 %) (Huggins *et al.*, 2004).

5.9.5 Coal gasification

Coal gasification refers to the heating of the fossil fuel under low oxygen conditions to produce hydrogen as H_2 and other combustible gases (Lu, Granatstein and Rose, 2004), 5402. The process uses steam and/or low levels of air or oxygen as O_2 to avoid combustion. Coal gasification is performed in gasifier units or through *underground coal gasification* (UCG). UCG involves the in situ heating of subsurface coal seams. Although UCG avoids the costs and efforts of mining and transporting coal, the process produces arsenic-bearing byproducts that may contaminate local groundwaters (Shuqin *et al.*, 2004).

Synthetic gas resulting from coal gasification may be combusted as a fuel or used in the manufacturing of commercial products, such as methanol (Quinn *et al.*, 2006). The gasification of coal mostly results in the production of carbon monoxide (CO), H_2, N_2, carbon dioxide (CO_2), water, and hydrogen sulfide (H_2S). Methane (CH_4) concentrations are variable (0.1–10 %) (Lu, Granatstein and Rose, 2004), 5402. Possible gaseous forms of arsenic in synthetic gas include: As^0, As_4, As_2, AsO, AsS, and arsine (AsH_3) (Diaz-Somoano, López-Antón and Martínez-Tarazona, 2004), 1241; (López-Antón *et al.*, 2007; Liu *et al.*, 2006).

In the subsurface, pressure increases by about 1 000 000 Pa with every 100 m of depth (Liu *et al.*, 2006), 211. In oxygen-steam UCG at 1500 Pa, As_2 begins to form at about 250 °C and is the dominant volatile arsenic species at approximately 300–900 °C (Liu *et al.*, 2006), 211. Gaseous AsS occurs above 500 °C and becomes the dominant form of volatile arsenic from about 900–1300 °C (Liu *et al.*, 2006), 211. Gaseous AsO and As^0 develop above 800 °C and become dominant above 1300 °C. In air-blown gasification at the same pressure, gaseous As_4 may occur at 200–400 °C (Liu *et al.*, 2006), 211. With increasing pressure from 1500 to 10 000 Pa, gaseous AsS decreases and AsH_3 gas becomes more abundant (Liu *et al.*, 2006), 211–212.

Arsenic impurities may be removed from synthetic gas with various sorbents, such as: zinc ferrite ($ZnFe_2O_4$) or, under carefully controlled temperatures, mixtures of copper(II) oxide and carbon (Quinn *et al.*, 2006; Diaz-Somoano, López-Antón and Martínez-Tarazona, 2004). Specifically, zinc ferrite may capture As_4 vapors through the following reaction (Diaz-Somoano, López-Antón and Martínez-Tarazona, 2004):

$$4ZnFe_2O_4 + As_4 + 8H_2S + 8H_2 \rightarrow 4FeAs + 4ZnS + 4FeS + 16H_2O \qquad (5.2)$$

5.9.6 Oil shale utilization

The extraction of oil from *oil shales* and other fine-grained *sedimentary rocks* often produces high-arsenic wastewaters and solid wastes with significant water-soluble arsenic ((Fruchter *et al.*, 1980; Jones, Chapman and Jung, 1990). Specifically, Fruchter *et al.* (1980), 1380) obtained wastewaters with 4.1–17.4 mg L^{-1} of arsenic from the extraction of oil from Green River Formation (western United States) samples that contained about 49 mg kg^{-1} of arsenic. While most crude oils contain much less than 1 mg kg^{-1} of arsenic (Chapter 3), the Green River Formation samples produced an oil with a concentration of 31 mg kg^{-1}. In the process, about 5.3 % of the arsenic in the raw shale transferred into the oil ((Fruchter *et al.*, 1980), 1379). Arsenic in shale oil may be removed under reducing conditions with a mixture of nickel and molybdenum sulfides in alumina (Committee on Medical and Biologic Effects of Environmental Pollutants (1997), 75).

References

Arai, Y., Lanzirotti, A., Sutton, S.R. *et al.* (2006) Spatial and temporal variability of arsenic solid-state speciation in historically used lead arsenate contaminated soils. *Environmental Science and Technology*, **40**(3), 673–79.

Azcue, J.M. and Nriagu, J.O. (1994) Arsenic: historical perspectives, in *Arsenic in the Environment: Part I: Cycling and Characterization* (eds J.O. Nriagu), John Wiley & Sons, Ltd, New York, pp. 1–15.

Bashkin, V.N. and Wongyai, K. (2002) Environmental fluxes of arsenic from lignite mining and power generation in northern Thailand. *Environmental Geology*, **41**(8), 883–88.

Ben Zirar, S., Gibaud, S., Camut, A. and Astier, A. (2007) Pharmacokinetics and tissue distribution of the antileukaemic organoarsenicals arsthinol and melarsoprol in mice. *The Journal of Organometallic Chemistry*, **692**(6 Special Issue), 1348–52.

Boyle, R.W. and Jonasson, I.R. (1973) Geochemistry of arsenic and its use as indicator element in geochemical prospecting. *The Journal of Geochemical Exploration*, **2**(3), 251–96.

Brooks, W.E. (2008) *Arsenic*, US Geological Survey Mineral Commodity Summaries, January.

Brown, B.L., Slaughter, A.D. and Schreiber, M.E. (2005) Controls on roxarsone transport in agricultural watersheds. *Applied Geochemistry*, **20**(1), 123–33.

Campos, V. (2002) Arsenic in groundwater affected by phosphate fertilizers at São Paulo, Brazil. *Environmental Geology*, **42**(1), 83–87.

Cancès, B., Juillot, F., Morin, G. *et al.* (2005) XAS evidence of As(V) association with iron oxyhydroxides in a contaminated soil at a former arsenical pesticide processing plant. *Environmental Science and Technology*, **39**(24), 9398–405.

Chein, H., Hsu, Y.-D., Aggarwal, S.G. *et al.* (2006) Evaluation of arsenical emission from semiconductor and opto-electronics facilities in Hsinchu, Taiwan. *Atmospheric Environment*, **40**(10), 1901–907.

Christensen, I., Pedersen, A., Ottosen, L. and Ribeiro, A. (2004) Electrodialytic remediation of CCA-treated wood in larger scale, *Environmental Impacts of Preservative-Treated Wood*, Florida Center for Environmental Solutions, Conference, February 8–11, Gainesville, Orlando, FL, pp. 227–37.

Committee on Medical and Biologic Effects of Environmental Pollutants (1997) *Arsenic*, National Academy of Sciences, Washington, DC.

Cortinas, I., Field, J.A., Kopplin, M. *et al.* (2006) Anaerobic biotransformation of roxarsone and related N-substituted phenylarsonic acids. *Environmental Science and Technology*, **40**(9), 2951–57.

Corwin, D.L., David, A. and Goldberg, S. (1999) Mobility of arsenic in soil from the Rocky Mountain Arsenal area. *Journal of Contaminant Hydrology*, **39**(1-2), 35–58.

Cox, C. (1991) Chromated copper arsenate. *The Journal of Pesticide Reform*, **11**(1), 2–6.

Craig, P.J., Eng, G. and Jenkins, R.O. (2003) Occurrence and pathways of organometallic compounds in the environment – general considerations, in *Organometallic Compounds in the Environment*, 2nd edn (ed. P.J. Craig), John Wiley & Sons, Ltd, West Sussex, 1–55.

Craig, J.R., Vaughan, D.J. and Skinner, B.J. (2001) *Resources of the Earth: Origin, Use, and Environmental Impact*, Prentice Hall, Upper Saddle River, NJ, p. 520.

Craw, D., Koons, P.O., Horton, T. and Chamberlain, C.P. (2002) Tectonically driven fluid flow and gold mineralisation in active collisional orogenic belts: comparison between New Zealand and western Himalaya. *Tectonophysics*, **348**(1-3), 135–53.

Cullen, W.R. and Reimer, K.J. (1989) Arsenic speciation in the environment. *Chemical Reviews*, **89**(4), 713–64.

Davis, A., Kempton, J.H., Nicholson, A. and Yare, B. (1994) Groundwater transport of arsenic and chromium at a historical tannery, Woburn, Massachusetts, USA. *Applied Geochemistry*, **9**(5), 569–82.

Diaz-Somoano, M., López-Antón, M.A. and Martínez-Tarazona, M.R. (2004) Retention of arsenic and selenium during hot gas desulfurization using metal oxide sorbents. *Energy and Fuels*, **18**(5), 1238–42.

Dubey, B., Solo-Gabriele, H.M. and Townsend, T.G. (2007) Quantities of arsenic-treated wood in demolition debris generated by hurricane Katrina. *Environmental Science and Technology*, **41**(5), 1533–36.

Dubey, B. and Townsend, T. (2004) Arsenic and lead leaching from the waste derived fertilizer ironite. *Environmental Science and Technology*, **38**(20), 5400–404.

Embrick, L.L., Porter, K.M., Pendergrass, A. and Butcher, D.J. (2005) Characterization of lead and arsenic contamination at Barber Orchard, Haywood County, NC. *The Microchemical Journal*, **81**(1), 117–21.

Fetter, C.W. (1993) *Contaminant Hydrology*, Prentice Hall, Upper Saddle River, NJ, p. 458.

Folkes, D.J., Kuehster, T.E. and Litle, R.A. (2001) Contributions of pesticide use to urban background concentrations of arsenic in Denver, Colorado, U.S.A. *Environmental Forensics*, **2**(2), 127–39.

Francesconi, K.A. and Kuehnelt, D. (2004) Determination of arsenic species: a critical review of methods and applications, 2000–2003. *The Analyst*, **129**(5), 373–95.

Frankenberger, W.T., Jr. and Arshad, M. (2002) Volatilization of arsenic, in *Environmental Chemistry of Arsenic* (ed. W.T. Frankenberger Jr.), Marcel Dekker, New York, 363–80.

Fruchter, J.S., Wilkerson, C.L., Evans, J.C. and Sanders, R.W. (1980) Elemental partitioning in an aboveground oil shale retort pilot plant. *Environmental Science and Technology*, **14**(11), 1374–81.

Galbreath, K.C. and Zygarlicke, C.J. (2004) Formation and chemical speciation of arsenic-, chromium-, and nickel-bearing coal combustion PM2.5. *Fuel Processing Technology*, **85**(6-7), 701–26.

Garbarino, J.R., Bednar, A.J., Rutherford, D.W. *et al.* (2003) Environmental fate of roxarsone in poultry litter. I. Degradation of roxarsone during composting. *Environmental Science and Technology*, **37**(8), 1509–14.

Goessler, W. and Kuehnelt, D.D. (2002) Analytical methods for the determination of arsenic and arsenic compounds in the environment, in *Environmental Chemistry of Arsenic* (ed. W.T. Frankenberger Jr., Marcel Dekker, New York, 27–50.

Goodarzi, F. (2006) Characteristics and composition of fly ash from Canadian coal-fired power plants. *Fuel*, **85**(10-11), 1418–27.

Goodarzi, F. and Huggins, F.E. (2005) Speciation of arsenic in feed coals and their ash byproducts from Canadian power plants burning sub-bituminous and bituminous coals. *Energy and Fuels*, **19**(3), 905–15.

Gorby, M.S. (1994) Arsenic in human medicine, in *Arsenic in the Environment: Part II: Human Health and Ecosystem Effects* (ed. J.O. Nriagu), John Wiley & Sons, Ltd, New York, 1–16.

Guo, R., Yang, J. and Liu, Z. (2004) Thermal and chemical stabilities of arsenic in three Chinese coals. *Fuel Processing Technology*, **85**(8-10), 903–12.

Han, F.X., Kingery, W.L., Selim, H.M. *et al.* (2004) Arsenic solubility and distribution in poultry waste and long-term amended soil. *Science of the Total Environment*, **320**(1), 51–61.

Han, F.X., Su, Y., Monts, D.L. *et al.* (2003) Assessment of global industrial-age anthropogenic arsenic contamination. *Naturwissenschaften*, **90**(9), 395–401.

He, B., Liang, L. and Jiang, G. (2002) Distributions of arsenic and selenium in selected Chinese coal mines. *Science of the Total Environment*, **296**(1-3), 19–26.

Helsen, L. and Van den Bulck, E. (2004) Review of thermochemical conversion processes as disposal technologies for chromated copper arsenate (CCA) treated wood waste, *Environmental Impacts of Preservative-Treated Wood*, Florida Center for Environmental Solutions, Conference, February 8–11, Gainesville, Orlando, FL, pp. 277–94.

Helsen, L. and Van Den Bulck, E. (2005) Review of disposal technologies for chromated copper arsenate (CCA) treated wood waste, with detailed analyses of thermochemical conversion processes. *Environmental Pollution*, **134**(2), 301–14.

Hetherington, L.E., Brown, T.J., Idoine, N.E. *et al.* (2008) *European Mineral Statistics 2002-06*, British Geological Survey, Keyworth, Nottingham.

Hingston, J.A., Collins, C.D., Murphy, R.J. and Lester, J.N. (2001) Leaching of chromated copper arsenate wood preservatives: a review. *Environmental Pollution*, **111**, 53–66.

Hirsch, M.E., Sterling, R.O., Huggins, F.E. and Helble, J.J. (2000) Speciation of combustion-derived particulate phase arsenic. *Environmental Engineering Science*, **17**(6), 315–27.

Hower, J.C., Robertson, J.D., Thomas, G.A. *et al.* (1996) Characterization of fly ash from Kentucky power plants. *Fuel*, **75**(4), 403–11.

Hower, J.C., Trimble, A.S., Eble, C.F. *et al.* (1999) Characterization of fly ash from low-sulfur and high-sulfur coal sources: partitioning of carbon and trace elements with particle size. *Energy Sources*, **21**(6), 511–25.

Huggins, F.E., Huffman, G.P., Linak, W.P. and Miller, C.A. (2004) Quantifying hazardous species in particulate matter derived from fossil-fuel combustion. *Environmental Science and Technology*, **38**(6), 1836–42.

Huggins, F.E., Shah, N., Zhao, J. *et al.* (1993) Nondestructive determination of trace element speciation in coal and coal ash by XAFS spectroscopy. *Energy and Fuels*, **7**(4), 482–89.

Illman, B. and Yang, V. (2004) Bioremediation and degradation of CCA-treated wood waste, *Environmental Impacts of Preservative-Treated Wood*, Florida Center for Environmental Solutions, Conference, February 8–11, Gainesville, Orlando, FL, pp. 259–69.

Jadhav, R.A. and Fan, L.-S. (2001) Capture of gas-phase arsenic oxide by lime: kinetic and mechanistic studies. *Environmental Science and Technology*, **35**(4), 794–99.

Jankowski, J., Ward, C.R., French, D. and Groves S. (2006) Mobility of trace elements from selected Australian fly ashes and its potential impact on aquatic ecosystems. *Fuel*, **85**(2), 243–56.

Jones, D.R., Chapman, B.M. and Jung, R.F. (1990) Column leaching of unretorted and retorted oil shales and claystone form the rundle deposit: water leaching. *Water Research*, **24**(2), 131–41.

Jones, D.E.H. and Ledingham, K.W.D. (1982) Arsenic in Napoleon's wallpaper. *Nature*, **299**(5884), 626–27.

Kapička, A., Petrovský, E., Ustjak, S. and Macháčková, K. (1999) Proxy mapping of fly-ash pollution of soils around a coal-burning power plant: a case study in the Czech Republic. *The Journal of Geochemical Exploration*, **66**(1-2), 291–97.

Keimowitz, A.R., Zheng, Y., Chillrud, S.N. *et al.* (2005) Arsenic redistribution between sediments and water near a highly contaminated source. *Environmental Science and Technology*, **39**(22), 8606–613.

Khan, B.I., Jambeck, J., Solo-Gabriele, H.M. *et al.* (2006a) Release of arsenic to the environment from CCA-treated wood. 2. Leaching and speciation during disposal. *Environmental Science and Technology*, **40**(3), 994–99.

Khan, B.I., Solo-Gabriele, H.M., Dubey, B.K. *et al.* (2004) Arsenic speciation of solvent-extracted leachate from new and weathered CCA-treated wood. *Environmental Science and Technology*, **38**(17), 4527–34.

Khan, B.I., Solo-Gabriele, H.M., Townsend, T.G. and Cai, Y. (2006b) Release of arsenic to the environment from CCA-treated wood. 1. Leaching and speciation during service. *Environmental Science and Technology*, **40**(3), 988–93.

Kost, D.A., Bigham, J.M., Stehouwer, R.C. *et al.* (2005) Chemical and physical properties of dry flue gas desulfurization products. *Journal of Environmental Quality*, **34**(2), 676–86.

Leist, M., Casey, R.J. and Caridi, D. (2000) The management of arsenic wastes: problems and prospects. *Journal of Hazardous Materials*, **76**(1), 125–38.

Lett, R.E., Bobrowsky, P., Cathro, M. and Yeow, A. (1998) Geochemical pathfinders for massive sulphide deposits in the southern Kootenay Terrane, in *Geological Fieldwork 1997*, British Columbia Geological Division, Victoria, BC, 15-1–15-9.

Li, Y., Tong, H., Zhuo, Y. *et al.* (2007) Simultaneous removal of SO_2 and trace As_2O_3 from flue gas: mechanism, kinetics study, and effect of main gases on arsenic capture. *Environmental Science and Technology*, **41**(8), 2894–900.

Lin, X., Alber, D. and Henkelmann, R. (2004) Elemental contents in Napoleon's hair cut before and after his death: Did Napoleon die of arsenic poisoning? *Analytical and Bioanalytical Chemistry*, **379**(2), 218–20.

Lin, X. and Henkelmann, R. (2003) Contents of arsenic, mercury and other trace elements in Napoleon's hair determined by INAA using the k_0-method. *The Journal of Radioanalytical and Nuclear Chemistry*, **257**(3), 615–20.

Liu, S., Wang, Y., Yu, L. and Oakey, J. (2006) Thermodynamic equilibrium study of trace element transformation during underground coal gasification. *Fuel Processing Technology*, **87**(3), 209–15.

Llorens, J.F., Fernández-Turiel, J.L. and Querol, X. (2001) The fate of trace elements in a large coal-fired power plant. *Environmental Geology*, **40**(4-5), 409–16.

Loebenstein, J.R. (1994) *The Materials Flow of Arsenic in the United States*, United States Department of the Interior, Bureau of Mines, Information Circular 9382, p. 12.

López-Antón, M.A., Diaz-Somoano, M., Fierro, J.L.G. and Martínez-Tarazona, M.R. (2007) Retention of arsenic and selenium compounds present in coal combustion and gasification flue gases using activated carbons. *Fuel Processing Technology*, **88**(8), 799–805.

Lu, D.Y., Granatstein, D.L. and Rose, D.J. (2004) Study of mercury speciation from simulated coal gasification. *Industrial and Engineering Chemistry Research*, **43**(17), 5400–404.

Lugli, A., Lugli, A.K. and Horcic, M. (2005) Napoleon's autopsy: new perspectives. *Human Pathology*, **36**(4), 320–24.

Lugli, A., Zlobec, I., Singer, G. *et al.* (2007) Napoleon Bonaparte's gastric cancer: A clinicopathologic approach to staging, pathogenesis, and etiology. *Nature Clinical Practice Gastroenterology and Hepatology*, **4**(1), 52–57.

Lund, U. and Fobian, A. (1991) Pollution of two soils by arsenic, chromium and copper, Denmark. *Geoderma*, **49**(1-2), 83–103.

Mahuli, S., Agnihotri, R., Chauk, S. *et al.* (1997) Mechanism of arsenic sorption by hydrated lime. *Environmental Science and Technology*, **31**(11), 3226–31.

Maki, T., Takeda, N., Hasegawa, H. and Ueda, K. (2006) Isolation of monomethylarsonic acid-mineralizing bacteria from arsenic contaminated soils of Ohkunoshima Island. *Applied Organometallic Chemistry*, **20**(9), 538–44.

Mandal, B.K. and Suzuki, K.T. (2002) Arsenic round the world: a review. *Talanta*, **58**(1), 201–35.

Mardon, S.M. and Hower, J.C. (2004) Impact of coal properties on coal combustion by-product quality: examples from a Kentucky power plant. *The International Journal of Coal Geology*, **59**(3-4), 153–69.

Mari, F., Bertol, E., Fineschi, V. and Karch, S.B. (2004) Channelling the emperor: What really killed Napoleon? *Journal of the Royal Society of Medicine*, **97**(8), 397–99.

Mastalerz, M., Hower, J.C., Drobniak, A. *et al.* (2004) From in-situ coal to fly ash: a study of coal mines and power plants from Indiana. *The International Journal of Coal Geology*, **59**(3-4), 171–92.

McCarthy, G.J., Hassett, D.J. and Manz, O.E.. (1986) Technical basis for codisposal of gasification and combustion ash from the plants at Beulah, North Dakota, in *Fly Ash and Coal Conversion By-Products: Characterization, Utilization and Disposal. II* (eds G.J. McCarthy, F.P. Glasser and D.M. Roy), Materials Research Society Symposia Proceedings, Materials Research Society, Pittsburgh, PA, Vol. **65**, 301–10.

McConnell, J.W. (1999) *Results of Geochemical Mapping Employing High-Density Lake-Sediment and Water Sampling in Central Labrador. Geological Survey, Report 99-1*, Current Research Newfoundland Department of Mines and Energy, 17–39.

Meharg, A.A. (2005) *Venomous Earth: How Arsenic Caused the World's Worst Mass Poisoning*, Macmillan, New York, p. 192.

Mehrabi, B., Yardley, B.W.D. and Cann, J.R. (1999) Sediment-hosted disseminated gold mineralisation at Zarshuran, NW Iran. *Mineralium Deposita*, **34**(7), 673–96.

Merwin, I., Pruyne, L.P.T., Ebel, J.G., Jr. *et al.* (1994) Persistence, phytotoxicity, and management of arsenic, lead and mercury residues in old orchard soils of New York State. *Chemosphere*, **29**(6), 1361–67.

Morrell, J.J. and Huffman, J. (2004) Copper, chromium, and arsenic levels in soils surrounding posts treated with chromated copper arsenate (CCA). *Wood and Fiber Science*, **36**(1), 119–28.

Morrell, J.J., Keefe, D. and Baileys, R.T. (2003) Copper, zinc, and arsenic in soil surrounding Douglas-fir poles treated with ammoniacal copper zinc arsenate (ACZA). *Journal of Environmental Quality*, **32**(6), 2095–99.

Mukherjee, A.B. and Kikuchi R. (1999) Coal ash from thermal power plants in Finland, in *Biogeochemistry of Trace Elements in Coal and Coal Combustion Byproducts* (eds K.S. Sajwan, A.K. Alva and R.F. Keefer), Kluwer Academic/Plenum Publishers, New York, 59–76.

Mukhopadhyay, P.K., Lajeunesse, G. and Crandlemire, A.L. (1996) Mineralogical speciation of elements in an eastern Canadian feed coal and their combustion residues from a Canadian power plant. *The International Journal of Coal Geology*, **32**(1-4), 279–312.

Mukhopadhyay, P.K., Lajeunesse, G., Crandlemire, A.L. and Finkelman, R.B. (1999) Mineralogy and geochemistry of selected coal seams and their combustion residues from the Sydney area, Nova Scotia, Canada. *The International Journal of Coal Geology*, **40**(2-3), 253–54.

Nico, P.S., Fendorf, S.E., Lowney, Y.W. *et al.* (2004) Chemical structure of arsenic and chromium in CCA-treated wood: implications of environmental weathering. *Environmental Science and Technology*, **38**(19), 5253–60.

Nico, P.S., Ruby, M.V., Lowney, Y.W. and Holm, S.E. (2006) Chemical speciation and bioaccessibility of arsenic and chromium in chromated copper arsenate-treated wood and soils. *Environmental Science and Technology*, **40**(1), 402–408.

Nriagu, J.O. (2002) Arsenic poisoning through the ages, in *Environmental Chemistry of Arsenic* (ed. W.T. FrankenbergerJr.), Marcel Dekker, New York, pp. 1–26.

Ölmez, E.I., Kut, D., Bilge, A.N. and Ölmez, I. (2004) Regional elemental signatures related to combustion of lignites. *The Journal of Radioanalytical and Nuclear Chemistry*, **259**(2), 227–31.

Oremland, R.S. and Stolz, J.F. (2003) The ecology of arsenic. *Science*, **300**(5621), 939–44.

Oskoui, K. (2004) Recovery and reuse of the wood and chromated copper arsenate (CCA) from CCA treated wood – a technical paper, *Environmental Impacts of Preservative-Treated Wood*, Florida Center for Environmental Solutions, Conference, February 8–11, Gainesville, Orlando, FL, pp. 238–44.

Ottosen, L., Pedersen, A. and Christensen, I. (2004) Characterization of residues from thermal treatment of CCA impregnated wood. Chemical and electrochemical extraction, *Environmental Impacts of Preservative-Treated Wood*, Florida Center for Environmental Solutions, Conference, February 8–11 Gainesville, Orlando, FL, pp. 295–311.

Papanicolaou, C., Kotis, T., Foscolos, A. and Goodarzi, F. (2004) Coals of greece: a review of properties, uses and future perspectives. *The International Journal of Coal Geology*, **58**(3), 147–69.

Pedersen, A.J., Kristensen, I.V., Ottosen, L.M. *et al.* (2005) Electrodialytic remediation of CCA-treated waste wood in pilot scale. *Engineering Geology*, **77**(3-4 Special Issue), 331–38.

Peryea, F.J. and Kammereck, R. (1997) Phosphate-enhanced movement of arsenic out of lead arsenate-contaminated topsoil and through uncontaminated subsoil. *Water, Air, and Soil Pollution*, **93**(1-4), 243–54.

Peters, S.C., Blum, J.D., Karagas, M.R. *et al.* (2006) Sources and exposure of the New Hampshire population to arsenic in public and private drinking water supplies. *Chemical Geology*, **228**(1-3 Special Issue), 72–84.

Petrides, B. (1999) The fatal voyage of the Polaris. *American History*, 23–28.

Pouschat, P. and Zagury, G.J. (2006) *In vitro* gastrointestinal bioavailability of arsenic in soils collected near CCA-treated utility poles. *Environmental Science and Technology*, **40**(13), 4317–23.

Quinn, R., Dahl, T.A., Diamond, B.W. and Toseland, B.A. (2006) Removal of arsine from synthesis gas using a copper on carbon adsorbent. *Industrial and Engineering Chemistry Research*, **45**(18), 6272–78.

Reisinger, H.J., Burris, D.R. and Hering, J.G. (2005) Remediating subsurface arsenic contamination with monitored natural attenuation. *Environmental Science and Technology*, **39**(22), 458A–64A.

Rutherford, D.W., Bednar, A.J., Garbarino, J.R. *et al.* (2003) Environmental fate of roxarsone in poultry litter. Part II. Mobility of arsenic in soils amended with poultry litter. *Environmental Science and Technology*, **37**(8), 1515–20.

Sarahney, H., Wang, J. and Alshawabkeh, A. (2005) Electrokinetic process for removing Cu, Cr, and As from CCA-treated wood. *Environmental Engineering Science*, **22**(5), 642–50.

Seames, W.S. and Wendt, J.O.L. (2000) Partitioning of arsenic, selenium, and cadmium during the combustion of Pittsburgh and Illinois #6 coals in a self-sustained combustor. *Fuel Processing Technology*, **63**(2), 179–96.

Seyferth, D. (2001) Cadet's fuming arsenical liquid and the cacodyl compounds of Bunsen. *Organometallics*, **20**, 1488–98.

Shadegg, S. (1970) *Clare Boothe Luce: A Biography*, Simon and Schuster, New York.

Shah, P., Strezov, V., Stevanov, C. and Nelson, P.F. (2007) Speciation of arsenic and selenium in coal combustion products. *Energy and Fuels*, **21**(2), 506–12.

Shen, Z.-X., Chen, G.-Q., Ni, J.-H. *et al.* (1997) Use of arsenic trioxide (As_2O_3) in the treatment of acute promyelocytic leukemia (AFL): II. Clinical efficacy and pharmacokinetics in relapsed patients. *Blood*, **89**(9), 3354–60.

Shuqin, L., Limei, Z., Xuedong Y. and Li Y. (2004) Leaching experiment of trace elements in lignite residue from underground gasification. *Chemical Journal on Internet*, **6**(11), http://www.chemistrymag.org/cji/2004/06b081pe.htm, accessed October 13, 2008.

Smoot, L.D. and Smith, P.J. (1985) *Coal Combustion and Gasification*, Plenum Press, New York.

Spongberg, A.L. and Becks, P.M. (2000) Inorganic soil contamination from cemetery leachate. *Water, Air, and Soil Pollution*, **117**(1-4), 313–27.

Sterling, R.O. and Helble, J.J. (2003) Reaction of arsenic vapor species with fly ash compounds: kinetics and speciation of the reaction with calcium silicates. *Chemosphere*, **51**(10), 1111–19.

Stook, K., Tolaymat, T., Ward, M. *et al.* (2005) Relative leaching and aquatic toxicity of pressure-treated wood products using batch leaching tests. *Environmental Science and Technology*, **39**(1), 155–63.

Taerakul, P., Sun, P., Golightly, D.W. *et al.* (2006) Distribution of arsenic and mercury in lime spray dryer ash. *Energy and Fuels*, **20**(4), 1521–27.

Townsend, T., Dubey, B. and Solo-Gabriele, H. (2004) Assessing potential waste disposal impact from preservative treated wood products, *Environmental Impacts of Preservative-treated Wood*, Florida Center for Environmental Solutions, Conference, February 8–11, Gainesville, Orlando, FL, pp. 169–88.

Townsend, T., Solo-Gabriele, H., Tolaymat, T. *et al.* (2003) Chromium, copper, and arsenic concentrations in soil underneath CCA-treated wood structures. *Soil and Sediment Contamination*, **12**(6), 779–98.

US Environmental Protection Agency (US EPA) (2002a) Zinc fertilizers made from recycled hazardous secondary materials. *Federal Register*, **67**(142), 48393–415.

US Environmental Protection Agency (US EPA) (2002b) *Proven Alternatives for Aboveground Treatment of Arsenic in Groundwater*, EPA-542-S-02-002, Office of Solid Wastes and Emergency (5102G).

U.S. Government Printing Office, US Code of Federal Regulations (CFR), Superintendent of Documents, Washington, DC.

Van der Hoek, E.E., Bonouvrie, P.A. and Comans, R.N.J. (1994) Sorption of As and Se on mineral components of fly ash: relevance for leaching processes. *Applied Geochemistry*, **9**(4), 403–12.

Wasson, S.J., Linak, W.P., Gullett, B.K. *et al.* (2005) Emissions of chromium, copper, arsenic, and PCDDs/Fs from open burning of CCA-treated wood. *Environmental Science and Technology*, **39**(22), 8865–76.

Weider, B. and Fournier, J.H. (1999) Activation analyses of authenticated hairs of Napoleon bonaparte confirm arsenic poisoning. *The American Journal of Forensic Medicine and Pathology*, **20**(4), 378–82.

Welch, A.H., Westjohn, D.B., Helsel, D.R. and Wanty, R.B. (2000) Arsenic in ground water of the United States: occurrence and geochemistry. *Ground Water*, **38**(4), 589–604.

Williams, A.G.B., Scheckel, K.G., Tolaymat, T. and Impellitteri, C.A. (2006) Mineralogy and characterization of arsenic, iron, and lead in a mine waste-derived fertilizer. *Environmental Science and Technology*, **40**(16), 4874–79.

Young, R.J., Jr. (2000) Arsenic and no lace: the bizarre tale of a Philadelphia murder ring. *Pennsylvania History*, **67**(3), 397–414.

Yudovich, Ya.E. and Ketris, M.P. (2005) Arsenic in coal: a review. *The International Journal of Coal Geology*, **61**(3-4), 141–96.

Zhang, J.-Y., Ren, D.-Y., Zhong, Q. *et al.* (2000) Restraining of arsenic volatility using lime in coal combustion. *The Journal of Fuel Chemistry and Technology*, **28**(3), 200.

Zhu, X.-H., Shen, Y.-L., Jing, Y.-K. *et al.* (1999) Apoptosis and growth inhibition in malignant lymphocytes after treatment with arsenic trioxide at clinically achievable concentrations. *Journal of the National Cancer Institute*, **91**(9), 772–78.

6

Major Occurrences of Elevated Arsenic in Groundwater and Other Natural Waters

ABHIJIT MUKHERJEE[1], ALAN E. FRYAR[2], and BETHANY M. O'SHEA[3]

[1] *Bureau of Economic Geology, Jackson School of Geosciences, University of Texas at Austin*
[2] *University of Kentucky, Department of Earth and Environmental Sciences*
[3] *Columbia University, Lamont-Doherty Earth Observatory, Geochemistry Division*

6.1 Introduction

Arsenic contamination of drinking water is a serious and widespread problem. (Smedley and Kinniburgh, 2002), 520 remarked that 'As and F are now recognized as the most serious inorganic contaminants in drinking water on a worldwide basis', yet arsenic was not routinely analyzed by water-quality laboratories until within the last two decades. The World Health Organization (WHO) has recommended a maximum arsenic concentration of $10 \mu g \, L^{-1}$ in drinking water and this value has been adopted, at least provisionally, as a regulatory standard by the European Union, the United States, Japan, and many other nations ((Smedley and Kinniburgh, 2002), Appendix E). However, the former WHO and United States limit of $50 \mu g \, L^{-1}$ remains the standard in many developing countries, in part because of financial and analytical limitations in measuring lower concentrations ((Smedley and Kinniburgh, 2002), Appendix E). Significant clusters of chronic illnesses associated with arsenic ingestion have been reported from Bangladesh, India, Taiwan, Chile, and Argentina. In these and other locations, exposure frequently results from drinking *groundwater* containing elevated concentrations of naturally occurring (*geogenic*) arsenic.

This chapter focuses on cases of severe and extensive arsenic contamination of groundwater at various locations around the world. With the exception of mining, the discussions include instances in which human activities have induced or exacerbated the contamination. Very localized cases of groundwater pollution resulting from the use of arsenical compounds (such as pesticides) or arsenic-contaminated products are also excluded. Readers interested in contamination associated with mining and the use of compounds containing arsenic should consult Chapters 3 and 5 and their references.

6.2   Arsenic speciation and mobility in natural waters

Dissolved arsenic in groundwater primarily exists as inorganic As(V) (*arsenate*) and inorganic As(III) (*arsenite*) (Chapters 2 and 3). Arsenic *speciation* is largely controlled by changes in the *redox potential* (Eh) of the host solution. Under oxidizing conditions (Eh generally $>100\,$mV) and circumneutral pH, $H_2AsO_4^-$, and $HAs_2O_4^{2-}$ usually dominate. However, As(III) oxidation is often sluggish and the species may persist in oxidizing groundwaters ((Boyle, Turner and Hall, 1998), Chapter 3). Under reducing conditions ($<100\,$mV) and circumneutral pH, $H_3AsO_3^0$ is the most common inorganic As(III) form. At Eh values lower than $-250\,$mV and in the presence of *sulfide* (HS^-, S^{2-}, and H_2S), compounds such as As_2S_3 can form. Furthermore, the speciation of arsenic under natural conditions may be very complex (Seyler and Martin, 1989). Arsenic exhibits a relatively slow *redox* transformation (Masscheleyn, Delaune and Patrick, 1991), so that both As(III) and As(V) often occur in the same redox environment. Biologically *catalyzed* arsenic redox reactions, which may result in the presence of a variety of *organoarsenicals*, can also be orders of magnitude faster than chemical *oxidation* and *reduction* ((Smedley and Kinniburgh, 2002), Chapter 3).

Inorganic As(V) generally dominates in surface waters, but its occurrence can be affected by stratification, *anoxic* bottom *sediments*, microbial interactions, and anthropogenic inputs (Chapter 3). In groundwaters, the As(III)/As(V) ratio can vary greatly. For example, the presence of *natural organic matter* (NOM) can have a profound effect on arsenic movement in groundwater, effectively hindering As(III) from reaching *sorption* equilibrium (Redman, Macalady and Ahmann, 2002). As(V) and As(III) have different *adsorption isotherms* (Chapter 2), which implies that they travel through geologic media at different velocities and with increased separation along the flow path. In hydrologic systems at circumneutral pH, inorganic As(III) is an uncharged molecule (i.e. $H_3AsO_3^0$; Chapter 2). Therefore, its mobility is enhanced, as it is less likely to sorb strongly to *mineral* surfaces (Korte and Fernando, 1991). In contrast, As(V) anions sorb more strongly at the pH of most natural waters, so the *distribution coefficient* (K_D) of inorganic As(V) is commonly greater than that of inorganic As(III) ((Smedley and Kinniburgh, 2002; Ferguson and Anderson, 1974), Chapters 2 and 7). Experiments by Gulens, Champ and Jackson (1979) showed that As(III) moves five to six times faster than As(V) under *acidic* and oxidizing conditions. At near neutral to neutral pH, As(V) was more mobile than under acidic conditions, but was still less mobile than As(III). Under alkaline and reducing conditions, both As(V) and As(III) moved rapidly (Gulens, Champ and Jackson, 1979).

6.3   Immobilization of arsenic in hydrologic systems

There are several natural processes that can remove arsenic from groundwaters and other natural waters. These processes were introduced in Chapter 3 and include: (1) *precipitation* and association with sulfides, (2) sorption on *clay minerals*, and (3) *carbonate* associations. This section discusses these processes in further detail. Additional discussions occur in Chapter 7, where some of the processes are utilized in treatment technologies for removing arsenic from water.

6.3.1   Precipitation, coprecipitation, and association with sulfides

Reduced iron (Fe(II)) and sulfide present in water can precipitate or *coprecipitate* to form pyrite (FeS_2) or other sulfide minerals, often in the presence of organic matter. Groundwater may be enriched in Fe(II) by *reductive dissolution* of iron *(oxy)(hydr)oxides* via microbial activity of dissimilatory (respiratory) bacteria like *Shewanella* BrY ((Cumming *et al.*, 1999), Chapter 3). Certain bacteria (e.g. *Desulfovibrio desulfuricans*) use sulfur as an electron sink during the oxidation of organic matter and reduce *sulfate* to sulfide. The sulfide then combines with Fe(II) in *anaerobic* environments (both marine and non-marine) (Fitzpatrick, Fritsch and Self, 1996). *Authigenic* pyrite is common where marine incursion has occurred in fine-grained sediments (Craw and Chappel, 1999). Belzile and Lebel (1986) showed an association of arsenic with *sedimentary* pyrite from the Laurentian Trough of Canada. Extractable iron strongly correlated with arsenic, which indicates that arsenic may be present in the lattice sites of pyrite (Chapters 2 and 3). Belzile (1988) noted that the distribution of arsenic is greatly affected by the formation of pyrite in coastal offshore and *estuarine* sediments. He proposed that arsenic is incorporated in the lattice sites of pyrite at a constant iron to arsenic (Fe/As) ratio of 1000, which falls within the range of 290 to 4600 suggested by Boyle and Jonasson (1973) for rocks containing pyrite. Belzile's (1988) constant Fe/As ratio could reflect the maximum arsenic retention capacity for *framboidal* pyrite in coastal environments, but may not be valid for other environments. Boyle and Jonasson (1973) found that the concentration of arsenic in pyrite may be as high as $5600\,mg\,kg^{-1}$. More recent work (Presser and Swain, 1990; Nesbitt, Muir and Pratt, 1995; Strawn *et al.*, 2002) suggests that arsenic forms a solid solution with reduced sulfur in the crystal structure of pyrite. Savage *et al.* (2000) proposed the term *arsenian* pyrite for pyrite crystals in which arsenic substitutes for sulfur with a molecular formula of $Fe(As_x S_{1-x})$ as suggested by Zacariáš *et al.* (2004). The substitution of arsenic causes arsenian pyrite to be more prone to *chemical weathering* than pure pyrite.

As discussed in Chapter 3, Bostick and Fendorf (2003) conducted a detailed study of arsenic kinetics relative to pyrite in anoxic environments. These authors suggested that As(III) sorbs onto iron sulfides by strong *inner-sphere complexation* (Chapter 2). They also hypothesized that As(III) reduction leads to the formation on FeS and FeS_2 of an intermediate surface precipitate with an arsenopyrite-like structure (Reaction 3.54). This *hypothesis* seems to be reasonable for explaining the immobilization of arsenic phases in sulfate-reducing environments. Bostick, Chen and Fendorf (2004) showed that the initial formation of FeAsS on pyrite would eventually produce more immobile and stable arsenic sulfides (As_2S_3), thereby leading to complete sequestration of arsenic in strongly reducing environments. However, it should be noted that such sequestration would not only depend on the redox state of the environment, but also on the amount of available iron and sulfide. For example, Mukherjee (2006) showed that in spite of the likelihood of sulfate-reducing conditions in the iron-rich deeper groundwater of West Bengal, India, arsenic is still detectable, probably because sulfur is insufficient for pervasive precipitation of pyrite. Moreover, arsenic sequestration may not automatically accompany iron sulfide precipitation, as shown by Van Geen *et al.* (2004) for sediment microcosms amended with acetate and incubated for 60 days.

Other metal sulfides, such as galena (PbS) and sphalerite (ZnS), may affect the mobility of arsenic in anoxic environments. However, immobilization depends on surface complexation rather than precipitation. In contrast to iron (oxy)(hydr)oxides (discussed later), As(III) adsorption on galena and sphalerite increases with pH (Bostick, Fendorf and Manning, 2003). Surface complexation does not occur by isomorphic substitution of lead or zinc, or by a *ligand* exchange mechanism. Instead, multinuclear, inner-surface arsenic-thiosulfide complexes probably form on galena or sphalerite surfaces (Bostick, Fendorf and Manning, 2003).

6.3.2 Arsenic sorption on metal (oxy)(hydr)oxides

Arsenic is most prone to form surface complexes by *adsorption* on metal (mostly iron and manganese) (oxy)(hydr)oxides, followed by clays and feldspars (Lin and Puls, 2003). As discussed in Chapters 3 and 7, iron (oxy)(hydr)oxides are groups of Fe(III) $\pm$ Fe(II) (hydrous) oxides, (hydrous) hydroxides, and (hydrous) oxyhydroxides. Individual compounds, such as *ferrihydrite*, often have highly variable and poorly defined compositions (see Section 3.9.2). Iron and manganese (oxy)(hydr)oxides (both crystalline and amorphous) are common in various geologic environments, such as surface coatings on sediments and as mineral grains, including dust particles. Due to their high sorption capacities ((Dudas, 1987), Chapter 2), (oxy)(hydr)oxide compounds are often very efficient in sorbing arsenic. However, the amount of sorption depends strongly on the *valence state* of the arsenic, the pH of the surrounding water, and the physical and chemical properties of the (oxy)(hydr)oxides.

Of the metal *sorbents*, amorphous to poorly crystalline iron (oxy)(hydr)oxides are most efficient at sorption because of their large *surface areas* (Chapters 2, 3, and 7). However, as these compounds crystallize into hematite, magnetite, or other minerals, their surface areas decrease. Although the affinity of the iron (oxy)(hydr)oxides to sorb arsenic may not always change very much as a result of crystallization (Dixit and Hering, 2003), the reduction of surface area may lead to the release of surface-complexed arsenic (O'Shea, 2006). Smedley and Kinniburgh (2002) provide a detailed list of sorption studies dealing with metal (oxy)(hydr)oxides (Table 6.1).

6.3.3 Arsenic sorption on clay minerals

The sorption of arsenic on clay minerals depends on the clay mineralogy, surface area, surface charge, and the availability of sorption sites, as well as on other chemical species that may be already present as surface complexes. In studying the sorption kinetics of arsenic in relation to clay minerals, like chlorite, kaolinite, illite, mixed illite-montmorillonite, and halloysite, Lin and Puls (2000) concluded that As(V) is preferentially sorbed relative to As(III) on clay surfaces. Sorption is largely affected by the pH of the system. As(V) sorption decreased at pH 7.5 when compared to experimental pH values of 5.5 and 6.5, whereas As(III) increased at pH 7.5. This difference can be attributed to the deprotonation of arsenic (H_3AsO_4) and arsenious (H_3AsO_3) acids (Chapter 2). As(V) species are more likely to be negatively charged and, therefore, more sorbable onto any positively charged clay surfaces than $H_3AsO_3^0$ at pH 5.5–7.5 ((Lin and Puls, 2000), Chapter 7).

Lin and Puls (2000) also observed that with aging of the clay minerals, the bonding strength between arsenic and clay mineral surfaces increases. This is probably because the surface-bonded arsenic slowly diffuses into internal micropores and the clay slowly dehydrates with time. Additionally, as the clay ages, the sorbed As(III) oxidizes to As(V), which exhibits a higher sorption affinity than As(III).

The sorption of arsenic on clay minerals is probably another example of inner-sphere complexation, since the *ionic strength* of the solution seems to have little effect on As(III) sorption ((Manning and Goldberg, 1997), Chapter 2). Maximum As(III) sorption on clay minerals occurs at pH 7.5–9.5, and illite, with its high surface area, sorbs more As(III) than kaolinite (Lin and Puls, 2000; Manning and Goldberg, 1997). At high pH, As(III) may also oxidize to As(V) in the presence of MnO_2 (Manning and Goldberg, 1997).

6.3.4 Carbonate interactions

The mobility of arsenic in groundwater and other natural waters may also be affected by the presence of carbonate phases. Cheng *et al.* (1999) suggested that As(III) is incorporated within the calcite ($CaCO_3$) crystal lattice at or near the surface of the mineral by a dissolution-precipitation mechanism. Maximum adsorption of arsenic on calcite occurs at pH values of 10 (Goldberg and Glaubig, 1988) to 10.5 (Romero, Amrienta and Carrilo-Chavez, 2004), which are more alkaline than most natural waters ((Krauskopf and Bird, 1995), 225). This phenomenon may be explained by *outer-sphere complexation* with calcite having a *zero point of charge* ranging from pH 7 to 10.8 ((Romero, Amrienta and Carrilo-Chavez, 2004), Chapter 2).

Table 6.1 *A summary of some of the studies on the sorption of arsenic species on metal (oxy)(hydr)oxides. (Reprinted from P.L. Smedley and D.G. Kinniburgh, A review of the source, behavior and distribution of arsenic in natural waters, Applied Geochemistry, 17, 517–568, Copyright 2002, with permission from Elsevier.)*

Metal (Oxy)(hydr)oxide	Comment	Reference(s)
'Amorphous' aluminum hydroxide	As(V) on precipitated Al(OH)$_3$ (pH 3–10). 'Adsorption' exceeded 15 mol kg^{-1} at pH 5. Fitted data to pH-dependent *Langmuir* isotherm	Anderson, Ferguson and Gavis (1976)
Alumina	On natural alumina. adsorption was As(V) > As(III) > MMAA[a] = DMAA[b] (pH > 6). Maximum adsorption at pH 5 for As(V) and pH 7 for As(III). As(V) but not As(III) adsorption decreased rapidly above pH 6. Log K_D (l kg^{-1}) at micromolar concentrations (pH 7) was 2.5–3.5 for As(V) and about 1.5 for As(III). FA[c] decreased adsorption.	Xu, Allard and Grimvall (1991)
Alumina, hematite, quartz, and kaolin	As(V) adsorption on natural, low surface area alumina, hematite, quartz, and kaolin (0.12–5 m^2 g^{-1}) at pH 3–10. Adsorption decreases with pH; alumina = kaolin > hematite >> quartz. Gives K_D values and isotherms at low concentrations. Some SO$_4{}^{2-}$ competition especially below pH 7. FA (>10 mg L^{-1}) generally reduced adsorption at pH 5–7 but not above pH 7, where FA is not adsorbed.	Xu, Allard and Grimvall (1988)
Aluminum oxides	As(V) and As(III) adsorption on activated alumina: pH dependence, kinetics, and column breakthrough. Regeneration by desorbing with NaOH. Modeling with pH-dependent Langmuir isotherm (for As) and surface complexation model (for protons)	Ghosh and Yuan (1987)
Birnessite, cryptomelane, and pyrolusite	Studied adsorption of As(III) and As(V) and kinetics of As(III) oxidation in presence of various MnO$_2$. As(III) adsorption (per unit weight of oxide): cryptomelane > birnessite > pyrolusite whereas for As(V): cryptomelane > pyrolusite > birnessite (not detectable). No isotherms given	Oscarson *et al.* (1981); Oscarson *et al.* (1983)
Goethite	An EXAFS[d] and XANES[e] study of As(III) adsorption to a synthetic goethite suggested *bidentate* inner-sphere binding. One plot of As(III) and As(V) pH adsorption envelopes. As(III) data fitted to constant capacitance SCM[f]	Manning, Fendorf and Goldberg (1998)
Goethite	Batch adsorption of As(V) on synthetic goethite. Used Mo blue analysis for As. Shows pH edge at about pH 9. Data fitted Langmuir isotherm presumably at constant pH (up to 60 mg L^{-1} As)	Matis *et al.* (1997)
Goethite	Successfully applied the CD-MUSIC[g] surface complexation model to literature data for anion adsorption to goethite including As(V)-P competition. CD-MUSIC is the most promising of the SCMs for modeling complex natural systems	Hiemstra and van Riemsdijk (1999)

(continued overleaf)

Table 6.1   *(continued)*

Goethite	As(V) adsorption on synthetic goethite primarily for a study of impact on flocculation and electrokinetics. No isotherms. Final pH varied but not defined	Matis *et al.* (1999)
Goethite	EXAFS study of As(V) and Cr(VI) adsorption on goethite. *Monodentate* binding favored at low surface coverage of As(V), bidentate at high surface coverage	Fendorf *et al.* (1997)
Goethite, hematite, and lepidocrocite	Batch adsorption of As(V), As(III), MMAA and DMAA on natural minerals (coarse-grained and very low He–Ar surface area). As adsorption: generally goethite > lepidocrocite >> hematite (pH 2–12, maximum often pH 5–8). At pH 7 on goethite, As(III) > MMAA > DMAA > As(V) (?). FA (up to 50 mg L^{-1}) tended to reduce As adsorption	Bowell (1994)
Granular 'ferric hydroxide' (akageneite)	As(V) isotherms given in the submillimolar concentration range; SO_4 competition significant at millimolar (mM) concentrations below pH 7 only; phosphate competition at 'natural' groundwater concentrations	Driehaus, Jekel and Hildebrandt (1998)
HFO[h]	Adsorption isotherms for arsenite and arsenate over free concentration range from 10^{-7} to 10^{-3} M (pH 4–10). Fitted to Langmuir isotherm at low concentrations and linear isotherm at higher concentrations. Dzombak and Morel (1990) fitted this data to their diffuse double layer model	Pierce and Moore (1982)
HFO	Sorption of As(V) and As(III) on HFO at As concentrations of environmental significance (low micromolar range) and pH 4–9. Compared results with Dzombak and Morel (1990) model predictions with generally reasonable agreement. SO_4 decreased adsorption of As(V) and As(III), especially at low pH, while Ca increased As(V) adsorption at high pH. 1 mM bicarbonate did not affect either As(V) or As(III) adsorption greatly	Wilkie and Hering (1996)
HFO	A wide angle X-ray scattering (and EXAFS) study of two-line ferrihydrite coprecipitated with varying amounts of As(V) suggested that the As reduced crystallite size because of the formation of strongly bound inner-sphere complex between As(V) and edge sharing $Fe(O,OH)_6$ octahedra. Saturation at As/Fe mol ratio of 0.68	Waychunas *et al.* (1996)

Table 6.1   *(continued)*

Metal (Oxy)(hydr)oxide	Comment	Reference(s)
HFO	As(III) and As(V) adsorption and OH$^-$ release/uptake on synthetic two-line ferrihydrite. As(V) at pH 9.2 released up to 1 mol OH$^-$ per mol As sorbed whereas As(III) released <0.25 mol As per mol Fe. At pH 4.6, OH$^-$ release was much less for As(V) adsorption and under these conditions there was a net release of H$^+$ by arsenite. These differences reflect the mechanism of As adsorption and influence the pH dependence of adsorption	Jain, Raven and Loeppert (1999)
HFO	Kinetics and pH dependence of As(V) and As(III) adsorption on HFO (202 m^2 g^{-1}). Found very high As(V) and As(III) loadings (up to 4–5 mol As kg^{-1}) at the highest concentrations. pH adsorption envelopes at various total As loadings	Raven, Jain and Loeppert (1998)
Manganese oxides	As(III) and As(V) removal by MnO$_2$(s) is similar, up to ~5 mmol As mol^{-1} Mn at micromolar As equilibrium solution concentrations. Freundlich isotherm obeyed. As(III) oxidized to As(V). Rapid oxidation (minutes) and adsorption of As(III). Monitored Mn release and effect of pH, Ca, phosphate, and sulfate	Driehaus, Seith and Jekel (1995)

[a] MMAA, *monomethylarsonic acid*, CH$_3$AsO(OH)$_2$
[b] DMAA, *dimethylarsinic acid*, (CH$_3$)$_2$AsO(OH)
[c] FA, *fulvic acid*
[d] EXAFS, *extended X-ray absorption fine structure*
[e] XANES, *X-ray absorption near-edge structure*
[f] SCM, *surface complexation model*
[g] CD-MUSIC, *charge distribution-multisite complexation model*
[h] HFO, 'hydrous ferric oxide', an iron (oxy)(hydr)oxide; Chapter 3.

6.4   Mobilization of arsenic in water

6.4.1   Competitive anion exchange

Arsenic *oxyanions* compete with more abundant anions, such as phosphate (PO$_4^{3-}$, HPO$_4^{2-}$, etc.), sulfate (SO$_4^{2-}$), carbonate (CO$_3^{2-}$), bicarbonate (HCO$_3^-$), and chloride (Cl$^-$), for sorption sites on metal (oxy)(hydr)oxides and clay minerals. Competitive anion exchange is an important mechanism by which arsenic and other *trace elements* are mobilized or remain mobile in water. Of these anions, phosphate may be most important because of its wide use in fertilizer, thereby potentially increasing its concentration (by a factor several times above natural background) in shallow groundwater (Chapters 3 and 5). Competitive exchange with arsenic has been found to be highest for phosphate, followed by CO$_3^{2-}$, SO$_4^{2-}$, and Cl$^-$ (Goh and Lim, 2005). Experimental studies have shown that a 1 mM KH$_2$PO$_4$ solution can leach as much as 150 mg kg^{-1} of arsenic from iron-coated sediments (Manning and Martens, 1997). Most likely, As(V) was desorbed from the surfaces of the iron coatings and replaced by the sorption of H$_2$PO$_4^-$ or HPO$_4^{2-}$.

High concentrations of carbonate (CO_3^{2-}) and bicarbonate (HCO_3^-) in groundwater have also been found to hinder the retention of sorbed arsenic on (oxy)(hydr)oxide compounds. Experiments have confirmed that carbonate ions compete with arsenic for sorption sites on ferrihydrite (Appelo *et al.*, 2002) and MnOOH (Anawar, Akai and Sakugawa, 2004). CO_3^{2-} and HCO_3^- may originate from either NOM or buried organic wastes from manufacturing and other human activities.

6.4.2   Effect of natural organic matter (NOM)

Natural organic matter (NOM) often exhibits high surface areas and has an affinity for forming complexes with arsenic (Mukhopadhyay and Sanyal, 2004). NOM affects arsenic mobility in water by forming water-soluble complexes with arsenic, inhibiting arsenic sorption on the surfaces of minerals and other solid compounds, and displacing arsenic that has already been sorbed (Chapter 3). In particular, NOM significantly interferes with the sorption of arsenic on hematite and goethite, two common iron (oxy)(hydr)oxides (Redman, Macalady and Ahmann, 2002; Grafe, Eick and Grossl, 2001). *Humic* and *fulvic acids* (FAs) in NOM may form stable complexes on the surfaces of clays, quartz, or metal (oxy)(hydr)oxides, thereby inhibiting the sorption of arsenic (Kaiser *et al.*, 1997; Bauer and Blodau, 2006). At near-neutral pH, humic acid sorbed onto kaolinite controls the amount of sorbed arsenic (Saada *et al.*, 2003). Decomposing NOM can also catalyze the dissolution of host minerals, thereby promoting arsenic mobilization. For example, NOM promotes the reductive dissolution of iron (oxy)(hydr)oxides and concurrently inhibits iron oxidation and precipitation ((Schwertmann, Fischer and Papendorf, 1968), Chapter 3). Harvey *et al.* (2002) and McArthur *et al.* (2004) suggested that the reduction of FeOOH and other metal (oxy)(hydr)oxides in Bengal basin sediments is mostly caused by the infiltration of NOM-rich water from the surface and by NOM existing in the sediments as remnants of paleovegetation.

6.4.3   Effect of pH

The sorption of arsenic to metal (oxy)(hydr)oxides is generally pH-dependent (McArthur *et al.*, 2004; Welch, Lico and Hughes, 1988; Robertson, 1989). As(V) tends to be strongly sorbed by (oxy)(hydr)oxide minerals at near-neutral to acidic pH conditions (Chapter 2). At alkaline pH, mineral surfaces become increasingly negatively charged, thus promoting arsenic desorption (O'Shea, 2006). In many oxidizing systems, pH is positively correlated with dissolved arsenic (see (Robertson, 1989)). The presence of HCO_3^- and phosphate in a solution with increasing pH can significantly modify the strongly nonlinear adsorption isotherm of arsenic, thereby affecting its mobilization in groundwater. Among the processes that may result from changing pH is proton uptake by mineral *weathering* and *ion exchange* reactions (Smedley and Kinniburgh, 2002). Furthermore, the dissolution of carbonate minerals, such as calcite, dolomite, and siderite, may take place with decreasing pH, which may in turn mobilize arsenic that had entered the carbonate lattices (Romero, Amrienta and Carrilo-Chavez, 2004) at higher pH.

Elevated arsenic concentrations in *oxic aquifers* in Arizona (US) were linked to pH-dependent desorption (Robertson, 1989). Similar results exist for *metamorphic* aquifers in New England (US), where moderately alkaline waters (pH 7.5–9.3) were found to have elevated concentrations of arsenic (Robinson and Ayotte, 2006). Conversely, (BGS (British Geological Survey), 1989) suggested that arsenic concentrations of $<4\,\mu g\,L^{-1}$ in water of the Lincolnshire Limestone (UK) cannot be explained by pH values of 7.0–9.5. McArthur *et al.* (2004) commented that the observations of pH increases with arsenic mobilization by Welch, Lico and Hughes (1988) and Robertson (1989) are not by themselves sufficient to prove that arsenic is mobilized by increasing pH. Arsenic may be mobilized by extended *residence times*, evaporation, and/or weathering, any of which could lead to both increases in pH and dissolved arsenic concentrations.

6.4.4 Redox-dependent mobilization

Arsenic can be liberated into solution both by oxidation and reduction. The electrochemical evolution sequence in natural *hydrologic systems*, which is commonly bacterially mediated, begins with the oxidation of NOM coupled to the reduction of dissolved O_2, resulting in the production of dissolved inorganic carbon (Chapelle, 2001). Arsenic phases that have formed complexes with NOM may be released at this stage. However, a more significant mechanism by which arsenic is liberated under oxic conditions is the *aerobic oxidation of iron sulfides* ((Smedley and Kinniburgh, 2002), Chapter 3). Arsenic released as a result of pyrite oxidation may be adsorbed by the oxidized iron (O'Shea, 2006). Van Geen *et al.* (2004) observed the slow release of arsenic from microcosms in reduced sediments exposed to low concentrations of O_2. They proposed that the gradual oxidation of As(III) could have been catalyzed by Fe(II) or manganese oxides.

Following consumption of dissolved O_2, the *thermodynamically* favored electron acceptor is nitrate (NO_3^-). Nitrate reduction can be coupled to anaerobic oxidation of metal sulfides (Appelo and Postma, 1999), which may include arsenic-rich phases. The release of sorbed arsenic may also be coupled to the reduction of Mn(IV) (oxy)(hydr)oxides, such as birnessite (δ-MnO_2) (Scott and Morgan, 1995). The electrostatic bond between the sorbed arsenic and the host mineral is dramatically weakened by an overall decrease of net positive charge so that surface-complexed arsenic could dissolve. However, arsenic liberated by these redox reactions may reprecipitate as a mixed As(III)–Mn(II) solid phase (Tournassat *et al.*, 2002) or resorb as surface complexes by iron (oxy)(hydr)oxides (McArthur *et al.*, 2004). The most widespread arsenic occurrence in natural waters probably results from reduction of iron (oxy)(hydr)oxides under anoxic conditions, which are commonly associated with rapid sediment accumulation and burial (Smedley and Kinniburgh, 2002). In anoxic alluvial aquifers, iron is commonly the dominant redox-sensitive *solute* with concentrations as high as $30\,mg\,L^{-1}$ (Smedley and Kinniburgh, 2002). However, the reduction of As(V) to As(III) may lag behind Fe(III) reduction (Islam *et al.*, 2004).

O'Shea (2006) has summarized the redox steps involved in the mobilization of arsenic in natural hydrologic systems (Table 6.2). It should be noted that in natural environments, discrete redox zones are not always observable and redox disequilibrium can occur, thus complicating arsenic mobilization and resorption (Mukherjee, 2006).

6.4.5 Complex and colloid formation

Mobilization of arsenic can be promoted by the presence of other solutes and their interaction with arsenic to form complexes, as detailed above for NOM. As(V) has been documented to form complexes with calcium and magnesium in brackish water and seawater (Cullen and Reimer, 1989). Arsenic commonly forms complexes with sulfide, such as $As_3S_3(SH)_3^0$ (Bostick, Fendorf and Manning, 2003) (Chapter 3). These complexes may exist in slightly sulfidic environments. Arsenic–fluoride complexes (e.g. AsO_3F^{2-} and $HAsO_3F^-$) can also form in high-fluoride waters ((Apambire and Hess, 2000; Bundschuh, Bonorino and Viero, 2000), Chapter 3).

Colloids are well known to have high mobility because of their small size. They also have a high affinity to sorb ions from solution because of their large surface areas. Sadiq (1997) and Csalagovits (1999) reported that arsenic enrichment in *soils* and sediments results from sorption to colloids, particularly iron (oxy)(hydr)oxides. Chen *et al.* (1994) noted that As(V) sorption can significantly affect the surface characteristics of soil colloids, such as the *isoelectric point*. In laboratory column experiments, As(V) sorbed onto Fe_2O_3 colloids traveled more than 21 times faster than dissolved As(V) through *sand* and gravel (Puls and Powell, 1992).

Table 6.2 Processes involved in the liberation of sorbed arsenic or sequestering with changes in the redox state of the hydrologic system (after (O'Shea, 2006))

Redox condition	Electron acceptor	Processes releasing or sequestering arsenic	Possible reaction	Reference(s)
Slightly oxidizing	Oxygen consumption	Decomposition of organic matter and release of bound arsenic	Humic acid $\equiv$ As $+ 2O_2 + H^+ \rightarrow$ $2CO_2 + 2H_2O + As_{(aq)}$	Mukhopadhyay and Sanyal (2004)
		Oxidation of arsenic-bearing pyrite with adsorption onto iron oxides and/or other metal (oxy)(hydr)oxides	$Fe(As,S)_2 + 15/4O_2 + 7/2H_2O \rightarrow Fe(OH)_3 \equiv$ As $+ 2SO_4^{2-} + 4H^+$	Kim, Nriagu and Haack (2000)
Slightly reducing	Nitrate reduction	Nitrate reduction by pyrite oxidation (note that Appelo and Postma, 1999 referred to pure rather than arsenian pyrite)	$8Fe(As,S)_2 + 13NO_3^- + 25H_2O + 10H^+ \rightarrow$ $8Fe^{2+} + 8HAsO_4^{2-} + 8SO_4^{2-} + 13NH_4^+$	Appelo and Postma (1999)
Slightly reducing	Manganese reduction	Manganese oxide reduction and release of sorbed arsenic	$MnO_2 \equiv HAsO_2 + 2H^+ \rightarrow$ $H_3AsO_{4(aq)} + Mn^{2+}$	Chaillou et al. (2003)
Reducing	Iron reduction	Fe(III) reduction on oxide surfaces changes net charge leading to arsenic desorption	$FeAsO_4^- + 2e^- + 5H^+ \rightarrow$ $Fe^{2+} + H_3AsO_3 + H_2O$	
		Iron oxide reductive dissolution and release of sorbed arsenic catalyzed by NOM degradation	$8FeOOH \equiv$ As $+ CH_3COOH + 14H_2CO_3 \rightarrow$ $8Fe^{2+} + HCO_3^- + 12H_2O + As_{(aq)}$	Nickson et al. (2000)
Reducing	Arsenate reduction	As(V) reduction to As(III) and change in sorption properties	$AsO_4^{3-} + 2H^+ + 2e^- \rightarrow$ $AsO_3^{3-} + H_2O$	Bose and Sharma (2002)
Strongly reducing	Sulfate reduction	Precipitation of insoluble sulfides (e.g. pyrite) and sequestration of arsenic as sorbed phases	$Fe_2O_3 + 1/2O_2 + 8CH_2O + 4SO_4^{2-} + As_{(aq)} \rightarrow 2Fe(As,S)_2 + 8HCO_3^- + 4H_2O$	Zheng et al. (2004)

6.5   Natural occurrences of elevated arsenic around the world

6.5.1   Introduction

Elevated concentrations of arsenic in groundwater and other natural waters with geogenic sources occur in many areas around the world (Appendix D). Table 6.3 summarizes some of the conditions at several of the most severely arsenic-affected locations. The first cases of widespread arsenic poisoning in contemporary times were discovered about half a century ago (e.g. Córdoba province in Argentina and Taiwan (Blackfoot disease)) ((Smedley and Kinniburgh, 2002), Table 6.3). However, the current interest in arsenic-contamination stems largely from its gradual discovery in the densely populated Bengal basin of Bangladesh and eastern India during the 1980s and 1990s. A proliferation of studies in the last two decades has shown geogenic groundwater and surface-water contamination by arsenic in parts of all the continents other than Antarctica. More than 100 million people may be at risk from the adverse health effects of ingesting detrimental levels of arsenic for prolonged periods. Most of the affected areas are in Asia (Bangladesh, India, Nepal, Pakistan, mainland China, Taiwan, Cambodia, Vietnam, and Iran), but large areas also occur in the Americas (Chile, Argentina, Mexico, and various parts of the United States) ((Smedley and Kinniburgh, 2002; Rahman *et al.*, 2006), Table 6.3; Appendix D).

Most of the major contaminated areas are marked by sediments of *Quaternary* age (less than about 1.75 million years old). As noted by Smedley and Kinniburgh (2002), these areas can be subdivided on the basis of geomorphic and geologic similarities. For example, in Asia, most of the affected areas occur in the flood and *delta* plains of the major Himalayan rivers, for example, the Ganges–Brahmaputra–Meghna system in India and Bangladesh, the Indus plain in Pakistan, the Irrawaddy delta of Myanmar, the Red River delta in Vietnam, and the Mekong delta of Cambodia and Laos. Some acutely contaminated areas are located in arid and semi-arid regions (e.g. Inner Mongolia, the Atacama Desert of northern Chile, and Nevada in the United States; Table 6.3). Arsenic contamination associated with *geothermal* waters and *igneous rocks* occurs in northern Chile, in the Chaco-Pampean aquifers of Argentina, in California, Nevada, and Yellowstone National Park in the western United States, and in the Donargarh rift belt of central India. Contamination in Ghana and northern Chile may be related to arsenic-enriched sulfide mineralization.

Arsenic-contaminated groundwaters also show a wide variation in depth. In some areas (e.g. the Bengal basin, Nepal, Myanmar, and Ghana), contaminated groundwater mainly occurs in wells drawing from depths of a few meters (m) to ~80 m below ground level. Surface or near-surface water supplies (e.g. shallow dug wells) and deeper wells are less contaminated. However, rivers are contaminated in northern Chile and Inner Mongolia, and artesian groundwaters are contaminated at depths as great as 280 m in southwestern Taiwan and as great as 660 m in Xinjiang, China (Smedley and Kinniburgh, 2002).

Smedley and Kinniburgh (2002) proposed a classification of contaminated aquifers based on the hydrogeochemical conditions prevailing in them – more precisely, the redox processes involved in the liberation and mobilization of arsenic in solution. Although this classification may be useful for understanding the cause of the contamination in some regions, in many cases (e.g. the Bengal basin, Cambodia, and Ghana) the inferred mobilization processes are still hypothetical and have yet to be universally accepted. Thus, the classification should be viewed as based on the most probable conditions.

At present, most contaminated aquifers in Quaternary delta plains are thought to contain arsenic mobilized under anoxic conditions by bacterial activity, which is driven by high concentrations of NOM (Smedley and Kinniburgh, 2002; Smedley, 2005). Arsenic sorbed on metal (oxy)(hydr)oxides is liberated by reductive dissolution (Chapter 3). The best-known example is the Bengal basin, where arsenic is widely present in *Holocene* (less than 11 500 years old) sediments, while the underlying, oxidized aquifers with *Pleistocene* (11 500–1.75 million years old) sediments are thought to have very low concentrations of arsenic. Similar conditions may occur in Cambodia, Vietnam, Myanmar, and elsewhere

Table 6.3 *A summary of some arsenic-affected areas of the world (see also Appendix D). Data compiled from reviews by Smedley and Kinniburgh (2002), Smedley (2005), and Rahman et al. (2006), including references therein, and from Tseng et al. (1968), Nicolli et al. (1989), Fredericks and Arsenic Secretariat (2004), Ghosh (2005), Haque (2005), Mukherjee (2006), Panthi, Sharma and Mishra (2006), Ramanathan, Bhattacharya and Tripathi (2006), and Farooqi, Masuda and Firdous (2007). Population exposed represents the minimum estimated number of people drinking water with $\geq 50\ \mu g L^{-1}$ of arsenic. Refer to the text and Appendix B glossary for definitions of terms. (Modified from P.L. Smedley and D.G. Kinniburgh, A review of the source, behavior and distribution of arsenic in natural waters, Applied Geochemistry, **17**, 517–568, Copyright 2002, with permission from Elsevier.)*

Region/country	Population exposed	Arsenic concentration ranges ($\mu g\ L^{-1}$)	Aquifer type	Groundwater conditions	Year(s) of discovery
Bengal basin, Bangladesh	35 million	<0.5–4730	Holocene alluvial/deltaic sediments with abundant solid organic matter	Strongly reducing, neutral pH, high alkalinity, slow groundwater	1993
Bengal basin, West Bengal, India	15 million	<1–4100	(same as above)	(same as above)	1978
Middle Ganges plain, India	1.4 million in Bihar	<3–1861 in Bihar	Holocene flood plain	Similar to Bengal basin, but more oxic?	2002
Terai, Nepal	550 000	<10–2620	Quaternary Himalayan sediments	Similar to Bengal basin	1999
Indus valley, Pakistan	2.36 million in Punjab	<10–1900	Holocene alluvial plain	Mixed oxidizing-reducing (?)	2000
Mekong delta, Cambodia	73 000	<1–1340	Holocene alluvial/deltaic sediments	Similar to Bengal basin	2000
Red River delta, Vietnam	100 000	1–3050	Holocene alluvial/deltaic sediments	Reducing with high Fe, Mn, NH$_4$ and alkalinity	2001

Location	Population	Concentration	Aquifer	Conditions	Date
Tianshan plain, Xinjiang, China	523 diagnosed	40–750	Holocene alluvial plain	Reducing, deep wells (up to 660 m) are artesian	1980s
Yellow River plains, Inner Mongolia, China	300 000 in Huhhot basin	<1–2400	Holocene alluvial/lacustrine sediments	Strongly reducing, neutral pH, high alkalinity. Deep groundwaters, often artesian, some have high humic acid concentrations	1990s
Taiwan	100 000?	10–1820	Sediments and black shale	Strongly reducing, artesian conditions, some groundwaters contain humic acids	1930s
Danube basin, Hungary and Romania	29 000	<2–4000	Quaternary alluvial plain	Reducing groundwater, some artesian, some high in humic acids	1941
Basin and Range, Arizona, USA	up to 1300		Alluvial basins, some evaporites	Oxidizing, high pH. Arsenic (mainly As[V]) correlates positively with Mo, Se, V, and F	
Tulare basin, San Joaquin Valley, California, USA		<1–2600	Holocene and older basin-fill sediments	Internally drained basin. Mixed redox conditions. Proportion of As(III) increases with well depth. High salinity in some shallow groundwaters. High Se, U, B, and Mo	

(continued overleaf)

Table 6.3 (continued)

Region/country	Population exposed	Arsenic concentration ranges (µg L^{-1})	Aquifer type	Groundwater conditions	Year(s) of discovery
Southern Carson Desert, Nevada, USA		up to 2600	Holocene mixed aeolian, alluvial, lacustrine sediments, some thin volcanic ash beds	Largely reducing, some high pH. Some with high salinity due to evaporation. Associated with high U, P, Mn, DOC, and Fe to a lesser extent	
Lagunera, Mexico	400 000	8–620	Volcanic sediments	Oxidizing, neutral to high pH, arsenic mainly as As(V)	1960s
Antofagasta, Atacama Desert, Chile	500 000	100–1000	Quaternary(?) volcanogenic sediment	Generally oxidizing. Arid conditions, high salinity, and B. Also high-arsenic river waters fed by geothermal springs.	1960s
Chaco-Pampean plain, Argentina	1.2 million	<1–11 500	Holocene and earlier loess with rhyolitic volcanic ash	Oxidizing, neutral to high pH, high alkalinity, groundwaters often saline. Arsenic mainly as As(V), accompanied by high B, V, Mo, and U. Also high arsenic in some river waters	1955

(Smedley, 2005). Arsenic (mainly as As(V)) can also be mobilized in oxic aquifers by the oxidation of sulfide minerals present in the aquifer matrix or by dissolution of arsenic-enriched *silicates* (Smedley and Kinniburgh, 2002). Such oxic arsenic-contaminated aquifers occur both in temperate humid regions (e.g. New England and the upper Midwest) and semi-arid to arid regions (e.g. Arizona) of the United States (Welch *et al.*, 2000), as well as in the Atacama Desert of northern Chile and the Chaco-Pampean plain of Argentina (Smedley and Kinniburgh, 2002). Contamination may also occur under mixed reducing-oxidizing (*redox*) conditions, such as in parts of the southwestern United States (Smedley and Kinniburgh, 2002).

The rest of this chapter provides descriptions of some regions with significant geogenic arsenic contamination (mainly in groundwater). The geochemical processes that have been proposed to control arsenic concentrations in these areas are reviewed, but in contrast to Smedley and Kinniburgh (2002), these processes are not used as a basis for classifying the areas.

6.5.2 Bengal basin, India and Bangladesh

Although its areal extent ($\sim200\,000\,km^2$) is not as large as that of some other arsenic-affected regions, the Bengal basin, which covers most of Bangladesh and parts of the Indian states of West Bengal, Assam, and Tripura (Figure 6.1), has received the most attention. According to Smedley (2005), about 4.2 million wells have been tested in Bangladesh with a large, but unknown, number of wells tested in West Bengal. The reason for such prominence is the fact that almost half of the ~120 million residents of the Bengal basin are thought to be at risk. As many as 30–35 million people in Bangladesh reside in areas with groundwater arsenic concentrations of $>50\,\mu g\,L^{-1}$ (Smedley and Kinniburgh, 2002), along with another ~15 million in West Bengal ((Mukherjee, 2006), Table 6.3). Gaus *et al.* (2003) estimated that the number of people in Bangladesh that are drinking water with $>10\,\mu g\,L^{-1}$ of arsenic may reach 57 million. Of these, at least 1 million are likely to be affected by arsenicosis, as calculated by Yu, Harvey and Harvey (2003) based on *dose–response* data from Guha Mazumder *et al.* (1998). The contamination has been termed as the *greatest mass poisoning in human history* (Smith, Lingas and Rahman, 2000).

The Bengal basin, the world's largest *fluvio*-deltaic basin (Coleman, 1981; Alam *et al.*, 2003), formed by the *sedimentation* of the Ganges, Brahmaputra, and Meghna (GBM) rivers, along with their numerous tributaries and distributaries. Indiscriminate use of surface waters for the disposal of sewage and industrial waste has rendered them impotable. This pollution and the introduction of high-yielding dry-season rice (Boro) increased the demand for irrigation (Harvey *et al.*, 2005), which led to a shift from surface water to groundwater usage in both West Bengal and Bangladesh during the early 1970s. Consequently, several million wells (ranging from hand-pumped to motor-driven) were installed to meet drinking, irrigation, and industrial water demands (Smith, Lingas and Rahman, 2000; BGS/DPHE (British Geological Survey/Department of Public Health Engineering [Bangladesh]), 2001; Harvey *et al.*, 2005; Horneman *et al.*, 2004).

Arsenic concentrations exceeding $50\,\mu g\,L^{-1}$ were discovered in 1978 (Guha Mazumder *et al.*, 1998; Table 6.3) in some wells in the North 24 Parganas district of West Bengal, followed by identification of arsenicosis by 1984 (Garai *et al.*, 1984). Soon, groundwater in other parts of West Bengal and, in 1993, Bangladesh (Swartz *et al.*, 2004), was found to have arsenic concentrations exceeding the current WHO safe drinking-water guideline of $10\,\mu g\,L^{-1}$ (Appendix E). Of the ~11 million tube wells providing water in Bangladesh (Smedley, 2005), about 25 % (McArthur *et al.*, 2004) to 33 % (Horneman *et al.*, 2004) are probably contaminated with arsenic. (BGS/DPHE [Bangladesh], 2001) estimated that 1.5–2.5 million wells may already be affected. Arsenic-contamination is reportedly present in 59 of the 64 districts

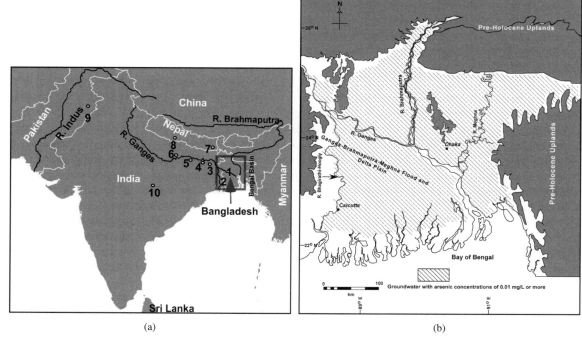

(a)   (b)

Figure 6.1   (a) Sites of geogenic arsenic contamination in the Indian subcontinent: (1) Bengal basin, Bangladesh; (2) Western Bengal basin, West Bengal, India; (3) Sahebganj, Jharkhand, India; (4) Bhagalpur, Bihar, India; (5) Bhojpur, Bihar, India; (6) Ballia and Gazipur, Uttar Pradesh, India; (7) Rajnandgaon, Chattisgarh, India; (8) Jhapa, Nepal; (9) Nawalparasi, Nepal; (10) Kalalanwala, Punjab, Pakistan; (b) Map of the arsenic-contaminated parts of the Bengal basin (modified from Mukherjee (2006)); (c) Map of the arsenic-contaminated areas of Bangladesh (based on data from (BGS/DPHE (British Geological Survey/Department of Public Health Engineering Bangladesh), 2001)).

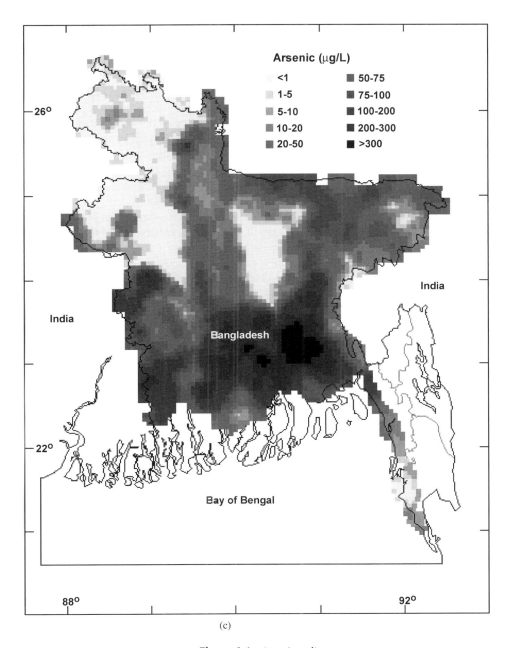

(c)

Figure 6.1   *(continued)*

(upazilas) of Bangladesh. The worst-affected districts of West Bengal are Maldah, Murshidabad, Nadia, North and South 24 Parganas (including parts of Calcutta), and minor adjoining areas of Bardhaman, Hoogly, and Howrah. Arsenic concentrations in groundwater of the Bengal basin vary widely. In West Bengal, reported concentrations range from $<5-4100\,\mu g\,L^{-1}$ (BGS/DPHE, 2001; CGWB (Central Ground Water Board [India]), 1999; Ghosh and Mukherjee, 2002)). In Bangladesh, the highest reported concentration is $4730\,\mu g\,L^{-1}$ (Rahman *et al.*, 2006).

Since 1990, numerous studies by governmental and nongovernmental agencies and researchers have tried to understand the causes and effects of arsenic contamination in the Bengal basin. These studies have led to a detailed understanding of the water-sediment chemistry and the proposition of several hypotheses on the occurrence and fate of arsenic in groundwater on both sides of the India–Bangladesh border (e.g. (Harvey *et al.*, 2002; BGS/DPHE Saha, 1991; Arsenic Investigation Project by Public Health Engineering Directorate of Government of West Bengal (AIP/PHED), 1995; Bhattacharya, Chatterjee and Jacks, 1997; Nickson *et al.*, 1998; Acharyya *et al.*, 1999; McArthur *et al.*, 2001; Dowling, Poreda and Basu, 2003; Van Geen *et al.*, 2003; Polizzotto *et al.*, 2005a)). It has been established that the arsenic-contaminated aquifers occur in modern flood and delta plains of the GBM. These aquifers are of Quaternary age and are comprised of *micaceous* sand, *silt*, and *clay* derived from the Himalayas and basement complexes of eastern India. They are bounded by the River Bhagirathi–Hoogly (a distributary of the Ganges) in the west, the rivers Ganges and Padma in the north, the flood plain of the River Meghna (tributary of the Padma), and the River Brahmaputra in the northeast (Acharyya *et al.*, 1999; 2000). Acharyya (2002) noted a strong correlation between the spatial distribution of the arsenic-affected areas and the paleogeomorphology and Quaternary stratigraphy of the Bengal basin. These authors suggested that the arsenic-affected areas of the GBM alluvial system were incised during the Pleistocene, exposing the already-deposited sediments to oxidation. Rapid sea level rises during the Holocene (including higher than present mean sea levels between 7000 and 5500 years ago) resulted in back-flooded and swamped incised channels with *peat* and other organic matter among fine-grained sediments. These Holocene sediments have been identified as enriched in arsenic.

Heterogeneity in arsenic concentrations exists both horizontally and with depth. Uncontaminated and contaminated wells may exist within a few meters of each other (Smedley, 2005). Weinman *et al.* (2006) observed arsenic concentrations of $<10-300\,\mu g\,L^{-1}$ within a 100-m lateral distance, as a result of differing *fluvial* sediment depositional regimes (i.e. a high-energy braided system versus a low-energy meandering system). Recent work by Van Geen *et al.* (2006) suggests that a variability between undetectable concentrations and $\geqslant 600\,\mu g\,L^{-1}$ of arsenic within $4\,km^2$ may be inversely related to the *permeability* of the surface soil, which in turn helps to determine the rate of groundwater recharge. At a regional scale, arsenic concentrations are lower in the northern alluvial fans and Pleistocene uplands (Barind and Madhupur areas). Vertically, the concentrations of dissolved arsenic are low in the first few meters, then increase rapidly over a narrow depth interval (Smedley, 2005). At Chapai Nawabganj, Bangladesh, the arsenic concentration increased from $17\,\mu g\,L^{-1}$ at 10 m depth to $\sim 400\,\mu g\,L^{-1}$ at 40 m depth ((BGS/DPHE), 2001). Shallow dug wells are thought to be safe from arsenic contamination ((Chakraborti, Basu and Biswas, 2001), $<10\,\mu g\,L^{-1}$). Smedley (2005) suggested that low concentrations in dug wells are a combined manifestation of lower arsenic at shallow depths, interaction with recently recharged groundwater, and aeration by exposure to atmospheric O_2. It has been thought that river water in the basin is also low in arsenic, but two recent studies from West Bengal have shown that elevated arsenic may be present in the water of smaller streams. Stüben *et al.* (2003) reported arsenic in water from the River Gobra in Murshidabad district and Mukherjee and Fryar (2008) documented arsenic concentrations up to $101\,\mu g\,L^{-1}$ from the Jalangi and Ichamati rivers.

It has been suggested that most of the arsenic-contaminated water exists at depths of 80–100 m or less below land surface (e.g. (BGS/DPHE), 2001; Chakraborti, Basu and Biswas, 2001)), and the number of polluted wells is insignificant at depths greater than 100–150 m. The depth of maximum contamination has been suggested to be between 15–30 m (Smedley, 2005) and 30–40 m, and the contamination could be associated with a zone of mixing between shallower and deeper groundwaters induced by pumping (Harvey *et al.*, 2002; Harvey *et al.*, 2005; Klump *et al.*, 2006). In the floodplains of Bangladesh, gray Holocene sediments are thought to be underlain at depths greater than ~150 m by brown Pleistocene sediments with lower arsenic concentrations (e.g (BGS/DPHE), 2001; Van Geen *et al.*, 2003)). Groundwater in the brown sediments may also contain more dissolved O_2 than the overlying Holocene sediments (Zheng *et al.*, 2004). BGS/DPHE (BGS/DPHE), 2001) found that of 335 groundwater samples collected at depths >150 m in Bangladesh, only 5 % had arsenic concentrations of $\geq 10\,\mu g\,L^{-1}$, whereas a recent survey by the Bangladesh Arsenic Mitigation Water Supply Project showed that only 3 % of more than 7000 deep wells have arsenic concentrations of $>50\,\mu g\,L^{-1}$ (Smedley, 2005). Hence, various studies (e.g. (Van Geen *et al.*, 2003; JICA (Japan International Cooperation Agency), 2002; Zheng *et al.*, 2005)) have suggested that the use of deeper aquifers may be a possible strategy for mitigation of arsenic exposure. However, the depths of the older, low-arsenic aquifers vary spatially. The extent of contamination in the deep aquifer in West Bengal was not publicized until Mukherjee (2006) showed that >60 % of the deep wells may have $\geq 10\,\mu g\,L^{-1}$ of arsenic. The disparity between the deep aquifers in Bangladesh and West Bengal may stem from differences in the stratigraphic and sedimentological evolution of the Bengal stable shelf basin (mostly West Bengal) and Bengal Foredeep basin (mostly Bangladesh), and in land-use patterns.

There has been much speculation about the primary source of arsenic in the groundwater of the basin. Concentrations of solid-phase arsenic in the soil and sediment are not greatly enriched, being typically $<10\,mg\,kg^{-1}$ in sandy sediments and $<100\,mg\,kg^{-1}$ in clay and peat (Harvey *et al.*, 2006). Several studies have confirmed that the contamination is natural and is intensified by anthropogenic interferences (Acharyya *et al.*, 1999; Acharyya *et al.*, 2000; Ray, 1999). Various workers have indicated that arsenic is associated with the clays in the basin subsurface (Swartz *et al.*, 2004; Chakraborti, Basu and Biswas, 2001; JICA, 2002; Anawar *et al.*, 2003). However, others (e.g. (Breit *et al.*, 2000; Foster *et al.*, 2003; Datta *et al.*, 2005)) have found that arsenic is mostly enriched in the iron-oxide-coated sand and *mica* grains in the aquifers. Some recent studies have also suggested that the contaminated aquifer sediments are gray (indicative of reducing conditions), in contrast to the uncontaminated yellow or brown sediments (indicative of oxidizing conditions) (McArthur *et al.*, 2004; Horneman *et al.*, 2004; von Brömssen *et al.*, 2007). Of the several hypotheses about the sources of arsenic, the most plausible are as follows:

1. Arsenic is transported by the River Ganges and its tributaries from the Gondwana *coal* seams in the Rajmahal trap area west of the basin. The arsenic concentration of the coal is as high as 200 mg kg^{-1} (Saha, 1991).
2. Arsenic is transported by the north Bengal tributaries of the Bhagirathi and Padma rivers from near the Gorubathan base-metal deposits in the eastern Himalayas (Ray, 1999).
3. Arsenic is a biogenic deposit in the palaeochannels of the Bhagirathi and Padma rivers under *euxinic* conditions (P. Chakrabarty, West Bengal State Remote Sensing Board 1999, personal communication to authors).
4. Arsenic is transported with the fluvial sediments from the Himalayas (e.g. (McArthur *et al.*, 2004)). This is the most widely accepted hypothesis at present.

Four main mechanisms have been identified by various workers to explain arsenic mobilization in groundwater of the Bengal basin. These are:

1. *Oxidation of pyrite:* Several workers (Mallick and Rajgopal, 1995; Das *et al.*, 1996; Mandal *et al.*, 1998) have proposed that arsenic is released by oxidation of arsenian pyrite within the alluvial sediments. Such oxidation could be caused by excessive pumping and *water-table* drawdown, which could lead to aeration of previously anoxic sediments. Chakraborti, Basu and Biswas (2001) and Sikdar, Sarkar and Palchoudhury (2001) noted water-table declines in the Bengal basin as a result of increased agricultural pumping beginning in the 1970s. Coincidentally, arsenic contamination in Bengal basin groundwater was first detected in 1978. Mallick and Rajgopal (1995) pointed out that pumping could induce mixing of oxidized and reduced waters.

   The presence of pyrite within sediments in West Bengal has been widely reported. These detections include arsenian pyrite in fluvial sediments within arsenic-affected areas (Chakraborti, Basu and Biswas, 2001; Das *et al.*, 1996; Mandal *et al.*, 1998). Das *et al.* (1996) found that pyrite collected at a depth of $79-82\,\text{m}$ in clays from Maldah had an arsenic content of $122.47\,\text{mg kg}^{-1}$. Pyrite framboids have also been reported in Bangladesh sediments (Nickson *et al.*, 2000; Nickson *et al.*, 1998; AAN/RGAG/NIPSOM (Asia Arsenic Network [AAN], Research Group of Applied Geology [RGAG], and National Institute for Preventive and Social Medicine [NIPSOM]), 1999). However, McArthur *et al.* (2001) suggested that the generally low SO_4^{2-} concentrations in groundwater did not indicate pyrite oxidation and Harvey *et al.* (2002) inferred an inverse relationship between dissolved arsenic and SO_4^{2-} in pore water of the Holocene aquifer in Bangladesh. Rather, the presence of pyrite may indicate that sediments are not oxidized enough to mobilize arsenic and that pyrite is forming via biogenic mineralization (AAN/RGAG/NIPSOM, 1999), thus sequestering arsenic from solution.

2. *Competitive ion exchange:* Acharyya *et al.* (1999; 2000), hypothesized that arsenic anions sorbed to aquifer sediments (mostly to iron ((oxy)(hydr)oxides) are displaced into solution by competitive exchange with phosphate from the use of fertilizers. However, (McArthur *et al.*, 2001) argued that dissolved phosphate is contributed primarily by sources other than fertilizers (e.g. reductive dissolution of iron (oxy)(hydr)oxides, degradation of human waste, and oxidation of peat) and that competitive exchange may be minor. DPHE (2000) data from site-specific studies at Laxmipur and Faridpur in Bangladesh showed that phosphate concentrations increase with depth, which is inconsistent with a surface input. Furthermore, the locations of high-arsenic groundwater do not match with the areas where phosphatic fertilizers are used (McArthur *et al.*, 2001; Ravenscroft, McArthur and Hoque, 2001). Mukherjee and Fryar (2008) did not detect any phosphate in deeper groundwater in West Bengal.

3. *Reductive dissolution of iron (oxy)(hydr)oxides:* Bhattacharya, Chatterjee and Jacks (1997) first proposed that the main mechanism for mobilization of arsenic sorbed to iron (oxy)(hydr)oxides is reductive dissolution. This hypothesis has since been generally accepted (Harvey *et al.*, 2002; McArthur *et al.*, 2004; Nickson *et al.*, 2000; Harvey *et al.*, 2005; McArthur *et al.*, 2001; Harvey *et al.*, 2006; Ravenscroft, McArthur and Hoque, 2001). McArthur *et al.* (2001) stated that arsenic liberation via reductive dissolution of metal (oxy)(hydr)oxides is common in nature and the good correlation between arsenic and HCO_3^- in Bangladesh groundwater is probably an indication of reduction.

   In general, Bengal basin groundwater has very low Eh, O_2, NO_3^-, and SO_4^{2-}, high iron and manganese, and some detectable NH_4^+, H_2S, and methane (CH_4) (e.g. (Bhattacharya, Chatterjee and Jacks, 1997; Ahmed *et al.*, 1998; Dowling *et al.*, 2002)). These results are all indicative of reducing conditions. For groundwater samples collected from different depths at sites in Bangladesh and West

Bengal, Swartz *et al.* (2004) and Mukherjee and Fryar (2008) found that Eh typically ranges between 15 and 120 mV. Mukherjee and Fryar (2008) calculated that the observed Eh values are within the range of Fe(III) and As(V) reduction.

4. *Reduction and oxidation:* A hypothesis combining aspects of the aforementioned mechanisms (1) and (3) has been proposed by Zheng *et al.* (2004) and elaborated upon by Mukherjee and Fryar (2008). According to this hypothesis, arsenic is primarily mobilized via reduction of iron (oxy)(hydr)oxides, but local oxidation of pyrite as a result of O_2 influxes is possible. Oxidized iron and sulfur species may be rereduced, thus resequestering arsenic, as suggested by the stable isotopic signature of SO_4^{2-} ($\delta^{34}S_{SO4}$) in groundwater. However, Mukherjee and Fryar (2008) commented that arsenic may not be completely immobilized under strongly reducing conditions, but rather may be retained in the solution as a result of partial redox equilibrium.

Polizzotto *et al.* (2005a; 2006), who conducted spectroscopic analyses and batch experiments on soil and sediment samples from Munshiganj, Bangladesh, offered a variation on the fourth hypothesis. These authors showed that arsenic in aquifer sediments primarily resides in *detrital* and authigenic sulfide grains with a lesser amount weakly sorbed to sands. In contrast, arsenic occurs at higher concentrations in surface soils, where it mainly resides in iron (oxy)(hydr)oxides and detrital sulfides. Polizzotto *et al.* (2005a, 2006), hypothesized that water-table fluctuations between dry and wet seasons result in alternating oxidation of sulfides and reduction of iron (oxy)(hydr)oxides in surface soils with residual arsenic and sulfur species being transported into the shallow aquifer. Leaching of SO_4^{2-} and competition with other ions (such as Fe^{2-}) would limit arsenic sequestration in authigenic sulfides.

Several workers have observed a moderate to strong correlation between dissolved arsenic and iron, which they attributed to reductive dissolution of iron (oxy)(hydr)oxides and mobilization of sorbed arsenic (Nickson *et al.*, 2000; Nickson *et al.*, 1998; McArthur *et al.*, 2001; Dowling *et al.*, 2002) (local correlations up to $r^2 = 0.8$ to 0.9); (Stüben *et al.*, 2003). However, (Swartz *et al.*, 2004) found that arsenic in sediments from a depth of 165 m at Munshiganj, Bangladesh, does not correlate well with iron. Mukherjee (2006) also showed at the regional scale that iron correlates only weakly with arsenic ($r^2 = 0.26$), although Fe(II) yields a better correlation with As(III) ($r^2 = 0.32$). The weak correlation suggests that iron reduction and arsenic release may not be simultaneous. Similarly, Islam *et al.* (2004) showed that sorbed arsenic is released from Bengal basin sediments after Fe(III).

In summary, reduction appears to be the major mechanism for arsenic mobilization in Bengal basin groundwater, although local oxidation of reduced species, resorption, and precipitation of authigenic phases may be complicating factors. In various studies at local and regional scales, arsenic in Bengal basin groundwater has been found to correlate with manganese (Stüben *et al.*, 2003), ammonium (NH_4^+) (Dowling *et al.*, 2002; Ahmed *et al.*, 2004), CH_4 (McArthur *et al.*, 2004; McArthur *et al.*, 2001; Ravenscroft, McArthur and Hoque, 2001; Dowling *et al.*, 2002), dissolved organic carbon (DOC) (Ahmed *et al.*, 2004), and zinc (Stüben *et al.*, 2003). The strongly reducing environment in groundwater, which may result in coliberation or production of the above-mentioned redox-sensitive solutes, has been offered as an explanation. However, Mukherjee and Fryar (2008) did not observe most of these correlations in his regional-scale study of groundwater in the western Bengal basin. He commented that the lack of these correlations suggests that arsenic is mobilized by multiple processes. Moreover, the lack of correlation between arsenic and pH indicates that pH-dependent sorption reactions may not play a significant role in regulating arsenic mobility in the deeper groundwater of the western Bengal basin. One unsolved question is the biogeochemical trigger for the reduction sequence that is thought to liberate arsenic. Cumming *et al.* (1999) experimentally showed that reductive dissolution of iron (oxy)(hydr)oxides is enhanced by microbial activity of dissimilatory bacterial strains like *Shewanella* BrY. The microcosm studies by Islam *et al.* (2004) showed that anaerobic metal-reducing bacteria could play a catalytic role in mobilization of arsenic from sediments collected from West Bengal.

The addition of a substrate may stimulate dissimilatory microbial reduction (Van Geen *et al.*, 2004), probably by the same group of microorganisms that are responsible for SO_4^{2-} reduction (Inskeep, McDermott and Fendorf, 2002; Oremland *et al.*, 2002). Yamazaki *et al.* (2000), McArthur *et al.* (2001, 2004), Ravenscroft, McArthur and Hoque (2001), and Ravenscroft *et al.* (2005) suggested that peat within aquifer and *aquitard* sediments drives microbial metabolism within basin sediments. Peat has been widely documented in many areas near the active delta. Yamazaki *et al.* (2000) reported a high average concentration of arsenic (121 mg kg^{-1}) from NOM-rich muddy sediments in comparison to NOM-poor sediments (0.7–23 mg kg^{-1} arsenic) from Samta village in Bangladesh. McArthur *et al.* (2004) argued that CH_4 in groundwater of the southern Bengal basin (North 24 Parganas district, West Bengal) probably results from decomposition of embedded peat layers.

In contrast, peat is unlikely to be the redox driver in older flood plain and delta areas with high-arsenic concentrations, where no peat layers have been documented (Mukherjee, 2006). Such areas include Munshiganj, Bangladesh (Harvey *et al.*, 2002), and Murshidabad and Nadia districts, West Bengal (Mukherjee, 2006), where elevated CH_4 concentrations have been reported. As an alternative explanation, Yamazaki *et al.* (2000) and Harvey *et al.* (2002, 2005, 2006) proposed that anthropogenic, dissolved NOM (from human waste or agricultural activity) infiltrates through paddy fields and the bottoms of surface-water bodies. Pumping may accelerate the downward movement of both NOM and arsenic in groundwater from the shallow subsurface Harvey *et al.* (2002, 2005, 2006). Although this scenario is plausible in light of the large population and land use in the basin, infiltration of anthropogenic NOM may not explain the existence of arsenic in deeper groundwater within the basin, because most of the dissolved NOM should be oxidized at relatively shallow depths under natural vertical hydraulic gradients.

6.5.3 Middle Ganges Plain, India

In India, arsenic contamination was thought to be restricted to the shallow alluvial aquifers along and east of the Bhagirathi–Hoogly River, the main distributary of the Ganges in West Bengal. However, several recent surveys have identified drinking water wells with elevated arsenic concentrations in various parts of the middle Ganges plain, adjoining the river and upstream from the Bengal basin, in the states of Jharkhand, Bihar, and Uttar Pradesh ((Ramanathan, Bhattacharya and Tripathi, 2006; Chakraborti *et al.*, 2003; Chakraborti *et al.*, 2004; Ahamed *et al.*, 2006), Table 6.3; Appendix D).

The Sahibgunj district of Jharkhand is situated on the southern and western banks of the Ganges at the transition from the middle to lower Ganges plain. On the opposite bank of the Ganges is the Maldah district of West Bengal, which has already been declared one of the most affected areas of the Bengal basin. Among the 698 groundwater samples collected from shallow tube wells in 17 villages of Sahibgunj district by Chakraborti *et al.* (2004), 33 % had $\geqslant 10\,\mu g\,L^{-1}$ of arsenic and 25 % had $\geqslant 50\,\mu g\,L^{-1}$. In studying three blocks (smaller administrative units) of the district, Bhattacharjee *et al.* (2005) found a maximum arsenic concentration of $620\,\mu g\,L^{-1}$ with a mean of $63\,\mu g\,L^{-1}$ for Sahibgunj block. The shallow dug wells were found to have low arsenic. The cause of the pollution has been speculated to be similar to that of the Bengal basin (reductive dissolution of metal (oxy)(hydr)oxides), as indicated by high concentrations of dissolved iron and manganese.

In Bihar state, north of Jharkhand, arsenic contamination was first discovered in Semria Ojha Patti village of Bhojpur district in June 2002 by Chakraborti *et al.* (2003) (Table 6.3). Bhojpur is one of the westernmost districts of Bihar, where the Ganges enters the district from Uttar Pradesh. Chakraborti *et al.* (2003) analyzed 206 drinking water tube wells (about 95 % of the wells in the village). Fifty-seven percent of the tested wells had $\geq 50\,\mu g\,L^{-1}$ of arsenic and another 25 % had $10–50\,\mu g\,L^{-1}$. Twenty percent of the tested wells had concentrations of $>300 \mu g L^{-1}$ with a maximum of $1654 \mu g L^{-1}$. Chakraborti *et al.* (2003) commented that the percentage of the contaminated wells and affected children in this village probably overshadows the condition in the Bengal basin. Iron concentrations reported from this study (average $2482 \mu g L^{-1}$) were higher than previous estimates in the middle Ganges plain (Acharyya *et al.*, 1999), although arsenic and iron were not well correlated in contrast to various studies in Bangladesh.

A recent survey of arsenic in the groundwater of Bihar included analyses of more than 28 000 samples along a 10-km zone following the main channel of the Ganges. The survey found severe contamination in several districts (Ghosh, 2005). Mukherjee *et al.* (2007) found a strong geologic-geomorphic control on the arsenic distribution in the shallow aquifers of middle Ganges plain, stretching from the Himalayan piedmonts to the Indian cratons. The study showed that while $\sim 80\%$ of the collected groundwater samples from the aquifers of younger alluvium are contaminated, aquifers in the older sediments were almost uncontaminated. In Patna district, more than 1050 wells were contaminated with a maximum arsenic concentration of $724 \mu g L^{-1}$ in Maner block. In Bhojpur district, 47.7 % of the 6292 wells tested were contaminated with the highest reading of $1851 \mu g L^{-1}$ in Barhara block. In Raghopur block of Vaishali district, almost 100 % of the tested wells were contaminated and arsenic concentrations ranged from ~ 10 to $300 \mu g L^{-1}$. The worst contamination was observed in Bhagalpur district adjoining the contaminated districts of Maldah in West Bengal and Sahibganj in Jharkhand. In Bhagalpur most of the hand-pumped wells and even dug wells were found to have elevated arsenic concentrations with values exceeding $500 \mu g L^{-1}$ in several villages of Nathnagar block. The estimated numbers of people at risk from arsenic exposure were over 550 000 in Patna district, 360 000 in Bhojpur district, and almost 480 000 in Vaishali district (Ghosh, 2005).

Immediately upstream along the Ganges (to the west) from Bhojpur in Bihar are the districts of Ballia, Ghazipur, and Varanasi in Uttar Pradesh state, where widespread arsenic contamination has recently been identified. Ahamed *et al.* (2006) reported that 46.5 % of 4780 wells tested in Ballia, Varanasi, and Gazipur had arsenic concentrations of $\geq 10 \mu g L^{-1}$. Twenty-seven percent of samples had concentrations of $\geq 50 \mu g L^{-1}$ and 10 % were $\geq 300 \mu g L^{-1}$. The worst-affected area is the southeastern part of Ballia, along the Ganges, where $\sim 96\%$ of the active wells in Chayan Chapra village had arsenic concentrations of $\geq 50 \mu g L^{-1}$ and almost 82 % of the wells in the village exceeded $300 \mu g L^{-1}$ of arsenic. The sampled tube wells were 6–60 m below ground level with a mean depth of about 26 m. Ahamed *et al.* (2006) found a qualitative correlation between the age of the sampled tube wells and the concentration of arsenic. The oldest wells in the area were up to about 30 years old with a mean age of 6.5 years. Ramanathan, Bhattacharya and Tripathi (2006), who studied 61 wells in Ballia and Ghazipur districts, found that arsenic concentrations in shallow (<35 m) to deep (>50 m) wells were highly variable, ranging from 6 to $259 \mu g L^{-1}$. Concentrations of NO_3^- (<1–120 mg L^{-1}) and SO_4^{2-} (15–379 mg L^{-1}) tended to be elevated relative to Bengal basin groundwater, which suggest a different redox environment. The observed Eh range (~ 120–700 mV) indicates oxic to slightly iron-reducing conditions, which could limit microbial reduction of NO_3^-, SO_4^{2-}, and Fe(III). Ramanathan, Bhattacharya and Tripathi (2006) also noted that the most contaminated depths are 10–20 and 30 m in Ballia, and 10–20 and 30–60 m in Ghazipur. The authors suggested that these are the depths of mixing between oxidized and reduced waters.

One obvious similarity among these contaminated areas is that they are located in the active flood plain of the Ganges and its tributaries. Chakraborti *et al.* (2004) commented that the flood plains of many of the rivers that originate from the Himalayas and Tibetan plateau have been found to be contaminated, which suggests a common source of arsenic-enriched sediments. The mechanisms for the mobilization of arsenic in the middle Ganges plain have not yet been studied, but because of geomorphic and geologic similarities, it is plausible that the processes responsible for groundwater contamination in the Bengal basin are also the cause of contamination in Jharkhand, Bihar, and Uttar Pradesh.

6.5.4   Donargarh rift belt, Chattisgarh, central India

Natural arsenic contamination of groundwater has been reported from some villages in Rajnandgaon district of Chattisgarh state in central India ((Acharyya, 2002; Chakraborti *et al.*, 1999), Appendix D). The total population of this district is about 1.5 million (Rahman *et al.*, 2006). The areas of contamination are generally restricted to the early *Proterozoic* Donargarh-Kotri rift belt (Acharyya, 2002). Eight of the 22 villages from where 146 groundwater samples were collected had arsenic $\geq 10 \,\mu g \, L^{-1}$ and four had arsenic $\geq 50 \,\mu g \, L^{-1}$ with a maximum concentration of $880 \,\mu g \, L^{-1}$ (Rahman *et al.*, 2006). The most affected areas are the villages of Kaurikasa, Joratarai, and Sonsayatola in the northern part of the rift belt, but Gurwandi and a few other villages in the southern part are also affected (Acharyya, 2002). In contrast to most other arsenic-affected areas in India, arsenic was also detected in shallow dug wells with a maximum concentration of $520 \,\mu g \, L^{-1}$ (Rahman *et al.*, 2006). However, the Seonath River, which passes through the most affected area, has been found to have $< 10 \,\mu g \, L^{-1}$ of arsenic, suggesting that the contamination is far more localized than has been documented in eastern India (Acharyya, 2002).

The source of the arsenic seems to have been *hydrothermal* fluids that disseminated arsenic within *felsic* volcanic rocks and some *granites*. Weathered samples of exposed volcanic rocks and soils had up to $250 \, mg \, kg^{-1}$ of arsenic with average concentrations in the range of $10-25 \, mg \, kg^{-1}$. *Regolith* samples from *rhyolites* in the affected village of Kaurikasa had up to $800 \, mg \, kg^{-1}$ of arsenic and the peak arsenic solid-phase concentrations in the village of Sonsayatola were $>260 \, mg \, kg^{-1}$ (Acharyya, 2002). Arsenic is probably liberated by weathering of the host rocks in the rift zone, but concentrations in groundwater may be limited because most of the arsenic in the sediments is bound to metal (oxy)(hydr)oxides (Acharyya *et al.*, 2005). However, under reducing conditions in the subsurface of Rajnandgaon, arsenic appears to be mobilized by reductive dissolution of the iron (oxy)(hydr)oxides, as in the Bengal basin.

6.5.5   Terai alluvial plain, Nepal

The Terai region of southern Nepal is located in the foothills of the Himalayas and is the source of sediment for many of the north Indian rivers. About 47 % of the population of Nepal resides in the region (Rahman *et al.*, 2006). The *unconfined* Quaternary aquifers, which are on the order of 300 m thick (Khadka, 1993), are tapped by an estimated 400 000 (Panthi, Sharma and Mishra, 2006) to 800 000 wells (Smedley, 2005). They supply water to ~11 million people, or about 90 % of the residents of Terai (Rahman *et al.*, 2006).

Elevated arsenic was first discovered in 1999 in the Jhapa, Morang, and Sunsari districts of eastern Terai, which border the Indian state of West Bengal (Table 6.3; Appendix D). About 8 % of the 268 groundwater samples had arsenic $>10 \,\mu g \, L^{-1}$ with two samples exceeding $50 \,\mu g \, L^{-1}$ (Panthi, Sharma and Mishra, 2006). Subsequent surveys of groundwater quality in Terai have revealed the presence of arsenic in some tube wells with depths <50 m. However, most of the samples had $< 10 \,\mu g \, L^{-1}$ of arsenic. Arsenic-related health problems have been detected in some of the affected areas. The most comprehensive survey was done by the National Arsenic Steering Committee of Nepal (NASC), which analyzed 25 058 water samples. NASC reported that 69 % of groundwater samples had arsenic concentrations of $<10 \,\mu g \, L^{-1}$, while 8 % exceeded $50 \,\mu g \, L^{-1}$ (Panthi, Sharma and Mishra, 2006; Tuinhof and Nanni, 2003; Shrestha *et al.*, 2004). The highest reported concentration was $2620 \,\mu g \, L^{-1}$ from Rupandehi district (Shrestha *et al.*, 2004; Tandukar, 2001). Nawalparasi represents the worst-affected district (Panthi, Sharma and Mishra, 2006), but major arsenic contamination has also been reported in the districts of Rautahat, Bara, Parsa, Kapilbastu, Rupandehi, Banke, Kanchanpur, and Kailali in the central and western Terai. Tandukar (2001) and Tandukar *et al.* (2006) reported the highest average concentrations of arsenic in the catchment of the Bagmati River. The Nepal Red Cross Society analyzed 2200 groundwater samples from 17 of the 20 districts of Nepal and reported more than 8 % of the samples having $>10 \,\mu g \, L^{-1}$

of arsenic. The highest arsenic concentrations have been found at depths of 10–30 m with much lower concentrations below 50 m depth (Tuinhof and Nanni, 2003; Shrestha *et al.*, 2004), which indicate that the deeper portions of the surficial aquifer and underlying aquifers may still be uncontaminated (Smedley, 2005). Fifty-five percent of water samples collected from streams that drain the Terai, *sedimentary rocks* of the Siwalik Group, and carbonate and low-grade metamorphic rocks of the Lesser Himalayas had $\geqslant 10\,\mu g\,L^{-1}$ of arsenic (Emerman *et al.*, 2005). Groundwater in Nawalparasi is anoxic and microbially mediated reductive dissolution similar to that observed in the Bengal basin is thought to be the primary mechanism of arsenic mobilization (Brikowski *et al.*, 2006).

6.5.6   Indus alluvial system, Pakistan

The Indus alluvial plain of Punjab and Sindh provinces in Pakistan is of Quaternary age and is similar to the Ganges alluvial system of India and Bangladesh. The sediments of the plain have their provenance in the western Himalayas and are brought down by the River Indus and its tributaries. The alluvial deposits are widespread and thick (Smedley, 2005) and mostly form unconfined aquifers. However, in contrast to the Bengal basin, the aquifers are relatively oxic (Mahmood *et al.*, 1998; Tasneem, 1999). The area also differs in having a more arid climate and greater proportion of Pleistocene deposits with greater apparent connectivity between the river systems and the aquifers (Smedley, 2005).

The occurrence of arsenic in the groundwater of Pakistan has mainly been identified by district-scale surveys (Appendix D; Table 6.3). In 1999–2000, more than 300 wells were tested in six districts representing different geologic settings in Punjab, including the main Indus alluvial aquifer in Gujarat district (Smedley, 2005; Haque, 2005). Ninety percent of samples had arsenic concentrations of $<10\,\mu g\,L^{-1}$ and only six samples (2 %) had concentrations of $>50\,\mu g\,L^{-1}$ (Iqbal, 2001). This survey revealed the existence of several arsenic hotspots in the Indus alluvial plain with concentrations ranging from 10 to $>200\,\mu g\,L^{-1}$, particularly in southern Punjab. Naseem *et al.* (2001) reported elevated arsenic in the groundwater of Kasur district, which was not included in this survey. A more detailed survey in 2002–2003 focused on Bahawalpur, Rahim Yar Khan, and Multan districts in southern Punjab. Of 2395 laboratory analyses, 22.7 % had arsenic concentrations of $>10\,\mu g\,L^{-1}$ and 2.8 % were $>50\,\mu g\,L^{-1}$ (Haque, 2005). Phase II of the more detailed survey, conducted in 2003–2004, involved 'blanket testing' of all water sources in these three districts and Sargodha district, as well as screening of four more districts (Dera Ghazi Khan, Jhang, Layyah, and Muzaffargarh). Laboratory analyses showed that 65 to 85 % of samples in Bahawalpur, Rahim Yar Khan, and Multan districts had arsenic $>10\,\mu g\,L^{-1}$, and 27–40 % of samples had $>50\,\mu g\,L^{-1}$ (Haque, 2005). Nickson *et al.* (2005) reported arsenic concentrations as high as 906 $\mu g\,L^{-1}$ from Muzaffargarh. Overall, 11.32 million residents of Punjab (24 % of the province's population) are estimated to have drinking water with arsenic $>10\,\mu g\,L^{-1}$ and 2.36 million are exposed to concentrations of $>50\,\mu g\,L^{-1}$ (Haque, 2005). The worst-affected district appears to be Multan with an estimated 2.62 million (74 % of the population) exposed to concentrations of $>10\,\mu g\,L^{-1}$ (Haque, 2005).

Results from Sindh province, down the Indus valley (to the south) from Punjab, indicate additional arsenic hotspots in groundwater (Appendix D). In 2001, the provincial Public Health Engineering Department found arsenic contamination in five of nine surveyed districts with 16 % of samples exceeding $50\,\mu g\,L^{-1}$ (Haque, 2005). In follow-up sampling conducted between 2002 and 2004, almost 15 000 (22 %) of field tests yielded arsenic concentrations of $>10\,\mu g\,L^{-1}$ and 4317 (6 %) were $>50\,\mu g\,L^{-1}$ (Haque, 2005). The worst-affected district was Dadu with 29 % of samples $>10\,\mu g\,L^{-1}$ and 10 % $>50\,\mu g\,L^{-1}$ (Haque, 2005).

Because of the generally oxic conditions in the Indus alluvial aquifers, the mechanisms of arsenic mobilization may differ from those observed in the Ganges basin. Elevated arsenic concentrations in the shallow Quaternary aquifers of the urban Thal Doab area (Muzaffargarh district, Punjab) are thought to have resulted from sewage pollution (Nickson *et al.*, 2005). Such pollution could induce anaerobic conditions and thus reductive release of arsenic via iron (oxy)(hydr)oxide dissolution. In urban areas, arsenic increased up to $170\,\mu g\,L^{-1}$ with depth, but rural areas in the same region had $< 25\,\mu g\,L^{-1}$ of arsenic, probably because of sorption by oxidized sediments in the absence of severe human pollution. In the Kalalanwala area (Kasur district, Punjab), both As(V) and F^- are elevated in groundwater with maximum concentrations at depths of 24–27 m. The highest arsenic concentration recorded from this area was $\sim 1900\,\mu g\,L^{-1}$ (Farooqi, Masuda and Firdous, 2007). Arsenic may be released into groundwater under alkaline pH conditions. Local coal combustion in brick fields may be another arsenic source (Farooqi, Masuda and Firdous, 2007).

6.5.7 Irrawaddy delta, Myanmar

As reviewed by Smedley (2005), preliminary studies have recently indicated that the delta of the Irrawaddy River in southern Myanmar may have high concentrations of arsenic (Appendix D). There are $\sim 400\,000$ drinking-water wells in Myanmar, most of which have been installed since 1990. The Ayeyarwaddy and Bago regions and the states of Mon and Shan are considered to be the worst affected. A reconnaissance study of 1912 wells in the southern delta area showed that $\sim 22\,\%$ of the wells have arsenic concentrations of $\geqslant 50\,\mu g\,L^{-1}$. Studies by the (Water Resources Utilization Department (WRUD), 2001) in the western coastal town of Sittway and around the southern towns of Hianthada and Kyaunkone indicated that about 10–13 % of the wells have arsenic $\geqslant 50\,\mu g\,L^{-1}$ with 35 % of the wells in Sittway being $\geqslant 10\,\mu g\,L^{-1}$. Shallower wells (30–50 m deep) were reported to be far more contaminated than deep (55–70 m) wells, but two of the three deep wells in Kyaungkone were found to be contaminated. In contrast, analyses of 125 samples from Nyaungshwe in Shan state found only 4 % of the wells to have arsenic concentrations of $\geqslant 50\,\mu g\,L^{-1}$ (UNDP/UNCHS (United Nations Development Programme and United Nations Centre for Human Settlements), 2001). More recent surveys by (WRUD, 2001) have shown that 15 % of the wells sampled have arsenic concentrations of $\geqslant 50\,\mu g\,L^{-1}$, including 8 % of the dug wells. However, because field test kits were used, the accuracy of the results is questionable (Smedley, 2005).

6.5.8 Mekong plain and delta, Cambodia, Vietnam, and Laos

The Mekong River originates in the Tibetan Plateau and flows for 4300 km through China, Myanmar, Laos, Thailand, Cambodia, and Vietnam to the South China Sea (Figure 6.2). The Mekong divides into two branches in southern Cambodia, the Mekong (main channel) to the east and the Bassac River to the west. The delta occupies $\sim 62\,000\,km^2$, mostly in Vietnam (Berg, Giger and Tran, 2006). The delta plain is bounded by Pleistocene uplands to the north and the Gulf of Thailand to the west (Nguyen, Ta and Tateishi, 2000). During the past few years, a large number of tube wells (probably >40 000 in Cambodia alone; (Fredericks and Arsenic Secretariat, 2004)) have been drilled in the late *Cenozoic* aquifers. This activity has stemmed from growing concerns about bacterial pollution of surface water and near-surface dug wells.

Most of the arsenic surveys have been conducted in the southern Mekong basin in Cambodia with some in Vietnam and Laos (Appendix D; Table 6.3). The WHO and the government of Cambodia initially conducted a reconnaissance screening of ~ 100 samples from drinking-water wells in 13 provinces of Cambodia. About 9 % of the wells were found to have arsenic $\geq 10\,\mu g\,L^{-1}$ and five of the 13 investigated provinces were declared to be contaminated. The regions with the highest concentrations (as high as $500\,\mu g\,L^{-1}$ of arsenic) were observed around the city of Phnom Penh, mostly in the Kien Svay and Ta Khman districts of Kandal province (Feldman and Rosenboom, 2001). The reason why the highest concentrations were observed around Phnom Penh is not clear, but Smedley (2005) speculated that they may be related to higher rates of pumping or large inputs of NOM from anthropogenic sources in and

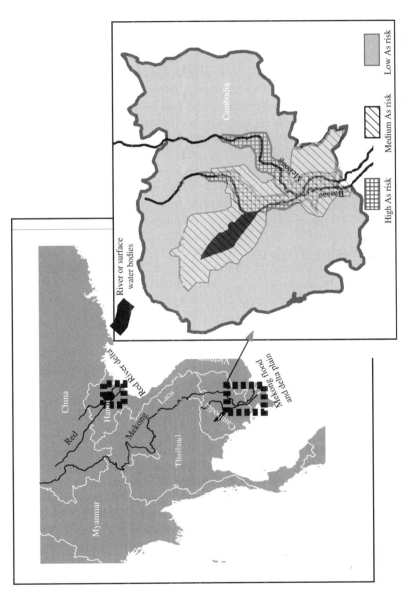

Figure 6.2 The location of the arsenic-contaminated areas of the Mekong delta in Cambodia and Vietnam. (Reproduced by permission of David Fredericks, Copyright 2004 Fredericks and Arsenic Secretariat.)

around the urban area. At present, seven provinces have been found to be arsenic-contaminated, mostly in the floodplains of the Mekong, Bassac, and Tonle Sap rivers (Figure 6.2). A survey of 5000 wells (Halperin, 2003) undertaken by the United Nations Children's Fund (UNICEF) in conjunction with nongovernment organizations concluded that 19 % of the wells occurring in the risk zones have arsenic $\geq 50 \mu g L^{-1}$ and about 50 % have concentrations of $\geq 10 \mu g L^{-1}$. Hazard maps developed by Polya *et al.* (2005) based on an investigation of $\sim$1000 wells indicate that the highest risk areas are around the present-day channels of the Mekong and Bassac rivers and that the arsenic-enriched water is hosted in aquifers comprised of Holocene sediments (Figure 6.2). Polya *et al.*, (2005) further observed that most of the contaminated wells are >16 m deep, although high concentrations also occur at shallower depths in the Holocene aquifers and in preHolocene (late Cenozoic) sediments. These authors could not definitively conclude if the shallow groundwater contamination resulted from enhanced exploitation for domestic use and irrigation.

Polizzotto *et al.* (2005b) conducted a site-specific study in Kandal province near Phnom Penh in the floodplain between the Mekong and Bassac rivers. They measured arsenic concentrations at different depths: shallow (8–12 m in the surface clay), intermediate (20–25 m, gray Holocene sand), and deep (37–60 m, gray Holocene sand). In general, higher concentrations were observed in the wells around the Mekong than the Bassac. The highest values (up to $1210 \mu g L^{-1}$) were mostly detected in the deep wells. Mean arsenic concentrations along the Mekong were $56 \mu g L^{-1}$ in shallow wells, $521 \mu g L^{-1}$ in intermediate-depth wells, and $716 \mu g L^{-1}$ in deep wells. Along the Bassac, mean concentrations were $110 \mu g L^{-1}$ in shallow wells, $127 \mu g L^{-1}$ in intermediate-depth wells, and $157 \mu g L^{-1}$ in deep wells. Polizzotto *et al.* (2005b) concluded that groundwater arsenic concentrations are spatially, but not temporally variable, and arsenic is positively correlated with increasing pH. Buschmann *et al.* (2006) collected 131 groundwater samples from tube wells in the same general area. Like Polizzotto *et al.* (2005b), Buschmann *et al.* (2006) found that arsenic is mostly restricted to sediments along the river channels with a range of $1–1340 \mu g L^{-1}$ (Table 6.3). Arsenic concentrations were $\geq 10 \mu g L^{-1}$ in 48 % of the sampled wells with a mean of $232 \mu g L^{-1}$ in the highly affected zones. Buschmann *et al.* (2006) concluded that arsenic concentrations are related to the microtopography of the floodplain because the concentrations decrease away from the center of the shallow valley.

Little is known about arsenic occurrences elsewhere in the Mekong basin. The thin upper sediments of Holocene age in Vietnam yield little groundwater (Smedley, 2005) that is mostly impotable because of high concentrations of sodium and chloride (Pham *et al.*, 2002). Consequently, most of the drinking water is pumped from arsenic-free deep aquifers at 150–250 m depth (Berg, Giger and Tran, 2006). Out of 99 groundwater samples from Holocene, Pleistocene, and *Pliocene* sediments in the Mekong delta of Vietnam, only one sample from the Pliocene sediments of the Ben Tre area was found to have arsenic $\geq 50 \mu g L^{-1}$ (Smedley, 2005). Fengthong, Dethoudom and Keosavanh, (2002) analyzed 200 samples from five regions of Laos and found only one sample with $>50 \mu g L^{-1}$ of arsenic, although some others had concentrations of $\geq 10 \mu g L^{-1}$. A UNICEF study of 680 drinking-water samples in Laos reported only 1 % of the wells had arsenic $>50 \mu g L^{-1}$ (Smedley, 2005).

Groundwater in the southern Mekong basin tends to be iron-reducing with high concentrations of iron, manganese, and ammonia (NH_3) (Berg, Giger and Tran, 2006). Polizzotto *et al.* (2005b) noted that groundwater and aquifer sediments of the Mekong are chemically similar to those of the Bengal basin; that is, low concentrations of O_2, NO_3^-, and SO_4^{2-} along with the occasional presence of HS^-. Buschmann *et al.* (2006) suggested that arsenic mobilization primarily results from reductive dissolution and to a lesser extent from pH values ≥ 7. In the Mekong delta in Vietnam, acid sulfate soils have developed (Husson *et al.*, 2000). Oxidation of pyrite, which is abundant in these soils (Minh *et al.*, 1998), can occur as water levels decline. Such conditions may result in the release of some arsenic sequestered in pyrite. The soils reportedly have arsenic concentrations in the range of $6–41$ mg kg^{-1} soil (Gustafsson and Tin, 1994).

6.5.9   Red River delta, Vietnam

In Vietnam, arsenic contamination has mostly been identified in the Red River delta (Bac Bo Plain) (Appendix D). The river flows for 1150 km from southern China to the Gulf of Tonkin. The river basin is bounded by the Truong Ginang and Chau Giang river basins to the north, the Ma River basin to the south and the Mekong basin to the west. The total area of the plain is around 17 000 km². The surficial Holocene sediments are about 10–40 m thick in the center of the plain and thin to about 1–3 m at its margins (Tong, 2002). The Pleistocene aquifers are as much as 100 m thick in the central and southeastern parts of the plain (Tong, 2002). The Quaternary deposits are composed of sand, silt, clay, and some *conglomerate* (Smedley, 2005).

Perhaps 150 000 tube wells, most of which are private, exist in the Red River delta (Smedley and Kinniburgh, 2002). Groundwater is withdrawn both from shallow, hand-pumped tube wells at depths of <45 m and from public-supply wells, which tap the Pleistocene aquifer beneath the city of Hanoi at depths of ∼30–70 m (Smedley, 2005). In recent years, the rate of groundwater abstraction in and around Hanoi has increased dramatically with an annual water table decline of about 1 m (Tong, 2002) and a seasonal decline of about 30 m (Trafford *et al.*, 1996).

In 1999, UNICEF, EAWAG (Swiss Federal Institute of Aquatic Science and Technology) and Hanoi National University carried out one of the first extensive surveys of arsenic concentrations in the Red River delta (Smedley and Kinniburgh, 2002). Out of 1228 samples from seven provinces, 12.5 % had arsenic >50 µg L^{-1}. In surveying arsenic contamination in four districts within a 1000-km² area around Hanoi, Berg *et al.* (2001) found that most of the sampled wells had arsenic concentrations of ≥ 10 µg L^{-1}. The mean concentration was 159 µg L^{-1} and significant numbers of wells had concentrations of ≥50 µg L^{-1}. Arsenic occurrences were spatially variable. The highest concentration (3050 µg L^{-1}) was observed from the southeastern plain (mean 140 µg L^{-1}) (Table 6.3). In the south, on both sides of the river, concentrations ranged from 1000–3000 µg L^{-1} with an average of 430 µg L^{-1}. Areas in the north and west had lower concentrations (≤230 µg L^{-1}) with a mean of 32 µg L^{-1} in the north and 67 µg L^{-1} in the west. In Hanoi itself, concentrations as high as 300 µg L^{-1} have been recorded. Tong (2002) reported arsenic hotspots in groundwater south, west, and east of the city. High concentrations were observed mainly in shallow wells, but some deep wells had concentrations up to 440 µg L^{-1}. Both the Holocene and Pleistocene aquifers were contaminated (Tong, 2001).

The reason for spatial variation in arsenic concentrations is not yet well understood for the Red River delta. Arsenic is probably derived from upstream weathering and *erosion* and is probably transported with iron (oxy)(hydr)oxides in alluvial sediments (Berg, Giger and Tran, 2006). In five sediment cores up to 40 m deep, arsenic concentrations were much higher in the clays (2–33 mg kg^{-1}) than in the sands (< 1–5 mg kg^{-1}) (Berg, Giger and Tran, 2006). Although no correlation was observed between extractable arsenic from the sediments and dissolved arsenic, extractable arsenic showed a strong correlation with extractable iron, which reinforces the idea of arsenic mobilization through reductive dissolution. Smedley (2005) suggested that the spatial variation may be a function of sediment thickness, chemistry, hydraulic connectivity, and the age of the aquifers. For example, the lower arsenic concentrations to the north of Hanoi may be related to the low thickness of the Pleistocene aquifers.

Arsenic concentrations have also been reported to fluctuate seasonally. Berg *et al.* (2001) noted that the highest arsenic concentrations occurred during the transition from the rainy to the dry season, which is between September and December. Tong (2002) found that concentrations are generally higher in the dry season. However, Berg, Giger and Tran (2006) observed that arsenic concentrations were highest during the latter part of the rainy season, which is from June to November. These authors commented that the variation in concentrations is probably related to the annual water table fluctuation of ∼10 m, which corresponds to the seasonality of the Red River's discharge. The elevated arsenic is probably released from the upper aquifer under reducing conditions in response to waterlogging during the rainy season.

Groundwater in the Quaternary aquifers of the Red River delta is strongly anoxic with 50–100 % of arsenic occurring as As(III) (Berg, Giger and Tran, 2006). Trafford *et al.* (1996), Berg, Giger and Tran (2006), and Berg *et al.* (2001) concluded that subsurface anoxia is probably microbially driven by abundant NOM, as indicated by sedimentary organic carbon concentrations that may be >15 %. Peat layers are occasionally encountered within 10 m of land surface. Berg, Giger and Tran (2006) suggested that the low dissolved O_2 and high bicarbonate alkalinity, which can exceed 800 mg L^{-1}, are manifestations of NOM. Elevated N_2 concentrations (10–48 mg L^{-1}) are also consistent with anoxia and the highest N_2 concentrations occurred in wells with the highest arsenic concentrations. Like Harvey *et al.* (2002) for Bangladesh, Smedley (2005) speculated that elevated arsenic in groundwater around Hanoi may somehow be related to the introduction of excess organic matter via wastewater recharge. At present, the increase of arsenic in the shallow aquifers cannot be definitively related to increased groundwater abstraction (Berg, Giger and Tran, 2006).

6.5.10   Yellow River plains, Inner Mongolia, China

Endemic arsenicosis has been reported in multiple studies of the alluvial plains of the Yellow River in Inner Mongolia, China (as summarized by (Smedley and Kinniburgh, 2002; Smedley, 2005; Xie *et al.*, 2006)). The first cases were reported in 1990 from Zhi Ji Liang village near Huhhot, the capital of Inner Mongolia (Luo, Zhang and Ma, 1997). Since then, groundwater containing elevated arsenic has been found in more than 655 villages in 11 counties (Rahman *et al.*, 2006). The most contaminated aquifers exist in the Tumet plain, including the area near Huhhot, the Hetao plain, and the Ba Men region (Luo, Zhang and Ma, 1997; Ma, Xia and Wu, 1999). More than 300 000 people living in these areas are thought to be at risk from arsenic poisoning ((Ma, Xia and Wu, 1999), Table 6.3).

Huhhot is located in the foothills of the Daqing Mountains on the north bank of the Yellow River (Xie *et al.*, 2006). The 4800-km^2 Huhhot alluvial basin (HAB), which consists of Quaternary fluvio-*lacustrine* sediments derived from the surrounding highlands, is thought to be enriched in arsenic (Luo, Zhang and Ma, 1997). The northern portion of the HAB is a geomorphic depression at the foot of alluvial fan complexes. The central portion is characterized by fluvio-lacustrine deposits of the Daheihe basin. Alluvial fans and lake deposits of the Yellow River are located in the southwest (Xie *et al.*, 2006).

The proportion of deep groundwater contaminated with arsenic appears to be greater in the HAB than in the aforementioned areas in the Indian subcontinent. Of 305 sampled shallow wells (4–30 m depth) reported by Luo, Zhang and Ma (1997) and Xie *et al.* (2006), 20.7 % had ≥50 µg L^{-1} of arsenic. In contrast, 54.6 % of the 33 sampled deeper wells (90–400 m) had ≥50 µg L^{-1} of arsenic. Of the 73 samples collected by Smedley *et al.* (2003), 25 % of the shallow wells and 57 % of the deeper wells had ≥50 µg L^{-1} of arsenic. The deeper aquifers are contaminated over an area of ~1000 km^2, while the contaminated portions of the shallow aquifer occur mostly along the northern margin of the HAB (Xie *et al.*, 2006). However, the maximum concentration recorded in shallow groundwater (>1800 µg L^{-1}) greatly exceeds that in deep groundwater (360 µg L^{-1}) (Xie *et al.*, 2006). The dug wells, which are generally 3–5 m deep and are most common near the Yellow River, were thought to be generally uncontaminated with maximum concentrations of no more than 20 µg L^{-1} (Xie *et al.*, 2006). However, some recent reports have noted that dug wells can have ≥50 µg L^{-1} and as much as 560 µg L^{-1} of arsenic (Smedley, 2005; Smedley *et al.*, 2003).

In the Quaternary alluvial aquifers of the Hetao plain, 96 % of the sampled wells in the Wuyuan area and 69 % of the sampled wells in the Alashan area had arsenic ≥50 µg L^{-1}. The highest arsenic concentrations were in 15- to 30-m deep tube wells with a peak value of 1350 µg L^{-1} (Smedley, 2005). In the Ba Men area, ~30 % of the sampled wells had arsenic ≥ 50 µg L^{-1} with a maximum concentration of 1800 µg L^{-1}

(Ma, Xia and Wu, 1999). The contamination in this area extends over $\sim$6000 km^2 and is mostly found in 10- to 35-m deep tube wells that tap Quaternary lacustrine sediments (Smedley, 2005).

Arsenic contamination in Inner Mongolia is thought to be geogenic (Guo, 1997) and is generally restricted to strongly reducing groundwaters in lowland areas. These waters are typically low not only in O_2, SO_4^{2-}, and NO_3^-, but also in iron and manganese (Xie *et al.*, 2006), unlike most other arsenic-contaminated reducing environments. As(III) constituted 52–75 % of the total arsenic in both shallow and deep wells. The similarity to geochemical conditions in the Bengal basin (Smedley, 2005) led Xie *et al.* (2006) to infer that arsenic in Inner Mongolia is probably mobilized by reduction. Arsenic concentrations are uniformly low in aerobic groundwater, which occurs mostly along the basin margins (Smedley, 2005). The contaminated dug wells also have reduced water and high concentrations of DOC (up to 11.4 mg L^{-1}) (Smedley, 2005; Smedley *et al.*, 2003). Arsenic contamination in deep aquifers of Inner Mongolia may stem from slow rates of groundwater flushing partly caused by limited recharge in the arid inland region (Smedley, 2005; Smedley *et al.*, 2003).

6.5.11 Taiwan

Blackfoot disease, a consequence of elevated arsenic in drinking water, has been known from the coastal plains of southwestern Taiwan since the 1930s ((Tseng *et al.*, 1968; Chi and Blackwell, 1968; Chen, 1994; Tseng, 1997), Appendix D). Shen and Chin (1964) reported groundwater contamination mostly at depths of 100–200 m, but as deep as 300 m beneath the Chianan plain. Surveys have shown that arsenic concentrations in this region commonly fall between 400 and 900 µg L^{-1}, but concentrations as high as 1820 µg L^{-1} have been noted ((Tseng *et al.*, 1968; Chen, 1994; Yeh, 1963; Kuo, 1968; Thornton and Farago, 1997), Table 6.3). Since the 1950s, treatment of surface water has limited Blackfoot disease in the Chianan plain (Smedley, 2005; Chen, 1994), yet groundwater is still extensively used on the southern Choushui River alluvial fan in Yun-Lin county on the northern margin of the plain (Liu *et al.*, 2006). TSC (TSC (Taiwan Sugar Company), 2003) monitored arsenic concentrations in 103 drinking-water wells on the fan between 1999 and 2003. About 33 % of the wells had arsenic concentrations of >50 µg L^{-1}. The geometric mean was 262 µg L^{-1} with a maximum of $\sim$600 µg L^{-1}. Depths of the contaminated wells on the fan (20–70 m) are shallower than in the Chianan plain. Hsu, Froines and Chen (1997) also reported arsenic contamination of shallow wells (16–40 m) in northeastern Taiwan with an average concentration of 135 µg L^{-1} and a maximum >600 µg L^{-1} for 377 samples.

Elevated arsenic is associated with sediments of marine origin in southwestern Taiwan (Yen, Long and Lu, 1980; Lee, 2000; Chen, 2003). Bi (1994) reported an average solid-phase arsenic concentration of 9.8 mg kg^{-1} for core samples from Ghaigang in the Chianan plain with a maximum of 26 mg kg^{-1} at a depth of 105 m below ground level. In analyzing 655 core samples from 13 wells in the southern Choushui River alluvial fan, Liu *et al.* (2006) obtained a much higher maximum concentration (590 mg kg^{-1} at 54 m depth). Liu *et al.* (2006) found a stronger correlation between dissolved arsenic and arsenic in aquitard sediments ($r^2 = 0.51$, $p < 0.01$) than in the aquifers ($r^2 = 0.21$, $p > 0.05$), which suggests that aquitard sediments are the source of the arsenic in the groundwater. Aquitard sediments, which are mostly clay and silt with some fine sand, are thought to have been deposited during repeated Holocene marine transgressions between 3000 and 9000 years ago (Liu *et al.*, 2006). These authors attributed elevated arsenic in sediments to *bioaccumulation* and biotransformation by marine organisms. The mechanism of arsenic mobilization is unclear, but positive correlations among arsenic, alkalinity, and total organic carbon (TOC) in groundwater (Liu, Lin and Kuo, 2003) suggest that microbial oxidation of organic matter is involved. Although arsenic was thought to have been liberated from metal sulfides in black-*shale* aquitards (Thornton and Farago, 1997), the predominance of As(III) relative to As(V) (Rahman *et al.*, 2006; Chen, 1994) and other redox indicators suggest reductive mobilization (Smedley, 2005).

6.5.12   Coastal aquifers of Australia

In the past few years, geogenic arsenic contamination has been reported in coastal aquifers of northern New South Wales (NSW) (O'Shea, 2006; Smith, Jankowski and Sammut, 2003; Smith, Jankowski and Sammut (2006); O'Shea, Jankowski and Sammut, 2007) and Western Australia (WA) ((Appleyard, Angeloni and Watkins, 2006), Appendix D). At Stuarts Point, NSW, arsenic-contaminated groundwater occurs in Holocene sands. Total aquifer thickness varies from a few meters to approximately 40 m depth. Two distinct arsenic peaks have been identified at different depths: at 10–11 m, total arsenic ranged from 52 to 85 $\mu g \, L^{-1}$ and As(V) dominated over As(III), while at 25 m below the surface, total arsenic was as high as 337 $\mu g \, L^{-1}$, and As(III) was the predominant arsenic species (Smith, Jankowski and Sammut, 2003). Although no major health concern has been reported from this area (drinking water is now being treated to remove arsenic), arsenic could accumulate in crops irrigated with contaminated water. Smith, Jankowski and Sammut (2006) have suggested two possible causes for the groundwater contamination. Water-table declines could result in the oxidation of arsenian pyrite present in surficial estuarine clays. In addition, dissolution of marine fossil shells within Late Pleistocene barrier sands could increase alkalinity and promote desorption of arsenic from metal (oxy)(hydr)oxides. More recent work by O'Shea, Jankowski and Sammut (2007) suggests the arsenic present within the Stuarts Point sand and clay sediments was predominantly derived from mineralized arsenian stibnite (Sb_2S_3) deposits in the hinterland. Erosion of these deposits during the Holocene transported arsenic toward the coast where sediments were deposited to form the Stuarts Point aquifer. A detailed groundwater study was conducted by O'Shea (2006), who found 39 % of 227 groundwater samples exceeded 10 $\mu g \, L^{-1}$ of arsenic. Increased arsenic was found in deeper parts of the aquifer (generally >20 m depth) and correlated with zones of seawater intrusion. Redox cycling involving the reductive dissolution of iron (oxy)(hydr)oxides and the precipitation of arsenian pyrite controls arsenic mobilization in the groundwater.

Appleyard, Angeloni and Watkins (2006) reported arsenic concentrations up to 7000 $\mu g \, L^{-1}$ from an unconfined sandy aquifer in the metropolitan area of Perth, WA. A combination of reduced rainfall, groundwater withdrawals for irrigation and water supply, and dewatering for construction during the past ~30 years has resulted in oxidation of arsenic-bearing pyrite in sediments near the water table. Arsenic is then released to groundwater tapped by shallow domestic wells for uses, such as irrigation of gardens. During the same period, concentrations of arsenic and other solutes (e.g. iron, calcium, and $SO_4{}^{2-}$) have risen in public-supply wells screened in the lower half of the 25 to 60-m thick surficial aquifer. Appleyard, Angeloni and Watkins (2006) inferred from well logs that sediments in the lower part of the aquifer were originally rich in Fe(III) compounds (i.e. Fe(III)-reducing conditions did not occur). These authors speculated that inputs of NOM from septic tanks, landfill leachate, or the oxidation of peat had progressively driven anoxia into deeper groundwater; thus, liberating arsenic by reductive dissolution of iron (oxy)(hydr)oxides or competitive sorption. The coexistence of $NO_3{}^-$, $NH_4{}^+$, $SO_4{}^{2-}$, and sulfides indicates that the lower aquifer is not at redox equilibrium, perhaps because of limited microbial activity (Appleyard, Angeloni and Watkins, 2006). Although arsenic concentrations in the deep supply wells have not exceeded 15 $\mu g \, L^{-1}$, the potential for additional release of arsenic associated with induced anoxia is a concern (Appleyard, Angeloni and Watkins, 2006).

6.5.13   Sedimentary basins and basement complexes of West Africa

In Africa, natural arsenic contamination of groundwater was apparently first identified in the Ashanti and Upper East regions of Ghana by Smedley (1996) (Appendix D; Chapter 3). The highest values observed were 64 $\mu g \, L^{-1}$ in Obuasi, Ashanti, and 141 $\mu g \, L^{-1}$ in the Bolgatanga area of the Upper East region. The main aquifers in these two areas occur below a regolith cover with thicknesses varying from < 10 m to >40 m. The source of the arsenic is thought to be arsenic-bearing pyrite in the aquifers. Smedley (1996) speculated that the lateral and vertical distributions of arsenic are related to sulfide oxidation and arsenic sorption. Arsenic concentrations are highest at depths of 20–40 m in Bolgatanga and 40–70 m in Obuasi. Groundwater at shallow depths within the regolith is acidic and oxidizing; thus, promoting immobilization of arsenic as Fe(III) arsenates or via sorption and/or *coprecipitation* with metal (oxy)(hydr)oxides in the aquifer matrix (Chapter 3). Up to 3.6 wt % of sorbed arsenic has been reported from the iron-rich sediments of this zone. At medium depths ($\sim$40 m), water-rock interactions increase because of the greater residence time of groundwater, which results in pH increases to near-neutral or slightly alkaline values. Moreover, the Eh (200–300 mV) of the water at this depth is still not sufficiently reducing to prevent sulfide oxidation and As(V) mobility. Hence, the liberated arsenic is conserved in solution. At greater depths with lower Eh values (< 200 mV), the sulfides are no longer oxidized, thereby limiting the mobility of arsenic.

A recent preliminary survey by Gbadebo (2005) documented arsenic contamination in groundwater in Nigeria. The geology of Nigeria is marked by igneous and metamorphic basement complexes, which are exposed in the central part of the country and are overlain by sedimentary basins elsewhere, notably in the Niger River delta of southern Nigeria. Gbadebo (2005) collected 82 groundwater samples for arsenic and associated hydrochemical parameters from shallow dug wells (with depths of 0.5–12.5 m below surface) and 10 boreholes (40–100 m deep). The depth to water in the main aquifers of the studied areas is typically on the order of 30 m. Arsenic was detected in many of the dug well samples, as well as in most of the borehole samples. Concentrations were typically higher in the boreholes than in the dug wells. The highest concentrations ($\sim$1100–3100 $\mu g\,L^{-1}$) occurred in boreholes from basement complexes in southwestern and northern Nigeria and from sedimentary rocks in northern Nigeria. Gbadebo (2005) concluded that elevated arsenic concentrations in the basement complexes are probably related to deep groundwater flow through crystalline rocks.

6.5.14 Western USA

Out of $\sim$30 000 analyses of groundwater in the United States, about 10 % had arsenic concentrations $\geqslant$10 $\mu g\,L^{-1}$ (Welch *et al.*, 2000) and $\sim$5 % had concentrations of $>$20 $\mu g\,L^{-1}$ (Welch *et al.*, 1999). Concentrations tend to be higher in the west (including the Interior Plains, Intermontane Plateau, Rocky Mountains, and Pacific Mountains physiographic provinces) than in the east, although recent studies have identified hotspots of contamination in several Midwestern and northeastern states (e.g. (Welch *et al.*, 2000; Peters *et al.*, 1999; Kelly *et al.*, 2005), Appendix D).

The Intermontane Plateau and Pacific Mountain regions generally have arid climates. This favors evaporative concentration of arsenic in shallow groundwater, such as in the hydrologically closed basins of eastern California, eastern Oregon, Nevada, western Utah, and Arizona ((Welch *et al.*, 2000), Table 6.3). In Nevada, at least 1000 private wells had arsenic concentrations of $>$50 $\mu g\,L^{-1}$ (Fontaine, 1994) and concentrations as high as 2600 $\mu g\,L^{-1}$ have been reported from the southern Carson Desert (Welch and Lico, 1998). Likewise, in the San Joaquin Valley of California, arsenic concentrations were as high as 2600 $\mu g\,L^{-1}$ ((Fujii and Swain, 1995), Table 6.3). About 7 % of 467 wells in the Basin and Range province of Arizona had arsenic $\geqslant$50 $\mu g\,L^{-1}$ (Robertson, 1989). In western Millard County, Utah, more than 250 wells were found to have 180–210 $\mu g\,L^{-1}$ of arsenic (Southwick, Western and Beck, 1983). Goldblatt, Van Denburgh and Marsland (1963) reported arsenic concentrations of 50–1700 $\mu g\,L^{-1}$ in western Oregon.

Arsenic in groundwater of the western Interior Plains and Intermontane Plateaus is generally thought to be liberated from iron (oxy)(hydr)oxides in aquifer sediments (Welch *et al.*, 2000). However, in contrast to other regions where arsenic appears to be mobilized primarily by reduction of iron (oxy)(hydr)oxides (e.g. south and east Asia), elevated arsenic occurs in both anoxic and oxic groundwaters in the western United

States (Smedley and Kinniburgh, 2002; Welch *et al.*, 2000). For example, in the Carson Desert of Nevada, arsenic-rich groundwater has very low dissolved O_2 and high concentrations of iron, manganese, and organic carbon ((Welch and Lico, 1998), Table 6.3). Conversely, high-arsenic concentrations in Arizona occur in aquifers up to 600 m deep with dissolved O_2 as high as 7 mg L^{-1} and where As(V) predominates ((Robertson, 1989), Table 6.3). Sorbed arsenic in the sediments of these aquifers can reach up to 88 mg kg^{-1}. About 86 % of arsenic in the groundwater of Millard County, Utah, occurs as As(V) (Southwick, Western and Beck, 1983). In the Tulare basin of California, arsenic has been detected in both oxidized and reduced groundwaters with reduction increasing with depth, as indicated by As(III) enrichment ((Fujii and Swain, 1995), Table 6.3). Mobilization of arsenic may be promoted by desorption under alkaline pH conditions, which may result from evaporative concentration of solutes or weathering of felsic volcanic rocks (Welch *et al.*, 2000). In addition, a correlation between arsenic and phosphate in groundwater (Welch *et al.*, 2000) suggests that competitive anion exchange can affect arsenic concentrations in the western United States.

Elevated arsenic in western US groundwater can also originate from geothermal sources (Appendix D). Arsenic concentrations up to 27 000 $\mu g\,L^{-1}$ in California geothermal waters and 2700–3500 $\mu g\,L^{-1}$ in Nevada and Utah *hot springs* have been reported (Smedley and Kinniburgh, 2002; Welch, Lico and Hughes, 1988). In the vicinity of the Yellowstone geothermal system in Wyoming, Idaho, and Montana, elevated arsenic concentrations occur in thermal waters (Smedley and Kinniburgh, 2002), in the Madison River (up to 370 $\mu g\,L^{-1}$; (Nimick *et al.*, 1998)), and in groundwater of the Madison and upper Missouri River valleys (Welch *et al.*, 2000).

6.5.15 New England, USA

Consistent with many arsenic occurrences in the western United States, the New England states in the north-eastern United States have reported elevated arsenic under oxidizing groundwater conditions (Appendix D). The sporadic occurrence of arsenic in the Goose River Basin in Maine has been investigated by Sidle, Wotten and Murphy (2001), Sidle *et al.* (2001), Sidle (2002, 2003), and Sidle and Fischer (2003). Arsenic is predominantly present as $H_2AsO_4^-$ and $HAsO_4^{2-}$. Sulfur and oxygen isotopes indicate that oxidizing conditions (dissolved O_2 of about 8.1 mg L^{-1}) promote the oxidation of arsenian pyrite along fractured zones in igneous and metamorphic aquifers ((Sidle, 2002), Chapter 3).

Domestic bedrock wells in New England commonly exhibit higher arsenic concentrations (up to 180 $\mu g\,L^{-1}$) than surficial wells (Peters *et al.*, 1999). Domestic bedrock wells in New Hampshire alone supply water to approximately 120 000 people (Peters *et al.*, 2006). Ayotte *et al.* (2003, 1999) found that nearly 30 % of wells in bedrock (primarily in Maine and New Hampshire) exceed 10 $\mu g\,L^{-1}$ of arsenic and that this association is related to a group of calc-silicate metamorphic rocks containing up to 40 mg kg^{-1} arsenic in whole rock samples. Utsunomiya *et al.* (2003) used nanoscale mineralogical techniques to determine that arsenic in New Hampshire bedrock is associated with the minerals arsenopyrite (FeAsS), westerveldite (FeAs), and magnetite (Fe_3O_4). Peters and Blum (2003) subsequently found that the arsenic distribution in groundwater of the same region was largely controlled by pH-dependent sorption/desorption reactions with iron (oxy)(hydr)oxides. More recently, Weldon and MacRae (2006) correlated the presence of the microbial genus *Geobacter* with total arsenic concentrations in Maine groundwater, which indicates that some *Geobacter* species may be involved in iron reduction and subsequent release of arsenic in these aquifers. Further research on the arsenic distribution and occurrence in the groundwater of Maine is currently the focus of several research groups (Clark *et al.*, 2006; O'Shea, Clark and Jankowski, 2006; Yang *et al.*, 2006).

6.5.16 Northern Chile

Arsenic-related health problems, including keratosis, carcinoma, and skin pigmentation, were reported from northern Chile in the early 1960s ((Rahman *et al.*, 2006; Borgono and Greiber, 1971), Chapter 4; Appendix D: Table 6.3). Elevated arsenic was found in both surface water and groundwater in Administrative Region II (the Second Region) of northern Chile, including the cities of Antofagasta, Calama, and Tocopilla (Cáceres, Gruttner and Contreras, 1992). From 1959 to 1970, about 1 300 000 inhabitants of Antofagasta unknowingly were exposed to groundwater with arsenic concentrations up to $800 \mu g L^{-1}$ (Borgono and Greiber, 1971). Smith *et al.* (1998) estimated that about 7 % of deaths in Antofagasta between 1989 and 1993 were caused by prolonged ingestion of drinking water with $\sim 500 \mu g L^{-1}$ of arsenic. In the region of Antofagasta, arsenic concentrations of $<100 \mu g L^{-1}$ in groundwater and surface water are unusual (Smedley and Kinniburgh, 2002). Groundwater samples collected from 23 locations near Calama had arsenic in the range of $100-800 \mu g L^{-1}$ (Rahman *et al.*, 2006). Although the affected aquifers are known to consist of volcanic rocks and sediments, geochemical studies of the affected areas are sparse (Smedley and Kinniburgh, 2002).

The Second Region (area $126 500 km^2$) is drained by the 440-km-long Rio Loa river. Arsenic concentrations of $100-1000 \mu g L^{-1}$ (mean $440 \mu g L^{-1}$) have been found in tributaries of the Rio Loa (Karcher *et al.*, 1999). The Rio Loa itself has an average arsenic concentration of $1400 \mu g L^{-1}$ and elevated arsenic has been found to be associated with high concentrations of boron and lithium (Romero *et al.*, 2003). Contaminated ground and surface waters in the Rio Loa watershed tend to be oxidized with As(V) as the dominant arsenic species (Rahman *et al.*, 2006; Thornton and Farago, 1997). Arsenic in stream water was thought to have been derived from mine drainage, but recent studies have indicated a primarily geologic origin. The source of the arsenic and associated solutes present in elevated concentrations has been attributed to the Rio Salado, a tributary to the Rio Loa, which is fed by the El Tatio geothermal springs in the Andean cordillera. Arsenic concentrations as high as $50 000 \mu g L^{-1}$ have been reported from El Tatio (Smedley and Kinniburgh, 2002; Ellis and Mahon, 1977). Near the confluence of the Rio Salado with the Rio Loa, 70 km downstream of El Tatio, the arsenic concentration decreases to about $1200 \mu g L^{-1}$, but the concentration subsequently increases farther downstream to the Pacific Ocean. Romero *et al.* (2003) thus concluded that geothermal water cannot be the only cause of abnormal arsenic concentrations in the Rio Loa and associated streams. One possible explanation is evaporative concentration in the hyperarid climate of the Atacama Desert, as well as a lack of dilution by uncontaminated waters from other sources. However, it is also likely (as in the western United States) that arsenic is liberated by weathering of volcanic rocks in the Rio Loa watershed. Romero *et al.* (2003) found that arsenic concentrations in sediments from the Rio Loa and its tributaries ranged from 26 to $2000 mg kg^{-1}$ with an average of $\sim 320 mg kg^{-1}$. Arsenic in these alluvial sediments was associated with metal (oxy)(hydr)oxides (mostly iron and manganese). About 20 % of the total arsenic was very *labile* (i.e. in the exchangeable and carbonate phases). Resorption of the mobilized arsenic is hindered by the near-neutral to alkaline pH and high salinity of the solutions (Romero *et al.*, 2003).

The $9800-km^2$ watershed of the Rio Elqui river, which is located $\sim 800 km$ south of the Rio Loa, is the water supply for more than 200 000 people (Dittmar, 2004). Stream water, groundwater, and sediments associated with the river have high concentrations of arsenic (Maturana *et al.*, 2001). The Rio Elqui river originates in the high Andes and drains through hydrothermal deposits enriched in arsenic (Oyarzun *et al.*, 2004). The affected waters are mostly oxidizing with dissolved O_2 concentrations up to 6.7 mg L^{-1}. Dittmar (2004) estimated that about 2 tons year^{-1} of arsenic enter the Rio Elqui system. Oyarzun *et al.* (2004) hypothesized that at least some of the arsenic sorbed onto goethite, which is common in the stream-bed sediments. These authors also concluded that the arsenic input in the Rio Elqui system is related to both mining and weathering of natural ore deposits. They detected high concentrations of arsenic in Early Holocene sediments deposited in lacustrine (119–2344 mg kg^{-1} arsenic) and fluvial (55–485 mg kg^{-1} arsenic) settings. The arsenic was mostly associated with iron (oxy)(hydr)oxides, although complex ions (mostly $H_2AsO_4^-$, but also some $HAsO_4^{2-}$ depending on pH) were thought to be mobile. Like Romero *et al.* (2003), Oyarzun *et al.* (2004) noted that heavy rainfall may remobilize arsenic-contaminated sediments. In particular, such conditions could produce debris flows from the Andean highlands as a result of El Niño episodes (Oyarzun *et al.*, 2004).

6.5.17   Chaco and Pampa plains of Argentina

One of the world's largest areas of arsenic-contaminated groundwater ($\sim$1 000 000 km^2) occurs in the Chaco and Pampa plains of Argentina (Smedley and Kinniburgh, 2002) (Figure 6.3; Table 6.3; Appendix D). At least 1.2 million rural residents of this region are thought to be drinking water with arsenic concentrations of $\geq$50 µg L^{-1} ((Bundschuh, Bonorino and Viero, 2000; Martin, 1999), Table 6.3) and arsenicosis has been reported in several locations (Hopenhayn-Rich *et al.*, 1996). Groundwater arsenic concentrations exceeding 1000 µg L^{-1} have been found in Córdoba, La Pampa, Santiago del Estero, and Tucumán provinces (Smedley and Kinniburgh, 2002; Bundschuh *et al.*, 2004) with a maximum of 11 500 µg L^{-1} in Córdoba ((Nicolli *et al.*, 1989), Table 6.3). The major aquifers consist of Cenozoic fluvial and alluvial sediments in the Chaco and *loess* deposits in the Pampa (Bundschuh *et al.*, 2004). High-arsenic groundwater occurs in loess or alluvium interbedded with volcanic ash (Smedley and Kinniburgh, 2002; Bundschuh *et al.*, 2004). Arsenic concentrations are highly variable both laterally and vertically (Smedley *et al.*, 2002; Bhattacharya *et al.*, 2006) and arsenic is commonly associated with various other trace elements, such as boron and fluoride ((Smedley *et al.*, 2002; Bhattacharya *et al.*, 2006; Farías *et al.*, 2003), Figure 6.3). High concentrations of arsenic are also associated with slow rates of groundwater flow, high pH, and high total dissolved solids (TDS) (Bundschuh *et al.*, 2004; Smedley *et al.*, 2002). Temporal variations of arsenic concentrations, which have been reported in some studies, may be linked to enhanced recharge, such as from irrigation return flow (Bundschuh *et al.*, 2004).

In the Rio Dulce alluvial aquifers of Santiago del Estero province in the Chaco region, the average arsenic concentration is as high as 743 µg L^{-1} and arsenic occurs primarily as As(V) (Bhattacharya *et al.*, 2006). The aquifers consist of loess and fluvial sand, silt, and clay (Bundschuh *et al.*, 2004). More than 48 % of 63 shallow wells studied by Bundschuh *et al.* (2004) showed elevated arsenic concentrations with a maximum of 4800 µg L^{-1} of arsenic. The deeper aquifers are mostly free of contamination. Correlations of dissolved arsenic with boron, fluoride, molybdenum, *silica*, and vanadium indicate that arsenic was originally liberated by dissolution of rhyolitic glass in volcanic ash layers (Bundschuh *et al.*, 2004; Bhattacharya *et al.*, 2006). Arsenic is subsequently sorbed onto iron and aluminum (oxy)(hydr)oxides. At alkaline pH conditions, which could result from the dissolution of carbonates and silicates and from cation exchange, arsenic desorption is promoted (Bhattacharya *et al.*, 2006). Relatively high DOC values (average 6.6 mg L^{-1}) and seasonally variable As(III)/total arsenic ratios suggest that arsenic may also be locally liberated under anoxic conditions induced by irrigation return flow (Bhattacharya *et al.*, 2006).

Smedley *et al.* (2002, 2005) observed geochemical conditions similar to those of the Chaco in groundwater and sediments of La Pampa province. Solute chemistry was spatially heterogeneous over distances of a few kilometers with particularly high concentrations close to depressions that act as discharge zones (Smedley *et al.*, 2002). Arsenic concentrations in groundwater varied from < 4 to 5300 µg L^{-1} with 95 % of samples having arsenic >10 µg L^{-1} and 73 % having arsenic >50 µg L^{-1} (Smedley *et al.*, 2002). The groundwater is neutral to alkaline (pH 7.0–8.8) and oxidizing with As(V) as the dominant form of dissolved arsenic Smedley *et al.* (2005). Smedley *et al.* (2002) concluded that weathering reactions, evaporative concentration, and slow groundwater flow had enabled the accumulation of arsenic and other solutes in groundwater. Elevated TDS concentrations were mainly due to evaporative concentration in the semi-arid

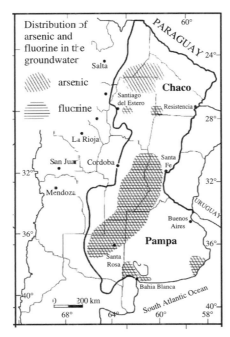

Figure 6.3 *Arsenic-contaminated areas of Chaco-Pampean plain of Argentina. (Reprinted from J. Bundschuh, B. Farias, R. Martin, A. Storniolo, P. Bhattacharya, J. Cortes, G. Bonorino, and R. Albouy, Groundwater arsenic in the Chaco-Pampean plain, Argentina: Case study from Robles county, Santiago del Estero province, Applied Geochemistry,* **19**, *231–243, Copyright 2004, with permission from Elsevier.)*

climate. Correlations of arsenic with pH, HCO_3^-, and various trace elements point to weathering of silicates in the loess and volcanic ash, followed by desorption of arsenic from iron and manganese oxides at high pH (Smedley *et al.*, 2002). Competitive sorption to iron (oxy)(hydr)oxides by other oxyanions, particularly vanadate, would further enhance arsenic mobility ((Smedley *et al.*, 2005), Chapter 3).

References

AAN/RGAG/NIPSOM (Asia Arsenic Network [AAN], Research Group of Applied Geology [RGAG], and National Institute for Preventive and Social Medicine [NIPSOM]) (1999) *Arsenic Contamination in Groundwater in Bangladesh: Interim Report of the Research at Samta Village*, Asia Arsenic Network (AAN), Research Group of Applied Geology (RGAG), and National Institute for Preventive and Social Medicine (NIPSOM).

Acharyya, S.K. (2002) Arsenic contamination in groundwater affecting major parts of southern West Bengal and parts of western Chhattisgarh: source and mobilization process. *Current Science*, **82**(6), 740–44.

Acharyya, S.K., Chakraborty, P., Lahiri, S. *et al.* (1999) Arsenic poisoning in the Ganges delta. *Nature*, **401**(6753), 545–47.

Acharyya, S.K., Lahiri, S., Raymahashay, B.C. and Bhowmik, A. (2000) Arsenic toxicity of groundwater in parts of the Bengal basin in India and Bangladesh: the role of Quaternary stratigraphy and Holocene sea-level fluctuation. *Environmental Geology*, **39**(10), 1127–37.

Acharyya, S.K., Shah, B.A., Ashyiya, I.D. and Pandey, Y. (2005) Arsenic contamination in groundwater from parts of Ambagarh-Chowki block, Chhattisgarh, India: Source and release mechanism. *Environmental Geology*, **49**(1), 148–58.

Ahmed, K.M., Bhattacharya, P., Hasan, M.A. *et al.* (2004) Arsenic enrichment in groundwater of the alluvial aquifers in Bangladesh: an overview. *Applied Geochemistry*, **19**(2), 181–200.

Ahmed, K.M., Hoque, M., Hasan, M.K. *et al.* (1998) Occurrence and origin of water well CH_4. gas in Bangladesh. *Journal of the Geological Society of India*, **51**, 697–708.

Ahamed, S., Kumar Sengupta, M., Mukherjee, A. *et al.* (2006) Arsenic groundwater contamination and its health effects in the state of Uttar Pradesh (UP) in upper and middle Ganga plain, India: a severe danger. *Science of the Total Environment*, **370**(2-3), 310–22.

AIP/PHED (1995) *Prospective Plan for Arsenic Affected Districts of West Bengal*, Arsenic Investigation Project (AIP) and Public Health Engineering Directorate (PHED), Government of West Bengal.

Alam, M., Alam, M.M., Curray, J.R. *et al.* (2003) An overview of the sedimentary geology of the Bengal Basin in relation to the regional tectonic framework and basin-fill history. *Sedimentary Geology*, **155**(3-4), 179–208.

Anawar, H.M., Akai, J., Komaki, K. *et al.* (2003) Geochemical occurrence of arsenic in groundwater of Bangladesh: sources and mobilization processes. *Journal of Geochemical Exploration*, **77**(2-3), 109–31.

Anawar, H.M., Akai, J. and Sakugawa, H. (2004) Mobilization of arsenic from subsurface sediments by effect of bicarbonate ions in groundwater. *Chemosphere*, **54**(6), 753–62.

Anderson, M.A., Ferguson, J.F. and Gavis, J. (1976) Arsenate adsorption on amorphous aluminum hydroxide. *Journal of Colloid and Interface Science*, **54**, 391–99.

Apambire, W.B. and Hess, J.W. (2000) The aqueous geochemistry of fluoride in the upper regions of Ghana. *Geological Society of America Abstracts with Programs*, **32**(7), 7.

Appelo, C.A.J. and Postma, D. (1999) *Geochemistry, Groundwater and Pollution*, Balkema, Rotterdam.

Appelo, C.A.J., Van Der Weiden, M.J.J., Tournassat, C. and Charlet, L. (2002) Surface complexation of ferrous iron and carbonate on ferrihydrite and the mobilization of arsenic. *Environmental Science and Technology*, **36**(14), 3096–103.

Appleyard, S.J., Angeloni, J. and Watkins, R. (2006) Arsenic-rich groundwater in an urban area experiencing drought and increasing population density, Perth, Australia. *Applied Geochemistry*, **21**(1), 83–97.

Ayotte, J.D., Montgomery, D.L., Flanagan, S.M. and Robinson, K.W. (2003) Arsenic in groundwater in eastern New England: occurrence, controls, and human health implications. *Environmental Science and Technology*, **37**(10), 2075–83.

Ayotte, J.D., Nielsen, M.G., Robinson, G.R., Jr. and Moore, R.B. (1999) *Relation of Arsenic, Iron, and Manganese in Ground Water to Aquifer Type, Bedrock Lithogeochemistry, and Land Use in the New England Coastal Basins. U.S. Geological Survey Water-Resources Investigations Report 99-4162*, USGS, Pembroke, NH.

Bauer, M. and Blodau, C. (2006) Mobilization of arsenic by dissolved organic matter from iron oxides, soils and sediments. *Science of the Total Environment*, **354**(2-3), 179–90.

Belzile, N. (1988) The fate of arsenic in sediments of the Laurentian trough. *Geochimica et Cosmochimica Acta*, **52**(9), 2293–302.

Belzile, N. and Lebel, J. (1986) Capture of arsenic by pyrite in near-shore marine sediments. *Chemical Geology*, **54**(3-4), 279–81.

Berg, M., Giger, W., Tran, H.C. (2006) Extent and severity of arsenic pollution in Vietnam and Cambodia, in *Managing Arsenic in the Environment: From Soil to Human Health* (eds R. Naidu, E. Smith, G. Owens *et al.*), CSIRO Publishing, Collingwood.

Berg, M., Tran, H.C., Nguyen, T.C. *et al.* (2001) Arsenic contamination of groundwater and drinking water in Vietnam: a human health threat. *Environmental Science and Technology*, **35**(13), 2621–26.

BGS (British Geological Survey) (1989) *Trace Element Occurrence in British Groundwaters*. Research Report SD/89/3, British Geological Survey (BGS), Keyworth.

BGS/DPHE (British Geological Survey/Department of Public Health Engineering [Bangladesh]) (2001) *Arsenic Contamination of Groundwater in Bangladesh* (eds D.G. Kinniburgh and P.L. Smedley), Report WC/00/19, British Geological Survey, Keyworth.

Bhattacharjee, S., Chakravarty, S., Maity, S. *et al.* (2005) Metal contents in the groundwater of Sahebgunj district, Jharkhand, India, with special reference to arsenic. *Chemosphere*, **58**(9), 1203–17.

Bhattacharya, P., Chatterjee, D. and Jacks, G. (1997) Occurrence of arsenic-contaminated groundwater in alluvial aquifers from delta plains, eastern India: options for safe drinking water supply. *International Journal of Water Resources Development*, **13**(1), 79–92.

Bhattacharya, P., Claesson, M., Bundschuh, J. *et al.* (2006) Distribution and mobility of arsenic in the Río Dulce alluvial aquifers in Santiago del Estero province, Argentina. *Science of the Total Environment*, **358**(1-3), 97–120.

Bi, R.L. (1994) *A Preliminary Study on the Arsenic Enrichment of Groundwater in Chianan Area, Taiwan*, Department of Geosciences, National Taiwan University, Taipei.

Borgono, J.M. and Greiber, R. (1971) Epidemiological study of arsenic poisoning in the city of Antofagasta. *Revista Medica Chile*, **99**, 702–07.

Bose, P. and Sharma, A. (2002) Role of iron in controlling speciation and mobilization of arsenic in subsurface environment. *Water Research*, **36**(19), 4916–25.

Bostick, B.C., Chen, C. and Fendorf, S. (2004) Arsenite retention mechanisms within estuarine sediments of Pescadero, CA. *Environmental Science and Technology*, **38**(12), 3299–304.

Bostick, B.C. and Fendorf, S. (2003) Arsenite sorption on troilite (FeS) and pyrite (FeS$_2$). *Geochimica et Cosmochimica Acta*, **67**(5), 909–21.

Bostick, B.C., Fendorf, S. and Manning, B.A. (2003) Arsenite adsorption on galena (PbS) and sphalerite (ZnS). *Geochimica et Cosmochimica Acta*, **67**(5), 895–907.

Bowell, R.J. (1994) Sorption of arsenic by iron oxides and oxyhydroxides in soils. *Applied Geochemistry*, **9**(3), 279–86.

Boyle, R.W. and Jonasson, I.R. (1973) Geochemistry of arsenic and its use as indicator element in geochemical prospecting. *Journal of Geochemical Exploration*, **2**(3), 251–96.

Boyle, D.R., Turner, R.J.W. and Hall, G.E.M. (1998) Anomalous arsenic concentrations in groundwaters of an island community, Bowen Island, British Columbia. *Environmental Geochemistry and Health*, **20**(4), 199–212.

Breit, G.N., Foster, A.L., Sanzolone, R.F. *et al.* (2000) Arsenic cycling in eastern Bangladesh: the role of phyllosilicates. Abstracts with Programs. *The Geological Society of America*, **32**(7), 192.

Brikowski, T.H., Leybourne, M., Shrestha, S.D. *et al.* (2006) Geochemical indicators of groundwater arsenic mobilization mechanisms in the Ganges floodplain of Nepal. Abstracts with Programs. *The Geological Society of America*, **38**(1), 8.

Bundschuh, J., Bonorino, G. and Viero. A.P. (2000) Arsenic and other trace elements in sedimentary aquifers in the Chaco-Pampean plain, in *Arsenic in Groundwater of Sedimentary Aquifers*, Pre-Congress Workshop, 31st International Geological Congress (eds P. Bhattacharya and A.H. Welch), Groundwater Arsenic Research Group, Division of Land and Water Resources, Royal Institute of Technology, Stockholm, pp. 27–32.

Bundschuh, J., Farias, B., Martin, R. *et al.* (2004) Groundwater arsenic in the Chaco-Pampean plain, Argentina: case study from Robles County, Santiago del Estero Province. *Applied Geochemistry*, **19**(2), 231–43.

Buschmann, J., Berg, M., Stengel, C. and Sampson, M. (2006) Arsenic and manganese contamination in Cambodia: relation to micro-topography. Abstracts with Programs. *The Geological Society of America*, **38**(7), 438.

Cáceres, L., Gruttner, E. and Contreras, R. (1992) Water recycling in arid regions-Chilean case. *Ambio*, **21**, 138–44.

CGWB (Central Ground Water Board [India]) (1999) *High Incidence of Arsenic in Groundwater in West Bengal*, Central Ground Water Board, Ministry of Water Resources, Government of India, Faridabad.

Chaillou, G., Schäfer, J., Anschutz, P. *et al.* (2003) The behaviour of arsenic in muddy sediments of the Bay of Biscay (France). *Geochimica et Cosmochimica Acta*, **67**(16), 2993–3003.

Chakraborti, D., Basu, G.K. and Biswas, B.K. (2001) Characterization of arsenic-bearing sediments in the Gangetic delta of West Bengal, India, in *Arsenic Exposure and Health Effects IV* (eds W.R. Chappell, C.O. Abernathy and R.L. Calderon), Elsevier Science, Oxford.

Chakraborti, D., Biswas, B.K., Roy Chowdhury, T. *et al.* (1999) Arsenic groundwater contamination and sufferings of people in Rajnandgaon district, Madhya Pradesh, India. *Current Science*, **77**, 502–4.

Chakraborti, D., Mukherjee, S.C., Pati S. *et al.* (2003) Arsenic groundwater contamination in middle Ganga Plain, Bihar, India: A future danger? *Environmental Health Perspectives*, **111**(9), 1194–201.

Chakraborti, D., Sengupta, M.K., Rahman, M.M. *et al.* (2004) Groundwater arsenic contamination and its health effects in the Ganga-Meghna-Brahmaputra Plain. *Journal of Environmental Monitoring*, **6**(6), 74N–83N.

Chapelle, F.H. (2001) *Ground-Water Microbiology and Geochemistry*, John Wiley & Sons Ltd, New York.

Chen, T. (1994) Arsenate adsorption in soil and its mechanism. *China Environmental Science*, **5**, 85–91.

Chen, C.C. (2003) *Accumulation and Release of Arsenic in Sediments from Hsindong and Jinhu in Chianan Plain, Taiwan*, Department of Geosciences, National Taiwan University, Taipei.

Chen, S.-L., Dzeng, S.R., Yang, M.-H. *et al.* (1994) Arsenic species in groundwaters of the blackfoot disease area, Taiwan. *Environmental Science and Technology*, **28**(5), 877–81.

Cheng, L., Fenter, P., Sturchio, N.C. *et al.* (1999) X-ray standing wave study of arsenite incorporation at the calcite surface. *Geochimica et Cosmochimica Acta*, **63**(19–20), 3153–57.

Chi, I.C. and Blackwell, R.Q. (1968) A controlled retrospective study of blackfoot disease, an endemic peripheral gangrene disease in Taiwan. *American Journal of Epidemiology*, **88**, 7–24.

Clark, G., O'Shea, B., Ongley, L. and Lev, S. (2006) The distribution of arsenic in ground and surface waters of the Woodbury Hills near Monmouth, Maine. Abstracts with Programs. *The Geological Society of America*, **38**(7), 422.

Coleman, J.M. (1981) *Deltas: Processes of Deposition and Models of Exploration*, Burgess, Minneapolis, MN.

Craw, D. and Chappel, D.A. (1999) Evolution and sulphide mineral occurrences of an incipient nonmarine sedimentary basin, New Zealand. *Sedimentary Geology*, **129**(1-2), 37–50.

Csalagovits, I. (1999) Arsenic-bearing artesian waters of hungary. *Annual Report Hungarian Geological Institute*, 1992–93 (in Hungarian), **2**, 85–92,

Cullen, W.R. and Reimer, K.J. (1989) Arsenic speciation in the environment. *Chemical Reviews*, **89**(4), 713–64.

Cumming, D.E., Caccavo, F., Jr., Fendorf, S. and Rosenzweig, R.F. (1999) Arsenic mobilization by the dissimilatory Fe(III)-reducing bacterium Shewanella alga BrY. *Environmental Science and Technology*, **33**(5), 723–29.

Das, D., Samanta, G., Mandal, B.K. *et al.* (1996) Arsenic in groundwater in six districts of West Bengal, India. *Environmental Geochemistry and Health*, **18**(1), 5–15.

Datta, S., Mallioux, B., Jung, H. *et al.* (2005). Mineralogy and mobility of Fe and As in Meghna River sediments in Bangladesh. Abstract B24B-04. *Eos Transactions American Geophysical Union*, **86** (52).

Dittmar, T. (2004) Hydrochemical processes controlling arsenic and heavy metal contamination in the Elqui River system (Chile). *Science of the Total Environment*, **325**(1-3), 193–207.

Dixit, S. and Hering, J.G. (2003) Comparison of arsenic(V) and arsenic(III) sorption onto iron oxide minerals: implications for arsenic mobility. *Environmental Science and Technology*, **37**(18), 4182–89.

Dowling, C.B., Poreda, R.J. and Basu, A.R. (2003) The groundwater geochemistry of the Bengal basin: weathering, chemsorption, and trace metal flux to the oceans. *Geochimica et Cosmochimica Acta*, **67**(12), 2117–36.

Dowling, C.B., Poreda, R.J., Basu, A.R. *et al.* (2002) Geochemical study of arsenic release mechanisms in the Bengal Basin groundwater. *Water Resources Research*, **38**(9), 12.1–12.18.

DPHE (Department of Public Health Engineering [Bangladesh]) (2000) *Groundwater Studies for Arsenic Contamination in Bangladesh, Supplemental Data to Final Report, Rapid Investigation Phase*, Department of Public Health Engineering (DPHE), Government of Bangladesh, in collaboration with British Geological Survey and Sir Mott MacDonald Ltd.

Driehaus, W., Jekel, M. and Hildebrandt, U. (1998) Granular ferric hydroxide - a new adsorbent for the removal of arsenic from natural water. *Journal of Water Supply Research and Technology - Aqua*, **47**, 30–35.

Driehaus, W., Seith, R. and Jekel, M. (1995) Oxidation of arsenate(III) with manganese oxides in water treatment. *Water Research*, **29**(1), 297–305.

Dudas, M.J. (1987) Accumulation of native arsenic in acid sulphate soils in Alberta. *Canadian Journal of Soil Science*, **67**(2), 317–31.

Dzombak, D.A. and Morel, F.M.M. (1990) *Surface Complexation Modeling: Hydrous Ferric Oxide*, John Wiley & Sons Ltd, New York.

Ellis, A.J. and Mahon, W.A.J. (1977) *Chemistry and Geothermal Systems*, Academic Press, New York.

Emerman, S.H., Bhattarai, T.N., Adhikari, D. *et al.* (2006) Arsenic and other heavy metals in the rivers of Nepal. Abstracts with Programs. *The Geological Society of America*, **38**, 242.

Farías, S.S., Casa, V.A., Vázquez, C. *et al.* (2003) Natural contamination with arsenic and other trace elements in ground waters of Argentine Pampean plain. *Science of the Total Environment*, **309**(1-3), 187–99.

Farooqi, A., Masuda, H. and Firdous, N. (2007) Toxic fluoride and arsenic contaminated groundwater in the Lahore and Kasur districts, Punjab, Pakistan and possible contaminant sources. *Environmental Pollution*, **145**(3), 839–49.

Feldman, P.R. and Rosenboom, J.W. (2001) *Cambodia Drinking Water Quality Assessment*, World Health Organization, in cooperation with the Ministry of Rural Development and the Ministry of Industry, Mines, and Energy, Government of Cambodia.

Fendorf, S., Eick, M.J., Grossl, P. and Sparks, D.L. (1997) Arsenate and chromate retention mechanisms on goethite. 1. Surface structure. *Environmental Science and Technology*, **31**(2), 315–20.

Fengthong, T., Dethoudom, S. and Keosavanh, O. (2002) Drinking water quality in the Lao People's Democratic Republic, in *Seminar on the Environmental and Public Health Risks Due to Contamination of Soils, Crops, Surface and Groundwater from Urban, Industrial and Natural Sources in South East Asia*, UN/ESCAP, Hanoi.

Ferguson, J.F. and Anderson, M.A. (1974) Chemical form of arsenic in water supplies and their removal, in *Chemistry of Water Supply, Treatment and Distribution* (ed A.J. Rubin), Ann Arbor Science, Ann Arbor, MI.

Fitzpatrick, R.W., Fritsch, E. and Self, P.G. (1996) Interpretation of soil features produced by ancient and modern processes in degraded landscapes: V. Development of saline sulfidic features in non-tidal seepage areas. *Geoderma*, **69**(1-2), 1–29.

Fontaine, J.A. (1994) Regulating arsenic in Nevada drinking water supplies: past problem, future challenges, in *Arsenic Exposure and Health Effects* (eds W.R. Chappell, C.O. Abernathy and R.L. Cothern), Science and Technology Letters, Northwood.

Foster, A.L., Perkins, R.B., Breit, G.N. *et al.* (2003) High arsenic accumulation in iron phases in near-surface sediments from eastern Bangladesh. Abstracts with Programs. *The Geological Society of America*, **35**(6), 47.

Fujii, R. and Swain, W.C. (1995) *Areal Distribution of Selected Trace Elements, Salinity, and Major Ions in Shallow Ground Water, Tulare Basin, Southern San Joaquin Valley, California . U.S. Geological Survey Water Resources Investigation Report 95-4048*, USGS, Sacramento, CA.

Fredericks, D. and Arsenic Secretariat (2004) *Situation Analysis: Arsenic Contamination of Groundwater in Cambodia*, report prepared for the Arsenic Inter-Ministerial Sub-Committee, Water and Sanitation Sector Coordinating Committee, Government of Cambodia.

Garai, R., Chakraborty, A.K., Dey, S.B. and Saha, K.C. (1984) Chronic arsenic poisoning from tube-well water. *Journal of the Indian Medical Association*, **82**(1), 34–35.

Gaus, I., Kinniburgh, D.G., Talbot, J.C. and Webster, R. (2003) Geostatistical analysis of arsenic concentration in groundwater in Bangladesh using disjunctive kriging. *Environmental Geology*, **44**(8), 939–48.

Gbadebo, A.M. (2005) Occurrence and fate of arsenic in the hydrogeological systems of Nigeria. Abstracts with Programs. *The Geological Society of America*. **37**(7), 375.

Ghosh, A. (2005) *Arsenic Hot Spots Detected in the State of Bihar (India) a Serious Health Hazard for Estimated Human Population of 5.5 Lakhs*. International Conference on Environmental Management, Hyderabad.

Ghosh, A.R. and Mukherjee, A. (2002) Arsenic contamination of groundwater and human health impacts in Burdwan District, West Bengal, India. Abstracts with Programs. *The Geological Society of America*, **34**(2), 107.

Ghosh, M.M. and Yuan, J.R. (1987) Adsorption of inorganic arsenic and organioarsenicals on hydrous oxides. *Environmental Progress*, **6**(3), 150–57.

Goh, K.-H. and Lim, T.-T. (2005) Arsenic fractionation in a fine soil fraction and influence of various anions on its mobility in the subsurface environment. *Applied Geochemistry*, **20**(2), 229–39.

Goldberg, S. and Glaubig, R.A. (1988) Anion sorption on a calcareous, montmorillonitic soil - arsenic. *Soil Science Society of America Journal*, **52**(5), 1297–300.

Goldblatt, E.L., Van Denburgh, S.A. and Marsland, R.A. (1963) *The Unusual and Widespread Occurrence of Arsenic in Well Waters of Lane County, Oregon*, Lane County Health Department Report.

Grafe, M., Eick, M.J. and Grossl, P.R. (2001) Adsorption of arsenate (V) and arsenite (III) on goethite in the presence and absence of dissolved organic carbon. *Soil Science Society of America Journal*, **65**(6), 1680–87.

Guha Mazumder, D.N., Haque, R., Ghosh, N. *et al.* (1998) Arsenic levels in drinking water and the prevalence of skin lesions in West Bengal, India. *International Journal of Epidemiology*, **27**(5), 871–77.

Gulens, J., Champ, D.R. and Jackson, R.E. (1979) Influence of redox environments on the mobility of arsenic in groundwater, in *Chemical Modeling in Aqueous Systems* (ed. E.A. Jenne), American Chemical Society.

Guo, X.-J. (1997) 96% of well water is undrinkable. *Newsletter Asia Arsenic Network*, **2**, 7–9.

Gustafsson, J.F. and Tin, N.T. (1994) Arsenic and selenium in some vietnamese acid sulphate soils. *Science of the Total Environment*, **151**(2), 153–58.

Halperin, A. (2003) Arsenic found in rural Mekong River wells. *The Cambodian Daily*, 11.

Haque, I. (2005) *National Action Plan for Arsenic Mitigation, 2005–09 (Pakistan)*. Final draft report prepared for the Ministry of Science and Technology, Pakistan.

Harvey, C.F., Ashfaque, K.N., Yu, W. *et al.* (2006) Groundwater dynamics and arsenic contamination in Bangladesh. *Chemical Geology*, **228**(Special Issue 1-3), 112–36.

Harvey, C.F., Swartz, C.H., Badruzzaman, A.B.M. *et al.* (2002) Arsenic mobility and groundwater extraction in Bangladesh. *Science*, **298**(5598), 1602–06.

Harvey, C.F., Swartz, C.H., Badruzzaman, A.B.M. *et al.* (2005) Groundwater arsenic contamination on the Ganges delta: biogeochemistry, hydrology, human perturbations, and human suffering on a large scale. *Comptes Rendus Geoscience*, **337**(1-2), 285–96.

Hiemstra, T. and van Riemsdijk, W.H. (1999) Surface structural ion adsorption modeling of competitive binding of oxyanions by metal (hydr)oxides. *Journal of Colloid and Interface Science*, **210**(1), 182–93.

Hopenhayn-Rich, C., Biggs, M.L., Fuchs, A. *et al.* (1996) Bladder cancer mortality associated with arsenic in drinking water in Argentina. *Epidemiology*, **7**(2), 117–24.

Horneman, A., van Geen, A., Kent, D.V. *et al.* (2004) Decoupling of As and Fe release to Bangladesh groundwater under reducing conditions. Part I: evidence from sediment profiles. *Geochimica et Cosmochimica Acta*, **68**(17), 3459–73.

Hsu, K.H., Froines, J.R. and Chen, C.J. (1997) Studies of arsenic ingestion from drinking water in northeastern Taiwan: chemical speciation and urinary metabolites, in *Arsenic Exposure and Health Effects* (eds C.O. Abernathy, R.L. Calderon and W.R. Chappell), Chapman and Hall, London.

Husson, O., Verburg, P.H., Phung, M.T. and Van Mensvoort, M.E.F. (2000) Spatial variability of acid sulphate soils in the Plain of Reeds, Mekong delta, Vietnam. *Geoderma*, **97**(1-2), 1–19.

Inskeep, W.P., McDermott, T.R. and Fendorf, S. (2002) Arsenic (V)/(III) cycling in soils and natural waters: chemical and microbiological processes, in *Environmental Chemistry of Arsenic* (ed. W.T. Frankenberger, Jr.), Marcel Dekker, New York, pp. 183–215.

Iqbal, S.Z. (2001) *Arsenic Contamination in Pakistan. UN-ESCAP Report*, Expert Group Meeting on Geology and Health, Bangkok.

Islam, F.S., Gault, A.G., Boothman, C. *et al.* (2004) Role of metal-reducing bacteria in arsenic release from Bengal delta sediments. *Nature*, **430**(6995), 68–71.

Jain, A., Raven, K.P. and Loeppert, R.H. (1999) Arsenite and arsenate adsorption on ferrihydrite: surface charge reduction and net OH- release stoichiometry. *Environmental Science and Technology*, **33**(8), 1179–84.

JICA (Japan International Cooperation Agency) (2002). *The Study on the Ground Water Development of Deep Aquifers for Safe Drinking Water Supply to Arsenic Affected Areas in Western Bangladesh*. Draft final report, Japan International Cooperation Agency, Kokusai Kogyo and Mitsui Mineral Development Engineering.

Kaiser, K., Guggenberger, G., Haumaier, L. and Zech, W. (1997) Dissolved organic matter sorption on subsoils and minerals studied by 13C-NMR and DRIFT spectroscopy. *European Journal of Soil Science*, **48**(2), 301–10.

Karcher, S., Cáceres, L., Jekel, M. and Contreras, R. (1999) Arsenic removal from water supplies in northern Chile using ferric chloride coagulation. *Journal of the Chartered Institution of Water and Environmental Management*, **13**(3), 164–69.

Kelly, W.R., Holm, T.R., Wilson, S.D. and Roadcap, G.S. (2005) Arsenic in glacial aquifers: sources and geochemical controls. *Ground Water*, **43**(4), 500–10.

Khadka, M.S. (1993) The groundwater quality situation in alluvial aquifers of the Kathmandu Valley, Nepal. *Journal of Australian Geology and Geophysics*, **14**, 207–11.

Kim, M.-J., Nriagu, J. and Haack, S. (2000) Carbonate ions and arsenic dissolution by groundwater. *Environmental Science and Technology*, **34**(15), 3094–100.

Klump, S., Kipfer, R., Cirpka, O.A. *et al.* (2006) Groundwater dynamics and arsenic mobilization in Bangladesh assessed using noble gases and tritium. *Environmental Science and Technology*, **40**(1), 243–50.

Korte, N.E. and Fernando, Q. (1991) A review of arsenic (III) in groundwater. *Critical Reviews in Environmental Control*, **21**(1), 1–39.

Krauskopf, K.B. and Bird, D.K. (1995) *Introduction to Geochemistry*, 3rd edn, McGraw-Hill, Boston, MA, 647 pp.

Kuo, T.-L. (1968) Arsenic content of artesian well water in endemic area of chronic arsenic poisoning. *Reports of the Institute of Pathology, National Taiwan University*, **20**, 7–13.

Lee, C.W. (2000) *Geochemical Characteristics of Porewater and Sediments from Kang-wei and Sin-wen, Southwestern Taiwan*, Department of Geosciences, National Taiwan University, Taipei.

Lin, Z. and Puls, R.W. (2000) Adsorption, desorption and oxidation of arsenic affected by clay minerals and aging process. *Environmental Geology*, **39**(7), 753–59.

Lin, Z. and Puls, R.W. (2003) Potential indicators for the assessment of arsenic natural attenuation in the subsurface. *Advances in Environmental Research*, **7**(4), 825–34.

Liu, C.-W., Lin, K.-H. and Kuo, Y.-M. (2003) Application of factor analysis in the assessment of groundwater quality in a blackfoot disease area in Taiwan. *Science of the Total Environment*, **313**(1-3), 77–89.

Liu, C.-W., Wang, S.-W., Jang, C.-S. and Lin, K.-H. (2006) Occurrence of arsenic in ground water in the Choushui River alluvial fan, Taiwan. *Journal of Environmental Quality*, **35**(1), 68–75.

Luo, Z.D., Zhang, Y.M., Ma, L. (1997) Chronic arsenicism and cancer in Inner Mongolia — consequences of well-water arsenic level greater than 50mg/L, in *Arsenic Exposure and Health Effects* (eds C.O. Abernathy, R.L. Calderon and W.R. Chappell), Chapman and Hall, London.

Ma, H.Z., Xia, Y.J., Wu, K.G. (1999) Human exposure to arsenic and health effects in Bayingnormen, Inner Mongolia, in *Arsenic Exposure and Health Effects* (eds W.R. Chappell, C.O. Abernathy and R.L. Calderon), Elsevier, Amsterdam.

Mahmood, S.N., Naseem, S., Siddiqui, I. and Khan F.A. (1998). Studies on physico-chemical nature of ground water of Korangi/Landhi (Karachi). *Journal of Chemical Society of Pakistan*, **19**, 42–8.

Mallick, S. and Rajgopal, N.R. (1995) Groundwater development in the arsenic affected alluvial belt of West Bengal — some questions. *Current Science*, **70**(11), 956–58

Mandal, B.K., Chowdhury, T.R., Samanta, G. *et al.* (1998) Impact of safe water for drinking and cooking on five arsenic-affected families for 2 years in West Bengal, India. *Science of the Total Environment*, **218**(2-3), 185–201.

Manning, B.A. and Goldberg, S. (1997) Adsorption and stability of arsenic(III) at the clay mineral-water interface. *Environmental Science and Technology*, **31**(7), 2005–11.

Manning, B.A., Fendorf, S.E. and Goldberg, S. (1998) Surface structures and stability of arsenic(III) on goethite: spectroscopic evidence for inner-sphere complexes. *Environmental Science and Technology*, **32**(16), 2383–88.

Manning, B.A. and Martens, D.A. (1997) Speciation of arsenic(III) and arsenic(V) in sediment extracts by high- performance liquid chromatography-hydride generation atomic absorption spectrophotometry. *Environmental Science and Technology*, **31**(1), 171–77.

Martin, A. (1999) *Hidrogeología de la Provincia de Santiago del Estero*, Ediciones del Rectorado, Universidad Nacional de Tucumán, Argentina.

Masscheleyn, P.H., Delaune, R.D. and Patrick, W.H., Jr. (1991) Effect of redox potential and pH on arsenic speciation and solubility in a contaminated soil. *Environmental Science and Technology*, **25**(8), 1414–19.

Matis, K.A., Zouboulis, A.I., Malamas, F.B. *et al.* (1997) Flotation removal of As(V) onto goethite. *Environmental Pollution*, **97**(3), 239–45.

Matis, K.A., Zouboulis, A.I., Zamboulis, D. and Valtadorou, A.V. (1999) Sorption of As(V) by goethite particles and study of their flocculation. *Water, Air, and Soil Pollution*, **111**(1-4), 297–316.

Maturana, H., Oyarzún, J., Pasieczna, A. and Paulo, A. (2001) Geoquímica de los sedimentos del Río Elqui (Coquimbo, Chile): Manejo de relaves y cierre de minas, in *Proceedings of 7th Argentinian Congress of Economic Geology, Salta, Argentina*, Asociación Geológica Argentina, Buenos Aires.

McArthur, J.M., Banerjee, D.M., Hudson-Edwards, K.A. *et al.* (2004) Natural organic matter in sedimentary basins and its relation to arsenic in anoxic ground water: the example of West Bengal and its worldwide implications. *Applied Geochemistry*, **19**(8), 1255–93.

McArthur, J.M., Ravenscroft, P., Safiulla, S. and Thirlwall, M.F. (2001) Arsenic in groundwater: testing pollution mechanisms for sedimentary aquifers in Bangladesh. *Water Resources Research*, **37**(1), 109–17.

Michael, H.A, and Voss, C.I. (2008) Evaluation of sustainability of deep groundwater as an arsenic-safe resource in the Bengal Basin. *Proceedings of National Academy of Science*, **105**(25), 8531–8536.

Minh, L.Q., Tuong, T.P., Van Mensvoort, M.E.F. and Bouma, J. (1998) Soil and water table management effects on aluminum dynamics in an acid sulphate soil in Vietnam. *Agriculture, Ecosystems and Environment*, **68**, 255–62.

Mukherjee, A. (2006) *Deeper groundwater chemistry and flow in the arsenic affected western Bengal basin, West Bengal, India*. Ph.D. dissertation, University of Kentucky, Lexington, 248 pp.

Mukherjee, A. and Fryar, A.E. (2008) Deeper groundwater chemistry and geochemical modeling of the arsenic affected western Bengal basin, West Bengal. *applied Geochemistry*, **23**(4), 863–892.

Mukherjee, A., Scanlon, B.R., Chaudhary, S., *et al.* (2007) Regional hydrogeochemical study of groundwater arsenic contamination along transects from the Himalayan alluvial deposits to the Indian Shield, Central Gangetic Basin, India. *Geological Society of America, Abstracts with Programs*, **39**(6), 519.

Mukherjee, A., von Brömssen, M., Scanlon, B.R., *et al.* (2008) Hydrogeochemical comparison and effects of overlapping redox zones on groundwater arsenic near the western (Bhagirathi sub-basin, India) and eastern (Meghna sub-basin, Bangladesh) of the Bengal basin. *Journal of Contaminant Hydrology*, **99**(1-4), 31–48.

Mukhopadhyay, D. and Sanyal, S.K. (2004) Complexation and release isotherm of arsenic in arsenic-humic/fulvic equilibrium study. *Australian Journal of Soil Research*, **42**(7), 815–24.

Naseem, M., Farooqi, A., Masih, D. and Anwar, M. (2001) *Invstigation of Toxic Elements in the Ground Water of Kalalanwala Area Near Lahore, Punjab, Pakistan*. Proceedings of GEOSAS-III, Lahore.

Nesbitt, H.W., Muir, I.J. and Pratt, A.R. (1995) Oxidation of arsenopyrite by air and air-saturated, distilled water, and implications for mechanism of oxidation. *Geochimica et Cosmochimica Acta*, **59**(9), 1773–86.

Nguyen, V.L., Ta, T.K.O. and Tateishi, M. (2000) Late Holocene depositional environments and coastal evolution of the Mekong River delta, southern Vietnam. *Journal of Asian Earth Sciences*, **18**(4), 427–39.

Nickson, R., McArthur, J., Burgess, W. *et al.* (1998) Arsenic poisoning of Bangladesh groundwater. *Nature*, **395**(6700), 338.

Nickson, R.T., McArthur, J.M., Ravenscroft, P. *et al.* (2000) Mechanism of arsenic release to groundwater, Bangladesh and West Bengal. *Applied Geochemistry*, **15**(4), 403–13.

Nickson, R.T., McArthur, J.M., Shrestha, B. *et al.* (2005) Arsenic and other drinking water quality issues, Muzaffargarh district, Pakistan. *Applied Geochemistry*, **20**(1), 55–68.

Nicolli, H.B., Suriano, J.M., Gomez Peral, M.A. *et al.* (1989) Groundwater contamination with arsenic and other trace elements in an area of the Pampa, province of Córdoba, Argentina. *Environmental Geology and Water Sciences*, **14**(1), 3–16.

Nimick, D.A., Moore, J.N., Dalby, C.E. and Savka, M.W. (1998) The fate of geothermal arsenic in the Madison and Missouri rivers, Montana and Wyoming. *Water Resources Research*, **34**(11), 3051–67.

O'Shea, B.M. (2006) *Delineating the source, geochemical sinks and aqueous mobilisation processes of naturally occurring Arsenic in a Coastal Sandy Aquifer, Stuarts Point, New South Wales, Australia*. Ph.D. thesis, University of New South Wales, Sydney.

O'Shea, B.M., Clark, G. and Jankowski, J. (2006) A comparison of arsenic occurrence and geochemistry in two groundwater environments. *Geochimica et Cosmochimica Acta*, **70**(Supplement 1 18), A–464.

O'Shea, B., Jankowski, J. and Sammut, J. (2007) The source of naturally occurring arsenic in a coastal sand aquifer of eastern Australia. *Science of the Total Environment*, **379**(2-3), 151–66.

Oremland, R.S., Newman, D.K., Kail, B.W. and Stolz, J.F. (2002) Bacterial respiration of arsenate and its significance in the environment, in *Environmental Chemistry of Arsenic* (ed. W.T. Frankenberger, Jr.), Marcel Dekker, New York.

Oscarson, D.W., Huang, P.M., Defosse, D. and Herbillion, A. (1981) Oxidative power of Mn(IV) and Fe(III) oxides with respect to As(III) in terrestrial and aquatic environments. *Nature*, **291**, 50–51.

Oscarson, D.W., Huang, P.M., Liaw, W.K. and Hammer, U.T. (1983) Kinetics of oxidation of arsenite by various manganese dioxides. *Soil Science Society of America Journal*, **47**, 644–48.

Oyarzun, R., Lillo, J., Higueras, P. *et al.* (2004) Strong arsenic enrichment in sediments from the Elqui watershed, northern Chile: industrial (gold mining at El Indio-Tambo district) vs. geologic processes. *Journal of Geochemical Exploration*, **84**(2), 53–64.

Panthi, S.R., Sharma, S. and Mishra, A.K. (2006) Recent status of arsenic contamination in groundwater of Nepal — a review. *Kathmandu University Journal of Science, Engineering and Technology*, **2**, 1–11.

Peters, S.C. and Blum, J.D. (2003) The source and transport of arsenic in a bedrock aquifer, New Hampshire, USA. *Applied Geochemistry*, **18**(11), 1773–87.

Peters, S.C., Blum, J.D., Karagas, M.R. *et al.* (2006) Sources and exposure of the New Hampshire population to arsenic in public and private drinking water supplies. *Chemical Geology*, **228**, 72–84.

Peters, S.C., Blum, J.D., Klaue, B. and Karagas, M.R. (1999) Arsenic occurrence in New Hampshire drinking water. *Environmental Science and Technology*, **33**(9), 1328–33.

Pham, V.N., Boyer, D., Le Mouël, J.-L. and Kim Thoa Nguyen, T. (2002) Hydrogeological investigation in the Mekong delta around Ho-Chi-Minh City (South Vietnam) by electric tomography. *Comptes Rendus Geoscience*, **334**(10), 733–40.

Pierce, M.L. and Moore, C.B. (1982) Adsorption of arsenite and arsenate on amorphous iron hydroxide. *Water Research*, **16**(7), 1247–53.

Polizzotto, M.L., Harvey, C.F., Li, G. *et al.* (2006) Solid-phases and desorption processes of arsenic within Bangladesh sediments. *Chemical Geology*, **228**(Special Issue 1-3), 97–111.

Polizzotto, M.L., Harvey, C.F., Sutton, S.R. and Fendorf, S. (2005a) Processes conducive to the release and transport of arsenic into aquifers of Bangladesh. *Proceedings of the National Academy of Sciences of the United States of America*, **102**(52), 18819–23.

Polizzotto, M.L., Kocar, B.D., Sampson, M. *et al.* (2005b) Arsenic contamination in the flood plain aquifer of the Mekong delta, Cambodia. Abstracts with Programs. *The Geological Society of America*, **37**(7), 376.

Polya, D.A., Gault, A.G., Diebe, N. *et al.* (2005) Arsenic hazard in shallow Cambodian groundwaters. *Mineralogical Magazine*, **69**(5), 807–23.

Presser, T.S. and Swain, W.C. (1990) Geochemical evidence for Se mobilization by the weathering of pyritic shale, San Joaquin Valley, California, USA *Applied Geochemistry*, **5**(5-6), 703–17.

Puls, R.W. and Powell, R.M. (1992) Transport of inorganic colloids through natural aquifer material: implications for contaminant transport. *Environmental Science and Technology*, **26**(3), 614–21.

Rahman, M.M., Sengupta, M.K. and Chowdhury, U.K. (2006) Arsenic contamination incidents around the world, in *Managing Arsenic in the Environment: From Soil to Human Health* (eds R. Naidu, E. Smith, G. Owens *et al.*), CSIRO Publishing, Collingwood, pp. 3–30.

Ramanathan, A., Bhattacharya, P. and Tripathi, P. (2006) Arsenic in groundwater of the aquifers of the central Gangetic plain of Uttar Pradesh, India. Abstracts with Programs. *The Geological Society of America*, **38**(7), 241.

Raven, K.P., Jain, A. and Loeppert, R.H. (1998) Arsenite and arsenate adsorption on ferrihydrite: kinetics, equilibrium, and adsorption envelopes. *Environmental Science and Technology*, **32**(3), 344–49.

Ravenscroft, P., Burgess, W.G., Ahmed, K.M. *et al.* (2005) Arsenic in groundwater of the Bengal Basin, Bangladesh: Distribution, field relations, and hydrogeological setting. *Hydrogeology Journal*, **13**(5-6), 727–51.

Ravenscroft, P., McArthur, J.M. and Hoque, B. (2001) Geochemical and palaeohydrological controls on pollution of groundwater by arsenic, in *Arsenic Exposure and Health Effects IV* (eds W.R. Chappell, C.O. Abernathy and R.L. Calderon), Elsevier Science Ltd, Oxford, pp. 53–77.

Ray, A.K. (1999). Chemistry of arsenic and arsenic minerals relevant to contamination of groundwater and soil from subterranean source. *Everyman's Science*, **35**(1).

Redman, A.D., Macalady, D.L. and Ahmann, D. (2002) Natural organic matter affects arsenic speciation and sorption onto hematite. *Environmental Science and Technology*, **36**(13), 2889–96.

Robertson, F.N. (1989) Arsenic in groundwater under oxidizing conditions, south-west United-States. *Environmental Geochemistry and Health*, **11**(3-4), 171–85.

Robinson, J.G.R. and Ayotte, J.D. (2006) The influence of geology and land use on arsenic in stream sediments and ground waters in New England, USA. *Applied Geochemistry*, **21**(9), 1482–97.

Romero, L., Alonso, H., Campano, P. *et al.* (2003) Arsenic enrichment in waters and sediments of the Rio Loa (second region, Chile). *Applied Geochemistry*, **18**(9), 1399–416.

Romero, M.A., Amrienta, M.A. and Carrilo-Chavez, A. (2004) Arsenic sorption by carbonate-rich aquifer material, a control on arsenic mobility at Zimpian, Mexico. *Environmental Contamination and Toxicology*, **47**, 1–13.

Saada, A., Breeze, D., Crouzet, C. *et al.* (2003) Adsorption of arsenic (V) on kaolinite and on kaolinite-humic acid complexes: role of humic acid nitrogen groups. *Chemosphere*, **51**(8), 757–63.

Sadiq, M. (1997) Arsenic chemistry in soils: an overview of thermodynamic predictions and field observations. *Water, Air, and Soil Pollution*, **93**(1-4), 117–36.

Saha, A.K. (1991) Genesis of arsenic in groundwater in parts of West Bengal. *Center for Studies on Man and Environment, Calcutta, Annual Volume*.

Savage, K.S., Tingle, T.N., O'Day, P.A. *et al.* (2000) Arsenic speciation in pyrite and secondary weathering phases, Mother Lode Gold District, Tuolumne County, California. *Applied Geochemistry*, **15**(8), 1219–44.

Schwertmann, U., Fischer, W.R. and Papendorf, H. (1968) *The Influence of Organic Carbon Compounds on the Formation of Iron Oxides*. 9th International Congress of Soil Science, Adelaide, pp. 645–55.

Scott, M.J. and Morgan, J.J. (1995) Reactions at oxide surfaces. 1. Oxidation of As(III) by synthetic birnessite. *Environmental Science and Technology*, **29**, 1898–905.

Seyler, P. and Martin, J.M. (1989) Biogeochemical processes affecting arsenic species distribution in a permanently stratified lake. *Environmental Science and Technology*, **23**, 1258–63.

Shen, Y.S. and Chin, C.S. (1964) Relation between blackfoot disease and the pollution of drinking water by arsenic in Taiwan, in *Advances in Water Pollution Research*, Pergamon Press, New York, pp. 173–90.

Shrestha, R.R., Shrestha, M.P., Upadhyay, N.P. *et al.* (2004) *Groundwater Arsenic Contamination in Nepal: A New Challenge for Water Supply Sector*, Environment and Public Health Organization, Kathmandu.

Sidle, W.C. (2002) $^{18}O_{SO4}$ and $^{18}O_{H2O}$ as prospective indicators of elevated arsenic in the Goose River ground-watershed, Maine. *Environmental Geology*, **42**(4), 350–59.

Sidle, W.C. (2003) Identifying discharge zones of arsenic in the Goose River Basin, Maine. *Journal of the American Water Resources Association*, **39**, 1067–77.

Sidle, W.C. and Fischer, R.A. (2003) Detection of ^{3}H and ^{85}Kr in groundwater from arsenic-bearing crystalline bedrock of the Goose River Basin, Maine. *Environmental Geology*, **44**(7), 781–89.

Sidle, W.C., Wotten, B. and Murphy, E. (2001) Provenance of geogenic arsenic in the Goose River Basin, Maine, USA. *Environmental Geology*, **41**(1-2), 62–73.

Sikdar, P.K., Sarkar, S.S. and Palchoudhury, S. (2001) Geochemical evolution of groundwater in the quaternary aquifer of Calcutta and Howrah, India. *Journal of Asian Earth Sciences*, **19**, 579–94.

Smedley, P.L. (1996) Arsenic in rural groundwater in Ghana. *Journal of African Earth Sciences*, **22**(4), 459–70.

Smedley, P. (2005) *Arsenic Occurrence in Groundwater in South and East Asia: Scale, Causes and Mitigation. Towards a More Effective Operational Response: Arsenic Contamination of Groundwater in South and East Asian Countries II*. Technical Report, World Bank Report No. 31303, World Bank, Washington, DC.

Smedley, P.L. and Kinniburgh, D.G. (2002) A review of the source, behaviour and distribution of arsenic in natural waters. *Applied Geochemistry*, **17**(5), 517–68.

Smedley, P.L., Kinniburgh, D.G., Macdonald, D.M.J. *et al.* (2005) Arsenic associations in sediments from the loess aquifer of La Pampa, Argentina. *Applied Geochemistry*, **20**(5), 989–1016.

Smedley, P.L., Nicolli, H.B., Macdonald, D.M.J. *et al.* (2002) Hydrogeochemistry of arsenic and other inorganic constituents in groundwaters from La Pampa, Argentina. *Applied Geochemistry*, **17**(3), 259–84.

Smedley, P.L., Zhang, M., Zhang, G. and Luo, Z. (2003) Mobilisation of arsenic and other trace elements in fluvio-lacustrine aquifers of the Huhhot Basin, Inner Mongolia. *Applied Geochemistry*, **18**(9), 1453–77.

Smith, A.H., Goycolea, M., Haque, R. and Biggs, M.L. (1998) Marked increase in bladder and lung cancer mortality in a region of northern Chile due to arsenic in drinking water. *American Journal of Epidemiology*, **147**, 660–69.

Smith, J.V.S., Jankowski, J. and Sammut, J. (2003) Vertical distribution of As(III) and As(V) in a coastal sandy aquifer: Factors controlling the concentration and speciation of arsenic in the Stuarts Point groundwater system, northern New South Wales, Australia. *Applied Geochemistry*, **18**(9), 1479–96.

Smith, J.V.S., Jankowski, J. and Sammut, J. (2006) Natural occurrences of inorganic arsenic in the Australian coastal groundwater environment, in *Managing Arsenic in the Environment: From Soil to Human Health* (eds R. Naidu, E. Smith, G. Owens *et al.*), CSIRO Publishing, Collingwood, pp. 129–53.

Smith, A.H., Lingas, E.O. and Rahman, M. (2000) Contamination of drinking-water by arsenic in Bangladesh: a public health emergency. *Bulletin of the World Health Organization*, **78** (9), 1093–103.

Southwick, J.W., Western, A.E., Beck, M.M. (1983) An epidemiological study of arsenic in drinking water in Millard County, Utah, in *Arsenic: Industrial, Biomedical, Environmental Perspectives* (eds W.H. Leaderer and J.F. Robert), Van Nostrand Reinhold, New York, pp. 210–25.

Strawn, D., Doner, H., Zavarin, M. and McHugo, S. (2002) Microscale investigation into the geochemistry of arsenic, selenium, and iron in soil developed in pyritic shale materials. *Geoderma*, **108**(3-4), 237–57.

Stüben, D., Berner, Z., Chandrasekharam, D. and Karmakar, J. (2003) Arsenic enrichment in groundwater of West Bengal, India: geochemical evidence for mobilization of As under reducing conditions. *Applied Geochemistry*, **18**(9), 1417–34.

Swartz, C.H., Keon Blute, N., Badruzzaman, B. *et al.* (2004) Mobility of arsenic in a Bangladesh aquifer: inferences from geochemical profiles, leaching data, and mineralogical characterization. *Geochimica et Cosmochimica Acta*, **68**(22), 4539–57.

Tandukar, E.N. (2001) *Scenario of Arsenic Contamination in Groundwater in Nepal*, Department of Water Supply and Sewerage.

Tandukar, N., Bhattacharya, P., Neku, A. and Mukherjee, A.B. (2006) Extent and severity of arsenic occurrence in groundwater of Nepal, in *Managing Arsenic in the Environment: From Soil to Human Health* (eds R. Naidu, E. Smith, G. Owens *et al.*), CSIRO Publishing, Collingwood, pp. 541–52.

Tasneem, M.A. (1999) Impact of agricultural and industrial activities on groundwater quality in Kasur area. *The Nucleus, Quarterly Journal of the Pakistan Atomic Energy Commission*, **36**.

Thornton, I. and Farago, M. (1997) The geochemistry of arsenic, in *Arsenic Exposure and Health Effects* (eds C.O. Abernathy, R.L. Calderon and W.R. Chappell), Chapman and Hall, London, pp. 1–16.

Tong, N.T. (2001) *Report on Investigated Results for Arsenic Groundwater Level in Ha Noi City*, Geological and Mineral Survey of Vietnam.

Tong, N.T. (2002) *Arsenic Pollution in Groundwater in the Red River Delta*, Geological Survey of Vietnam, Northern Hydrogeological-Engineering Geological Division.

Tournassat, C., Charlet, L., Bosbach, D. and Manceau, A. (2002) Arsenic(III) oxidation by birnessite and precipitation of manganese(II) arsenate. *Environmental Science and Technology*, **36**(3), 493–500.

Trafford, J.M., Lawrence, A.R., Macdonald, D.M.J. *et al.* (1996) *The Effect of Urbanization on the Groundwater Quality Beneath the City of Hanoi, Vietnam*. British Geological Survey Technical Report WC/96/22, British Geological Survey, Keyworth.

TSC (Taiwan Sugar Company) (2003) *Groundwater Quality by the Taiwan Groundwater Monitoring Network (5/5)*, Taiwan Sugar Company (TSC), Taiwan Water Resources Bureau, Taipei.

Tseng, W.P. (1997) Effects and dose-response relationship of skin cancer and blackfoot disease with arsenic. *Environmental Health Perspectives*, **19**, 109–19.

Tseng, W.P., Chu, H.M., How, S.W. *et al.* (1968) Prevalence of skin cancer in an endemic area of chronic arsenicism in Taiwan. *Journal of National Cancer Institute*, **40**, 453–63.

Tuinhof, A. and Nanni, M. (2003) *Arsenic Testing and Finalization of Groundwater Legislation, Nepal*, World Bank Group.

UNDP/UNCHS (United Nations Development Programme and United Nations Centre for Human Settlements) (2001) *Water Quality Testing in 11 Project Townships*, United Nations Development Programme and United Nations Centre for Human Settlements.

Utsunomiya, S., Peters, S.C., Blum, J.D. and Ewing, R.C. (2003) Nanoscale mineralogy of arsenic in a region of New Hampshire with elevated As-concentrations in the groundwater. *American Mineralogist*, **88**, 1844–52.

Van Geen, A., Ahmed, K.M., Seddique, A.A. and Shamsudduha, M. (2003) Community wells to mitigate the arsenic crisis in Bangladesh. *Bulletin of the World Health Organization*, **81**, 632–38.

Van Geen, A., Aziz, Z., Horneman, A. *et al.* (2006) Preliminary evidence of a link between surface soil properties and the arsenic content of shallow groundwater in Bangladesh. *Journal of Geochemical Exploration*, **88**(1-3), 157–61.

Van Geen, A., Rose, J., Thoral, S. *et al.* (2004) Decoupling of As and Fe release to Bangladesh groundwater under reducing conditions. Part II: evidence from sediment incubations. *Geochimica et Cosmochimica Acta*, **68** (17), 3475–86.

von Brömssen, M., Jakariya, M., Bhattacharya, P. *et al.* (2007) Targeting low-arsenic aquifers in Matlab Upazila, southeastern Bangladesh. *Science of the Total Environment*, **379**(2-3), 121–32.

Waychunas, G.A., Fuller, C.C., Rea, B.A. and Davis, J.A. (1996) Wide angle X-ray scattering (WAXS) study of 'two-line' ferrihydrite structure: effect of arsenate sorption and counterion variation and comparison with EXAFS results. *Geochimica et Cosmochimica Acta*, **60**(10), 1765–81.

Weinman, B., Goodbred, S., Zheng, Y. *et al.* (2006) Fluvio-deltaic processes and geomorphic development at the scale of 100 to 1000 M: their importance in governing the heterogeneity of groundwater arsenic in Araihazar, Bangladesh. Abstracts with Programs. *The Geological Society of America*, **38**(7), 180.

Welch, A.H., Helsel, D.R., Focazio, M.J. and Watkins, S.A. (1999) Arsenic in ground water supplies of the United States, in *Arsenic Exposure and Health Effects* (eds W.R. Chappell, C.O. Abernathy and R.I., Calderon), Elsevier, Amsterdam, pp. 9–17.

Welch, A.H. and Lico, M.S. (1998) Factors controlling As and U in shallow ground water, southern Carson Desert, Nevada. *Applied Geochemistry*, **13**(4), 521–39.

Welch, A.H., Lico, M.S. and Hughes, J.L. (1988) Arsenic in ground water of the western United States. *Ground Water*, **26**, 333–47.

Welch, A.H., Westjohn, D.B., Helsel, D.R. and Wanty, R.B. (2000) Arsenic in ground water of the United States: occurrence and geochemistry. *Ground Water*, **38**(4), 589–604.

Weldon, J.M. and MacRae, J.D. (2006) Correlations between arsenic in Maine groundwater and microbial populations as determined by fluorescence in situ hybridization. *Chemosphere*, **63**, 440–48.

Wilkie, J.A. and Hering, J.G. (1996) Adsorption of arsenic onto hydrous ferric oxide: effects of adsorbate/adsorbent ratios and co-occurring solutes. *Colloids And Surfaces A-Physicochemical and Engineering Aspects*, **107**, 97–110.

WRUD (Water Resource Utilization Department) (2001) *Preliminary Study on Arsenic Contamination in Selected areas of Myanmar*. Report of the Water Resource Utilization Department, Ministry of Agriculture and Irrigation, Myanmar.

Xie, Z.M., Zhang, Y.M. and Naidu, R. (2006) Extent and severity of arsenic poisoning in China, in *Managing Arsenic in the Environment: From Soils to Human Health* (eds R. Naidu, E. Smith, G. Owens *et al.*), CSIRO Publishing, pp. 541–52.

Xu, H., Allard, B. and Grimvall, A. (1988) Influence of pH and organic substances on the adsorption of As(V) on geologic materials. *Water, Air, and Soil Pollution*, **40**, 293–305.

Xu, H., Allard, B. and Grimvall, A. (1991) Effects of acidification and natural organic materials on the mobility of arsenic in the environment. *Water, Air, and Soil Pollution*, **57**, 269–78.

Yamazaki, C., Ishiga, H., Dozen, K. *et al.* (2000) Geochemical compositions of sediments of Ganges delta of Bangladesh - arsenic release from peat? *Earth Science*, **54**, 81–93.

Yang, Q., Jung, H.B., Culbertson, C. *et al.* (2006) Geochemical characteristics of groundwater from domestic wells in greater Augusta area, Maine, USA. Abstracts with Programs. *The Geological Society of America*, **38**(7), 422.

Yeh, S. (1963) Relative incidence of skin cancer in Chinese in Taiwan: with special reference to arsenical cancer. *National Cancer Institute Monographs*, **10**, 81–170.

Yen, F.S., Long, C.N. and Lu, T.H. (1980) An environmental model of high arsenic concentration in the groundwater aquifer of Taiwan. *Journal of Taiwan Environmental Sanitation*, **12**, 66–80.

Yu, W., Harvey, C.M. and Harvey, C.F. (2003) Arsenic in groundwater in Bangladesh: a geostatistical and epidemiological framework for evaluating health effects and potential remedies. *Water Resources Research*, **39**, 1146.

Zachariáš, J., Frýda, J., Paterova, B. and Mihaljevìc, M. (2004) Arsenopyrite and As-bearing pyrite from the Roudný deposit, Bohemian Massif. *Mineralogical Magazine*, **68**(1), 31–46.

Zheng, Y., Stute, M., van Geen, A. *et al.* (2004) Redox control of arsenic mobilization in Bangladesh groundwater. *Applied Geochemistry*, **19**(2), 201–14.

Zheng, Y., van Geen, A., Stute, M. *et al.* (2005) Geochemical and hydrogeological contrasts between shallow and deeper aquifers in two villages of Araihazar, Bangladesh: implications for deeper aquifers as drinking water sources. *Geochimica et Cosmochimica Acta*, **69**(22), 5203–18.

7

Waste Treatment and Remediation Technologies for Arsenic

University of Kentucky Center for Applied Energy Research

7.1 Introduction

Arsenic occurs in a wide variety of natural materials, wastes, commercial products, and at numerous contaminated sites (Chapters 3, 5, and 6). In some cases, the element is in high enough concentrations to threaten human health or the environment. High-arsenic materials are most often encountered in two situations: (1) as commercial products or wastes from industrial, mining, utility, agricultural, or other human activities and (2) in *soils*, *sediments*, gases, and water at sites that have been naturally enhanced in arsenic or artificially contaminated (Chapters 3, 5, and 6). In both situations, the arsenic concentrations of liquids, solids, or gases may have to be treated to reduce potential environmental and health threats. Considering the large number of arsenic wastes, arsenic-contaminated sites, and the millions of people threatened by arsenic-contaminated *groundwater*, low-cost technologies are desperately needed to effectively treat arsenic in liquids, solids, and gases.

Waste management refers to the proper collection, handling, treatment, transportation, and disposal of wastes or other byproducts from mining, utility, agricultural, municipal, industrial, or other operations so that they are not environmental or human health threats. Although many countries have recently implemented extensive environmental and health and safety regulations to deal with arsenic and other contaminants (Appendix E), waste management has often been inadequate, improper, or even nonexistent. The lack of suitable waste management has resulted in numerous sites that have seriously contaminated soils, sediments, and water.

While waste management attempts to prevent site contamination, *remediation* deals with sites that have already been contaminated. Remediation refers to the restoration of a site through the treatment of its contaminated soils, sediments, *aquifers*, air, water, previously discarded wastes, and/or other materials so that they no longer pose a threat to the environment or human health. Not all sites require remediation. In some cases, contaminants are not present in sufficient concentrations to pose a significant threat as defined by applicable regulations or *risk assessment* studies. In other cases, adequate remediation may not be technically or economically feasible, and the contaminated area or feature (such as an aquifer; Chapters 3 and 6) can only be isolated from further use. If remediation is necessary, soils, sediments, groundwater, surface water, and other contaminated materials may be treated in place (*in situ remediation*), excavated and treated on-site (*ex situ remediation*), or removed for off-site treatment and disposal (also *ex situ* remediation). Each treatment technology for water, solids, or gases employs one of the following approaches: (1) concentration and removal, (2) dilution and dispersion, (3) conversion to less toxic species, or (4) encapsulation (Leist, Casey and Caridi, 2000, 127).

Selecting an arsenic-treatment technology or group of technologies for a waste management or remediation project depends on several key factors. Among these, one must first consider the medium that requires treatment; that is, water, gases, soil, ash, chromated copper arsenate (CCA)-treated wood, and so on. Arsenic *speciation*, initial arsenic concentrations, regulatory requirements, and target treatment levels must also be addressed. Additional considerations include available treatment and disposal options, the volume and general chemistry of material to be treated, how a remediated site will ultimately be used (for example, landfill, agricultural use, urban development, etc.), and costs. Due to variations in arsenic speciation and large differences in the chemical and physical properties of groundwaters, sediments, soils, and wastes at different locations (Chapters 2, 3, and 6), no single technology will adequately meet the needs of every project. Furthermore, successful remediation or waste management often requires a combination of two or more treatment technologies. For example, the remediation of inorganic *arsenite* (inorganic As(III); e.g. $H_3AsO_3^0$, $H_2AsO_3^-$, $HAsO_3^{2-}$, and/or AsO_3^{3-}) in groundwater may include oxidation to inorganic *arsenate* (inorganic As(V); e.g. $H_3AsO_4^0$, $H_2AsO_4^-$, $HAsO_4^{2-}$, and/or AsO_4^{3-}), followed by *precipitation coprecipitation*, and finally *filtration* (US Environmental Protection Agency (US EPA), 2002a, 2.5). Unless stated otherwise, the abbreviations As(III) and As(V) in this chapter only refer to inorganic trivalent and pentavalent arsenic, respectively.

This chapter reviews the major technologies that are used to treat arsenic in water, solid materials, and gases. Many of these technologies have important applications in both waste management and site remediation. For more detailed information on individual technologies, readers should consult US EPA (2002a, 2002b), Dambies (2004), Clifford and Ghurye (2002), and the other references in this chapter.

7.2 Treatment technologies for arsenic in water

7.2.1 Introduction

Arsenic contamination may occur in a variety of aqueous solutions, including groundwaters, surface waters, drinking water, industrial wastewaters, *mine drainage*, and *leachates* from landfills. In recent years, arsenic contamination has been especially widespread and problematic in the groundwaters of Bangladesh and West Bengal, India (Chakraborti *et al.*, 2002; Alaerts and Khouri, 2004; Chapter 6). As discussed in Chapters 3 and 6, the distributions and concentrations of arsenic in surface waters and groundwaters are fundamentally linked to the mineralogical, chemical, and physical characteristics of the soils, sediments, and rocks in contact with them. Considering that arsenic contamination can originate from *geologic materials*, the remediation of these materials is usually necessary to reduce arsenic concentrations in associated waters.

Groundwater is generally less accessible to treatment methods than surface water, which means that groundwater treatment is usually more difficult and expensive. Alaerts and Khouri (2004) identified several factors that affect the costs and feasibility of treating arsenic in groundwater. As expected, the lowering of arsenic drinking water standards from 50 to $10\,\mu g\,L^{-1}$ in many countries (Appendix E) has resulted in increased demands for additional technologies and funding for treating groundwaters that are utilized for human consumption. In other words, the natural laws of chemistry and physics require more effort and costs to lower the concentration of a dissolved contaminant from 50 to $10\,\mu g\,L^{-1}$ than from 100 to $50\,\mu g\,L^{-1}$.

The physical and chemical characteristics of each groundwater will affect the selection of reliable treatment options. For instance, *acidic* waters are usually easier to treat than alkaline ones (Alaerts and Khouri, 2004). The number and location of contaminated groundwater sites are other factors that strongly influence treatment and costs. In general, treatment is easier and costs are lower when processing a groundwater at a central plant in an urban setting than when dealing with scattered individual wells in rural areas. Trained personnel are also critical in maintaining the effectiveness of many treatment systems. In some cases, groundwater contamination is so severe that affordable and effective remediation is not possible. The contaminated aquifer can only be isolated as much as possible from its surroundings and alternative sources of drinking water must be found. Although some surface waters may be readily treated to replace arsenic-contaminated groundwaters (Yokota *et al.*, 2001), many of the surface waters in rural areas are too contaminated with chemicals and dangerous microorganisms for inexpensive treatment technologies (Alaerts and Khouri, 2004; Adeel, 2002, 70). Other suitable alternatives might include bottled water, pipelines, or uncontaminated aquifers. If uncontaminated aquifers are found below or laterally from a contaminated aquifer, great care must be taken to ensure that drilling and pumping new wells will not unwittingly contaminate the clean water source.

Methods for treating arsenic in water may be divided into several broad categories, which are named after the process that is primarily responsible for removing the contaminant. These categories include *sorption, ion exchange*, precipitation/coprecipitation, various *separation technologies* (including: filtration, *membranes*, and even magnetic methods), biological methods, and *natural remediation*. No water treatment technology will work in every situation. The effectiveness of a given technology depends on the types and concentrations of the arsenic species, *pH*, the presence of other dissolved species (such as calcium, phosphate (PO_4^{3-}), or organic contaminants, which may enhance or interfere with the treatment technology), the volume of water that requires treatment, costs, and the target value for arsenic treatment (that is, whether the water must be made fit for human consumption or simply meet *wastewater discharge standards*, which are usually not as stringent) (Leist, Casey and Caridi, 2000, 130–131; Chen *et al.*, 1999; Newcombe, Hart and Möller, 2006; Sahai *et al.*, 2007; Huang *et al.*, 2007, Appendix E). Multiple technologies (such as preoxidation, coprecipitation, and filtering) may also be required to attain treatment goals. US EPA (2002b) provides a review of several *ex situ* precipitation/coprecipitation, sorption, ion exchange, and filtration technologies for removing arsenic from groundwater. Detailed discussions include costs, operational requirements, reliability, capabilities of achieving the new US $10\,\mu g\,L^{-1}$ arsenic drinking water standard (*maximum contaminant level* (MCL); Appendix E), and any pretreatment steps. Another report, US EPA (2002a), includes additional reviews of *in situ* technologies for treating arsenic in water, as well as soils, sediments, and solid wastes.

7.2.2 Preoxidation of As(III) in water

Inorganic As(V) is usually the prominent form of arsenic in aerated surface waters. Depending on *oxidation-reduction* (*redox*) conditions and biological activity, groundwaters and deeper surface waters may contain As(V) and/or the more toxic As(III) forms (US EPA, 2002b, 2, 19; Chapter 3). Considering that many wastewaters and the vast majority of natural waters have pH values below 9 (Krauskopf and Bird, 1995, 225), any dissolved inorganic As(III) in them should primarily exist as $H_3AsO_3^0$ (Chapter 2). In most cases, the lack of a charge on $H_3AsO_3^0$ makes As(III) difficult to remove from water with sorption and ion-exchange technologies (Hug and Leupin, 2003, 2734). To effectively treat these waters, As(III) is usually oxidized to As(V) prior to treatment (Clifford and Ghurye, 2002, 218, 230). As(V) mostly exists as the *oxyanions* $H_2AsO_4^-$ and $HAsO_4^{2-}$ in pH 4–9 waters (Dambies, 2004, 604–605; Chapter 2). The charges on the As(V) oxyanions allow them to be removed from most waters by sorption, anion exchange, or precipitation/coprecipitation (Table 7.1).

By themselves, air and pure oxygen are generally too slow to effectively oxidize As(III) in water (Bissen and Frimmel, 2003; Burkitbaev, 2003; Hering and Kneebone, 2002, 173; Bisceglia *et al.*, 2005). Specifically, the *half-life* of naturally occurring As(III) (arsenite) in groundwater samples from two locations in Michigan, USA, was about four to nine days when saturated with air and two to five days with pure oxygen (Kim and Nriagu, 2000, 78). In comparison, saturating the groundwaters with ozone (O_3) resulted in an As(III) half-life of only 4.2–4.5 minutes (Kim and Nriagu, 2000, 78).

As(III) oxidation in the presence of air may be enhanced with radiation (Hug *et al.*, 2001), electrochemical methods (Maldonado-Reyes, Montero-Ocampo and Solorza-Feria, 2007; Arienzo *et al.*, 2002; Licht and Yu, 2005), chemicals (Hug and Leupin, 2003; Dodd *et al.*, 2006), and/or bacteria (Ehrlich, 2002; De, 2005, 684). Specifically, García *et al.* (2004) used solar energy to accelerate the oxidation of As(III) in water as part of a treatment technology. After oxidation, the resulting As(V) was coprecipitated with iron *(oxy)(hydr)oxides* (see Section 7.2.3.5.2). Nakajima *et al.* (2005) also noted that titanium dioxide (TiO_2) photocatalysts in the presence of sunlight may oxidize As(III) and convert monomethylarsonate (MMA(V); $(CH_3)AsO(OH)_2$; Chapter 2) and dimethylarsinate (DMA(V); $(CH_3)_2AsO(OH)$; Chapter 2) to As(V) in water.

Majumder and Chaudhuri (2005) developed a tartrate citrate photocatalyst for oxidizing As(III) with sunlight. The fruit of the tree, *Tamarindus indica*, provided tartrate citrate that photooxidized and removed $250–260 \mu g L^{-1}$ of As(III) in water to less than $50 \mu g L^{-1}$. The material is inexpensive and might have wide applications in treating domestic waters in India and other developing countries.

Besides sunlight, ultraviolet, and gamma radiation may accelerate the oxidation of As(III) in aerated water. Using ionizing (gamma) radiation, Burkitbaev (2003) significantly increased the oxidation rate of As(III) in the presence of molecular oxygen (O_2). Ultraviolet radiation from high-pressure mercury lamps (190–254 nm wavelengths) may also effectively oxidize As(III) to As(V). However, low-pressure mercury lamps with dominant wavelengths of 254 nm were generally ineffective (Jekel, 1994, 123). The oxidation of As(III) with ultraviolet radiation can be further enhanced with iron compounds (Leist, Casey and Caridi, 2000, 134; Dutta *et al.*, 2005, 1827), *humic acids* (Buschmann *et al.*, 2005), or titanium dioxides (TiO_2) (Zhang and Itoh, 2006; Dutta *et al.*, 2005; Xu, Kamat and O'Shea, 2005; Lee and Choi, 2002; Ferguson, Hoffmann and Hering, 2005). Hug *et al.* (2001) also found that 90 % of $500 \mu g L^{-1}$ of As(III) in water with initial Fe(II) concentrations of $0.06–5 mg L^{-1}$ can be oxidized within 2–3 hours after exposure to a $90 W m^{-3}$) ultraviolet A lamp. The addition of citrate to the water further enhanced As(III) oxidation in the Hug *et al.* (2001) study.

Effective chemical *oxidants* for As(III) in water include chlorine (Cl_2), ozone, sodium hypochlorite (NaOCl), chlorine dioxide (ClO_2), potassium permanganate ($KMnO_4$), iron(VI) compounds (such as K_2FeO_4), manganese (oxy)(hydr)oxides (including birnessite (approximately, $Na_4Mn_{14}O_{27} \cdot 9H_2O$), cryptomelane ($KMn_8O_{16}$), manganite (MnOOH), and pyrolusite (β-MnO_2)), and manganese oxide-coated sands (Clifford and Ghurye, 2002, 231; Thirunavukkarasu *et al.*, 2005; Licht and Yu, 2005; Dodd *et al.*, 2006; Bissen and Frimmel, 2003; Dutta *et al.*, 2005, 1827; Foster, 2003, 58; Tournassat *et al.*, 2002; Tani *et al.*, 2004, 6618; Manning *et al.*, 2002a). By itself, hydrogen peroxide (H_2O_2) is generally slow at oxidizing As(III) under neutral and acidic pH conditions (Hug and Leupin, 2003, 2734). The half-life of As(III) in $1 mM H_2O_2$ at pH 7.5 is 2.1 days (Hug and Leupin, 2003, 2734). H_2O_2 tends to only react with $H_2AsO_3^-$ and $HAsO_3^{2-}$, but not with the more abundant $H_3AsO_3^0$ (Hug and Leupin, 2003, 2734; Pettine, Campanella and Millero, 1999). The oxidation of As(III) with H_2O_2 vastly improves in the presence of Fe(II)

Table 7.1 Treatment methods that have the potential to reduce inorganic arsenic concentrations in water to below 10 μg L^{-1}. For some inorganic As(III) treatment methods, sorbents (such as, manganese (oxy)(hydr)oxides) may actually oxidize the As(III) to inorganic As(V) before sorbing it.

Inorganic species	Method	Reference(s)
As(III) (without separate preoxidation methods to convert to As(V))		
	Bacteria (modified) bioaccumulation	Kostal *et al.* (2004)
	Carbon powder sorbents modified with L-cysteine methyl ester	Xiao *et al.* (2006)
	FIBAN-As (Ion-exchange polymeric fibers impregnated with iron (oxy)(hydr)oxides)	Vatutsina *et al.* (2007)
	Hydrated stannic (Sn(IV)) oxide sorbent	Manna and Ghosh (2007)
	Iron(III)-impregnated granular activated carbon sorbent	Gu, Fang and Deng (2005)
	Iron (oxy)(hydr)oxide-coated sand filtration	Joshi and Chaudhuri (1996), Ko *et al.* (2007)
	Iron-rich manganese ore sorbent	Chakravarty *et al.* (2002)
	Iron and manganese (oxy)(hydr)oxide sorbents	Deschamps, Ciminelli and Höll (2005)
	Manganese (oxy)(hydr)oxide-coated sand filtration	Bajpai and Chaudhuri (1999)
	Manganese (oxy)(hydr)oxide sorbents and manganese-oxidizing bacteria	Katsoyiannis, Zouboulis and Jekel (2004)
	Permeable reactive barriers with lime, iron oxides, and limestone	Blowes *et al.* (2000)
	Siderite (coprecipitation and possibly sorption)	Guo, Stüben and Berner (2007b)
	Water hyacinth (*Eichhornia crassipes*) roots (dried) sorbent	Al Rmalli *et al.* (2005)
	Wood (maple) ash sorbent	Rahman, Wasiuddin and Islam (2004)
	Zerovalent iron sorbent	Nikolaidis, Dobbs and Lackovic (2003)
	Zirconium-loaded anion-exchange resins	Suzuki *et al.* (2001)
As(V)		
	Aluminum (oxy)(hydr)oxide coprecipitation followed by filtration	Hering *et al.* (1997)
	ArsenX^{np} sorbent	Sylvester *et al.* (2007)
	Bauxsol sorbent	Genç-Fuhrman, Tjell and McConchie (2004)
	Calcinated bauxite (modified) sorbent	Ayoob, Gupta and Bhakat (2007)
	Cationic surfactant and ultrafiltration	Gecol, Ergican and Fuchs (2004)
	Cerium-iron oxide sorbent	Zhang *et al.* (2005)
	Coconut coir pith anion exchanger	Anirudhan and Unnithan (2007)

(continued overleaf)

Table 7.1   (continued)

Inorganic species	Method	Reference(s)
	Electrochemical (co)precipitation ± filtration	Sagitova et al. (2005), Wang, Bejan and Bunce (2003), Maldonado-Reyes, Montero-Ocampo and Solorza-Feria (2007)
	Fern (living Pteris sp.) bioaccumulation	Huang et al. (2004)
	FIBAN-As (Ion-exchange polymeric fibers impregnated with iron (oxy)(hydr)oxides)	Vatutsina et al. (2007)
	Hematite ore deposit sorbents	Zhang et al. (2004)
	Hydrotalcite sorbent	Gillman (2006)
	Iron(III)-impregnated granular activated carbon sorbent	Gu, Fang and Deng (2005)
	Iron-loaded anion-exchange polymer	Cumbal and Sengupta (2005), Greenleaf, Lin and Sengupta (2006)
	Iron (oxy)(hydr)oxide-coated sand filtration	Joshi and Chaudhuri (1996), Ko et al. (2007)
	Iron (oxy)(hydr)oxide coprecipitation followed by filtration	Han et al. (2003), Hering et al. (1997)
	Iron and manganese (oxy)(hydr)oxide sorbents	Deschamps, Ciminelli and Höll (2005)
	Manganese (oxy)(hydr)oxide-coated sand filtration	Bajpai and Chaudhuri (1999)
	Manganese (oxy)(hydr)oxide sorbents and manganese-oxidizing bacteria	Katsoyiannis, Zouboulis and Jekel (2004)
	Permeable reactive barriers with lime, iron oxides, and limestone	Blowes et al. (2000)
	Polyaniline in activated carbon sorbent	Yang, Wu and Chen (2007)
	Portland cement (hardened, granular) sorbent	Kundu et al. (2004)
	Reverse osmosis	Shih (2005), Saitúa et al. (2005)
	Siderite (coprecipitation and possibly sorption)	Guo, Stüben and Berner (2007b), Wang and Reardon (2001)
	Strong-base ion-exchange resins	Korngold, Belayev and Aronov (2001)
	Titanium dioxide (granular anatase) sorbents	Bang et al. (2005b)
	Water hyacinth (E. crassipes) roots (dried) sorbent	Al Rmalli et al. (2005)
	Wood (maple) ash sorbent	Rahman, Wasiuddin and Islam (2004)
	Zerovalent iron sorbent	Su and Puls (2001a)
	Zirconium-loaded anion-exchange resins	Suzuki et al. (2001)

ions (Fenton's reagent) (Hug and Leupin, 2003; Jekel, 1994, 124; Krishna *et al.*, 2001), which creates As(V) and Fe(III). As(V) can then sorb or coprecipitate with iron (oxy)(hydr)oxides, which are removed through filtration.

In some circumstances, water chemistry may complicate or interfere with the preoxidation of As(III). Humic acids and other *natural organic matter* often consume oxidants (Floch and Hideg, 2004, 76–77). Chlorine oxidants may react with organic matter and produce carcinogenic trihalomethane (THM) byproducts (Jekel, 1994, 123–124; Dutta *et al.*, 2005, 1827). If ozone is applied as an oxidant to water containing bromide (Br^-), carcinogenic bromate (BrO_3^-) may form. The oxidation of soils, sediments, or water could also transform metals into more toxic chemical species (such as, the conversion of Cr(III) to Cr(VI)). Furthermore, permanganates may leave residual dissolved manganese in treated water (Dutta *et al.*, 2005, 1827), which may exceed drinking water recommendations and standards. The removal of manganese or other byproducts from water would further increase treatment costs. To avoid the formation of potentially toxic byproducts during As(III) oxidation, the chemistry of the water must be carefully determined before selecting an appropriate oxidant and proper dosage.

Although highly oxidizing Fe(VI) compounds (such as K_2FeO_4) are less toxic than chlorine and may be used to effectively oxidize As(III), they are usually expensive to produce (Licht and Yu, 2005, 8071). Recently, Licht and Yu (2005) developed an inexpensive electrochemical system, which synthesizes FeO_4^{2-} from an iron anode. Once produced, the FeO_4^{2-}-bearing solution is immediately available for oxidizing contaminants, including As(III). The oxidation reaction for As(III) is shown in the following reaction (Licht and Yu, 2005, 8074):

$$FeO_4^{2-} + 3/2AsO_2^- - OH^- + H_2O \rightarrow Fe(OH)_3\downarrow + 3/2AsO_4^{3-} \quad (7.1)$$

Rather than using expensive and possibly toxic chemicals, certain bacteria and other microorganisms have the ability to oxidize As(III) in water, wastes, soils, and sediments (Ehrlich, 2002; Santini, Vanden Hoven and Macy, 2002; Anderson *et al.*, 2002). *Microbacterium lacticum* (Mokashi and Paknikar, 2002) and *Alcaligenes faecalis* (Anderson *et al.*, 2002) are examples of bacteria that are known to oxidize As(III). Some bacteria use As(III) as a source of energy, whereas others simply oxidize it as a means of detoxifying their environments (Ehrlich, 2002, 313).

7.2.3 Sorption and ion-exchange technologies

7.2.3.1 Introduction

With water treatment technologies, *adsorption* refers to the removal of contaminants by causing them to attach onto the surfaces of solid materials (*adsorbents* or *sorbents*) (Chapters 2 and 3). Sometimes the adsorbed *solute* is called the *adsorbate* (Krauskopf and Bird, 1995, 145). Adsorption usually involves ion exchange (Eby, 2004, 345). For example, adsorbing arsenic oxyanions will replace other ions on the surface of the sorbent (Chapter 2).

Absorption is the assimilation of a chemical species into the interior of a solid material. Absorption may include the migration of solutes into internal pores (Fetter, 1993, 117) or the migration or exchange of atoms within the crystalline structure of a mineral (Krauskopf and Bird, 1995, 150). Some researchers use the generic term '*sorption*' to refer to a treatment method where adsorption and/or absorption are involved or if adsorption and absorption cannot be distinguished.

7.2.3.2 *Applications and limitations of sorption and ion-exchange technologies*

To sorb contaminants from water, sorbents are sometimes suspended in a reactor. More commonly, however, 0.3–0.6 mm diameter sorbents or ion-exchange granules, fibers, or other materials are packed into columns or filters (Clifford and Ghurye, 2002, 219; Greenleaf, Lin and Sengupta, 2006). The materials are large enough to facilitate *permeability* and water flow while still providing sufficient *surface areas* for numerous sorption and ion-exchange sites. Other desirable properties for sorbents and ion-exchange media include (1) an ability to rapidly and effectively remove large amounts of both As(III) and As(V) before regeneration or disposal, (2) capable of being regenerated, (3) high durability in water, and (4) reasonable costs. Nevertheless, few, if any, sorption or ion-exchange systems adequately achieve all of these goals.

Selecting an appropriate sorption or ion-exchange technology strongly depends on the characteristics of the contaminated water, including its total dissolved solids (TDS) content, pH, redox conditions, microbial activity, the presence of any organic or inorganic species that might interfere with treatment, and the concentrations of the arsenic species. For example, very acidic waters may corrode *alumina* sorbents, thus decreasing the number of active sorption sites and reducing the effective life spans of the sorbents (US EPA, 2002b, 25; Prasad, 1994, 141).

For arsenic, sorption onto inorganic solids is generally more convenient than chemical precipitation/coprecipitation methods and less expensive than ion-exchange resins or membrane filtration (Kim *et al.*, 2004, 924). In contrast to precipitation/coprecipitation, which may produce large volumes of sludge that are difficult to dewater (Deliyanni *et al.*, 2003a, 155), sorption and ion-exchange columns usually produce relatively little waste and can often be regenerated. Sorption and ion-exchange technologies also require less technical training and can be applied to small-scale operations, including small villages and even individual household wells (US Environmental Protection Agency (US EPA), 2002a, 1.1; Yokota *et al.*, 2001; Singh, 2007). On the other hand, organic materials and *silica* (SiO_2) may foul the systems (US EPA, 2002b, 27). Therefore, sorption and ion-exchange technologies are most often used with low TDS waters where arsenic is the only significant contaminant or as a 'polishing step' after precipitation/coprecipitation (US Environmental Protection Agency (US EPA), 2002a, 1.1). The costs of sorption and ion-exchange systems would depend on the concentration and speciation of the arsenic contaminants, regeneration and disposal requirements for the spent media, water flow rates, and the overall chemistry of the water, which controls fouling. To meet the lower $10 \mu g L^{-1}$ drinking water standard (MCL) for arsenic, sorption and ion-exchange systems in the United States may have to be modified to include (1) larger systems with a variety of sorbents and ion-exchange media, (2) systems that allow for more frequent replacement or regeneration of sorbents and ion-exchange materials, (3) systems that can operate at decreased water flow rates, and/or (4) the addition of one or more additional technologies to supplement the systems (US EPA, 2002b, 27; Boccelli, Small and Dzombak, 2006). Implementing such changes could significantly raise costs and even require the adoption of a completely different technology or waste management approach.

7.2.3.3 *Interferences*

Compared to precipitation/coprecipitation methods, sorbents and ion-exchange media are more vulnerable to chemical interferences that hinder the removal of arsenic from water (US EPA, 2002b, 5). Some interfering chemicals directly compete with arsenic species for sorption and ion-exchange sites. Two prime examples are phosphate (PO_4^{3-}) and *silicate* (SiO_4^{4-}), which have the same tetrahedral structure as arsenate (AsO_4^{3-}). Due to these similarities, phosphate and silicate may desorb As(V) from *clay*, iron, aluminum, and other sorbents over a wide range of pH values, or at least hinder the sorption of As(V) onto these materials (Clifford and Ghurye, 2002, 227; Zhang *et al.*, 2004; Stollenwerk, 2003, 85, 91; Su and Puls, 2003, 2582; Violante *et al.*, 2006; Lafferty and Loeppert, 2005, 2120; Holm, 2002; Smith and Edwards, 2005; McNeill, Chen and Edwards, 2002, 146; Liu *et al.*, 2007a; Zeng, Fisher and Giammar, 2008). While *carbonates* ($H_2CO_3^0$, HCO_3^-, and/or CO_3^{2-}) present in natural waters often have little or no effect on As(V) sorption, evidence suggests that they may interfere with As(III) sorption due to their similar trigonal molecular structures (Stollenwerk, 2003, 88). Dissolved organic materials may also compete with arsenic for sorption sites (Stollenwerk, 2003, 89). Specifically, *fulvic acid* is known to interfere with the sorption of As(V) onto aluminum compounds, especially under acidic conditions, and with the sorption of As(III), As(V), MMA(V), and DMA(V) on iron compounds at pH values of 4−8 (Stollenwerk, 2003, 89). Furthermore, humic acids significantly inhibit the sorption of As(V) and As(III) on goethite at pH conditions of 6−9 and 3−8, respectively (Grafe, Eick and Grossl, 2001). In contrast, the presence of nitrogen-rich (*amine*) humic acids on the surfaces of kaolinite ($Al_2Si_2O_5(OH)_4$) often improves As(V) sorption from water (Saada *et al.*, 2003). Unlike kaolinite, the amine humic acids have abundant positive surface charges under pH 7 conditions, which attract As(V) oxyanions (Saada *et al.*, 2003).

Some chemicals, such as silica, have the ability to *polymerize*, change surface charges, or otherwise modify surface chemistries to the point that arsenic sorption and ion exchange are much less favorable (Stollenwerk, 2003, 85, 89). For instance, if present in high enough concentrations in water, sulfate may accumulate on sorbents and dramatically increase the number of negative surface charges (Leist, Casey and Caridi, 2000, 131). The negative surface charges would then repel arsenic oxyanions and thus hinder their sorption and ion exchange.

7.2.3.4 *Regeneration and disposal of spent sorbents and ion-exchange materials*

Depending on their chemistry and costs, exhausted sorbents and ion-exchange media may be regenerated and reused rather than discarded. To increase the time between regeneration, treatment systems often use a series of two or more sorption or ion-exchange columns. The first column in the series is regenerated or replaced first and the second column then advances to the first position while a fresh column is added to the end of the series (US EPA, 2002b, 25). Nonregenerable sorbents and ion-exchange media tend to be more expensive. On the other hand, the disposal costs of nonregenerable products may be less than potentially hazardous wastewaters and other byproducts that often result from regeneration.

Eventually, even a multiple reuse sorbent or ion-exchange resin can no longer be regenerated and will require disposal. The disposal options and costs for spent materials will greatly depend on whether or not the materials are hazardous. For landfilling in the United States, the toxicity characteristic leaching procedure (TCLP; Appendix E) determines the toxic *hazardousness* of a spent sorbent or resin under federal regulations. However, local and state regulations in the United States may require that the TCLP be supplemented with the California Waste Extraction Test (*CWET*) and/or other leaching tests before the materials can be legally landfilled (Appendix E). Furthermore, a number of studies indicate that the TCLP may seriously underestimate the leaching of arsenic from spent materials in landfills (Ghosh *et al.*, 2006). Therefore, additional studies beyond the regulatory requirements may be needed to adequately predict any environmental or health threats from the landfilling of spent sorbents and ion-exchange resins.

Jing *et al.* (2005a) compared results from both the TCLP and CWET to evaluate the ability of arsenic to leach from five spent sorbents from a field site in southern New Jersey. The sorbents had been used to treat raw water with 39−53 μg L^{-1} of As(V) and they included granular 'ferric hydroxide' (actually mostly akaganéite, β-FeO(OH)), granular 'ferric oxide' (goethite), titanium dioxide (anatase), activated alumina, and modified (iron-impregnated) activated alumina. In all cases, the TCLP solutions contained less than 180 μg L^{-1} of arsenic, which is far below the 5 mg L^{-1} limit (Appendix E). In contrast to the TCLP, CWET uses a stronger citrate leaching solution and has a lower liquid to solid ratio. As a result, the arsenic concentrations in the CWET leachates were consistently more than ten times higher than the TCLP results and were as high as $6.65\,mg\,L^{-1}$ (Jing *et al.*, 2005a). Although the spent sorbents were considered nonhazardous under US federal regulations, the CWET suggested that the sorbents could release deleterious concentrations of arsenic if the disposal environment of a landfill was more severe than the conditions modeled by the TCLP. The critical question then arises as to how accurate either leaching test models the actual landfill leaching environment. Rather than performing additional leaching studies, the dilemma might be resolved by focusing on reducing the leachability of the spent sorbents, perhaps through *solidification/stabilization* (see below discussions in Section 7.3.2.2).

7.2.3.5   *Review of various sorption and ion-exchange technologies*

While Chapters 2 and 3 provide additional details on the mechanisms by which arsenic species sorb or undergo ion exchange onto solid surfaces, the chemistry and effectiveness of some of the more widely used sorbents and ion-exchange media will be reviewed in the following sections. Since costs are often very site specific and fluctuate with market conditions and regulatory changes, they are generally not discussed in this chapter. US EPA (2002a), US EPA (2002b) contains some recent examples of cost analyses for various arsenic sorption and ion-exchange technologies. In particular, US EPA (2002b) reviews the performance and costs of seven full-scale ion-exchange and 23 pilot- and full-scale sorption technologies for the *ex situ* removal of arsenic from water, including water from US Superfund (National Priorities List (*NPL*), Appendix E) sites.

7.2.3.5.1   Zerovalent (elemental) iron   In the presence of water and oxygen, zerovalent (elemental) iron (Fe(0)) rapidly oxidizes or 'rusts'. Depending on pH, the amount of molecular oxygen (O_2), the presence of bacteria, and other aqueous conditions, the oxidation of zerovalent iron may produce a wide variety of compounds, including lepidocrocite (γ-FeO(OH)), sulfate ($4Fe(OH)_2 \cdot 2FeOOH \cdot FeSO_4 \cdot 4H_2O$) and carbonate (possibly $Fe(II)_4Fe(III)_2(OH)_{12}(CO_3 \cdot 2H_2O)$) 'green rusts', akaganéite, mackinawite (Fe_9S_8), magnetite (Fe_3O_4), maghemite (γ-Fe_2O_3), goethite, and amorphous ferrous sulfide (FeS) (Gu *et al.*, 1999, 2175–2176). These oxidation products are largely responsible for sorbing and/or coprecipitating arsenic from water.

While iron (oxy)(hydr)oxides typically sorb more As(V) than As(III), zerovalent iron is generally more effective in treating As(III) (Cheng *et al.*, 2005, 7665). Unlike most sorption/ion-exchange technologies, As(III) does not require preoxidation before treatment with zerovalent iron (Lien and Wilkin, 2005). Although As(III) may spontaneously oxidize in the presence of zerovalent iron (Cheng *et al.*, 2005, 7662) and the resulting As(V) may be sorbed by iron oxidation products, the removal of As(III) by zerovalent iron cannot be adequately explained by this process. The exact sorption, ion-exchange, and coprecipitation mechanisms that are responsible for removing arsenic with zerovalent iron are complex and depend on the characteristics of the water and the iron (Su and Puls, 2001a; Lien and Wilkin, 2005; Bang *et al.*, 2005a; Yu *et al.*, 2006). Among the many possible oxidation products, carbonate green rust could have an important role in the removal of As(III). As(III) is capable of substituting for carbonate in the crystalline structure of the compounds (Lien and Wilkin, 2005, 383, 385). Besides sorption and ion exchange, arsenic could also react under certain redox and pH conditions to form precipitates on zerovalent iron surfaces, such as symplesite ($Fe_3(AsO_4)_2$) or realgar (As_4S_4) (Lien and Wilkin, 2005, 383).

The removal of As(III) and As(V) from water greatly increases as zerovalent iron oxidizes. In a series of batch system experiments in Bang *et al.* (2005a), zerovalent iron at pH 7 and under a nitrogen atmosphere removed about 36 % of $0.5\,mg\,L^{-1}$ of As(V) and 56 % of $0.5\,mg\,L^{-1}$ of As(III) in 120 hours. When exposed to air, the removal values increased to 72 % for As(V) and 55 % for As(III) after only 90 minutes of mixing at pH 7 (Bang *et al.*, 2005a, 765–766). Kanel *et al.* (2005, 1297) further suggests that As(III) is initially sorbed (over 0–24 hours) onto amorphous Fe(II) and Fe(III) compounds, magnetite, and/or maghemite that form on the surfaces of the zerovalent iron. As oxidation proceeds, crystalline magnetite and lepidocrocite develop and become important sorption sites for any remaining As(III). Eventually, the sorbed arsenic may be buried and trapped under fresh layers of oxidized iron precipitates (Kanel *et al.*, 2005, 1297).

In the absence of dissolved oxygen, zerovalent iron can still oxidize to Fe(II) and produce magnetite and other iron oxides, which can sorb As(III) and As(V) from water (Bang *et al.*, 2005a, 766; Manning *et al.*, 2002b). According to Manning *et al.* (2002b, 5455) and Su and Puls (2001a, 1490), water would be the primary oxidant for elemental iron as shown in the following *anaerobic* reaction:

$$Fe^0 + 2H_2O \rightarrow Fe^{2+} + H_2 + 2OH^- \tag{7.2}$$

Fe^{2+} can then react with OH^- to form $Fe(OH)_2$. Over hours to days, $Fe(OH)_2$ further oxidizes to form magnetite or maghemite (Manning *et al.*, 2002b, 5455; Farrell *et al.*, 2001, 2030).

Fe^{2+} can also form under more *aerobic* conditions (Su and Puls, 2001a, 1490):

$$2Fe^0 + 2H_2O + O_2 \rightarrow 2Fe^{2+} + 4OH^- \tag{7.3}$$

However, in the presence of O_2, Fe^{2+} oxidizes to Fe^{3+}. At pH conditions above about 3.5, Fe^{3+} then rapidly precipitates as iron (oxy)(hydr)oxides (Bednar *et al.*, 2005, 58; Chapters 2 and 3).

Sulfate, bicarbonate (HCO_3^-), and microbes in contaminated water can substantially influence the oxidation rate of zerovalent iron, which affects the removal of arsenic (Gu *et al.*, 1999). Bicarbonate in the presence of sulfate is especially corrosive to zerovalent iron (Gu *et al.*, 1999). Corrosion products included ferrous carbonate ($FeCO_3$) and hydrogen gas, which can stimulate microbial growth and increase the reduction of sulfate. Nevertheless, in the studies described in Gu *et al.* (1999), it took about two months for sulfate reduction to produce significant concentrations of sulfide. Gu *et al.* (1999) demonstrated that chemicals and microbes in groundwater could significantly corrode in-ground zerovalent iron treatment barriers and possibly interfere with their operations (also see Section 7.2.5).

The effectiveness of zerovalent iron in removing arsenic from water also greatly depends on the properties of the iron. As(III) removal is especially effective with high surface area 1–120 nm spheres of zerovalent iron (Kanel *et al.*, 2005). Provided that interfering anions (such as, carbonate, silicate, and phosphate) are insignificant, colloidal spheres of zerovalent iron could be injected into arsenic-contaminated soils, sediments, and aquifers for possible *in situ* remediation (Kanel *et al.*, 2005, 1291).

Different brands of zerovalent iron or the presence of alloys of nickel and other metals with zerovalent iron may have significant effects on arsenic removal. When compared with pure iron, nanometer-sized alloys of zerovalent nickel–iron increased the sorption rate of As(V) by 2.5 times (Jegadeesan, Mondal and Lalvani, 2005). In another study, Su and Puls (2001a) compared the abilities of zerovalent iron from four different manufacturers (Aldrich, Fisher, Master Builders, and Peerless) to sorb As(III) and As(V) from water. A total of 1 g of iron was added to 41.6 ± 0.17 ml of 0.01 M sodium chloride (NaCl) batch solutions containing either $2\,mg\,L^{-1}$ of As(V), As(III), or a 50–50 % mixture of the two. The iron was allowed to react with the arsenic in the dark at 23 °C for up to five days. Except for the Aldrich samples, dissolved arsenic concentrations declined to levels below $10\,\mu g\,L^{-1}$ in four days. The effectiveness of the four iron products was highly variable (Fisher > Peerless $\sim$ Master Builders > Aldrich). The reason(s) for the variations among the four different products is unknown, but they cannot be readily explained by initial differences in surface area (Su and Puls, 2001a, 1489).

Using columns and water initially containing $50\,mg\,L^{-1}$ of As(III), Lien and Wilkin (2005) estimated that the As(III) removal capacity of Peerless zerovalent iron was about 6.5–8.9 mg As(III)/g Fe(0) under anaerobic conditions (Table 7.2). As a comparison, Kanel *et al.* (2005) synthesized their own zerovalent iron. Using batch instead of column methods, Kanel *et al.* (2005) obtained a maximum As(III) removal capacity of 3.5 mg As(III)/g Fe(0) at 25 °C and in the presence of an unknown amount of dissolved oxygen (Table 7.2).

Su and Puls (2001b) investigated possible interferences from different oxyanions on the sorption of As(V) and As(III) on Peerless zerovalent iron. Using chloride as a relatively nonreactive standard, Su and Puls (2001b) found that phosphate had the greatest effect in decreasing arsenic sorption on zerovalent iron. This observation is consistent with other batch (Kanel *et al.*, 2005) and column (Su and Puls, 2003) studies. Like As(V), phosphate can form *inner sphere complexes* (Chapter 2) or produce distinct precipitate layers on zerovalent iron surfaces (Su and Puls, 2001b, 4564).

Silicate, chromate, molybdenate, and humic acids also may create significant interferences with the sorption of As(V) and As(III) on zerovalent iron (Su and Puls, 2001b; Giasuddin, Kanel and Choi, 2007). Interferences from borate and sulfate, however, were negligible with As(V) and only minor with As(III). Some interferences with As(III) and As(V) sorption occurred with carbonate and nitrate (Su and Puls, 2001b). Similarly, Farrell *et al.* (2001) concluded that carbonate and nitrate could hinder the efforts of zerovalent iron to lower As(V) concentrations in water to below $5 \mu g L^{-1}$.

Several researchers have sought to develop inexpensive arsenic treatment systems utilizing zerovalent iron. The systems could then be commercialized for residential and other point-of-use applications in developing countries, such as Bangladesh. To be widely used, any treatment system must be inexpensive, nontoxic, fast, convenient, effective over a wide range of As(III) and As(V) concentrations (at least $10–1000 \mu g L^{-1}$), have a high potential to be regenerated, and not depend on electricity to function.

Leupin, Hug and Badruzzaman (2005) tested the removal of arsenic from Bangladeshi groundwater with filter columns containing sand and zerovalent iron. Each column consisted of 2.5 g of iron in 100–150 g of sand. The groundwater initially contained $440 \mu g L^{-1}$ of As(III), $1.8 mg L^{-1}$ of phosphorus (P), $4.7 mg L^{-1}$ of iron, $19 mg L^{-1}$ of silica, and $6 mg L^{-1}$ of dissolved organic carbon. Using a flow rate of only 1 L hour^{-1} and without As(III) preoxidation, Leupin, Hug and Badruzzaman (2005) were able to lower the arsenic concentrations in 75–90 L of groundwater to below $50 \mu g L^{-1}$. In this system, the removal of arsenic from the water was probably accomplished through spontaneous oxidation of the As(III) and subsequent sorption of the As(V) onto various iron oxidation products (Leupin, Hug and Badruzzaman, 2005; Leupin and Hug, 2005).

On a larger scale, Nikolaidis, Dobbs and Lackovic (2003) evaluated a zerovalent iron treatment system for the removal of As(III) from municipal landfill leachate at a Superfund site in Maine, United States. The composition of the leachate included $300 \pm 50 \mu g L^{-1}$ of arsenic (96–99 % As(III)), $65 \pm 5 mg L^{-1}$ of Fe(II), $180 \pm 20 mg L^{-1}$ of sulfate, $8 \pm 3 mg L^{-1}$ of dissolved organic carbon, a pH range of 6.0–6.3, and an *Eh* of $-50 mV$ (Nikolaidis, Dobbs and Lackovic, 2003, 1418, 1421). They found that over at least an eight-month period, iron filings could reduce As(III) concentrations in the groundwater to below $10 \mu g L^{-1}$ (Table 7.1). TCLP leachates of the spent zerovalent iron were two orders of magnitude below the $5 mg L^{-1}$ standard and would be considered nontoxic under US EPA regulations (Nikolaidis, Dobbs and Lackovic, 2003). The TCLP results are consistent with previous studies, which indicate that arsenic strongly sorbs onto iron filings and does not readily desorb (Nikolaidis, Dobbs and Lackovic, 2003; Ramaswami, Tawachsupa and Isleyen, 2001; Lackovic, Nikolaidis and Dobbs, 2000).

Zerovalent iron also removes methylated arsenate (i.e. MMA(V) and DMA(V)) from water. Cheng *et al.* (2005) used laboratory and field studies at a Superfund site in New Jersey, United States, to evaluate the removal of MMA(V) and DMA(V) from water with iron filings. After 20 hours in the dark at pH 4–7, batch experiments using 1 g of iron filings removed more than 95 % of $1 mg L^{-1}$ of MMA(V) from 25 ml of water. In contrast, less than 30 % of the DMA(V) was removed under the same conditions (Cheng *et al.*, 2005, 7663–7664). The addition of $2 mg L^{-1}$ of phosphorus in the form of potassium dihydrogen phosphate (KH_2PO_4) reduced the ability of 1 g of iron fillings to remove $1 mg L^{-1}$ of MMA(V) by about 20 %. The

Table 7.2 *Inorganic arsenic removal capacities of various sorbents and ion-exchange media in water. Ambient conditions indicate about one atmosphere pressure and room temperature or about 20–25 °C.*

Sorbent or ion-exchange material	Type of water	Temperature (°C)	Initial pH	Inorganic arsenic species	Initial arsenic concentration (mg L^{-1})	Batch sorbent dosage (sorbent/batch solution) or column	Initial surface area (m^2 g^{-1}) of sorbent or ion-exchange material	Maximum removal capacity (mg As g^{-1} sorbent or ion-exchange material)	References
Akaganéite-loaded cellulose beads	Laboratory solutions	25 ± 0.5	7.0 ± 0.1	As(III)	75–7500	1 ml/50 ml	?	99.6	Guo and Chen (2005)
Akaganéite-loaded cellulose beads	Laboratory solutions	25 ± 0.5	7.0 ± 0.1	As(V)	75–7500	1 ml/50 ml	?	33.2	Guo and Chen (2005)
Akaganéite	Laboratory solutions	25	7.5	As(V)	5–20	0.5 g L^{-1}	330	120	Deliyanni et al. (2003a)
Akaganéite	Groundwater: Arizona	Ambient?	7.6–8.9	As(V)	0.014–0.053	Column	?	0.02–0.28	Westerhoff et al. (2005)
Akaganéite	Laboratory solutions	Ambient?	8.6	As(V)	0.100	Column	?	0.99–1.5	Westerhoff et al. (2005)
Algae powder: *Lessonia nigrescens*	Wastewater: copper smelter	20	2.5	As(V)	50–600	2 g/500 ml	?	45.2	Hansen, Ribeiro and Mateus (2006)
Alumina (activated)	Laboratory solutions	25 ± 0.5	6.9 ± 0.1	As(III)	1.09	0.1–0.5 g/100 ml	115–118	0.2	Lin and Wu (2001)
Alumina (activated)	Laboratory solutions	25 ± 0.5	6.9 ± 0.1	As(III)	4.90	0.1–0.5 g/100 ml	115–118	0.9	Lin and Wu (2001)
Alumina (activated)	Laboratory solutions	25 ± 0.5	5.2 ± 0.1	As(V)	2.85	0.1–0.5 g/100 ml	115–118	3	Lin and Wu (2001)
Alumina (activated)	Laboratory solutions	25 ± 0.5	5.2 ± 0.1	As(V)	11.52	0.1–0.5 g/100 ml	115–118	11	Lin and Wu (2001)
Alumina (mesoporous)	Laboratory solutions	Ambient?	?	As(III)	7–1500	0.1 g/20 ml	307	47	Kim et al. (2004)
Alumina (mesoporous)	Laboratory solutions	Ambient?	5	As(V)	1500	0.1 g/20 ml	307	121	Kim et al. (2004)
Aluminum-coated limestone	Laboratory solutions	22–25	?	As(V)	5.0	Column	?	0.150	Ohki et al. (1996)

(continued overleaf)

Table 7.2 (continued)

Sorbent or ion-exchange material	Type of water	Temperature (°C)	Initial pH	Inorganic arsenic species	Initial arsenic concentration (mg L^{-1})	Batch sorbent dosage (sorbent/batch solution) or column	Initial surface area (m^2 g^{-1}) of sorbent or ion-exchange material	Maximum removal capacity (mg As g^{-1} sorbent or ion-exchange material)	References
ArsenX^{np}	Groundwater: New Mexico and Arizona, United States	Ambient?	7.5–8.2	As(V)	0.011–0.023	Column	?	8.17	Sylvester et al. (2007)
Calcium alginate	Laboratory solutions	20	5–6	As(V)	6	0.5 g L^{-1}	313	6.75	Lim and Chen (2007)
Carbon (activated, impregnated with copper)	Laboratory solutions	22	8	As(III)	1–100	Column	1200	31	Rajaković (1992)
Carbon (activated, impregnated with copper)	Laboratory solutions	22	8	As(V)	1–100	Column	?	17	Rajaković (1992)
Carbon (granular activated)	Laboratory solutions	22	8	As(V)	1–100	Column	?	20	Rajaković (1992)
Carbon (activated, prepared from extract of olive wastes)	Laboratory solutions	Ambient?	7.0	As(III)	5–20	0.250 g/25 ml	1030	1.39	Budinova et al. (2006)
Carbon (black, modified with sulfuric acid)	Laboratory solutions	20 ± 1	~4	As(V)	100	50 mg/50 ml	62.1	150–160	Borah et al. (2008)
Carbon (granular activated, untreated with iron)	Laboratory solutions	25 ± 1	4.70	As(V)	0.105, 1.031	0.090 g/30.0 ml	600–1000	0.0378	Gu, Fang and Deng (2005)
Carbon (granular activated with iron)	Laboratory solutions	25 ± 1	4.70	As(V)	0.105, 1.031	0.090 g/30.0 ml	?	2.96	Gu, Fang and Deng (2005)

Carbon (granular activated with iron + O_2)	Laboratory solutions	25 ± 1	As(V)	0.105, 1.031	0.090 g/30.0 ml	?	1.92	Gu, Fang and Deng (2005)
Carbon (granular activated with iron + H_2O_2)	Laboratory solutions	25 ± 1	As(V)	0.105, 1.031	0.090 g/30.0 ml	?	3.94	Gu, Fang and Deng *et al.* (2005)
Carbon (granular activated with iron + NaClO)	Laboratory solutions	25 ± 1	As(V)	0.105, 1.031	0.090 g/30.0 ml	?	6.57	Gu, Fang and Deng (2005)
Carbon (mesoporous, Fe[III] containing)	As-spiked Columbia, Missouri, USA tap water	25 ± 1	As(III)	4.925	0.090 g/30.00 ml	401	5.96	Gu and Deng (2007)
Carbon (mesoporous, Fe[III] containing)	As-spiked Columbia, Missouri, USA tap water	25 ± 1	As(V)	4.910	0.090 g/30.00 ml	401	5.15	Gu and Deng (2007)
Chitosan sorbent	Laboratory solutions	25 ± 2	As(III)	10	1 g/200 ml	?	1.83	Chen and Chung (2006)
Chitosan sorbent	Laboratory solutions	25 ± 2	As(V)	10	1 g/200 ml	?	1.94	Chen and Chung (2006)
Coconut coir pith anion exchanger	Laboratory solutions	30 ± 1	As(V)	1	2 g L^{-1}	175	0.50	Anirudhan and Unnithan (2007)
Coconut coir pith anion exchanger	Laboratory solutions	30 ± 1	As(V)	5–100	2 g L^{-1}	175	12.51	Anirudhan and Unnithan (2007)
Copper(II) oxide (nanostructured)	Laboratory solutions	25	As(III)	11.5	0.03 g/20 ml	87	4.7	Cao *et al.* (2007)
Drinking water treatment residuals (aluminum rich)	Laboratory solutions	23 ± 2	As(V)	3750–15 000	100 g L^{-1}	?	15	Makris, Sarkar and Datta (2006)

Table 7.2 (continued)

Sorbent or ion-exchange material	Type of water	Temperature (°C)	Initial pH	Inorganic arsenic species	Initial arsenic concentration (mg L^{-1})	Batch sorbent dosage (sorbent/batch solution) or column	Initial surface area (m^2 g^{-1}) of sorbent or ion-exchange material	Maximum removal capacity (mg As g^{-1} sorbent or ion-exchange material)	References
Drinking water treatment residuals (iron rich)	Laboratory solutions	23 ± 2	6.0 ± 0.3	As(III)	3750–15 000	100 g L^{-1}	?	15	Makris, Sarkar and Datta (2006)
Fe(III)-coated sand	As-spiked tap water	22 ± 1	?	As(III)	0.100	?	?	0.136	Thirunavukkarasu et al. (2005)
Fe(III)-loaded cellulose sponge	Laboratory solutions	Ambient	9.0	As(III)	100–10000	?	?	18	Muñoz, Gonzalo and Valiente (2002)
Fe(III)-loaded cellulose sponge	Laboratory solutions	Ambient	4.5	As(V)	100–10000	?	?	137	Muñoz, Gonzalo and Valiente (2002)
Fe(III)-loaded chelating resin (Lewatit TP 207)	Laboratory solutions	20	1.7	As(V)	1000	?	?	58	Rau, Gonzalo and Valiente (2000)
thiol resin	Laboratory solutions	Ambient?	2	As(V)	15 000	?	?	124	Hrubý et al. (2003)
Fe(III)-loaded thiol polymer	Laboratory solutions	Ambient?	2	As(V)	15 000	?	?	283	Hrubý et al. (2003)
Feathers treated with thioglycolate	Laboratory solutions	25 ± 1	5	As(III)	?	2.0 g L^{-1}	?	19.9	Teixeira and Ciminelli (2005)
Ferrihydrite ('two line')	Laboratory solutions	Ambient	3.0	As(V)	750	?	226	210	Carlson et al. (2002)
FIBAN-As (Ion-exchange polymeric fibers impregnated with iron [oxy][hydr] oxides)	Laboratory solutions	20 ± 2	9	As(III)	500	?	?	110	Vatutsina et al. (2007)

Material									Reference
FIBAN-As (Ion-exchange polymeric fibers impregnated with iron [oxy][hydr] oxides)	Laboratory solutions	20 + ?	5.64	As(V)	500	?	?	121	Vatutsina et al. (2007)
Fungus: Penicillium purpurogenum	Laboratory solutions	20	5	As(III)	10–750	100 mg/50 ml	?	35.6	Say, Yilmaz and Denizli (2003b)
Gibbsite	Laboratory solutions	25 ± 0.5	5.5	As(III)	10–1000	10 g/100 ml	13.5	3.3	Ladeira and Ciminelli (2004)
Gibbsite	Laboratory solutions	25 ± 0.5	5.5	As(V)	10–1000	10 g/100 ml	13.5	4.6	Ladeira and Ciminelli (2004)
Goethite	Laboratory solutions	Ambient?	9	As(III)	0–30	1.6 g L^{-1}	39	22	Lenoble et al. (2002)
Goethite	Laboratory solutions	Ambient?	9	As(V)	0–60	1.6 g L^{-1}	39	4	Lenoble et al. (2002)
Goethite	Laboratory solutions	25 ± 0.5	5.5	As(III)	10–1000	10 g/100 ml	12.7	7.5	Ladeira and Ciminelli (2004)
Goethite	Laboratory solutions	25 ± 0.5	5.5	As(V)	10–1000	10 g/100 ml	12.7	12.4	Ladeira and Ciminelli (2004)
Goethite	Laboratory solutions	25–45	5	As(V)	10	1 g L^{-1}	132	~25	Matis et al. (1999)
Goethite	Laboratory solutions	Ambient	2	As(V)	10	1 g L^{-1}	?	11.4	Mohapatra et al. (2006)
Goethite	Laboratory solutions	Ambient	4	As(V)	10	1 g L^{-1}	?	8.97	Mohapatra et al. (2006)
Goethite	Laboratory solutions	Ambient	7	As(V)	10	1 g L^{-1}	?	3.32	Mohapatra et al. (2006)
Goethite (Co[II]-doped)	Laboratory solutions	Ambient	2	As(V)	10	1 g L^{-1}	40–130	16.32	Mohapatra et al. (2006)
Goethite (Co[II]-doped)	Laboratory solutions	Ambient	4	As(V)	10	1 g L^{-1}	40–130	11.86	Mohapatra et al. (2006)

(continued overleaf)

Table 7.2 (continued)

Sorbent or ion-exchange material	Type of water	Temperature (°C)	Initial pH	Inorganic arsenic species	Initial arsenic concentration (mg L⁻¹)	Batch sorbent dosage (sorbent/ batch solution) or column	Initial surface area (m² g⁻¹) of sorbent or ion-exchange material	Maximum removal capacity (mg As g⁻¹ sorbent or ion-exchange material)	References
Goethite (Co(II)-doped)	Laboratory solutions	Ambient	7	As(V)	10	1 g L⁻¹	40–130	5.72	Mohapatra et al. (2006)
Goethite (Cu(II)-doped)	Laboratory solutions	Ambient	2	As(V)	10	1 g L⁻¹	40–160	19.55	Mohapatra et al. (2006)
Goethite (Cu(II)-doped)	Laboratory solutions	Ambient	4	As(V)	10	1 g L⁻¹	40–160	17.36	Mohapatra et al. (2006)
Goethite (Cu(II)-doped)	Laboratory solutions	Ambient	7	As(V)	10	1 g L⁻¹	40–160	13.25	Mohapatra et al. (2006)
Goethite (Ni(II)-doped)	Laboratory solutions	Ambient	2	As(V)	10	1 g L⁻¹	40–160	18.9	Mohapatra et al. (2006)
Goethite (Ni(II)-doped)	Laboratory solutions	Ambient	4	As(V)	10	1 g L⁻¹	40–160	15.59	Mohapatra et al. (2006)
Goethite (Ni(II)-doped)	Laboratory solutions	Ambient	7	As(V)	10	1 g L⁻¹	40–160	7.55	Mohapatra et al. (2006)
Hematite iron ore	Laboratory solutions	Ambient?	4.5–6.5	As(V)	1	5 g L⁻¹	10.2	0.4	Zhang et al. (2004)
Hydrotalcite (calcinated)	Laboratory solutions	25	4.2–5.4	As(V)	0.020–0.200	0.12 g L⁻¹	198	5.609	Yang et al. (2005)
Hydrotalcite (uncalcinated)	Laboratory solutions	25	4.2–5.4	As(V)	0.020–0.200	0.12 g L⁻¹	47	4.545	Yang et al. (2005)
Iron-modified light-expanded clay aggregates	Laboratory solutions	Ambient	6.0	As(V)	0.1–100	10 mg ml⁻¹	?	3.12	Haque et al. (2008)
Iron (oxy)(hydr)oxides (amorphous)	Laboratory solutions	Ambient?	<9	As(III)	0–20	1.6 g L⁻¹	200	28	Lenoble et al. (2002)
Iron (oxy)(hydr)oxides (amorphous)	Laboratory solutions	Ambient?	<9	As(V)	0–60	1.6 g L⁻¹	200	7	Lenoble et al. (2002)

Material	Water type	Temperature	pH	Species	Concentration	Dose			Reference
Iron (oxy)(hydr)oxides (granular, akaganéite?)	Laboratory solutions	24 ± 0.5	7.00 ± 0.02	As(V)	0.100	?	230–252	4	Badruzzaman, Westerhoff and Knappe (2004)
Iron (oxy)(hydr)oxide-coated sand	Laboratory solutions	27 ± 2	7.5	As(III)	0.100–0.800	$20\,g\,L^{-1}$	?	0.02857	Gupta, Saini and Jain (2005)
Iron and manganese (oxy)(hydr)oxides	Laboratory solutions	25 ± 0.5	3.0	As(III)	0.100–100	10 g/100 ml	40.8	14.7	Deschamps, Ciminelli and Höll (2005)
Iron and manganese (oxy)(hydr)oxides	Arsenic spiked tap water and mine drainage	25 ± 0.5	3.0	As(V)	0.100–100	10 g/100 ml	40.0	0.5	Deschamps, Ciminelli and Höll (2005)
Iron- and manganese-(oxy)(hydr)oxide sea nodules	Laboratory solutions	Ambient?	5.9–6.1	As(III)	0.34		?	0.74	Maity et al. (2005)
Iron- and manganese-(oxy)(hydr)oxide sea nodules	Laboratory solutions	Ambient?	2.0–2.2	As(V)	0.78		?	0.74	Maity et al. (2005)
Kaolinite	Laboratory solutions	22	?	As(III)	7–75	1 g/10 ml	15	0.04	Li et al. (2007)
Kaolinite	Laboratory solutions	22	?	As(V)	7–150	1 g/10 ml	15	0.14	Li et al. (2007)
Kaolinite treated with hexadecyltrimethylammonium bromide	Laboratory solutions	22	?	As(III)	7–150	1 g/10 ml	?	0.3	Li et al. (2007)

Table 7.2 (*continued*)

Sorbent or ion-exchange material	Type of water	Temperature (°C)	Initial pH	Inorganic arsenic species	Initial arsenic concentration (mg L^{-1})	Batch sorbent dosage (sorbent/batch solution) or column	Initial surface area (m^2 g^{-1}) of sorbent or ion-exchange material	Maximum removal capacity (mg As g^{-1} sorbent or ion-exchange material)	References
Kaolinite treated with hexadecyltrimethylammonium bromide	Laboratory solutions	22	?	As(V)	7–150	1 g/10 ml	?	0.7	Li *et al.* (2007)
Manganese-substituted Fe (III) oxyhydroxide (Mn$_{0.13}$Fe$_{0.87}$OOH)	Laboratory solutions	30–60	7.0	As(III)	0.2–25.0	?	101	4.58	Lakshmipathiraj *et al.* (2006)
Manganese-substituted iron oxyhydroxide (Mn$_{0.13}$Fe$_{0.87}$OOH)	Laboratory solutions	30–60	7.0	As(V)	1.0–50.0	?	101	5.72	Lakshmipathiraj *et al.* (2006)
Methacryloylamido phenylalanine beads	Synthetic wastewater	20	7.0	As(III)	37	?	19.1	8.0	Say *et al.* (2003a)
Methacryloylamido phenylalanine beads	Laboratory solution	20	6.0	As(III)	750	?	19.1	204.1	Say *et al.* (2003a)
Mica (biotite)	Laboratory solutions	24 ± 1	4.1–6.2	As(III)	0.97	4.21 g L^{-1}	8.34	0.214	Chakraborty *et al.* (2007)
Mica (biotite)	Laboratory solutions	24 ± 1	4.6–5.6	As(V)	0.97	4.35 g L^{-1}	8.34	0.241	Chakraborty *et al.* (2007)
Mica (muscovite)	Laboratory solutions	24 ± 1	4.2–5.5	As(III)	0.97	4.1 g L^{-1}	14.28	0.231	Chakraborty *et al.* (2007)
Mica (muscovite)	Laboratory solutions	24 ± 1	4.2–5.5	As(V)	0.97	4.1 g L^{-1}	14.28	0.235	Chakraborty *et al.* (2007)

Material	Solution	Temperature	pH	As species	Concentration range	Dose		Capacity	Reference
Montmorillonite (loaded with iron [oxy][hydr]oxides)	Laboratory solutions	Ambient?	<9	As(III)	0–60	1.6 g L^{-1}	165	13	Lenoble et al. (2002)
Montmorillonite (loaded with iron [oxy][hydr]oxides)	Laboratory solutions	Ambient?	<9	As(V)	0–60	1.6 g L^{-1}	165	4	Lenoble et al. (2002)
Montmorillonite (loaded with titanium)	Laboratory solutions	Ambient?	<9	As(III)	0–60	1.6 g L^{-1}	249	13	Lenoble et al. (2002)
Montmorillonite (loaded with titanium)	Laboratory solutions	Ambient?	<9	As(V)	0–60	1.6 g L^{-1}	249	3	Lenoble et al. (2002)
Mycan/HDTMA (fungal biomass modified with hexadecyltrimethylammonium bromide)	Laboratory solution	20–25	3	As(V)	1–300	1 g L^{-1}	?	57.85	Loukidou et al. (2003)
Pisolite (manganese and iron [oxy][hydr] oxide mining wastes)	Arsenic-spiked river water and laboratory solutions	Ambient?	6.5 ± 0.2	As(V)	50	1.0 g/100 ml	61.40	1.42	Pereira, Dutra and Martins (2007)
Pisolite (manganese and iron [oxy][hydr] oxide mining wastes)	Arsenic-spiked river water and laboratory solutions	Ambient?	6.5 ± 0.2	As(V)	50	Column	61.40?	1.51	Pereira, Dutra and Martins (2007)
Pisolite (activated; manganese and iron [oxy][hydr] oxide mining wastes)	Laboratory solutions	Ambient?	6.5 ± 0.2	As(V)	50	1.0 g/100 ml	90.45	3.17	Pereira, Dutra and Martins (2007)
Pisolite (activated; manganese and iron [oxy][hydr] oxide mining wastes)	Laboratory solutions	Ambient?	6.5 ± 0.2	As(V)	50	Column	90.45?	3.51	Pereira, Dutra and Martins (2007)

(continued overleaf)

Table 7.2 (continued)

Sorbent or ion-exchange material	Type of water	Temperature (°C)	Initial pH	Inorganic arsenic species	Initial arsenic concentration (mg L⁻¹)	Batch sorbent dosage (sorbent/batch solution) or column	Initial surface area ($m^2 g^{-1}$) of sorbent or ion-exchange material	Maximum removal capacity (mg As g^{-1} sorbent or ion-exchange material)	References
Portland cement (hardened, granular)	Laboratory solutions	30 ± 2	4–5	As(V)	0.2	15 g L⁻¹	15.38	3.98	Kundu et al. (2004)
Schwertmannite (Fe₈O₈[OH]₆SO₄ to Fe₈O₈[OH]₄.₅[SO₄]₁.₇₅)	Laboratory solutions	Ambient	3.0	As(V)	750	?	180	175	Carlson et al. (2002)
Seeds (shelled and powdered) of Moringa oleifera tree	Laboratory solutions	Ambient?	7.5	As(III)	25	2.0 g/200 ml	?	1.50	Kumari et al. (2006)
Seeds (shelled and powdered) of Moringa oleifera tree	Laboratory solutions	Ambient?	2.5	As(V)	25	2.0 g/200 ml	?	2.14	Kumari et al. (2006)
Siderite	Laboratory solutions	20 ± 2	?	As(III)	0.250–2.000	0.1 g/50 ml	?	1.040	Guo, Stüben and Berner (2007b)
Siderite	Laboratory solutions	20 ± 2	?	As(V)	0.250–2.000	0.1 g/50 ml	?	0.520	Guo, Stüben and Berner (2007b)
Silica (mesoporous) impregnated with copper(II)	Laboratory solutions	Ambient?	5–6	As(V)	1–4063	0.1 or 0.02 g/10 ml	?	75	Fryxell et al. (1999)
Silica (mesoporous, MCM-41) impregnated with iron (oxy)(hydr)oxides	Laboratory solutions	25	6	As(V)	<1500	50 mg/100 ml	310	120	Yoshitake, Yokoi and Tatsumi (2003)

Material	Water/solution	Temperature	pH	Arsenic species	Concentration	Dose			Reference
Silica (mesoporous, MCM-48) impregnated with iron (oxy)(hydr)oxides	Laboratory solutions	25	6	As(V)	<1500	50 mg/100 ml	352	187	Yoshitake, Yokoi and Tatsumi (2003)
Slag iron-titanium dioxide sorbent with photocatalytic oxidation	Laboratory solution	Ambient?	3.0	As(III)	100	2 g L^{-1}	163	0.0256	Zhang and Itoh (2006)
Soil (oxisol)	Laboratory solutions	25 ± 0.5	5.5	As(III)	10–1000	10 g/100 ml	35.7	2.6	Ladeira and Ciminelli (2004)
Soil (oxisol)	Laboratory solutions	25 ± 0.5	5.5	As(V)	10–1000	10 g/100 ml	35.7	3.2	Ladcira and Ciminelli (2004)
Titanium dioxide (99 % anatase)	Laboratory solutions	22 ± 3	9	As(III)	0–110	?	334	~26	Dutta et al. (2004)
Titanium dioxide (99 % anatase)	Laboratory solutions	22 ± 3	4	As(V)	0–110	?	334	~24	Dutta et al. (2004)
Titanium dioxide (~80 % anatase, ~20 % rutile)	Laboratory solutions	22 ± 3	9	As(III)	0–110	?	~55	~10	Dutta et al. (2004)
Titanium dioxide (~80 % anatase, ~20 % rutile)	Laboratory solutions	22 ± 3	4	As(V)	0–110	?	~55	~4	Dutta et al. (2004)
Titanium dioxide (70 % anatase, 30 % rutile)	Laboratory solutions	25	3	As(V)	5–30	1 g L^{-1}	50	8.0	Jézéquel and Chu (2006)
Titanium dioxide (70 % anatase, 30 % rutile)	Laboratory solutions	25	7	As(V)	5–30	1 g L^{-1}	50	2.7	Jézéquel and Chu (2006)
Titanium dioxide (granular anatase)	Arsenic-spiked New Jersey, US groundwater	Ambient?	7.0 ± 0.1	As(III)	0.4–80	1.0 g L^{-1}	250.7	32.4	Bang et al. (2005b)
Titanium dioxide (granular anatase)	Arsenic-spiked New Jersey, US groundwater	Ambient?	7.0 ± 0.1	As(V)	0.4–80	1.0 g L^{-1}	250.7	41.4	Bang et al. (2005b)

(continued overleaf)

Table 7.2 (continued)

Sorbent or ion-exchange material	Type of water	Temperature (°C)	Initial pH	Inorganic arsenic species	Initial arsenic concentration (mg L^{-1})	Batch sorbent dosage (sorbent/ batch solution) or column	Initial surface area (m^2 g^{-1}) of sorbent or ion-exchange material	Maximum removal capacity (mg As g^{-1} sorbent or ion-exchange material)	References
Titanium dioxide-loaded resin	Laboratory solutions	Ambient?	7	As(III)	10	200 mg/20 ml	208.8	9.7	Balaji and Matsunaga (2002)
Titanium dioxide-loaded resin	Laboratory solutions	Ambient?	4	As(V)	10	200 mg/20 ml	208.8	4.7	Balaji and Matsunaga (2002)
Titanium dioxide (nanometer) on silica gel	Laboratory solutions	Ambient?	10	As(III)	5–60	Column	?	4.22	Liang and Liu (2007)
Zeolite (clinoptilolite) treated with hexadecyltrimethylammonium bromide	Synthetic soil leachates spiked with arsenic	25	12.5	As(V)	500	5–25 g/100 ml	?	5.4	Sullivan, Bowman and Legiec (2003)
Zeolite (clinoptilolite) treated with hexadecyltrimethylammonium bromide	Laboratory solutions	22	7–9	As(V)	7–150	1 g/10 ml	?	0.5	Li et al. (2007)
Zeolite (clinoptilolite-heulandite)-rich tuff modified with lanthanum (III) nitrate solution	Laboratory solutions	18	3	As(V)	0.1	5–120 mg/10 ml	?	0.0754	Macedo-Miranda and Olguin (2007)

Zeolite (clinoptilolite-heulandite)-rich tuff modified with iron(III) chloride solution	Laboratory solutions	18	6	As(V)	0.1	5–120 mg/10 ml	?	0.0536	Macedo-Miranda and Olguín (2007)
Zerovalent iron (Fe(0))	Laboratory solutions	Ambient?	7.25 ± 0.25	As(III)	50	Column	2.53 ± 0.44	6.8–8.9 (average = 7.5)	Lien and Wilkin (2005)
Zerovalent iron (Fe(0))	Laboratory solutions	25	7	As(III)	0–0.15	20 mg/20 ml	24.4	3.5	Kanel et al. (2005)
Zirconium-loaded resin	Laboratory solutions	25	8	As(III)	?	?	?	79	Suzuki et al. (2001)
Zirconium-loaded resin	Laboratory solutions	25	4.5	As(V)	?	?	?	54	Suzuki et al. (2001)
Zirconium-loaded lysine diacetic acid chelating resin	Laboratory solutions	Ambient	9	As(III)	10	200 mg/20 ml	?	89	Balaji, Yokoyama and Matsunaga (2005)
Zirconium-loaded lysine diacetic acid chelating resin	Laboratory solutions	Ambient	4	As(V)	10	200 mg/20 ml	?	49	Balaji, Yokoyama and Matsunaga (2005)
Zirconium-loaded phosphoric acid chelating resin	Laboratory solutions	Ambient?	1.14	As(V)	190	Column	?	50 (per dry weight resin)	Zhu and Jyo (2001)

removal of inorganic As(III), As(V), and DMA(V) were not substantially affected by the presence of the phosphate (<5 % decrease) (Cheng *et al.*, 2005, 7664).

In the field, a 3-l cartridge of 1 : 1 (by volume) quartz sand and iron filings removed >85 % of $1-1.5\,mg\,L^{-1}$ of arsenic (~30 % organic arsenic) over four months from 16 000 L of groundwater at the Vineland, New Jersey, United States, Superfund site. Cheng *et al.* (2005, 7666) speculated that MMA(V) formed stable *bidentate complexes* (Chapter 2) with iron compounds, whereas DMA(V) could not.

7.2.3.5.2 Iron (oxy)(hydr)oxides

Iron (oxy)(hydr)oxides are groups of Fe(III) ± Fe(II) (hydrous) oxides, (hydrous) hydroxides, and (hydrous) oxyhydroxides, which are often very effective in removing arsenic from water. The compounds may be synthesized (e.g., Mayo *et al.*, 2007) or collected from rocks, sediments, or soils (Ndur and Norman, 2003). As discussed in Chapters 2 and 3, many iron (oxy)(hydr)oxides are amorphous and have poorly defined chemical compositions. Common iron (oxy)(hydr)oxides include hematite (Fe_2O_3), maghemite, magnetite, goethite, akaganéite, lepidocrocite, limonite ($FeO(OH)\cdot nH_2O$, where $n>0$), *ferrihydrites* (variable compositions), and the 'green rust' groups. Compared to their crystalline counterparts, amorphous iron (oxy)(hydr)oxides generally have higher surface areas (Cumbal and Sengupta, 2005, 6508), which are more favorable for sorption and ion exchange.

The sorption properties of iron and other metal (oxy)(hydr)oxides are largely controlled through ion exchanges involving OH_2^+, OH, and O^- surface functional groups (Stollenwerk, 2003, 73; Chapter 2). In contrast to zerovalent iron, iron (oxy)(hydr)oxides are usually more effective in sorbing As(V) than As(III) (Roberts *et al.*, 2004, 307). The treatment of As(III) in aqueous solutions may improve if amorphous manganese(IV) (oxy)(hydro)oxides are mixed with the iron (oxy)(hydr)oxides. The Mn(IV) (oxy)(hydro)oxides would tend to oxidize the As(III) and the resulting As(V) could sorb onto the iron (oxy)(hydr)oxides (Deschamps, Ciminelli and Höll, 2005; Zhang *et al.*, 2007a).

The sorption of As(V) onto iron (oxy)(hydr)oxides generally decreases as pH conditions rise from 3 to 10 (Su and Puls, 2001a, 1489). $H_2AsO_4^-$ is the dominant inorganic As(V) form at about pH 2–7 (Chapter 2). Under these acidic conditions, the surfaces of most iron (oxy)(hydr)oxides tend to have net positive charges as indicated by their *isoelectric points* and *zero points of charge* (ZPC) (Chapter 2). These positive surface charges readily attract $H_2AsO_4^-$ (Su and Puls, 2001a, 1489). When pH is alkaline, As(III) sorption on iron (oxy)(hydr)oxides increases as As(V) sorption declines. Maximum As(III) sorption generally occurs around pH 9 (Su and Puls, 2001a, 1489), where As(III) oxyanions start to become prominent and yet iron (oxy)(hydr)oxide surfaces still have some residual positive charges.

The ability of ferrihydrite to effectively sorb or ion-exchange arsenic depends on several factors, including the age, surface area, and exact composition of the compound. Freshly precipitated ferrihydrites are usually poorly crystalline and have high surface areas, which are ideal for sorption (Stollenwerk, 2003, 77). As the ferrihydrites age and crystallize, the surface area and number of sorption/ion-exchange sites decrease (Stollenwerk, 2003, 79). Aging ferrihydrite for just six days can decrease As(V) sorption by 20 % (Stollenwerk, 2003, 79–80). Depending on the pH, composition, and temperature of any surrounding solutions, ferrihydrite may eventually transform into crystalline hematite and/or goethite (Stipp *et al.*, 2002, 322).

The sorption of As(III) and/or As(V) on ferrihydrites also depends on the pH and arsenic concentrations of the solutions. At relatively high arsenic concentrations (26.7 mmol As L^{-1} or 2000 mg As L^{-1}) and during the first 2 hours of contact, As(III) reacted faster with 'two-line' ferrihydrites (Chapter 3) than As(V) at both pH 4.6 and 9.2 conditions. However, at pH conditions of 4.6 and at lower arsenic concentrations (0.534 mmol As L^{-1} (40 mg As L^{-1})), As(V) sorbed faster (Raven, Jain and Loeppert, 1998, 344–346). Raven, Jain and Loeppert (1998) also found that maximum sorption of As(III) on 'two-line' ferrihydrite was 0.60 mol As mol^{-1} Fe (810 mg As g^{-1} Fe) at pH 4.6 and 0.58 mol As mol^{-1} Fe (770 mg As g^{-1} Fe) at pH 9.2. In contrast, the sorption of As(V) was only 0.25 moles As mol^{-1} Fe (340 mg As g^{-1} Fe) at pH 4.6

and 0.16 moles As mol^{-1} Fe (210 mg As g^{-1} Fe) at pH 9.2. Raven, Jain and Loeppert (1998) concluded that the unusually high removal capacity for As(III) was probably due to the formation of Fe(III) arsenite compounds and not just sorption onto iron (oxy)(hydr)oxide surfaces.

Another important factor is how pH may affect the competition between As(III) and As(V) for sorption/ion-exchange sites on ferrihydrites. As(III) has little effect on As(V) sorption at pH values below 6. However, As(V) sorption may significantly decrease above pH 6 because of preferential sorption of As(III). In turn, As(V) may interfere with As(III) sorption on ferrihydrites at pH conditions of 4–6, but has little effect on As(III) sorption at pH values above 9 (Stollenwerk, 2003, 79). As discussed in Chapters 2 and 3, these changes in sorption with pH are largely related to the presence or absence of charges on dissolved As(III) and As(V) species, and the surface charge of the ferrihydrite.

In addition to pH, the sorption of arsenic on ferrihydrites is also affected by other chemicals in the water. In many situations, lime (CaO) is used to neutralize sulfuric acid (H$_2$SO$_4$) mine drainage, which may contain ferrihydrite precipitates and arsenic. The addition of lime can enhance the removal of As(V) with ferrihydrite by providing abundant Ca^{2+} that sorb onto ferrihydrites and increase the number of positive surface charges. The positive surface charges would readily attract and sorb As(V) oxyanions (Stollenwerk, 2003, 89; Jia and Demopoulos, 2005, 9526; Wilkie and Hering, 1996, 107).

Although ferrihydrites and other iron (oxy)(hydr)oxides are good to excellent sorbents of arsenic, granular forms tend to disintegrate into impermeable sludges in water treatment columns (Zeng, 2003, 4352). To improve structural integrity, ferrihydrites and other iron compounds may be (1) coated on sands, *diatomite*, or cement granules; (2) impregnated into porous materials (such as *activated carbon*, ion-exchange fibers, mesoporous silica, mesoporous carbon, *zeolites*, or biopolymers); or (3) mixed into permeable gels and resins (Vatutsina *et al.*, 2007; Ko *et al.*, 2007; Greenleaf, Lin and Sengupta, 2006; Gu and Deng, 2007; Thirunavukkarasu *et al.*, 2005; Muñoz, Gonzalo and Valiente, 2002; Gupta, Saini and Jain, 2005; Yoshitake, Yokoi and Tatsumi, 2003; Yuan *et al.*, 2002; Berg *et al.*, 2006; Jang *et al.*, 2006; Jang *et al.*, 2007a; Gu, Deng and Yang, 2007; Kundu and Gupta, 2006; Kundu and Gupta, 2007; Payne and Abdel-Fattah, 2005; Doušová *et al.*, 2006; Vaughan, Reed and Smith, 2007; Murugesan, Sathishkumar and Swaminathan, 2006; Dupont, Jolly and Aplincourt, 2007; Solozhenkin, Zouboulis and Katsoyiannis, 2007). For a series of sorption experiments, Thirunavukkarasu *et al.* (2005) thermally coated quartz sands with iron compounds (ferrihydrite?) precipitated from a Fe(III) nitrate nonahydrate (Fe(NO$_3$)$_3$ · 9H$_2$O) solution. The sand sorbed up to 95.8 % of 100 µg L^{-1} of As(III). The sorption capacity was about 0.136 mg As g^{-1} sand (Thirunavukkarasu *et al.*, 2005, 215; Table 7.2).

Vaishya and Gupta (2004) modeled the removal of 0.5–4.0 mg L^{-1} of As(V) from water with quartz sand that had been coated with barium sulfate and unspecified iron 'oxides' (ferrihydrite?) produced from Fe(III) nitrate nonahydrate. Maximum As(V) removal occurred under acidic to near neutral conditions (Vaishya and Gupta, 2004, 663), which is similar to other studies with iron (oxy)(hydr)oxide compounds. As(V) removals were not significantly affected by the *ionic strength* (0.1 M and less as NaNO$_3$) and chloride (up to 300 mg L^{-1}) concentrations of the water. Small concentrations (up to 10 mg L^{-1}) of silica however, dramatically reduced arsenic removal efficiency, probably by blocking active surface sites. In contrast, removal efficiency improved in the presence of Mg^{2+} and Ca^{2+}, which may have involved the precipitation of calcium arsenates (Vaishya and Gupta, 2004, 661).

Lafferty and Loeppert (2005) investigated the sorption and desorption of MMA(V), MMA(III) (monomethylarsonous acid; (CH$_3$)As(OH)$_2$; Chapter 2), DMA(V), DMA(III) (dimethylarsinous acid; (CH$_3$)$_2$As(OH); Chapter 2), inorganic As(V), and inorganic As(III) on goethite and 'two-line' ferrihydrite. As(III) was strongly sorbed by goethite and ferrihydrite, but MMA(III) and DMA(III) were not substantially sorbed at pH 3–11 conditions. Under all pH conditions, MMA(V) and As(V) were sorbed in higher concentrations on goethite and ferrihydrite than DMA(V). All of the arsenic species were more effectively desorbed from the iron compounds with phosphate than sulfate (Lafferty and Loeppert, 2005,

2120). Increasing the number of methyls in the arsenic species resulted in decreased sorption at low arsenic concentrations and an increase in the ability of the species to desorb (Lafferty and Loeppert, 2005, 2120). Like As(V), maximum sorption of MMA(V) and DMA(V) occurs under low pH conditions and decreases with increasing pH (Lafferty and Loeppert, 2005, 2120).

Ferrihydrites and amorphous varieties of other iron (oxy)(hydr)oxides are usually more effective than alumina in sorbing As(III) and As(V) (Jekel, 1994, 128–129). However, if the iron compounds are not regenerated, their costs are often similar to alumina (Clifford and Ghurye, 2002, 223–224). Iron (oxy)(hydr)oxide sorbents are typically regenerated with sodium hydroxide (NaOH) followed by rinsing with sulfuric acid (US EPA, 2002b, 25).

Although ferrihydrites are usually more effective than goethite in removing As(III) and As(V) from water (Stollenwerk and Colman, 2003, 365), goethite is still widely utilized in arsenic treatment. Like other iron (oxy)(hydr)oxides, the sorption of arsenic on goethite strongly depends on the properties and chemistry of the water, including temperature, TDS content, whether phosphate and other anions that might interfere with arsenic sorption are present, and pH, which influences the types of exchangeable functional groups (O^-, OH_2^+, etc.) on the surfaces of goethite (Luengo, Brigante and Avena, 2007; Chapter 2). Changing the compositional chemistry of goethite (α-FeOOH) may also affect arsenic sorption. Specifically, the sorption of As(V) may significantly improve if Cu(II), Ni(II), or Co(II) are doped into goethite through coprecipitation (Mohapatra *et al.*, 2006). Besides sorption, As(V) may also be removed from acidic waters through the precipitation of iron arsenates on goethite surfaces (Matis *et al.*, 1999, 301).

Matis *et al.* (1999) investigated the sorption of As(V) with aqueous suspensions of fine-grained synthetic goethite. At initial concentrations of $10 \, mg \, L^{-1}$ of As(V) and at a pH of about 5, removals on the order of 100 % were obtained with 1.0 g goethite/l of water with a contact time of about 1 hour (Matis *et al.*, 1999, 313). Matis *et al.* (1999, 304, 306) also found that prewetting goethite with water to form a paste was more effective in sorbing As(V) than using dry goethite. The greater effectiveness of the paste may be due to surface hydrolysis.

Under pH 7–12 conditions, As(V) sorption on goethite increased as the ionic strength of the aqueous solutions was raised with 0.001, 0.01, or 0.05 M potassium nitrate (KNO_3) (Matis *et al.*, 1999, 307). However, ionic strength did not affect As(V) removals under acidic conditions. Potassium nitrate probably suppressed negative charges on the goethite particles under alkaline conditions, which would promote reactions between the goethite surfaces and arsenic oxyanions (Matis *et al.*, 1999, 307, 309).

Many commercial 'Fe(III) hydroxide' granular sorbents are actually poorly crystalline akaganéite, an Fe(III) oxyhydroxide, β-FeO(OH) (Clifford and Ghurye, 2002, 223; Westerhoff *et al.*, 2005). The poor crystallinity of akaganéite results in high surface areas, which are very effective in sorbing arsenic (Clifford and Ghurye, 2002, 223). Granular akaganéite, which has relatively low surface areas, tends to be less effective in removing As(V) than the nanocrystalline (typically, 3–6 nm) form, except at high ion strengths (0.1 M potassium nitrate) where both forms have about the same effectiveness (Deliyanni *et al.*, 2003b, 3973).

Deliyanni *et al.* (2003a) studied the sorption of As(V) with synthetic akaganéite nanocrystals. Nearly 100 % removal of $20 \, mg \, L^{-1}$ As(V) was achieved in 24 hours with 2 g akaganéite/l water at pH conditions below 8.5. Under alkaline conditions, As(V) sorption improved as the ionic strength of the solution increased from 0.001 to 0.1 M with potassium nitrate. The K^+ probably suppressed the negative charges on the surfaces of the akaganéite, which allowed for greater sorption of As(V) oxyanions. By varying temperatures at 25, 35, and 45 °C and akaganéite dosages between 0.5 and $2 \, g \, L^{-1}$, Deliyanni *et al.* (2003a, 159) obtained a maximum sorption capacity of 120 mg As(V) g^{-1} akaganéite at 25 °C (Table 7.2). Compared to the sorption capacity of goethite (about $25–40 \, mg \, g^{-1}$ sorbent) and most other iron sorbents from the literature, this value is relatively high and is probably due to the high surface area of the nanocrystalline akaganéite ($330 \, m^2 \, g^{-1}$). Iron arsenate precipitates may also have formed on the surfaces of the akaganéite particles. The akaganéite can be regenerated three or four times with NaOH followed by a sulfuric acid rinse before it needs to be replaced (Deliyanni *et al.*, 2003a, 162).

Both As(III) and As(V) were effectively removed from water with cotton cellulose beads loaded with iron (oxy)(hydr)oxides, mostly as akaganéite (Guo and Chen, 2005; Guo *et al.*, 2007a). Finely powdered iron (oxy)(hydr)oxides are too impermeable for sorption columns. To improve permeability, Guo and Chen (2005) developed high-surface area cotton cellulose beads that can be loaded with as much as 50 wt % iron (oxy)(hydr)oxides. At pH 7.0 and with a bead iron content of 220 mg ml^{-1} (mass of iron/volume of beads), the sorption capacities for the materials were 59.6 and 33.2 mg As(V) g^{-1} FeOOH (Guo and Chen, 2005; Table 7.2). As(V) removal was favored under acidic conditions, whereas As(III) could be removed over a pH range of 5–11 (Guo and Chen, 2005). Sulfate, chloride, nitrate, hydrocarbonate ions, and sodium chloride caused no appreciable interferences with As(V) and As(III) sorption on the beads (Guo *et al.*, 2007a, 431). However, phosphate suppressed both As(III) and As(V) sorption and silicate significantly hindered the sorption of As(III) (Guo and Chen, 2005). Once exhausted, the beads can be regenerated with 2 M NaOH (Guo and Chen, 2005).

Westerhoff *et al.* (2005) performed a series of laboratory studies to evaluate the removal of arsenic from three Arizona (United States) groundwaters by sorption with granular 'Fe(III) hydroxide' (akaganéite) columns. The groundwaters contained 13–53 µg L^{-1} of As(V) and no detectable As(III) (Westerhoff *et al.*, 2005, 265). Removal capacities for the columns using a pH 8.9 groundwater with an initial As(V) concentration of 51 µg L^{-1} were 0.02–0.28 mg As g^{-1} akaganéite. In contrast, model waters with pH values of 8.6 and initial As(V) concentrations of 10–108 µg L^{-1} gave removal capacities of 0.99–1.5 mg As g^{-1} akaganéite (Table 7.2). Silica, phosphate, and other inorganics were probably responsible for the lower As(V) removal capacities with the groundwater. Due to the ability of akaganéite and other iron (oxy)(hydr)oxides to strongly sorb arsenic, spent materials generally pass most leaching tests and are often landfilled (Westerhoff *et al.*, 2005, 262).

'Green rust' refers to a group of *metastable* Fe(II) and Fe(III) hydrated hydroxides that also contain carbonate, chloride, and/or sulfate (Stipp *et al.*, 2002, 322; Refait, Abdelmoula and Génin, 1998; Trolard *et al.*, 1997; Génin *et al.*, 1998; Randall, Sherman and Ragnarsdottir, 2001; Ona-Nguema *et al.*, 2002; Doušová *et al.*, 2005, 32). Sulfate green rust, approximately Fe(II)$_4$Fe(III)$_2$(OH)$_{12}$SO$_4$ · 3H$_2$O, may form as an intermediate product from the oxidation of pyrite and other iron sulfides. Eventually, the green rust converts to more stable iron (oxy)(hydr)oxides, such as goethite, lepidocrocite, and magnetite (Randall, Sherman and Ragnarsdottir, 2001, 1015). Green rust may also form during the *reductive dissolution* of iron (oxy)(hydr)oxides (Randall, Sherman and Ragnarsdottir, 2001, 1015) and the oxidation of zerovalent iron (Su and Puls, 2003, 2586).

Several varieties of green rusts effectively sorb As(V) from water (Su and Puls, 2004). Randall, Sherman and Ragnarsdottir (2001) investigated the sorption of As(V) onto sulfate green rust and the behavior of sorbed arsenic during the transformation of the rust to lepidocrocite. Instead of substituting for sulfate or otherwise incorporating into the green rust structure, *extended X-ray absorption fine structure* (EXAFS) spectrometry indicates that arsenic is adsorbed onto the surface of the rust. During the transformation to lepidocrocite, the As(V) remains preferentially bound to any remaining green rust and does not reduce to As(III). The As(V) only transfers and sorbs onto lepidocrocite once the green rust has disappeared.

Hematite is a common mineral in rocks and soils, including many iron ore deposits. The mineral is usually more crystalline than ferrihydrite and goethite, which results in a lower surface area and less sorption sites on hematite (Stollenwerk, 2003, 80). The sorption capacity of As(III) or As(V) typically has the following order: ferrihydrite > goethite > hematite (Stollenwerk, 2003, 80).

Zhang *et al.* (2004) estimated the sorption capacity of an Australian hematite iron ore as 0.4 mg As(V) g^{-1} ore (Table 7.2). The presence of silicate and phosphates in the ores hindered As(V) sorption, whereas sorption was slightly enhanced by water-soluble sulfate and chloride. Zhang *et al.* (2004) found that the presence of $2 mg L^{-1}$ of phosphate in water required six times as much ore ($30 g L^{-1}$) to reduce As(V) concentrations from $1 mg L^{-1}$ to below $10 \mu g L^{-1}$ (Zhang *et al.*, 2004, 522). Dissolved silicate decreased As(V) sorption by up to $\sim30\%$ at pH 7, but the interferences could be minimized by reducing the pH to 5. Organic materials in water may also interfere with the sorption of As(V) and As(III) onto hematite (Redman, Macalady and Ahmann, 2002). Specifically, negatively charged functional groups on humic acids may compete with arsenic oxyanions for sorption sites on hematite (Ko, Kim and Kim, 2004).

Prasad (1994, 139) found that As(V) sorption attained equilibrium in contact with hematite in about 35 minutes. Using an initial As(V) concentration of 1.3×10^{-4} mol L^{-1} (9.7 mg As/l) at 20°C, maximum As(V) removal with hematite occurred at a pH of 4.2 (Prasad, 1994, 141). At this pH, 200-μm hematite removed 81% of the As(V), whereas hematite with a mean particle size of only 100 μm removed 97% (Prasad, 1994, 139–140). The percentage As(V) removal by hematite decreased with increasing temperatures of 20–40°C, which is consistent with an *exothermic* adsorption mechanism (Prasad, 1994, 141).

Maximum adsorption of As(III) on hematite occurs around pH 7 (Su and Puls, 2001a, 1489). As pH conditions rise from 5 to 7, the number of positive charges on the hematite surfaces decreases (Su and Puls, 2001a, 1489). However, under acidic to neutral conditions, low sorbing $H_3AsO_3{}^0$ is the dominant As(III) species (Wolthers *et al.*, 2005, 3484; Chapter 2). The adsorbable oxyanion, $H_2AsO_3{}^-$, only becomes dominant once pH conditions exceed 9 (Chapter 2). Therefore, maximum adsorption of As(III) on hematite is expected to be around pH 7–9, where a few positive charges and $H_2AsO_3{}^-$ may coexist.

Sometimes the sorption of arsenic from water is improved if hematite is combined with another sorbent, such as alumina. Hossain *et al.* (2005) field tested several sorption, ion-exchange, and coprecipitation technologies for removing arsenic from groundwater in West Bengal, India. Most of the technologies were unable to consistently remove arsenic to below the World Health Organization (WHO) recommended value of $10 \mu g L^{-1}$. However, sorption media containing granular hematite, quartz sand, and activated aluminum typically reduced arsenic concentrations to $5 \mu g L^{-1}$ and below. Nevertheless, most of the groundwater treatment technologies had serious problems with maintenance, clogging, and/or a lack of user friendliness (Hossain *et al.*, 2005, 4305). Therefore, Dipankar Chakraborti, leader of the Hossain *et al.* team, concluded that sanitizing and purifying low-arsenic rain and surface waters are better alternatives (Ball, 2005).

7.2.3.5.3 Manganese (oxy)(hydr)oxides Common manganese (oxy)(hydr)oxides include pyrolusite (β-MnO_2), manganite ($MnO(OH)$), cryptomelane (KMn_8O_{16}), and birnessite (approximately $Na_4Mn_{14}O_{27} \cdot 9H_2O$) (Foster, 2003, 59; Klein, 2002, 394). Manganese (oxy)(hydr)oxides are capable of both oxidizing As(III) in water and sorbing the resulting As(V) (Manning *et al.*, 2002a; Stollenwerk, 2003, 81). Below pH 9 under natural conditions and in water treatment systems, MnO_2 may oxidize As(III) through the following reactions (Nesbitt, Canning and Bancroft, 1998):

$$2MnO_2 + H_3AsO_3{}^0 + H_2O \rightarrow 2MnOOH + H_3AsO_4{}^0 \tag{7.4}$$

$$2MnOOH + H_3AsO_3{}^0 + 4H^+ \rightarrow 2Mn^{2+} + H_3AsO_4{}^0 + 3H_2O \tag{7.5}$$

Depending on the pH of the water, $H_3AsO_4^0$ may quickly dissociate into $H_2AsO_4{}^-$ and/or $HAsO_4{}^{2-}$ (Chapter 2).

The oxidation rate of As(III) in the presence of manganese and water may be substantially enhanced by manganese-oxidizing bacteria, such as *Leptothrix ochracea* (Katsoyiannis, Zouboulis and Jekel, 2004). Katsoyiannis, Zouboulis and Jekel (2004) found that the bacteria are important in oxidizing Mn(II) to Mn(IV), Fe(II) to Fe(III), and As(III) to As(V). The oxidation of Mn(II) leads to the precipitation of Mn(IV) (oxy)(hydr)oxides, which then *abiotically* oxidize additional As(III) and significantly sorb the As(V) that results from both abiotic and *biotic* oxidation.

Like aluminum and iron (oxy)(hydr)oxides, amorphous or poorly crystalline manganese (oxy)(hydr)oxide sorbents have higher surface areas and are usually more effective sorbents than crystalline varieties (Stollenwerk, 2003, 82). Potentially important sources of manganese (oxy)(hydr)oxides include marine nodules and *mine tailings*. The iron–manganese tailings tend to have high surface areas and *porosities* (Bai and Wiltshire, 2005), which could be effective in the removal of arsenic and other inorganic contaminants from water.

Like other sorbents, the isoelectric points of manganese (oxy)(hydr)oxides are important in determining their abilities to sorb arsenic. For example, birnessite has a low isoelectric point ($\sim$2.5), which means that the material is negatively charged at the pH values of most groundwaters and would not appreciably adsorb As(V) oxyanions (Stollenwerk, 2003, 81). However, well-crystallized birnessite can react with As(V) oxyanions and precipitate them. The resulting precipitates form on the surfaces of the birnessite and have a composition that approximates krautite ($MnHAsO_4 \cdot H_2O$) (Tournassat *et al.*, 2002). The studies in Tournassat *et al.* (2002) utilized a 0.011 M (824 mg L^{-1}) As(III) solution containing 2.7 g L^{-1} of suspended birnessite. The birnessite oxidized the As(III) and then precipitated the resulting As(V) as Mn(II) arsenates during 583 hours (Tournassat *et al.*, 2002, 497).

7.2.3.5.4 Aluminum (oxy)(hydr)oxides

Al^{3+} has an identical charge and similar radius to Fe^{3+}, which means that aluminum (oxy)(hydr)oxides share some chemical properties with their Fe(III) analogs. Like iron(III) hydroxides, aluminum hydroxides are often amorphous. The most common crystalline varieties are bayerite (α-Al(OH)$_3$, monoclinic; $P2_1/a$ *space group*) and gibbsite (γ-Al(OH)$_3$, monoclinic; $P2_1/n$ space group). Diaspore (α-AlO(OH)) and boehmite (γ-AlO(OH)) are the major aluminum oxyhydroxides. Diaspore has the same crystalline structure as goethite (orthorhombic; Pbnm space group), whereas boehmite is analogous to lepidocrocite (orthorhombic; Amam space group).

Corundum is aluminum oxide, α-Al$_2$O$_3$, which has a hexagonal crystalline structure that is analogous to hematite. However, water treatment systems most often use activated alumina, which is typically produced by thermally dehydrating aluminum (oxy)(hydr)oxides to form amorphous, cubic (γ), and/or other *polymorphs* of corundum (Clifford and Ghurye, 2002, 220; Hlavay and Polyák, 2005; Mohan and Pittman, 2007). When compared with corundum, amorphous alumina tends to have higher surface areas, greater numbers of sorption sites, and better sorption properties.

Activated alumina is usually ineffective in removing As(III) from water (Lin and Wu, 2001; Table 7.2; Jekel, 1994, 127) and any As(III) should be preoxidized to As(V) (Clifford and Ghurye, 2002, 224). As(V) is routinely removed from water with columns of granular activated alumina. Although Al(III) and Fe(III) have similar properties, alumina and other aluminum compounds generally remove less As(V) from water than their iron(III) counterparts (Stollenwerk, 2003, 81). On the other hand, granular iron (oxy)(hydr)oxides tend to disintegrate into impermeable sludges in water (Zeng, 2003, 4352). Furthermore, if iron compounds are not regenerated, the costs of aluminum and iron (oxy)(hydr)oxides are similar (Clifford and Ghurye, 2002, 223–224).

To effectively sorb As(V) with activated alumina, pH must be carefully controlled (Clifford and Ghurye, 2002, 221). The optimal pH conditions for sorbing As(V) onto alumina are around 5.5–6.0 (Clifford and Ghurye, 2002, 221). If pH conditions exceed 8, the alumina approaches its zero point of charge (about 8.4–9.1; Chapter 2; Lin and Wu, 2001, 2052) and its surfaces have relatively few positive charges to attract oxyanions (Dambies, 2004, 607). Under acidic conditions, the alumina surfaces acquire excess positive charges through the adsorption of H$^+$ (protonation). Once the alumina surfaces have excess H$^+$, they can attract As(V) oxyanions, such as H$_2$AsO$_4^-$ (Clifford and Ghurye, 2002, 221). Chloride (Cl$^-$) anions from pH adjustments with hydrochloric acid (HCl) are also attracted toward the adsorbed H$^+$, but they are usually readily displaced by As(V) oxyanions. As(V) oxyanions adsorb onto protonated alumina surfaces

($Al\text{-}OH_2^+$) through the following exchange reaction (Clifford and Ghurye, 2002, 220):

$$\vdash Al\text{-}OH_2^+ + H_2AsO_4^- \rightarrow \vdash Al\text{-}H_2AsO_4 + H_2O \tag{7.6}$$

Lin and Wu (2001, 2053) found that As(V) sorption on alumina was fairly constant at pH values below 6. Their sorption capacity was about $3\,mg$ As(V) g^{-1} activated alumina with an initial As(V) concentration of $2.85\,mg\,L^{-1}$ (Table 7.2). In general, the sorption of As(V) onto activated alumina is rapid during the first 24 hours and then slows down (Leist, Casey and Caridi, 2000, 130).

Activated alumina often has poor porosity, which may lead to inadequate sorption of arsenic (Kim et al., 2004, 924). Kim et al. (2004) developed mesoporous alumina, which mostly consists of the cubic forms η- and γ-Al_2O_3. Through the sorption of H^+, the surfaces of mesoporous alumina are positively charged in the pH 4–7.8 range (Kim et al., 2004, 926). The uniform and well-connected pore spaces in the alumina have good surface areas, which allow for greater and faster arsenic sorption. Specifically, the maximum uptake of As(V) with mesoporous alumina was about seven times higher ($121\,mg$ As(V) g^{-1} alumina) than conventional activated alumina. The sorptive capacity with As(III) was also high, $47\,mg$ As(III) g^{-1} alumina (Kim et al., 2004; Table 7.2).

Chloride, fluoride, phosphate, silica, organic materials, and sulfate may interfere with As(V) sorption on activated alumina (US EPA, 2002b, 27; Dambies, 2004, 607, 610; Clifford and Ghurye, 2002, 224, 227; Jekel, 1994, 127). In general, sulfate interferes more than chloride (Clifford and Ghurye, 2002, 224). In one example, 15 meq L^{-1} of chloride ($532\,mg\,L^{-1}$) suppressed the sorption of As(V) onto activated alumina by 16 %, whereas 15 meq L^{-1} of sulfate ($720\,mg\,L^{-1}$) reduced As(V) sorption by 50 % (Clifford and Ghurye, 2002, 224, 226). Silica typically interferes at pH 6 and higher by competing with arsenic for sorption sites on activated alumina (Clifford and Ghurye, 2002, 227). At a pH of 7.5, the sorption capacity of As(V) on an activated alumina dropped from 0.55 to $0.15\,mg\,g^{-1}$ from the addition of $15\,mg\,L^{-1}$ of silica (Clifford and Ghurye, 2002, 227).

Depending on the chemistry of the water, activated alumina is typically used for one to three months before regeneration or disposal is required (Jekel, 1994, 128). Alumina regeneration begins with a wash of $0.25–1.0$ normal (N) NaOH (Clifford and Ghurye, 2002, 221, 229). As shown in the following reaction, hydroxides from the *base* replace arsenic oxyanions adsorbed onto the surface aluminums:

$$\vdash Al\text{-}H_2AsO_4 + OH^- \rightarrow \vdash Al\text{-}OH + H_2AsO_4^- \tag{7.7}$$

The alumina is then rinsed with $0.1–0.5\,N$ H_2SO_4 or HCl (Clifford and Ghurye, 2002, 222, 229). The acids remove surface hydroxides and reprotonate the alumina surfaces for use in future arsenic removal.

In many cases, disposal of spent activated alumina is more cost effective than regeneration. Besides producing hazardous effluents, regeneration may fail to release as much as 20–30 % of the sorbed arsenic (Leist, Casey and Caridi, 2000, 128). With repeated regenerations, unextractable arsenic and other firmly sorbed species may accumulate and eventually eliminate the effectiveness of the sorbent. Simply increasing the normality of the NaOH and regenerating acids may not improve regeneration. Strongly alkaline and acidic regenerating solutions could corrode the alumina and produce arsenic- and aluminum-rich sludges that would require disposal (US EPA, 2002b, 25).

7.2.3.5.5   Combining aluminum compounds with other sorbents

The sorption of arsenic from water can often be greatly improved by combining sorbents, such as amending aluminum (oxy)(hydr)oxides with manganese (Kunzru and Chaudhuri, 2005), iron (Clifford and Ghurye, 2002, 222–223; Masue, Loeppert and Kramer, 2007; Zhang et al., 2007b), or both of these metals (Dhiman and Chaudhuri, 2007). Alcan AAFS-50 is an example of an iron-doped alumina sorbent. According to the manufacturers, Alcan has two to five times the arsenic sorption capacity of alumina and is less sensitive to pH (Clifford and Ghurye, 2002, 222–223). Although Alcan cannot be regenerated, supposedly it is nonhazardous and can be readily landfilled under current US regulations (Clifford and Ghurye, 2002, 222).

Hlavay and Polyák (2005) investigated the surface properties of an iron(III) hydroxide-coated alumina sorbent. Amorphous iron(III) hydroxide was coated onto a mixture of amorphous and various crystalline forms of Al_2O_3. The iron concentration of the sorbent was 56.1 mmol g^{-1}. Depending on pH, batch experiments indicated that the sorbent was effective with As(V) and to some extent As(III). In a pH 5.6 solution with an ionic strength of 0.1 mmol L^{-1} and containing 1 mmol L^{-1} (75 mg As(V) L^{-1}) of As(V), a dosage of 2.5 g sorbent 100 ml^{-1} of solution sorbed 0.3 mmol As(V) g^{-1} sorbent (22 mg As(V) g^{-1} sorbent). In comparison, the sorption capacity of As(III) was only about 0.06 mmol g^{-1} (4.5 mg As(III) g^{-1} sorbent) under the same conditions (Hlavay and Polyák, 2005, 76).

Activated Bauxsol may effectively sorb As(V) from water (Genç-Fuhrman, Tjell and McConchie, 2004; Table 7.1). This product is made from 'red mud', a byproduct of aluminum production. Red mud contains a variety of aluminum, iron, and other compounds, including hematite, boehmite, gibbsite, brucite ($Mg(OH)_2$), calcite ($CaCO_3$), diaspore, ferrihydrite, gypsum ($CaSO_4 \cdot 2H_2O$), hydrocalumite ($Ca_2Al(OH)_7 \cdot 2H_2O$), hydrotalcite ($Mg_6Al_2(CO_3)(OH)_{16} \cdot 4H_2O$), and portlandite (hydrated lime, $Ca(OH)_2$) (Munro, Clark and McConchie, 2004, 183–184). To produce Bauxsol, the hydroxides and hydroxylcarbonates in the original red mud are neutralized with seawater (Genç-Fuhrman, Tjell and McConchie, 2004). The material is then activated by washing with HCl, ammonia, and distilled water, and finally *calcinated* at 500 °C for 2 hours. The treatment process produces a more pH neutral product.

In water, Bauxsol can achieve almost 100 % As(V) removal if the pH is adjusted to 4.5. Even at pH 7 and with Bauxsol dosages of only 0.4 g L^{-1} water, As(V) concentrations up to 330 µg L^{-1} were lowered within 3 hours to below 10 µg L^{-1} (Genç-Fuhrman, Tjell and McConchie, 2004, 2433). Like many other sorption technologies, arsenic removal with Bauxsol is best achieved by preoxidizing any As(III) to As(V).

7.2.3.5.6 Hydrotalcite   Hydrotalcite ($Mg_6Al_2(CO_3)(OH)_{16} \cdot 4H_2O$) and other double-layered hydroxide compounds may effectively sorb As(V) from water. Hydrotalcites remove As(V) through sorption and/or exchanges with interlayer anions. The minerals may also be heated (calcinated) until most carbonate anions leave the interlayers. The calcinated hydrotalcites are then placed in water, where the vacant interlayers can sorb anions, including As(V) oxyanions (Yang *et al.*, 2005, 6805).

Depending on the types of hydrotalcites and the chemistry of the arsenic-contaminated waters, As(V) sorption capacities can be very good, perhaps more than 500 mg As(V) g^{-1} of hydrotalcite (Lazaridis *et al.*, 2002, 321–322), or relatively poor (4.545 mg As(V)/uncalcinated hydrotalcite and 5.609 mg As(V)/calcinated hydrotalcite; (Yang *et al.*, 2005, 6809; Table 7.2). Even with excellent As(V) sorption abilities, hydrotalcites are rather expensive compared to aluminum (oxy)(hydr)oxides (Doušová *et al.*, 2003, 266). Although some hydrotalcites can remove As(III) to less than 20 µg L^{-1} in water (Jiang *et al.*, 2007), most of them, as with many other sorbents, are not very effective with As(III). Therefore, any As(III) should be oxidized before treatment with these compounds (Yang *et al.*, 2005; Bhaumik, Samanta and Mal, 2005).

Doušová *et al.* (2003) found that after mixing for 24–72 hours at 20 °C, calcinated synthetic hydrotalcite and calcinated natural boehmite were much more effective in removing As(V) from water than oxihumolite (i.e. humic-rich *low-rank coals*) (Doušová *et al.*, 2003, 259). Dosages of 1 g L^{-1} of hydrotalcite or 2.6 g L^{-1} of boehmite removed 70.5 and 97.1 %, respectively, of 2 mmol L^{-1} (150 mg As L^{-1}) of As(V) (Doušová *et al.*, 2003, 265). The arsenic probably bonded with oxygens on the hydrotalcite and boehmite. Infrared spectra detected As–O bonds in both inorganic sorbents (Doušová *et al.*, 2003, 261, 263). In contrast, As–O bonds were absent in the infrared spectra of the oxihumolite (Doušová *et al.*, 2003, 263).

Yang *et al.* (2005) investigated the removal of trace concentrations of arsenic and selenium from water with calcinated and uncalcinated hydrotalcites. Calcinated hydrotalcites had higher As(V) sorption capacities than the uncalcinated varieties, which was possibly due to calcination increasing surface areas and/or producing interlayer vacancies that could be filled by As(V) oxyanions (Yang *et al.*, 2005, 6809; Table 7.2). As long as the pH was above 4, the calcinated hydrotalcite structures did not decompose and the pH of the starting aqueous solutions had little effect on As(V) sorption. In contrast, the sorption properties of the uncalcinated varieties were much more sensitive to pH.

With solutions containing $20 \, \mu g \, L^{-1}$ of As(V), nitrate concentrations up to $1000 \, mg \, L^{-1}$ had no effect on As(V) sorption with calcinated hydrotalcites. As(V) sorption on calcinated hydrotalcites was lowered by 14 and 20 % from $500 \, mg \, L^{-1}$ of sulfate and $500 \, mg \, L^{-1}$ of carbonate ($CO_3^{2-} + HCO_3^{-}$), respectively. Phosphate can completely stop As(V) sorption on calcinated hydrotalcites, but only at unusually high concentrations of greater than $500 \, mg \, L^{-1}$ (Yang *et al.*, 2005, 6811).

Hydrotalcite is often too fine grained to produce treatment columns with suitable permeability. As an alternative, the sorbent may be mixed with contaminated water in a tank (Lazaridis *et al.*, 2002). The spent sorbent is then separated from the treated water by *flocculation*, flotation, or other separation methods (see Section 7.2.4). Lazaridis *et al.* (2002) investigated the use of *surfactants* with dispersed-air flotation to separate spent hydrotalcites from treated water. At ionic strengths of 0.1 M using KNO_3, effective flotation and separation could be obtained by using a mixture of dodecylpyridinium chloride, sodium dodecylhydrogen sulfate, and a cetyltrimethyl ammonium bromide frother (Lazaridis *et al.*, 2002, 322, 323).

7.2.3.5.7 Titanium dioxides The most common forms of titanium dioxide (TiO_2) are rutile, anatase, and brookite. Brookite is orthorhombic, whereas rutile and anatase are tetragonal ($P4_2/mnm$ and $I4_1/amd$ space groups, respectively) (Klein, 2002, 383–384). Titanium dioxides may sorb both As(III) and As(V) from water. The compounds are also important photocatalysts in the oxidation of As(III), MMA(V), and DMA(V) to inorganic As(V) in water (Nakajima *et al.*, 2005; Xu, Cai and O'Shea, 2007).

Dutta *et al.* (2004) investigated the sorption of As(V) and As(III) in water with TiO_2 suspensions (Table 7.2). Two German commercial titanium dioxides were tested: mesoporous Hombikat UV100 (99 % anatase, surface area of $334 \, m^2 \, g^{-1}$) and nonporous Degussa P25 (~80 % anatase and ~20 % rutile, surface area of only $\sim 55 \, m^2 \, g^{-1}$) (Dutta *et al.*, 2004, 271). Like most results from iron and aluminum (oxy)(hydr)oxides, As(V) sorption was greater than As(III) sorption on the titanium dioxides at pH 4, but As(III) sorption became greater than As(V) sorption at pH 9 once substantial As(III) oxyanions formed (Chapter 2). In accordance with its much larger surface area and porosity, the sorption capacity of Hombikat UV100 at pH 4 was $\sim 320 \, \mu mol \, As(V) \, g^{-1} \, TiO_2$ ($\sim 24 \, mg \, As(V) \, g^{-1} \, TiO_2$) at equilibrium As(V) concentrations of $100–500 \, \mu M$ ($7.5–37 \, mg \, L^{-1}$), whereas the value for Degussa P25 was only $\sim 50 \, \mu mol$ As(V) $g^{-1} \, TiO_2$ ($\sim 4 \, mg \, As(V) \, g^{-1} \, TiO_2$) at the same pH and As(V) concentrations (Dutta *et al.*, 2004, 273; Table 7.2). At pH 9, the sorption capacity of Hombikat UV100 for As(III) was $\sim 350 \, \mu mol \, As(III)$ $g^{-1} \, TiO_2$ ($\sim 26 \, mg \, As(III) \, g^{-1} \, TiO_2$) for equilibrium As(III) concentrations of $\sim 400 \, \mu M$ ($\sim 30 \, mg \, L^{-1}$). The results for Degussa P25 were, again, much lower with $\sim 60–100 \, \mu mol \, As(III) \, g^{-1} \, TiO_2$ ($\sim 5–8 \, mg$ As(III) $g^{-1} \, TiO_2$) at $\sim 400 \, \mu M$ ($\sim 30 \, mg \, L^{-1}$) and only increasing to $\sim 130 \, \mu mol \, As(III) \, g^{-1} \, TiO_2$ ($\sim 10 \, mg$ As(III) $g^{-1} \, TiO_2$) at $\sim 540 \, \mu M$ ($\sim 40 \, mg \, L^{-1}$) (Dutta *et al.*, 2004, 273).

In another sorption study with Degussa P25 (Table 7.2), Jézéquel and Chu (2006) found that carbonate and less than $10 \, mmol \, L^{-1}$ of $CaCl_2$, $MgCl_2$, $NaCl$, or $NaNO_3$ did not appreciably interfere with the sorption of As(V) on the Degussa product. However, $0.267 \, mmol \, L^{-1}$ of phosphate significantly diminished As(V) sorption and $0.267 \, mmol \, L^{-1}$ of sulfate was a moderate hindrance.

Hydrous titanium dioxides ($TiO_2 \cdot nH_2O$, where $n > 0$) may also be important sorbents for arsenic species. Manna, Dasgupta and Ghosh (2004) investigated the removal of As(III) with synthetic crystalline hydrous titanium dioxide. At pH 7, the adsorption capacity of the compound was $72–75 \, mg \, g^{-1}$. Approximately

70 % of the sorption occurred within 30 minutes. Like many other sorbents, crystalline hydrous titanium dioxide may be regenerated with 1 M NaOH.

Jing *et al.* (2005b) investigated the adsorption mechanisms of MMA(V) and DMA(V) on nanocrystalline TiO_2 (anatase). The anatase had an initial surface area of $329 \, m^2 \, g^{-1}$ (Jing *et al.*, 2005b, 15). Jing *et al.* (2005b) concluded that both arsenic species form inner sphere complexes. However, MMA(V) is bidentate with TiO_2 surfaces, whereas DMA(V) is *monodentate* (Chapter 2). In their studies with Degussa P25 TiO_2 (~80 % anatase and ~20 % rutile), Xu, Cai and O'Shea (2007, 5473) found that the sorption capacities of inorganic arsenic on the TiO_2 were about 1.5–2.0 times greater than MMA(V) and 3–4 times greater than DMA(V).

Dutta *et al.* (2005), Xu *et al.* (2005), and Lee and Choi (2002) discuss some of the details and controversies associated with deciphering the mechanisms that are responsible for the oxidation of As(III) by TiO_2 photocatalysts. One controversy deals with the role of pH in the photooxidation process. Lee and Choi (2002, 3874) concluded that the initial oxidation rate of As(III) was much faster at pH 9 than pH 3. In contrast, Dutta *et al.* (2005, 1829) found no significant pH effect on As(III) oxidation rates. In the Lee and Choi (2002) pH 3 and 9 experiments, a dosage of $1.5 \, g \, TiO_2 \, L^{-1}$ required up to 2 hours for complete oxidation of $500 \, \mu M$ ($37 \, mg \, L^{-1}$) of As(III). In contrast, Dutta *et al.* (2005, 1829) only used $0.1 \, g \, TiO_2 \, L^{-1}$ to completely oxidize $200 \, \mu M$ ($15 \, mg \, L^{-1}$) of As(III) in pH 3, 4, 7, and 9 solutions within 15 minutes. Dutta *et al.* (2005, 1829) attributed their differences with Lee and Choi (2002) to variations in reactor design and TiO_2 dosages.

7.2.3.5.8 Activated and impregnated activated carbon   Activated carbon is a largely amorphous form of porous and high surface area graphite, which is used to sorb a wide variety of contaminants from air and water. A number of organic sources may be utilized to produce activated carbon, including: wood, coal, *peat*, crop residues, bones, or *petroleum* byproducts (Mohan and Pittman, 2007). After charring the source materials to remove volatiles, the resulting carbon is usually 'activated' with heat and chemicals (such as acids, bases, steam, or metal salts; (Mohan and Pittman, 2007)) to increase its surface area and improve sorption. Although activated carbon will sorb a wide variety of organic and inorganic contaminants from water, the pure compound is generally ineffective in removing As(III) (Dambies, 2004, 608; Mondal, Majumder and Mohanty, 2007). Depending upon the composition and surface area of the activated carbon and the chemistry of the water, the carbon may (Mokashi and Paknikar, 2002) or may not (Daus, Wennrich and Weiss, 2004) extensively remove As(V) from water. Mohan and Pittman (2007) list and discuss specific details on the arsenic removal capacities and other sorptive properties of various activated carbons mentioned in the literature. A typical sorption capacity for a granular activated carbon (GAC) is about 20 mg As(V) g^{-1} (Rajaković, 1992; Table 7.2). Even if a particular activated carbon is effective in removing As(V) in a given situation, the carbon is often too expensive (Shih, 2005, 86) and regeneration of the material may not be possible (US EPA, 2002b, 25).

Powdered varieties of activated carbon generally have higher surface areas and more sorption sites than granular forms. However, water treatment with powders often produces impermeable sludges that are difficult to dewater. As an alternative, GAC may be impregnated with iron and other materials and then placed into permeable treatment columns. Impregnating activated carbon with inorganic materials, such as calcium chloride ($CaCl_2$) or iron, copper, or zirconium salts, can greatly improve the sorption of As(III) and/or As(V) (US EPA, 2002b, 25; (Rajaković, 1992; Mohan and Pittman, 2007; Mondal, Majumder and Mohanty, 2007; Zhang *et al.*, 2007c; Schmidt *et al.*, 2008). The addition of polyanilines (($C_6H_7N_4)_n$) or other organic compounds to activated carbon may also substantially enhance the sorption of As(V) (Yang, Wu and Chen, 2007; Table 7.1). Under acidic conditions, the presence of polyanilines increases the number of positive surface charges on activated carbons, which allows for the greater sorption of As(V) oxyanions (Yang, Wu and Chen, 2007).

Gu, Fang and Deng (2005) evaluated the ability of Fe(III)-impregnated GAC to sorb As(III) and As(V) from water (Table 7.2). The Fe(III) was primarily amorphous and was prepared by impregnating Fe(II) chloride into GAC followed by chemical oxidation with sodium hypochlorite (NaClO). Arsenic was most effectively removed with GACs made from steam-activated *lignite* containing about 6 % iron. Maximum removal of As(V) occurred under acidic conditions. Above pH 8.5, phosphate and silicate could significantly interfere with As(V) removal, although sulfate, chloride, and fluoride had little effect. Column studies indicated that both As(III) and As(V) could be reduced to below $10 \mu g \, L^{-1}$ from groundwaters containing $50 \mu g \, L^{-1}$ of arsenic (Gu, Fang and Deng, 2005; Table 7.1).

The addition of zirconium to activated carbon may substantially increase the removal of arsenic from water (Daus, Wennrich and Weiss, 2004; Schmidt *et al.*, 2008). (Daus, Wennrich and Weiss, 2004) used batch and column tests to evaluate the ability of five materials (activated carbon, zirconium-loaded activated carbon, zerovalent iron, granulated Fe(III) hydroxide, and a commercial product, 'Absorptionsmittel 3') to sorb As(III) and As(V) from water. The GAC had grain sizes between 1.0 and 1.5 mm. The material was primarily chosen as a comparison with the zirconium-loaded sample. The zirconium-loaded activated carbon contained 28 mg zirconium g^{-1} activated carbon and was produced by shaking activated carbon in a solution of zirconyl nitrate ($ZrO(NO_3)_2$). The zerovalent iron (Fe(0)) primarily had particle sizes of 1.2–1.7 mm. Absorptionsmittel 3 is a mixture of calcite, brucite, fluorite, and iron hydroxides. The granular iron hydroxides consisted of mostly amorphous Fe(III) hydroxide coatings on sand grains (particle sizes of 3–4 mm) (Daus, Wennrich and Weiss, 2004, 2950).

The batch tests consisted of adding 0.5 g of sorbent to 50 ml of pH 7 carbonate-buffered water containing $10 \, mg \, L^{-1}$ sulfate as an ammonium salt, $14 \, mg \, L^{-1}$ chloride as an ammonium salt, and $500 \mu g \, L^{-1}$ each of As(III) and As(V) (Daus, Wennrich and Weiss, 2004, 2949). Although activated carbon has a reputation of being ineffective in adsorbing As(III) (Dambies, 2004, 608), the activated carbon in the batch experiments was more effective in sorbing As(III) than the other four products. In the presence of activated carbon, the As(III) concentrations in the batch samples decreased from 500 to $\sim 200 \mu g \, L^{-1}$ in about 30 minutes. The activated carbon also tended to oxidize the As(III) even under anaerobic (no more than 0.5 v % oxygen) batch conditions (Daus, Wennrich and Weiss, 2004, 2951).

For As(V), the zirconium-loaded activated carbon was the most effective sorbent, whereas the activated carbon was largely ineffective (Daus, Wennrich and Weiss, 2004, 2950). The zirconium-loaded activated carbon reduced the As(V) concentrations to below $50 \mu g \, L^{-1}$ in about 30 minutes (Daus, Wennrich and Weiss, 2004, 2951). After 24 hours, the As(V) concentrations in the Absorptionsmittel 3 and Fe(III) hydroxide batch leachates were below $50 \mu g \, L^{-1}$ (Daus, Wennrich and Weiss, 2004, 2951).

The arsenic sorption capacities of activated carbons may significantly increase because of natural impurities in the source materials of the carbons. For example, activated carbons produced from lignites and *subbituminous coals* usually contain more inorganic materials than those made from *bituminous coals*. The higher inorganic contents allow activated carbons produced from lignite and subbituminous coals to more effectively remove As(V) from water than carbons made from bituminous specimens (Leist, Casey and Caridi, 2000, 130).

The modification of carbon powders with organic compounds may substantially improve the sorption of As(III). Xiao *et al.* (2006) chemically modified carbon powders with L-cysteine methyl ester ($(CH_3)OOCCH(CH_2SH)NH_2$) (Table 7.1). The modified materials are capable of reducing As(III) in water from about $70 \mu g \, L^{-1}$ to below $10 \mu g \, L^{-1}$ within 20 minutes at a dosage of 40 mg of carbon in 20 ml of water (Xiao *et al.*, 2006, 619–620). However, the experimental dosages are quite high and the materials are very expensive (about £1 per g or about $US 4 to treat 1 L of water with a dosage of $2 \, g \, L^{-1}$). Unless the dosages can be substantially reduced, the process is probably too costly for widespread applications in Bangladesh and other developing nations.

7.2.3.5.9 Biomass Crop residues, such as sorghum biomass or waste rice husks, may be effective and economical materials for treating As(V) in water (Cano-Aguilera *et al.*, 2005; Amin, Kaneco and Kitagawa, 2006). At pH 4.5, $10\,mg\,ml^{-1}$ of sorghum biomass can remove about 90 % of $1\,mg\,L^{-1}$ of As(V) from water in about 12 hours. If $0.1\,mM$ of $FeSO_4$ or $Fe(NO_3)_2$ are added to the mixtures, As(V) removal increases to more than 95 % in only about 1 hour (Cano-Aguilera *et al.*, 2005, 59). In contrast, magnesium sulfate decreases the sorption of As(V) on sorghum biomass by 21 %. The decrease may be due to sulfate competing with As(V) for the same sorption and ion-exchange sites on the biomass (Cano-Aguilera *et al.*, 2005). Humic acids also severely interfere with As(V) sorption on sorghum biomass. After 24 hours, only about 80 % of $1\,mg\,L^{-1}$ As(V) was removed from an aqueous mixture containing $10\,mg\,ml^{-1}$ biomass and $2\,mg\,ml^{-1}$ humic acids (Cano-Aguilera *et al.*, 2005, 59).

Biomass may also sorb As(III) from water. Teixeira and Ciminelli (2005) removed considerable As(III) with ground chicken feathers treated with ammonium thioglycolate. *X-ray absorption near edge structure* (XANES) spectra indicate that the adsorbed arsenic is still in the +3 *valence state* and that each atom is bound to three sulfur atoms associated with reduced cysteine amino acids ($HO_2CCH(NH_2)CH_2SH$) in the feathers. At pH 5 and biomass dosages of $2.0\,g\,L^{-1}$, the sorption capacity of the material was as high as $0.265\,mmol\,As(III)\,g^{-1}$ biomass ($19.9\,mg\,As(III)\,g^{-1}$ biomass; Table 7.2). The presence of $0.01\,mol\,L^{-1}$ of phosphate had only minor effects on the sorption capacity, which was $0.260\,mmol\,As(III)\,g^{-1}$ biomass ($19.5\,mg\,As(III)\,g^{-1}$ biomass) (Teixeira and Ciminelli, 2005, 898).

7.2.3.5.10 Metal-loaded gels and ion-exchange resins A number of *polymers* (including resins) and metal-loaded gels have been developed to treat arsenic in water. Typically, the polymers or gels are packed into columns for treatment applications. In some cases, the polymers alone will sorb arsenic. Specifically, $150–200\,\mu m$ porous beads containing methacryloylamidophenylalanine can sorb up to $8.0\,mg\,As(III)\,g^{-1}$ beads from a $0.5\,mmol\,L^{-1}$ ($37\,mg\,As\,L^{-1}$) As(III) solution (Say *et al.*, 2003a; Table 7.2). In other cases, polymers and gels only act as support media for iron and other powdered sorbents (Sylvester *et al.*, 2007; Balaji and Matsunaga, 2002; Zeng, 2003) or the sorption and ion-exchange properties of the polymers are enhanced by impregnation with inorganic compounds (Dambies, 2004).

Ion-exchange resins consist of natural or synthetic organic polymers and possibly inorganic components. Like many sorbents, the resins contain ionic functional groups, which provide sites for ion exchange (US EPA, 2002b, 30). When compared with precipitation/coprecipitation, chemicals in water are more likely to interfere with the removal of arsenic with ion-exchange methods (US EPA, 2002b, 5). However, for waters with less than $500–1000\,mg\,L^{-1}$ of TDS and no more than $120\,mg\,L^{-1}$ sulfate, ion exchange is one of the better technologies for removing arsenic (Clifford and Ghurye, 2002, 231). Like other water treatment technologies, ion-exchange methods are usually more effective with As(V) than As(III). Arsenic removal may be enhanced by preoxidizing any As(III). In ion-exchange systems, pH controls the surface properties of the polymers and the charges of dissolved arsenic species (Dambies, 2004, 609). Therefore, pH must be carefully monitored and controlled to achieve optimal arsenic removal with the resins.

Ion-exchange media are typically divided into four groups: strong acid, weak acid, strong base, and weak base (US EPA, 2002b, 30). The acid groups exchange cations, whereas the base resins exchange anions, including arsenic oxyanions. For arsenic, weak-base resins tend to only be effective over a narrow pH range. Therefore, strong-base resins are better alternatives (US EPA, 2002b, 30).

Strong-base anion-exchange resins contain amine, quaternary ammonium (NR_4^+), or other functional groups within their polymer matrices (Dambies, 2004, 605; Clifford and Ghurye, 2002, 232). Quaternary ammonium strong-base resins are typically divided into two types. Type I resins have three methyls ($—CH_3$) attached onto each nitrogen of the ammonium group, whereas Type II often have two methyls and one ethanolamine (C_2H_7NO) on each nitrogen (Dambies, 2004, 605). Overall, strong-base resins are more selective for sulfate than $HAsO_4^{2-}$, but $HAsO_4^{2-}$ is more selective than monovalent and most other anions, as summarized in the following order: $SO_4^{2-} > HAsO_4^{2-} > CO_3^{2-} \sim NO_3^- > Cl^- > H_2AsO_4^- \sim HCO_3^- \gg H_3AsO_3^0$ (Clifford and Ghurye, 2002, 232). The removal effectiveness of $H_3AsO_3^0$ is very low because of its absence of charge (Clifford and Ghurye, 2002, 232). The resins are most effective in removing As(V) at about pH 7–9, where $HAsO_4^{2-}$ is the dominant As(V) oxyanion.

Several strong-base resins are very effective in treating As(V) in water (Table 7.1). Amberlite IRA-458, a Type I resin, can remove 90 % of $1 \, mg \, L^{-1}$ of As(V) in 5 minutes (Dambies, 2004, 606). Korngold, Belayev and Aronov (2001) successfully removed As(V) to below $10 \, \mu g \, L^{-1}$ with a commercial Type I resin, Purolite A-505 by Purolite Co. and another strong base resin, Relite A-490 by Mitsubishi Co. Like other ion-exchange resins, sulfate and chloride may significantly interfere with As(V) removal. The TDS of the water should also be below $1000 \, mg \, L^{-1}$ (Korngold, Belayev and Aronov, 2001, 84).

Base resins also may be classified by the type of anions that are exchanged ('selected') for the contaminant (US EPA, 2002b, 30). Nitrate-, chloride-, and sulfate-selected resins are typically used to remove arsenic oxyanions (US EPA, 2002b, 30; Clifford and Ghurye, 2002). Chloride resins remove arsenic by releasing two Cl^- for $HAsO_4^{2-}$ ((Clifford and Ghurye, 2002), 232). Although the resins are generally effective, the treated water may be acidic and corrosive (US EPA, 2002b, 32).

Although base resins are generally ineffective in removing arsenic from high sulfate and high TDS water (Clifford and Ghurye, 2002, 231), researchers have developed other types of polymers and resins that are less susceptible to interferences from sulfate and other anions. Hrubý *et al.* (2003) describe the preparation of ion-exchange resins containing *thiol* (-SH) and quaternary ammonium groups. They found that for a $0.2 \, mol \, L^{-1}$ ($15 \, g \, As \, L^{-1}$) As(V) solution, the highest sorption with an iron-loaded thiol resin occurred at a pH of 2 and was $3.78 \, mmol$ ($283 \, mg$) As(V) g^{-1} dry resin (Hrubý *et al.*, 2003, 2168–2169; Table 7.2).

Besides having the potential for achieving very low treatment levels, ion-exchange resins have the advantage of being relatively insensitive to water flow variations. Most of them are also easily regenerated. The potential disadvantages of ion-exchange resins often resemble the problems associated with aluminum and other inorganic sorbents, including disposal of large volumes of spent materials after regeneration is no longer possible, variable removal rates depending on the presence of interfering species, increased ineffectiveness with high TDS waters, and possible fouling from organic materials and other chemicals (US EPA, 2002b, 30; Clifford, Subramonian and Sorg, 1986). Sulfate interferences, fouling and competing ions, pH requirements, and the frequency of bed regeneration will all affect the costs of ion-exchange systems (Leist, Casey and Caridi, 2000, 130; US EPA, 2002b, 32).

US EPA (2002b) reviews the performance and costs of seven full-scale ion-exchange systems for removing arsenic from water. The authors concluded that ion-exchange systems may need modification to meet the new US arsenic drinking water standard (MCL), which could include using larger amounts of ion-exchange materials, changing the type or composition of the materials, increasing the frequency of material replacement or regeneration, decreasing water flow rates, and adding another treatment technology (US EPA, 2002b, 33).

Cation-loaded polymers and gels are generally prepared by mixing or passing aqueous metal or other cation solutions through the materials until they become saturated. Useful metals and other cations in removing arsenic from water include: iron(III), aluminum(III), copper(II), zirconium(IV), manganese(IV), lanthanum(III), titanium(IV), or molybdenum(VI) (Dambies, 2004; Balaji and Matsunaga, 2002; Seko *et al.*, 2004; Seko, Tamada and Yoshii, 2005; Lenoble *et al.*, 2004; Jang *et al.*, 2003). Some of these cation-loaded materials are also used in analytical laboratories to concentrate or separate arsenic species in solutions prior to analysis (e.g. lanthanum(III) resins) (Dambies, 2004, 619).

Most metal-loaded polymers are less affected by chloride and sulfate interferences than strong-base resins (Dambies, 2004, 608). However, As(V) sorption with iron-loaded polymers may still be seriously hindered by fluoride, sulfate, chloride, or phosphate in the water. Furthermore, chloride and sulfate may interfere with As(III) sorption (Dambies, 2004, 616). Iron and other metal-loaded polymers must also be carefully designed so that they will not release metals and other potential toxins into water during treatment (Dambies, 2004, 608). Loss of functional groups, iron, aluminum, or other chemicals will not only lower water quality, but also degrade the polymer.

Zeng (2003) produced sorbent granules by coprecipitating amorphous iron (oxy)(hydr)oxides with a silica sol binder. Batch sorption studies were performed by adding 100 mg of the sorbent granules to 100 ml solutions containing either 2000 $\mu g \, L^{-1}$ of As(III) or As(V). The mixtures were stirred for 24 hours. Without a silica binder, the Fe(III) (oxy)(hydr)oxides removed about 100 % of the As(V) at pH 6.5 and 99.5 % of As(III) at pH 7.3. In contrast, the silica gel without iron loading removed only 6.8 % of the As(V) and no detectable As(III) at a pH of 6.9. The addition of silica reduced the capacity of the iron (oxy)(hydr)oxides to sorb As(III) and As(V), but granules with Si/Fe molar ratios of 0.33 and 0.5 still removed more than 95 % of the arsenic over a pH range of 7.3–7.4 for As(V) and 7.3–7.6 for As(III). The granules also showed little cracking in water, which indicates that they have suitable strength and durability for column applications (Zeng, 2003).

ArsenX^{np} is a sorbent that consists of porous (300–1200 μm diameter) polymeric beads that have been impregnated with iron (oxy)(hydr)oxide nanoparticles (Sylvester *et al.*, 2007; Sarkar *et al.*, 2007). The high porosity, surface area, and iron content of the beads allow them to reduce As(V) concentrations to below 10 $\mu g \, L^{-1}$ in water (Tables 7.1 and 7.2). Sulfate, chloride, fluoride, and bicarbonate do not substantially interfere with the removal of As(V). However, like other iron-based sorbents, dissolved silica at pH > 7.5 and phosphate may interfere (Sylvester *et al.*, 2007, 106). ArsenX^{np} is regenerated with a proprietary process, which involves rinsing the sorbent with a warm caustic solution and then neutralizing the surface charges on the beads (Sylvester *et al.*, 2007, 109).

Cumbal and Sengupta (2005) produced an effective arsenic removal system by dispersing high-surface area iron (oxy)(hydr)oxide nanoparticles within a base (anion exchange) polymer for permeable matrix support (Table 7.1). The polymer consisted of a quaternary ammonium functional group within a styrene-divinylbenzene matrix. The addition of the nanoparticles improved arsenic removal when compared with the polymer alone.

Column studies with the nanoparticle base polymer used solutions containing either 100 $\mu g \, L^{-1}$ of As(III) or As(V) (Cumbal and Sengupta, 2005). Even after more than 10 000 bed volumes of water had passed through the system, arsenic concentrations in the treated water still remained below 10 $\mu g \, L^{-1}$. The nanoparticle system releases less than 5 $\mu g \, L^{-1}$ of iron into treated water and can be regenerated with an aqueous mixture of 2 % NaOH and 3 % NaCl (Cumbal and Sengupta, 2005).

Zirconium-loaded resins can have strong affinities for As(III) even at circumneutral pH conditions (Dambies, 2004, 618, 622; Tables 7.1 and 7.2; Suzuki *et al.*, 2001). Specifically, Amberlite™ XAD™-7 resin loaded with hydrous zirconium(IV) oxides ($ZrO_2 \cdot nH_2O$, where $n>0$) has sorption capacities of 1.06 mmol g^{-1} (79.4 mg As g^{-1}) for As(III) (pH of 8.5), 0.72 mmol g^{-1} (54 mg As g^{-1}) for As(V) (pH of 4.5), 0.61 mmol g^{-1} (46 mg As g^{-1}) for MMA(V) (pH of 4.5), and 0.19 mmol g^{-1} (14 mg As g^{-1}) for DMA(V) (pH of 4.5) (Suzuki *et al.*, 2001, 109; Table 7.2). Zirconium(IV) oxide resins effectively remove arsenic without significant interferences from sulfate, nitrate, and chloride. However, fluoride and phosphate may interfere (Dambies, 2004, 618; Suzuki *et al.*, 2001). The resins can be readily regenerated with 1 M NaOH followed by rinsing with an acetate buffer (Suzuki *et al.*, 2001).

As ion-exchange resins become exhausted, chromatographic peaking may occur. That is, the arsenic previously sorbed in the resins may begin to leak out of them. In the worst situations, continued use of the resins may actually increase the arsenic concentrations of the water (Clifford and Ghurye, 2002, 235). Chromatographic peaking is avoided by replacing the resins before extensive exhaustion (Clifford and Ghurye, 2002, 235). Like sorption columns, ion-exchange systems may include a series of two or more columns to reduce the frequency of replacement and regeneration.

Although ion exchange may be very effective in treating arsenic in water, resin regeneration creates wastewaters that may be hazardous (Shih, 2005, 86). Depending on costs, the effectiveness of regeneration, and whether or not the spent sorbents and resins are hazardous, the materials may be regenerated or discarded. Chloride-selected strong-base resins, for example, are typically regenerated with NaCl solutions (Clifford and Ghurye, 2002, 232). In many cases, 0.5 N NaCl is more effective in regenerating the resins than 1.0 or 2.0 N NaCl (Clifford and Ghurye, 2002, 235). The NaCl solutions may also be reused in future regenerations if fresh NaCl is added (Clifford and Ghurye, 2002, 235–236). Resin columns have been regenerated as many as 18 times with recycled NaCl solutions (Clifford and Ghurye, 2002, 235–236). Arsenic may be removed from the spent regeneration solutions through coprecipitation using iron(III) chloride ($FeCl_3$) or aluminum sulfate ($Al_2(SO_4)_3$) salts (Clifford and Ghurye, 2002, 236). If the resulting iron and aluminum sludges pass the TCLP test, they may be legally landfilled in most of the United States (Clifford and Ghurye, 2002, 237).

7.2.4 Precipitation/coprecipitation

7.2.4.1 Introduction

Precipitation refers to the process of adding one or more chemical reagents to water so that dissolved contaminants are transformed (precipitated) into insoluble solids (*precipitates*) (US EPA, 2002b, 17). The precipitates can then be collected and removed from the water by filtration, flotation, centrifugation, or other methods. For arsenic in water, precipitation typically involves reactions between arsenic oxyanions and dissolved cations. A common example is the precipitation of calcium arsenates from the addition of lime to a wastewater containing dissolved As(V).

In coprecipitation, a *minor* or *trace element* (such as arsenic) adsorbs onto or absorbs within the precipitates of other chemical species. Although often difficult to distinguish, sorption involves the incorporation of contaminants onto or within preexisting solids (sorbents), whereas coprecipitation occurs as or shortly after the host solids precipitate from the solution. For example, the addition of Fe(III) chloride to aerated water would readily precipitate Fe(III) (oxy)(hydr)oxides. Any arsenic in the water could then coprecipitate with the Fe(III) (oxy)(hydr)oxides. Coprecipitation might also involve arsenic-bearing *colloids* or fine-grained particles becoming trapped (absorbed) in the interiors of precipitating compounds (US EPA, 2002b, 17; Yuan *et al.*, 2003). Additionally, arsenic could coprecipitate by substituting into the crystalline structures of precipitating compounds, such as As(III) partially substituting for carbonate in the developing crystalline structures of 'green rust' (Lien and Wilkin, 2005, 383, 385).

7.2.4.2 Applications and limitations of precipitation/coprecipitation

When compared with ion exchange and other water treatment technologies, the major advantages of precipitation/coprecipitation are relatively low costs and the ability of the process to treat waters with high TDS. That is, with a good knowledge of the chemistry of the water and by applying the correct dosage, chemical species that frequently interfere with sorption and ion exchange (such as phosphate) can often be managed with precipitation/coprecipitation (US EPA), 2002a, 1.1; US EPA, 2002b, 19). The favorable economics and versatility of precipitation/coprecipitation technologies often result in their use to treat arsenic at US Superfund sites (US EPA), 2002a, 2.3). On the other hand, effective removal of arsenic with precipitation/coprecipitation requires an accurate assessment of the initial arsenic concentration of the water, its speciation, and strict control of pH, chemical dosages, and other water characteristics. The process may also produce voluminous and potentially toxic sludges that will require disposal (Mohan and Pittman, 2007).

Precipitation/coprecipitation technologies are most effective in large-scale water treatment plants where skilled technicians are available to monitor and operate the systems (US EPA), 2002a, 1:1). The costs of precipitation/coprecipitation would depend on the types of treatment chemicals and their dosages, treatment goals, and sludge recovery and disposal requirements (US EPA, 2002b, 20). For specific examples, US EPA (2002b) reviews the effectiveness and available cost data of 69 precipitation/coprecipitation technologies that are used to remove arsenic from water.

With the lowering of the US arsenic drinking water standard (MCL) from 50 to $10\,\mu g\,L^{-1}$ (Appendix E), water treatment facilities may have had to modify their procedures to comply with the new standard. Boccelli, Small and Dzombak (2005) identifies some of the factors associated with Fe(III) chloride coprecipitation that might require modification while still maintaining reasonable costs. Besides the influent arsenic concentration and Fe(III) chloride dose, other critical factors in improving arsenic removal may include pH, the calcium and major anion concentrations of the influent, the volume of sludge, equilibrium sorption constants, and filter efficiency.

7.2.4.3 *General design of precipitation/coprecipitation systems for arsenic treatment in water*

Figure 7.1 shows a diagram of a typical precipitation/coprecipitation system for treating dissolved arsenic in water (US EPA, 2002b, 18). In most cases, precipitation/coprecipitation is more effective if any inorganic As(III) is oxidized to As(V) before treatment. The pH of the water may also require adjustment for optimal performance (US EPA, 2002b, 17). To achieve optimal performance during oxidation, pH adjustment, and subsequent treatment steps, the operating conditions of the system may be monitored or modeled with computer software (e.g. Pal, Ahammad and Bhattacharya, 2007).

After pH adjustment, the water enters a precipitation/coprecipitation tank, where chemical reagents (such as iron or aluminum salts) are carefully added to form the precipitates. The resulting precipitates are often colloidal or are otherwise too fine grained to readily settle out of solution. They may also have repulsive surface charges that prevent them from agglomerating and settling. As shown in Figure 7.1,

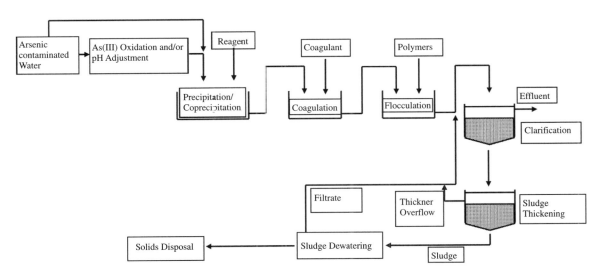

Figure 7.1 *A typical precipitation/coprecipitation system. (Modified after US EPA (2002b, 18) with the permission of the United States Environmental Protection Agency.)*

after precipitation/coprecipitation, the water and its suspended particles are transferred into one or more *coagulation* and flocculation tanks. The definitions of coagulation and flocculation are sometimes confused and often used interchangeably in the literature (Faust and Aly, 1998, 217). According to the general consensus, *coagulants* refer to organic compounds, iron or aluminum salts, or other chemicals that are added to neutralize repulsive surface charges on suspended particles (Spellman, 2003, 24, 473–474). Once the surface charges diminish, the suspended particles will begin to agglomerate and form fragile *flocs* (Spellman, 2003, 474). After the flocs form, they may be filtered out of solution. Alternatively, the flocs may undergo flocculation, which refers to the gentle mixing of the water to promote additional agglomeration and the separation of the flocs from the water by either floating or settling (Spellman, 2003, 26, 475; Pacheco and Torem, 2002). Mixing should be great enough to promote particle collisions and additional agglomeration, but not so severe that the particles are torn apart. Polymers may be added to increase bridging between the particles and add weight to the agglomerates, which further advance settling (Spellman, 2003, 475; Montgomery, 1985, 128).

During the *clarification* or *sedimentation* step (Figure 7.1), the main objective is to allow gravity to settle the enlarged particles (Spellman, 2003, 476). Suspended particles may also be separated from the effluent by filtering (Spellman, 2003, 476). Once the effluent is essentially free of precipitates (that is, 'clear'), it may undergo additional treatment or be discharged.

The settled precipitates form a wet *sludge* that may require thickening and dewatering (Figure 7.1). Thickening and dewatering reduce the mass of the sludge, which can significantly lessen transportation and disposal costs. The main objective of thickening is to remove excess water from the sludge, but still allow it to be pumped. Thickening may involve additional gravity settling, flotation, or mechanical methods (American Water Works Association, 1998, 499, 501). The most common mechanical method involves porous and moving gravity belts, which absorb water from the sludge (American Water Works Association, 1998, 501).

During the dewatering step, the sludge is transformed from a viscous liquid to a damp solid. Dewatering methods may include filter presses, belt presses, centrifuges, vacuum filtration, and/or drying (American Water Works Association, 1998, 502–516). If the dewatered solids fail to pass regulatory leaching tests (such as the TCLP, Appendix E), they would have to undergo additional treatment (see discussions in Section 7.3.2.2) before disposal.

7.2.4.4 Review of various precipitation/coprecipitation reagents and methods

7.2.4.4.1 Lime In the past, lime (CaO) was extensively used to remove As(V) from water through the precipitation of calcium arsenates (Bothe and Brown, 1999). Specifically, about 90 % of $0.4\,mg\,L^{-1}$ of As(V) may be precipitated with lime at pH values above 10.5. In contrast, the precipitation of As(III) with lime is much less effective (Jekel, 1994, 126).

Bothe and Brown (1999) concluded that lime precipitates As(V) from aqueous solutions as hydroxyl and hydrated calcium arsenates ($Ca_4(OH)_2(AsO_4)_2 \cdot 4H_2O$, $Ca_5(AsO_4)_3OH$ (arsenate apatite), and/or $Ca_3(AsO_4)_2 \cdot 3^2/_3H_2O$) rather than anhydrous tricalcium orthoarsenate ($Ca_3(AsO_4)_2$). Although calcium arsenates may readily precipitate, Robins and Tozawa (1982) warn that the compounds may not have long-term stability, which could lead to disposal and environmental problems. Besides dissolving under acidic conditions, the presence of carbonate, bicarbonate, or carbon dioxide may decompose calcium arsenates. When calcium arsenates react with carbon dioxide, calcium carbonate forms and the arsenic could be released into the environment (Ghimire *et al.*, 2003, 4946; Jing, Korfiatis and Meng, 2003, 5055–5056). Although arsenic- and lime-rich sludges are potentially unstable, lime may still be used for pH adjustment in some precipitation/coprecipitation systems. Iron salts or other reagents would then be primarily responsible for precipitating/coprecipitating arsenic (Leist, Casey and Caridi, 2000, 128).

7.2.4.4.2   Iron salts   Coprecipitation with Fe(III) salts is one of the most popular and effective techniques for removing As(V) from water (Table 7.1). In water, Fe(III) chlorides and sulfates ($Fe_2(SO_4)_3$) react and precipitate iron (oxy)(hydr)oxides. Fe(III) salts can readily coprecipitate about 99 % of $0.1-1$ mg L^{-1} As(V) under pH conditions below 7.2 (Jekel, 1994, 124). Unless the arsenic is oxidized before treatment, only about $50-60$ % of As(III) can be removed with Fe(III) salts (Jekel, 1994, $124-125$).

Although both Fe(III) sulfates and chlorides are effective in coprecipitating As(V) from water, the sulfates may produce less turbidity and corrosion (Han *et al.*, 2003; Floch and Hideg, 2004, 76). As(V) coprecipitation with iron (oxy)(hydr)oxides may be further improved by filtering out the precipitates with membranes or sand. Han *et al.* (2003) were able to reduce arsenic concentrations to below $2\,\mu g\,L^{-1}$ with Fe(III) doses of 6 mg L^{-1} followed by membrane filtration (Table 7.1). Using Fe(III) sulfate coprecipitation followed by sand filtration, Yuan *et al.* (2003) found that 0.25 mM of Fe(III) could inexpensively remove about 98 % of 1 mg L^{-1} of As(V) from household water supplies. Dosages of aluminum sulfate (0.25 mM as Al(III)) achieved about 95 % arsenic removal.

Hering *et al.* (1997) used laboratory batch experiments to compare the coprecipitation of As(V) and As(III) with Fe(III) chloride and alum (aluminum sulfate). When separately added to waters spiked with $20\,\mu g\,L^{-1}$ of As(V), 4.9 mg L^{-1} of Fe(III) chloride and 40 mg L^{-1} $Al_2(SO_4)_3 \cdot 18H_2O$ were able to lower the arsenic concentrations to below $5\,\mu g\,L^{-1}$ at neutral pH (Hering *et al.*, 1997, 802; Table 7.1). Below pH 8, As(V) coprecipitation with alum is fairly insensitive to water composition (Hering *et al.*, 1997). Under pH $8-9$, however, coprecipitation with Fe(III) chloride in the presence of 4 mg L^{-1} of natural organic matter may remove less than 50 % of the $20\,\mu g\,L^{-1}$ of As(V) (Hering *et al.*, 1997, 805). When coprecipitating As(V) with Fe(III) (oxy)(hydr)oxides produced from Fe(III) chloride or other Fe(III) salts, interferences from natural organic matter (Hering *et al.*, 1997, 805), phosphate (Leist, Casey and Caridi, 2000, 131), or silicate (Liu *et al.*, 2007b) may be reduced or even eliminated by the addition of calcium.

In the studies described in Hering *et al.* (1997), As(III) coprecipitation from Fe(III) chloride or alum was less successful and reinforces the belief that As(III) should be oxidized before treatment. As(III) coprecipitation from Fe(III) chloride was adversely affected by the presence of sulfate (pH $4-5$) and natural organic matter (pH $4-9$) (Hering *et al.*, 1997, 800). In comparison, alum was unable to significantly coprecipitate As(III) under any conditions.

Some As(III)-contaminated groundwaters contain natural Fe(II). When exposed to air, Fe(II) readily oxidizes in water and arsenic may coprecipitate with the resulting iron (oxy)(hydr)oxides (Guo, Stüben and Berner, 2007b; Roberts *et al.*, 2004; Appelo and de Vet, 2003). Although most Bangladeshi and other arsenic-contaminated groundwaters have insufficient natural Fe(II) to effectively coprecipitate arsenic to below $10\,\mu g\,L^{-1}$ (Kim and Nriagu, 2000; Roberts *et al.*, 2004), As(III) oxidation and coprecipitation may be improved by supplementing the natural Fe(II) with multiple doses of artificial Fe(II) sulfate solutions (Roberts *et al.*, 2004). Roberts *et al.* (2004) found that $500\,\mu g\,L^{-1}$ of As(III) in the presence of 3 mg L^{-1} of phosphorus using KH_2PO_4 and 30 mg L^{-1} of silicon using $Na_2SiO_3 \cdot 9H_2O$ could be lowered to below $50\,\mu g\,L^{-1}$ (as total dissolved arsenic) with $20-25$ mg L^{-1} of Fe(II) in multiple doses over several hours as opposed to single dosages, which require $50-55$ mg L^{-1} of Fe(II). Multiple doses of Fe(II) over several hours provide enough time for the As(III) to partially oxidize and more effectively coprecipitate.

Electrochemical coprecipitation with iron electrodes may also be effective in removing As(V) from water (Table 7.1). Recently, Sagitova *et al.* (2005) developed an electrochemical unit for treating household water. The unit oxidizes As(III) to As(V). A sacrificial anode creates dissolved iron, which coprecipitates As(V) with iron (oxy)(hydr)oxides. Filtration of the iron (oxy)(hydr)oxides can lower the arsenic concentrations of the water to below $10\,\mu g\,L^{-1}$.

Wang, Bejan and Bunce (2003) found that As(III) and As(V) could be removed from pH $2.0-2.1$ synthetic *acid mine drainage* by raising the pH with electrochemical methods and coprecipitating the arsenic with iron (oxy)(hydr)oxides. In the presence of 260 mg L^{-1} of Fe(III), $40-42$ mA of current lowered

$8 \, mg \, L^{-1}$ of As(V) to below $5 \, \mu g \, L^{-1}$. The final pH values of the solutions ranged from 4.7 to 5.7. As(III) concentrations at pH 6.0–7.0 were reduced from $8 \, mg \, L^{-1}$ to $19–28 \, \mu g \, L^{-1}$ at 42 mA with $260 \, mg \, L^{-1}$ of Fe(III). With $300 \, mg \, L^{-1}$ of Fe(II) in the synthetic acid mine drainage, $8 \, mg \, L^{-1}$ of As(V) was lowered to below $5 \, \mu g \, L^{-1}$ at an approximate pH of 4 and with currents of 25–32 mA. With $300 \, mg \, L^{-1}$ of Fe(II) and at a pH of 5.2, As(III) concentrations were reduced from $8 \, mg \, L^{-1}$ to $18 \, \mu g \, L^{-1}$ using 35 mA. The studies demonstrate that electrical currents and coprecipitation with iron (oxy)(hydr)oxides can remove As(V) to below the US drinking water standard (MCL) of $10 \, \mu g \, L^{-1}$ and that As(III) removals may almost meet the standard (Wang, Bejan and Bunce, 2003; Table 7.1).

7.2.4.4.3 Aluminum salts   Like Fe(III) salts, aluminum sulfate ($Al_2(SO_4)_3$) and other aluminum salts react with water and precipitate (oxy)(hydr)oxides. Overall, aluminum salts are ineffective in coprecipitating As(III) (Hering and Kneebone, 2002, 166; Floch and Hideg, 2004, 76). They can effectively coprecipitate As(V), although Fe(III) sulfates and chlorides are usually better (Leist, Casey and Caridi, 2000, 129; Hering and Kneebone, 2002, 166; Jekel, 1994, 126; Deliyanni *et al.*, 2003b, 3968; Cheng *et al.*, 1994). The superiority of coprecipitation with Fe(III) salts may be due to higher sorption site densities on iron (oxy)(hydr)oxides when compared with the aluminum precipitates (Boccelli, Small and Dzombak, 2005, 6501). Furthermore, arsenic coprecipitation is hindered by the tendency of aluminum hydroxides to be more soluble in water than Fe(III) hydroxides (Hering *et al.*, 1997, 802).

7.2.5 Permeable reactive barriers

7.2.5.1 *Introduction and applications of permeable reactive barriers*

Permeable reactive barriers (PRBs) are *in situ* remediation devices, which are inserted into the subsurface to intercept contaminated groundwaters flowing from landfills, mines, or other sites. Once contaminants come into contact with the barriers, they may be sorbed, undergo ion exchange, biodegrade, precipitate, coprecipitate, or filter out. PRBs are effective in treating a wide variety of metals, organic contaminants, and anions, including arsenic species (Blowes *et al.*, 2000). If possible, the barriers should be designed to last for decades. The costs associated with PRBs would include their design, installation, long-term monitoring, and possible recovery and disposal after they are exhausted.

Zerovalent iron is often used in PRBs to remove arsenic from groundwater (Lien and Wilkin, 2005, 378). Other barriers for removing arsenic consist of lime, portlandite, steel industry byproducts, iron compounds, iron (oxy)(hydr)oxide-coated sands, and/or limestone (Ko *et al.*, 2007; Blowes *et al.*, 2000, 129–130; Ahn *et al.*, 2003). Mixtures of lime, iron (oxy)(hydr)oxides, and limestone may remove dissolved As(III) and As(V) to less than $3 \, \mu g \, L^{-1}$ (Blowes *et al.*, 2000, 129–130; Table 7.1).

In subsurface environments, microorganisms may colonize and reduce sulfate to sulfide on PRBs made of zerovalent iron or iron compounds. Microbial reduction of sulfate may result in the reductive dissolution of iron (oxy)(hydr)oxides in aquifers and PRBs, which could release arsenic into surrounding groundwaters (Köber *et al.*, 2005a, 7650; Chapters 2, 3, and 6). However, Köber *et al.* (2005b) concluded that moderate to high sulfide concentrations ($0.01–100 \, mg \, L^{-1}$) do not interfere with the ability of zerovalent iron to remove arsenic from water. In laboratory treatment columns, zerovalent iron in the presence of sulfide was still able to lower $2–200 \, mg \, L^{-1}$ of As(V) or As(III) in water to below $10 \, \mu g \, L^{-1}$. Substantial As(V) and As(III) were removed in the presence of sulfides by sorption onto iron (oxy)(hydr)oxides (Köber *et al.*, 2005b, 8043). *X-ray absorption spectroscopy* (XAS) further suggested that additional arsenic coprecipitated with iron sulfides (Köber *et al.*, 2005b, 8043). At the same time, the formation of arsenopyrite and zerovalent or elemental arsenic (As(0)) was excluded by XAS (Köber *et al.*, 2005b, 8043).

7.2.6   Filtration, membranes, and other separation technologies

7.2.6.1   *Introduction to filtration*

In water treatment, filtration refers to the physical separation of particles, colloids, or other contaminants from water by passing the liquid through permeable or semipermeable materials. Typically, the physical separation involves particles larger than a given size failing to pass through the filtering medium and instead collecting on the surface or becoming trapped within the medium (Montgomery, 1985, 152). Chemical processes, such as sorption and ion exchange, may also remove contaminants during filtration.

Pressure or vacuums may be applied to filtering systems to hasten the removal of contaminants. Other systems only use gravity, which involves percolating contaminated water through artificial beds or columns containing sand and other materials. In the field, artificial or natural sediment layers routinely filter water as it migrates into the subsurface (Jekel, 1994, 129). Although many filtration systems will not remove dissolved arsenic, filters can physically remove iron (oxy)(hydr)oxides and other particles that coprecipitate and sorb arsenic.

7.2.6.2   *Types of membranes*

Pressure filtration involves the use of pressure to force contaminated water through a semipermeable membrane or other type of barrier (US EPA), 2002a, 1.3; US EPA, 2002b, 35). As water passes through the membrane, the membrane captures or otherwise separates out the contaminants from the water. Depending on the type of membrane, pH, temperature, pressure, arsenic speciation, the presence of dissolved organics, and other characteristics of the water, pressure filtration can reduce arsenic concentrations to below $10 \mu g \, L^{-1}$ (Shih, 2005, 95; Brandhuber and Amy, 1998, 9). Although the membranes themselves usually require no chemical additives and are fairly simple to use (Shih, 2005), 86), chemicals may be required to precipitate, coprecipitate, or sorb arsenic so that filters can remove it. That is, filtration is often used as a 'polishing step' after water has been treated with precipitation, coprecipitation, sorption, or other technologies.

The four major types of pressure filtration processes are *microfiltration, ultrafiltration, nanofiltration,* and *reverse osmosis* (US EPA, 2002b, 35). Microfiltration is generally used to remove suspended particles with diameters that are greater than about $0.1 \, \mu m$. For recovering particles as small as $0.01 \, \mu m$, ultrafilters are available. Nanofiltration removes particles that are larger than about $0.001 \, \mu m$ and reverse osmosis can capture particles as small as $0.0001 \, \mu m$ (Shih, 2005, 94). Nanofiltration and reverse osmosis require high pressures ($0.34-6.9 \, MPa$ or $50-1000 \, psi$) to function, which can lead to significant energy consumption and costs. In comparison, microfiltration and ultrafiltration operate under relatively low pressures ($0.034-0.69 \, MPa$ ($5-100 \, psi$)) (US EPA, 2002b, 35; Shih, 2005, 90).

Ultrafiltration and especially microfiltration remove contaminants through mechanical sieving. In contrast, capillary flow and solution diffusion are important processes in nanofiltration and reverse osmosis. *Donnan exclusion* may also occur with nanofiltration (Brandhuber and Amy, 1998, 2–3). In Donnan exclusion, charges on the nanofilter repel chemical species of the same charge and hinder their movement through the membrane. Many nanofilters have negative surface charges that are capable of filtering out arsenic oxyanions (Brandhuber and Amy, 1998, 3; Urase, Oh and Yamamoto, 1998).

Brandhuber and Amy (1998) and Shih (2005) provide concise summaries of arsenic removal with different types of membranes. Microfiltration and ultrafiltration are used to remove particles that contain sorbed, coprecipitated, or precipitated arsenic (US EPA, 2002b, 35; Shih, 2005, 93, 94). Both reverse osmosis and nanofiltration can effectively remove As(V) oxyanions from water and, in some cases, $H_3AsO_3^0$ can also be treated without preoxidation (Brandhuber and Amy, 1998, 9; Uddin *et al.*, 2007a; Xia *et al.*, 2007; Uddin *et al.*, 2007b).

Compared with sorption, precipitation, and coprecipitation technologies, high-pressure filtration is fairly expensive. The costs of pressure filtration technologies depend on the type of membrane, the chemistry of the initial stream, any associated treatment technologies, temperature, pH, arsenic speciation, and the presence of organic compounds, colloids, or suspended material that could create fouling (US EPA, 2002b, 36–37). Due to the recent lowering of arsenic drinking water standards in the United States and other countries, additional expenses associated with pressure filtration are likely to occur. The costs of modifying filtration treatment systems to meet the new standards would depend on any changes in membrane type, decreases in water flow rates, and adding or changing any supplementary treatment technologies (US EPA, 2002b, 37).

7.2.6.3 *Separation technologies*

7.2.6.3.1 Microfiltration Microfilters are too coarse to remove dissolved or most colloidal species containing arsenic. However, larger particulates with sorbed, coprecipitated, or precipitated arsenic can be removed (Shih, 2005, 93; Ergican, Gecol and Fuchs, 2005, 11). Microfiltration after precipitation/coprecipitation with Fe(III) salts is one of the most effective and popular strategies for removing As(V) from water. Specifically, a combination of iron (oxy)(hydr)oxide coprecipitation using Fe(III) chloride followed by microfiltration can reduce As(V) concentrations in water from 40 to less than $2\,\mu g\,L^{-1}$ (Shih, 2005, 93–94, Table 7.1; Ghurye, Clifford and Tripp, 2004).

7.2.6.3.2 Ultrafiltration Like microfiltration, ultrafiltration primarily removes contaminants through physical sieving and the filter pores are too coarse to remove dissolved arsenic (Shih, 2005, 94). In some cases, electrically charging ultrafilters (*electrodialysis*) improves As(V) removal (Shih, 2005, 94; Weng *et al.*, 2005). As(V) removal with ultrafiltration may also improve in the presence of dissolved organic carbon (Shih, 2005, 94). The sorption of dissolved organic carbon onto the membrane changes the surface charges and increases the repulsion of arsenic oxyanions.

Cationic surfactants with or without electrical charging of the membranes may increase the removal of As(V) with ultrafilters (Gecol, Ergican and Fuchs, 2004; Pookrod, Haller and Scamehorn, 2004; Beolchini *et al.*, 2006; Rivas, Del Carmen Aguirre and Pereira, 2007a; Rivas *et al.*, 2007b). Gecol, Ergican and Fuchs (2004) found that up to $221\,\mu g\,L^{-1}$ of As(V) in water could be reduced to below $10\,\mu g\,L^{-1}$ by using $10\,mM$ of the cationic surfactant cetylpyridinium chloride followed by ultrafiltration with regenerated cellulose or polyethersulfone membranes. Arsenic oxyanions bind onto the highly charged surfaces of *micelles* created by the surfactant. The arsenic-bearing micelles are then removed by ultrafiltration (Table 7.1).

In a study related to Gecol, Ergican and Fuchs (2004), Ergican, Gecol and Fuchs (2005) investigated the effects of $0–4.1\,mg\,L^{-1}$ bicarbonate ($HCO_3{}^-$), $0–0.3\,mg\,L^{-1}$ $HPO_4{}^{2-}$, $0–90\,mg\,L^{-1}$ $H_4SiO_4^0$, and $0–400\,mg\,L^{-1}$ sulfate ($SO_4{}^{2-}$) on the removal of up to $105\,\mu g\,L^{-1}$ of As(V) with cetylpyridinium chloride and polyethersulfone ultrafilters. Except for a sample containing $105\,\mu g\,L^{-1}$ arsenic(V) and $400\,mg\,L^{-1}$ of sulfate, the membrane and surfactant reduced arsenic concentrations to below $10\,\mu g\,L^{-1}$.

Like microfiltration, ultrafiltration may be used as a polishing step to remove arsenic from water. Floch and Hideg (2004) developed an extensive treatment system that lowered up to $300\,\mu g\,L^{-1}$ of arsenic in Hungarian groundwaters to below $10\,\mu g\,L^{-1}$. As(III) in the waters was initially oxidized with potassium permanganate. After oxidation, the water underwent precipitation/coprecipitation with Fe(III) sulfate. Finally, the water passed through a ZW-1000 (ZeeWeed (Zenon), nominal $0.02\,\mu m$) ultrafilter.

7.2.6.3.3 Nanofiltration Compared to microfiltration and ultrafiltration, nanofiltration and reverse osmosis are more expensive and susceptible to fouling (Shih, 2005, 95). Most of the expenses result from the high densities of the membranes, which require high pressures ($0.34–6.9\,MPa$) and a considerable amount of energy to operate (Shih, 2005, 90; Ergican, Gecol and Fuchs, 2005, 11). Nanofiltration can be done at lower pressures (as low as 0.1 MPa ($\sim$15 psi)) to save energy. However, the membranes will not capture as much arsenic (Urase, Oh and Yamamoto, 1998).

The ability of a nanofilter to remove arsenic from water depends on the concentration and speciation of the arsenic, pH, the presence of dissolved organic matter and salts, pressure conditions, and the properties of the individual membrane. Nanofiltration removes arsenic and other contaminants through charge and size exclusion (Shih, 2005, 92; Košutic *et al.*, 2005). Therefore, the effectiveness of a particular nanofilter would depend on its electrical charge and permeability. Charge exclusions of oxyanions result in As(V) being more effectively removed with nanofiltration than $H_3AsO_3^0$ (Shih, 2005, 92–93; Xia *et al.*, 2007). Although the removal of As(III) and DMA(V) with nanofiltration can increase at higher pH values, As(V) removal tends to remain consistently greater (Urase, Oh and Yamamoto, 1998; Oh, Lee and Yamamoto, 2004, 171).

Increasing pressure with nanofiltration may increase the removal of arsenic from water, including As(III) (Shih, 2005, 93). Sato *et al.* (2002) evaluated the removal of 50 μg L^{-1} of As(V) or As(III) with three types of nanofilters (ES-10, an aromatic polyamid and two polyvinyl alcohols, NTR-729HF and NTR-7250) at pressures of 0.3–1.1 MPa (44–160 psi). The removal was consistently higher for As(V) ($>$85 %) than As(III) over the entire pressure range and for all membranes (Sato *et al.*, 2002, 3374–3375). As(V) removal improved slightly ($<$4 %) with increasing pressure from 0.3 to 1.1 MPa. Maximum As(III) removals increased from 62 to 82 % over the pressure range (Sato *et al.*, 2002, 3375). On the other hand, Saitúa *et al.* (2005) found that changes in transmembrane pressure, temperature, and crossflow velocity had little effect on the removal of 100–382 μg L^{-1} of As(V) with a composite polyamide (192-NF-300 by Osmonic Inc.) nanofilter.

Chloride, sulfate, and other anions may affect the removal of arsenic by nanofiltration. The effects often depend on the composition of the nanofilter. Specifically, increasing NaCl concentrations actually improves arsenic removals with polyamide thin-film composite filters. On the other hand, NaCl solutions may interfere with the removal of arsenic with sulfonated polysulfone thin-film composite nanofilters (Shih, 2005, 92).

7.2.6.3.4 Reverse osmosis

Although expensive, reverse osmosis is a suitable technology for reducing As(V) concentrations to below 10 μg L^{-1} (Shih, 2005, 90, Table 7.1; Saitúa *et al.*, 2005, 174). Reverse osmosis membranes typically contain cellulose acetate, polyamides, polyvinyl alcohol, or other synthetic materials (Shih, 2005, 90–91). The process can usually remove As(V) at 98–99 %. At pH $<$ 9, As(III) primarily exists as uncharged $H_3AsO_3^0$ (Chapter 2), which is not as effectively removed by reverse osmosis (about 46–75 %) (Jekel, 1994, 129). Preoxidation of As(III) is not always practical with this technology because many reserve osmosis membranes are sensitive to oxidants (Shih, 2005, 90). In most cases, reverse osmosis is simply too expensive to use for arsenic removal unless desalination is also required (Jekel, 1994, 129). However, some systems that operate with bicycle pumps have been developed for removing arsenic from groundwater in nonelectrified areas of developing countries (Shih, 2005, 91).

7.2.6.3.5 Magnetic separation

Magnetite is a highly magnetic compound and, like many other iron (oxy)(hydr)oxides, it can sorb As(III) and As(V) from water. Nanoparticulate forms of magnetite have high surface areas that are also especially good sorbents of arsenic. For arsenic removal, magnetite nanoparticles are best utilized when they are homogeneously dispersed in contaminated solutions and magnetically recovered after they are spent (Mayo *et al.*, 2007, 71). Mayo *et al.* (2007) found that decreasing magnetite particle sizes from 300 to 12 nm increases their sorption capacities for both As(III) and As(V) by nearly 200 times. The ability of magnetite to sorb As(V) from water may be further improved by doping magnetite with trace elements, such as cerium(IV) (Zhang *et al.*, 2005).

High-arsenic *geothermal* waters are sometimes used to heat cleaner river water for greenhouses, indoor swimming pools, and other facilities (Okada *et al.*, 2004, 1576). If arsenic could be removed from the geothermal waters, they could be directly utilized for recreation, business, agriculture, drinking, and washing (Gallup, 2007). Recently, Okada *et al.* (2004) used a superconducting magnet as part of a treatment system that reduced arsenic concentrations from 3400 (mostly as As(III)) to $15 \mu g \, L^{-1}$ in geothermal waters from Kakkonda, Japan. Before magnetic treatment, the arsenic was entirely oxidized to As(V) with hydrogen peroxide (Okada *et al.*, 2004, 1577). Fe(III) sulfate was then added to the water at pH 4 to coprecipitate the As(V) with paramagnetic iron (oxy)(hydr)oxides (Okada *et al.*, 2004, 1576, 1577). Magnetic wire meshes were used to capture the precipitates. Although the arsenic concentration of the treated geothermal water was still slightly above the $10 \mu g \, L^{-1}$ WHO recommendation for human consumption (Appendix E), the concentration was well below the Japanese $100 \mu g \, L^{-1}$ limit for discharging waters into the environment.

Magnetic separation has the potential to be widely used as an effective method for recovering spent sorbents from treated water. To facilitate magnetic separation, magnetic materials may be inserted into nonmagnetic sorbents. For arsenic sorption technologies, magnetite has been combined with activated charcoal sorbents (Nakahira *et al.*, 2007a). In another example, Fe(III) and Ni(II) concentrations were increased in hydrotalcite and other double-layered hydroxide clays (Nakahira, Kubo and Murase, 2007b). The increased concentrations improved both the arsenic sorptive properties of the clays and their ability to be recovered by magnetic separation.

The recovery of spent nonmagnetic sorbents may also be enhanced by coating them onto magnetic particles and utilizing magnetic separation. Specifically, Lim and Chen (2007) coated synthetic magnetite particles with a calcium alginate sorbent $((Ca_{0.5}C_6H_7O_6)_n)$. The particles took less than 25 hours to reach equilibrium in aqueous solutions containing $6 \, mg \, L^{-1}$ of As(V) (Lim and Chen, 2007, 5772–5773). The maximum As(V) sorption capacity of the calcium alginate was $6.75 \, mg \, g^{-1}$ (Table 7.2). The presence of magnetite cores allowed the spent calcium alginate particles to be readily recovered.

7.2.7 Biological treatment and bioremediation

7.2.7.1 Introduction

Biological treatment of water refers to the use of living organisms (such as plants, fungi, or bacteria) or biological materials (e.g. bones, biomass, hair, seeds, leaves, or wood) to sorb or otherwise treat contaminants. As examples, human hair, produce and crop wastes, fungal biomass, algae, and chitosan may be used to sorb arsenic from water (Anirudhan and Unnithan, 2007; Hansen, Ribeiro and Mateus, 2006; Budinova *et al.*, 2006; Chen and Chung, 2006; Say, Yilmaz and Denizli, 2003b; Kumari *et al.*, 2006; Murugesan, Sathishkumar and Swaminathan, 2006; Ghimire *et al.*, 2003; Wasiuddin, Tango and Islam, 2002; Kamala *et al.*, 2005). Living bacteria, fungi, plants, and other biological organisms may also treat arsenic in surface waters, groundwater, soils, sediments, and wastewaters (e.g. Macy and Santini, 2002, 297, 299; Leblanc *et al.*, 2002). Furthermore, as mentioned earlier in this chapter, some bacteria can oxidize As(III). Once oxidized, the resulting As(V) could be treated with nonbiological methods, such as precipitation/coprecipitation or sorption (Mokashi and Paknikar, 2002). At the same time, biological treatment with fungi and bacteria must be carefully managed to avoid substantially *methylating* inorganic arsenic into highly toxic methylarsine (Gosio) gases (Chapters 4 and 5).

Bioremediation uses living organisms or biological materials to decontaminate sediments, soils, aquifers, groundwaters, and surface waters at field sites. The methods may be in or *ex situ*. For arsenic, bioremediation includes the use of organisms or biological materials to change redox, pH, or other ambient conditions so that arsenic is less mobile in the environment. Living organisms (such as plants and fungi) may also extract arsenic contaminants from soils, sediments, water, wood, and wastes. The plants or fungi may then be harvested to remove the arsenic from the site.

7.2.7.2   *Applications with bacteria and fungi*

Rather than using expensive and possibly deleterious oxidizing chemicals, Katsoyiannis *et al.* (2002) utilized bacteria and achieved As(III) removal levels of about 80 % in an arsenic-spiked groundwater. Fixed-bed upflow bioreactors were used to oxidize 2.8 mg L^{-1} of naturally dissolved iron, 0.6 mg L^{-1} of manganese, and possibly As(III) in the spiked groundwater. The bioreactors were most effective with water that had been spiked with 35–60 μg L^{-1} of As(III) (Katsoyiannis *et al.*, 2002, 329). The bacteria in the system (*Gallionella ferruginea* and *L. ochracea*) precipitated iron and manganese 'oxides' ((oxy)(hydr)oxides?). Most likely, the bulk of the arsenic was removed through bacterial oxidation to As(V) followed by coprecipitation with the iron (oxy)(hydr)oxides.

In other studies with arsenic-spiked groundwater, Katsoyiannis and Zouboulis (2004) were able to further optimize arsenic removal with *G. ferruginea* and *L. ochracea*. The As(III) concentrations in the groundwater were lowered from 200 μg L^{-1} to below 10 μg L^{-1} in the presence of Fe(II) and oxygen. They also confirmed that bacteria catalytically oxidized the Fe(II) to Fe(III) and As(III) to As(V). Oxidation produced iron (oxy)(hydr)oxide precipitates, which removed the As(V) from the water (Katsoyiannis and Zouboulis, 2004).

Rather than oxidizing As(III), some bacteria precipitate/coprecipitate, sorb, or otherwise directly remove As(III) from water. In their studies, Kostal *et al.* (2004) *overexpressed* the ArsR protein in *Escherichia coli* bacteria cells. The protein has a high affinity for As(III), which allowed the cells to remove nearly 100 % of 50 μg L^{-1} of As(III) from water (Kostal *et al.*, 2004; Table 7.1).

Acidithiobacillus ferrooxidans can remove As(III) from water without oxidizing it. Duquesne *et al.* (2003) concluded that this species of bacteria was probably responsible for coprecipitating As(III) from acid mine drainage at Carnoulès in southeastern France. The bacteria naturally lowered arsenic levels at the site by two to three orders of magnitude from initial concentrations that averaged 250 mg L^{-1} (Duquesne *et al.*, 2003, 6165). In laboratory studies, *A. ferrooxidans* was only able to remove arsenic from synthetic aqueous solutions if they were grown on Fe(II). Without oxidizing the arsenic, the bacteria coprecipitated As(III) with Fe(III), which was produced from their energy metabolism. The coprecipitated As(III) was primarily associated with poorly ordered schwertmannite (Chapter 3). Details on the association between As(III) and schwertmannite are unknown. Duquesne *et al.* (2003, 6171) speculated that $H_3AsO_3^0$ might be linked to FeO_6 octahedra in the schwertmannite structure.

Schwertmannite may also be associated with the biological precipitation and coprecipitation of As(V). Through field and laboratory investigations, Ohnuki *et al.* (2004) studied the coprecipitation of As(V) from pH 4.7 mine water by iron-rich bacterial colonies of *Gallionella sp.* They found that As(V) was associated with iron and sulfur in schwertmannite precipitates within the colonies.

Loukidou *et al.* (2003) investigated the removal of As(V) from wastewaters with chemically modified dead fungal (*P. chrysogenum*) biomasses (Mycan). To improve performance, the biomass was pretreated with a surfactant (either hexadecyltrimethylammonium bromide (HDTMA-Br) or dodecylamine) or a cationic polyelectrolyte (Magnafloc-463). In general, As(V) sorption increased through the lowering of pH, probably because of interactions between $H_2AsO_4^-$ and protonated *carboxyl* and amine groups in the biomass (Loukidou *et al.*, 2003, 4549). Compared with a sample of activated alumina (12.34 mg As(V) g^{-1} alumina), the modified fungus had a maximum sorption capacity of 57.85 mg As(V) g^{-1} using HDTMA Br, 56.08 mg As(V) g^{-1} with Magnafloc-463, and 33.31 mg As(V) g^{-1} with dodecylamine (Loukidou *et al.*, 2003, 4550; Table 7.2).

Kang, Kim and Kim (2007) investigated the removal of As(V) from water with a methylated microbial activated sludge. The sludge originated from a sewage treatment plant. After washing and drying, the sludge was methylated with methanol containing 0.1 N HCl (Kang, Kim and Kim, 2007, 314). The methylation process neutralized negative charges in the sludge, which allowed the sludge to more readily sorb As(V) oxyanions from pH 3–9 water. Methylating the sludge increased the sorption of As(V) from 23 up to $266\,\mu g\,g^{-1}$ dry mass.

7.2.7.3 *Phytoremediation and water treatment with plants*

Phytoremediation uses living plants, plant parts, or plant extracts to treat contaminated sites. Certain plants have the ability to *bioaccumulate* arsenic and detoxify their surroundings. Although inexpensive, phytoremediation is typically very slow and its success is often hindered by drought and pests (US EPA), 2002a, 1.2). Furthermore, during the long remediation process, chemical and physical conditions at a site may change and arsenic phytoremediation could become ineffective. Finally, to permanently remediate a site, the plants should be harvested and undergo treatment or disposal through incineration, gasification, *pyrolysis*, or contaminant extraction (Mulligan *et al.*, 2001, 202).

Growing plants may remove arsenic contaminants in soils, sediments, and water by either sorbing the contaminants on iron coatings on their roots or through bioaccumulation within the plant. In a greenhouse study, Hu *et al.* (2005) found that arsenic (probably as As(V)) sorbs onto the iron plaque of the roots of rice (*Oryza sativa* L.). The plaque typically consists of ferrihydrite, goethite, and minor amounts of siderite ($FeCO_3$) (Hu *et al.*, 2005, 170). The addition of phosphorus to the soils hindered the development of arsenic-sorbing plaque. While phosphorus often reduces the bioaccumulation of arsenic in plants, in this case, the reduced plaque allowed more arsenic to migrate into the plants.

Phytoremediation with living plants may be improved through genetic engineering and a thorough understanding of arsenic metabolism and detoxification in plants (Montes-Bayón *et al.*, 2004). As an initial step, Montes-Bayón *et al.* (2004) studied arsenic metabolism in *Brassica juncea* (Indian mustard). They found that some of the arsenic was associated with thiol groups in the plant.

The common Mediterranean aquatic plant, *Apium nodiflorum* (Vlyssides, Barampouti and Mai, 2005) and ferns, such as *Salvinia natans* (Mukherjee and Kumar, 2005) and *Pteris vittata* (Pickering *et al.*, 2006; Fitz *et al.*, 2003; Embrick *et al.*, 2005; Baldwin and Butcher, 2007; Wei *et al.*, 2007), have widespread applications in removing As(V) from water, sediments, and soils. In a study with living ferns, Huang *et al.* (2004) examined the removal of As(V) from water with three species (*P. vittata*, *Pteris cretica*, and *Nephrolepis exaltata*). The ferns were suspended in water with $20–500\,\mu g\,L^{-1}$ of radiolabeled $^{73}As(V)$. *P. vittata* removed 98.6 % of $200\,\mu g\,L^{-1}$ of arsenic within 24 hours. In the presence of $20\,\mu g\,L^{-1}$ of As(V), *P. vittata* lowered the arsenic concentration to $7.2\,\mu g\,L^{-1}$ in 6 hours and $0.4\,\mu g\,L^{-1}$ in 24 hours. Both *P. vittata* and *P. cretica* were capable of removing As(V) to below $10\,\mu g\,L^{-1}$ (Table 7.1). In contrast, *N. exaltata* was not very effective under the same conditions (Huang *et al.*, 2004). In a related study, Baldwin and Butcher (2007) found that arsenic preferentially accumulated in the stems and leaves of living *P.* cv *Mayii* when compared with its roots. Furthermore, living specimens of *P. vittata* showed no statistical differences in their ability to accumulate As(III) and As(V). The ferns accumulated about $1000\,\mu g\,g^{-1}$ of As(III) or As(V) from aqueous solutions containing $500\,\mu M$ ($37\,mg\,L^{-1}$) of arsenic (Baldwin and Butcher, 2007).

Living trees and flowering shrubs may remediate sites through the bioaccumulation of arsenic in their needles, leaves, and other body parts. Pratas *et al.* (2005) evaluated the accumulation of arsenic in plants at old mine sites in Portugal. They found elevated levels of arsenic in the old needles of *Pinus pinaster*, *Calluna vulgaris*, and *C. tridentatum* and in leaves from *C. ladanifer*, *Erica umbellate*, and *Quercus ilex* subsp. *ballota*. The accumulation of arsenic and trace metals in plants may also have important applications in ore prospecting (i.e. as *pathfinder elements*. Chapter 5). That is, elevated trace element concentrations in plants may indicate the presence of valuable ore deposits in surrounding rocks, soils, and sediments.

Considerable interest exists in using *wetlands* to treat arsenic-bearing wastewaters from electric utilities and other industries (Goodrich-Mahoney, 1996; Mitzman, 1999; Fox and Doner, 2003). In a recent study in Indiana, United States, Ye *et al.* (2003) used wetlands to remove arsenic from wastewaters at a *coal gasification* plant. About 51 % of the total arsenic entered the wetland sediments, approximately 17 % remained in the surface water, 2 % was taken up by plants, 28 % exited the system, and 2 % was unaccounted. Cattail, *Thalia sp.*, and rabbitfoot grass were highly tolerant of the arsenic, boron, and selenocyanate. They exhibited no evidence of growth retardation.

7.2.8   Natural remediation

At some field sites, natural chemical, physical, and biological processes are capable of suitably treating contaminants in waters, sediments, and soils without human intervention. In other words, the sites are 'self-cleaning'. *Natural remediation* refers to allowing already existing biological, chemical, and physical processes to biodegrade, precipitate, sorb, or otherwise reduce the toxicity and mobility of contaminants at a site with little or no human intervention. For arsenic, the most effective natural remediation processes would be sorption and precipitation/coprecipitation, including sorption onto iron (oxy)(hydr)oxides under oxidizing conditions and coprecipitation with sulfides under strongly reducing conditions (Reisinger, Burris and Hering, 2005, 459A, 461A).

As long as it is fast and effective, allowing nature to cleanup a site is a very attractive idea and can be economical. On the other hand, natural remediation is often too slow to meet immediate environmental and human health concerns. Furthermore, regulators will often demand conclusive scientific evidence that natural remediation is effective at a site, contaminants are not spreading further in the subsurface, and any cocontaminants are not increasing arsenic mobility and toxicity. Regulators could also require long-term monitoring of the sites to ensure that natural remediation continues. The costs of the long-term monitoring could easily exceed *ex situ* remediation methods that would permanently remove the contaminants. The public may also view natural remediation as irresponsibly 'walking away' from environmental problems (Reisinger, Burris and Hering, 2005, 463A).

On a case by case basis, regulators may permit natural remediation at a site. Currently, the US EPA is developing federal guidelines on natural remediation. For their contaminated sites, the US Department of Energy (DOE) has already adopted policies for using natural remediation (US DOE), 1999). Reisinger, Burris and Hering (2005) further discusses the monitoring and US regulatory requirements that must be met before adopting natural remediation at arsenic-contaminated sites.

7.3   Treatment technologies for arsenic in solids

7.3.1   Introduction

As discussed in Chapter 3, at least trace amounts of arsenic commonly occur in rocks, soils, sediments, sludges and spent sorbents from water treatment systems, coal ashes, industrial wastes, and many other natural and artificial solids. Depending upon whether they are considered regulatory hazards (Appendix E), solid materials may require treatment before disposal (waste management) or remediation if they are located at a contaminated site. For solids, arsenic treatment may involve reducing the arsenic concentrations in the materials so that they are no longer hazardous (for example, soil washing). However, because arsenic cannot be destroyed, eventually the element will require permanent disposal in a manner that does not pose a threat to human health or the environment. Most often, the ultimate solution involves changing the chemical and physical properties of the arsenic-bearing materials (e.g. solidification/stabilization) so that the arsenic is essentially immobile and can safely undergo long-term disposal. Depending on the regulations, materials may be landfilled after treatment. Otherwise, they would be sent to a hazardous waste disposal facility.

A large variety of solid materials may be treated with solidification/stabilization as part of waste management and site remediation programs. Other technologies for remediating and/or managing arsenic in solids include site isolation, physical separation, bioremediation, washing, heating methods, and electrical (*electrokinetic* and *electrodialytic*) methods. Some technologies (such as *in situ vitrification*) are specifically designed to treat large volumes of contaminated soils and sediments, and would have few, if any, applications in waste management.

7.3.2 Review of various treatment technologies for arsenic in inorganic solids

7.3.2.1 Site isolation

Landfills are best described as disposal options rather than treatment technologies. However, most landfills will eventually leak. *Site isolation technologies* may be used to segregate old landfills or other contaminated sites from their surroundings until funding or technologies become available to permanently remediate them. Isolation technologies use clay, grout, plastic, steel, cement, or other impermeable ($<1 \times 10^{-7}$ m s^{-1} fluid flow) caps and barriers to physically prevent water and contaminants from migrating between a landfill and its surrounding environment (Mulligan *et al.*, 2001, 193). In particular, caps on landfills and contaminated areas are designed to minimize the infiltration of rainwater, which can leach and mobilize contaminants. Vertical barriers or walls are often emplaced as slurries around contaminated sites. They should extend to impermeable bedrock to prevent pristine groundwaters from coming into contact with the contaminants (Mulligan *et al.*, 2001, 197). As mentioned in Section 7.2.5.1, PRBs containing zerovalent iron or other materials may be inserted into the subsurface to treat contaminated groundwaters (Mulligan *et al.*, 2001, 199).

7.3.2.2 Solidification/stabilization

Solidification/stabilization refers to reducing the mobility of a contaminant in soils, other solids, or even liquid wastes by mixing them with Portland cement, lime, cement kiln dust, clays, slags, polymers, water treatment sludges, iron-rich gypsum, coal *flyash*, and/or other *binders* (Leist, Casey and Caridi, 2000, 132; (US EPA), 2002a, 4.1; Mulligan *et al.*, 2001, 193; Mendonça *et al.*, 2006). *Solidification* specifically deals with the physical encapsulation of contaminants, whereas *stabilization* refers to the immobilization of contaminants through the formation of chemical bonds between the contaminants and the binders (Leist, Casey and Caridi, 2000, 132–133; Mulligan *et al.*, 2001, 193).

The physical processes of solidification and the chemistry in stabilization are both important in reducing the leaching of arsenic and other hazardous contaminants. Solidification typically decreases the surface areas and permeability of the contaminated materials, thereby hindering the leaching and migration of contaminants from the wastes and into the environment. Stabilization may involve altering the chemistry of the arsenic species so that they are less toxic, water soluble, or otherwise mobile (US EPA), 2002a, 1.3). If the solidified/stabilized materials are identified as nonhazardous by leaching (e.g. the TCLP in the United States) and other regulatory tests, they are usually landfilled.

In situ solidification/stabilization refers to the injection and mixing of binders into contaminated soils and sediments. Although the process can be effective, obstructions, deep contamination, shallow water tables, rocks, and coarse particles usually hinder the process. Furthermore, *in situ* solidification/stabilization is often too difficult to evaluate for short- and long-term effectiveness. As alternatives, contaminated soils and sediments are usually excavated, treated, and disposed of on-site or transported to a landfill (Mulligan *et al.*, 2001, 197). Typically, cement or coal fly ashes are used to solidify and stabilize arsenic-contaminated soils and sediments. The effectiveness of the solidification/stabilization techniques are evaluated with laboratory leaching tests and field monitoring of disposal sites. The exact costs of solidification/stabilization largely depend on the speciation of the arsenic, the volume of the wastes, the types of binders that are used, any required waste pretreatment, and the chemical and physical properties of the solidified/stabilized materials and its disposal environment (US EPA), 2002a, 4.2, 4.4).

Prior to solidification/stabilization, a waste may undergo pretreatment to increase the effectiveness of the binder and decrease long-term leachability. To maximize the solidification/stabilization of arsenic-bearing wastes, any inorganic As(III) is usually preoxidized (Jing, Liu and Meng, 2005c, 1242). Other pretreatment procedures may include removing organic or other contaminants that might undermine the long-term stability of the binders, adjusting pH, or otherwise modifying the chemical and physical properties of the waste so that its arsenic is in less water-soluble and more stable forms (US EPA), 2002a, 4.2).

When binders are selected, they must be compatible with the waste. Considering that the chemistry of wastes tends to be highly variable and arsenic may occur in a variety of forms, no single solidification/stabilization recipe will work in all situations (Leist, Casey and Caridi, 2000, 134–135). Although Portland cement is often used in binding arsenic-bearing solid wastes (US EPA), 2002a, 4.1), this binder is not always effective or efficient. In particular, Fernández-Jiménez *et al.* (2005) evaluated alkali-activated metakaolin and flyash matrices as possibly more effective alternatives to Portland cement. Miller *et al.* (2000) found that the effectiveness of Type I Portland cement as a binder for arsenic-contaminated sandy soils improved if the soils were treated with $FeSO_4 \cdot 7H_2O$ before mixing with the cement. However, the addition of arsenic-bearing Fe(III) (oxy)(hydr)oxide sludges to fresh cement may extend curing times and inhibit solidification/stabilization (Leist, Casey and Caridi, 2000, 134; Phenrat, Marhaba and Rachakornkij, 2007). Modifying the composition of the cement with flyash, lime, or slag will not always improve solidification/stabilization (Leist, Casey and Caridi, 2000, 132–133). Furthermore, the use of flyash and cement mixtures usually increases the volume of the wastes much more than solidification/stabilization with silicate or metal hydroxides (Leist, Casey and Caridi, 2000, 133). Higher volumes would result in higher landfilling costs. Due to these concerns and limitations, the US EPA only recommends the solidification/stabilization of arsenic-contaminated materials on a case by case basis (US EPA), 1999, C.1; Akhter *et al.*, 2000).

With solidification/stabilization using cement, arsenic is probably immobilized by sorbing onto the surfaces of compounds, replacing sulfate in ettringite ($Ca_6Al_2(SO_4)_3(OH)_{12} \cdot 26H_2O$), and/or reacting with calcium to form sparsely soluble arsenates and arsenites (Jing, Liu and Meng, 2005c, 1242; Phenrat, Marhaba and Rachakornkij, 2005). The long-term stability of the solidified/stabilized arsenic would depend on several factors, including arsenic speciation, the pH and redox properties of the mixture and disposal environment, the presence of any organic compounds that might weaken the binder, the long-term compatibility of the chemicals in the binders and the waste, the presence of fine-grained particulates that could weaken bonds between waste and binder particles, how the binders were prepared, and how thoroughly the binders were mixed with the waste (US EPA), 2002a, 4.2).

Jing, Korfiatis and Meng (2003) investigated the behavior of As(V) in an iron (oxy)hydroxide sludge that had been stabilized with cement. The sludge was from a groundwater remediation site in Tacoma, Washington, United States. EXAFS analyses indicated that As(V) formed *bidentate–mononuclear complexes* on the surfaces of the iron (oxy)hydroxides (Chapter 2). However, once encapsulated in cement, highly alkaline (pH ~13) porewaters tended to desorb As(V) from the iron (oxy)hydroxides (Phenrat, Marhaba and

Rachakornkij, 2007, 602–603). The released As(V) then reacted with calcium to form calcium arsenates. As previously mentioned, the long-term stability of calcium arsenates depends on ambient conditions. In the presence of carbon dioxide from air or carbonate in water, calcium arsenates may decompose over time and release arsenic (Robins and Tozawa, 1982; Ghimire *et al.*, 2003, 4946; Jing, Korfiatis and Meng, 2003, 5055–5056).

Once a waste is solidified/stabilized, it must pass regulatory tests (such as the TCLP in the United States, Appendix E) before it can be classified as nonhazardous and undergo disposal in secured landfills (Leist, Casey and Caridi, 2000, 131). The TCLP uses acetic acid ± sodium acetate buffering solutions, which are designed to model leaching from rotting garbage in municipal waste landfills. However, because solidified/stabilized arsenic wastes are unlikely to undergo disposal in municipal landfills, researchers often use other types of leaching procedures to better predict the release of arsenic from groundwater interactions with the wastes. In particular, Leist, Casey and Caridi (2003) used *sequential batch leaching* tests to evaluate the ability of cement and other compounds to successfully encapsulate sodium arsenite, sodium arsenate, arsenic trioxide, and arsenic pentoxide. The dosages of the arsenic compounds in the binders were about 10 %. Calcium-rich stabilization mixtures were most effective in hindering the leaching of arsenic. Although Fe(II) frequently enhances the solidification/stabilization of arsenic (Leist, Casey and Caridi, 2000, 134), Leist, Casey and Caridi (2003, 359) found that the addition of Fe(II) to their cement mixtures failed to improve the solidification/stabilization of the contaminant.

Wastes from copper mines and smelters often contain very high concentrations of arsenic. Using fly-ashes from a copper refinery, which contained 23–47 wt % As(III), Dutré and Vandecasteele (1998) found that mixing the ashes with cement and lime reduced their leachability with the German DIN (Deutsches Institut für Normung) 38 414 *batch leaching* test from approximately $5 \, g \, L^{-1}$ to about $5 \, mg \, L^{-1}$. The DIN test consists of placing 100 g of dried waste in 1 L of distilled and deionized water, and then shaking the mixture for 24 hours at room temperature. The relatively low As(III) concentrations in some of the DIN leachates (approximate pH of 12.5) of the solidified/stabilized wastes are explained by the precipitation of calcium hydrogen arsenite ($CaHAsO_3$) in the presence of portlandite (Dutré and Vandecasteele, 1998).

Shih and Lin (2003) investigated the solidification/stabilization of arsenic-rich flyash from an abandoned copper smelter in northern Taiwan. The flyashes (2–40 % total arsenic, mostly as As(III)) were collected from three *flue gas* discharge tunnels. Extremely high cement dosages (cement/waste mass ratio of greater than 6) were required to stabilize the wastes so that they would pass the US TCLP for arsenic ($<5 \, mg \, L^{-1}$; Appendix E). (The TCLP is often used in research outside of the United States.) Cement dosages could be reduced and the mixtures would still pass the TCLP for both arsenic and lead if municipal waste incinerator flyash was added. Lime alone was able to stabilize arsenic and pass the TCLP; however, the leachates exceeded the TCLP lead standard of $5 \, mg \, L^{-1}$. The immobilization of arsenic in lime may be due to the formation of sparsely water-soluble calcium arsenites and arsenates, such as: $CaHAsO_3 \cdot nH_2O$ or $Ca_3(AsO_4)_2 \cdot nH_2O$, where $n \geq 0$ (Shih and Lin, 2003, 692).

7.3.2.3   *Physical separation*

In some sediments, soils, and granular wastes, contaminants may occur in specific size fractions or sorb onto high-surface area fine-grained particles. The volume of contaminated materials may be significantly reduced by physically separating or *sorting* the contaminated size fractions from the cleaner matrix. Physical separation may include sieving, flotation in water or other liquids, gravity settling in liquids, water washing, or magnetic separation (US EPA), 2002a, 1.3, 6.1; Mulligan *et al.*, 2001, 198).

7.3.2.4 Washing

Washing consists of using aqueous solutions or hot fluids to remove contaminants or reduce their concentrations in solid materials. The process is one of the more common physical separation and leaching methods for treating granular solid wastes and contaminated soils and sediments. Washing is also used to remove arsenic and other contaminants from ore deposits prior to smelting and from coal before combustion (Mihajlovic *et al.*, 2007; Chapter 5). Chemicals that can promote the leaching or extraction of arsenic from solid materials include oxidants, EDTA (ethylenediaminetetracetic acid; Lee *et al.*, 2004; Englehardt *et al.*, 2007), supercritical carbon dioxide (Wang and Guan, 2005), solutions containing citric acid ($C_6H_8O_7$) and KH_2PO_4 (Lee *et al.*, 2007), or a series of washes with HCl and NaOH (Jang, Hwang and Choi, 2007b). In soils, arsenic is often associated with humic materials, which are more soluble in highly alkaline solutions than acids (Sullivan, Bowman and Legiec, 2003, 2388). Chemical similarities also allow phosphate solutions to desorb As(V) from mineral surfaces and soils (Hering and Kneebone, 2002, 166; Stollenwerk, 2003, 85-86, 91; Lee *et al.*, 2007; Jackson and Miller, 2000).

Kahakachchi, Uden and Tyson (2004) investigated the ability of various liquids to extract As(III), As(V), DMA(V), and MMA(V) from spiked soils. The extractants included deionized water, a citrate buffer, an ammonium dihydrogen phosphate buffer, 1 M phosphoric acid, 5 % acetic acid, household vinegar, 0.1 M NaOH, and even Coca Cola®. After eight days, the highest extractions for As(III), MMA(V), and As(V) were achieved with NaOH at 46, 100, and 84 %, respectively. A 10 mM citrate buffer was most effective with DMA(V) with about 85 % removal after eight days.

Contaminated materials may be washed in vats or the solutions may be applied *in situ* with sprinklers, trenches, and drains (Mulligan *et al.*, 2001, 202). Soil flushing is an *in situ* washing technology, which removes arsenic and other contaminants from soils and sediments without excavating them (US EPA), 2002a, 1.3). The technology consists of spraying leaching solutions onto contaminated soils and sediments or injecting treatment solutions into the subsurface through wells. The leaching solutions may include water, acids, NaOH, organic solutions, or other liquids. The contaminants are leached from the soils and sediments through dissolution or *emulsification* (US EPA), 2002a, 8.1).

Effective washing requires an appropriate understanding of the environment surrounding the contaminated materials and the physical, geological, and chemical properties of the materials, especially the speciation of the arsenic. Specifically, soil flushing is most effective with homogeneous soils and sediments that contain few or no other contaminants besides arsenic. That is, soils and sediments that contain large rocks, a high water table, and complex mixtures of organic contaminants and metals may not be suitable for *in situ* washing methods. Weather is another factor that affects both *in situ* and *ex situ* washing of soils and sediments. Cold temperatures may freeze *in situ* leaching solutions and frozen soils and sediments may be difficult to excavate for *ex situ* methods (US EPA), 2002a, 6.2).

Serious environmental and technical problems may result from the improper application of flushing solutions. If chemicals precipitate in soils and sediments, they may clog pores, decrease soil/sediment permeability, and reduce the effectiveness of the treatment technique. Furthermore, hazardous solutions may leak from the treatment site and contaminate valuable groundwaters in surrounding areas. To prevent the spread of contaminants, barriers are used to control the subsurface flow of the flushing solutions (US EPA), 2002a, 8.1).

Recovery of flushing solutions may involve pumping and treating or capturing the spent solutions in trenches around the site. The recovered liquids usually require treatment, which may involve precipitation/coprecipitation, ion-exchange resins, or sorption. Like many other treatment technologies, the treatment of washing solutions ultimately transfers arsenic into a smaller and more manageable volume.

In treating spent washing solutions, zeolite sorbents are often economically competitive with activated carbon and ion-exchange resins (Sullivan, Bowman and Legiec, 2003, 2387). Sullivan, Bowman and Legiec (2003) investigated the use of surfactant-modified zeolites to remove $500\,mg\,L^{-1}$ of As(V) (as As_2O_5) from humate-rich pH 12 soil washing solutions. The zeolite consisted of a New Mexico clinoptilolite treated with HDTMA-Br. The batch sorption tests used liquid to zeolite ratios of 40:1 to 4:1. A maximum sorption capacity of $72.0\,mmol$ of As(V) kg^{-1} of zeolite ($5.4\,mg\,As(V)\,g^{-1}$ zeolite) occurred at a temperature of $25\,°C$ (Sullivan, Bowman and Legiec, 2003; Table 7.2).

7.3.2.5 *Biological treatment*

Like water, arsenic in contaminated soils, sediments, and even solid wastes may be treated with plants, fungi, bacteria, or other biological organisms. The applications, limitations, and advantages of biological treatment methods with solid materials are often similar to those with water. To be exact, many bioremediation methods are designed to simultaneously treat contaminants in soils, sediments, and water (e.g. phytoremediation).

Bacteria may be used to extract or immobilize arsenic in solid materials. In bioleaching, bacteria in aqueous solutions extract or flush arsenic from contaminated soils, sediments, and solid wastes. Like chemical washing, bioleaching solutions may be mixed with the solids or applied *in situ* onto contaminated soils and sediments. Bacteria may also be involved in producing sulfides and immobilizing arsenic in very reducing subsurface environments (Chapter 3). The arsenic may sorb or coprecipitate with Fe(II) sulfides (Wolthers *et al.*, 2005) or precipitate as arsenic sulfides (such as amorphous As_2S_3, orpiment (As_2S_3), and realgar (As_4S_4)) (Köber *et al.*, 2005a, 7650).

7.3.2.6 *Pyrometallurgical treatment (arsenic volatilization and recovery)*

Pyrometallurgical treatment uses heat from incinerators or furnaces to extract or concentrate metals and other inorganic contaminants from soils, sediments, or solid wastes. Common incinerators for arsenic-contaminated materials include rotary kilns, arc furnaces, and fluidized bed systems (Mulligan *et al.*, 2001, 199). Pyrometallurgical methods are primarily used with solid materials that contain exceptionally high concentrations (usually wt % levels) of inorganic contaminants (US EPA), 2002a, 1.3). Depending on the volume of any contaminated soils and sediments, the materials may be excavated and transported to an incinerator, or mobile on-site incinerators may be used. Most pyrometallurgical technologies essentially treat contaminated geologic materials and solid wastes as ore deposits.

Pyrometallurgical technologies volatilize arsenic from solid materials. These technologies are often part of ore smelting and refining waste management operations (US EPA), 2002a, 7.1). Prior to heating, *reductants* or fluxing agents are usually mixed with the solid materials to promote melting and arsenic volatilization (Mulligan *et al.*, 2001, 198). The volatilized arsenic is then captured either by filtration or *scrubbing* (Smith *et al.*, 1995, 98). Although the recovered arsenic might be sold, few, if any, remediation and waste management operations recover any substantial costs by selling captured volatile contaminants (such as arsenic, lead, or mercury). In 2006, the average price of $\geq 99\,\%$ As(0) was only $US 1.21/kg (Hetherington *et al.*, 2008, 147; Chapter 5). Instead, most of the recovered arsenic simply undergoes disposal, usually through solidification/stabilization in a volume that is much smaller than the original waste.

7.3.2.7 *Vitrification*

Vitrification refers to the melting of soils, sediments, and solid wastes to primarily incinerate organic contaminants and encapsulate arsenic and other inorganic species into melts. The melts then cool into impermeable and chemically resistant glass. Unlike pyrometallurgical technologies, vitrification attempts to minimize the volatilization of arsenic by incorporating as much of it as possible into the slags. To reduce volatilization, the wastes may be pretreated to transform the arsenic into less volatile compounds. Specifically, arsenic in flue dust or other solid wastes may be stabilized by heating them in the presence of lime and air, which produces less volatile calcium arsenates and arsenites (Leist, Casey and Caridi, 2000, 132; (US EPA), 2002a, 5.1). The calcium arsenates and arsenites are then encapsulated into molten iron silicates or other melts (Leist, Casey and Caridi, 2000, 132). Despite problems with volatilization, the US EPA considers vitrification to be the Best Demonstrated Available Technology (*BDAT*) for treating arsenic in soils (US EPA), 1999, Chapter 1).

Vitrification technologies may be *ex situ* or *in situ*. In *ex situ* vitrification, any soils, sediments, or buried wastes are excavated. After possible pretreatment to reduce arsenic volatilization, the materials are placed in a furnace and melted at temperatures as high as 2000 °C. The heat may be generated by fossil fuels, electricity, plasma torches, or microwaves (US EPA), 2002a, 5.1).

In situ vitrification refers to the use of electrical currents to melt, incinerate, or encapsulate contaminants in unexcavated soils and sediments. The method consists of placing electrodes into the contaminated materials (Mulligan *et al.*, 2001, 198). As the electricity is applied, the electrodes sink up to 6 m into the melting soils and sediments (US EPA), 2002a, 5.1). Temperatures may reach 1600–2000 °C (Buelt, Timmerman and Westsik, 1989). As much as 1000 tons of material may be melted at once with the technology (US EPA), 2002a, 5.1). The melt eventually cools into a glass, which encapsulates most of the arsenic. Any volatilized arsenic may be captured in collection hoods (Buelt, Timmerman and Westsik, 1989). Nevertheless, because *in situ* vitrification is energy intensive and expensive, its use with arsenic-rich materials has been limited (US EPA), 2002a, 3.1).

Before vitrification technologies can be applied, the mineralogical, physical, and chemical properties of the geologic materials must be well understood, especially the concentrations and speciation of the contaminants. Specifically, chlorinated organics can produce corrosive flue gases and even toxic dioxin fumes that are difficult to treat (US EPA), 2002a, 5.2). The moisture content of the materials must also be considered. Excess steam can overwhelm the vitrification process (US EPA), 2002a, 5.3). The technology is also most effective with silicate-rich soils and sediments. High alkali and low silicate soils and other materials will not produce a chemically inert glass. Furthermore, *in situ* vitrification is not suitable for contaminated materials that have large volumes of coarse particles, extremely high concentrations (>15 wt %) of metals, subsurface air pockets, high water tables, or buried drums of volatile liquids (US EPA), 2002a, 5.3).

Highly arsenic-contaminated soils and sediments may not always produce inert slags. The solubility limit of arsenic in silicate glass is only about 1–3 wt % (US EPA), 2002a, 5.2). Furthermore, excessive water, chloride, fluoride, sulfide, and sulfate in the slags may increase the leachability of arsenic. The addition of silicate sand during melting may counteract these effects (US EPA), 2002a, 5.2, 5.3).

US EPA (2002a, 5.2) summarizes an *in situ* vitrification project at the Parsons Chemical Superfund Site in Grand Ledge, Michigan, United States. The soil at the old agrichemical facility contained 8.4–10.1 mg kg^{-1} of arsenic. A total of 3000 cubic yards of soil was treated in eight melts. Like other encapsulation technologies, the TCLP (Appendix E) was used to evaluate the effectiveness of the vitrification technology. TCLP leachates of the vitrified materials contained <0.004–0.0305 mg L^{-1} of arsenic, which is far below the 5 mg L^{-1} TCLP limit. Volatilized arsenic was <0.000269 mg m^{-3} or <0.59 mg released per hour (US EPA), 2002a, 5.2).

7.3.2.8 Electrokinetic methods

Electrokinetic methods refer to *in situ* and, in some cases, *ex situ* technologies that remove contaminants from wet soils, sediments, or other solid materials by passing electric currents through them. Unlike *in situ* vitrification, the currents in electrokinetic methods are too low to melt the materials. Instead, the electric currents cause contaminant ions and charged particles in aqueous solutions within solid materials to migrate toward electrodes, where they may be collected or otherwise treated (Mulligan *et al.*, 2001, 193, 199–200). Several processes are important in removing contaminants with electrokinetic methods, including electromigration (movement of charged chemicals), electroosmosis (movement of fluids), electrophoresis (movement of charged particles), and electrolysis (chemical reactions resulting from electrical fields) (Mulligan *et al.*, 2001, 200).

Electrokinetic methods are often *in situ* and involve inserting the electrodes directly into contaminated soils and sediments. When activated, the electrodes dissociate water, which produces oxidizing conditions and an acid front (perhaps at pH < 2) at the positively charged anode (Acar and Alshawabkeh, 1993, 2638):

$$2H_2O - 4e^- \rightarrow O_2 + 4H^+ \tag{7.8}$$

Anions, including arsenic oxyanions and OH^-, migrate toward the anode. The negatively charged cathode attracts metal cations and creates reducing conditions and an alkaline (perhaps pH > 12) front in the surrounding waters (Acar and Alshawabkeh, 1993, 2638):

$$2H_2O + 2e^- \rightarrow H_2 + 2OH^- \tag{7.9}$$

Depending on the properties of the contaminant, common collection methods at the electrodes include electroplating, precipitation/coprecipitation, sorption, and ion exchange. Water around the electrodes in *in situ* applications may also be pumped and treated (US EPA), 2002a, 1.3; Mulligan *et al.*, 2001, 200). Unlike soil washing, *in situ* electrokinetic methods are often very effective with clays and other low-permeability, fine-grained materials (US EPA, 2002a, 1.2; Smith *et al.*, 1995, 118; Cundy and Hopkinson, 2005). In the presence of electric fields, the lower permeability materials may allow ion diffusion of the contaminants (Smith *et al.*, 1995, 118). On the other hand, buried metal objects, building foundations, rock formations, and other obstacles may interfere with the effectiveness of *in situ* electrokinetic methods (Mulligan *et al.*, 2001, 200). Many of these electrokinetic techniques are also energy intensive and complicated (Cundy and Hopkinson, 2005), which hinders their commercialization.

Kim, Kim and Kim (2005) evaluated the treatment of arsenic in two fine-grained soils with an *ex situ* electrokinetic technology, which consisted placing the samples in a three-compartment chamber with a platinum anode and titanium cathode on opposite ends. One soil consisting of a Korean kaolinite was spiked with 1500 mg kg^{-1} of As(V). The second soil sample, which contained 3210 mg kg^{-1} of arsenic, was collected from the abandoned Myungbong gold mine in southern South Korea. The soils were treated with NaOH or KH$_2$PO$_4$ electrolyte solutions. Deionized water was used with control samples to establish a baseline.

KH$_2$PO$_4$ was most effective in removing arsenic from the kaolinite (88.75 % removal). The removal was probably due to ion exchange between the phosphate and As(V). In contrast, NaOH was best in removing arsenic from the mine soil (65.67 % removal). In the mine soil, the oxidation of pyrite had produced iron (oxy)(hydr)oxides. The arsenic in the soil probably sorbed onto the iron (oxy)(hydr)oxides or coprecipitated with them (Kim, Kim and Kim, 2005, 451). The addition of NaOH increased the pH of the mine soil, which may have mobilized the arsenic through dissolution of the iron (oxy)(hydr)oxides. Kim, Kim and Kim (2005, 452) concluded that optimizing arsenic removal with the electrokinetic technology requires information on the properties of the soils, including pH, arsenic speciation, and electrical properties.

7.3.2.9 *Electrodialytic methods*

Electrodialytic methods are *ex situ* procedures that combine electrokinetic processes with electrodialysis membranes to remove contaminants from wet solids and aqueous solutions (Pedersen *et al.*, 2005, 332). The methods were originally developed to remediate copper and other heavy metals in contaminated soils (Ottosen *et al.*, 1997; Ottosen, Hansen and Hansen, 2000). However, in recent years, electrodialytic methods have been modified to also remove arsenic from ash, water, and treated wood (Wang, Bejan and Bunce, 2003; Christensen *et al.*, 2004; Ottosen, Pedersen and Christensen, 2004). The procedure uses an electrodialytic cell, which is typically divided into three compartments (Figure 7.2). The chambers on the ends of the cell are the electrode (anode and cathode) compartments, while samples are placed in the middle compartment. In the middle sample compartment, solids are typically mixed with an electrolytic solution, which may be water, dilute oxalic acid (H$_2$C$_2$O$_4$), or 0.01 M sodium nitrate (NaNO$_3$). An anion-exchange membrane separates the anode compartment from the sample chamber. On the other end of the cell, a cation-exchange membrane is located between the sample chamber and the cathode compartment (Figure 7.2). The electrode compartments are filled with circulating solutions, such as 0.01 M NaNO$_3$ (Christensen *et al.*, 2004, 232). During sample treatment, direct electric current causes contaminants to collect on the ion-exchange membranes (Christensen *et al.*, 2004). The membranes may also allow

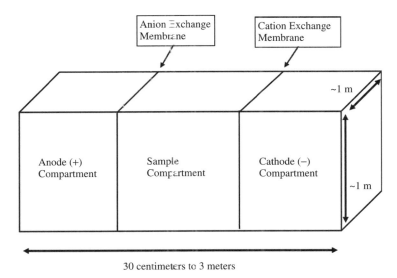

Figure 7.2 *Schematic diagram of an electrodialytic system for treating arsenic and other contaminants in solid samples (based on descriptions in Christensen et a l., (2004)).*

arsenic and other contaminants to pass from the sample into the electrode compartments, where they can be collected. However, the ion-exchange membranes prevent contaminants from freely migrating between the different compartments (Christensen *et al.*, 2004, 229).

Ottosen, Pedersen and Christensen (2004) investigated the removal of arsenic from ashes with an electrodialytic method. The ashes resulted from the incineration of wood that had been treated with CCA preservatives. Using the electrodialytic method over five days, Ottosen, Pedersen and Christensen (2004, 300) were able to extract about 92 % of the arsenic from a 40-g mixture of *bottom ash* and flyash. The ash mixture initially contained 35 000 mg kg^{-1} of arsenic (Ottosen, Pedersen and Christensen, 2004, 300). Arsenic was found in both the anode and cathode compartments. In the cathode compartment, arsenic was associated with copper (Ottosen, Pedersen and Christensen, 2004, 307).

7.3.3   Review of various treatment technologies for chromated copper arsenate (CCA)-treated wood

7.3.3.1   *Introduction*

CCA is a group of wood preservatives that were first developed in India in the 1930s (Morrell and Huffman, 2004, 119; Chapter 5). Since the 1950s, the preservatives have been widely used to protect wood from microorganisms, fungi, wood-feeding insects, and marine borers (Pedersen *et al.*, 2005, 332; Cox, 1991, 2). By 1993, about 80 % of US lumber were treated with CCA (Sarahney, Wang and Alshawabkeh, 2005, 642). However, in recent years, there have been concerns over the toxicity and environmental impacts of arsenic, chromium, and copper from the preservatives. Norway and Denmark have even banned CCA wood preservatives (Ottosen, Pedersen and Christensen, 2004, 296) and bans are expected in additional countries in the near future. In the United States, a voluntary agreement between the US EPA and manufacturers discontinued the use of CCA in virtually all consumer products after 2003 (Wasson *et al.*, 2005, 8865; Townsend *et al.*, 2003, 781). Even with the implementation of bans, the life expectancy of CCA-treated wood is at least 30 years and the preservatives are common in fenceposts, playground equipment, decking, and other outdoor wood in many countries (Christensen *et al.*, 2004, 228; Cox, 1991, 2). The challenge is then to find suitable alternative preservatives and develop effective and economical technologies that will continue to deal with CCA-treated wood wastes for many decades to come (Illman and Yang, 2004, 259; Helsen and Van den Bulck, 2005).

Although TCLP leachates of CCA-treated wood often exceed the 5 mg L^{-1} limit for arsenic (Townsend, Dubey and Solo-Gabriele, 2004, 171–172; Stook *et al.*, 2005), spent CCA wood is normally exempted from being a hazardous waste under US federal regulations (40 Code of Federal Regulations (*CFR*) 261.4(b)(9); Appendix E). However, some US states (such as Minnesota) have adopted stricter disposal regulations and individual landfill operators may not accept the material (Oskoui, 2004, 241). Depending upon state and local regulations, CCA-treated wood is often landfilled with construction and demolition wastes in the United States (Townsend, Dubey and Solo-Gabriele, 2004; Oskoui, 2004, 241). Although there are concerns about arsenic leaching from landfills and contaminating groundwater, Saxe *et al.* (2007) found no evidence of significant arsenic migrating from unlined construction and demolition landfills in Florida, United States, through 2003, despite the wet climate and that approximately 13 million kg of arsenic in CCA-treated wood had been landfilled in the state by the year 2000.

The volume and toxicity of CCA-treated wood wastes will probably begin to curtail landfilling in the near future and promote the development of alternative waste management technologies that involve incineration, biological treatment, and/or extraction. Helsen and Van den Bulck (2005) provide a concise review of landfilling alternatives for CCA-treated wood, especially thermal methods. Most of the alternatives are still undergoing commercialization and their future marketability will greatly depend on whether any local landfill operators are still able and willing to accept CCA-treat materials and, if so, the landfilling costs. The following sections discuss some of the landfill alternatives for treating and disposing of CCA wood waste.

7.3.3.2   Solidification/stabilization

In some circumstances, CCA-treated wood may be mixed with cement and utilized in construction (Gong Kamdem and Harcihandran, 2004). Wood-cement composites have compressive strengths similar to concrete. However, the composites are far more capable of sustaining plastic deformation than concrete. Therefore, they may be useful in circumstances where compressive strength and energy dissipation are required (Gong, Kamdem and Harcihandran, 2004, 276).

7.3.3.3   Extraction

CCA-treated wood may be detoxified with acidic, organic, or other extracting solutions, provided that the use and disposal of the solutions are cost effective (Helsen and Van den Bulck, 2004, 281; Kakitani *et al.*, 2007). Aclean is an example of a proprietary organic and biodegradable solution or 'lixivent' that can extract 88–96 % of the arsenic from CCA-treated wood (Oskoui, 2004). The wood is ground or chipped to about 10–20 mm and placed in a reaction tank. Typically, the water to solids ratio is 0.25:1. Heating to 40 °C increases the effectiveness of the lixivent (Oskoui, 2004, 241). Mixing and extraction continues for 2 hours. Afterwards, the wood is allowed to settle for at least 1 hour. In their studies, the spent lixivent contained up to $3000 \, \mathrm{mg \, L^{-1}}$ of arsenic, $2000 \, \mathrm{mg \, L^{-1}}$ of copper, and $3000 \, \mathrm{mg \, L^{-1}}$ of chromium (Oskoui, 2004, 241). A screw press removed any remaining liquid from the wood and produced a cake. The Aclean-treated wood passed the US TCLP (an average of $0.737 \, \mathrm{mg \, L^{-1}}$ of arsenic). Therefore, the material could be landfilled or possibly reused in pulp for paper or materials for composites (Oskoui, 2004). The CCA was removed from the extracting solutions with a zirconium-based ion-exchange system (Oskoui, 2004, 241). Like other waste management options for CCA-treated wood, the cost effectiveness of the Aclean technology would strongly depend on whether or not inexpensive landfills are locally available and if they are willing to accept wood waste containing CCA (Oskoui, 2004, 241).

Kartal and Kose (2003) investigated the removal of arsenic from CCA-treated wood with common chelating agents (EDTA, nitrilotriacetic acid ($C_5H_9NO_6$), and oxalic acid). Dual mixtures of EDTA and oxalic acid and nitrilotriacetic and oxalic acids removed about 100 % of the arsenic from sawdust of the wood. However, only about 80 % of the arsenic was removed from wood chips with the same dual solutions. Once the arsenic, copper, chromium, and any other toxins have been removed from the wood, it could be used as a biofuel.

7.3.3.4   Biological treatment

Although CCA preservatives are designed to protect wood from fungi and other organisms, several researchers are developing strains that are resistant to CCA so that they could be used to detoxify wood wastes. Specifically, Illman and Yang (2004) isolated and cultured several CCA-resistant fungi, including two strains of *Meruliporia incrassata* and two of *Antrodia radiculosa*. To completely remove arsenic and metals, the new strains must be able to thoroughly penetrate the interior of the wood. The development of CCA-resistant organisms also has raised concerns that natural fungi and other organisms might eventually develop resistances to wood preservatives (Illman and Yang, 2004, 260).

Kartal *et al.* (2004) also investigated the use of fungi to remove arsenic from CCA-treated wood. Three brown-rot fungi (*Fomitopsis palustris*, *Coniophora puteana*, and *Laetiporus sulphureus*) were initially cultivated in an oxalic acid broth. When placed on sawdust of CCA-treated wood, *F. palustris* and *L. sulphureus* removed 100 and 85 %, respectively, of the arsenic in 10 days. *C. puteana* was far less effective and only removed 18 % of the arsenic. *F. palustris* and *L. sulphureus* were able to accumulate more oxalic acid than *C. puteana*, which allowed these fungi to release more of the acid onto the wood substrate and extract more of the arsenic.

Using a chemical extracting solution and bacteria, Clausen and Kenealy (2004) were able to removed 81–97 % of the arsenic from CCA-coated samples of southern yellow pine. The wood was initially extracted for 18 hours with a solution of 0.8 % (pH 1.52) oxalic acid. The extraction was then followed by seven to nine days of bioleaching with *Bacillus licheniformis*. The method was most effective with flaked wood because it is thin and has a relatively high surface area (Clausen and Kenealy, 2004, 79). Currently, this chemical extraction and biological leaching method is not cost effective (Clausen and Kenealy, 2004, 72, 78–79). However, Clausen (2004) found that recycling the oxalic acid could somewhat reduce the costs. Furthermore, future bans on the landfilling of CCA-treated wood may improve the marketability of this method.

7.3.3.5 Electrodialytic and electrokinetic methods

Incineration of CCA-treated wood is not allowed in Denmark. As an alternative, a research group at the Technical University of Denmark developed an electrodialytic method to remove arsenic, copper, and chromium from wood (Christensen *et al.*, 2004; Figure 7.2; Pedersen *et al.*, 2005). The method is especially effective in removing >95 % of the arsenic, but costs may be high (Helsen and Van den Bulck, 2004, 282).

A pilot-scale system was developed to detoxify 100 kg samples of CCA-treated wood (Christensen *et al.*, 2004). The process consists of placing wood chips within the sample compartment (Figure 7.2). The sample compartment contains an electrolytic solution, which may be water, dilute oxalic acid, or 0.01 M sodium nitrate. The electrode compartments are filled with circulating 0.01 M sodium nitrate (Christensen *et al.*, 2004, 232). During operation, ion-exchange membranes allow the arsenic and metals to pass from the sample into the electrode compartments, where they may be collected.

In the study described in Christensen *et al.* (2004, 230), the arsenic content of the wood before electro-dialysis was 837 ± 114 mg kg^{-1} (95 % confidence level) based on 95 samples. The electrodialytic process was more effective if the wood was first broken down into <2 cm chips and soaked in phosphoric acid followed by oxalic acid (Christensen *et al.*, 2004, 236). The soaking probably leaches a significant amount of the arsenic and metals from the wood, which allows the electrodialytic process to remove most of the remaining arsenic, copper, and chromium (Christensen *et al.*, 2004, 235–236). The most efficient results for all three contaminants, which included >95 % removal of arsenic, used 100 kg of wood chips with a 60-cm spacing between the electrodes. The electrodialysis lasted for 21 days (Christensen *et al.*, 2004, 231).

Sarahney *et al.* (2005) investigated the application of an *ex situ* electrokinetic process to remove arsenic, copper, and chromium from CCA-treated wood chips. According to Sarahney *et al.* (2005), oxalic acid and oxalic acid mixed with EDTA enhanced the treatment process for arsenic. The oxalic acid and EDTA mixture removed 88 % of the arsenic from the wood.

7.3.3.6 Incineration

Incineration has the advantages of reducing the volume of wood waste, producing heat for energy generation, and possibly allowing for the recovery and reuse of copper, chromium, or arsenic. Although CCA-treated wood is often purposely or inadvertently burned (Wasson *et al.*, 2005; Solo-Gabriele and Townsend, 1999), the process is either restricted or completely illegal in many areas. Burning unavoidably volatilizes some of the arsenic and could produce a toxic ash (Townsend, Dubey and Solo-Gabriele,

2004, 174; Morrell, 2004, 204; Rogers *et al.*, 2007). The fraction of arsenic volatilized during pyrolysis or combustion ranges from 8–95 % (Helsen and Van den Bulck, 2004, 285). Helsen and Van den Bulck (2004) discuss some of the factors that affect arsenic volatilization from the incineration of CCA-treated wood, including temperature, *residence time*, effects of an oxidizing or reducing atmosphere, the presence of chlorine and sulfur, gas flow rate, the types of organics in the wood, and how the CCA was impregnated into the wood. Additionally, Lundholm *et al.* (2007) found that the addition of peat during the incineration of CCA-treated wood can marginally reduce the volatilization of arsenic. Peats often contain calcium, iron, and other inorganic compounds that may sorb volatile arsenic during combustion (Lundholm *et al.*, 2007, 6534).

Volatilized arsenic from the combustion of CCA-treated wood usually consists of arsenic oxides (probably As_4O_6, Chapter 5) and As(III) and As(V) compounds on particulate matter (Helsen and Van den Bulck, 2004, 286). Specifically, Lundholm *et al.* (2007) identified KH_2AsO_4 and As_2O_3 on flyash particles from the combustion of CCA-treated wood. For coal combustion systems, sorbent injection can capture arsenic in flue gases. Similar injection systems might also be effective with CCA-treated wood combustors. Effective sorbents of arsenic in flue gas include hydrated lime ($Ca(OH)_2$, portlandite), calcium carbonate, limestone, flyash, and to some extent activated carbon (Helsen and Van den Bulck, 2004, 287, 289; López-Antón *et al.*, 2007; Gupta *et al.*, 2007; Chapter 5). On the other hand, submicron particles, which may contain significant arsenic, routinely pass through flue gas treatment systems (Ottosen, Pedersen and Christensen, 2004, 296).

Ashes from the combustion of CCA-treated wood often fail the German DIN 38 414 leaching test for arsenic and chromium and the US TCLP for at least arsenic (Helsen and Van den Bulck, 2004, 287–288; Townsend, Dubey and Solo-Gabriele, 2004, 171–172). One possible solution to reduce the arsenic concentrations in the ashes might be coincineration, which involves diluting CCA-treated wood with untreated wood, municipal waste, coal, or other low-arsenic fuels. However, Helsen and Van den Bulck (2004, 289) advise against coincineration because potentially toxic concentrations of arsenic could still substantially accumulate in flyashes and bottom ashes. Specifically, an ash produced from the combustion of wood in which only 5 % has been treated with CCA is still likely to fail the US TCLP for arsenic (Townsend, Dubey and Solo-Gabriele, 2004, 174). Even if an ash passes the TCLP, the presence of significant arsenic and other contaminants still might make it unacceptable for reuse or disposal in a landfill. Some nations also have regulations against the coincineration of CCA-treated wood. Except for the Netherlands, which currently allows some blending of coal with up to 40 % CCA-treated wood, coincineration is not a legal option in the European Union (Helsen and Van den Bulck, 2004, 289). Danish regulations, in particular, require that CCA-treated wood be removed from other waste streams and treated separately.

To deal with the hazardous status of ashes from CCA-treated wood, the arsenic must either be extracted or the ashes would need to be encapsulated through solidification/stabilization. Hypothetically, the recovery and reuse of arsenic from the incineration of CCA-treated wood could reduce arsenic mining and imports. However, arsenic use has declined in recent years (Chapter 5) and there is little economic incentive to incinerate solid wastes and recover any volatile arsenic (Leist, Casey and Caridi, 2000, 126, 127).

7.3.3.7   *Low-temperature pyrolysis*

The volatilization of arsenic during the thermal destruction of CCA-treated wood may be reduced by utilizing low-temperature pyrolysis. Low-temperature pyrolysis uses temperatures of approximately 300–400 °C with a limited air supply (Helsen and Van den Bulck, 2004, 286, 290; Helsen and Van den Bulck, 2003). Pyrolysis includes slow and flash methods (Helsen and Van den Bulck, 2004). Flash pyrolysis, which produces an oil byproduct, is not effective with CCA-treated wood because only 5–18 % of the arsenic collects in the oil, yet the arsenic concentrations are high enough to make the oil useless (Helsen and Van den Bulck, 2004, 288). Slow pyrolysis shows more promise with CCA-treated wood. However, as the name indicates, the process involves slow decomposition and reactions. (Helsen and Van den Bulck, 2004, 290) concluded that additional research is required before any pyrolysis technique for CCA-treated wood is commercialized.

Although arsenic is less volatile during low-temperature pyrolysis than combustion, some arsenic still volatilizes during the process. The volatilization of arsenic during pyrolysis chiefly results from the reduction of As(V) to As_4O_6 ('As_2O_3') and other As(III) oxides (Helsen and Van den Bulck, 2003; Hata *et al.*, 2003; Helsen *et al.*, (2003)). To minimize arsenic volatilization, the characteristics of the wood must be known and pyrolysis operations must be carefully controlled at temperatures below 320 °C (Helsen and Van den Bulck, 2003).

7.4 Treatment technologies for arsenic in gases

The residence time of arsenic in the atmosphere is about 7–10 days (Matschullat, 2000, 300). About 60 % of anthropogenic arsenic emissions to the global atmosphere originate from copper smelters and coal combustion, which represent about 12 800 and 6240 tons arsenic/year, respectively (Matschullat, 2000, 301). Other emitters of gaseous arsenic include coal gasification (Chapter 5) and natural sources, such as volcanoes, wind-blown particles, seasalt spray, wild fires, and biological activity (Nriagu, 1989; Chapter 3). The vast majority (approximately 89–98.6 %) of atmospheric arsenic occurs with particulates rather than existing as vapors. Both gaseous and particulate arsenic are inhalation hazards and may contaminate surface soils, sediments, and waters near their points of origin (Shih and Lin, 2003; Chapter 4; Chein *et al.*, 2006; Hedberg *et al.*, 2005; Martley, Gulson and Pfeifer, 2004).

In most coal combustion systems, a large portion of the arsenic in the original coal is captured by flyash particles or condenses onto the particles (López-Antón *et al.*, 2006, 3947). Arsenic sorption is especially favored on submicron flyash particles with high surface areas (Helble, 2000, 128). After sorbing arsenic, the submicron particles often readily pass through *baghouses* and other flue gas treatment systems, and into the atmosphere.

Gaseous arsenic in coal combustion and smelting operations primarily exists as arsenic(III) oxide (As_4O_6) (Shih and Lin, 2003; Jadhav and Fan, 2001, 794; Cotton *et al.*, 1999, 400), although $AsCl_3$ may be prominent from the combustion of chlorine-rich coals at 527–927 °C (Urban and Wilcox, 2006). As_4O_6 is more volatile than any zerovalent arsenic that might be present and gaseous As(V) pentoxide (As_2O_5) only forms under pressurized conditions (Jadhav and Fan, 2001, 794, 797; Mahuli *et al.*, 1997, 3226). Arsine (AsH_3) gas does not develop under oxidizing conditions in combustors and smelters. However, the gas is used and released by semiconductor and optoelectronics facilities. Once in the atmosphere, arsine oxidizes to As_4O_6, which subsequently deposits on atmospheric particles (Chein *et al.*, 2006, 1901–1902).

In most developed countries, coal combustion and smelting facilities are required to treat gaseous and particulate arsenic in flue gases before the gases are released into the atmosphere (Chapter 5). Potentially effective sorbents for arsenic in flue gases include hydrated lime (portlandite), lime, calcium carbonate, limestone, and flyash (Helsen and Van den Bulck, 2004, 287, 289; Gupta *et al.*, 2007; Chapter 5; Jadhav and Fan, 2001; Taerakul *et al.*, 2006). The injection of hydrated lime is especially effective and probably removes volatile As_4O_6 through the formation of calcium arsenates at least over a temperature range of 600–1000 °C (Mahuli *et al.*, 1997; Chapter 5).

'Lime' (portlandite) spray dryers are often used to remove sulfur dioxide from flue gases at coal combustion facilities. Calcium in the spray and flyash may also simultaneously remove arsenic. The resulting arsenic-bearing flyash and calcium sulfites and sulfates are then collected in baghouses or with *electrostatic precipitators* (ESPs) (Taerakul *et al.*, 2006). Although ESPs are widely used at coal combustion facilities, their average capture efficiency for gaseous arsenic is relatively low (96.1 % in a study with 19 American facilities) when compared with particulate arsenic (99.2 %; (Helble, 2000, 129).

Taerakul *et al.* (2006) investigated the distribution of arsenic in lime spray dryer byproducts and bituminous coal flyashes from a power plant in Ohio, USA. The arsenic preferentially concentrated in the calcium-enriched fractions ($47.1-66.2\,mg\,kg^{-1}$) when compared with the flyash- and unburned carbon-enriched materials ($30.4-39.6\,mg\,kg^{-1}$). These results suggest that arsenic was captured by the spray dryer. Due to the low operating temperature of the spray dryer ($140-190\,^\circ C$), the arsenic was captured by aqueous droplets of portlandite rather than reacting with solid lime to form calcium arsenates (Taerakul *et al.*, 2006, 1524). Drying of the droplets precipitated portlandite, hannebachite ($CaSO_3 \cdot 0.5H_2O$), and ettringite, which sorbed the arsenic. The precipitates were subsequently removed from the flue gas with a baghouse.

References

Acar, Y.B. and Alshawabkeh, A.N. (1993) Principles of electrokinetic remediation. *Environmental Science and Technology*, **27**(13), 2638–47.

Adeel, Z. (2002) The disaster of arsenic poisoning of groundwater in south Asia — A focus on research needs and UNU's role. *Global Environmental Change*, **12**(1) 69–72.

Ahn, J.S., Chon, C.-M., Moon, H.-S. and Kim, K.-W. (2003) Arsenic removal using steel manufacturing byproducts as permeable reactive materials in mine tailing containment systems. *Water Research*, **37**(10), 2478–88.

Akhter, H., Cartledge, F.K., Miller, J. and McLearn, M. (2000) Treatment of arsenic-contaminated soils. I: soil characterization. *Journal of Environmental Engineering*, **126**(11), 999–1003.

Al Rmalli, S.W., Harrington, C.F., Ayub, M. and Haris, P.I. (2005) A biomaterial based approach for arsenic removal from water. *Journal of Environmental Monitoring*, **7**(4), 279–82.

Alaerts, G.J. and Khouri, N. (2004) Arsenic contamination and groundwater: mitigation strategies and policies. *Hydrogeology Journal*, **12**(1), 103–14.

American Water Works Association (1998) *Water Treatment Plant Design*, McGraw-Hill, New York, p. 806.

Amin, M.N., Kaneco, S., Kitagawa, T. *et al.* (2006) Removal of arsenic in aqueous solutions by adsorption onto waste rice husk. *Industrial and Engineering Chemistry Research*, **45**(24), 8105–10.

Anderson, G.L., Ellis, P.J., Khun, P. and Hille, R. (2002) Oxidation of arsenite by Alcaligenes faecalis, in *Environmental Chemistry of Arsenic* (ed. W.T. Frankenberger Jr.), Marcel Dekker, New York, pp. 343–61.

Anirudhan, T.S. and Unnithan, M.R. (2007) Arsenic(V) removal from aqueous solutions using an anion exchanger derived from coconut coir pith and its recovery. *Chemosphere*, **66**(1), 60–66.

Appelo, C.A.J. and de Vet, W.W.J.M. (2003) Modeling in situ iron removal from groundwater with trace elements such as As, in *Arsenic in Ground Water* (eds A.H. Welch and K.G. Stollenwerk), Kluwer Academic Publishers, Boston, MA, pp. 381–401.

Arienzo, M., Adamo, P., Chiarenzelli, J. *et al.* (2002) Retention of arsenic on hydrous ferric oxides generated by electrochemical peroxidation. *Chemosphere*, **48**(10), 1009–18.

Ayoob, S., Gupta, A.K. and Bhakat, P.B. (2007) Analysis of breakthrough developments and modeling of fixed bed adsorption system for As(V) removal from water by modified calcined bauxite (MCB). *Separation and Purification Technology*, **52**(3), 430–38.

Badruzzaman, M., Westerhoff, P. and Knappe, D.R. (2004) Intraparticle diffusion and adsorption of arsenate onto granular ferric hydroxide (GFH). *Water Research*, **38**(18), 4002–12.

Bai, Z. and Wiltshire, J.C. (2005) Composition and useful properties of tailings of marine manganese nodules and crusts. *Marine Georesources and Geotechnology*, **23**(1-2), 13–24.

Bajpai, S. and Chaudhuri, M. (1999) Removal of arsenic from ground water by manganese dioxide-coated sand. *Journal of Environmental Engineering*, **125**(8), 782–84.

Balaji, T. and Matsunaga, H. (2002) Adsorption characteristics of As(III) and As(V) with titanium dioxide loaded amberlite XAD-7 resin. *Analytical Sciences*, **18**(12), 1345–49.

Balaji, T., Yokoyama, T. and Matsunaga, H. (2005) Adsorption and removal of As(V) and As(III) using Zr-loaded lysine diacetic acid chelating resin. *Chemosphere*, **59**(8), 1169–74.

Baldwin, P.R. and Butcher, D.J. (2007) Phytoremediation of arsenic by two hyperaccumulators in a hydroponic environment. *Microchemical Journal*, **85**(2), 297–300.

Ball, P. (2005) Arsenic-free water still a pipedream. *Nature*, **436**(7049), 313.

Bang, S., Johnson, M.D., Korfiatis, G.P. and Meng, X. (2005a) Chemical reactions between arsenic and zero-valent iron in water. *Water Research*, **39**(5), 763–70.

Bang, S., Patel, M., Lippincott, L. and Meng, X. (2005b) Removal of arsenic from groundwater by granular titanium dioxide adsorbent. *Chemosphere*, **60**(3), 389–97.

Bednar, A.J., Garbarino, J.R., Ranville, J.F. and Wildeman, T.R. (2005) Effects of iron on arsenic speciation and redox chemistry in acid mine water. *Journal of Geochemical Exploration*, **85**(2), 55–62.

Beolchini, F., Pagnanelli, F., De Michelis, I. and Vegliò, F. (2006) Micellar enhanced ultrafiltration for arsenic(V) removal: effect of main operating conditions and dynamic modeling. *Environmental Science and Technology*, **40**(8), 2746–52.

Berg, M., Luzi, S., Trang, P.T.K. *et al.* (2006) Arsenic removal from groundwater by household sand filters: comparative field study, model calculations, and health benefits. *Environmental Science and Technology*, **40**(17), 5567–73.

Bhaumik, A., Samanta, S. and Mal, N.K. (2005) Efficient removal of arsenic from polluted ground water by using a layered double hydroxide exchanger. *Indian Journal of Chemistry, Section A: Inorganic, Physical, Theoretical and Analytical Chemistry*, **44**(7), 1406–9.

Bisceglia, K.J., Rader, K.J., Carbonaro, R.F. *et al.* (2005) Iron(II)-catalyzed oxidation of arsenic(III) in a sediment column. *Environmental Science and Technology*, **39**(23), 9217–22.

Bissen, M. and Frimmel, F.H. (2003) Arsenic — a review. Part II: oxidation of arsenic and its removal in water treatment. *Acta Hydrochimica et Hydrobiologica*, **31**(2), 97–107.

Blowes, D.W., Ptacek, C.J., Benner, S.G. *et al.* (2000) Treatment of inorganic contaminants using permeable reactive barriers. *Journal of Contaminant Hydrology*, **45**(1–2), 123–37.

Boccelli, D.L., Small, M.J. and Dzombak, D.A. (2005) Enhanced coagulation for satisfying the arsenic maximum contaminant level under variable and uncertain conditions. *Environmental Science and Technology*, **39**(17), 6501–7.

Boccelli, D.L., Small, M.J. and Dzombak, D.A. (2006) Effects of water quality and model structure on arsenic removal simulation: an optimization study. *Environmental Engineering Science*, **23**(5), 835–50.

Borah, D., Satokawa, S., Kato, S. and Kojima, T. (2008) Surface-modified carbon black for As(V) removal. *Journal of Colloid and Interface Science*, **319**(1), 53–62.

Bothe, J.V., Jr. and Brown, P.W. (1999) Arsenic immobilization by calcium arsenate formation. *Environmental Science and Technology*, **33**(21), 3806–11.

Brandhuber, P. and Amy, G. (1998) Alternative methods for membrane filtration of arsenic from drinking water. *Desalination*, **117**(1-3), 1–10.

Budinova, T., Petrov, N., Razvigorova, M. *et al.* (2006) Removal of As(III) from aqueous solution by activated carbons prepared from solvent extracted olive pulp and olive stones. *Industrial and Engineering Chemistry Research*, **45**(6), 1896–901.

Buelt, J.L., Timmerman, C.L. and Westsik, J.H., Jr. (1989) *In-Situ Vitrification: Test Results for a Contaminated Soil-Melting Process*. Report PNL-SA-15767, US Department of Energy, p. 32.

Burkitbaev, M. (2003) Radiation-stimulated oxidation reactions of oxo anions in aqueous solutions. *High Energy Chemistry*, **37**(4), 216–19.

Buschmann, J., Canonica, S., Lindauer, U. *et al.* (2005) Photoirradiation of dissolved humic acid induces arsenic(III) oxidation. *Environmental Science and Technology*, **39**(24), 9541–46.

Cano-Aguilera, I., Haque, N., Morrison, G.M. *et al.* (2005) Use of hydride generation-atomic absorption spectrometry to determine the effects of hard ions, iron salts and humic substances on arsenic sorption to sorghum biomass. *Microchemical Journal*, **81**(1), 57–60.

Cao, A.-M., Monnell, J.D., Matranga, C. *et al.* (2007) Hierarchical nanostructured copper oxide and its application in arsenic removal. *Journal of Physical Chemistry C*, **111**(50), 18624–28.

Carlson, L., Bigham, J.M., Schwertmann, U. *et al.* (2002) Scavenging of As from acid mine drainage by schwertmannite and ferrihydrite: a comparison with synthetic analogues. *Environmental Science and Technology*, **36**(8), 1712–19.

Chakraborti, D., Rahman, M.M., Paul, K. *et al.* (2002) Arsenic calamity in the Indian subcontinent: What lessons have been learned? *Talanta*, **58**(1), 3–22.

Chakraborty, S., Wolthers, M., Chatterjee, D. and Charlet, L. (2007) Adsorption of arsenite and arsenate onto muscovite and biotite mica. *Journal of Colloid and Interface Science*, **309**(2), 392–401.

Chakravarty, S., Dureja, V., Bhattacharyya, G. *et al.* (2002) Removal of arsenic from groundwater using low cost ferruginous manganese ore. *Water Research*, **36**(3), 625–32.

Chein, H., Hsu, Y.-D., Aggarwal, S.G. *et al.* (2006) Evaluation of arsenical emission from semiconductor and opto-electronics facilities in Hsinchu, Taiwan. *Atmospheric Environment*, **40**(10), 1901–7.

Chen, C.-C. and Chung, Y.-C. (2006) Arsenic removal using a biopolymer chitosan sorbent. *Journal of Environmental Science and Health, Part A: Toxic/Hazardous Substances and Environmental Engineering*, **41**(4), 645–58.

Chen, H.-W., Frey, M.M., Clifford, D. *et al.* (1999) Arsenic treatment considerations. *Journal of American Water Works Association*, **91**(2–3), 74–85.

Cheng, R.C., Liang, S., Wang, H-C. and Beuhler, M.D. (1994) Enhanced coagulation for arsenic removal. *Journal of American Water Works Association*, **86**(9), 79–90.

Cheng, Z., Van Geen, A., Louis, R. *et al.* (2005) Removal of methylated arsenic in groundwater with iron filings. *Environmental Science and Technology*, **39**(19), 7662–66.

Christensen, I., Pedersen, A., Ottosen, L. and Riberic, A. (2004) Electrodialytic remediation of CCA-treated wood in larger scale, in *Environmental Impacts of Preservative-Treated Wood, Conference*, Gainesville, FL, February 8-11, Florida Center for Environmental Solutions, Orlando, FL, pp. 227–37.

Clausen, C. (2004) Improving the two-step remediation process for CCA-treated wood: Part I. Evaluating oxalic acid extraction. *Waste Management*, **24**(4), 401–5.

Clausen, C. and Kenealy, W. (2004) Scaled-up remediation of CCA-treated wood, in *Environmental Impacts of Preservative-Treated Wood, Conference*, Gainesville, FL, February 8-11, Florida Center for Environmental Solutions, Orlando, FL, pp. 71–80.

Clifford, D.A. and Ghurye, G.L. (2002) Metal-oxide adsorption, ion exchange, and coagulation-microfiltration for arsenic removal from water, in *Environmental Chemistry of Arsenic* (ed. W.T. Frankenberger Jr.), Marcel Dekker, New York, pp. 217–45.

Clifford, D., Subramonian, S. and Sorg, T.J. (1986) Removing dissolved inorganic contaminants from water. *Environmental Science and Technology*, **20**(11), 1072–80.

Code of Federal Regulations (CFR). US Government Printing Office, Superintendent of Documents, Washington, DC.

Cotton, F.A., Wilkinson, G., Murillo, C.A. and Bochmann, M. (1999) *Advanced Inorganic Chemistry*, 6th edn, John Wiley & Sons, Inc., New York.

Cox, C. (1991) Chromated copper arsenate. *Journal of Pesticide Reform*, **11**(1), 2–6.

Cumbal, L. and Sengupta, A.K. (2005) Arsenic removal using polymer-supported hydrated iron(III) oxide nanoparticles: Role of Donnan membrane effect. *Environmental Science and Technology*, **39**(17), 6508–15.

Cundy, A.B. and Hopkinson, L. (2005) Electrokinetic iron pan generation in unconsolidated sediments: implications for contaminated land remediation and soil engineering. *Applied Geochemistry*, **20**(5), 841–48.

Dambies, L. (2004) Existing and prospective sorption technologies for the removal of arsenic in water. *Separation Science and Technology*, **39**(3), 603–27.

Daus, B., Wennrich, R. and Weiss, H. (2004) Sorption materials for arsenic removal from water: a comparative study. *Water Research*, **38**(12), 2948–54.

De, M. (2005) Arsenic — India's health crisis attracting global attention. *Current Science*, **88**(5), 683–84.

Deliyanni, E.A., Bakoyannakis, D.N., Zouboulis, A.I. and Matis, K.A. (2003a) Sorption of As(V) ions by akagane'ite-type nanocrystals. *Chemosphere*, **50**(1), 155–63.

Deliyanni, E.A., Bakoyannakis, D.N., Zouboulis, A.I. and Peleka, E. (2003b) Removal of arsenic and cadmium by akaganeite fixed-beds. *Separation Science and Technology*, **38**(16), 3967–81.

Deschamps, E., Ciminelli, V.S.T. and Höll, W.H. (2005) Removal of As(III) and As(V) from water using a natural Fe and Mn enriched sample. *Water Research*, **39**(20), 5212–20.

Dhiman, A.K. and Chaudhuri, M. (2007) Iron and manganese amended activated alumina — a medium for adsorption/oxidation of arsenic from water. *Journal of Water Supply: Research and Technology - AQUA*, **56**(1), 69–74.

Dodd, M.C., Vu, N.D., Ammann, A. *et al.* (2006) Kinetics and mechanistic aspects of As(III) oxidation by aqueous chlorine, chloramines, and ozone: relevance to drinking water treatment. *Environmental Science and Technology*, **40**(10), 3285–92.

Doušová, B., Grygar, T., Martaus, A. *et al.* (2006) Sorption of AsV on aluminosilicates treated with FeII nanoparticles. *Journal of Colloid and Interface Science*, **302**(2), 424–31.

Doušová, B., Kolous?ek, D., Kovanda, F. *et al.* (2005) Removal of As(V) species from extremely contaminated mining water. *Applied Clay Science*, **28**(Spec. Iss. 1-4), 31–42.

Doušová, B., Machovč, V., Koloušek, D. *et al.* (2003) Sorption of As(V) species from aqueous systems. *Water, Air, and Soil Pollution*, **149**(1-4), 251–67.

Dupont, L., Jolly, G. and Aplincourt, M. (2007) Arsenic adsorption on lignocellulosic substrate loaded with ferric ion. *Environmental Chemistry Letters*, **5**(3), 125–29.

Duquesne, K., Lebrun, S., Casiot, C. *et al.* (2003) Immobilization of arsenite and ferric iron by Acidithiobacillus ferrooxidans and its relevance to acid mine drainage. *Applied and Environmental Microbiology*, **69**(10), 6165–73.

Dutré, V. and Vandecasteele, C. (1998) Immobilization mechanism of arsenic in waste solidified using cement and lime. *Environmental Science and Technology*, **32**(18), 2782–87.

Dutta, P.K., Pehkonen, S.O., Sharma, V.K. and Ray, A.K. (2005) Photocatalytic oxidation of arsenic (III): evidence of hydroxyl radicals. *Environmental Science and Technology*, **39**(6), 1827–34.

Dutta, P.K., Ray, A.K., Sharma, V.K. and Millero, F.J. (2004) Adsorption of arsenate and arsenite on titanium dioxide suspensions. *Journal of Colloid and Interface Science*, **278**(2), 270–75.

Eby, G.N. (2004) *Principles of Environmental Geochemistry*, Thomson Brooks/Cole, Pacific Grove, CA, p. 514.

Ehrlich, H.L. (2002) Bacterial oxidation of As(III) compounds, in *Environmental Chemistry of Arsenic* (ed. W.T. Frankenberger Jr.), Marcel Dekker, New York, pp. 313–27.

Embrick, L.L., Porter, K.M., Pendergrass, A. and Butcher, D.J. (2005) Characterization of lead and arsenic contamination at Barber Orchard, Haywood County, NC. *Microchemical Journal*, **81**(1), 117–21.

Englehardt, J.D., Meeroff, D.E., Echegoyen, L. *et al.* (2007) Oxidation of aqueous EDTA and associated organics and coprecipitation of inorganics by ambient iron-mediated aeration. *Environmental Science and Technology*, **41**(1), 270–76.

Ergican, E., Gecol, H. and Fuchs, A. (2005) The effect of co-occurring inorganic solutes on the removal of arsenic (V) from water using cationic surfactant micelles and an ultrafiltration membrane. *Desalination*, **181**(1-3), 9–26.

Farrell, J., Wang, J.P., O'Day, P. and Conklin, M. (2001) Electrochemical and spectroscopic study of arsenate removal from water using zero-valent iron media. *Environmental Science and Technology*, **35**(10), 2026–32.

Faust, S.D. and Aly, O.M. (1998) *Chemistry of Water Treatment*, Ann Arbor Press, Chelsea, MA.

Ferguson, M.A., Hoffmann, M.R. and Hering, J.G. (2005) TiO2-photocatalyzed As(III) oxidation in aqueous suspensions: reaction kinetics and effects of adsorption. *Environmental Science and Technology*, **39**(6), 1880–86.

Fernández-Jiménez, A., Palomo, A., Macphee, D.E. and Lachowski, E.E. (2005) Fixing arsenic in alkali-activated cementitious matrices. *Journal of the American Ceramic Society*, **88**(5), 1122–26.

Fetter, C.W. (1993) *Contaminant Hydrogeology*, Prentice Hall, Upper Saddle River, NJ, p. 458.

Fitz, W.J., Wenzel, W.W., Zhang, H. *et al.* (2003) Rhizosphere characteristics of the arsenic hyperaccumulator Pteris vittata L. and monitoring of phytoremoval efficiency. *Environmental Science and Technology*, **37**(21), 5008–14.

Floch, J. and Hideg, M. (2004) Application of ZW-1000 membranes for arsenic removal from water sources. *Desalination*, **162**(1-3), 75–83.

Foster, A.L. (2003) Spectroscopic investigation of arsenic species in solid phases, in *Arsenic in Ground Water* (eds A.H. Welch and K.G. Stollenwerk), Kluwer Academic Publishers, Boston, MA, pp. 27–65.

Fox, P.M. and Doner, H.E. (2003) Accumulation, release, and solubility of arsenic, molybdenum, and vanadium in wetland sediments. *Journal of Environmental Quality*, **32**, 2428–35.

Frankenberger, W. T., Jr. and Arshad, M. (2002) Volatilization of arsenic, in *Environmental Chemistry of Arsenic* (ed. W.T. Frankenberger Jr.) Marcel Dekker, New York, pp. 363–80.

Fryxell, G.E., Liu, J., Hauser, T.A. *et al.* (1999) Design and synthesis of selective mesoporous anion traps. *Chemistry of Materials*, **11**(8), 2148–54.

Gallup, D.L. (2007) Treatment of geothermal waters for production of industrial, agricultural or drinking water. *Geothermics*, **36**(5), 473–83.

García, M.G., D'Hiriart, J., Giullitti, J. *et al.* (2004) Solar light induced removal of arsenic from contaminated groundwater: the interplay of solar energy and chemical variables. *Solar Energy*, **77**(5), 601–13.

Gecol, H., Ergican, E. and Fuchs, A. (2004) Molecular level separation of arsenic (V) from water using cationic surfactant micelles and ultrafiltration membrane. *Journal of Membrane Science*, **241**(1), 105–19.

Genç-Fuhrman, H., Tjell, J.C. and McConchie, D. (2004) Adsorption of arsenic from water using activated neutralized red mud. *Environmental Science and Technology*, **38**(8), 2428–34.

Génin, J.-M.R., Bourrié, G., Trolard, F. *et al.* (1998) Thermodynamic equilibria in aqueous suspensions of synthetic and natural Fe(II)-Fe(III) green rusts: occurrences of the mineral in hydromorphic soils. *Environmental Science and Technology*, **32**(8), 1058–68.

Ghimire, K.N., Inoue, K., Yamaguchi, H. *et al.* (2003) Adsorptive separation of arsenate and arsenite anions from aqueous medium by using orange waste. *Water Research*, **37**(20), 4945–53.

Ghosh, A., Mukiibi, M., Sáez, A.E. and Ela, W.P. (2006) Leaching of arsenic from granular ferric hydroxide residuals under mature landfill conditions. *Environmental Science and Technology*, **40**(19), 6070–75.

Ghurye, G., Clifford, D. and Tripp, A. (2004) Iron coagulation and direct microfiltration to remove arsenic from groundwater. *Journal of American Water Works Association*, **96**(4), 143–52.

Giasuddin, A.B.M., Kanel, S.R. and Choi, H. (2007) Adsorption of humic acid onto nanoscale zerovalent iron and its effect on arsenic removal. *Environmental Science and Technology*, **41**(6), 2022–27.

Gillman, G.P. (2006) A simple technology for arsenic removal from drinking water using hydrotalcite. *Science of the Total Environment*, **366**(2–3), 926–31.

Gong, A., Kamdem, D. and Harcihandran, R. (2004) Compression tests on wood-cement particle composites made of CCA-treated wood removed from service, in *Environmental Impacts of Preservative-Treated Wood, Conference*, Gainesville, FL, February 8-11, Florida Center for Environmental Solutions, Orlando, FL, pp. 270–76.

Goodrich-Mahoney, J.W. (1996) Constructed wetland treatment systems applied research program at the Electric Power Research Institute. *Water, Air, and Soil Pollution*, **90**(1-2), 205–17.

Grafe, M., Eick, M.J. and Grossl, P.R. (2001) Adsorption of arsenate (V) and arsenite (III) on goethite in the presence and absence of dissolved organic carbon. *Soil Science Society of America Journal*, **65**, 1680–87.

Greenleaf, J.E., Lin, J.-C. and Sengupta, A.K. (2006) Two novel applications of ion exchange fibers: arsenic removal and chemical-free softening of hard water. *Environmental Progress*, **25**(4), 300–11.

Gu, Z. and Deng, B. (2007) Use of iron-containing mesoporous carbon (IMC) for arsenic removal from drinking water. *Environmental Engineering Science*, **24**(1), 113–21.

Gu, Z., Deng, B. and Yang, J. (2007) Synthesis and evaluation of iron-containing ordered mesoporous carbon (FeOMC) for arsenic adsorption. *Microporous and Mesoporous Materials*, **102**(1-3), 265–73.

Gu, Z., Fang, J. and Deng, B. (2005) Preparation and evaluation of GAC-based iron-containing adsorbents for arsenic removal. *Environmental Science and Technology*, **39**(10), 3833–43.

Gu, B., Phelps, T.J., Liang, L. *et al.* (1999) Biogeochemical dynamics in zero-valent iron columns: implications for permeable reactive barriers. *Environmental Science and Technology*, **33**(13), 2170–77.

Guo, X. and Chen, F. (2005) Removal of arsenic by bead cellulose loaded with iron oxyhydroxide from groundwater. *Environmental Science and Technology*, **39**(17), 6808–18.

Guo, X., Du, Y., Chen, F. *et al.* (2007a) Mechanism of removal of arsenic by bead cellulose loaded with iron oxyhydroxide (β-FeOOH): EXAFS study. *Journal of Colloid and Interface Science*, **314**(2), 427–33.

Guo, H., Stüben, D. and Berner, Z. (2007b) Adsorption of arsenic(III) and arsenic(V) from groundwater using natural siderite as the adsorbent. *Journal of Colloid and Interface Science*, **315**(1), 47–53.

Gupta, V.K., Saini, V.K. and Jain, N. (2005) Adsorption of As(III) from aqueous solutions by iron oxide-coated sand. *Journal of Colloid and Interface Science*, **288**(1), 55–60.

Gupta, H., Thomas, T.J., Park, A.-H.A. *et al.* (2007) Pilot-scale demonstration of the OSCAR process for high-temperature multipollutant control of coal combustion flue gas, using carbonated fly ash and mesoporous calcium carbonate. *Industrial and Engineering Chemistry Research*, **46**(14), 5051–60.

Han, B., Zimbron, J., Runnells, T.R. *et al.* (2003) New arsenic standard spurs search for cost-effective removal techniques. *Journal of American Water Works Association*, **95**(10), 109–18.

Hansen, H.K., Ribeiro, A. and Mateus, E. (2006) Biosorption of arsenic(V) with Lessonia nigrescens. *Minerals Engineering*, **19**, 486–90.

Haque, N., Morrison, G., Cano-Aguilera, I. and Gardea-Torresdey, J.L. (2008) Iron-modified light expanded clay aggregates for the removal of arsenic(V) from groundwater. *Microchemical Journal*, **88**(1), 7–13.

Hata, T., Bronsveld, P.M., Vystavel, T. *et al.* (2003) Electron microscopic study on pyrolysis of CCA (chromium, copper and arsenic oxide)-treated wood. *Journal of Analytical and Applied Pyrolysis*, **68-69**, 635–43.

Hedberg, E., Gidhagen, L. and Johansson, C. (2005) Source contributions to PM10 and arsenic concentrations in central Chile using positive matrix factorization. *Atmospheric Environment*, **39**(3), 549–61.

Helble, J.J. (2000) Model for the air emissions of trace metallic elements from coal combustors equipped with electrostatic precipitators. *Fuel Processing Technology*, **63**(2), 125–47.

Helsen, L. and Van den Bulck, E. (2003) Metal retention in the solid residue after low-temperature pyrolysis of chromated copper arsenate (CCA)-treated wood. *Environmental Engineering Science*, **20**(6), 569–80.

Helsen, L. and Van den Bulck, E. (2004) Review of thermochemical conversion processes as disposal technologies for chromated copper arsenate (CCA) treated wood waste, in *Environmental Impacts of Preservative-Treated Wood, Conference*, Gainesville, FL, February 8-11, Florida Center for Environmental Solutions, Orlando, FL, pp. 277–94.

Helsen, L. and Van den Bulck, E. (2005) Review of disposal technologies for chromated copper arsenate (CCA) treated wood waste, with detailed analyses of thermochemical conversion processes. *Environmental Pollution*, **134**(2), 301–14.

Helsen, L., Van den Bulck, E., Van Bael, M.K. and Mullens, J. (2003) Arsenic release during pyrolysis of CCA treated wood waste: current state of knowledge. *Journal of Analytical and Applied Pyrolysis*, **68-69**, 613–33.

Hering, J.G., Chen, P.-Y., Wilkie, J.A. and Elimelech, M. (1997) Arsenic removal from drinking water during coagulation. *Journal of Environmental Engineering*, **123**(8), 800–7.

Hering, J.G. and Kneebone, P.E. (2002) Biogeochemical controls on arsenic occurrence and mobility in water supplies, in *Environmental Chemistry of Arsenic* (ed. W.T. Frankenberger Jr.), Marcel Dekker, New York, pp. 155–81.

Hetherington, L.E., Brown, T.J., Idoine, N.E. *et al.* (technical support) (2008) *European Mineral Statistics 2002-06*, British Geological Survey, Keyworth, Nottingham.

Hlavay, J. and Polyák, K. (2005) Determination of surface properties of iron hydroxide-coated alumina adsorbent prepared for removal of arsenic from drinking water. *Journal of Colloid and Interface Science*, **284**(1), 71–77.

Holm, T.R. (2002) Effects of CO_3^{2-}/bicarbonate, Si, and PO_4^{3-} on arsenic sorption to HFO. *Journal of American Water Works Association*, **94**(4), 174–81.

Hossain, M.A., Sengupta, M.K., Ahamed, S. *et al.* (2005) Ineffectiveness and poor reliability of arsenic removal plants in West Bengal, India. *Environmental Science and Technology*, **39**(11), 4300–6.

Hrubý, M., Korostyatynets, V., Beneš, M.J. and Matějka, Z. (2003) Bifunctional ion exchange resin with thiol and quaternary ammonium groups for the sorption of arsenate. *Collection of Czechoslovak Chemical Communications*, **68**(11), 2159–70.

Hu, Y., Li, J.-H., Zhu, Y.-G. *et al.* (2005) Sequestration of as by iron plaque on the roots of three rice (Oryza sativa L.) cultivars in a low-P soil with or without P fertilizer. *Environmental Geochemistry and Health*, **27**(2), 169–76.

Huang, C., Pan, J.R., Lee, M. and Yen, S. (2007) Treatment of high-level arsenic-containing wastewater by fluidized bed crystallization process. *Journal of Chemical Technology and Biotechnology*, **82**(3), 289–94.

Huang, J.W., Poynton, C.Y., Kochian, L.V. and Elless, M.P. (2004) Phytofiltration of arsenic from drinking water using arsenic-hyperaccumulating ferns. *Environmental Science and Technology*, **38**(12), 3412–17.

Hug, S.J., Canonica, L., Wegelin, M. *et al.* (2001) Solar oxidation and removal of arsenic at circumneutral pH in iron containing waters. *Environmental Science and Technology*, **35**(10), 2114–21.

Hug, S.J. and Leupin, O. (2003) Iron-catalyzed oxidation of arsenic(III) by oxygen and by hydrogen peroxide: pH-dependent formation of oxidants in the Fenton reaction. *Environmental Science and Technology*, **37**(12), 2734–42.

Illman, B. and Yang, V. (2004) Bioremediation and degradation of CCA-treated wood waste, in *Environmental Impacts of Preservative-Treated Wood, Conference*, Gainesville, FL, February 8-11, Florida Center for Environmental Solutions, Orlando, FL, pp. 259–69.

Jia, Y. and Demopoulos, G.P. (2005) Adsorption of arsenate onto ferrihydrite from aqueous solution: influence of media (sulfate vs nitrate), added gypsum, and pH alteration. *Environmental Science and Technology*, **39**(24), 9523–27.

Jackson, B.P. and Miller, W.P. (2000) Effectiveness of phosphate and hydroxide for desorption of arsenic and selenium species from iron oxides. *Soil Science Society of America Journal*, **64**(5), 1616–22.

Jadhav, R.A. and Fan, L.-S. (2001) Capture of gas-phase arsenic oxide by lime: kinetic and mechanistic studies. *Environmental Science and Technology*, **35**(4), 794–99.

Jang, M., Min, S.-H., Kim, T.-H. and Park, J.K. (2006) Removal of arsenite and arsenate using hydrous ferric oxide incorporated into naturally occurring porous diatomite. *Environmental Science and Technology*, **40**(5), 1636–43.

Jang, M., Min, S.-H., Park, J.K. and Tlachac, E.J. (2007a) Hydrous ferric oxide incorporated diatomite for remediation of arsenic contaminated groundwater. *Environmental Science and Technology*, **41**(9), 3322–28.

Jang, M., Hwang, J.S. and Choi, S.I. (2007b) Sequential soil washing techniques using hydrochloric acid and sodium hydroxide for remediating arsenic-contaminated soils in abandoned iron-ore mines. *Chemosphere*, **66**(1), 8–17.

Jang, M., Shin, E.W., Park, J.K. and Choi, S.I. (2003) Mechanisms of arsenate adsorption by highly-ordered nano-structured silicate media impregnated with metal oxides. *Environmental Science and Technology*, **37**(21), 5062–70.

Jegadeesan, G., Mondal, K. and Lalvani, S.B. (2005) Arsenate remediation using nanosized modified zerovalent iron particles. *Environmental Progress*, **24**(3), 289–96.

Jekel, M.R. (1994) Removal of arsenic in drinking water treatment, in *Arsenic in the Environment: Part I: Cycling and Characterization* (ed. J.O. Nriagu). John Wiley & Sons, Inc., New York, pp. 119–32.

Jézéquel, H. and Chu, K. (2006) Removal of arsenate from aqueous solution by adsorption onto titanium dioxide nanoparticles. *Journal of Environmental Science and Health, Part A: Toxic/Hazardous Substances and Environmental Engineering*, **41**(8), 1519–28.

Jiang, J.-Q., Xu, Y., Quill, K. *et al.* (2007) Laboratory study of boron removal by Mg/Al double-layered hydroxides. *Industrial and Engineering Chemistry Research*, **46**(13), 4577–83.

Jing, C., Korfiatis, G.P. and Meng, X. (2003) Immobilization mechanisms of arsenate in iron hydroxide sludge stabilized with cement. *Environmental Science and Technology*, **37**(21), 5050–56.

Jing, C., Liu, S., Patel, M. and Meng, X. (2005a) Arsenic leachability in water treatment adsorbents. *Environmental Science and Technology*, **39**(14), 5481–87.

Jing, C., Meng, X., Liu, S. *et al.* (2005b) Surface complexation of organic arsenic on nanocrystalline titanium oxide. *Journal of Colloid and Interface Science*, **290**(1), 14–21.

Jing, C., Liu, S. and Meng, X. (2005c) Arsenic leachability and speciation in cement immobilized water treatment sludge. *Chemosphere*, **59**(9), 1241–47.

Joshi, A. and Chaudhuri, M. (1996) Removal of arsenic from ground water by iron oxide-coated sand. *Journal of Environmental Engineering*, **122**(8), 769–71.

Kahakachchi, C., Uden, P.C. and Tyson, J.F. (2004) Extraction of arsenic species from spiked soils and standard reference materials. *Analyst*, **129**(8), 714–18.

Kakitani, T., Hata, T., Katsumata, N. *et al.* (2007) Chelating extraction for removal of chromium, copper, and arsenic from treated wood with bioxalate. *Environmental Engineering Science*, **24**(8), 1026–37.

Kamala, C.T., Chu, K.H., Chary, N.S. *et al.* (2005) Removal of arsenic(III) from aqueous solutions using fresh and immobilized plant biomass. *Water Research*, **39**(13), 2815–26.

Kanel, S.R., Manning, B., Charlet, L. and Choi, H. (2005) Removal of arsenic(III) from groundwater by nanoscale zero-valent iron. *Environmental Science and Technology*, **39**(5), 1291–98.

Kang, S.-Y., Kim, D.-W. and Kim, K.-W. (2007) Enhancement of As(V) adsorption onto activated sludge by methylation treatment. *Environmental Geochemistry and Health*, **29**(4), 313–18.

Kartal, S.N. and Kose, C. (2003) Remediation of CCA-C treated wood using chelating agents. *Holz als Roh - und Werkstoff*, **61**(5), 382–87.

Kartal, S.N., Munir, E., Kakitani, T. and Imamura, Y. (2004) Bioremediation of CCA-treated wood by brown-rot fungi Fomitopsis palustris, Coniophora puteana, and Laetiporus sulphureus. *Journal of Wood Science*, **50**(2), 182–88.

Katsoyiannis, I.A. and Zouboulis, A.I. (2004) Application of biological processes for the removal of arsenic from groundwaters. *Water Research*, **38**(1), 17–26.

Katsoyiannis, I., Zouboulis, A., Althoff, H. and Bartel, H. (2002) As(III) removal from groundwaters using fixed-bed upflow bioreactors. *Chemosphere*, **47**(3), 325–32.

Katsoyiannis, I.A., Zouboulis, A.I. and Jekel, M. (2004) Kinetics of bacterial As(III) oxidation and subsequent As(V) removal by sorption onto biogenic manganese oxides during groundwater treatment. *Industrial and Engineering Chemistry Research*, **43**(2), 486–93.

Kim, Y., Kim, C., Choi, I. *et al.* (2004) Arsenic removal using mesoporous alumina prepared via a templating method. *Environmental Science and Technology*, **38**(3), 924–31.

Kim, S.-O., Kim, W.-S. and Kim, K.-W. (2005) Evaluation of electrokinetic remediation of arsenic-contaminated soils. *Environmental Geochemistry and Health*, **27**(5-6), 443–53.

Kim, M-J. and Nriagu, J. (2000) Oxidation of arsenite in groundwater using ozone and oxygen. *Science of the Total Environment*, **247**(1), 71–79.

Klein, C. (2002) *Mineral Science* (after J.D. Dana), 22nd edn, John Wiley & Sons, Inc., New York, p. 641.

Ko, I., Davis, A.P., Kim, J.-Y. and Kim, K.-W. (2007) Arsenic removal by a colloidal iron oxide coated sand. *Journal of Environmental Engineering*, **133**(9), 891–98.

Ko, I., Kim, J.-Y. and Kim, K.-W. (2004) Arsenic speciation and sorption kinetics in the As-hematite-humic acid system. *Colloids and Surfaces A: Physicochemical and Engineering Aspects*, **234**(1-3), 43–50.

Köber, R., Daus, B., Ebert, M. *et al.* (2005a) Compost-based permeable reactive barriers for the source treatment of arsenic contaminations in aquifers: column studies and solid-phase investigations. *Environmental Science and Technology*, **39**(19), 7650–55.

Köber, R., Welter, E., Ebert, M. and Dahmke, A. (2005b) Removal of arsenic from groundwater by zerovalent iron and the role of sulfide. *Environmental Science and Technology*, **39**(20), 8038–44.

Korngold, E., Belayev, N. and Aronov, L. (2001) Removal of arsenic from drinking water by anion exchangers. *Desalination*, **141**(1), 81–84.

Kostal, J., Yang, R., Wu, C.H. *et al.* (2004) Enhanced arsenic accumulation in engineered bacterial cells expressing ArsR. *Applied and Environmental Microbiology*, **70**(8), 4582–87.

Košutic, K., Furač, L., Sipos, L. and Kunst, B. (2005) Removal of arsenic and pesticides from drinking water by nanofiltration membranes. *Separation and Purification Technology*, **42**(2), 137–44.

Krauskopf, K.B. and Bird, D.K. (1995) *Introduction to Geochemistry*, 3rd edn, McGraw-Hill, Boston, p. 647.

Krishna, M.V.B., Chandrasekaran, K., Karunasagar, D. and Arunachalam, J. (2001) A combined treatment approach using Fenton's reagent and zero valent iron for the removal of arsenic from drinking water. *Journal of Hazardous Materials*, **84**(2-3), 229–40.

Kumari, P., Sharma, P., Srivastava, S. and Srivastava, M.M. (2006) Biosorption studies on shelled Moringa oleifera Lamarck seed powder: removal and recovery of arsenic from aqueous system. *International Journal of Mineral Processing*, **78**(3), 131–39.

Kundu, S. and Gupta, A.K. (2006) Investigations on the adsorption efficiency of iron oxide coated cement (IOCC) towards As(V)- kinetics, equilibrium and thermodynamic studies. *Colloids and Surfaces A: Physicochemical and Engineering Aspects*, **273**(1-3), 121–28.

Kundu, S. and Gupta, A.K. (2007) As(III) removal from aqueous medium in fixed bed using iron oxide-coated cement (IOCC): experimental and modeling studies. *Chemical Engineering Journal*, **129**(1-3), 123–31.

Kundu, S., Kavalakatt, S.S., Pal, A. *et al.* (2004) Removal of arsenic using hardened paste of Portland cement: batch adsorption and column study. *Water Research*, **38**(17), 3780–90.

Kunzru, S. and Chaudhuri, M. (2005) Manganese amended activated alumina for adsorption/oxidation of arsenic. *Journal of Environmental Engineering*, **131**(9), 1350–53.

Lackovic, J.A., Nikolaidis, N.P. and Dobbs, G.M. (2000) Inorganic arsenic removal by zero-valent iron. *Environmental Engineering Science*, **17**(1), 29–39.

Ladeira, A.C.Q. and Ciminelli, V.S.T. (2004) Adsorption and desorption of arsenic on an oxisol and its constituents. *Water Research*, **38**(8), 2087–94.

Lafferty, B.J. and Loeppert, R.H. (2005) Methyl arsenic adsorption and desorption behavior on iron oxides. *Environmental Science and Technology*, **39**(7), 2120–27.

Lakshmipathiraj, P., Narasimhan, B.R.V., Prabhakar, S. and Bhaskar Raju, G. (2006) Adsorption studies of arsenic on Mn-substituted iron oxyhydroxide. *Journal of Colloid and Interface Science*, **304**, 317–22.

Lazaridis, N.K., Hourzemanoglou, A. and Matis, K.A. (2002) Flotation of metal-loaded clay anion exchangers. Part II: the case of arsenates. *Chemosphere*, **47**(3), 319–24.

Leblanc, M., Casiot, C., Elbaz-Poulichet, F. and Personne', C. (2002) Arsenic removal by oxidizing bacteria in a heavily arsenic-contaminated acid mine drainage system (Carnoule's, France). *Geological Society Special Publication* 198, 267–74.

Lee, H. and Choi, W. (2002) Photocatalytic oxidation of arsenite in TiO2 suspension: kinetics and mechanisms. *Environmental Science and Technology*, **36**(17), 3872–78.

Lee, S.W., Kim, J.Y., Lee, J.U. *et al.* (2004) Removal of arsenic in tailings by soil flushing and the remediation process monitoring. *Environmental Geochemistry and Health*, **26**(4), 403–9.

Lee, M., Paik, I.S., Do, W. *et al.* (2007) Soil washing of As-contaminated stream sediments in the vicinity of an abandoned mine in Korea. *Environmental Geochemistry and Health*, **29**(4), 319–29.

Leist, M., Casey, R.J. and Caridi, D. (2000) The management of arsenic wastes: problems and prospects. *Journal of Hazardous Materials*, **76**(1), 125–38.

Leist, M., Casey, R.J. and Caridi, D. (2003) The fixation and leaching of cement stabilized arsenic. *Waste Management*, **23**(4), 353–59.

Lenoble, V., Bouras, O., Deluchat, V. *et al.* (2002) Arsenic adsorption onto pillared clays and iron oxides. *Journal of Colloid and Interface Science*, **255**(1), 52–58.

Lenoble, V., Chabroullet, C., Al Shukry, R. *et al.* (2004) Dynamic arsenic removal on a MnO2-loaded resin. *Journal of Colloid and Interface Science*, **280**(1), 62–67.

Leupin, O.X. and Hug, S.J. (2005) Oxidation and removal of arsenic (III) from aerated groundwater by filtration through sand and zero-valent iron. *Water Research*, **39**(9), 1729–40.

Leupin, O.X., Hug, S.J. and Badruzzaman, A.B.M. (2005) Arsenic removal from Bangladesh tube well water with filter columns containing zerovalent iron filings and sand. *Environmental Science and Technology*, **39**(20), 8032–37.

Li, Z., Beachner, R., McManama, Z. and Hanlie, H. (2007) Sorption of arsenic by surfactant-modified zeolite and kaolinite. *Microporous and Mesoporous Materials*, **105**(3), 291–97.

Liang, P. and Liu, R. (2007) Speciation analysis of inorganic arsenic in water samples by immobilized nanometer titanium dioxide separation and graphite furnace atomic absorption spectrometric determination. *Analytica Chimica Acta*, **602**(1), 32–36.

Licht, S. and Yu, X. (2005) Electrochemical alkaline Fe(VI) water purification and remediation. *Environmental Science and Technology*, **39**(20), 8071–76.

Lien, H-L. and Wilkin, R.T. (2005) High-level arsenite removal from groundwater by zero-valent iron. *Chemosphere*, **59**(3), 377–86.

Lim, S.F. and Chen, J.P. (2007) Synthesis of an innovative calcium-alginate magnetic sorbent for removal of multiple contaminants. *Applied Surface Science*, **253**(Spec. Iss. 13), 5772–75.

Lin, T.-F. and Wu, J.-K. (2001) Adsorption of arsenite and arsenate within activated alumina grains: equilibrium and kinetics. *Water Research*, **35**(8), 2049–57.

Liu, R., Qu, J., Xia, S. *et al.* (2007a) Silicate hindering in situ formed ferric hydroxide precipitation: inhibiting arsenic removal from water. *Environmental Engineering Science*, **24**(5), 707–15.

Liu, R., Li, X., Xia, S. *et al.* (2007b) Calcium-enhanced ferric hydroxide co-precipitation of arsenic in the presence of silicate. *Water Environment Research*, **79**(11), 2260–64.

López-Antón, M.A., Diaz-Somoano, M., Fierro, J.L.G. and Martínez-Tarazona, M.R. (2007) Retention of arsenic and selenium compounds present in coal combustion and gasification flue gases using activated carbons. *Fuel Processing Technology*, **88**, 799–805.

López-Antón, M.A., Di'az-Somoano, M., Spears, D.A. and Marti'nez-Tarazona, M.R. (2006) Arsenic and selenium capture by fly ashes at low temperature. *Environmental Science and Technology*, **40**(12), 3947–51.

Loukidou, M.X., Matis, K.A., Zouboulis, A.I. and Liakopoulou-Kyriakidou, M. (2003) Removal of As(V) from wastewaters by chemically modified fungal biomass. *Water Research*, **37**(18), 4544–52.

Luengo, C., Brigante, M. and Avena, M. (2007) Adsorption kinetics of phosphate and arsenate on goethite. A comparative study. *Journal of Colloid and Interface Science*, **311**(2), 354–60.

Lundholm, K., Boström, D., Nordin, A. and Shchukarev, A. (2007) Fate of Cu, Cr, and As during combustion of impregnated wood with and without peat additive. *Environmental Science and Technology*, **41**(18), 6534–40.

Macedo-Miranda, M.G. and Olguín, M.T. (2007) Arsenic sorption by modified clinoptilolite-heulandite rich tuffs. *Journal of Inclusion Phenomena and Macrocyclic Chemistry*, **59**(1-2), 131–42.

Macy, J.M. and Santini, J.M. (2002) Unique modes of arsenate respiration by Chrysiogenes arsenatis and Desulfomicrobium sp. str. Ben-RB, in *Environmental Chemistry of Arsenic* (ed. W.T. Frankenberger Jr.), Marcel Dekker, New York, pp. 297–312.

Mahuli, S., Agnihotri, R., Chauk, S. *et al.* (1997) Mechanism of arsenic sorption by hydrated lime. *Environmental Science and Technology*, **31**(11), 3226–31.

Maity, S., Chakravarty, S., Bhattacharjee, S. and Roy, B.C. (2005) A study on arsenic adsorption on polymetallic sea nodule in aqueous medium. *Water Research*, **39**, 2579–90.

Majumder, A. and Chaudhuri, M. (2005) Solar photocatalytic oxidation and removal of arsenic from ground water. *Indian Journal of Engineering and Materials Sciences*, **12**(2), 122–28.

Makris, K.C., Sarkar, D. and Datta, R. (2006) Evaluating a drinking-water waste by-product as a novel sorbent for arsenic. *Chemosphere*, **64**(5), 730–41.

Maldonado-Reyes, A., Montero-Ocampo, C. and Solorza-Feria, O. (2007) Remediation of drinking water contaminated with arsenic by the electro-removal process using different metal electrodes. *Journal of Environmental Monitoring*, **9**(11), 1241–47.

Manna, B., Dasgupta, M. and Ghosh, U.C. (2004) Crystalline hydrous titanium (IV) oxide (CHTO): an arsenic (III) scavenger from natural water. *Journal of Water Supply: Research and Technology - AQUA*, **53**(7), 483–95.

Manna, B. and Ghosh, U.C. (2007) Adsorption of arsenic from aqueous solution on synthetic hydrous stannic oxide. *Journal of Hazardous Materials*, **144**(1–2), 522–31.

Manning, B.A., Fendorf, S.E., Bostick, B. and Suarez, D.L. (2002a) Arsenic(III) oxidation and arsenic(V) adsorption reactions on synthetic birnessite. *Environmental Science and Technology*, **36**(5), 976–81.

Manning, B.A., Hunt, M.L., Amrhein, C. and Yarmoff, J.A. (2002b) Arsenic(III) and arsenic(V) reactions with zerovalent iron corrosion products. *Environmental Science and Technology*, **36**(24), 5455–61.

Martley, E., Gulson, B.L. and Pfeifer, H.-R. (2004) Metal concentrations in soils around the copper smelter and surrounding industrial complex of Port Kembla, NSW, Australia. *Science of the Total Environment*, **325**(1-3), 113–27.

Masue, Y., Loeppert, R.H. and Kramer, T.A. (2007) Arsenate and arsenite adsorption and desorption behavior on coprecipitated aluminum:iron hydroxides. *Environmental Science and Technology*, **41**(3), 837–42.

Matis, K.A., Zouboulis, A.I., Zamboulis, D. and Valtadorou, A.V. (1999) Sorption of As(V) by goethite particles and study of their flocculation. *Water, Air, and Soil Pollution*, **111**(1-4), 297–316.

Matschullat, J. (2000) Arsenic in the geosphere — a review. *Science of the Total Environment*, **249**(1-3), 297–312.

Mayo, J.T., Yavuz, C., Yean, S. *et al.* (2007) The effect of nanocrystalline magnetite size on arsenic removal. *Science and Technology of Advanced Materials*, **8**, 71–75.

McNeill, L.S., Chen, H. and Edwards, M. (2002) Aspects of arsenic chemistry in relation to occurrence, health and treatment, in *Environmental Chemistry of Arsenic* (ed. W.T. Frankenberger Jr.), Marcel Dekker, New York, pp. 141–53.

Mendonça, A.A., Brito Galvão, T.C., Lima, D.C. and Soares, E.P. (2006) Stabilization of arsenic-bearing sludges using lime. *Journal of Materials in Civil Engineering*, **18**(2), 135–39.

Mihajlovic, I., Strbac, N., Zivkovic, Z. *et al.* (2007) A potential method for arsenic removal from copper concentrates. *Minerals Engineering*, **20**(1), 26–33.

Miller, J., Akhter, H., Cartledge, F.K. and McLearn, M. (2000) Treatment of arsenic-contaminated soils. II: treatability study and remediation. *Journal of Environmental Engineering*, **126**(11), 1004–12.

Mitzman, S. (1999) *The application of PHREEQCi, a geochemical computer program, to aid in the management of a wastewater treatment wetland*. Master's Thesis, Texas A&M University, College Station, p. 163.

Mohan, D. and Pittman, C.U., Jr. (2007) Arsenic removal from water/wastewater using adsorbents-A critical review. *Journal of Hazardous Materials*, **142**(1-2), 1–53.

Mohapatra, M., Sahoo, S.K., Anand, S. and Das, R.P. (2006) Removal of As(V) by Cu(II)-, Ni(II)-, or Co(II)-doped goethite samples. *Journal of Colloid and Interface Science*, **298**, 6–12.

Mokashi, S.A. and Paknikar, K.M. (2002) Arsenic (III) oxidizing Microbacterium lacticum and its use in the treatment of arsenic contaminated groundwater. *Letters in Applied Microbiology*, **34**(4), 258–62, Erratum: **35**(2), 171.

Mondal, P., Majumder, C.B. and Mohanty, B. (2007) Removal of trivalent arsenic (As(III)) from contaminated water by calcium chloride (CaCl2)-impregnated rice husk carbon. *Industrial and Engineering Chemistry Research*, **46**(8), 2550–57.

Montes-Bayón, M., Meija, J., LeDuc, D.L. *et al.* (2004) HPLC-ICP-MS and ESI-Q-TOF analysis of biomolecules induced in Brassica juncea during arsenic accumulation. *Journal of Analytical Atomic Spectrometry*, **19**(1), 153–58.

Montgomery, J.M. (1985) *Water Treatment Principles and Design*, John Wiley & Sons, Inc., New York.

Morrell, J. (2004) Disposal of treated wood, in *Environmental Impacts of Preservative-Treated Wood, Conference*, Gainesville, FL, February 8-11, Florida Center for Environmental Solutions, Orlando, FL, pp. 196–209.

Morrell, J.J. and Huffman, J. (2004) Copper, chromium, and arsenic levels in soils surrounding posts treated with chromated copper arsenate (CCA). *Wood and Fiber Science*, **36**(1), 119–28.

Mukherjee, S. and Kumar, S. (2005) Adsorptive uptake of arsenic (V) from water by aquatic fern Salvinia natans. *Journal of Water Supply: Research and Technology - AQUA*, **54**(1), 47–53.

Mulligan, C.N., Yong, R.N. and Gibbs, B.F. (2001) Remediation technologies for metal-contaminated soils and groundwater: an evaluation. *Engineering Geology*, **60**(1-4), 193–207.

Muñoz, J.A., Gonzalo, A. and Valiente, M. (2002) Arsenic adsorption by Fe(III)-loaded open-celled cellulose sponge. Thermodynamic and selectivity aspects. *Environmental Science and Technology*, **36**(15), 3405–11.

Munro, L.D., Clark, M.W. and McConchie, D. (2004) A Bauxsol^{TM}-based permeable reactive barrier for the treatment of acid rock drainage. *Mine Water and the Environment*, **23**(4), 183–94.

Murugesan, G.S., Sathishkumar, M. and Swaminathan, K. (2006) Arsenic removal from groundwater by pretreated waste tea fungal biomass. *Bioresource Technology*, **97**(3), 483–87.

Nakahira, A., Nagata, H., Takimura, M. and Fukunishi, K. (2007a) Synthesis and evaluation of magnetic active charcoals for removal of environmental endocrine disrupter and heavy metal ion. *Journal of Applied Physics*, **101**(9), 09J114. doi: 10.1063/1.2713430.

Nakahira, A., Kubo, T. and Murase, H. (2007b) Synthesis of LDH-type clay substituted with Fe and Ni ion for arsenic removal and its application to magnetic separation. *IEEE Transactions on Magnetics*, **43**(6), 2442–44.

Nakajima, T., Xu, Y.-H., Mori, Y. *et al.* (2005) Combined use of photocatalyst and adsorbent for the removal of inorganic arsenic(III) and organoarsenic compounds from aqueous media. *Journal of Hazardous Materials*, **120**(1-3), 75–80.

Ndur, S.A. and Norman, D.J. (2003) Sorption of arsenic onto laterite: a new technology for filtering rural water. *Abstracts with Programs - Geological Society of America*, **35**(6), 413.

Nesbitt, H.W., Canning, G.W. and Bancroft, G.M. (1998) XPS study of reductive dissolution of 7Å-birnessite by H3AsO3, with constraints on reaction mechanism. *Geochimica et Cosmochimica Acta*, **62**(12), 2097–110.

Newcombe, R.L., Hart, B.K. and Möller, G. (2006) Arsenic removal from water by moving bed active filtration. *Journal of Environmental Engineering*, **132**(1), 5–12.

Nikolaidis, N.P., Dobbs, G.M. and Lackovic, J.A. (2003) Arsenic removal by zero-valent iron: field, laboratory and modeling studies. *Water Research*, **37**(6), 1417–25.

Nriagu, J.O. (1989) A global assessment of natural sources of atmospheric trace metals. *Nature*, **338**(6210), 47–49.

Oh, J.-I., Lee, S.-H. and Yamamoto, K. (2004) Relationship between molar volume and rejection of arsenic species in groundwater by low-pressure nanofiltration process. *Journal of Membrane Science*, **234**(1-2), 167–75.

Ohki, A., Nakayachigo, K., Naka, K. and Maeda, S. (1996) Adsorption of inorganic and organic arsenic compounds by aluminium-loaded coral limestone. *Applied Organometallic Chemistry*, **10**(9), 747–52.

Ohnuki, T., Sakamoto, F., Kozai, N. *et al.* (2004) Mechanisms of arsenic immobilization in a biomat from mine discharge water. *Chemical Geology*, **212**(Spec. Iss. 3-4), 279–90.

Okada, H., Kudo, Y., Nakazawa, H. *et al.* (2004) Removal system of arsenic from geothermal water by high gradient magnetic separation-HGMS reciprocal filter. *IEEE Transactions on Applied Superconductivity*, **14**(2), 1576–79.

Ona-Nguema, G., Abdelmoula, M., Jorand, F. *et al.* (2002) Iron(II,III) hydroxycarbonate green rust formation and stabilization from lepidocrocite bioreduction. *Environmental Science and Technology*, **36**(1), 16–20.

Oskoui, K. (2004) Recovery and reuse of the wood and chromated copper arsenate (CCA) from CCA treated wood — a technical paper, in *Environmental Impacts of Preservative-Treated Wood, Conference*, Gainesville, FL, February 8-11, Florida Center for Environmental Solutions, Orlando, FL, pp. 238–44.

Ottosen, L.M., Hansen, H.K. and Hansen, C.B. (2000) Water splitting at ion-exchange membranes and potential differences in soil during electrodialytic soil remediation. *Journal of Applied Electrochemistry*, **30**(11), 1199–207.

Ottosen, L.M., Hansen, H.K., Laursen, S. and Villumsen, A. (1997) Electrodialytic remediation of soil polluted with copper from wood preservation industry. *Environmental Science and Technology*, **31**(6), 1711–15.

Ottosen, L., Pedersen, A. and Christensen, I. (2004) Characterization of residues from thermal treatment of CCA impregnated wood. Chemical and electrochemical extraction, in *Environmental Impacts of Preservative-Treated Wood, Conference*, Gainesville, FL, February 8-11, Florida Center for Environmental Solutions, Orlando, FL, pp. 295–311.

Pacheco, A.C.C. and Torem, M.L. (2002) Influence of ionic strength on the removal of As5 + by adsorbing colloid flotation. *Separation Science and Technology*, **37**(15), 3599–610.

Pal, P., Ahammad, Z. and Bhattacharya, P. (2007) ARSEPPA: a visual basic software tool for arsenic separation plant performance analysis. *Chemical Engineering Journal*, **129**(1-3), 113–22.

Payne, K.B. and Abdel-Fattah, T.M. (2005) Adsorption of arsenate and arsenite by iron-treated activated carbon and zeolites: effects of pH, temperature, and ionic strength. *Journal of Environmental Science and Health, Part A: Toxic/Hazardous Substances and Environmental Engineering*, **40**(4), 723–49.

Pedersen, A.J., Kristensen, I.V., Ottosen, L.M. *et al.* (2005) Electrodialytic remediation of CCA-treated waste wood in pilot scale. *Engineering Geology*, **77**(Spec. Iss. 3-4), 331–38.

Pereira, P.A.L., Dutra, A.J.B. and Martins, A.H. (2007) Adsorptive removal of arsenic from river waters using pisolite. *Minerals Engineering*, **20**(1), 52–59.

Pettine, M., Campanella, L. and Millero, F.J. (1999) Arsenite oxidation by H2O2 in aqueous solutions. *Geochimica et Cosmochimica Acta*, **63**(18), 2727–35.

Phenrat, T., Marhaba, T.F. and Rachakornkij, M. (2005) A SEM and X-ray study for investigation of solidified/stabilized arsenic-iron hydroxide sludge. *Journal of Hazardous Materials*, **118**(1-3), 185–95.

Phenrat, T., Marhaba, T.F. and Rachakornkij, M. (2007) XRD and unconfined compressive strength study for a qualitative examination of calcium-arsenic compounds retardation of cement hydration in solidified/stabilized arsenic-iron hydroxide sludge. *Journal of Environmental Engineering*, **133**(6), 595–607.

Pickering, I.J., Gumaelius, L., Harris, H.H. *et al.* (2006) Localizing the biochemical transformations of arsenate in hyperaccumulating fern. *Environmental Science and Technology*, **40**(16), 5010–14.

Pookrod, P., Haller, K.J. and Scamehorn, J.F. (2004) Removal of arsenic anions from water using polyelectrolyte-enhanced ultrafiltration. *Separation Science and Technology*, **39**(4), 811–31.

Prasad, G. (1994) Removal of arsenic(V) from aqueous systems by adsorption onto some geologic materials, in *Arsenic in the Environment: Part I: Cycling and Characterization* (ed. J.O. Nriagu), John Wiley & Sons, Inc., New York, pp. 133–54.

Pratas, J., Prasad, M.N.V., Freitas, H. and Conde, L. (2005) Plants growing in abandoned mines of Portugal are useful for biogeochemical exploration of arsenic, antimony, tungsten and mine reclamation. *Journal of Geochemical Exploration*, **85**(3), 99–107.

Rahman, M.H., Wasiuddin, N.M. and Islam, M.R. (2004) Experimental and numerical modeling studies of arsenic removal with wood ash from aqueous streams. *Canadian Journal of Chemical Engineering*, **82**(5), 968–77.

Rajaković, L.V. (1992) The sorption of arsenic onto activated carbon impregnated with metallic silver and copper. *Separation Science and Technology*, **27**(11), 1423–33.

Ramaswami, A., Tawachsupa, S. and Isleyen, M. (2001) Batch-mixed iron treatment of high arsenic waters. *Water Research*, **35**(18), 4474–79.

Randall, S.R., Sherman, D.M. and Ragnarsdottir, K.V. (2001) Sorption of As(V) on green rust (Fe4(II)Fe2(III)(OH) 12SO4 - 3H2O) and lepidocrocite (γ -FeOOH): surface complexes from EXAFS spectroscopy. *Geochimica et Cosmochimica Acta*, **65**(7), 1015–23.

Rau, I., Gonzalo, A. and Valiente, M. (2000) Arsenic(V) removal from aqueous solutions by iron(III) loaded chelating resin. *Journal of Radioanalytical and Nuclear Chemistry*, **246**(3), 597–600.

Raven, K.P., Jain, A. and Loeppert, R.H. (1998) Arsenite and arsenate adsorption on ferrihydrite: kinetics, equilibrium, and adsorption envelopes. *Environmental Science and Technology*, **32**(3), 344–49.

Redman, A.D., Macalady, D.L. and Ahmann, D. (2002) Natural organic matter affects arsenic speciation and sorption on hematite. *Environmental Science and Technology*, **36**(13), 2889–96.

Refait, Ph., Abdelmoula, M. and Génin, J.-M.R. (1998) Mechanisms of formation and structure of green rust one in aqueous corrosion of iron in the presence of chloride ions. *Corrosion Science*, **40**(9), 1547–60.

Reisinger, H.J., Burris, D.R. and Hering, J.G. (2005) Remediating subsurface arsenic contamination with monitored natural attenuation. *Environmental Science and Technology*, **39**(22), 458A–464A.

Rivas, B.L., Del Carmen Aguirre, M. and Pereira, E. (2007a) Cationic water-soluble polymers with the ability to remove arsenate through an ultrafiltration technique. *Journal of Applied Polymer Science*, **106**(1), 89–94.

Rivas, B.L., Del Carmen Aguirre, M., Pereira, E. *et al.* (2007b) Capability of cationic water-soluble polymers in conjunction with ultrafiltration membranes to remove arsenate ions. *Polymer Engineering and Science*, **47**(8), 1256–61.

Roberts, L.C., Hug, S.J., Ruettimann, T. *et al.* (2004) Arsenic removal with iron(II) and iron(III) in waters with high silicate and phosphate concentrations. *Environmental Science and Technology*, **38**(1), 307–15.

Robins, R.G. and Tozawa, K. (1982) Arsenic removal from gold processing waste waters: the potential ineffectiveness of lime. *CIM Bulletin*, **75**(840), 171–74.

Rogers, J.M., Stewart, M., Petrie, J.G. and Haynes, B.S. (2007) Deportment and management of metals produced during combustion of CCA-treated timbers. *Journal of Hazardous Materials*, **139**(3), 500–5.

Saada, A., Breeze, D., Crouzet, C. *et al.* (2003) Adsorption of arsenic (V) on kaolinite and on kaolinite-humic acid complexes role of humic acid nitrogen groups. *Chemosphere*, **51**(8), 757–63.

Sagitova, F., Bejan, D., Bunce, N.J. and Miziolek, R. (2005) Development of an electrochemical device for removal of arsenic from drinking water. *Canadian Journal of Chemical Engineering*, **83**(5), 889–95.

Sahai, N., Lee, Y.J., Xu, H. *et al.* (2007) Role of Fe(II) and phosphate in arsenic uptake by coprecipitation. *Geochimica et Cosmochimica Acta*, **71**(13), 3193–210.

Saitúa, H., Campderro's, M., Cerutti, S. and Pacilla, A.P. (2005) Effect of operating conditions in removal of arsenic from water by nanofiltration membrane. *Desalination*, **172**(2), 173–80.

Santini, J.M., Vanden Hoven, R.N. and Macy, J.M. (2002) Characteristics of newly discovered arsenite-oxidizing bacteria, in *Environmental Chemistry of Arsenic* (ed. W.T. Frankenberger Jr.), Marcel Dekker, New York, pp. 329–42.

Sarahney, H., Wang, J. and Alshawabkeh, A. (2005) Electrokinetic process for removing Cu, Cr, and As from CCA-treated wood. *Environmental Engineering Science*, **22**(5), 642–50.

Sarkar, S., Blaney, L.M., Gupta, A. *et al.* (2007) Use of ArsenXnp, a hybrid anion exchanger, for arsenic removal in remote villages in the Indian subcontinent. *Reactive and Functional Polymers*, **67**(Spec. Iss. 12), 1599–611.

Sato, Y., Kang, M., Kamei, T. and Magara, Y. (2002) Performance of nanofiltration for arsenic removal. *Water Research*, **36**(13), 3371–77.

Saxe, J.K., Wannamaker, E.J., Conklin, S.W. *et al.* (2007) Evaluating landfill disposal of chromated copper arsenate (CCA) treated wood and potential effects on groundwater: evidence from Florida. *Chemosphere*, **66**(3), 496–504.

Say, R., Emir, S., Garipcan, B. *et al.* (2003a) Novel methacryloylamidophenylalanine functionalized porous chelating beads for adsorption of heavy metal ions. *Advances in Polymer Technology*, **22**(4), 355–64.

Say, R., Yilmaz, N. and Denizli, A. (2003b) Biosorption of cadmium, lead, mercury, and arsenic ions by the fungus penicillium purpurogenum. *Separation Science and Technology*, **38**(9), 2039–53.

Schmidt, G.T., Vlasova, N., Zuzaan, D. *et al.* (2008) Adsorption mechanism of arsenate by zirconyl-functionalized activated carbon. *Journal of Colloid and Interface Science*, **317**(1), 228–34.

Seko, N., Basuki, F., Tamada, M. and Yoshii, F. (2004) Rapid removal of arsenic(V) by zirconium(IV) loaded phosphoric chelate adsorbent synthesized by radiation induced graft polymerization. *Reactive and Functional Polymers*, **59**(3), 235–41.

Seko, N., Tamada, M. and Yoshii, F. (2005) Current status of adsorbent for metal ions with radiation grafting and crosslinking techniques. *Nuclear Instruments and Methods in Physics Research, Section B: Beam Interactions with Materials and Atoms*, **236**(1-4), 21–29.

Shih, M-C. (2005) An overview of arsenic removal by pressure-driven membrane processes. *Desalination*, **172**(1), 85–97.

Shih, C.-J. and Lin, C.-F. (2003) Arsenic contaminated site at an abandoned copper smelter plant: waste characterization and solidification/stabilization treatment. *Chemosphere*, **53**(7), 691–703.

Singh, A.K. (2007) Approaches for removal of arsenic from groundwater of northeastern India. *Current Science*, **92**(11), 1506–15.

Smith, S.D. and Edwards, M. (2005) The influence of silica and calcium on arsenate sorption to oxide surfaces. *Journal of Water Supply: Research and Technology - AQUA*, **54**(4), 201–11.

Smith, L.A., Means, J.L., Chen, A. *et al.* (1995) *Remedial Options for Metals-Contaminated Sites*, CRC Lewis Publishers, Boca Raton, FL.

Solo-Gabriele, H. and Townsend, T. (1999) Disposal practices and management alternatives for CCA-treated wood waste. *Waste Management and Research*, **17**(5), 378–89.

Solozhenkin, P.M., Zouboulis, A.I. and Katsoyiannis, I.A. (2007) Removal of arsenic compounds by chemisorption filtration. *Journal of Mining Science*, **43**(2), 212–20.

Spellman, F.R. (2003) *Handbook of Water and Wastewater Treatment Plant Operations*, Lewis Publishers, Boca Raton, FL, p. 661.

Stipp, S.L.S., Hansen, M., Kristensen, R. *et al.* (2002) Behaviour of Fe-oxides relevant to contaminant uptake in the environment. *Chemical Geology*, **190**(1-4), 321–37.

Stollenwerk, K.G. (2003) Geochemical processes controlling transport of arsenic in groundwater: a review of adsorption, in *Arsenic in Ground Water* (eds A.H. Welch and K.G. Stollenwerk), Kluwer Academic Publishers, Boston, MA, pp. 67–100.

Stollenwerk, K.G. and Colman, J.A. (2003) Natural remediation potential of arsenic-contaminated ground water, in *Arsenic in Ground Water* (eds A.H. Welch and K.G. Stollenwerk), Kluwer Academic Publishers, Boston, MA, pp. 351–79.

Stook, K., Tolaymat, T., Ward, M. *et al.* (2005) Relative leaching and aquatic toxicity of pressure-treated wood products using batch leaching tests. *Environmental Science and Technology*, **39**(1), 155–63.

Su, C. and Puls, R.W. (2001a) Arsenate and arsenite removal by zerovalent iron: kinetics, redox transformation, and implications for in situ groundwater remediation. *Environmental Science and Technology*, **35**(7), 1487–92.

Su, C. and Puls, R.W. (2001b) Arsenate and arsenite removal by zerovalent iron: effects of phosphate, silicate, carbonate, borate, sulfate, chromate, molybdate, and nitrate, relative to chloride. *Environmental Science and Technology*, **35**(22), 4562–68.

Su, C. and Puls, R.W. (2003) In situ remediation of arsenic in simulated groundwater using zerovalent iron: laboratory column tests on combined effects of phosphate and silicate. *Environmental Science and Technology*, **37**(11), 2582–87.

Su, C. and Puls, R.W. (2004) Significance of iron (II,III) hydroxycarbonate green rust in arsenic remediation using zerovalent iron in laboratory column tests. *Environmental Science and Technology*, **38**(19), 5224–31.

Sullivan, E.J., Bowman, R.S. and Legiec, I.A. (2003) Sorption of arsenic from soil-washing leachate by surfactant-modified zeolite. *Journal of Environmental Quality*, **32**(6), 2387–91.

Suzuki, T.M., Tanco, M.L., Tanaka, D.A.P. *et al.* (2001) Adsorption characteristics and removal of oxo-anions of arsenic and selenium on the porous polymers loaded with monoclinic hydrous zirconium oxide. *Separation Science and Technology*, **36**(1), 103–111.

Sylvester, P., Westerhoff, P., Moller, T. *et al.* (2007) A hybrid sorbent utilizing nanoparticles of hydrous iron oxide for arsenic removal from drinking water. *Environmental Engineering Science*, **24**(1), 104–12.

Taerakul, P., Sun, P., Golightly, D.W. *et al.* (2006) Distribution of arsenic and mercury in lime spray dryer ash. *Energy and Fuels*, **20**(4), 1521–27.

Tani, Y., Miyata, N., Ohashi, M. *et al.* (2004) Interaction of inorganic arsenic with biogenic manganese oxide produced by a Mn-oxidizing fungus, strain KR21-2. *Environmental Science and Technology*, **38**(24), 6618–24.

Teixeira, M.C. and Ciminelli, V.S.T. (2005) Development of a biosorbent for arsenite: structural modeling based on X-ray spectroscopy. *Environmental Science and Technology*, **39**(3), 895–900.

Thirunavukkarasu, O.S., Viraraghavan, T., Subramanian, K.S. *et al.* (2005) Arsenic removal in drinking water — impacts and novel removal technologies. *Energy Sources*, **27**(1–2), 209–19.

Tournassat, C., Charlet, L., Bosbach, D. and Manceau, A. (2002) Arsenic(III) oxidation by birnessite and precipitation of manganese(II) arsenate. *Environmental Science and Technology*, **36**(3), 493–500.

Townsend, T., Dubey, B. and Solo-Gabriele, H. (2004) Assessing potential waste disposal impact from preservative treated wood products, in *Environmental Impacts of Preservative-Treated Wood, Conference*, Gainesville, FL, February 8-11, Florida Center for Environmental Solutions, Orlando, FL, pp. 169–88.

Townsend, T., Solo-Gabriele, H., Tolaymat, T. *et al.* (2003) Chromium, copper, and arsenic concentrations in soil underneath CCA-treated wood structures. *Soil and Sediment Contamination*, **12**(6), 779–98.

Trolard, F., Génin, J.-M.R., Abdelmoula, M. *et al.* (1997) Identification of a green rust mineral in a reductomorphic soil by Mössbauer and Raman spectroscopies. *Geochimica et Cosmochimica Acta*, **61**(5), 1107–11.

Uddin, M.T., Mozumder, M.S.I., Islam, M.A. *et al.* (2007a) Nanofiltration membrane process for the removal of arsenic from drinking water. *Chemical Engineering and Technology*, **30**(9), 1248–54.

Uddin, M.T., Mozumder, Md.S.I., Figoli, A. *et al.* (2007b) Arsenic removal by conventional and membrane technology: an overview. *Indian Journal of Chemical Technology*, **14**(5), 441–50.

Urase, T., Oh, J.-I. and Yamamoto, K. (1998) Effect of pH on rejection of different species of arsenic by nanofiltration. *Desalination*, **117**(1-3), 11–18.

Urban, D.R. and Wilcox, J. (2006) Theoretical study of the kinetics of the reactions Se + O2 → Se + O and As + HCl → AsCl + H. *Journal of Physical Chemistry A*, **110**(28), 8797–801.

US Department of Energy (US DOE) (1999) *Decision-Making Framework Guide for the Evaluation and Selection of Monitored Natural Attenuation Remedies at Department of Energy Sites*, Office of Environmental Restoration, p. 19.

US Environmental Protection Agency (US EPA) (1999) Presumptive Remedy for Metals-in-Soils Sites. EPA-540-F-98-054, Office of Solid Wastes and Emergency (5102G).

US Environmental Protection Agency (US EPA) (2002a) Arsenic Treatment Technologies for Soil, Waste, and Water. EPA-542-R-02-004, Office of Solid Wastes and Energency (5102G).

US Environmental Protection Agency (US EPA) (2002b) Proven Alternatives for Aboveground Treatment of Arsenic in Groundwater. EPA-542-S-02-002, Office of Solid Wastes and Emergency (5102G).

Vaishya, R.C. and Gupta, S.K. (2004) Modeling arsenic(V) removal from water by sulfate modified iron-oxide coated sand (SMIOCS). *Separation Science and Technology*, **39**(3), 645–66.

Vatutsina, O.M., Soldatov, V.S., Sokolova, V.I. *et al.* (2007) A new hybrid (polymer/inorganic) fibrous sorbent for arsenic removal from drinking water. *Reactive and Functional Polymers*, **67**(3), 184–201.

Vaughan, R.L., Jr., Reed, B.E. and Smith, E.H. (2007) Modeling As(V) removal in iron oxide impregnated activated carbon columns. *Journal of Environmental Engineering*, **133**(1), 121–24.

Violante, A., Ricciardella, M., Del Gaudio, S. and Pigna, M. (2006) Coprecipitation of arsenate with metal oxides: nature, mineralogy, and reactivity of aluminum precipitates. *Environmental Science and Technology*, **40**(16), 4961–67.

Vlyssides, A., Barampouti, E.M. and Mai, S. (2005) Heavy metal removal from water resources using the aquatic plant Apium nodiflorum. *Communications in Soil Science and Plant Analysis*, **36**(7-8), 1075–81.

Wang, J.W., Bejan, D. and Bunce, N.J. (2003) Removal of arsenic from synthetic acid mine drainage by electrochemical pH adjustment and coprecipitation with iron hydroxide. *Environmental Science and Technology*, **37**(19), 4500–6.

Wang, T. and Guan, Y. (2005) Extraction of arsenic-containing anions by supercritical CO2 with ion-pairing. *Chemical Engineering Journal*, **108**(1-2), 145–53.

Wang, Y. and Reardon, E.J. (2001) A siderite/limestone reactor to remove arsenic and cadmium from wastewaters. *Applied Geochemistry*, **16**(9–10), 1241–49.

Wasiuddin, N.M., Tango, M. and Islam, M.R. (2002) A novel method for arsenic removal at low concentrations. *Energy Sources*, **24**(11), 1031–41.

Wasson, S.J., Linak, W.P., Gullett, B.K. *et al.* (2005) Emissions of chromium, copper, arsenic, and PCDDs/Fs from open burning of CCA-treated wood. *Environmental Science and Technology*, **39**(22), 8865–76.

Wei, C.-Y., Wang, C., Sun, X. and Wang, W.-Y. (2007) Arsenic accumulation by ferns: a field survey in southern China. *Environmental Geochemistry and Health*, **29**(3), 169–77.

Weng, Y.-H., Lin, H.C.-H., Lee, H.-H. *et al.* (2005) Removal of arsenic and humic substances (HSs) by electro-ultrafiltration (EUF). *Journal of Hazardous Materials*, **122**(1-2), 171–76.

Westerhoff, P., Highfield, D., Badruzzaman, M. and Yoon, Y. (2005) Rapid small-scale column tests for arsenate removal in iron oxide packed bed columns. *Journal of Environmental Engineering*, **131**(2), 262–71.

Wilkie, J.A. and Hering, J.G. (1996) Adsorption of arsenic onto hydrous ferric oxide: effects of adsorbate/adsorbent ratios and co-occurring solutes. *Colloids and Surfaces A: Physicochemical and Engineering Aspects*, **107**, 97–110.

Wolthers, M., Charlet, L., van Der Weijden, C.H. *et al.* (2005) Arsenic mobility in the ambient sulfidic environment: sorption of arsenic(V) and arsenic(III) onto disordered mackinawite. *Geochimica et Cosmochimica Acta*, **69**(14), 3483–92.

Xia, S., Dong, B., Zhang, Q. *et al.* (2007) Study of arsenic removal by nanofiltration and its application in China. *Desalination*, **204**(Spec. Iss. 1-3), 374–79.

Xiao, L., Wildgoose, G.G., Crossley, A. *et al.* (2006) Removal of toxic metal-ion pollutants from water by using chemically modified carbon powders. *Chemistry-An Asian Journal*, **1**(4), 614–22.

Xu, T., Cai, Y. and O'Shea, K.E. (2007) Adsorption and photocatalyzed oxidation of methylated arsenic species in TiO2 suspensions. *Environmental Science and Technology*, **41**(15), 5471–77.

Xu, T., Kamat, P.V. and O'Shea, K.E. (2005) Mechanistic evaluation of arsenite oxidation in TiO2 assisted photocatalysis. *Journal of Physical Chemistry A*, **109**(40), 9070–75.

Yang, L., Shahrivari, Z., Liu, P.K.T. *et al.* (2005) Removal of trace levels of arsenic and selenium from aqueous solutions by calcined and uncalcined layered double hydroxides (LDH). *Industrial and Engineering Chemistry Research*, **44**(17), 6804–15.

Yang, L., Wu, S. and Chen, J.P. (2007) Modification of activated carbon by polyaniline for enhanced adsorption of aqueous arsenate. *Industrial and Engineering Chemistry Research*, **46**(7), 2133–40.

Ye, Z.H., Lin, Z.-Q., Whiting, S.N. *et al.* (2003) Possible use of constructed wetland to remove selenocyanate, arsenic, and boron from electric utility wastewater. *Chemosphere*, **52**(9), 1571–79.

Yokota, H., Tanabe, K., Sezaki, M. *et al.* (2001) Arsenic contamination of ground and pond water and water purification system using pond water in Bangladesh. *Engineering Geology*, **60**(1–4), 323–31.

Yoshitake, H., Yokoi, T. and Tatsumi, T. (2003) Adsorption behavior of arsenate at transition metal cations captured by amino-functionalized mesoporous silicas. *Chemistry of Materials*, **15**(8), 1713–21.

Yu, X., Amrhein, C., Zhang, Y. and Matsumoto, M.R. (2006) Factors influencing arsenite removal by zero-valent iron. *Journal of Environmental Engineering*, **132**(11), 1459–69.

Yuan, T., Hu, J.Y., Ong, S.L. *et al.* (2002) Arsenic removal from household drinking water by adsorption. *Journal of Environmental Science and Health, Part A: Toxic/Hazardous Substances and Environmental Engineering*, **37**(9), 1721–36.

Yuan, T., Luo, Q.-F., Hu, J.-Y. *et al.* (2003) A study on arsenic removal from household drinking water. *Journal of Environmental Science and Health, Part A: Toxic/Hazardous Substances and Environmental Engineering*, **38**(9), 1731–44.

Zeng, L. (2003) A method for preparing silica-containing iron(III) oxide adsorbents for arsenic removal. *Water Research*, **37**(18), 4351–58.

Zeng, H., Fisher, B. and Giammar, D.E. (2008) Individual and competitive adsorption of arsenate and phosphate to a high-surface-area iron oxide-based sorbent. *Environmental Science and Technology*, **42**(1), 147–52.

Zhang, F-S. and Itoh, H. (2006) Photocatalytic oxidation and removal of arsenite from water using slag-iron oxide-TiO2 adsorbent. *Chemosphere*, **65**(1), 125–31.

Zhang, G.-S., Qu, J.-H., Liu, H.-J. *et al.* (2007a) Removal mechanism of As(III) by a novel Fe-Mn binary oxide adsorbent: oxidation and sorption. *Environmental Science and Technology*, **41**(13), 4613–19.

Zhang, Q.L., Lin, Y.C., Chen, X. and Gao, N.Y. (2007b) A method for preparing ferric activated carbon composites adsorbents to remove arsenic from drinking water. *Journal of Hazardous Materials*, **148**(3), 671–78.

Zhang, Q.L., Gao, N.Y., Lin, Y.C. *et al.* (2007c) Removal of arsenic(V) from aqueous solutions using iron-oxide-coated modified activated carbon. *Water Environment Research*, **79**(8), 931–36.

Zhang, W., Singh, P., Paling, E. and Delides, S. (2004) Arsenic removal from contaminated water by natural iron ores. *Minerals Engineering*, **17**(4), 517–24.

Zhang, Y., Yang, M., Dou, X.-M. *et al.* (2005) Arsenate adsorption on an Fe–Ce bimetal oxide adsorbent: role of surface properties. *Environmental Science and Technology*, **39**(18), 7246–53.

Zhu, X. and Jyo, A. (2001) Removal of arsenic(V) by zirconium(IV)-loaded phosphoric acid chelating resin. *Separation Science and Technology*, **36**(14), 3175–89.

Appendix A

Common Physical and Chemical Constants and Conversions for Units of Measure

Constants

Avogadro's number $= 6.022\ 045 \times 10^{23}\ \text{mol}^{-1}$
$e \approx 2.718\ 28$
Faraday constant $(F) = 96\ 485\ \text{C mol}^{-1}$
Gas constant $(R) = 8.314\ 41\ \text{J mol}^{-1}\ \text{K}^{-1} = 1.987\ 17\ \text{cal mol}^{-1}\cdot\text{K}^{-1}$
Molar volume of ideal gas at 273.15 K and 1 atm $= 0.022\ 413\ \text{m}^3\ \text{mol}^{-1}$
Pi $(\pi) \approx 3.141\ 59$
Speed of light in a vacuum $(c) = 2.997\ 924\ 5\overline{\text{\ }} \times 10^8\ \text{m s}^{-1}$

Energy

To obtain	From	Multiple by
British thermal units (BTU, thermochemical)	Calories (cal, thermochemical)	3.968×10^{-3}
British thermal units (BTU, thermochemical)	Joules (J)	$9.484\ 52 \times 10^{-4}$
Calories (cal, thermochemical)	British thermal units (BTU, thermochemical)	252.0
Calories (cal, thermochemical)	Joules (J)	0.2390
Electron volts (eV)	Joules (J)	$6.241\ 509\ 48 \times 10^{18}$
Joules (J)	British thermal units (BTU, thermochemical)	1054.35
Joules (J)	Calories (cal, thermochemical)	4.184
Joules (J)	Electron volts (eV)	$1.602\ 176\ 53 \times 10^{-19}$

Arsenic   Edited by Kevin R. Henke
© 2009 John Wiley & Sons, Ltd

Length and Distance

To obtain:	From:	Multiple by:
Ångström (Å)	Meters (m)	1.0×10^{10}
Centimeters (cm)	Feet (ft)	30.48
Centimeters (cm)	Inches (in)	2.54
Feet (ft)	Centimeters (cm)	0.0328
Feet (ft)	Meter (m)	3.2808
Inches (in)	Centimeters (cm)	0.3937
Kilometers (km)	Miles (US, mi)	1.6093
Meters (m)	Ångström (Å)	1.0×10^{-10}
Meters (m)	Feet (ft)	0.3048
Meters (m)	Yards (yd)	0.9144
Miles (US, mi)	Kilometers (km)	0.6214
Yards (yd)	Meters (m)	1.0936

Mass and Weight

To obtain:	From:	Multiple by:
Atomic mass units (amu)	Daltons (Da)	1
Atomic mass units (amu)	Kilograms (kg)	$6.022\ 141 \times 10^{26}$
Grams (g)	Ounces (oz)	28.35
Grams (g)	US pounds (avoirdupois)	453.6
Kilograms (kg)	Atomic mass units (amu)	$1.660\ 539 \times 10^{-27}$
Kilograms (kg)	Metric tons (t)	1000
Kilograms (kg)	US pounds (avoirdupois)	0.4536
Metric tons (tonne)	Kilograms (kg)	0.001
Ounces (oz)	Grams (g)	0.0353
US pounds (avoirdupois)	Kilograms (kg)	2.2046

Pressure

To obtain:	From:	Multiple by:
Atmospheres (atm, standard)	Bars	0.986 9
Atmospheres (atm, standard)	Pascals (Pa)	9.869×10^{-6}
Atmospheres (atm, standard)	Pounds per square inch (psi)	0.068 05
Atmospheres (atm, standard)	Torr	0.001 316

To obtain:	From:	Multiple by:
Bars	Atmospheres (atm, standard)	1.013 25
Bars	Pascals (Pa)	1×10^{-5}
Bars	Pounds per square inch (psi)	0.0689 5
Bars	Torr	0.001 333
Pascals (Pa)	Atmospheres (atm, standard)	101 325
Pascals (Pa)	Bars	100 000
Pascals (Pa)	Pounds per square inch (psi)	6 894.76
Pascals (Pa)	Torr	133.3
Pounds per square inch (psi)	Atmospheres (atm, standard)	14.696
Pounds per square inch (psi)	Bars	14.504
Pounds per square inch (psi)	Pascals (Pa)	$1.450\ 4 \times 10^{-4}$
Pounds per square inch (psi)	Torr	0.0193 4
Torr	Atmospheres (atm, standard)	760
Torr	Bars	750
Torr	Pascals (Pa)	7.501×10^{-3}
Torr	Pounds per square inch (psi)	51.715

Temperature

Celsius ($^\circ$C) = Kelvin (K) $-$ 273.15
Celsius ($^\circ$C) = ($^\circ$F $-$ 32)/1.8
Fahrenheit ($^\circ$F) = 1.8($^\circ$C) + 32
Kelvin (K) = Celsius ($^\circ$C) + 273.15

Time

1 day = 1440 minutes
1 day = 86 400 seconds
1 hour = 3600 seconds
1 tropical year = 365.242 19 days
1 tropical year = $3.155\ 6925 \times 10^7$ seconds

Volume

To obtain:	From:	Multiple by:
Cubic centimeters (cm^3)	Cubic feet (ft^3)	28 317
Cubic centimeters (cm^3)	Cubic inches (in^3)	16.39
Cubic centimeters (cm^3)	Cubic meters (m^3)	1×10^6

To obtain:	From:	Multiple by:
Cubic centimeters (cm^3)	US gallons (gal, liquid)	3785
Cubic centimeters (cm^3)	Liters (L)	1000
Cubic centimeters (cm^3)	Milliliters (mL)	1
Cubic feet (ft^3)	Cubic centimeters (cm^3)	3.532×10^{-5}
Cubic feet (ft^3)	Cubic meters (m^3)	35.32
Cubic feet (ft^3)	Liters (L)	0.353 2
Cubic inches (in^3)	Cubic centimeters (cm^3)	0.061
Cubic inches (in^3)	US gallons (gal, liquid)	231
Cubic meters (m^3)	Cubic centimeters (cm^3)	1×10^{-6}
Cubic meters (m^3)	Cubic feet (ft^3)	0.028 317
Cubic meters (m^3)	US gallons (gal, liquid)	3.785×10^{-3}
Cubic meters (m^3)	Liters (L)	0.001
Cubic meters (m^3)	Cubic yards (yd^3)	0.764 555
Cubic yards (yd^3)	Cubic meters (m^3)	1.3080
Liters (L)	Cubic centimeters (cm^3)	0.001 Liters (L)
Liters (L)	Cubic feet (ft^3)	28.32
Liters (L)	Cubic meters (m^3)	1000
Liters (L)	US quarts (qt, liquid)	0.946 4
Liters (L)	US gallons (gal, liquid)	3.785 4
US gallons (gal, liquid)	Cubic centimeters (cm^3)	2.642×10^{-4}
US gallons (gal, liquid)	Cubic inches (in^3)	4.329×10^{-3}
US gallons (gal, liquid)	Cubic meters (m^3)	264.2
US gallons (gal, liquid)	Liters (L)	0.264 2
US quarts (qt, liquid)	Liters (L)	1.056 7

Table A.1 *Prefixes for metric measurements.*

Prefix	Abbreviation	Value
Tera	T	10^{12}
Giga	G	10^9
Mega	M	10^6
Kilo	k	1000
Hecto	h	100
Deca	da	10
Deci	d	0.10
Centi	c	0.01
Milli	m	0.001
Micro	μ	10^{-6}
Nano	n	10^{-9}
Pico	p	10^{-12}

Table A.2 *Atomic masses of selective elements.*

Element	Atomic Mass (atomic mass units, amu)
Arsenic (As)	74.92160
Aluminum (Al)	26.98154
Antimony (Sb)	121.757
Calcium (Ca)	40.078
Carbon (C)	12.011
Chlorine (C)	35.453
Chromium (Cr)	51.996
Copper (Cu)	63.546
Fluorine (F)	18.998403
Gallium (Ga)	69.72
Germanium (Ge)	72.61
Hydrogen (H)	1.00794
Iron (Fe)	55.847
Magnesium (Mg)	24.305
Manganese (Mn)	54.938
Nickel (Ni)	58.69
Nitrogen (N)	14.0067
Oxygen (O)	15.9994
Phosphorus (P)	30.97376
Potassium (K)	39.0983
Selenium (Se)	78.96
Silicon (Si)	28.0855
Sodium (Na)	22.98977
Sulfur (S)	32.066
Zinc (Zn)	65.39

Appendix B

Glossary of Terms

B.1 Introduction

The terms in this glossary originate from a wide variety of scientific and engineering disciplines, including chemistry, toxicology, geology, planetary geology, biology, waste management and treatment, and physics. The definitions of the terms were generally derived from the disciplines that most often use them. However, researchers in a given discipline may have diverse or even contradictory definitions for some terms (e.g. 'theory' or 'water table'). The definitions in this glossary are attempts at the best consensus. Other professionals and the general public may define some of the terms very differently. Italicized words in the definitions or words closely related to the italicized words are also defined in this glossary. Furthermore, Appendix A has conversions for many of the units of measure mentioned in this glossary.

B.2 Glossary

Abiotic: Nonliving (compare with biotic).

Absorbent: A solid that assimilates a gas, liquid, or *solute* into its interior (compare with adsorbent and sorbent).

Absorption: The assimilation of a *solute*, gas, or liquid into the interior of a solid material, an *absorbent*. Absorption may include the migration of solutes into internal pores or the migration or exchange of atoms within the crystalline structure of a *mineral*. Some researchers use the generic term '*sorption*' to refer to a natural or artificial process where both absorption and adsorption may be involved or if absorption and adsorption cannot be distinguished (compare with adsorption and sorption).

Achondrite: A type of stony (low in elemental metals) *meteorite*. Unlike *chrondrites*, they do not contain *chondrules*.

Acid: A substance that releases H^+ when it's dissolved in water (compare with base, strong acid, and weak acid).

Acid mine drainage: Water runoff from coal or ore mines that is rich in sulfuric acid and possibly toxic metals (such as lead, arsenic, and mercury). The sulfuric acid and metals result from the chemical weathering of pyrite (FeS_2) and other sulfide minerals in the mined rocks.

Arsenic   Edited by Kevin R. Henke
© 2009 John Wiley & Sons, Ltd

Action Level: A regulatory standard, especially used by the United States Occupational Safety and Health Administration (OSHA) for air contaminants, to indicate an unhealthy concentration of a substance. The US action level for arsenic in air is $5\,mg\,m^{-3}$ averaged over an eight-hour period (29 *Code of Federal Regulations* 1910.1018). If an action level is exceeded, efforts must be made to protect individuals from exposure to the contaminant, reduce the concentration, and possibly monitor or treat exposed individuals. In general, action levels are 50 % of the *permissible exposure limit* (PEL) (compare with threshold limit value).

Activated carbon: A largely amorphous form of porous and high surface area graphite, which is used to sorb a wide variety of contaminants from air and water. A number of organic sources may be utilized to produce activated carbon, including: wood, coal, peat, crop residues, bones, or petroleum byproducts. After charring the source materials to remove volatiles, the resulting carbon is usually 'activated' with heat and chemicals (such as acids, bases, steam, or metal salts) to increase its surface area and improve sorption.

Activity (chemistry): The effective concentration of an ion in a solution. In solutions with high ionic strengths, ions interact with each other and are not totally independent chemical units. The interactions affect the boiling point, freezing point, and other properties of the solution (Nebergall, Schmidt and Holtzclaw, 1976), 316. The effective concentration (activity) of an ion in a high ionic strength solution is usually less than the total number of moles in the solution. The activity of an ion is calculated by multiplying its molal concentration by its activity coefficient. In very dilute solutions, activities and molal concentrations are essentially equal.

Activity coefficient: A value (γ) when multiplied by the molal concentration of an ion in an aqueous solution gives the activity of the ion. Activity coefficients correct the molal concentrations of ions in solution for interferences from other ions (see (Faure, 1998), 113). In very dilute solutions, $\gamma \approx 1$ and activities and molal concentrations are essentially equal. Equations, such as the Debye-Hückel and Davies, are used to calculate activity coefficients for ions (for detailed discussions, see (Faure, 1998), 140–142; (Langmuir, 1997), 123–135). The Debye-Hückel is utilized with solutions that have ionic strengths below 0.1 molar (M) and the Davies equation is applied to solutions with ionic strengths up to about 0.5 M. Other equations are available for solutions with higher ionic strengths (Faure, 1998), 139–142; (Langmuir, 1997), 129–135; (Eby, 2004), 36–40. Geochemical computer models, such as MINTEQA2 (Allison, Brown and Novo-Gradac, 1991), readily calculate activities and activity coefficients by using the Debye-Hückel or Davies equations.

Activity product: For a reaction not at equilibrium, the activity product (Q) at a given temperature and pressure is derived from the ratio of the activities of the products to the activities of the reactants. For the following reaction:

$$aA + bB = cC + dD,$$

Q = [C]c[D]d/[A]a[B]b, where Q ≠ Keq, the equilibrium constant for the reaction.

Acute (toxicology): An adverse event that occurs following a one-time or short-time exposure to an agent (chemical or physical) over 24 hours. In some cases, urgent treatement may be required (compare with chronic).

Acute promyelocytic leukemia: A form of leukemia involving the infiltration of bone marrow by abnormal pre- and young leukocytes.

Adsorbate: A contaminant or other substance that is removed from its host liquid or gas through *adsorption* onto an *adsorbent*.

Adsorbent: A solid material that removes chemical species from liquids or gases by causing them to adhere onto its surface (compare with absorbent and sorbent).

Adsorption: The adherence of a solute, gas, or liquid onto the surface of a solid substance (the *adsorbent*). Also, adsorption refers to a treatment technology for liquids and gases that primarily uses adsorption to remove contaminants. Some researchers use the generic term '*sorption*' to refer to a natural or artificial process where both adsorption and absorption may be involved or if adsorption and absorption cannot be distinguished (compare with absorption and sorption).

Adsorption isotherm: An equation or distribution on a graph representing the concentration of an adsorbed chemical species as a function of its concentration in an associated aqueous solution. Adsorption isotherms are measured in systems that are at equilibrium and where temperature, pressure, and possibly other conditions are held constant. Isotherms are often described with linear, *Freundlich*, or *Langmuir* equations.

Aerobic: With O_2. Also refers to environments with O_2 and organisms that live in environments with O_2 (compare with anaerobic, anoxic, and oxic).

Aerodynamic diameter: Particles suspended in gases typically have irregular shapes. The aerodynamic behavior of an irregularly shaped particle may be modeled with an idealized sphere of uniform composition. The diameter of the sphere is the aerodynamic diameter (compare with mass median aerodynamic diameter).

A horizon (soil): An organic-rich soil layer or 'topsoil' in the upper portion of *soil profiles*. The A horizon is critical for the growth of terrestrial plants.

Aldehyde: An organic molecule with the general formula HC(O)R, where R represents an organic carbon chain. Formaldehyde is an example of an aldehyde (Figure B.1).

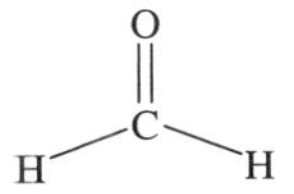

Figure B.1

Alkane: A hydrocarbon with the general formula of C_nH_{2n+2}, where n ≥ 1. Methane (CH_4) and ethane (C_2H_6) are alkanes.

Alkene: An organic molecule with a carbon to carbon double bond (C=C). Ethylene (C_2H_4) is an example of an alkene.

Alkene epioxidation reaction: A reaction which results in the formation a carbon–oxygen three-membered ring.

Alkyl: An organic group that forms from the removal of one hydrogen from an alkane. The methyl groups (–CH3) in methylarsenic species are alkyls.

Alpine glacier: A glacier that forms and flows in a mountain valley (compare with continental glacier).

Alumina: Aluminum oxide.

Amine: An organic group that is derived from ammonia (NH_3), where one or more of the hydrogens on the ammonia are replaced by *alkyl* or *aromatic* groups.

Amorphous: A noncrystalline solid substance. Unlike crystalline materials, the atoms of amorphous substances do not have regular (ordered) arrangements. Glass is an amorphous substance.

Anaerobic: Without O_2. In particular, environments that lack O_2 and microorganisms that live in environments without O_2 (compare with aerobic, anoxic, and oxic).

Anion exchange capacity: The ability of a solid substance to adsorb anions. The anion exchange capacity of a material represents the total positive charge on the surface of the material and is generally expressed in milli*equivalents* per 100 grams of material (compare with cation exchange capacity).

Anorthosite: An igneous rock almost entirely composed of calcium-rich *plagioclase*.

Anoxic: Without O_2 (compare with aerobic, anaerobic, and oxic).

Anthracite: A *metamorphic coal* that contains very little water in its chemical structure. Anthracites usually form at subsurface temperatures above 200 °C (Boggs, 1995), 279. The coal results from the *metamorphism* of *lignite, subbituminous,* or *bituminous coals.*

Apatite: A phosphate *mineral* (ideally, $Ca_5(PO_4)_3(F,Cl,OH)$) that may occur in *sedimentary, metamorphic,* and *igneous rocks. Arsenate* may substitute for phosphate.

Apoptosis: Programmed cell death.

Aquifer: A *rock, sediment,* or *soil* that is capable of transmitting enough water to supply wells.

Aquitard: A water-saturated, *impermeable rock, sediment,* or *soil* that is incapable of transmitting useful quantities of water (compare with aquifer, also see (Freeze and Cherry, 1979), 47).

Aragonite: A calcium carbonate mineral ($CaCO_3$) that is a *polymorph* of calcite.

Aromatic: A group of organic compounds that result from the substitution of an atom or group of atoms for one or more of the hydrogens on a *benzene* ring. Organic compounds also exist where nitrogen is present in one or more of the carbon positions in the ring.

Arsenate: A *mineral,* compound, or aqueous species containing $AsO_4{}^{3-}$, where the *valence* state of the arsenic is pentavalent (As(V)). Arsenate is often abbreviated "As(V)" in the literature, especially in documents dealing with arsenic treatment. The inorganic arsenic acid species, $H_2AsO_4{}^-$ and $HAsO_4{}^{2-}$, are the most common dissolved forms of arsenate in near *pH* neutral aqueous solutions (compare with arsenite and thioarsenate).

Arsenian: Arsenic-bearing, especially refers to pyrite, where arsenic partially substitutes for sulfur in the crystalline structure. The term may be used to describe a *mineral* that contains significant amounts of arsenic, although the arsenic is not a major component in the mineral and usually does not appear in its formula; for example, arsenian pyrite (FeS_2) (compare with arsenical).

Arsenical: An arsenic compound. Unlike *arsenian minerals,* arsenic is a major component in arsenicals (compare with organoarsenical).

Arsenide: An arsenic atom with a valence state of -3. A *mineral* or other compound where the major anion is As^{3-}.

Arsenite: A *mineral,* compound or aqueous species containing $AsO_3{}^{3-}$, where the *valence* state of the arsenic is trivalent (As(III)). Arsenite is often abbreviated "As(III)" in the literature, especially in documents dealing with arsenic treatment. The inorganic arsenious acid species, $H_3AsO_3{}^0$, is the most common dissolved form of arsenite in near *pH* neutral aqueous solutions (compare with arsenate and thioarsenite).

Arsenolysis: Cleavage of an *arsenate* group from a *molecule.*

Arsenosulfide: A *mineral* or other compound containing both *arsenide* and *sulfide* as the major anions. Arsenopyrite (FeAsS) is an arsenosulfide mineral (compare with thioarsenic).

Asteroid: A nonicy object or small planet up to 1020 km in diameter, most of which orbit the Sun in the inner solar system.

Asthenosphere: A plastic and partially molten layer in the *mantle* under the *lithosphere.* The asthenosphere usually extends from about 70 to 200 km below the Earth's surface.

Atomic mass: The number of *atomic mass units* of an atom.

Atomic mass unit (amu): 1/12 the mass of a carbon-12 atom, 1 *Da.*

Atomic number (Z): The number of protons in the nucleus of an atom. Each element has its own atomic number (compare with mass number, *A*).

Authigenic: A *rock* or *mineral* that formed in its current location rather than forming elsewhere and being transported to its current location.

Baghouse: Fabric filters used to remove fine-grained particles from *flue gas* in combustion facilities.

Basalt: An *extrusive mafic igneous rock*. An extrusive equivalent of a *gabbro*.

Base: A substance that releases OH^- when it's dissolved in water (compare with acid).

Batch leaching: A *leaching test* that estimates how readily contaminants could mobilize out of a solid waste or other solid material if the material comes into contact with natural waters. The test involves placing a given mass of the solid sample in a container with a specific volume of a liquid *leaching* solution. The leaching solution may be water, salt buffers, acidic solutions, basic solutions, or organic solvents. The mixture is agitated for a specific amount of time (usually, hours to days). Afterwards, the mixture is filtered and the liquid (the *leachate*) is analyzed for contaminants, such as arsenic. A common batch leaching method is the US Environmental Protection Agency's *toxicity characteristic leaching procedure* (TCLP), which is used to determine whether a solid or liquid waste has the *toxicity characteristic* of *hazardousness* (compare with leaching, leaching test, column leaching, sequential batch leaching, and serial batch leaching).

Best demonstrated available technology (BDAT): According to the US Environmental Protection Agency, the best commercial technology for treating a specific hazardous waste. For example, *vitrification* is the BDAT for treating arsenic in soils (see Chapter 7).

Benzene: An *aromatic hydrocarbon* consisting of six carbons in a ring with hybrid single and double bonds (C_6H_6). See Figure B.2.

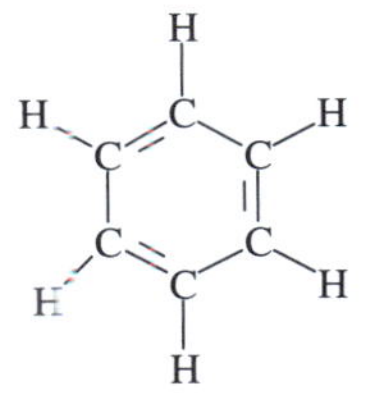

Figure B.2

B horizon (soil): A layer of *soil* generally in the middle section of most *soil profiles* that is rich in *clay minerals* and/or *(oxy)(hydr)oxides* of iron or aluminum.

Bidentate complexes: Refers to *Stern inner-sphere bidentate-mononuclear* or *bidentate-binuclear complexes*.

Bidentate–binuclear complex: A *Stern inner-sphere adsorption* complex that consists one adsorbed atom bonding to two separate metal oxides. As discussed in Chapter 2, an *arsenate* may adsorb onto two separate pairs of iron and oxygen atoms.

Bidentate-mononuclear complex: A *Stern inner-sphere adsorption* complex that consists of two atoms bridging between an atom of the adsorbed species and an *adsorbent* metal atom. Commonly, two oxygens may bridge between an iron atom in the adsorbent and adsorbed *arsenate* (see Chapter 2).

Big Bang: An event that began about 13.7 billion years ago, which explains the origin, expansion, and development of the *Universe*.

Binder: In site *remediation* and waste treatment, binders refer to cement, *clays*, or other cohesive solid materials that encapsulate and immobilize solid and even liquid wastes. Binders are key components in waste *solidification/stabilization*.

Bioaccumulate: Refers to a biological organism absorbing and retaining more of a toxic substance than it excretes.

Bioremediation: The use of plants, bacteria, or other living organisms to collect, immobilize, or biodegrade organics or other contaminants in water, *sediments*, or *soils*.

Biotic: Living or caused by living organisms (compare with abiotic).

Bituminous coal: A *sedimentary rock* composed of at least 50 wt% and 70 v% combustible organic materials that contains more moisture than *anthracites*, but less than *subbituminous coals*.

Body mass index: A measure of body mass relative to height that can approximate total body fat.

Bottom ash: The noncombustible and largely nonvolatile solid remains of *coal* or other solid fuels that accumulate in the high-temperature regions of a combustor (compare with flyash).

Brine: Water that is saturated or nearly saturated with sodium chloride and perhaps other salts. Brines contain considerably more dissolved solids than seawater.

Calcinate: Refers to heating a solid substance to remove moisture and decompose *carbonates* to oxides and carbon dioxide, but without melting any of the sample.

California Waste Extraction Test (CWET): In addition to complying with US federal regulations, the state of California has established a regulatory *batch leaching* test to evaluate the *toxicity characteristic* of *hazardousness* for solid, liquid, and mixed solid and liquid wastes in that state. For solid wastes, CWET uses a *pH* 5.0 ± 0.1 sodium citrate *leaching* solution. For chromium(VI), however, the leaching solution is deionized water. In general, the procedure involves leaching 50 g of solid waste in 500 mL of the leaching solution. Further details on CWET are in Appendix 2 of Title 22, Division 4.5, and Chapter 4 of the *Unofficial California Code of Regulations* maintained by the California Department of Toxic Substances Control (2006), which is separate from the official *California Code of Regulations*.

Calorie (nonnutritional): The amount of heat required to raise the temperature of 1 g of pure water by 1 °C (for conversions, see Appendix A).

Capillary fringe: A zone of *rocks*, *sediments*, or *soils* immediately above the *water table* and between the *saturated* and *unsaturated zones*. The pores in the capillary fringe are full of water, but the water is held by surface tension rather than flowing under atmospheric pressure like shallow *groundwater*. Rather than identify it as a separate zone, some scientists consider the capillary fringe as part of the saturated zone, while others group it with the unsaturated zone.

Capillary movement: The upward migration of water through small fractures in solids because of surface tension.

Carbonate: A group of *minerals* or aqueous chemical species containing CO_3^{2-}. Carbonates also refer to *rocks* (such as *limestones*) that mostly or almost entirely consist of calcite ($CaCO_3$) or other carbonate minerals.

Carboxyl: An organic group with a carbon–oxygen double bond and the formula –C(O)OH. See Figure B.3.

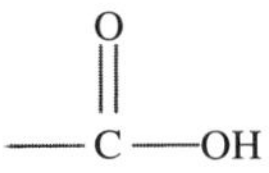

Figure B.3

Carboxylic acid: An organic *acid* characterized by the presence of *carboxyl*, –C(O)OH. Acetic acid ($CH_3C(O)OH$) is a carboxylic acid.

Cataclastic metamorphism: Localized *metamorphism* involving the mechanical crushing and grinding of *rocks* in faults during earthquakes (compare with contact and regional metamorphism).

Catalyst: A substance that increases the rate of a chemical reaction without being consumed by the reaction.

Cation exchange capacity: The ability of a solid substance (especially *clay minerals*) to *adsorb* cations. The cation exchange capacity of a material represents the total negative charge on the surface of the material and is generally expressed in milli*equivalents* per 100 g of material (compare with anion exchange capacity).

Cenozoic Era: A geologic time span that ranges from the extinction of the dinosaurs about 65 million years ago to the present.

Challenger mechanism: Also called *the Challenger scheme*, which was proposed by Frederick Challenger. The mechanism is a series of metabolic *reduction* and oxidative *methylation* reactions that begin with the reduction of inorganic *arsenate* to inorganic *arsenite* and ultimately ends with the formation of *trimethylarsine*. See Chapters 2 and 4 for details.

Challenger scheme: See Challenger mechanism.

Charge distribution multisite complexation model (CD-MUSIC): A *surface complexation model* for explaining ion *adsorption* on the surfaces of *adsorbents*. Hiemstra and van Riemsdijk (1999) used the model to explain the adsorption of *arsenate oxyanions* on goethite.

Chemical weathering: The conversion of *minerals* and other *rock* components into new, usually finer-grained materials through chemical reactions that typically involve water, natural *acids*, salts, carbon dioxide, and/or oxygen (compare with weathering and physical weathering).

Chemisorption: The formation of chemical bonds (*inner-sphere complexes*) between atoms on the surfaces of a solid (an *adsorbent*) and chemical species from an associated solution (compare physisorption).

Chondrite meteorite: A stony (low in elemental metals) *meteorite* containing *chondrules*.

Chondrules: Small spheroids of silicate materials common in certain *meteorites*. They probably condensed from the *Solar Nebula* as liquid droplets.

C horizon (soil): A layer of partially *weathered* bedrock in the lower most portion of a *soil profile*.

Chronic: Refers to a prolonged or long-term exposure to an agent (chemical or physical) that may result in an adverse health effect (compare with acute).

Clarification: Also called *sedimentation* (Spellman, 2003), 476. In water treatment, the removal of suspended particles from water through settling, flotation, and/or *filtration*.

Clastic sedimentary rock: A *sedimentary rock* largely consisting of cemented particles from preexisting *rocks* rather than chemical *precipitates* or materials excreted by organisms (e.g. coral reef deposits). *Sandstones* are examples of clastic sedimentary rocks.

Clay (mineral): A group of fine-grained and platy aluminum silicate *minerals*. Kaolinite ($Al_2Si_2O_5(OH)_4$), montmorillonite ((Al,Mg)$_8$(Si_4O_{10})$_4$(OH)$_8\cdot12H_2O$), and halloysite ($Al_2Si_2O_5(OH)_4\cdot2H_2O$) are common clay minerals.

Clay (size): A particle in a *sediment*, *soil*, or *sedimentary rock* that is less than 0.004 mm in diameter. Not all clay-sized particles are *clay minerals*.

Claystone: A *clastic* non*fissile sedimentary rock* composed of *clay-sized* particles (<0.004 mm in diameter) (compare with mudstone, shale, and siltstone).

Closed system: A location that can exchange heat with its surroundings, but not matter. A well-sealed landfill is an example of a closed system (compare with open and isolated systems).

Coagulant: An organic compound, iron or aluminum salt, or another chemical that causes *coagulation*.

Coagulation: According to the general consensus, coagulation refers to the addition of chemicals (*coagulants*) to neutralize repulsive surface charges on suspended particles in water that is being treated (Spellman, 2003), 24, 473–474. Once the surface charges diminish, the suspended particles will begin to agglomerate and form fragile *flocs* (Spellman, 2003), 474, which can be removed by *filtration* or *flocculation* (compare with flocculation; also see Chapter 7).

Coal: A *sedimentary* or *metamorphic rock* composed of at least 50 wt % and 70 v % combustible organic materials. In most cases, the organic carbon was derived from ancient terrestrial plants. Coals are potentially useful fuels (compare with lignite, subbituminous coal, bituminous coal, and anthracite).

Coal gasification: The conversion of *coal* to useful combustible gases (compare with underground coal gasification).

Coalification: The natural conversion of plant debris into *coal* through burial and *diagenesis*.

Colloid: Fine-grained materials (usually 2–10 000 Å in diameter) that remain indefinitely suspended in their host medium. For example, solid colloids may occur in water. *Double layers* often exist on colloids.

Column leaching: A laboratory method that involves passing water, *acids*, or other *leaching* solutions through a stationary solid waste or another solid sample to model the dissolution of contaminants from the sample. The leaching process uses a plastic or glass column. The column is periodically recharged (usually at the top) with the leaching solution. The solution passes through the solids and is collected at the bottom of the column, usually filtered, and then analyzed for contaminants. *Leachate* collection may occur over days, weeks, or even months. Column leaching tests provide estimates of the types and concentrations of contaminants that are likely to mobilize from a wetted solid sample and move into the environment. Unlike *batch leaching*, no agitation of the column occurs and the system is *open* and more closely resembles subsurface conditions in many natural environments and contaminated sites. Although the procedures tend to be more expensive, time-consuming, and labor intensive than batch leaching, column leaching methods have the advantages of allowing observers to study longer term chemical interactions between solid samples and leachates, to note changes in the *permeability* of solid samples with time, and to evaluate how chemical reactions may change once more soluble compounds are flushed out of the solids (compare with leaching, batch leaching, leaching test, sequential batch leaching, and serial batch leaching).

Comet: A small (typically 1–10 km in diameter) interplanetary body that mostly consists of ice, other frozen materials, and dust. Comets originate from the outer solar system (compare with asteroid and meteoroid).

Comprehensive metabolic panel: A group of clinical tests that will assess the state of biochemical reactions (anabolism and catabolism) in a patient.

Confined aquifer: An *aquifer* located between two *aquitards* and with one or more aquitards between it and the surface (compare with unconfined aquifer).

Conglomerate: A *clastic sedimentary rock* mostly composed of rounded or subrounded particles that are >2 mm in diameter.

Connate water: Water trapped in a *sedimentary rock* when its *sediments* were originally deposited. In most cases, the origin(s) of saline water in sedimentary rocks cannot be determined and they are simply identified as *formation waters* (compare with meteoric, juvenile, magmatic, and metamorphic waters).

Contact metamorphism: Localized *metamorphism* of *rocks* involving heat, fluids, and minor deformations resulting from the intrusion of a *magma* (compare with regional and cataclastic metamorphism).

Continent-continent convergence: A *convergent* boundary between two colliding *continental tectonic plates*. The Himalayas formed from the collision of the Indian continental plate with the Eurasian continental plate. Continent-continent convergence results in high mountains without substantial *subduction* and volcanism.

Continent-ocean convergence: A *convergent* boundary where an *oceanic tectonic plate* collides with and *subducts* underneath a *continental tectonic plate*. The volcanoes of the Andes Mountains result from the Nazca oceanic plate subducting under the continental South American plate.

Continental crust: The *felsic* to *intermediate* outer *rock* layers of the Earth that constitute large portions of the continents. The maximum thickness of the continental crust is about 65 km. Compared with continental crust, *oceanic crust* is much thinner and more *mafic*.

Continental glacier: Thick and extensive sheets of flowing ice covering significant portions of a continent. Modern continental glaciers exist in Greenland and Antarctica (compare with alpine glacier and glacier).

Continental plate: A *tectonic plate* largely consisting of *continental* rather than *oceanic crust* (compare with oceanic plate).

Convergent plate boundary: Areas where *tectonic plates* collide. These boundaries include: *continent–continent, continent–ocean,* and *ocean–ocean boundaries*.

Coordination number (chemistry): The number of atoms surrounding an atom in a solid material. Usually, the coordination refers to the number of anions surrounding a cation in a crystal.

Coprecipitation: Occurs in solutions and refers to a *minor* or *trace element* (such as arsenic) *adsorbing* onto or *absorbing* within the developing or fresh *precipitates* of other chemical species. Although sometimes difficult to distinguish, *sorption* involves the incorporation of contaminants onto or within preexisting solids (*sorbents*), whereas coprecipitation occurs as or shortly after the host solids precipitate from solution, such as arsenic coprecipitating with iron (*oxy*)(*hydr*)*oxides* in *acid mine drainage*. Coprecipitation might also involve arsenic-bearing *colloids* or other fine-grained particles becoming trapped (absorbed) in the interiors of precipitating compounds (compare with precipitation and sorption).

Core (Earth): A spherical nickel–iron body at the center of the Earth. The core is divided into inner and outer sections. The outer core is a molten iron–nickel layer at about 2900 to 5100 km below the Earth's surface. The inner core is the central shell of the Earth, which is solid and extends from a depth of about 5100 km to the very center of the Earth at a depth of 6370 km.

Covalent bond: The linking of two atoms, where *valence electrons* are more or less evenly shared between the two atoms (compare with ionic bond).

Cretaceous: A geologic *period* that lasted from about 144–65 million years ago.

Criterion continuous concentration (CCC): The highest concentration of a contaminant that aquatic life will not show any deleterious effects after exposure for four days (an 'extended period of time') (40 *Code of Federal Regulations* 131.38; compare with Criterion Maximum Concentration (CMC)).

Criterion maximum concentration (CMC): The highest concentration of a contaminant in a surface water that will not show any deleterious effects to aquatic life after exposure for a brief period of time (an average of 1 hour) (40 *Code of Federal Regulations* 131.38, 131.36; compare with criterion continuous concentration).

Crust (Earth): The outermost layer of the Earth, which includes the *oceanic* and *continental crusts*. The *oceanic crust* may be as thin as 5 km and the thickness of the continental crust may be as much as 65 km or so.

Curie: A unit of measure of radioactivity, where 1 curie $= 2.22 \times 10^{12}$ disintegrations/minute.

Cysteine: An amino *acid* with a *sulfhydryl* functional group.

Cysteinyl: A substance containing *cysteine*.

Cytosol: A substance of a cell exclusive of the nucleus, other organelles, and membranous components (e.g. endoplasmic reticulum).

Cytotoxin: A substance that is poisonous to a cell.

Cytotoxicant: A cellular poison.

Dalton (Da): A unit of mass equal to 1/12 of the mass of a carbon-12 atom (compare with atomic mass unit; Appendix A).

Delta: A fan-shaped accumulation of *sediments* that forms when a river empties into a sea, large lake, or ocean.

Demethylation: The removal of a methyl ($-CH_3$) from a chemical species (compare with methylation).

Density: The mass per unit volume of a substance. It's often measured in grams per cubic centimeter.

Detrital: Refers to loose *rock* and *sediments* produced by *physical weathering*.

Deuterium: A heavy hydrogen *isotope*, ^2H, where the *mass number* (A) is two. Unlike most hydrogen atoms, the nucleus of a deuterium atom contains a neutron.

Diagenesis: Any chemical reactions or physical changes resulting from water, heat, and pressure as a buried *sediment* or *soil* changes into a *sedimentary rock*.

Diatomite: A *sedimentary rock* formed by the burial and *diagenesis* of *sediments* consisting of abundant microscopic and *amorphous silica* shells from marine microorganisms.

Diffusion: The movement of dissolved materials in fluids because of a concentration gradient; that is, dissolved materials diffuse from areas of higher concentration to areas of lower concentration. Unlike *dispersion*, fluid flow is not involved with diffusion.

Dike: A layered *intrusive igneous rock* or *magma* that cuts across layers and other structures in its host rocks.

Dimethylarsinic acid: An organic pentavalent arsenical with the composition of $(CH_3)_2AsO(OH)$, abbreviated DMA(V), see Figure B.4 (compare with dimethylarsinous acid, monomethylarsonic acid, and monomethylarsonous acid).

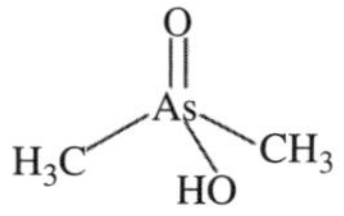

Figure B.4

Dimethylarsinous acid: An organic trivalent arsenical with the composition of $(CH_3)_2As(OH)$, abbreviated DMA(III), see Figure B.5 (compare with dimethylarsinic acid, monomethylarsonic acid, and monomethylarsonous acid).

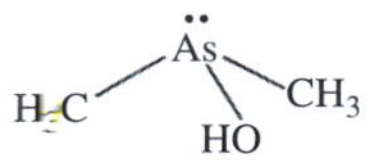

$$H_2C \overset{\overset{\displaystyle ..}{As}}{\underset{HO}{\diagdown}} CH_3$$

Figure B.5

Diorite: An *intermediate intrusive igneous rock* that is more *mafic* than a *granodiorite* and more *felsic* than a *gabbro*.

Dispersion: The movement and mixing of dissolved or suspended materials in liquids or gases by fluid flow (compare with diffusion).

Distribution coefficient (K_d): In a *system* at *equilibrium*, the distribution coefficient (K_d) is a constant that results from taking the ratio of the concentration of a chemical species in a solid or *adsorbed* onto it with the species concentration in the host solution of the solid.

Divergent plate boundary: Areas where adjacent *tectonic plates* are separating. A *spreading zone*.

Dolomite: A calcium–magnesium *carbonate mineral*, $CaMg(CO_3)_2$.

Dolostone: A *sedimentary carbonate rock* primarily consisting of *dolomite*.

Donnan exclusion: The presence of charges on a *membrane*, which repel ions of like charge in a solution and prevent the ions from passing through the membrane with the filtering solution. Donnan exclusion may remove arsenic *oxyanions* from solutions by preventing the oxyanions from passing through negatively charged *nanofilters*.

Dose–response relationship: A description of changes in effect (response) of an organism due to differing levels of exposure (dose) to a stressor such as a chemical or dosages of a drug.

Double layer: Two layers of ions associated with a charged particle in an aqueous solution. The *Stern layer* is attached onto the surface of the particle. The diffuse *Gouy layer* surrounds particle and its Stern layer.

Dunite: An *intrusive igneous rock* that almost entirely consists of *olivine*, $(Mg,Fe)_2SiO_4$.

Dysregulation: Impairment or disruption of a biochemical process that is normally regulated by an organism.

Eclogite: A high pressure and moderate to high-temperature *metamorphic rock* that forms in the Earth's mantle.

Eh: Reduction–oxidation (*redox*) potential. A value measured in millivolts or volts when compared with a $H_2 \rightarrow 2H^+ + 2e^-$ standard of 0.00 V at 25 °C and one bar pressure. Eh describes the *reduction* or *oxidation* of an element. Most natural waters are chemically complex and not at redox equilibrium. However, unless all redox reactions are at equilibrium, a single accurate Eh value cannot be obtained for a water sample (compare with oxidation and reduction).

Electrodialysis: Applying an electrical charge to an *ultrafilter* or another *membrane* to remove ions, including *arsenate oxyanions*, from a water undergoing treatment.

Electrodialytic method: Combines *electrokinetic* processes with *electrodialysis* membranes to remove contaminants from wet solids and aqueous solutions.

Electrokinetic method: Refers to an *in situ*, and sometimes an *ex situ*, technology that removes contaminants from wet soils and sediments by passing electric currents through them. Unlike *in situ vitrification*, the currents in electrokinetic methods are too low to melt the materials. Instead, the electric currents cause contaminant ions and charged particles in aqueous solutions within the solid materials to migrate towards the electrodes, where they may be collected or otherwise treated.

Electron capture: A form of radioactive decay, in which one of the electrons in an inner *orbital* of an unstable atom is capture by a proton in its nucleus. The proton converts into a neutron and releases a neutrino (compare with electron emission).

Electron emission (β-): A type of radioactive decay, where a neutron in the nucleus of an unstable atom converts into a proton and releases an electron and an antineutrino (compare with electron capture and positron emission).

Electron spin traps: Diamagnetic chemicals (e.g., nitrones) that can react with unstable free radicals to form more stable free radicals. These stable free radicals can then be detected in an electron spin resonance spectrometer.

Electronegativity: The tendency of an atom or ion to attract an electron. Electronegativity is described with a unitless scale that was initially developed by Pauling (1960). The electronegativities of the noble gases are usually undefined. Fluorine has the highest electronegativity, which is 4.0, and the lowest is francium with 0.7. Arsenic has electronegativities of 2.0–2.2 (see Chapter 2). Atoms with the highest electronegativities (such as fluorine) attract electrons and those with the lowest donate them. Thus, if two atoms with similar electronegativities (generally <1.7 difference) bond to each other, the bond would tend to be *covalent.* Two atoms with dissimilar electronegativity values would probably form *ionic bonds* (Langmuir, 1997), 101.

Electrostatic precipitator (ESP): An air pollution control device that uses electrostatic charges to remove particles from *flue gas.*

Emulsification: The development of an emulsion by mixing two or more *immiscible* liquids together, such as oil and water. The liquids do not dissolve into each other, but the emulsion consists of distinct microscopic droplets of one or more liquids dispersed into the most abundant liquid (compare with surfactant).

Endothermic: A chemical reaction that absorbs more energy from its surroundings than what it gives off (*compare with exothermic*).

Enthalpy: The 'heat content' of a substance or *system* (compare with entropy and Gibbs free energy).

Entropy: A measure of the amount of disorder in a substance or *system*. Entropy is usually measured in joules/mole degree Kelvin or calories/mole degree Kelvin (compare with enthalpy and Gibbs free energy).

Enzyme: A biomolecule that *catalyzes* a change in another *molecule* without being affected itself.

Eolian: Describes materials that have been transported by wind or landforms that are created by wind.

Eon: The largest division of geologic time. An eon includes one or more *eras*. The Phanerozoic Eon ranges from about 540 million years ago to the present.

Epidemiology: The study of the causes, distributions, and management of diseases within human populations.

Epilimnion: The upper surface layer of the water of a lake characterized by uniform temperature and high O_2 content (compare with hypolimnion and metalimnion).

Epoch: Usually, the smallest division of geologic time. *Periods* are divided into *epochs.* The *Holocene* epoch ranges from the end of the last *glaciation* about 11 500 years ago to the present.

Equilibrium (chemical): A condition where the forward and reserve of a chemical reaction occurs at the same rate.

Equilibrium constant: For a reaction at *equilibrium* at a given temperature and pressure, the equilibrium constant (K_{eq}) is derived from the ratio of the *activities* of the products to the activities of the reactants.

For the following reaction:

$$aA + bB = cC + dD,$$

$$K_{eq} = [C]^c [D]^d / [A]^a [B]^b$$

(compare with activity product and distribution coefficient).

Equivalent: The amount of a substance that will react with one mole of H^+ or OH^-. The related term equivalent mass ('weight') refers to a molar mass of a substance divided by the absolute value of its *valence state*. For *anion exchange capacity* and *cation exchange capacity* measurements, the results are usually reported in milliequivalents (1/1000 of an equivalent) per 100 g of material.

Era: A division of geologic time that includes several *periods*. The Cenozoic Era ranges from the extinction of the dinosaurs about 65 million years ago to the present.

Erosion: The removal and transportation of *rocks*, *soils*, or *sediments* by wind, liquid water, or ice.

Erythrocyte: A red blood cell.

Ester: An organic molecule with a carbon–oxygen double bond that has the general formula RC(O)OR', where R and R' represent possibly different organic carbon chains. See Figure B.6.

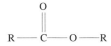

Figure B.6

Estuary: A semi-enclosed coastal body of water (e.g. bays, lagoons, inlets, and *fjords*), where fresh water (usually from rivers) mixes with seawater. Chesapeake Bay is an estuary.

Estuarine: A term used to describe structures, organisms, *sediments*, and other features associated with *estuaries*.

Euxinic: An aquatic environment with stagnant and restricted circulation. *Anaerobic* water or *sediments*.

Evaporite: Sediments or *sedimentary rocks* that primarily form through the chemical *precipitation* of salts from evaporating lakes or seas.

Exothermic: A chemical reaction that gives off more heat to its surroundings than what it absorbs from its surroundings (compare with endothermic).

Expression (epidemiology): Appearance of a specific genetic trait.

Ex situ: Offsite or to remove a material from its position (compare with *in situ*).

Ex situ remediation: Refers to the treatment of contaminants either by excavating contaminated *soils* and *sediments* or pumping contaminated *groundwater* to the surface for on- or off-site treatment and disposal (compare with *in situ* remediation).

Exsolution: The unmixing of an initially homogeneous substance into two or more separate crystalline substances. For example, at 1000 °C and one bar pressure, sodium and potassium can readily substitute for each other in the crystalline structure of the aluminum *silicate mineral* alkali feldspar to form $(Na,K)AlSi_3O_8$. At much lower temperatures, the sizes of the sodium and potassium atoms are too dissimilar for the crystalline structure to remain stable and the alkali feldspar separates (exsolves) into $NaAlSi_3O_8$ and $KAlSi_3O_8$ components (Klein, 2002), 143.

Extended X-ray absorption fine structure (EXAFS) spectrum: Part of an *X-ray absorption* spectrum that is used to identify the *coordination* of atoms, estimate bond lengths, and determine the *adsorption* complexes on the surfaces of *adsorbents*. EXAFS spectra may provide useful information on the *speciation* (*valence state*), surface complexes, and the coordination of arsenic on *adsorbents* (e.g. (Randall, Sherman and Ragnarsdottir, 2001; Ladeira, *et al.* (2001); Teixeira and Ciminelli (2005); Köber, *et al.* (2005)) (compare with X-ray absorption spectroscopy (XAS), X-ray absorption near edge structure (XANES) spectra, and X-ray absorption fine structure spectroscopy (XAFS)).

Extrusive: A variety of *igneous rocks* that erupts and cools on the Earth's surface. A volcanic rock. A rock that forms from *lava* or volcanic ash (compare with intrusive).

Felsic: Describes an *igneous rock*, *lava*, or *magma* that is relatively enriched in *silica* ($>66\,wt\%$ as SiO_2) and is generally more depleted in magnesium and iron than *intermediate* and *mafic* rocks (Hyndman, 1985), 42; (Press and Siever, 2001), 74.

Ferrihydrite: A group of iron *(oxy)(hydr)oxide* compounds with highly variable chemical compositions. Rather than determining the potentially complex chemical formula of every ferrihydrite sample, researchers often classify the compounds by their number of broad powder *X-ray diffraction* (XRD) peaks, which include the poorly crystalline 'two-line' and better crystalline 'six- or seven-line' varieties.

Filtration (water treatment): Refers to the physical separation of particles, colloids, or other contaminants from water by passing the liquid through *permeable* or semipermeable materials (compare with microfiltration, nanofiltration, reverse osmosis, and ultrafiltration).

First ionization potential: The amount of energy required to remove the most weakly held electron from a neutral atom in a vacuum and place the electron at rest an infinite distance away. The first ionization potentials are usually measured in electron volts. $1\,eV = 1.602\,176\,53 \times 10^{-19}$ J (Appendix A).

Fissile: The tendency of a *sedimentary rock* (especially *shales*) to break into platy fragments. Fissility is usually due to the alignment of platy *clay minerals* in the rock.

Fjord: A deep *glacially* carved channel partially submerged by seawater. Fjords typically have U-shaped profiles and are common in Alaska and Scandinavia.

Floc: Agglomerates of suspended particles in a water treatment tank created by the addition of *coagulants* to neutralize repulsive surface charges on the particles (compare with coagulation and flocculation).

Flocculation: Refers to the gentle mixing of *flocs* created by *coagulants* during water treatment to promote further agglomeration of the flocs and allow them to separate from the water by either floating or settling ((Spellman, 2003), 26, 475; also see Chapter 7; compare with coagulation).

Flue gas: Gas emitted by the combustion or other utilization of fuels in a facility.

Flue gas desulfurization: The injection of calcium or other *sorbents* to remove sulfur from *flue gases*.

Fluorescent probe: A chemical that can produce light when being acted upon by a source of energy.

Fluvial: A term used to describe structures, organisms, *sediments*, and other features associated with rivers.

Flyash: Fined-grained and noncombustible solid components of *coal* and other fuels that accumulate in *flue gas* (compare with bottom ash).

Foliation: Planar features in *metamorphic rocks* resulting from the parallel alignment of elongated *minerals* during *regional metamorphism*.

Formation water: Any saline waters in deep *sedimentary rocks*. If the origin of the waters can be determined, they may be identified as *meteoric* or *connate* (compare with magmatic, juvenile, and metamorphic waters).

Fossil fuel: Combustible, organic-rich materials (such as *petroleum* and *coal*) that are natural remains of ancient plants and animals.

Fourier transform infrared (FTIR) spectroscopy: An analytical method that uses infrared radiation to investigate the chemical characteristics of a sample. This method may be used to identify the *valence states* of arsenic on *adsorbents* and bonds between arsenic and other elements (e.g. (Goldberg and Johnston, 2001)) (compare with Raman spectroscopy).

Framboid: A microscopic cluster of spheroidal *minerals*, usually pyrite (FeS_2).

Freundlich isotherm: An equation for an *adsorption isotherm* in the form of $C_{ads} = K_f(C_{soln})^n$ (if $n = 1$, the distribution is linear, see Chapter 2 for details), which describes the distribution of a chemical species between an *adsorbent* and an associated solution under *equilibrium* conditions. Some adsorption isotherms are better described with the *Langmuir isotherm*.

Fulvic acids: A group of naturally occurring organic compounds of biological origin that are common in the *A horizons* of *soils* and other natural environments. While *humic acids* are only soluble in water at pH > 2, fulvic acids are water-soluble under all pH conditions (Drever, 1997), 113–114 (compare humic acid and humin).

Fusion (nuclear): The combining or 'fusing together' of lighter elements in the cores of stars and possibly future reactors to produce heavier elements and energy.

Gabbro: An *intrusive mafic igneous rock*. *Basalts* are *extrusive* equivalents of gabbros.

Genotoxin: A chemical or physical agent (e.g. radiation) that causes a change in genetic material (i.e. deoxyribonucleic acid, DNA).

Geochemical modeling: The use of computers to investigate and predict chemical reactions in natural waters, *soils*, *sediments*, and *rocks*.

Geogenic: Naturally occurring. Refers to the contamination of *groundwater*, *soils*, *sediments*, and *rocks* by natural sources.

Geologic material: A solid, inorganic, and natural material in the upper portion of the *crust* of the Earth including *rocks*, *soils*, and *sediments*.

Geothermal: Heat from within the Earth (compare with hydrothermal).

Geothermal water: If *hydrothermal* waters reach the Earth's surface as *hot springs* and *geysers*, they are identified as geothermal waters. Geothermal waters may also have economic benefits, such as generating electricity, promoting tourism, and utilization in spas.

Geyser: A spring that periodically erupts hot water and steam because of the boiling of *groundwater* from shallow *magmas* (compare with hydrothermal and geothermal waters).

Gibbs free energy: The difference between the *enthalpy* of a *system* and the product of its *entropy* and absolute temperature (K); that is, for a reaction: $\Delta G_R = \Delta H_R - T\Delta S_R$, where: ΔG^o_R = Gibbs free energy for the reaction, ΔH^o_R = enthalpy of the reaction, ΔS^o_R = entropy of the reaction, and T is temperature in Kelvin. Normally, Gibbs free energies are determined at one bar pressure and 298.15 K (25 °C).

Glaciation: A time period when *glaciers* more extensively covered continents and high mountains than at present. The last major glaciation ended about 11 500 years ago.

Glacier: A voluminous mass of ice that flows over a land surface.

Glutathione: An organic *acid* ($C_{10}H_{17}N_3O_6S$) that widely occurs in the tissues of plants and animals.

Gneiss: A high-temperature *regional metamorphic rock* consisting of alternating light and dark layers.

Gouy layer: A diffuse layer of ions surrounding a charged particle in an aqueous solution. The Gouy layer ions do not come into contact with the particle surface, but are separated from it by the *Stern layer*. In the Gouy layer, the imbalance between cations and anions exponentially decreases with distance away from the solid surface. The Gouy and Stern layers comprise the *double layer*.

Granite: An *intrusive felsic igneous rock*. *Rhyolites* are *extrusive* equivalents of granites.

Granodiorite: An *intrusive intermediate rock* with a composition between that of a *granite* and a *diorite*.

Granulocyte macrophage-colony stimulating factor: A biomolecule that is produced by several different tissues that stimulates the bone marrow to produce granulocytes and stem cells.

Gravel: A *sediment* or *sedimentary rock* particle greater than 2 mm in diameter. Gravels include pebbles, cobbles, and boulders.

Groundwater: Subsurface water in the *saturated zone*. If the groundwater is above 50 °C, it is usually classified as *hydrothermal* water.

Haber-Weiss reaction: A two step reaction *catalyzed* by iron that forms hydroxyl *radicals* from hydrogen peroxide (H_2O_2) and superoxide (O_2^-); a source for oxidative stress.

Half-life: The amount of time that it takes for one-half of a given number of atoms of a radioactive *isotope* to undergo radioactive decay. Each isotope has its own half-life, which may range from a fraction of a second to billions of years. For the mathematical details, (see Faure (1998), 281; Krauskopf and Bird (1995), 243–245; Dalrymple (1991), 79–86), or other geochemistry and geochronology textbooks.

Hazardousness: Having a hazardous property. A chemical, procedure, or event that creates a substantial threat to human health and life. The United States Environment Protection Agency (US EPA) classifies hazardous wastes by four characteristics, including: corrosivity (very alkaline (*basic*) ($pH \geq 12$) or *acidic* ($pH \geqslant 2$) wastes), ignitability, toxicity, and reactivity (40 *Code of Federal Regulations* 261.20–261.24) (see Appendix E; compare with toxicity characteristic and toxicity characteristic leaching procedure, TCLP).

Heat capacity: Without a change in phase and under constant pressure, the amount of heat (usually in joules or *calories*) required to raise the temperature of a given quantity (usually a *mole*) of a substance by 1 ° K.

Heat content: The amount of heat in *calories* or joules released by the combustion of a kilogram or other mass of a *coal*.

Heavy metal: A metallic element with an *atomic mass* greater than that of calcium (40.08 *atomic mass units*) (Drever, 1997), 175.

Hemolytic agent: A chemical or physical agent that lyses (destroys) red blood cells.

Hepatoma cell: A malignant liver cell.

Heterologous expression: The manifestation of a genetic trait not normally found in the organism.

High-rank coal: Refers to *anthracite*, *meta-anthracite*, and *bituminous coals*, which are relatively low in moisture, high in combustible carbon, and have high *heat contents* (compare low-rank coal, lignite, and subbituminous coal).

Holocene: An *epoch* of time lasting from the end of the last glaciation (about 11 500 years ago) to the present.

Homology: A state of being alike in certain attributes; in genetics, having identical chromosomes or chromosomal parts with respect to their content and construction.

Horizon (soil): A layer in a *soil profile* that is distinguishable from other soil layers (horizons) by its physical and/or chemical properties.

Hot spring: Surface water with a temperature above the temperature of a human body (about 37 °C; also see (Pentecost, Jones and Renaut, 2003); compare with geothermal water and geyser).

Humic acids: A group of naturally occurring organic compounds of biological origin that are common in the *A horizons* of *soils* and other natural environments. By definition, they are only soluble in water at $pH > 2$ (Drever, 1997), 113–114 (compare with humin and fulvic acid).

Humin: Naturally occurring, organic-rich substances of biological origin that are common in the *A horizons* of *soils*. By definition, humin is insoluble in water under all *pH* conditions (Drever, 1997), 113 (compare with humic acid and fulvic acid).

Hydrocarbon: Organic compounds mostly or entirely consisting of carbon and hydrogen. Hydrocarbons are used as fuels and to manufacture plastics, medicines, and other products (compare with petroleum, oil, and natural gas).

Hydrologic: Related to water.

Hydrosphere: The Earth's water, including liquid, vapor, and ice.

Hydrothermal: Water heated by the Earth to more than 50 °C.

Hydrothermal fluid: Hot aqueous gases or liquids that form in the Earth.

Hypolimnion: The cold and deep bottom waters of a temperature stratified lake below its *metalimnion* (compare with epilimnion and metalimnion).

Hypothesis: In a scientific investigation, a tentative explanation of observations in a laboratory or nature. Hypotheses are then tested with the *scientific method* to determine their validity. If a hypothesis is highly reliable at explaining and even predicting a phenomenon, it becomes a *theory*.

Igneous: A *rock, mineral*, or other feature that results from the melting of rocks. Also, temperature, pressure or other conditions associated with the melting of rocks.

Immiscible: Two substances (usually liquids) that will not dissolve into each other, such as oil and water (compare with miscible).

Inner-sphere complex: A type of *Stern layer*, where a chemical species bonds directly onto the surface of a solid material (*adsorbent*) in an aqueous solution. The formation of inner-sphere complexes is called *chemisorption*. Stern inner-sphere adsorption complexes are further divided into *monodentate*, *bidentate-mononuclear*, and *bidentate-binuclear* types (compare with outer-sphere complex).

In situ: In place or onsite (compare with *ex situ*).

In situ remediation: Treating *groundwater, soils*, and *sediments* in place to remove unwanted contaminants. *In situ* treatment technologies for groundwater may include the emplacement of *permeable reactive barriers* that remove the contaminants as groundwater flows through them. Chemicals or biological agents may also be injected into *aquifers* to precipitate or biodegrade contaminants, respectively. Cementitious materials may be mixed into contaminated soils and sediments to encapsulate the contaminants (compare with *ex situ* remediation).

In situ vitrification: Refers to the use of electrical currents to melt, incinerate, or encapsulate contaminants in unexcavated *soils* and *sediments* (compare with vitrification).

Intermediate (igneous): Describes a *magma, lava*, or *igneous rock* that contains more *silica* (52–66 wt % as SiO_2) and less iron and magnesium than a *mafic* magma, lava, or igneous rock, but less silica and more magnesium and iron than a *felsic* specimen (Hyndman, 1985), 42; (Press and Siever, 2001), 74.

Internal transition: A mode of radioactive decay, where an excited nucleus transfers energy to an electron and expels the electron from the atom. Internal transition is responsible for transforming certain arsenic *isomers* from higher to lower energy states (Table 2.1).

Intraperitoneal dose: Administration of a chemical into the peritoneal cavity.

Intratracheal instillation: Administration of a chemical into the lungs via the trachea.

Intrusive: A variety of *igneous rock* that cools and solidifies below the Earth's surface. *Magmas* solidify into intrusive rocks.

Ion exchange: A process where one or more ions from a solution replace one or more other chemicals on the surface of an *adsorbent* through *adsorption* and desorption. Most adsorption involves ion exchange. Ion exchange is also a water treatment technology, where *sorbents* remove contaminants from water and release nontoxic ions into the water.

Ionic bond: The loose linking together of two atoms in a solid substance, where one or more *valence electrons* are preferentially associated with one of the bonded atoms (compare with covalent bond).

Ionic character: How ionic a chemical bond is or how unevenly one or more electrons are shared between two bonded atoms. If the bond is strongly *covalent*, the ionic character is very low.

Ionic strength (I): A value usually given in *molal* that describes the overall effect of the concentrations and charges of ions in an aqueous solution. The ionic strength of a solution is equal to one-half the sum of the molal concentration of each ion multiplied by its charge squared, that is, $I = 0.5 \sum (m_i z_i^2)$.

Iron formation: A chemically precipitated *Precambrian sedimentary* or *metamorphosed* sedimentary *rock* consisting of thin layers rich in iron minerals (such as hematite, Fe_2O_3, and magnetite, Fe_3O_4) and usually *quartz* (compare with ironstone).

Iron meteorite: A type of *meteorite* that is largely or entirely composed of elemental nickel and iron. Iron meteorites may be the remains of the cores of large *asteroids* and small planets that once existed in the inner solar system.

Ironstone: A *Phanerozoic* iron-rich *sedimentary rock* (compare with iron formation).

Island arc: A chain of volcanic islands that are commonly associated with the *subduction* of an *oceanic plate* underneath another oceanic plate (*ocean–ocean convergence zone*). The Aleutian Islands of Alaska are an example of an island arc.

Isoelectric point: The *pH* of water in contact with a *mineral* or another solid substance, where the solid has a net surface charge of zero. When determining the isoelectric point, the surface charges on the solid are only controlled by the *adsorption* of OH^- or H^+. In contrast, if the solution in contact with the solid contains electrolytes, the *zero point of charge* describes the pH of the solution at which the net surface charge of the solid is zero.

Isolated system: An ideal situation where a location does not exchange heat or matter with its surroundings (compare with *closed* and *open systems*).

Isomers: Two *isotopes* that have the same *atomic* (Z) and *mass numbers* (A), but their nuclei are in different energy states. Isotope isomers are labeled '*m*', '*n*', '*p*', or '*q*' with increasing excitation energy (Audi *et al.*, 2003), 6. Also, isomers may refer to two substances (especially organic compounds) that have the same chemical formula, but a different arrangement of their atoms (compare with polymorph).

Isotherm: Constant temperature. Refers to data collected from laboratory experiments, where temperature was held constant (e.g. *adsorption isotherms*).

Isotopes: Two or more atoms of the same element that have different numbers of neutrons.

Juvenile water: Water and steam derived from the Earth's *mantle* and thought not to have been previously on the Earth's surface.

Karst: An area underlain with thick *limestones* or other *carbonate rocks*, which results in abundant sinkholes and caves.

Ketone: An organic *molecule* with the general formula RC(O)R' where R and R' represent organic carbon chains. Acetone is an example of a ketone (Figure B.7).

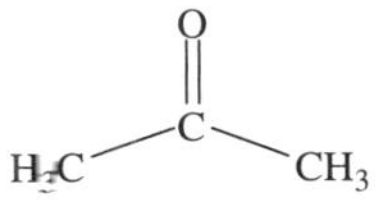

Figure B.7

Labile: Unstable. A material that undergoes change or is likely to undergo change.

Lacustrine: A term used to describe structures, organisms, *sediments*, and other features associated with lakes.

Laminar: Parallel flow in water or thin parallel layers in a *sediment* or *sedimentary rock*. *Shales* typically have laminar layering.

Langmuir isotherm: An equation for an *adsorption isotherm* in the form of $C_{ads} = C_{max} ((K_{Lang} C_{soln})/(1 + (K_{Lang} C_{soln})))$ (see Chapter 2 for details), which describes the distribution of a chemical species between an *adsorbent* and an associated solution under *equilibrium* conditions. Some adsorption isotherms are better described with linear or *Freundlich isotherms*.

Lava: Molten rock erupting at the Earth's surface.

Leachate: A contaminated liquid resulting from natural or laboratory aqueous solutions interacting with contaminated solids, such as wastes in a landfill or at a mine.

Leaching: The dissolution and extraction of chemical species from solid samples by water, *acids*, *bases*, organic solvents, salt solutions, or other liquids (compare with batch leaching, column leaching, sequential batch leaching, serial batch leaching, and leaching test).

Leaching test: Refers to laboratory procedures that use water, *acids*, *bases*, organic solvents, salt solutions, or other liquids to estimate the types and concentrations of contaminants that will dissolve out of a solid waste or another solid sample. Leaching procedures are useful in predicting whether the dumping or burial of a solid material could potentially release problematic levels of contaminants into *groundwater*, surface waters, or other natural environments. Leaching procedures include *batch*, *sequential batch*, *serial batch*, and *column* methods. Some batch leaching tests, such as the *toxicity characteristic leaching procedure* (TCLP), are used in regulations to identify solid and liquid wastes for the *toxicity characteristic* of *hazardousness*. Kosson *et al.* (2002) also discuss the purposes and limitations of leaching tests, and provide good background information on their design.

Le Châtelier's principle: When a chemical reaction at *equilibrium* is disturbed by a change in temperature, the concentration of a reactant or product, or other conditions, Le Châtelier's principle states that the reaction will counteract the disturbance to restore equilibrium. For example, the addition of more reactants will cause a reaction to produce more products.

Ligand: An anion or neutral *molecule* that can attach onto a cation.

Lignite: A *coal* that forms at relatively low *diagenetic* temperatures (around 65 °C) and contains more moisture than *subbituminous*, *bituminous*, and *anthracite* coals.

Limestone: A *sedimentary rock* mostly composed of calcite ($CaCO_3$).

Lipoic acid: An organic *acid*, $(SSCH_2CH_2)CH(CH_2)_4COOH$, which is important in human nutrition.

Lithosphere: The outermost, rigid layer of the Earth, which includes the *crust* and upper *mantle*. The thickness of the lithosphere usually ranges from about 50 km in ocean basins to 100 km in the high mountainous areas of the continents.

Loess: Sediment deposits of wind-blown *silt.*

Low-rank coal: Refers to *lignite* and *subbituminous* coals, which are relatively high in moisture, low in combustible carbon, and have low *heat contents* (compare high-rank coal).

Lung clearance: Removal of a substance from the lung.

Luster: The appearance of a *mineral* or other solid substance as light reflects off its surface, which may be metallic, resinous, glassy, pearly, and so on.

Mafic: Describes a *magma, lava,* or *igneous rock* that is generally richer in magnesium and iron than *intermediate* and *felsic* varieties, but has less *silica* (45–52 wt %) than intermediate and felsic magmas, lavas, and rocks (Hyndman, 1985), 42; (Press and Siever, 2001), 74.

Magma: Molten rock below the Earth's surface.

Magmatic water: Water in or originating from a *magma.*

Major element: An element in the Earth's *crusts,* a *rock,* an organism, or another sample that has a concentration above 1 wt % ($10\,000\,mg\,kg^{-1}$). The major elements in the Earth's crusts include: oxygen, silicon, aluminum, iron, calcium, sodium, potassium, and magnesium (compare with minor and trace element).

Mantle (Earth): An *ultramafic* layer in the Earth's interior that exists between the *crust* and *core.* The mantle is located at depths of about 40 to 2900 km.

Marble: A *metamorphic rock* chiefly consisting of calcite ($CaCO_3$). Marbles typically form from the metamorphism of *limestones* (compare with skarn).

Mass median aerodynamic diameter: A measure of particle size related to its mass. The *aerodynamic diameter* of an aerosol, where 50 % of the mass of the aerosol has particles with larger aerodynamic diameters and 50 % has smaller diameters.

Mass number (A): The total number of protons and neutrons in an atom (compare with atomic number, Z).

Maximum adsorption capacity: The maximum amount of an *adsorbate,* often measured in *molal,* that a given mass of an *adsorbent* can *adsorb.* A *Langmuir isotherm* may approach the maximum adsorption capacity.

Maximum contaminant levels (MCLs): Drinking water standards for various contaminants that are enforced by the United States Environmental Protection Agency (US EPA) that take into account the risks of human health effects and the costs of cleaning the water (Appendix E). The current MCL for arsenic is $10\,\mu g\,L^{-1}$ (40 *Code of Federal Regulations* 141.62).

Maximum contaminant level goals (MCLGs): Unenforceable levels recommended by the United States Environmental Protection Agency (US EPA) for drinking water contaminants that only take into account the risks of human health effects. The MCLG for arsenic is $0\,\mu g\,L^{-1}$ (Appendix E; 40 *Code of Federal Regulations* 141.51).

Membrane: A microporous sheet or layer of material that is used to separate solids from liquids and, in some cases, *solutes* from *solvents.*

Meta-anthracite: A term used by some geologists to refer to highly *metamorphosed anthracites.*

Metabolism: A biochemical process that alters the structure of a chemical.

Metabolite: A chemical that has been altered by *metabolism.*

Metalimnion: A moderately deep layer of water in a thermally stratified lake where the water temperature rapidly declines with depth (compare with epilimnion and hypolimnion).

Metalloid: An element, such as silicon or arsenic, that has properties that are intermediate between metals and nonmetals (such as the ability to conduct electricity), or shares properties with both metals and nonmetals.

Metamorphic: A *rock, mineral,* or other feature that results from metamorphism. Also, temperature, pressure or other conditions associated with metamorphism.

Metamorphic water: Water that is expelled from *minerals* during *metamorphism.*

Metamorphism: Sedimentary, igneous, or previously metamorphosed *rocks* that have been considerably altered by subsurface heat (usually above $200\,^{\circ}C$), pressure, and/or grinding, but not to the point of melting.

Metastable: A *mineral* or other material that is present in an environment where it should be unstable and absent. For example, metastable *biotic arsenite* may exist for extended periods of time in oxygenated seawater.

Meteoric water: Water recently originating from the atmosphere.

Meteorite: A *meteoroid* that has survived impacting the Earth's surface.

Meteoroid: A small (typically having a size of a few meters or less) essentially nonicy interplanetary object (compare with *asteroid* and *comet*).

Methylation: Refers to the addition of one or more methyls ($-CH_3$) onto a chemical species (compare with demethylation).

Methylthioarsenate: A chemical species containing pentavalent arsenic, *sulfide,* and methyl groups, as examples: $(CH_3)AsO_2S^{2-}$, $(CH_3)AsOS_2^{2-}$, $(CH_3)_2AsOS^-$, and $(CH_3)_2AsS_2^-$ (compare with thioarsenic, thioarsenate, and thioarsenite).

Methyltransferase: An *enzyme* that *catalyzes* the transfer of a methyl group from one *molecule* to another.

Mica: A group of platy *silicate minerals* with the general formula of $AB_{2-3}(Si,Al)_4O_{10}(OH,F)_2$, where $A = Ca^{2+}$, K^+, or Na^+ and $B = Mg^{2+}$, Fe^{2+}, Fe^{3+}, Li^+, or Al^{3+}. Muscovite ($KAl_2(AlSi_3O_{10})(OH)_2$) and biotite ($K(Mg,Fe)_3(AlSi_3O_{10})(OH)_2$) are common micas.

Micelles: Electrically charged *colloidal* particles that are usually composed of aggregates of organic *molecules.*

Microfiltration: Filtration for the treatment of water that removes suspended particles with diameters that are greater than about $0.1\,\mu m$ from water (compare with filtration, nanofiltration, reverse osmosis, and ultrafiltration).

Mid-oceanic ridge basalt: A *mafic* volcanic (*extrusive*) *rock* that erupts on the ocean floor in a *spreading zone.*

Mine drainage: Water runoff or *leachates* from mining wastes. They are often, but not always, *acidic* (compare with acid mine drainage).

Mine tailings: Rocks, minerals, sediments, soils, and other wastes that result from the mining of ore deposits or *coal.* Mine tailings often contain pyrite and other *sulfide* minerals, which *oxidize* in the presence of oxygen and water to form *acid mine drainage.*

Mineral: A naturally occurring and well-crystallized inorganic and nonbiological solid substance. Minerals are major components of most *rocks, sediments,* and *soils.*

Minor element: An element in the Earth's *crusts,* a *rock,* an organism, or another sample that has a concentration between 1000 and 10000 mg kg^{-} (compare with major and trace element).

Miscible: Two substances (usually liquids) that will readily mix and entirely dissolve into each other, such as ethanol and water (compare with immiscible).

Mitogen activated protein kinase (MAPK): An *enzyme* involved in cellular regulation; it is activated by an extracellular signal (mitogen) and *catalyzes* the addition of a phosphate to a biomolecule.

Mohorovičić Discontinuity: The boundary between the lower *crust* and the upper *mantle.*

Molal (m): A measure of concentration given in *moles* of *solute* per kilogram of *solvent* (compare with molar).

Molar (M): A measure of concentration given in *moles* of *solute* per liter of *solvent* (compare with molal).

Molar mass: The mass in grams of one *mole* of a substance. For example, one mole of elemental arsenic has a mass of 74.92160 g.

Mole: Avogadro's number (6.022×10^{23}, Appendix A) of a substance. One mole of carbon contains 6.022×10^{23} carbon atoms (compare with molar mass).

Molecule: Two or more atoms *covalently* bonded together to form an electrically neutral unit of matter. For many substances, a molecule is the smallest stable unit of matter that has the same chemistry as the substance (such as an H_2O molecule in liquid water).

Monodentate: A *Stern inner-sphere adsorption* complex where each adsorbed species attaches onto only one atom on the *adsorbent* surface.

Monomer: A small *molecule* that can combine with many other molecules with the same chemistry to form a *polymer.*

Monomethylarsonic acid: An organic pentavalent arsenical with the composition of $(CH_3)AsO(OH)_2$, abbreviated MMA(V), see Figure B.8. (compare with dimethylarsinous acid, dimethylarsinic acid, and monomethylarsonous acid).

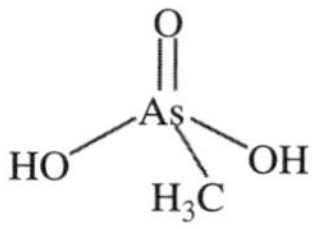

Figure B.8

Monomethylarsonous acid: An organic trivalent arsenical with the composition of $(CH_3)As(OH)_2$, abbreviated MMA(III), Figure B.9 (compare dimethylarsinous acid, dimethylarsinic acid, and monomethylarsonic acid).

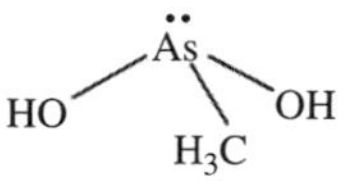

Figure B.9

Mudstone: A non*fissile clastic sedimentary rock* largely or entirely consisting of *silt-* and *clay*-sized particles (less than 0.063 mm in diameter) (compare with claystone, shale, and siltstone).

Multi-drug resistance protein-2 (MRP-2): A protein in the bile canaliculus of liver cells involved in the hepatobiliary transport of organic chemicals.

Multiple logistic regression analysis: A statistical model used to predict the probability of the occurrence of an event using several predictor variables.

Multivariate-adjusted odds ratio: The odds of a disease in the exposed group relative to the odds of disease in the unexposed group that are adjusted for other risk factors for that disease.

Mutagenesis: A process whereby a chemical or a physical agent (e.g. radiation) alters deoxyribonucleic acid (DNA).

Nanofiltration: Filtration for the treatment of water that removes suspended particles that are larger than about 0.001 μm (compare with Donnan exclusion, filtration, microfiltration, reverse osmosis, and ultrafiltration).

National priorities list (NPL): A list of US sites contaminated with hazardous wastes that is maintained by the United States Environmental Protection Agency (US EPA). The sites are eligible for long-term *remediation* under the US EPA *Superfund* program.

Natural gas: Gaseous *hydrocarbons* that are used as fuel. Natural gas is primarily methane, although other gaseous hydrocarbons (ethane and propane) may be present (compare with petroleum and oil).

Natural organic matter (NOM): Complex mixtures of organic compounds that result from the partial decay of plants and other dead organisms. NOM commonly occurs in natural waters, *soils*, and *sediments*.

Natural remediation: Refers to allowing already existing biological, chemical, and physical processes in nature to biodegrade, *precipitate*, *sorb*, or otherwise reduce the toxicity and mobility of contaminants at a field site with little or no human intervention.

Nebula: A 'cloud' of interstellar dust and gas, which may condense to form stars and even solar systems (compare with *Solar Nebula*).

Neutron activation analysis (NAA): A nuclear process whereby the elements in a material can be qualitatively and quantitatively determined.

Neutron emission: A type of radioactive decay, where a neutron is ejected from the nucleus of a decaying atom.

Neutron flux: A measure of the number of neutrons passing through a given area in a given amount of time.

Nicotinamide adenine dinucleotide phosphate (NADPH): A biochemical cofactor (a nonprotein chemical that can interact with an *enzyme*) that can donate electrons for the *reduction* of molecules.

Null genotype: An organism not containing the normal genotype (e.g. AS3MT knockout mice).

Nucleosynthesis: The creation of elements from the *fusion* of lighter elements during the *Big Bang* or in the interiors of stars.

Oceanic crust: The *mafic* outer *rock* layer of the Earth that covers most of the seafloor. The oceanic crust may be as thin as 5 km. Compared to the oceanic crust, the *continental crust* is much thicker and more *felsic*.

Oceanic plate: A *tectonic plate* largely consisting of *oceanic* rather than *continental crust* (compare with continental plate).

Ocean–ocean plate convergence: A *convergent plate tectonic* boundary consisting of an *oceanic plate* *subducting* under another *oceanic plate*.

Oil: Naturally occurring liquid *hydrocarbons*. Oil is used as a fuel and a lubricant, and in the manufacturing of plastics, medicines, and other products (compare with petroleum).

Oil shale: Fine-grained *sedimentary rocks* that are rich in heavy *oils*. Many of them are actually *siltstones*, fine-grained *limestones*, or impure *coals* rather than *shales*.

Oligomer: A *polymer* with only a few *monomer* units, such as a *tetramer*.

Olivine: A *silicate mineral*, $(Fe,Mg)_2SiO_4$, that is common in *ultramafic* and *mafic igneous rocks* and some *metamorphic rocks*.

Open system: A location that can exchange heat and matter with its surroundings (compare with *closed* and *isolated systems*). The Pacific Ocean is an example of an open system.

Orbital: A region around the nucleus of an atom where an electron is mostly likely to reside (compare with principal quantum number, shell, and subshell).

Organoarsenical: An organic compound that contains arsenic as a major component (compare with *arsenian* and *arsenical*).

Organohalide: An organic compound that contains a halide (fluoride, chloride, bromide, or iodide).

Organotriflate: An organic compound that contains a trifluoromethanesulfonate ($CF_3SO_3^-$) group.

Outer-sphere complex: A type of *Stern layer*, where an *adsorbate* indirectly attaches to an *adsorbent* surface through one or more water molecules. The formation of outer-sphere complexes is called *physisorption* (compare with inner-sphere complex).

Overexpress: Increasing the quantity of a protein encoded by a gene and synthesized by a cell by increasing the number of copies of the gene or enhancing the binding strength of the gene's promoter region (compare with expression).

Oxic: With O_2. An environment that contains O_2 (compare with aerobic, anaerobic, and anoxic).

Oxidant: Also called an *oxidizing agent*. A substance that causes another substance to be oxidized by accepting one or more electrons from the other substance (compare with reductant).

Oxidation: The loss of one or more electrons by an atom. Although oxygen as O_2 is often the *oxidant* in oxidation, oxidation may also occur without the presence of O_2. Other possible oxidants include nitrate (NO_3^-) and Fe(III). Oxidation and *reduction* occur simultaneously.

Oxidation–reduction: Also called *reduction–oxidation*. The transfer of one or more electrons between a pair of atoms. The atom accepting the electrons is *reduced* and the electron donor is oxidized.

Oxidizing agent: An oxidant.

Oxyanion: An anion, such as $HAsO_4^{2-}$, that contains oxygen.

(Oxy)(hydr)oxide compounds: Non*silicate*, crystalline, or *amorphous* solids that contain oxide (O^{2-}), hydroxide (OH^-), or oxyhydroxide (OOH^{3-}) as the major anion. Common cations include: aluminum, iron(II), iron(III), or manganese.

Pathfinder element: In ore prospecting, a relatively mobile, common, and easy to identify element in *soils*, *sediments*, *rocks*, or plants that may provide information on the location of a less common and more valuable element. Arsenic is sometimes a pathfinder element for gold.

Peat: An organic-rich *soil* consisting entirely or almost entirely of the remains of partially decayed terrestrial plants.

Pedosphere: The Earth's *soils*.

Pegmatite: An *intrusive igneous rock* that mostly consists of *minerals* that are larger than two centimeters long.

Peridotite: An *ultramafic rock* that exists in the upper *mantle*. In rare circumstances, peridotite fragments are brought to the Earth's surface by volcanic eruptions.

Period (geological): A commonly used division of geologic time. *Eras* are divided into periods. The *Quaternary* period extends from 1.75 million years ago to the present.

Permeability: The ability of a *rock, sediment,* or *soil* to transmit water. Permeable materials commonly include *sands* and *gravels.* Water does not readily flow through impermeable materials, such as most *clays* (compare with porosity).

Permeable reactive barrier: An *in situ remediation* wall, *membrane,* or layer, which is installed in the subsurface to intercept and remove contaminants from *groundwaters* flowing out of landfills, mines, or other sites. Once contaminants come into contact with the barriers, they may be *sorbed,* undergo *ion exchange,* biodegrade, *precipitate, coprecipitate,* or *filter* out.

Permissible exposure limit (PEL): According to United States Occupational Safety and Health Administration (OSHA) regulations, the highest concentration of an air contaminant that an individual is allowed to be exposed to. PELs include 'ceiling values', which are concentration limits that may not be exceeded at any time and '8-hour time weighted averages', which is the maximum average concentration that an individual may be exposed to over an 8-hour work day. The PEL for inorganic arsenic is $10\,\mathrm{mg\,m^{-3}}$ of air averaged over an 8-hour period (29 *Code of Federal Regulations* 1910.1018) (compare with action level and threshold limit value).

Petroleum: Naturally occurring *hydrocarbons,* which include gases (*natural gas*), liquids (*oil*), and solids.

pH: The negative $\log_{10}$ of the *molar* H^+ *activity* of an aqueous solution. *Acids* have pH values of 0–6.9, the pH values of *bases* are 7.1–14, and pH 7.0 is neutral.

Phanerozoic: A geologic *eon* that ranges from about 540 million years ago to the present.

Phenotype: Physical, morphological, or biochemical characteristics of an individual determined by the genotype and environment.

Phosphorite: A s*edimentary rock* that is rich in phosphate (PO_4^{3-}) *minerals* or other phosphate compounds.

Phosphorylytic: Cleavage or lysis of a phosphate group from a *molecule.*

Photic zone: The top layer of a lake, sea, or ocean where sufficient light penetrates to permit photosynthesis.

Phyllite: A fine-grained *foliated rock* that is more *metamorphosed* than a *slate,* but less than a *schist.* Phyllites have visible sheens from microscopic *micas.* However, individual mica crystals are not visible with the naked eye.

Physical weathering: The mechanical breakdown of *rocks* into smaller fragments, which then contribute to the formation of *soils* and *sediments* (compare with chemical weathering and weathering).

Physisorption: The formation of *outer-sphere complexes* (compare with chemisorption).

Phytoremediation: The use of living plants, plant parts, or plant extracts to treat contaminated sites. Certain plants have the ability to *bioaccumulate* arsenic and detoxify their surroundings.

Placer deposit: Sediments that contain significant concentrations of dense valuable *minerals,* such as gold or diamonds. Placers are usually stream deposits.

Plagioclase: A sodium and calcium aluminum *silicate mineral* that commonly forms in *igneous* and *metamorphic rocks,* and occasionally survives *weathering* to occur in *sediments, soils,* and *sedimentary rocks.* A *solid solution* exists in plagioclase, where $(CaAl)^{5+}$ and $(NaSi)^{5+}$ may substitute for each other to produce a composition that ranges from $NaAlSi_3O_8$ to $CaAl_2Si_2O_8$.

Planetesimal: A body in the early solar system that eventually accreted with other planetesimals to form the rocky planets (Mercury, Venus, Earth, and Mars). *Asteroids* are probably the remains of unaccreted planetesimals.

Plasma: Completely ionized gas, common in stars and the early *Big Bang.* In plasma, electrons are independent of nuclei.

Plate tectonics: A *theory* that states that the Earth's *lithosphere* is divided into several sections or plates that collide, separate, or slide past each other over geologic time.

Playa: A permanent or temporary desert lake.

Pleistocene: An *epoch* that lasted between 1.75 million years and 11 500 years ago.

Pliocene: An *epoch* that lasted between 5 and 1.75 million years ago.

Polymer: A larger *molecule* formed by five or more identical *monomers*.

Polymerize: The formation of *polymers*.

Polymorph: Two or more solid crystalline substances with the same chemistry, but different crystalline structures. Calcite (rhombohedral $CaCO_3$) and aragonite (orthorhombic $CaCO_3$) are polymorphs.

Pore: A small opening in a *rock*, *sediment*, or *soil*.

Porosity: The percentage or fraction of openings in a *soil*, *sediment*, or *rock* when compared with the total volume of the material. If the pores are not interconnected, they are not permeable (compare with permeability).

Porewater: Water within the *pores* (openings) of *soils*, *sediments*, or *rocks*.

Positron emission ($\beta+$): A type of radioactive decay where a proton converts into a neutron, positron, and neutrino (compare with electron emission).

Precambrian: A span of geologic time from the formation of the Earth about 4.5 billion years ago to about 540 million years ago.

Precipitate: Solid particles that settle out of a liquid or gas that are commonly the products of a chemical reaction.

Precipitation: Refers to dissolved species (such as arsenate *oxyanions*) in water or other liquids reacting with other dissolved species (such as Ca^{2+}, Fe^{3+}, or manganese cations) to form solid insoluble reaction products. Precipitation may result from evaporation, *oxidation*, *reduction*, changes in *pH*, or the mixing of chemicals into an aqueous solution. Also a water treatment technology that primarily uses precipitation to remove contaminants (compare with coprecipitation).

Principal quantum number: A number ($n = 1, 2, 3, \ldots$) that describes the location of an electron *shell* (energy level) around the nucleus of an atom (compare with orbital, shell, and subshell).

Proterozoic: A geologic time span between 2500 and 540 million years ago.

Proton decay: A *hypothetical* form of radioactive decay, where a proton decays into lighter subatomic particles.

Pulmonary toxicity: An adverse effect on the lungs.

Pyridine: An *aromatic* organic compound with a *benzene*-like ring structure and a composition of C_5H_5N. A nitrogen atom substitutes for one of the carbons in the benzene ring.

Pyrolysis: Heating an organic-rich sample to decompose it, which may involve low or no O_2 to avoid combustion.

Pyrometallurgical treatment: The use of heat in furnaces or incinerators to volatilize and remove arsenic or other inorganic contaminants from solid wastes. For millennia prior to waste treatment applications, this smelting technology was used to extract and concentrate valuable elements from ore deposits.

Pyroxene: A group of *silicate minerals* that are common in *mafic igneous* and some *metamorphic rocks*. The general formula for pyroxenes is XYZ_2O_6, where $X = Ca^{2+}$, Fe^{2+}, Mg^{2+}, and/or Na^+, $Y = Al^{3+}$, Fe^{2+}, Fe^{3+}, and/or Mg^{2+}, and $Z = Si^{4+}$ and possibly with some Al^{3+}.

QT interval: Duration of the ventricular systole or contraction of the heart.

Quartz: A very common *mineral* found in *sediments*, *soils*, *sedimentary rocks*, most *metamorphic rocks*, and *felsic igneous rocks*. The ideal formula for quartz is SiO_2.

Quartzite: A *metamorphosed* quartz-rich rock. *Sandstones* may metamorphose into quartzites.

Quaternary: A geologic time *period* that ranges from 1.75 million years ago to the present.

Radical: Atoms or *molecules* with unpaired electrons.

Raman spectroscopy: An analytical technique that involves the study of the vibrational and rotational properties of *molecules* in samples, which are used to identify chemical bonds. The method may be used to identify the *valence states* of arsenic in *adsorbents* (e.g. (Goldberg and Johnston, 2001)).

Rank (coal): A classification system that describes the amount of heating that a *coal* has experienced during *diagenesis* or *metamorphism*. Heating causes a coal to lose water and concentrate carbon.

Rate ratio: The incidence rate of developing a disease in the exposed group relative to the rate in the unexposed group.

Redox: Reduction–oxidation or oxidation–reduction.

Redox potential: The *Eh* of a sample.

Reducing agent: A *reductant*.

Reductant: Also called a *reducing agent*. A substance that causes another substance to be reduced by donating one or more electrons to it (compare with oxidant).

Reductase: An *enzyme* that *catalyzes* the *reduction* of a *molecule*.

Reduction: The acceptance of one or more electrons by an atom or ion. The addition of hydrogen to a compound. In geology, reduction refers to the removal of O_2 from a natural environment.

Reduction–oxidation: *Redox* or *oxidation–reduction*. The transfer of one or more electrons between a pair of atoms. The atom accepting the electrons is reduced and the electron donor is oxidized.

Reductive dissolution: The dissolution of iron or other elements in solid materials through their *reduction* to more water-soluble forms. For example, water-insoluble Fe(III) in iron *(oxy)(hydr)oxides* may reduce to water-soluble Fe(II), which causes the (oxy)(hydr)oxides to dissolve. In low *sulfide* subsurface environments, reductive dissolution is an important process in releasing arsenic that *coprecipitated* with iron (oxy)(hydr)oxides.

Regional metamorphism: The widespread *metamorphism* in the lower portion of the Earth's *crust* (compare with contact and cataclastic metamorphism).

Regolith: A surface layer of fragmented and rocky material on a moon or planet.

Remediation: Cleaning up contaminated water, *soils*, or *sediments*, and restoring a site to an environmentally acceptable condition.

Reservoir: A natural storage location where water, arsenic, or other substances may at least temporarily reside. Oceans, lakes, clouds, and *soils* are examples of reservoirs for water.

Residence time: The average amount of time that an atom or a *molecule* of a particular substance remains in a *reservoir*.

Resonance (chemistry): The migration of electrons between atoms in a *molecule* or ion (compare with resonance compound and ylide). Figure B.10 shows an example of resonance in inorganic *arsenate*.

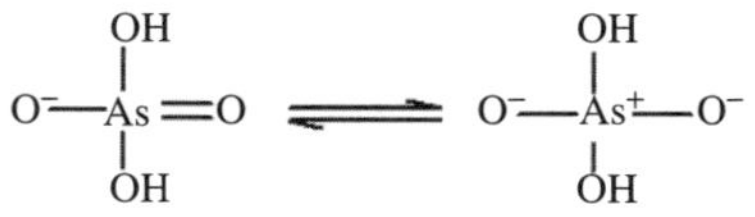

Figure B.10

Resonance compound: A *molecule* where electrons continuously migrate between atoms in the structure (compare with resonance and ylide).

Reverse osmosis: The use of pressure to force the *solvent* of a *solution* through a membrane, but not its *solutes* (compare with filtration, microfiltration, nanofiltration, and ultrafiltration).

Rhyolite: A *felsic extrusive rock*. Its *intrusive* equivalent is *granite*.

Risk assessment: An evaluation of the potential impacts of a chemical or physical hazard on human health or the environment. A risk assessment is the first step in managing and minimizing risks. Risk assessments often include identifying human health or environmental threats, possible exposure routes (e.g. inhalation, digestion, or contact with skin), the likely duration of any exposure, and the individuals that are at risk (e.g. workers, the general public, or both). A risk assessment may also involve defining the probability of an adverse effect and establishing 'safety limits' based on health standards.

Rock: A solid heterogeneous nonliving material consisting of two or more *minerals* and/or other naturally occurring solid substances.

r-process: Also called *the rapid process*. A rapid event during the last few minutes of the life of a massive star as it explodes into a *supernova*. The process involves the fusion of neutrons into nuclei to produce heavy elements, such as gold and uranium (compare with s-process).

Saltation: The movement of *sediment* particles through a series of bounces caused by flowing wind or liquid water.

Sand: A *sediment* or *sedimentary rock* particle between 0.063 and 2 mm in diameter.

Sandstone: A *clastic sedimentary rock* primarily consisting of *sand*-sized grains (0.063–2 mm in diameter). Usually, but not always, the major mineral in sandstone is *quartz*.

Saturated zone: Soils, *sediments*, and *rocks* whose *pores* are entirely filled with *groundwater* (compare with unsaturated zone and capillary fringe).

Schist: A strongly *foliated metamorphic* rock with visible *micas* or other platy or linear *minerals* (compare with slate and phyllite).

Scientific method: The research procedures for conducting a scientific investigation, which involves laboratory or field observations, the development of multiple *hypotheses* to explain the observations, and the testing of the hypotheses with computers, laboratory measurements and experiments, and/or field research. The investigation may result in none, one, several, or many of the hypotheses being verified as plausible explanations. If a hypothesis is repeatedly shown to be a reliable explanation and can actually make reasonably accurate predictions, it becomes a *theory*.

Scrubbing: The spraying of *sorbents* into *flue gas* to remove contaminants. Scrubbers are common in *coal*-combustion facilities (compare with flue gas desulfurization).

Sediment: Materials (including *rock* and *mineral* fragments (clasts), salt deposits, and fossils) produced by the *weathering* of rocks, evaporation of water, or accumulations of the remains of once-living organisms. Compared to most rocks, sediments are unconsolidated (loose). Often, sediments have been transported by wind, liquid water, or ice from their places of origin (compare with soils).

Sedimentary rock: A *rock* that formed from the natural cementing of *sediments* during burial, compaction, and interactions with groundwater (*clastic sedimentary rocks*) or that resulted from the burial and *diagenesis* of materials excreted from organisms or salts that *precipitated* from surface water.

Sedimentation (water treatment): Clarification

Sedimentation (geology): The accumulation of *sediment* in a natural environment, including: rivers, lakes, seas, oceans, and deserts.

Separation technology: A *waste treatment* or *remediation* method that physically removes contaminants from solids, liquids, or gases. Separation technologies include: *filtration*, sieving, washing, and magnetic methods.

Sequential batch leaching: A *batch leaching test* where a solid sample is *leached* in successive volumes of different types of leaching solutions (compare with serial batch leaching, leaching test, column leaching, and batch leaching).

Serial batch leaching: A *batch leaching test* where a solid sample is *leached* in successive volumes of fresh aliquots of the same leaching solution (compare with sequential batch leaching, leaching test, column leaching, and batch leaching).

Serine: An amino *acid* with a hydroxyl (–OH) functional group.

Serpentinite: A *metamorphic rock* derived from the hydration of iron- and magnesium-rich minerals, such as *olivine* and *pyroxenes*, in *peridotites* and other *ultramafic igneous* rocks.

Shale: A very fine-grained, *laminar*, and *fissile clastic sedimentary rock* consisting of 67 % or more *clay-sized* (<4 μm in diameter) particles (compare with claystone, mudstone, and siltstone).

Shell (electron): A region surrounding the nucleus of an atom where one or more electrons are expected to be found. Each shell corresponds to a *principal quantum number* (compare with orbital and subshell).

Silica: A silicon dioxide (SiO_2) compound or *mineral* (such as *quartz*). Silica may also refer to the silicon content of a *rock* or another sample, where the amount of silicon is listed as a chemical oxide (i.e. weight percent silicon dioxide or wt % SiO_2). For many rocks and other solid samples (e.g. *mafic igneous rocks*), the silicon dioxide in the sample does not exist as quartz or another SiO_2 mineral, but combines with other elements to form different *silicate* minerals (such as *olivine*, *micas*, or *pyroxenes*).

Silicate: A *mineral*, compound, or *rock* that contains *silica* (SiO_2) or SiO_4^{4-} as the major anion.

Silt: A *sediment* or *sedimentary rock* particle between 0.004 and 0.063 mm in diameter.

Siltstone: A *clastic sedimentary rock* primarily consisting of *silt*-sized particles (0.004–0.063 mm in diameter) (compare with claystone, mudstone, and shale).

Site isolation technology: A method that attempts to isolate wastes and contaminated *soils* and *sediments* from their surrounding environments. Site isolation technologies include the use of barriers and *clay* caps to isolate landfills from contaminating surrounding *groundwater*, surface water, air, sediments, and soils.

Skarn: A calcite- or *dolomite*-rich *rock* produced by *contact metamorphism* (compare with marble).

Slate: A very fine grained and cleavable *rock* that typically forms from the *metamorphism* of *shales* at low pressures and temperatures.

Sludge: A viscous mixture of fine-grained solids and water.

Smectite: A group of *clay minerals* that often contain iron, magnesium, calcium, sodium, and/or potassium in their crystalline structures. Common smectite clays include: montmorillonite $((Al,Mg)_8(Si_4O_{10})_4(OH)_8 \cdot 12H_2O)$, nontronite $(Na_{0.3}Fe_2(Al,Si)_4O_{10}(OH)_2 \cdot nH_2O$, where $n > 0)$ and beidellite $((Ca,Na)_{0.3}Al_2(OH)_2 (Al,Si)_4O_{10} \cdot 4H_2O)$.

Soil: Layered (having *horizons*) and usually fine-grained and often unconsolidated (loose) organic and inorganic materials that form from the *weathering* of *rocks* and biological activity. At least some of the materials in soils form in-place rather than being transported from distant locations by wind, ice, or liquid water (compare with sediment, A horizon, B horizon, and C horizon).

Soil profile: A vertical cross-section of a *soil* displaying all of its *horizons* down to its parent bedrock.

Solar Nebula: An ancient 'cloud' (*nebula*) of dust and gas that condensed to form our solar system about 4.5 billion years ago.

Solid solution: A group of *minerals* or other compounds with the same crystalline structure and whose chemical compositions vary somewhat due to substitutions between two or more different elements or pairs of elements. As examples, Mg^{2+} and Fe^{2+} substitute for each other to form a solid solution in *olivine* $((Fe,Mg)_2SiO_4)$ and $(CaAl)^{5+}$ and $(NaSi)^{5+}$ substitute for each other in *plagioclase*.

Solidification: The physical encapsulation of contaminants in *clays*, cement, or other solid *binders* so that they may undergo safe disposal in the environment (such as landfilling). In contrast, *stabilization* refers to the immobilization of contaminants through the formation of chemical bonds between the contaminants and the binders. Solidification and stabilization simultaneously occur in waste treatment and are commonly called *solidification/stabilization*.

Solidification/stabilization: Refers to reducing the mobility of a contaminant in *soils*, other solids, or even liquid wastes by mixing them with Portland cement, lime, cement kiln dust, *clays*, slags, polymers, water treatment *sludges*, iron-rich gypsum, *flyash*, and/or other *binders*. The process decreases the mobility of contaminants through physical encapsulation (*solidification*) and chemical bonding between the contaminants and the binders (*stabilization*).

Solubility product constant: An *equilibrium constant* that describes the dissolution of a solid in water.

Solute: A substance that will dissolve into a *solvent* or that has dissolved into a solvent.

Solvent: A liquid or gas that is capable of dissolving another substance (a *solute*). Liquid water is a common solvent.

Sorbent: A solid material that removes chemical species from liquids or gases through *adsorption* and/or *absorption* (compare with absorbent and adsorbent).

Sorption: Refers to *adsorption* and/or *absorption*. Some researchers use the generic term '*sorption*' to refer to a natural or artificial process where both adsorption and absorption may be involved or if adsorption and absorption cannot be distinguished. In particular, sorption is a treatment technology that primarily uses adsorption and/or absorption to remove contaminants from water, other liquids, or gases.

Sorting: A physical process done by nature or humans. Through sieving, flotation, or washing of *soils*, *sediments*, or other solid materials, humans separate (sort) finer-sized particles from coarser ones, heavier from lighter, or sort them on the basis of other differences in properties. Specifically, sorting may be used to separate contaminated materials with a narrow size range from a cleaner matrix of different size particles. Sorting also refers to the particle size distribution in soils, sediments, and *sedimentary rocks*. Well-sorted soils, sediments, and sedimentary rocks contain particles with very uniform sizes. Poorly sorted materials consist of particles with a wide variety of sizes.

Space group: A group of 230 known symmetry descriptions for the crystalline structures of solid substances. Bloss (1971) and other mineralogy textbooks list the 230 space groups and provide additional details.

Speciation (chemistry): The chemical species in a sample, which includes information on its *valence state* and specific chemistry. For example, the speciation of arsenic in a *groundwater* sample may include arsenic fluoride species, such as AsO_3F^{2-}. Also, an analytical method that identifies the valence state of a chemical species in a sample.

Spreading zone: A *divergent boundary* between two *tectonic plates* where *magma* from the *asthenosphere* rises and creates new *crust* on the adjacent plates. The Mid-Atlantic Ridge is an example of a spreading zone.

s-process: Also called the slow process. A slow phenomenon in massive stars where neutrons are fused into nuclei to produce heavier elements, including arsenic-75 (compare with r-process).

Squamous epithelium: A layer of flattened epithelial cells.

Stabilization: In waste treatment, the immobilization of contaminants through the formation of chemical bonds between the contaminants and encapsulating *binders*. In contrast, *solidification* deals with the physical encapsulation of contaminants (compare with solidification/stabilization).

Standard mortality ratio (SMR): The ratio of observed number of deaths to that expected.

Stern layer: A fixed layer of ions on the surface of a charged particle in an aqueous solution. The Stern layer consists of *inner-* and *outer-sphere* complexes. The Stern and *Gouy layers* comprise the *double layer* (compare with Gouy layer).

Stille reaction: A palladium *catalyzed* reaction that results in the formation of new carbon to carbon bonds by reacting an *organohalide* or *organotriflate* with an organotin compound.

Stone meteorite: A *meteorite* largely consisting of *silicate minerals*. Stony meteorites are divided into *achondrite* and *chondrite* varieties.

Stony iron meteorite: A *meteorite* consisting of a mixture of *silicates* and elemental iron–nickel. Its composition is intermediate between *iron* and *stony meteorites*.

Strong acid: An *acid*, such as HCl, where H^+ completely dissociates from its anion in water (compare with acid and weak acid).

Subbituminous coal: A *sedimentary rock* composed of at least 50 wt % and 70v % combustible organic materials that has less moisture than *lignite coal*, but more moisture than *bituminous coal*.

Subclinical disorder: Early stages in the development of a disease that does not manifest symptoms.

Subduction: A key process in *plate tectonics* that involves the movement of an *oceanic plate* and some of its associated *sediments* underneath a *continental plate* or another oceanic plate.

Sublimation: The transformation of a solid directly into vapor without melting. Ice may sublimate into water vapor.

Sublimation temperature: The temperature at which a solid substance converts into a vapor phase under normal atmospheric pressure.

Subshell: One or more orbitals in an electron *shell* of an atom. Subshells are designated s, p, d, and f. The s, p, d, and f subshells contain a maximum of 2, 6, 10, and 14 electrons, respectively (compare with orbital and principal quantum number).

Subsurface water: Water below the Earth's surface, including buried ice, *groundwater*, and *hydrothermal fluids*.

Sulfate: A group of *minerals* or aqueous chemical species containing SO_4^{2-}.

Sulfhydryl: A functional group containing −SH (compare with thiol).

Sulfide: A group of *minerals* or aqueous chemical species containing S^{2-}.

Supercritical: Refers to a fluid at certain temperatures and pressures that cause it to have properties of both a liquid and a gas.

Superfund: A program funded by the US federal government and managed by the US Environmental Protection Agency to *remediate* seriously contaminated sites in the United States.

Supernova: An exploding massive star that results from instabilities after exhausting its nuclear fuel. Supernovas can synthesize elements heavier than iron.

Supersaturate: A situation where a *solvent* contains more dissolved material (*solute*) than predicted by chemical *equilibrium*.

Surface area: The total area of the exposed surface of a solid particle, typically measured in square meters per gram ($m^2 g^{-1}$).

Surface complexation model: A computer code or *geochemical model* that provides an explanation and attempts to predict the partitioning of a chemical species between the surface of an *adsorbent* and the associated *solvent*. The models consider a number of factors, including *pH* and *ionic strength* (see (Langmuir, 1997), 369–395 for details; compare with charge distribution multisite complexation model).

Surfactant (water treatment): A compound that reduces surface tension between *immiscible* substances, like oil and water, which prevents them from separating. Examples include detergents and emulsifiers (compare with emulsification).

Suzuki reaction: A palladium *catalyzed* reaction that results in the formation of new carbon to carbon bonds by reacting an *organohalide* or *organotriflate* with an organoboronic acid.

System (chemistry): A portion of the *Universe* under study.

Tectonic plate: A section of the Earth's *lithosphere* (the *crusts* and underlying rigid upper *mantle*) that slowly moves on the underlying plastic *asthenosphere*. The Earth's lithosphere is divided into about 15 tectonic plates. Some plates consist entirely of *oceanic crust*, while others have substantial *continental crust*. The movement of tectonic plates over time is described by the *theory* of *plate tectonics*.

Tektite: Glassy spheroids that probably form from meteorite impacts. Most tektites are less than 1 mm in diameter.

Tenacity: The cohesiveness of a material or how it behaves under tension and compression. Common forms of tenacity include: brittle (breaks and powders easily), malleable (material can be hammered into sheets), sectile (material can be cut into thin layers with a knife), and ductile (the material can be drawn into a wire).

Tertiary: A span of geologic time that lasted from about 65 to 1.75 million years ago.

Tetramer: An *oligomer* consisting of only four *monomers*.

Theory: A highly reliable and well-tested scientific explanation of a natural phenomenon. While laypeople often define a 'theory'as a hunch, guess, or mere speculation, scientists use the term *hypothesis* to describe a suggested explanation or an untested idea.

Thermodynamics: A science that studies energy transformations in chemical reactions and heat transfers between *systems* and their surroundings. As Nebergall, Schmidt and Holtzclaw (1976), 461 state, thermodynamics may be useful in: (1) predicting whether two or more substances will react, (2) identifying the energy changes if the reaction occurs, and (3) predicting the final concentrations of the reactants and the products when the reaction reaches *equilibrium*. Thermodynamics, however, says nothing about the rates of any reactions or the detailed mechanisms on how they occur. Thermodynamics considers changes in *enthalpy*, *entropy*, *Gibbs free energy*, and *heat capacity* in chemical reactions. Thermodynamic data for arsenic and many of its compounds and aqueous and gaseous species are listed in Appendix C.

Thioarsenate: An aqueous species, such as $H_2AsO_3S^-$ and $H_2AsS_2O_2^-$, that contains pentavalent arsenic and *sulfide* (compare with arsenate, thioarsenic, thioarsenite, and methylthioarsenate).

Thioarsenic: An aqueous species, such as $HAs_3S_6^{2-}$ and $H_2AsO_3S^-$, that contains *sulfide* and arsenic, either as *arsenide* (As^{3-}), As^{3+}, or As^{5+} (compare with arsenosulfide, methylthioarsenate, thioarsenate, and thioarsenite).

Thioarsenite: An aqueous species, such as $H_2As_3S_6^-$, that contains trivalent arsenic and *sulfide* (compare with arsenite, thioarsenic, thioarsenate, and methylthioarsenate).

Threshold limit value (TLV): The maximum concentration of a chemical substance in air that a worker can be exposed to day after work day without suffering any harmful effects. Typically, TLVs are guidelines or recommendations rather than enforceable standards. Like *permissible exposure limits* (PELs), TLVs are often given as time weighted averages (TWAs) for an 8-hour work day (compare with action level and permissible exposure limit).

Thiol: An organic compound or chemical species containing a *sulfhydryl* (–SH) group. Methanethiol (CH_3SH) is an example of a thiol.

Till: A poorly *sorted sediment* deposited by *glaciers*.

Toxicity characteristic: One of the four properties (ignitability, corrosivity, and reactivity being the other three) that are frequently used to evaluate solid and liquid wastes for *hazardousness* under the regulations of the United States Environmental Protection Agency (US EPA) (40 *Code of Federal Regulations* 261). Toxic wastes are possibly harmful or fatal if humans inhale, ingest, and/or come into physical contact with them. Arsenic-bearing wastes often have this characteristic. The *toxicity characteristic leaching procedure* (TCLP) is used to evaluate liquid and solid wastes for the toxicity characteristic (also see Appendix E).

Toxicity characteristic leaching procedure (TCLP): A *batch leaching* method that is used to evaluate solid and liquid wastes for the *toxicity characteristic* of *hazardousness* under the regulations of the United States Environmental Protection Agency (US EPA) (40 *Code of Federal Regulations* 261; ((US EPA), 2007), Method 1311). The TCLP is designed to predict if contaminants are likely to mobilize from wastes and move into surrounding *groundwaters* if they are buried in an unlined municipal waste landfill (Greene, 1991). Liquid wastes containing less than 0.5% dry solids are filtered through 0.6–0.8 µm glass fiber filters. The filtrates are TCLP *leachates*. For an arsenic-bearing nonvolatile solid waste, the TCLP uses either a *pH* 2.88 ± 0.05 acetic acid leaching solution or a pH 4.93 ± 0.05 acetic acid and sodium hydroxide leaching solution, depending upon the pH of an aqueous slurry of the waste. At least 100 g of the solid waste is mixed with 20 times the mass of the leaching solution (i.e. 2.0 kg of solution for 100 g of waste). The mixture is then tumbled for 18 ± 2 hours at 23 ± 1 °C on an end-over-end stirrer. After tumbling, the liquid is separated from the sample with a 0.6–0.8 µm glass fiber filter. The filtrate is a TCLP leachate. If a TCLP leachate contains more than 5 mg L^{-1} of arsenic, the liquid or solid waste has the toxic characteristic of hazardousness and the waste is identified as hazardous under US federal regulations. Some wastes, such as chromated copper arsenate (CCA)-treated wood, are currently exempt from US federal hazardous waste regulations even though they often produce TCLP leachates that exceed the 5 mg L^{-1} limit for arsenic. Nevertheless, the wastes may still be classified as hazardous under US state and local regulations.

Trace element: An element in the Earth's *crusts*, a *rock*, an organism, or another sample that has a concentration below 1000 mg kg^{-1} (compare with minor and major elements).

Transforming growth factor-α (TGF-α): A cytokine or messenger protein that mediates signals of cell proliferation and survival.

Triexponential model: A mathematical equation that sums three exponents to describe data.

Trimethylarsine: An *organoarsenical* with a composition of $(CH_3)_3As$ (see Figure 2.2).

Trimethylarsine oxide (TMAO): An *organoarsenical* with a composition of $(CH_3)_3AsO$ (see Figure 2.2).

Triple-alpha process: The formation of a carbon-12 nucleus in the interior of a massive star through the essentially simultaneous *fusion* of three helium nuclei.

Tumor necrosis factor-α: A cytokine or messenger protein involved in systemic inflammation and other biological processes.

Ultrafiltration: Filtration for the treatment of water that removes suspended particles with diameters that are greater than about 0.01 μm (compare with filtration, microfiltration, nanofiltration, and reverse osmosis).

Ultramafic: Describes a *magma, lava,* or *igneous rock* that is generally richer in magnesium and iron and has less *silica* (<45 wt %) than *mafic* magmas, lavas, and igneous rocks (Hyndman, 1985), 42; (Press and Siever, 2001), 74.

Ultramafic breccia: An *ultramafic stony meteorite* that consists of angular fragments.

Unconfined aquifer: An *aquifer* that is not overlain by any *aquitards.* Precipitation and surface water may readily infiltrate into an unconfined aquifer (compare with confined aquifer).

Underground coal gasification: The production of recoverable and useful combustible gases from the *in situ* heating of subsurface *coal* deposits.

Universe: All existing matter, energy, and the space between them. The Universe began with the *Big Bang* about 13.7 billion years ago.

Unsaturated zone: Soils, sediments, and *rocks* whose *pores* contain both air and water (compare with saturated zone and capillary fringe).

Valence electron: An electron in the outermost *shell* of an atom. Valence electrons are responsible for chemical reactions.

Valence state: Also called the *oxidation number or oxidation state.* An integer (positive, negative, or zero) that describes the number of electrons that must be added or removed from an atom to give it a neutral charge. Typical valence states for arsenic are -3, 0, $+3$, and $+5$.

van der Waals force: A weak attraction between unbonded atoms or *molecules.* For example, weak van der Waals forces occur between layers of arsenic and sulfur in the crystalline structure of the *mineral* orpiment (As_2S_3).

van't Hoff equation: Is used to calculate the *equilibrium constant* of a reaction at a desired temperature between about 10 and 40 °C and a constant pressure of one bar. The equation is $\ln (K_{eq2}/K_{eq1}) = -\Delta H_R^o/R$ $(1/T_2 - 1/T_1)$. See Chapter 2 for details.

Vicinal dithol: Two *sulfhydryl* functional groups bonded to adjoining carbons (e.g. $HSCH_2CH_2SH$).

Vitrification: A treatment technology for contaminated *soils, sediments,* and solid wastes that involves melting them to incinerate organic contaminants and encapsulate arsenic and other inorganic species in the resulting melts (compare with *in situ* vitrification).

Waste management: The proper collection, handling, treatment, transportation, and disposal of unusable liquids, solids, or gases. Proper waste management minimizes the negative impacts of wastes on the environment. While waste management seeks to prevent negative environmental impacts, the purpose of *remediation* is to restore sites that have already been contaminated into an environmental acceptable condition.

Wastewater discharge standards: Regulatory limits on the volume, discharge rate, and concentrations of organic and inorganic chemicals in wastewaters that must not be exceeded if the untreated waters are to be discharged into the environment. If any of the standards are exceeded, the wastewater must be treated before discharging. Wastewater discharge standards for arsenic are generally higher than standards for drinking water because the regulations assume that the discharged waters will disperse into the environment rather than being directly utilized by humans. Different nations have their own wastewater discharge standards for arsenic and other contaminants (see Appendix E for available information).

Water table: A subsurface boundary located between the *capillary fringe* and the *saturated zone*.

Weak acid: An *acid* that does not completely dissociate (release all of its hydrogens) in *pH* neutral water, but selectively releases H^+ with increasing pH. Arsenic (H_3AsO_4) and arsenious (H_3AsO_3) acids are weak (compare with acid and strong acid).

Weathering: The physical or chemical conversion of surface *rocks* into *sediments*, *soils*, and dissolved and suspended materials in water through exposure of the rocks to liquid water, oxygen, carbon dioxide, wind, *acids*, salts, ice, biological activity, temperature fluctuations, and/or other factors at and near the Earth's surface (compare with chemical and physical weathering).

Wetland: Poorly drained, low relief areas in which *soils* and *sediments* are seasonally or continuously saturated or covered with water. Wetlands include bogs, swamps, and fens.

Wetting capacity: The amount or volume of water a chemical can hold.

Wittig reaction: A reaction which converts an *aldehyde* or *ketone* into an *alkene* via a reaction with a *ylide*.

X-ray absorption fine structure (XAFS) spectroscopy: An *X-ray absorption spectroscopy* (XAS) method. XAFS provides information on the physical and chemical properties of matter on an atomic scale. XAFS may include *X-ray absorption near edge structure* (XANES) and *extended X-ray absorption fine structure* (EXAFS) spectra.

X-ray absorption near edge structure (XANES) spectrum: An analysis from *X-ray absorption spectroscopy* (XAS) and, in particular, *X-ray absorption fine structure* (XAFS) spectroscopy. XANES can be used to identify the *valence state* of arsenic in solid samples (Teixeira and Ciminelli, 2005; Köber *et al.*, 2005).

X-ray absorption spectroscopy (XAS): Methods that use X-rays to investigate the physical and chemical properties of materials on an atomic scale. XAS includes *X-ray adsorption fine structure* (XAFS) spectroscopy and its *X-ray absorption near edge structure* (XANES) and *extended X-ray absorption fine structure* (EXAFS) spectra.

X-ray diffraction (XRD): Analytical techniques, which include powder and single-crystal methods, that use X-rays to study the crystalline properties of solid substances. Traditionally, powder XRD has been primarily used to identify crystalline substances in solid samples. Single-crystal methods provide detailed information on the crystal structure and chemistry of a single crystal, such as identification of the *space group* and the types and distributions of the atoms in the crystal.

X-ray photoelectron spectroscopy (XPS): A method that uses X-rays to identify *valence states* and provide semiquantitative concentrations for elements ($Z \geq 3$, lithium and above) on the surfaces of solid samples (e.g. (Nesbitt *et al.* (2003); Goodarzi and Huggins (2005)).

Ylide: A *resonating* organic *molecule* that can be commonly visualized as alternating between adjacent positive and negative charges and a double bond ($X^+ - Y^- \leftrightarrow X = Y$).

Zeolite: A group of aluminum *silicate minerals* that may form under low *metamorphic* temperatures and pressures.

Zero point of charge (ZPC): The zero point of charge of a *mineral* or another solid substance is the *pH* of an aqueous solution in contact and *equilibrium* with the solid when the solid has a net surface charge of zero. The ZPC depends on the composition of the solid and the concentration and chemistry of the electrolytes in the aqueous solution. In situations where the net surface charge of the solid is only controlled by the *adsorption* of OH^- or H^+, the zero point of charge is the *isoelectric point*.

References

Allison, J.D., Brown, D.S. and Novo-Gradac, K.J. (1991) *MINTEQA2/PRODEFA2: A Geochemical Assessment Model for Environmental Systems: Version 3.0 User's Manual*, US Environmental Protection Agency, Washington, DC.

Audi, G., Bersillon, O., Blachot, J. and Wapstra, A.H. (2003) The NUBASE evaluation of nuclear and decay properties. *Nuclear Physics A*, **729**(1), 3–128.

Bloss, F.D. (1971) *Crystallography and Crystal Chemistry: An Introduction*, Holt, Rinehart & Winston, Inc., New York, 545pp.

Boggs, S., Jr. (1995) *Principles of Sedimentology and Stratigraphy*, 2nd edn, Prentice Hall, Upper Saddle River, NJ.

California Department of Toxic Substances Control (2006) *Unofficial Code of California Regulations*, California Department of Toxic Substances Control, Sacramento, CA.

Dalrymple, G.B. (1991) *The Age of the Earth*, Stanford University Press, Stanford, CA, 474pp.

Drever, J.I. (1997) *The Geochemistry of Natural Waters: Surface and Groundwater Environments*, Prentice Hall, Upper Saddle River, NJ, 436pp.

Eby, G. N. (2004) *Principles of Environmental Geochemistry*, Brooks/Cole-Thomson Learning, Pacific Grove, CA, 514pp.

Faure, G. (1998) *Principles and Applications of Geochemistry*, 2nd edn, Prentice Hall, Upper Saddle River, NJ, 600pp.

Freeze, R.A. and Cherry, J.A. (1979) *Groundwater*, Prentice Hall, Englewood Cliffs, NJ, 604pp.

Goldberg, S. and Johnston, C.T. (2001) Mechanisms of arsenic adsorption on amorphous oxides evaluated using macroscopic measurements, vibrational spectroscopy, and surface complexation modeling. *Journal of Colloid and Interface Science*, **234**(1), 204–16.

Goodarzi, F. and Huggins, F.E. (2005) Speciation of arsenic in feed coals and their ash byproducts from Canadian power plants burning sub-bituminous and bituminous coals. *Energy and Fuels*, **19**(3), 905–15.

Greene, G.C. (1991) *TCLP: Where are we and what lies ahead?* American Environmental Laboratory, December: 9–14.

Hiemstra, T., and van Riemsdijk, W.H. (1999) Surface structural ion adsorption modeling of competitive binding of oxyanions by metal (hydr)oxides. *Journal of Colloid and Interface Science*, **210**(1), 182–93.

Hyndman, D.W. (1985) *Petrology of Igneous and Metamorphic Rocks*, 2nd edn, McGraw-Hill, New York, 786pp.

Klein, C. (2002) in *The 22nd Edition of the Manual of Mineral Science* (J.D. Dana), John Wiley & Sons, Inc., New York, 641pp.

Köber, R., Daus, B., Ebert, M. *et al.* (2005) Compost-based permeable reactive barriers for the source treatment of arsenic contaminations in aquifers: column studies and solid-phase investigations. *Environmental Science and Technology*, **39**(19), 7650–55.

Köber, R., Welter, E., Ebert, M. and Dahmke, A. (2005) Removal of arsenic from groundwater by zerovalent iron and the role of sulfide. *Environmental Science and Technology*, **39**(20), 8038–44.

Kosson, D.S., Van Der Sloot, H.A., Sanchez, F. and Garrabrants, A.C. (2002) An integrated framework for evaluating leaching in waste management and utilization of secondary materials. *Environmental Engineering Science*, **19**(3), 159–204.

Krauskopf, K.B. and Bird, D.K. (1995) *Introduction to Geochemistry*, 3rd edn, McGraw-Hill, Boston, MA.

Ladeira, A.C.Q., Ciminelli, V.S.T., Duarte, H.A., *et al.* (2001) Mechanism of anion retention from EXAFS and density functional calculations: arsenic (V) adsorbed on gibbsite. *Geochimica et Cosmochimica Acta*, **65**(8), 1211–17.

Langmuir, D. (1997) *Aqueous Environmental Geochemistry*, Prentice Hall, Upper Saddle River, NJ, 600pp.

Nebergall, W.H., Schmidt, F.C. and Holtzclaw, H.F., Jr. (1976) *College Chemistry with Qualitative Analysis*, 5th edn, D. C. Heath & Company, Lexington, MA, 1058pp.

Nesbitt, H.W., Schaufuß, A., Sciani, M. *et al.* (2003) Monitoring fundamental reactions at NiAsS surfaces by synchrotron radiation X-ray photoelectron spectroscopy: As and S air oxidation by consecutive reaction schemes. *Geochimica et Cosmochimica Acta*, **67**(5), 845–58.

Pauling, L. (1960) *Nature of the Chemical Bond*, 3rd edn, Cornell University Press, Ithaca, NY.

Pentecost, A., Jones, B. and Renaut, R.W. (2003) What is a hot spring? *Canadian Journal of Earth Sciences*, **40**(11), 1443–46.

Press, F. and Siever, R. (2001) *Understanding Earth*, 3rd edn, W. H. Freeman & Company, New York.

Randall, S.R., Sherman, D.M. and Ragnarsdottir, K. V. (2001) Sorption of As(V) on green rust (Fe_4(II)Fe_2(III)(OH)$_{12}$ SO_4-$3H_2O$) and lepidocrocite (γ-FeOOH): surface complexes from EXAFS spectroscopy. *Geochimica et Cosmochimica Acta*, **65**(7), 1015–23.

Spellman, F.R. (2003) *Handbook of Water and Wastewater Treatment Plant Operations*, Lewis Publishers, Boca Raton, FL, 661pp.

Teixeira, M.C. and Ciminelli, V.S.T. (2005) Development of a biosorbent for arsenite: structural modeling based on X-ray spectroscopy. *Environmental Science and Technology*, **39**(3), 895–900.

US Environmental Protection Agency (US EPA) (2007) *Test Methods for Evaluating Solid Waste, Physical/Chemical Methods*, SW-486, 6th revision, Office of Solid Waste, National Technical Information Service, Springfield, VA.

US Government Printing Office, US Code of Federal Regulations (CFR), Superintendent of Documents, Washington, DC.

Appendix C

Arsenic Thermodynamic Data

C.1 Introduction

Italicized terms in the text of this appendix or words derived from the italicized terms are defined in the glossary of Appendix B.

Table C.1 lists published *thermodynamic data* for arsenic, including: the more common elemental forms, selected compounds, and some aqueous and gaseous species. The table contains *Gibbs free energy* (ΔG_f), *enthalpy* (ΔH_f), *entropy* (S), and *heat capacity* (Cp) values at 1 bar pressure and mostly at 298.15 K (25 °C). For thermodynamic data of chemical species not containing arsenic, extensive tables occur in geochemistry textbooks and other references, including: (Barin, 1995; Robie, Hemingway and Fisher, 1979; Wagman *et al.*, 1982; Drever, 1997; Eby, 2004; Faure, 1998; Lide, 2007; Krauskopf and Bird, 1995). Thermodynamic data are important in examining the behavior of arsenic in mineral–water interactions, studying water quality issues, investigating the formation of ore deposits, developing and interpreting *Eh-pH* diagrams, and designing environmental *remediation* projects (Nordstrom and Archer, 2003, 1–2).

Not all published and widely accepted thermodynamic values are reliable. Nordstrom and Archer (2003) provide a detailed review of the controversies, uncertainties, and problems related to thermodynamic data for arsenic and its compounds and aqueous species. Many of the data are contradictory and the methods that produce the data are sometimes questionable or have not been thoroughly documented. Too often, data in the literature have been passed from reference to reference without critical evaluations. Some of the data have high measurement errors, were produced under undefined or poorly defined laboratory conditions, and involved unrepresentative sampling (Matschullat 2000, 298; Nordstrom and Archer, 2003). Furthermore, other questionable data originate from obscure documents or are written in languages that many individuals cannot read and properly interpret. Therefore, thermodynamic results must be accepted with a certain amount of caution. The table in this appendix includes thermodynamic data from various sources, which provide users with some idea of their variability. Although sometimes unavoidable, users

Arsenic Edited by Kevin R. Henke
© 2009 John Wiley & Sons, Ltd

are cautioned to avoid mixing data from different literature sources when performing calculations for a reaction. Using data from multiple sources may introduce serious errors (Wagman *et al.*, 1982; Eby, 2004, 474).

C.2   Modeling applications with thermodynamic data

Once high-quality laboratory data are available, they may be used in geochemical computer models, such as MINTEQA2 (Allison, Brown and Novo-Gradac, 1991) and PHREEQC (Parkhurst and Appelo, 1999). Langmuir (1997, 558–561) provides a brief review of the different types of geochemical models and their advantages and limitations. Geochemical models allow users to avoid tedious calculations. When properly utilized, they may: (1) derive *adsorption* models, (2) identify probable aqueous species and estimate their *activities*, (3) model the effects of pH, *reduction/oxidation* (*redox* conditions), temperature, *ionic strength*, and other factors on arsenic chemistry, and/or (4) identify possible *precipitates*. However, as discussed in Chapter 2, geochemical models are often incapable of accurately representing and predicting the complex behavior of arsenic in natural environments. Specifically, the models typically fail to predict the presence of *metastable* species, sluggish chemical reactions, and the effects of biological organisms.

C.3   Thermodynamic data

Table C.1 *Thermodynamic data for arsenic, its compounds, and its aqueous and gaseous species at 1 bar pressure. Note that the units of G_i and H_i are 1000 times larger than S_i and Cp_i. 1 kcal = 4.184 kJ. °C = K −273.15. (See Appendix A for other unit conversions). Phases include: amorphous (am), aqueous species (aq), gas, liquid (liq), and crystalline solids (xls).*

Chemical species	Phase	Temperature (K)	Gibbs free energy (G_i) (kJ mol^{-1})	Enthalpy (H_i) (kJ mol^{-1})	Entropy (S_i) (J mol^{-1}·K^{-1})	Heat Capacity (Cp_i) (J mol^{-1}·K^{-1})	Reference
Elemental							
As(0) (elemental arsenic, rhombohedral or 'gray')	xls	298.15	0	0	35.63	24.43	Nordstrom and Archer (2003)
As(0) (elemental arsenic, rhombohedral or 'gray')	xls	298.15	0	0	35.1	—	Wagman et al. (1982)
As(0) (elemental arsenic, rhombohedral or 'gray')	xls	298.15	0	0	35.69	24.65	Robie, Hemingway and Fisher (1979)
As(0) (elemental arsenic, rhombohedral or 'gray')	xls	298.15	0	0	35.6	—	Naumov, Ryzhenko and Khodakovsky (1974)
As(0) (elemental arsenic, rhombohedral or 'gray')	xls	298.15	0	0	35.706	24.652	Barin (1995)
As(0) (elemental arsenic, rhombohedral or 'gray')	xls	400	0	0	43.04	25.38	Robie, Hemingway and Fisher (1979)
As(0) (elemental arsenic, rhombohedral or 'gray')	xls	400	0	0	43.065	25.388	Barin (1995)
As(0) (elemental arsenic, rhombohedral or 'gray')	xls	500	0	0	48.77	25.96	Robie, Hemingway and Fisher (1979)
As(0) (elemental arsenic, rhombohedral or 'gray')	xls	500	0	0	48.790	25.945	Barin (1995)
As(0) (elemental arsenic, rhombohedral or 'gray')	xls	600	0	0	53.55	26.51	Robie, Hemingway and Fisher (1979)
As(0) (elemental arsenic, rhombohedral or 'gray')	xls	600	0	0	53.569	26.501	Barin (1995)
As(0) (elemental arsenic, rhombohedral or 'gray')	xls	700	0	0	57.68	27.05	Robie, Hemingway and Fisher (1979)
As(0) (elemental arsenic, rhombohedral or 'gray')	xls	700	0	0	57.696	27.054	Barin (1995)

(continued overleaf)

Table C.1   (continued)

Chemical species	Phase	Temperature (K)	Gibbs free energy (G_i) (kJ mol^{-1})	Enthalpy (H_i) (kJ mol^{-1})	Entropy (S_i) (J mol$^{-1} \cdot$K^{-1})	Heat Capacity (Cp_i) (J mol$^{-1} \cdot$K^{-1})	Reference
As(0) (elemental arsenic, rhombohedral or 'gray')	xls	800	0	0	61.33	27.60	Robie, Hemingway and Fisher (1979)
As(0) (elemental arsenic, rhombohedral or 'gray')	xls	800	0	0	61.344	27.580	Barin (1995)
As(0) (elemental arsenic, rhombohedral or 'gray')	xls	875	0	0	63.82	28.03	Robie, Hemingway and Fisher (1979)
As(0) (elemental arsenic, amorphous or 'black')	am	298.15	—	4.2	—	—	Wagman et al. (1982)
As(0) (elemental arsenic, amorphous or 'black')	am	298.15	—	13.6	—	—	Naumov, Ryzhenko and Khodakovsky (1974)
As$^+$	gas	298.15	—	1255.6	—	—	Wagman et al. (1982)
As^{2+}	gas	298.15	—	3059.8	—	—	Wagman et al. (1982)
As^{3+}	gas	298.15	—	5801.1	—	—	Wagman et al. (1982)
As^{4+}	gas	298.15	—	10644	—	—	Wagman et al. (1982)
As^{5+}	gas	298.15	—	16694	—	—	Wagman et al. (1982)
As^{6+}	gas	298.15	—	29045	—	—	Wagman et al. (1982)
As$_2$	gas	298.15	171.9	222.2	239.4	35.003	Wagman et al. (1982)
As$_2$	gas	298.15	143	194	242.2	—	Naumov, Ryzhenko and Khodakovsky (1974)
As$_4$	gas	298.15	92.4	143.9	314	—	Wagman et al. (1982)
As$_4$	gas	298.15	87.9	144	330	—	Naumov, Ryzhenko and Khodakovsky (1974)
As$_4$	gas	298.15	98.261	153.30	327.43	77.208	Barin (1995)

Gases and aqueous species of arsenic compounds

Formula	State	T (K)					Reference
Ag_3AsO_4	xls	298.15	−545.39	−634.29	275.83	173.86	Barin (1995)
$AlAs$	xls	298.15	—	−116.3	—	—	Wagman et al. (1982)
$AlAs$	xls	298.15	−115.20	−116.32	60.25	45.80	Barin (1995)
$AlAsO_4$	xls	298.15	−1333.1	−1431.1	145.60	118.34	Barin (1995)
$AlAsO_4$	xls	298.15	−1280.8	—	—	—	Ryu et al. (2002)
$AlAsO_4 \cdot 2H_2O$ (mansfieldite)	xls	298.15	−1709	—	—	—	Naumov, Ryzhenko and Khodakovsky (1974)
$AsBr_3$	gas	298.15	−161.67	−132.1	363.8	—	Pankratov and Uchaeva (2000)
$AsBr_3$	gas	298.15	−159	−130	363.87	79.16	Wagman et al. (1982)
$AsBr_3$	gas	298.15	−159.80	−130.00	363.99	70.992	Barin (1995)
$AsCl_3$	gas	298.15	−258.05	−270.3	328.8	—	Pankratov and Uchaeva (2000)
$AsCl_3$	gas	298.15	−248.9	−261.5	327.17	75.73	Wagman et al. (1982)
$AsCl_3$	gas	298.15	−248.66	−261.50	327.30	75.395	Barin (1995)
$AsCl_3$	liq	298.15	−259.4	−305.0	216.3	—	Wagman et al. (1982)
$AsCl_3$	liq	298.15	−259.08	−305.01	216.31	133.47	Barin (1995)
AsF_3	gas	298.15	−905.67	−785.8	289.0	—	Pankratov and Uchaeva (2000)
AsF_3	gas	298.15	−770.76	−785.76	298.10	65.61	Wagman et al. (1982)
AsF_3	gas	298.15	−770.65	−785.76	289.22	64.684	Barin (1995)
AsF_3	liq	298.15	−774.16	−821.3	181.21	126.57	Wagman et al. (1982)
AsF_3	liq	298.15	−774.01	−821.32	181.21	126.55	Barin (1995)
AsF_5	gas	298.15	−1172.5	−1236.7	—	—	Pankratov and Uchaeva (2000)

(*continued overleaf*)

Table C.1 (continued)

Chemical species	Phase	Temperature (K)	Gibbs free energy (G_i) (kJ mol^{-1})	Enthalpy (H_i) (kJ mol^{-1})	Entropy (S_i) (J mol^{-1}·K^{-1})	Heat Capacity (Cp_i) (J mol^{-1}·K^{-1})	Reference
AsF$_5$	gas	298.15	−1169.0	−1234.2	—	87.84	O'Hare (1993)
AsF$_5$	gas	298.15	−1169.6	−1236.8	317.26	97.541	Barin (1995)
AsH$_3$ (arsine)	gas	298.15	68.91	66.40	223.0	—	Pankratov and Uchaeva (2000)
AsH$_3$ (arsine)	gas	298.15	68.93	66.44	222.78	38.07	Wagman et al. (1982)
AsH$_3$ (arsine)	gas	298.15	68.83	66.4	223	—	Naumov, Ryzhenko and Khodakovsky (1974)
AsH$_3$ (arsine)	gas	298.15	69.109	66.442	222.78	38.072	Barin (1995)
AsI$_3$	xls	298.15	−59.4	−58.2	213.05	105.77	Wagman et al. (1982)
AsI$_3$	xls	298.15	−59.091	−58.158	213.05	105.77	Barin (1995)
AsI$_3$	gas	298.15	—	—	388.34	80.63	Wagman et al. (1982)
AsI$_3$	gas	298.15	−15.321	38.911	391.82	80.960	Barin (1995)
AsN	gas	298.15	167.97	196.27	225.6	30.42	Wagman et al. (1982)
AsO	gas	298.15	−84.715	−57.287	230.27	32.342	Barin (1995)
As$_2$O$_3$ (arsenolite, cubic)	xls	298.15	−576.34	−657.27	107.38	96.88	Nordstrom and Archer (2003)
As$_2$O$_3$ (arsenolite, cubic)	xls	298.15	−576	−657	107.4	—	Robie, Hemingway and Fisher (1979)
As$_2$O$_3$ (arsenolite, cubic)	xls	298.15	−588.3	−666.1	117	—	Naumov, Ryzhenko and Khodakovsky (1974)
As$_2$O$_3$ (arsenolite, cubic)	xls	298.15	−576.41	—	—	—	Davis et al. (1996)
As$_2$O$_3$ (arsenolite, cubic)	xls	298.15	−575.96	−656.97	107.41	96.878	Barin (1995)
As$_2$O$_3$ (claudetite, monoclinic)	xls	298.15	−576.53	−655.67	113.37	96.98	Nordstrom and Archer (2003)
As$_2$O$_3$ (claudetite, monoclinic)	xls	298.15	−576	−655	113.3	—	Robie, Hemingway and Fisher (1979)
As$_2$O$_3$ (claudetite, monoclinic)	xls	298.15	−589.1	−664.0	127	—	Naumov, Ryzhenko and Khodakovsky (1974)

Species	State	T (K)	ΔfG°	ΔfH°	S°	Cp°	Reference
As₂O₃ (claudetite, monoclinic)	xls	298.15	−576.64	−654.80	117.00	96.979	Barin (1995)
As₂O₃·SO₃	xls	298.15	—	−1194.32	—	—	Wagman et al. (1982)
As₂O₅	xls	298.15	−774.96	−917.59	105.44	115.9	Nordstrom and Archer (2003)
As₂O₅	xls	298.15	−782.0	−924.7	105	—	Naumov, Ryzhenko and Khodakovsky (1974)
As₂O₅	xls	298.15	−782.78	—	—	—	Davis et al. (1996)
As₂O₅	xls	298.15	−782.3	−924.87	105.4	116.52	Wagman et al. (1982)
As₂O₅	xls	298.15	−782.09	−924.87	105.40	116.54	Barin (1995)
As₂O₅	xls	400	−733.31	−924.59	142.62	136.54	Barin (1995)
As₂O₅	xls	500	−685.66	−922.93	174.77	151.78	Barin (1995)
As₂O₅·4H₂O	xls	298.15	—	−2104.6	—	—	Wagman et al. (1982)
As₂O₅·4H₂O	xls	298.15	−1770	—	—	—	Faure (1998)
(As₂O₅)₃·5H₂O	xls	298.15	—	−4248.4	—	—	Wagman et al. (1982)
As₄O₆	gas	298.15	−1092.2	−1196.2	409.35	173.60	Barin (1995)
AsS (α, realgar)	xls	298.15	−31.3	−31.8	62.9	47	Nordstrom and Archer (2003)
AsS (α, realgar)	xls	298.15	−35	−36	63.5	—	Naumov, Ryzhenko and Khodakovsky (1974)
AsS (β)	xls	298.15	−30.9	−31.0	63.5	47	Nordstrom and Archer (2003)
AsS(OH)(SH)⁻	aq	298.15	−245.11	—	—	—	Helz et al. (1995)
As₂S₃	am	298.15	−76.8	−66.9	200	—	Nordstrom and Archer (2003)
As₂S₃ (orpiment)	xls	298.15	−84.9	−85.8	163.8	163	Nordstrom and Archer (2003)
As₂S₃ (orpiment)	xls	298.15	−95.4	−96.2	164	—	Naumov, Ryzhenko and Khodakovsky (1974)
As₃S₄(SH)₂⁻	aq	298.15	−127.19	—	—	—	Helz et al. (1995)

(continued overleaf)

Table C.1 (continued)

Chemical species	Phase	Temperature (K)	Gibbs free energy (G_i) (kJ mol^{-1})	Enthalpy (H_i) (kJ mol^{-1})	Entropy (S_i) (J mol$^{-1} \cdot$K^{-1})	Heat Capacity (Cp_i) (J mol$^{-1} \cdot$K^{-1})	Reference
As$_2$Se$_3$	xls	298.15	−88.4	−86.1	7.8	—	O'Hare et al. (1990)
As$_2$Se$_3$	xls	298.15	—	−81.1	—	—	O'Hare (1993)
As$_2$Se$_3$	xls	298.15	−101.43	−102.51	194.56	121.39	Barin (1995)
As$_2$Se$_3$	am	298.15	—	−53.1	—	—	O'Hare (1993)
As$_2$Te$_3$	xls	298.15	−39.580	−37.656	226.35	127.49	Barin (1995)
Ba$_3$(AsO$_4$)$_2$	xls	298.15	−3113.40	—	—	—	Zhu et al. (2005)
Ba$_3$(AsO$_4$)$_2$	xls	298.15	−3192.2	−3421.7	309.62	257.21	Barin (1995)
BaHAsO$_4 \cdot$H$_2$O	xls	298.15	−1544.47	—	—	—	Zhu et al. (2005)
Be$_3$(AsO$_4$)$_2$	xls	298.15	−2525.5	−2738.1	207.28	232.52	Barin (1995)
BiAsO$_4$	xls	298.15	−619	—	—	—	Wagman et al. (1982)
BiAsO$_4$	xls	298.15	−695.43	−795.21	168.07	121.12	Barin (1995)
BiAsO$_4$ (rooseveltite)	xls	298.15	−613.58	—	—	—	Naumov, Ryzhenko and Khodakovsky (1974)
(CH$_3$)$_3$As	gas	298.15	—	12	—	—	Pankratov and Uchaeva (2000)
Ca(H$_2$AsO$_4$)$_2$	xls	298.15	−2054	—	—	—	Ryu et al. (2002)
Ca$_2$AsO$_4$OH	xls	298.15	−1988	—	—	—	Ryu et al. (2002)
Ca$_3$(AsO$_4$)$_2$	xls	298.15	−3061	—	—	—	Ryu et al. (2002)
Ca$_3$(AsO$_4$)$_2$	xls	298.15	−3063.1	−3298.7	226.00	249.79	Barin (1995)
Ca$_3$(AsO$_4$)$_2$	xls	298.15	−3063.0	−3298.7	226	—	Wagman et al. (1982)
Ca$_3$(AsO$_4$)$_2 \cdot$2.25H$_2$O	xls	298.15	−3611.50	—	—	—	Zhu et al. (2006)
Ca$_3$(AsO$_4$)$_2 \cdot$3H$_2$O	xls	298.15	−3787.87	—	—	—	Zhu et al. (2006)
Ca$_3$(AsO$_4$)$_2 \cdot$3.67H$_2$O	xls	296	−3945	—	—	—	Bothe and Brown (1999)
Ca$_3$(AsO$_4$)$_2 \cdot$4H$_2$O	xls	298.15	−4019	—	—	—	Naumov, Ryzhenko and Khodakovsky (1974)
Ca$_3$(AsO$_4$)$_2 \cdot$4.25H$_2$O	xls	296	−4085	—	—	—	Bothe and Brown (1999)

Formula		T					Reference
$Ca_3(AsO_4)_2 \cdot 6H_2O$	xls	298.15	−2732	—	—	—	Naumov, Ryzhenko and Khodakovsky (1974)
$Ca_3(AsO_4)_2 \cdot 8H_2O$	xls	298.15	−3530	—	—	—	Naumov, Ryzhenko and Khodakovsky (1974)
$Ca_4(OH)_2(AsO_4)_2 \cdot 4H_2O$	xls	296	−4941	—	—	—	Bothe and Brown (1999)
$Ca_4(OH)_2(AsO_4)_2 \cdot 4H_2O$	xls	298.15	−4928.86	—	—	—	Zhu et al. (2006)
$Ca_5(AsO_4)_3OH$	xls	296	−5087	—	—	—	Bothe and Brown (1999)
$Ca_5(AsO_4)_3OH$	xls	298.15	−5096.47	—	—	—	Zhu et al. (2006)
$Ca_5H_2(AsO_4)_4$	xls	298.15	−5636.7	—	—	—	Ryu et al. (2002)
$Ca_5I_2(AsO_4)_4 \cdot 9H_2O$ (ferrarisite)	xls	296	−7808	—	—	—	Bothe and Brown (1999)
$Ca_5H_2(AsO_4)_4 \cdot 9H_2O$ (guerinite)	xls	296	−7803	—	—	—	Bothe and Brown (1999)
$CaH(AsO_4) \cdot H_2O$	xls	296	−1533	—	—	—	Bothe and Brown (1999)
$CaHAsO_4^0$	aq	298.15	—	−1446.0	—	—	Wagman et al. (1982)
$Cd_3(AsO_4)_2$	xls	298.15	−1716.1	—	—	—	Wagman et al. (1982)
$Cd_3(AsO_4)_2$	xls	298.15	−1712.0	−1934.3	301.71	258.82	Barin (1995)
Cd_3As_2	xls	298.15	—	−41.8	—	—	Wagman et al. (1982)
Cd_3As_2	xls	298.15	−35.885	−41.840	206.83	125.30	Barin (1995)
$CdAs_2$	xls	298.15	—	−17.6	—	—	Wagman et al. (1982)
$CoAs$	xls	298.15	—	−40.6	—	—	Wagman et al. (1982)
$CoAs$ (modderite)	xls	298.15	−49	−51	59	—	Naumov, Ryzhenko and Khodakovsky (1974)
$CoAs_2$	xls	298.15	—	−61.5	—	—	Wagman et al. (1982)
$CoAs_2$ (safflorite)	xls	298.15	−97	−83	100	—	Naumov, Ryzhenko and Khodakovsky (1974)

(continued overleaf)

Table C.1   *(continued)*

Chemical species	Phase	Temperature (K)	Gibbs free energy (G_i) (kJ mol^{-1})	Enthalpy (H_i) (kJ mol^{-1})	Entropy (S_i) (J mol^{-1}·K^{-1})	Heat Capacity (Cp_i) (J mol^{-1}·K^{-1})	Reference
Co$_2$As	xls	298.15	—	−39.7	—	—	Wagman et al. (1982)
Co$_2$As$_3$	xls	298.15	—	−97.5	—	—	Wagman et al. (1982)
Co$_3$As$_2$	xls	298.15	—	−81.2	—	—	Wagman et al. (1982)
Co$_5$As$_2$	xls	298.15	—	−79.5	—	—	Wagman et al. (1982)
Co$_3$(AsO$_4$)$_2$	xls	298.15	−1620.8	—	—	—	Wagman et al. (1982)
Co$_3$(AsO$_4$)$_2$	xls	298.15	−1671.9	−1864.3	337.02	264.31	Barin (1995)
CrAsO$_4$	xls	298.15	−968.36	−1062.1	155.35	119.10	Barin (1995)
Cr$_3$(AsO$_4$)$_2$	xls	298.15	−2027.0	−2218.2	321.42	261.83	Barin (1995)
Cs$_3$AsO$_4$	xls	298.15	−1543.9	−1668.5	283.59	176.22	Barin (1995)
Cu$_3$As	xls	298.15	—	−11.7	—	—	Wagman et al. (1982)
Cu$_3$As	xls	298.15	−12.322	−11.715	137.24	93.080	Barin (1995)
Cu$_3$(AsO$_4$)	xls	298.15	−624.04	−710.36	255.98	176.36	Barin (1995)
—	xls	298.15	−1300.7	—	—	—	Wagman et al. (1982)
Cu$_3$(AsO$_4$)$_2$	xls	298.15	−1301.32	—	—	—	Davis et al. (1996)
Cu$_3$(AsO$_4$)$_2$	xls	298.15	−1316.0	−1522.6	298.61	258.21	Barin (1995)
Cu$_3$(AsO$_4$)$_2$·6H$_2$O	xls	298.15	−2733.04	—	—	—	Davis et al. (1996)
Cu$_3$AsS$_4$ (enargite)	xls	298.15	−206.74	−179.0	356.4	—	Castro and Baltierra (2005) and Kantar (2002)
Cu$_3$AsS$_4$ (enargite)	xls	298.15	—	—	257.6	190.4	Seal et al. (1996)
Cu$_3$AsS$_4$ (enargite)	xls	298.15	−437.26	—	—	—	Davis et al. (1996)
Fe$_3$(AsO$_4$)$_2$	xls	298.15	−1753.93	—	—	—	Davis et al. (1996)
Fe$_3$(AsO$_4$)$_2$·8H$_2$O (symplesite)	xls	298.15	−3687.26	—	—	—	Davis et al. (1996)
FeAs (westerveldite)	xls	298.15	−43.36	−43.51	62.5	50.36	Perfetti et al. (2008)
FeAs$_2$ (loellingite)	xls	298.15	−52.09	−43.5	127	—	Naumov, Ryzhenko and Khodakovsky (1974)

$FeAsO_4$	xls	298.15	−772.39	161.54	117.05	Barin (1995)
$FeAsO_4$	xls	298.15	−772.62	—	—	Ryu *et al.* (2002)
$FeAsO_4 \cdot 2H_2O$ (scorodite)	xls	298.15	−1280.1	—	—	Ryu *et al.* (2002)
$FeAsO_4 \cdot 2H_2O$ (scorodite)	xls	296	−1279.2	—	—	Krause and Ettel (1988)
$FeAsO_4 \cdot 2H_2O$ (scorodite)	xls	298.15	−1267.69	—	—	Davis *et al.* (1996)
$Fe_3(AsO_4)_2$	xls	298.15	−1766.0	339.91	264.64	Barin (1995)
FeAsS (arsenopyrite)	xls	298	−136.45	68.5	68.44	Perfetti *et al.* (2008)
FeAsS (arsenopyrite)	xls	298	−141.6	—	—	Pokrovski, Kara and Roux (2002)
FeAsS (arsenopyrite)	xls	298.15	−127.25	—	—	Davis *et al.* (1996)
$FeAsS_2$ (loellingite)	xls	298.15	−80.23	80.1	70.83	Perfetti *et al.* (2008)
$FeAsS_2$ (loellingite)	xls	298.15	−52.116	—	—	Davis *et al.* (1996)
GaAs (gallium arsenide)	xls	298.15	−67.8	64.18	46.23	Wagman *et al.* (1982)
GaAs (gallium arsenide)	xls	298.15	−70.374	64.183	46.858	Barin (1995)
GaAs (gallium arsenide)	xls	298.15	−67.8	64.2	—	Naumov, Ryzhenko and Khodakovsky (1974)
$GaAsO_4$	xls	298.15	−901.85	150.12	118.26	Barin (1995)
$H_3AsO_3^0$	aq	298.15	−639.8	212.4	85.0	Perfetti *et al.* (2008)
$H_3AsO_3^0$	aq	298.15	−640.03	195.8	—	Nordstrom and Archer (2003)
$H_3AsO_3^0$	aq	298.15	−639.78	200	197	Pokrovski *et al.* (1996)
$H_2AsO_3^-$	aq	298.15	−587.13	110.5	—	Wagman *et al.* (1982)
$H_2AsO_3^-$	aq	298.15	−587.66	112.79	—	Nordstrom and Archer (2003)
$HAsO_3^{2-}$	aq	298.15	−524	−15	—	Dove and Rimstidt (1985)

(continued overleaf)

Table C.1 (continued)

Chemical species	Phase	Temperature (K)	Gibbs free energy (G_i) (kJ mol^{-1})	Enthalpy (H_i) (kJ mol^{-1})	Entropy (S_i) (J mol^{-1}·K^{-1})	Heat Capacity (Cp_i) (J mol^{-1}·K^{-1})	Reference
$HAsO_3^{2-}$	aq	298.15	−524.3	—	—	—	Naumov, Ryzhenko and Khodakovsky (1974)
AsO_3^{3-}	aq	298.15	−448	−664	−187	—	Dove and Rimstidt (1985)
AsO_3^{3-}	aq	298.15	−447.7	—	—	—	Naumov, Ryzhenko and Khodakovsky (1974)
$H_3AsO_4^0$	aq	298.15	−766.3	−898.6	198.3	105.0	Perfetti et al. (2008)
$H_3AsO_4^0$	aq	298.15	−766.75	−903.45	183.07	—	Nordstrom and Archer (2003)
$H_2AsO_4^-$	aq	298.15	−753.17	−909.56	117	—	Wagman et al. (1982)
$H_2AsO_4^-$	aq	298.15	−753.65	−911.42	112.38	—	Nordstrom and Archer (2003)
$HAsO_4^{2-}$	aq	298.15	−713.73	−908.41	−11.42	—	Nordstrom and Archer (2003)
$HAsO_4^{2-}$	aq	298.15	−714.60	−906.34	—	—	Wagman et al. (1982)
AsO_4^{3-}	aq	298.15	−646.36	−890.21	−176.31	—	Nordstrom and Archer (2003)
$HAsO_3F^-$	aq	298.15	−1060.96	—	—	—	Wagman et al. (1982)
AsO_3F^{2-}	aq	298.15	−1027.45	—	—	—	Wagman et al. (1982)
$Hg_3(AsO_4)_2$	xls	298.15	−1033.6	−1271.0	323.47	261.72	Barin (1995)
InAs (indium arsenide)	xls	298.15	−53.6	−58.6	75.7	47.78	Wagman et al. (1982)
InAs (indium arsenide)	xls	298.15	−53.286	−58.601	75.701	47.780	Barin (1995)
InAs (indium arsenide)	xls	298.15	−52.7	−57.7	75.7	—	Naumov, Ryzhenko and Khodakovsky (1974)
$InAsO_4$	xls	298.15	−874.26	−978.30	154.85	119.32	Barin (1995)
KAs	xls	298.15	—	−102.9	—	—	Wagman et al. (1982)

Compound	State	T					Reference
KAs_2	xls	298.15	—	-127.2	—	—	Wagman et al. (1982)
K_3As	xls	298.15	—	-186.2	—	—	Wagman et al. (1982)
K_5As_4	xls	298.15	—	-452.7	—	—	Wagman et al. (1982)
KH_2AsO_4	xls	298.15	-1035.9	-1180.7	155.02	126.73	Wagman et al. (1982)
KH_2AsO_4	xls	298.15	-1041.6	-1186.2	155	—	Naumov, Ryzhenko and Khodakovsky (1974)
$K_3(AsO_4)$	xls	298.15	-1548.8	-1668.7	237.82	172.29	Barin (1995)
KUO_2AsO_4	xls	298.15	-2021.1	—	—	—	Naumov, Ryzhenko and Khodakovsky (1974)
$LaAsO_4$	xls	298.15	-1455.7	-1556.9	163.47	117.89	Barin (1995)
$Li_3(AsO_4)$	xls	298.15	-1595.0	-1702.4	173.13	161.84	Barin (1995)
$MgAs_4$	xls	298.15	—	-126	—	—	Wagman et al. (1982)
Mg_3As_2	xls	298.15	—	-371.5	—	—	Wagman et al. (1982)
$Mg_3(AsO_4)_2$	xls	298.15	—	-3092.8	—	—	Wagman et al. (1982)
$Mg_3(AsO_4)_2$	xls	298.15	-2775	—	—	—	Ryu et al. (2002)

(continued overleaf)

Table C.1 (continued)

Chemical species	Phase	Temperature (K)	Gibbs free energy (G_i) (kJ mol⁻¹)	Enthalpy (H_i) (kJ mol⁻¹)	Entropy (S_i) (J mol⁻¹·K⁻¹)	Heat Capacity (Cp_i) (J mol⁻¹·K⁻¹)	Reference
$Mg_3(AsO_4)_2$	xls	298.15	−2831.7	−3059.8	225.10	236.31	Barin (1995)
$Mg_3(AsO_4)_2 \cdot 10H_2O$	xls	298.15	−5147.2	—	—	—	Ryu et al. (2002)
$MgHAsO_4$	xls	298.15	−1187	—	—	—	Ryu et al. (2002)
$MgHAsO_4 \cdot 7H_2O$ (roesslerite)	xls	298.15	−2848	—	—	—	Ryu et al. (2002)
$MgNH_4AsO_4 \cdot 6H_2O$	xls	298.15	—	−3316.7	—	—	Wagman et al. (1982)
$MnAs$ (α)	xls	298.15	−61.636	−51.861	107.00	71.116	Barin (1995)
$MnAs$ (β)	xls	298.15	−59.865	−53.297	91.442	70.340	Barin (1995)
$MnAs$ (γ)	xls	298.15	−59.691	−56.902	77.069	70.164	Barin (1995)
$MnAs$	xls	298.15	—	−59	—	—	Wagman et al. (1982)
$MnAs$ (kaneite)	xls	298.15	−55.2	−57.3	60	—	Naumov, Ryzhenko and Khodakovsky (1974)
$Mn_3(AsO_4)_2$	xls	298.15	−2167.4	—	—	—	Ryu et al. (2002)
$Mn_3(AsO_4)_2$	xls	298.15	—	−2145.6	—	—	Wagman et al. (1982)
$Mn_3(AsO_4)_2$	xls	298.15	−2167.0	−2366.3	319.62	261.79	Barin (1995)
$Mn_3(AsO_4)_2 \cdot 8H_2O$	xls	298.15	−4055	—	—	—	Naumov, Ryzhenko and Khodakovsky (1974)
$MnHAsO_4$	xls	298.15	—	−1102.5	—	—	Wagman et al. (1982)
$MoAsO_4$	xls	298.15	−817.79	−910.69	163.01	120.06	Barin (1995)
$NaAs$	xls	298.15	−89.1	−96.2	62.8	—	Wagman et al. (1982)
$NaAs_2$	xls	298.15	−100.4	−106.7	100	—	Wagman et al. (1982)
Na_3As	xls	298.15	−187.4	−205	130	—	Wagman et al. (1982)
Na_3As	xls	298.15	−187.10	−205.02	130.00	97.769	Barin (1995)

Formula	State	T (K)					Reference
NaAsO$_2$	xls	298.15	—	-660.53	—	—	Wagman et al. (1982)
NaUO$_2$AsO$_4$·4H$_2$O	xls	298.15	-2944.6	—	—	—	Naumov, Ryzhenko and Khodakovsky (1974)
Na$_3$AsO$_4$	xls	298.15	—	-1540	—	—	Wagman et al. (1982)
Na$_3$AsO$_4$	xls	298.15	-1426.0	-1540.0	217.94	170.13	Barin (1995)
Na$_3$AsO$_4$·12H$_2$O	xls	298.15	—	-5092	—	—	Wagman et al. (1982)
NiAs (nickeline, niccolite)	xls	298.15		—	61	—	Naumov, Ryzhenko and Khodakovsky (1974)
NiAs	xls	298.15	-67.268	-73.291	45.380	56.001	Barin (1995)
Ni$_3$(AsO$_4$)$_2$	xls	298.15	-1579.3	—	—	—	Wagman et al. (1982)
Ni$_3$(AsO$_4$)$_2$	xls	298.15	-1659.4	-1849.2	344.80	265.41	Barin (1995)
Ni$_3$(AsO$_4$)$_2$·8H$_2$O (Annabergite)	xls	298.15	-3482	—	—	—	Naumov, Ryzhenko and Khodakovsky (1974)
Ni$_5$As$_2$	xls	298.15	-242.11	-251.10	190.63	215.98	Barin (1995)
Ni$_{11}$As$_8$	xls	298.15	-762.63	-774.04	468.88	552.00	Barin (1995)
NH$_4$H$_2$AsO$_4$	xls	298.15	-832.9	-1059.8	172.05	151.17	Wagman et al. (1982)

(continued overleaf)

Table C.1 *(continued)*

Chemical species	Phase	Temperature (K)	Gibbs free energy (G_i) (kJ mol^{-1})	Enthalpy (H_i) (kJ mol^{-1})	Entropy (S_i) (J mol^{-1}·K^{-1})	Heat Capacity (Cp_i) (J mol^{-1}·K^{-1})	Reference
$NH_4H_2AsO_4$	xls	298.15	−837.34	−1064.1	172	—	Naumov, Ryzhenko and Khodakovsky (1974)
$(NH_4)_2HAsO_4$	xls	298.15	—	−1181.6	—	—	Wagman et al. (1982)
$(NH_4)_3AsO_4$	xls	298.15	—	−1286.2	—	—	Wagman et al. (1982)
$(NH_4)_3AsO_4 \cdot 3H_2O$	xls	298.15	—	−2166.9	—	—	Wagman et al. (1982)
$Pb_3(AsO_4)_2$	xls	298.15	−1553.1	−1780.2	324.60	258.00	Barin (1995)
$Pb_3(AsO_4)_2$	xls	298.15	−1580.01	—	—	—	Davis et al. (1996)
$Pb_3(AsO_4)_2$	xls	298.15	−1579.3	—	—	—	Naumov, Ryzhenko and Khodakovsky (1974)
$Pb_5(AsO_4)_3Cl$ (mimetite)	xls	298.15	−2616.25	—	—	—	Davis et al. (1996)
$Pb_5(AsO_4)_3OH$	xls	298.15	−2659	—	—	—	Lee and Nriagu (2007)
$PbHAsO_4$	xls	298.15	−805.66	—	—	—	Lee and Nriagu (2007)
$Rb_3(AsO_4)$	xls	298.15	−1547.0	−1669.0	267.06	175.28	Barin (1995)
Re_3As_7	xls	298.15	−88.624	−95.395	336.81	248.64	Barin (1995)
$ReAsO_4$	xls	298.15	−678.39	−771.57	170.00	121.37	Barin (1995)
$Sn_3(AsO_4)_2$	xls	298.15	−1564.6	−1785.8	303.93	259.1	Barin (1995)
$Sr_3(AsO_4)_2$	xls	298.15	−3094.4	−3317.1	312.1	257.6	Barin (1995)
$Ti_3(AsO_4)_2$	xls	298.15	−2547.3	−2617.3	339.4	264.5	Barin (1995)

$TlAsO_4$	xls	298.15	−885.81	299.74	145.69	Barin (1995)
$YAsO_4$	xls	298.15	−1416.8	160.54	120.75	Barin (1995)
$Zn_3(AsO_4)_2$	xls	298.15	−1895.91	—	—	Davis et al. (1996)
$Zn_3(AsO_4)_2$	xls	298.15	−1895	—	—	Wagman et al. (1982)
$Zn_3(AsO_4)_2$	xls	298.15	−1915.6	281.92	255.28	Barin (1995)
$Zn_3(AsO_4)_2 \cdot 2.5H_2O$ (legrandite)	xls	298.15	−2611	—	—	Naumov, Ryzhenko and Khodakovsky (1974)
$ZnAs_2$	xls	298.15	—	−41.8	—	Wagman et al. (1982)
Zn_3As_2	xls	298.15	−125.46	−133.89	125.23	Barin (1995)
Zn_3As_2	xls	298.15	—	−32.2	—	Wagman et al. (1982)

References

Allison, J.D., Brown, D.S. and Novo-Gradac, K.J. (1991) *MINTEQA2/PRODEFA2: A Geochemical Assessment Model for Environmental Systems: Version 3.0 User's Manual*, U.S. Environmental Protection Agency, Washington, DC.

Barin, I. (1995) *Thermochemical Data of Pure Substances: Parts 1 and II*, 3rd edn. Weinheim, VCH Verlagsgesellschaft, 1885 pp.

Bothe, J.V. Jr. and Brown, P.W. (1999) The stabilities of calcium arsenates at 23 ± 1 °C. *Journal of Hazardous Materials*, **69**(2), 197–207.

Castro, S.H. and Baltierra, L. (2005) Study of the surface properties of enargite as a function of pH. *International Journal of Mineral Processing*, **77**(2), 104–15.

Davis, A., Ruby, M.V., Bloom, M. *et al.* (1996) Mineralogic constraints on the bioavailability of arsenic in smelter-impacted soils. *Environmental Science and Technology*, **30**(2), 392–99.

Dove, P.M. and Rimstidt, J.D. (1985) The solubility and stability of scorodite, $FeAsO_4.2H_2O$. *American Mineralogist*, **70**(7–8), 838–44.

Drever, J.I. (1997) *The Geochemistry of Natural Waters: Surface and Groundwater Environments*, Upper Saddle River, NJ, Prentice Hall, 436 pp.

Eby, G.N. (2004) *Principles of Environmental Geochemistry*, Brooks/Cole-Thomson Learning, Pacific Grove, CA, 514 pp.

Faure, G. (1998) *Principles and Applications of Geochemistry*, 2nd edn. Prentice Hall, Upper Saddle River, NJ, 600 pp.

Helz, G.R., Tossell, J.A., Charnock, J.M. *et al.* (1995) Oligomerization in As(III) sulfide solutions: theoretical constraints and spectroscopic evidence. *Geochimica et Cosmochimica Acta*, **59**(22), 4591–604.

Kantar, C. (2002) Solution and flotation chemistry of enargite. *Colloids and Surfaces A: Physicochemical and Engineering Aspects*, **210**(1), 23–31.

Krause, E. and Ettel, V.A. (1988) Solubility and stability of scorodite, $FeAsO_4.2H_2O$: new data and further discussion. *American Mineralogist*, **73**(7–8), 850–54.

Krauskopf, K.B. and Bird, D.K. (1995) *Introduction to Geochemistry*, 3rd edn. McGraw-Hill, Boston.

Langmuir, D. (1997) *Aqueous Environmental Geochemistry*, Upper Saddle River, NJ, Prentice Hall, 600 pp.

Lee, J.S. and Nriagu, J.O. (2007) Stability constants for metal arsenates. *Environmental Chemistry*, **4**(2), 123–33.

Lide, D.R. (ed) (2007) *CRC Handbook of Chemistry and Physics*, 88th edn. CRC Press, Boca Raton, FL.

Matschullat, J. (2000) Arsenic in the geosphere –A review. *Science of the Total Environment*, **249**(1–3), 297–312.

Naumov, G.B., Ryzhenko, B.N. and Khodakovsky, I.L. (1974) *Handbook of Thermodynamic Data*, (English translation). PB 226 722, Report No. USGS-WRD-74-001, U.S. Geological Survey, Menlo Park California, National Technical Information Service, Springfield, Virginia.

Nordstrom, D.K. and Archer, D.G. (2003) Arsenic thermodynamic data and environmental geochemistry, in *Arsenic in Ground Water*, (eds A.H. Welch and K.G. Stollenwerk), Kluwer Academic Publishers, Boston, pp. 1–25.

O'Hare, P.A.G. (1993) Calorimetric measurements of the specific energies of reaction of arsenic and of selenium with fluorine. Standard molar enthalpies of formation $\Delta_f H°_m$ at the temperature 2.98.15 K of AsF_5, SeF_6, As_2Se_3, As_4S_4, and As_2S_3. Thermodynamic properties of AsF_5 and SeF_6 in the ideal-gas state. Critical assessment of $\Delta_f H°_m$ (AsF_3, l)), and the dissociation enthalpies of As-F bonds. *Journal of Chemical Thermodynamics*, **25**, 391–402.

O'Hare, P.A.G., Lewis, B.M., Susman, S. and Volin, K.J. (1990) Standard molar enthalpies of formation and transition at the temperature 298.15 K and other thermodynamic properties of the crystalline and vitreous forms of arsenic sesquiselenide (As_2S_3). Dissociation enthalpies of As-Se bonds. *Journal of Chemical Thermodynamics*, **22**, 1191–206.

Pankratov, A.N. and Uchaeva, I.M. (2000) A semiempirical quantum chemical testing of thermodynamic and molecular properties of arsenic compounds. *Journal of Molecular Structure (Theochem)*, **498**, 247–54.

Parkhurst, D.L. and Appelo, C.A.J. (1999) User's Guide to PHREEQC (Version 2): A Computer Program for Speciation, Batch-reaction, One-dimensional transport, and Inverse Geochemical Calculations. U.S. Geological Survey Water-Resources Investigations Report 99-4259, 312 pp.

Perfetti, E., Pokrovski, G.S., Ballerat-Busserolles, K. *et al.* (2008) Densities and heat capacities of aqueous arsenious and arsenic acid solutions to 350 °C and 300 bar, and revised thermodynamic properties of As(OH)$_3$°(aq), AsO(OH)$_3$°(aq) and iron sulfarsenide minerals. *Geochimica et Cosmochimica Acta*, **72**(3), 713–31.

Pokrovski, G., Gout, R., Schott, J. *et al.* (1996) Thermodynamic properties and stoichiometry of As(III) hydroxide complexes at hydrothermal conditions. *Geochimica et Cosmochimica Acta*, **60**(5), 737–49.

Pokrovski, G.S., Kara, S. and Roux, J. (2002) Stability and solubility of arsenopyrite, FeAsS, in crustal fluids. *Geochimica et Cosmochimica Acta*, **66**(13), 2361–73.

Robie, R.A., Hemingway, B.S. and Fisher, J.R. (1979) *Thermodynamic Properties of Minerals and Related Substances at 298.15 K and 1 Bar (10^5 Pascals) Pressure and at Higher Temperatures*, Reprinted with corrections. Geological Survey Bulletin 1452, United States Printing Office, Washington, DC, 456 pp.

Ryu, J-H., Gao, S., Dahlgren, R.A. and Zierenberg, R.A. (2002) Arsenic distribution, speciation and solubility in shallow groundwater of Owens Dry Lake, California. *Geochimica et Cosmochimica Acta*, **66**(17), 2981–94.

Seal, R.R. II, Robie, R.A., Hemingway, B.S. and Evans, H.T. Jr (1996) Heat capacity and entropy at the temperatures 5 K to 720 K and thermal expansion from the temperatures 298 K to 573 K of synthetic enargite (Cu$_3$AsS$_4$). *Journal of Chemical Thermodynamics*, **28**(4), 405–12.

Wagman, D.D., Evans, W.H., Parker, V.B. *et al.* (1982) The NBS tables of chemical thermodynamic properties: selected values for inorganic and C1 and C2 organic substances in SI units. *Journal of Physical and Chemical Reference Data*, **11**(2), 1–392.

Zhu, Y., Zhang, X., Xie, Q. *et al.* (2005) Solubility and stability of barium arsenate and barium hydrogen arsenate at 25°C. *Journal of Hazardous Materials*, **120**(1-3), 37–44.

Zhu, Y.N., Zhang, X.H., Xie, Q.L. *et al.* (2006) Solubility and stability of calcium arsenates at 25°C. *Water, Air, and Soil Pollution*, **169**(1–4), 221–38.

Appendix D

Locations of Significant Arsenic Contamination

The tables and maps in this appendix list the locations of prominent areas with arsenic-contaminated groundwater, geothermal waters, and/or substantial arsenic-bearing rocks and mining wastes. Additionally, there are countless sites that have local waters, sediments, and soils that have been contaminated by arsenic from chemical manufacturing facilities, pesticide applications, and individual mines. These small-scale areas, which include hundreds of Superfund sites in the United States, are too numerous and poorly documented to list in this appendix. Appendix B in US Environmental Protection Agency (US EPA) (2002) lists the locations of Superfund sites where arsenic is a contaminant of concern.

The arsenic-contaminated areas are grouped into four categories according to the following key: (1) modern geothermal waters; (2) mine drainage, smelter wastes, and solutions and products from the weathering of igneous, metamorphic, or hydrothermally altered rocks; (3) coal and coal mining wastes; and (4) groundwater, drinking water, and sediment porewaters. The arsenic at some locations occurs in more than one category. In such cases, overlapping symbols are used on the maps.

Map Key

 Modern geothermal

 Weathering, mining, or smelting of igneous, metamorphic, or hydrothermally altered rocks

 Coal and coal mining

 Groundwater, drinking water, or porewaters

Abundant literature exists on arsenic contamination problems in some countries (e.g. Bangladesh). In such cases, the tables in this appendix simply list a few of the more recent key references, which provide overall and concise summaries of the problems. For other regions and countries (e.g. Laos), arsenic contamination has been identified as a significant problem, but further details are not readily available. In these situations, the reference(s) with the limited information is listed. No public information is readily available on arsenic contamination in North Korea, many of the republics of the former Soviet Union, and

Arsenic Edited by Kevin R. Henke
© 2009 John Wiley & Sons, Ltd

Table D.1 *Locations of significant arsenic contamination problems in Asia*

Number	Location	Arsenic-bearing medium	Reference(s)
1	Bangladesh	Groundwater	Chakraborti *et al.* (2002); Ravenscroft *et al.* (2005); Hossain (2006)
2	Cambodia	Groundwater	Polya *et al.* (2005); Buschmann *et al.* (2007); Berg, Giger and Tron (2006)
3	Caspian Sea	Mine runoff and other anthropogenic sources	De Mora *et al.* (2004)
4	China: Anhui	Groundwater	Xia and Liu (2004)
5	China: Beijing	Groundwater	Xia and Liu (2004)
6	China: Guangdong: Qingyuan County: Xinzhou gold deposit	Ore deposits	Zeng (1994)
7	China: Guizhou	Coal and coal mining	Dai *et al.* (2005); Dai, Zeng and Sun (2006); Zhang *et al.* (2004)
7	China: Guizhou	Ore deposits	Zhang *et al.* (2003)
8	China: Henan	Groundwater	Mandal and Suzuki (2002)
9	China: Hunan	Ore deposits and mining wastes	Lu and Zhang (2005); Yang and Blum (1999)
10	China: Inner Mongolia	Groundwater	Smedley *et al.* (2003); Zhang (2004)
11	China: Jilin	Groundwater	Lu *et al.* (1998); Xie, Zhang and Naidu (2006)
12	China: Liaoning	Groundwater	Mandal and Suzuki (2002)
13	China: Ningxia	Groundwater	Mandal and Suzuki (2002); Xia and Liu (2004)
14	China: Northeast	Coals	He, Liang and Jiang (2002); Guo, Yang and Liu (2004)
15	China: Qinghai	Groundwater	Mandal and Suzuki (2002); Xia and Liu (2004)
16	China: Shandong: Jinan	Groundwater	Liu (2000)
17	China: Shanxi	Groundwater	Guo and Wang (2005); Xie, Zhang and Naidu (2006); Wang, Lunshan and Wu (1998)
17	China: Shanxi: Datong Basin	Coal	He, Liang and Jiang (2002); Guo, Yang and Liu (2004)
18	China: Sichuan	Ore deposits	Wang and Zhang (2001)
19	China: Xinjiang	Groundwater	Lianfang and Jianzhong (1994)
20	China: Yunnan	Coal	Zhou and Ren (1992); Guo, Yang and Liu (2004)
21	India: Andhra Pradesh: Hyderabad: Nacharam Lake	Groundwater	Govil, Reddy and Gnaneswara Rao (1999); Rahman, Sengupta and Chowdhury (2006)
22	India: Andhra Pradesh: Visakhapatnam	Groundwater	Subba and Subba (1999)
23	India: Assam	Groundwater	Mukherjee, Bhattacharya and Jacks (2006)

23	India: Assam: Makum Coalfield	Coal	Mukherjee and Srivastava (2005)
24	India: Bihar and Jharkhand	Groundwater	Rahman, Sengupta and Chowdhury (2006); Mukherjee, Bhattacharya and Jacks (2006)
25	India: Chhattisgarh (Chattisgarh)	Groundwater	Patel *et al.* (2005); Pandey *et al.* (2006); Mukherjee, Bhattacharya and Jacks (2006)
26	India: Karnataka: Raichur district	Schist belt	Sahoo and Pandalai (2000)
27	India: Punjab, Haryana, and Himachal Pradesh	Groundwater	Mukherjee, Bhattacharya and Jacks (2006)
28	India: Rajasthan	Groundwater	Mukherjee, Bhattacharya and Jacks (2006)
28	India: Rajasthan	Ore deposits and mining wastes	Mukherjee, Bhattacharya and Jacks (2006)
29	India: Tamil Nadu: Chennai	Groundwater	Mukherjee, Bhattacharya and Jacks (2006)
30	India: Uttar Pradesh	Groundwater	Mukherjee, Bhattacharya and Jacks (2006)
31	India: West Bengal	Groundwater	Dowling *et al.* (2002); Nath *et al.* (2005); Burgess and Ahmed (2006); Smedley (2003); Chakraborti *et al.* (2002)
32	Indonesia: Ngada district: Bejawa geothermal area	Geothermal	Nasution and Muraoka (1999)
33	Indonesia: North Sulawesi: Ratatotok district	Ore deposits and mining wastes	Turner *et al.* (1994)
34	Indonesia: West Java: Ciwidey River	Volcanic deposits	Sriwana *et al.* (1998)
35	Iran: Kurdistan	Groundwater	Rahman, Sengupta and Chowdhury (2006)
36	Iran: Zarshuran	Ore deposits and mining wastes	Mehrabi, Yardley and Cann (1999)
37	Japan: Ehime Prefecture: Ichinokawa Mine	Ore deposits and mining wastes	Mitsunobu, Harada and Takahashi (2006)
38	Japan: Honshu: Gunma Prefecture	Ore deposits and mining wastes	Ohnuki *et al.* (2004)
38	Japan: Honshu: Gunma: Manza hot springs	Geothermal	Noguchi and Nakagawa (1994a)
38	Japan: Honshu: Narugo and Onikobe areas: Miyagi	Geothermal	Noguchi and Nakagawa (1994b)
38	Japan: Honshu: Niigata Plain and Shinji Lowland	Groundwater	Kubota, Yokota and Ishiyama (2001)

(continued overleaf)

Table D.1   (continued)

Number	Location	Arsenic-bearing medium	Reference(s)
38	Japan: Honshu: Okuaizu geothermal area	Geothermal	Seki (2000)
39	Japan: Honshu: Aomori Prefecture: Osoreyama hot springs	Geothermal	Noguchi and Nakagawa (1994c)
40	Japan: Kyushu	Geothermal	Yokoyama, Takahashi and Tarutani (1993)
40	Japan: Kyushu: Fukuoka Prefecture	Groundwater	Mandal and Suzuki (2002); Rahman, Sengupta and Chowdhury (2006)
41	Japan: Osaka Prefecture	Groundwater	Ito, Masuda and Kusakabe (2003); Masuda, Ibuki and Tonokai (1999)
42	Korea, South	Ore deposits and mining wastes	Lee, Lee and Lee (2001); Jung,Thornton and Chon (2002); Woo and Choi (2001); Park et al. (2006); Lee and Lee (2004)
43	Laos	Groundwater	Chakraborti et al. (2002); Fengthong, Dethoudom and Keosavanh (2002)
44	Malaysia: Peninsula	Ore deposits and mining wastes	Williams (2001)
45	Malaysia: Sabah	Ore deposits and mining wastes	Williams (2001)
46	Malaysia: Sarawak	Ore deposits and mining wastes	Williams (2001)
47	Myanmar: Ayeyarwaddy Division	Groundwater	Smedley (2003)
47	Myanmar: Bago	Groundwater	Smedley (2005)
48	Myanmar: Rakhine state: Sittway Township	Groundwater	Smedley (2003)
49	Nepal: Southern	Groundwater	Tandukar et al. (2006); Panthi, Sharma and Mishra (2006)
50	Pakistan: Punjab: Muzaffargarh district	Groundwater	Nickson et al. (2005)
50	Pakistan: Punjab: Bahawalpur, Rahim Yar Khan, and Multan districts	Groundwater	Haque (2005)
51	Pakistan: Punjab: Gujarat	Groundwater	Chakraborti et al. (2002)
51	Pakistan: Punjab: Kasur district	Groundwater	Naseem et al. (2001)
51	Pakistan: Punjab: Jhelum	Groundwater	Chakraborti et al. (2002)
52	Pakistan: Sindh: Larkana district	Drinking water	Asghar et al. (2006)
53	Philippines: Luzon	Ore deposits and mining wastes	Williams (2001)

54	Philippines: Mindanao	Ore deposits and mining wastes	Williams (2001)
54	Philippines: Mindanao Island	Geothermal	Mandal and Suzuki (2002)
55	Philippines: Palawan	Ore deposits and mining wastes	Williams (2001)
56	Russia: Altai Mountains	Coal and Coal Mining	Yudovich and Ketris (2005)
57	Russia: Kamchatka Peninsula	Geothermal	Tazaki *et al.* (2003); Nishikawa *et al.* (2006); Cleverley, Benning and Mountain (2003)
58	Russia: Kemerovo region	Ore deposits and mining wastes	Hozhina *et al.* (2001); Bortnikova *et al.* (2001); Giere, Sidenko and Lazareva (2003)
59	Russia: Ural Mountains	Ore deposits, smelting, and mining wastes	Williamson *et al.* (2004); Gelova (1977); Smedley and Kinniburgh (2002)
60	Sri Lanka	Groundwater	Mandal and Suzuki (2002)
61	Taiwan	Groundwater	Chen *et al.* (1994); Liu *et al.* (2006)
61	Taiwan	Smelter	Shih and Lin (2003)
62	Thailand: Mae Moh	Coal mining and combustion	Kouprianov *et al.* (2002); Bashkin and Wongyai (2002)
63	Thailand: Nakhon Si Thammarat Province: Ron Phibun	Groundwater	Williams *et al.* (1996); Milintawisamai *et al.* (1998)
63	Thailand: Nakhon Si Thammarat Province: Ron Phibun	Ore deposits and mining wastes	Williams *et al.* (1998)
64	Vietnam: Red River region	Groundwater	Berg *et al.* (2001); Berg, Giger and Tron (2006)
65	Vietnam–Cambodia: Mekong River region	Groundwater	Buschmann *et al.* (2007); Berg, Giger and Tron (2006)

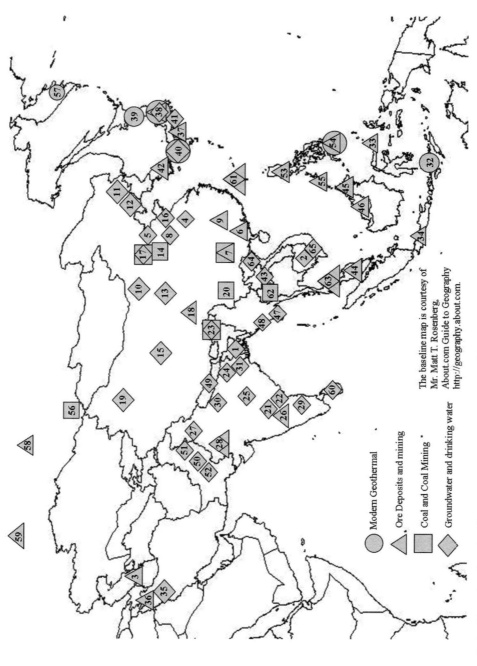

Figure D.1 Map showing significant arsenic contamination problems in Asia. (Map reproduced by permission of Mr. Matt T. Rosenberg, http://geography.about.com.)

Table D.2 *Locations of significant arsenic contamination problems in Australia, New Zealand, and the southwestern Pacific*

Number	Location	Arsenic-bearing medium	Reference(s)
66	Australia: New South Wales: Hillgrove	Ore deposits and mining wastes	Ashley *et al.* (2003; 2007)
67	Australia: New South Wales: Port Kembla	Smelter and industrial complex	Martley, Gulson and Pfeifer (2004)
68	Australia: New South Wales: Stuarts Point	Groundwater	Smith, Jankowski and Sammut (2003; 2006); O'Shea, Jankowski and Sammut (2007)
69	Australia: New South Wales: Tomakin Park gold mine	Ore deposits and mining wastes	Reith, McPhail and Christy (2005)
70	Australia: Queensland: Cloncurry district: Lorena gold mine	Ore deposits	Lawrence *et al.* (1999)
70	Australia: Queensland: Mammoth area	Ore deposits and mining water	Scott (1990)
71	Australia: Western Australia: Granny Smith gold deposit	Ore deposits	Lintern and Butt (1998)
72	Australia: Western Australia: Perth	Groundwater	Appleyard, Angeloni and Watkins (2006)
73	New Zealand: North Island: Kawerau geothermal region	Geothermal	Mroczek (2005)
73	New Zealand: North Island: Lake Ohakuri	Geothermal	Aggett and O'Brien (1985)
73	New Zealand: North Island: Waiotapu geothermal system	Geothermal	Hedenquist and Henley (1985); Phoenix *et al.* (2005)
74	New Zealand: North Island: Puhipuhi area	Ore deposits and mining wastes	Craw, Chappell and Reay (2000a)
75	New Zealand: South Island	Ore deposits and mining wastes	Craw, Koons and Chappell (2002b); Craw, Falconer and Youngson (2003)
75	New Zealand: South Island: Wangaloa coal mine	Coal and coal mining	Black and Craw (2001)
76	Papua New Guinea: Ambitle Island: Tatum Bay	Geothermal	Price and Pichler (2005)

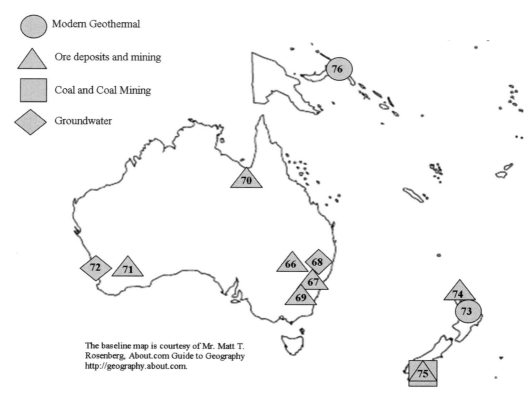

Figure D.2 Map showing significant arsenic contamination problems in Australia, New Zealand, and the southwestern Pacific. (Map reproduced by permission of Mr. Matt T. Rosenberg, http://geography.about.com.)

Table D.3 *Locations of significant arsenic contamination problems in Africa*

Number	Location	Arsenic-bearing medium	Reference(s)
77	Botswana	Groundwater	Huntsman-Mapila *et al.* (2006)
78	Ghana	Groundwater	Smedley (1996)
79	Ghana	Ore deposits and mining wastes	Duker, Carranza and Hale (2005), Buwell, Morley and Din (1994); Donkor *et al.* (2005); Serfor-Armah *et al.* (2006)
80	Nigeria	Groundwater	Gbadebo (2005)
81	South Africa	Ore deposits and mining wastes	Aucamp (2003); Robb, Freeman and Armstrong (2001)
82	Tanzania: Serengeti NP: Orangi River	Ore deposits and mining wastes	Bowell *et al.* (1995)
83	Tunisia	Ore deposits and mining wastes	Jdid *et al.* (1999)
84	Zimbabwe	Ore deposits and mining wastes	Meck, Love and Mapani (2006); Williams (2001)

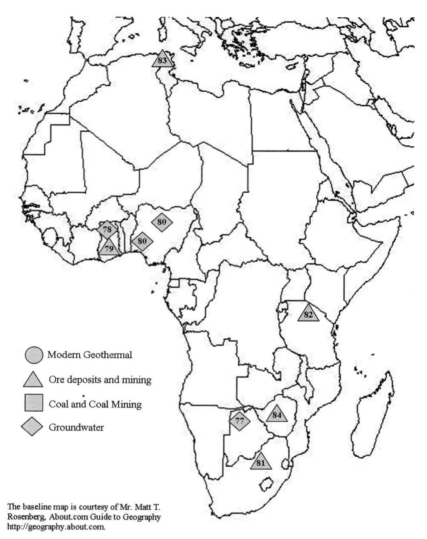

Figure D.3   *Map showing significant arsenic contamination problems in Africa. (Map reproduced by permission of Mr. Matt T. Rosenberg, http://geography.about.com.)*

Table D.4 *Locations of significant arsenic contamination problems in Europe and Turkey*

Number	Location	Arsenic-bearing medium	Reference(s)
85	Austria: East Alps	Ore deposits and mining wastes	Abrecht (1990); Bojar *et al.* (2001)
86	Belgium	Groundwater	Coetsiers and Walraevens (2006)
87	Bulgaria	Coal	Eskenazy (1995); Yudovich and Ketris (2005)
87	Bulgaria: Plovdiv-Assenovgrad area	Ore deposits, smelting and mining	Pentcheva, Velitchkova and Karadjov (2003); Atanassova, Kerestedjian and Satchanska (2003)
87	Bulgaria: Southwest	Geothermal	Criaud and Fouillac (1989)
88	Cyprus: Mathiatis Mine	Ore deposits and mining wastes	Hudson-Edwards and Edwards (2005)
89	Czech Republic: Bohemian Massif	Ore deposits and mining wastes	Filippi, Goliáš and Pertold (2004); Zachariáš *et al.* (2004)
89	Czech Republic: Krušné hory Mountains	Coal	Bouška and Pešek (1999)
89	Czech Republic: Sokolov Basin	Coal	Šlejkovec and Kanduč (2005)
90	Czech Republic: Ostrava-Karvina Basin	Coal and coal mining wastes	Yudovich and Ketris (2005)
90	Czech Republic–Poland	Mineralization	Samecka-Cymerman and Kempers (1994)
91	Finland: Ilomantsi	Ore deposits and mining wastes	Rasilainen, Nurmi and Bornhorst (1993)
92	Finland: Kittila	Groundwater	Loukola, Tanskanen and Lahermo (1999); Tanskanen *et al.* (2004); Loukola and Lahermo (2004)
93	Finland: Pirkanmaa	Groundwater	Juntunen, Vartiainen and Pullinen (2004)
93	Finland: Seinäjoki district	Tills	Lestinen *et al.* (1991)
93	Finland: Ylöjärvi: Paroistenjärvi Mine	Ore deposits and mining wastes	Carlson *et al.* (2002)
94	France: Corsica	Ore deposits and mining wastes	Migon and Mori (1999)
95	France: Douai area	Smelter	Sterckeman *et al.* (2002)
96	France: Massif Central	Geothermal	Criaud and Fouillac (1989)
96	France: Massif Central	Ore deposits and mining wastes	Lerouge, Bouchot and Guerrot (2000); Morin *et al.* (2003); Courtin-Nomade *et al.* (2003); Néel *et al.* (2003)
96	France: Rhône River	Ore deposits and mining wastes	Ollivier, Radakovitch and Hamelin (2006)
97	Germany: Bavaria	Groundwater	Heinrichs and Udluft (1999)

(continued overleaf)

Table D.4 *(continued)*

Number	Location	Arsenic-bearing medium	Reference(s)
98	Germany: Bitterfeld-Wolfen: Mulde River	Mining and chemical industry	Kalbitz and Wennrich (1998)
98	Germany: Dessau area	Groundwater	Riemann (1999)
98	Germany: Freiberg: Himmelfahrt Fundgrube Mine	Ore deposits and mining wastes	Zänker et al. (2002)
98	Germany: Koenigstein Mine	Ore deposits and mining wastes	Jakubick, Jenk and Kahnt (2002)
98	Germany: Lake Süßer See	Ore deposits and mining wastes	Becker et al. (2001)
98	Germany: Mansfeld region	Ore deposits and mining wastes	Wenrich et al. (2004)
98	Germany: Southeast Harz Forelands	Ore deposits and mining wastes	Jung, Knitzschke and Gerlach (1974)
99	Germany: Ebersdorf Coal Deposit	Coal deposits and mining wastes	Yudovich and Ketris (2005)
100	Greece: Central Macedonia	Groundwater	Meladiotis, Veranis and Nikolaidis (2002)
101	Greece: Crete	Coal and coal mining wastes	Yudovich and Ketris (2005); Gentzis et al. (1996)
102	Greece: Elassona Basin	Coal and coal mining wastes	Yudovich and Ketris (2005); Pentari, Foscolos and Perdikatsis (2004)
102	Greece: Northwestern	Coal and coal mining wastes	Yudovich and Ketris (2005)
100	Greece: Mygdonia region	Groundwater	Baker et al. (1995)
103	Greece: Eastern Attica: Lavrion area	Ore deposits and mining wastes	Komnitsas, Xenidis and Adam (1995)
104	Greece: Santorini	Geothermal	Varnavas and Cronan (1988)
105	Hungary: Great Hungarian Plain	Groundwater	Deak, Liebe and Deseo (1998); Rahman, Sengupta and Chowdhury (2006)
105	Hungary: Mátra Mountains	Ore deposits and mining wastes	Ódor et al. (1998)
105	Hungary: Pannonian Basin	Groundwater	Varsányi and Kovács (2006)
106	Hungary–Romania: Danube Basin	Groundwater	Smedley and Kinniburgh (2002)
107	Italy: Ischia and Phlegrean fields (Campanion Volcanic Province)	Geothermal	Aiuppa et al. (2006); Daniele (2004)
108	Italy: Sardinia: Baccu Locci streams	Ore deposits and mining wastes	Frau and Ardau (2003)
109	Italy: Southern Tuscany	Ore deposits and mining wastes	Baroni et al. (2004); Mascaro et al. (2001)
110	Netherlands	Groundwater	Huisman et al. (2002)
90	Poland–Czech Republic: Sudety Mountains	Mineralization	Samecka-Cymerman and Kempers (1994)
111	Poland: Lyublin Basin	Coals and coal mining wastes	Yudovich and Ketris (2005)
90	Poland: Sudety Mountains: Zloty Stok mining area	Ore deposits and mining wastes	Marsza lek and Wasik (2000)

112	Portugal: Castromil gold deposit	Ore deposits and mining wastes	Vallance et al. (2003)
113	Romania: Baia Mare region: Bozanta area	Groundwater	Frentiu et al. (2007)
114	Russia: Kola Peninsula: Nikel and Zapoljarnij	Smelter	Niskavaara, Reimann and Chekushin (1996)
115	Slovakia	Mining wastes	Klukanová and Rapant (1999)
115	Slovakia	Groundwater	Rapant and Krčmová (2007)
115	Slovakia: Cierna Lehota	Mineralized Shales	Pršek et al. (2005)
116	Spain: Anllóns River	Ore deposits and mining wastes	Rubinos et al. (2003)
116	Spain: Asturias and León	Ore deposits and mining wastes	Loredo et al. (2003, 1999, 2004); Crespo et al. (2000)
117	Spain: Aznalcóllar	Ore deposits and mining wastes	Hudson-Edwards et al. (2005)
118	Spain: Central	Groundwater	García-Sánchez, Moyano and Mayorga (2005)
119	Spain: Eastern Pyrenees	Ore deposits and mining wastes	Ruiz-Chancho, López Sánchez and Rubio (2007); Ayora and Casas (1986)
120	Spain: Madrid aquifer	Groundwater	Hernández-García and Custodio (2004)
121	Spain: Puertollano Basin	Coal and coal mining wastes	Yudovich and Ketris (2005)
122	Sweden: Kalix River and Bothnian Bay	Smelter	Widerlund and Ingri (1995)
123	Sweden: Kristineberg mining site	Ore deposits and mining wastes	Malmström, Gleisner and Herbert (2006); Öhlander et al. (2007)
124	Sweden: Rönnskär Smelter	Smelter	Mandal and Suzuki (2002)
123	Sweden: Västerbotten district: Adak Mine	Ore deposits and mining wastes	Routh et al. (2007)
123	Sweden: Vormbäcken River	Ore deposits and mining wastes	Sjöblom, Håkansson and Allard (2004)
125	Switzerland	Geothermal and groundwater	Pfeifer and Zobrist (2002)
126	Switzerland: Alps: Aar Massif: Grimsel Pass	Hydrothermal deposits	Hofmann et al. (2004)
125	Switzerland: Camignolo area	Groundwater	Temgoua and Pfeifer (2002)
126	Switzerland: Malcantone watershed	Ore deposits and mining wastes	Pfeifer et al. (2004)
127	Turkey	Geothermal	Baba and Ármannsson (2006); Dogan and Dogan (2007)

(continued overleaf)

Table D.4 *(continued)*

Number	Location	Arsenic-bearing medium	Reference(s)
128	Turkey: Central Anatolia: Beypazari Lignite	Coal	Querol *et al.* (1997)
129	Turkey: Emet	Borate ore deposits and mining wastes	Helvaci and Alonso (2000)
129	Turkey: Gediz: Gokler coalfield	Coal	Karayigit, Spears and Booth (2000)
129	Turkey: Kütahya region	Coal	Dogan and Dogan (2007)
129	Turkey: Kütahya region	Groundwater	Dogan and Dogan (2007)
129	Turkey: Kütahya region	Volcanics	Dogan and Dogan (2007)
129	Turkey: Kütahya: Igdeköy-Emet	Ore deposits and mining wastes	Çolak, Gemici and Tarcan (2003)
130	Ukraine–Russia: Donbas Basin	Coal deposits and mining wastes	Kizilshtein and Kholodkov (1999); Yudovich and Ketris (2005)
131	Ukraine: L'vov-Volynsk Basin	Coal deposits and mining wastes	Yudovich and Ketris (2005)
132	Ukraine	Groundwater	Panov *et al.* (1999)
133	United Kingdom	Groundwater	Burgess and Pinto (2005)
134	United Kingdom: Cornwall	Ore deposits, smelting, and mining wastes	Pirrie *et al.* (2002); Moon (2002); Camm *et al.* (2003); Willis-Richards and Jackson (1989)
135	United Kingdom: Scotland: Talnotry	Intrusive igneous rocks	Power *et al.* (2004)
136	United Kingdom: South Wales	Coal	Gayer *et al.* (1999)

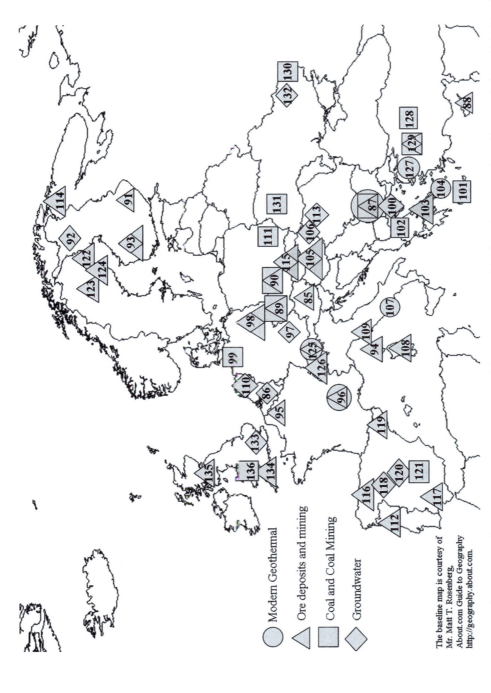

Figure D.4 Map showing significant arsenic contamination problems in Europe and Turkey. (Map reproduced by permission of Mr. Matt T. Rosenberg, http://geography.about.com.)

Table D.5 Locations of significant arsenic contamination problems in Canada

Number	Location	Arsenic-bearing medium	Reference(s)
137	Alberta	Weathered shale	Muloin and Dudas (2005)
138	Alberta: Northeastern	Ore deposits	McDonough (1997)
139	British Columbia	Geothermal	Hirner et al. (1998)
139	British Columbia: Bowen Island	Groundwater	Boyle, Turner and Hall (1998)
139	British Columbia: Sunshine Coast: Powell River	Groundwater	Mattu, Clarkson and Carmichael (2000)
140	British Columbia: Jack of Clubs Lake	Ore deposits and mining wastes	Azcue et al. (1995)
141	British Columbia: Southern Kootenay Terrace	Ore deposits and mining wastes	Lett et al. (1998)
142	British Columbia: Hat Creek	Coal	Goodarzi and Swaine (1993)
143	British Columbia: Vancouver Island	Coal	Goodarzi and Swaine (1993)
144	Labrador: Northern	Ore deposits and mining wastes	Skanes, Kerr and Sylvester (2004)
145	Manitoba: Northern	Ore deposits and mining wastes	Moncur et al. (2002)
145	Manitoba: Snow Lake	Ore deposits and mining wastes	Salzsauler, Sidenko and Sherriff (2005)
146	New Brunswick	Groundwater	Bottomley (1984)
147	New Brunswick: Mount Pleasant	Ore deposits and mining wastes	Petrunic and Al (2005)
148	Newfoundland: Burin Peninsula	Ore deposits	Hinchey, O'Driscoll and Wilton (2000); O'Brien, Dubé and O'Driscoll (1999)
149	Newfoundland: White Bay	Ore deposits and mining wastes	Kerr (2004)
150	Northwest Territories: Yellowknife region	Ore deposits and mining wastes	Mudroch et al. (1989); Halbert et al. (2003); Walker et al. (2005); Ollson et al. (2003)
151	Nova Scotia	Groundwater	Bottomley (1984); Rahman, Sengupta and Chowdhury (2006)
151	Nova Scotia: Caribou gold mining area	Ore deposits and mining wastes	Wong et al. (2002)
151	Nova Scotia: Cumberland Basin: Joggins Formation: Westphalian A coal	Coal	Hower et al. (2000)
151	Nova Scotia: Sydney and Stellarton Basins	Coal	Mukhopadhyay et al. (1998)
152	Nunavut	Ore deposits	Kerswill et al. (1999)
153	Ontario: Balmer Lake	Ore deposits and mining wastes	Martin and Pedersen (2002)
153	Ontario: Campbell Mine	Ore deposits and mining wastes	Penczak and Mason (1997)
154	Ontario: Cobalt area:	Ore deposits and mining wastes	Beauchemin and Kwong (2006)

154	Ontario: Kidd Creek Tailings	Ore deposits and mining wastes	Martin, Al and Cabri (1997)
155	Ontario: Kelly Lake	Ore deposits and mining wastes	Sadiq *et al.* (2002)
155	Ontario: Moira Mine	Ore deposits and mining wastes	Azcue and Nriagu (1995); Diamond (1995)
156	Ontario: Kitchener-Waterloo	Groundwater	Sanderson *et al.* (1994)
157	Ontario: Sudbury	Ore deposits and mining wastes	McGregor *et al.* (1998)
158	Quebec: Laurentian Trough	Porewaters	Belzile (1988)
159	Saskatchewan	Groundwater	Thompson *et al.* (1999)
160	Saskatchewan: McClean Lake	Ore deposits and mining wastes	Langmuir, Mahoney and Rowson (2002)
160	Saskatchewan: Northern	Ore deposits and mining wastes	Langmuir *et al.* (1999)
161	Saskatchewan: Rabbit Lake	Ore deposits and mining wastes	Pichler, Hendry and Hall (2001)
162	Yukon: Mount Nansen	Ore deposits and mining wastes	Kwong (2003)

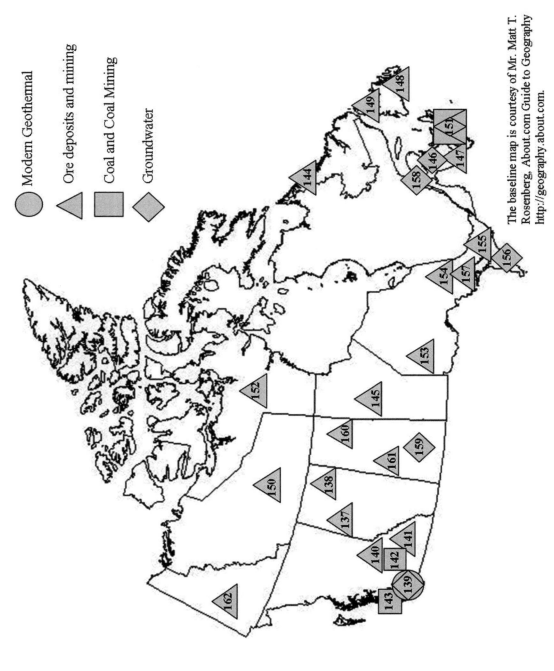

Figure D.5   Map showing significant arsenic contamination problems in Canada. (Map reproduced by permission of Mr. Matt T. Rosenberg, http://geography.about.com.)

Table D.6 *Locations of significant arsenic contamination problems in the United States of America*

Number	Location	Arsenic-bearing medium	Reference(s)
163	Alabama: Black Warrior Basin	Coal and coal mining	Diehl, Goldhaber and Hatch (2004); Goldhaber, Lee and Hatch (2003)
164	Alaska: Copper River Basin	Geothermal	Motyka *et al.* (1986)
165	Alaska: Ester Dome	Groundwater	Verplanck *et al.* (2003)
165	Alaska: Fairbanks	Groundwater	Mueller, Goldfarb and Verplanck (2001a); Mueller *et al.* (2001b)
165	Alaska: Fortymile River Watershed	Ore deposits	Crock *et al.* (1999)
166	Alaska: Southeast	Groundwater	Ryker (2001)
167	Appalachia (Eastern Kentucky, western West Virginia, southeastern Ohio, and northeastern Tennessee)	Groundwater	Shiber (2005)
168	Arizona: Avra Valley: Central Arizona Project	Groundwater	Lutz (2001)
169	Arizona: Copper Mountain Mine	Ore deposits and mining wastes	Wenrich and Silberman (1984)
170	Arizona: Near Kingman	Groundwater	Rösner (1998)
171	Arizona: Montezuma Well	Groundwater	Compton-O'Brien *et al.* (2003); Foust *et al.* (2004)
172	Arizona: Navajo Nation	Groundwater	Breit and Hiza (2002)
173	Arizona: Southwestern	Groundwater	Welch *et al.* (2000)
174	Arkansas: Eastern	Groundwater	Ryker (2001)
175	California: Argonaut Mine near Jackson	Ore deposits and mining wastes	Foster *et al.* (1998)
175	California: Jamestown Mine	Ore deposits and mining wastes	Savage, Ashley and Bird (2001)
175	California: Nevada County	Ore deposits and mining wastes	Ashley and Ziarkowski (1999)
175	California: Sierra Nevada Mountains	Ore deposits and mining wastes	Savage, Bird and Ashley (2000a); Ashley *et al.* (2000b); Ashley (2001)
175	California: Smartville Mine, east of Marysville	Ore deposits and mining wastes	Foster *et al.* (1998)
176	California: Crowley Lake	Geothermal	Kneebone and Hering (2000)
176	California: Eastern Owens Lake	Groundwater	Levy *et al.* (1997)
176	California: Hot Creek	Geothermal	Hering *et al.* (1999)
176	California: Hanford	Groundwater	Hering and Chiu (2000)
176	California: Kern River alluvial fan	Groundwater	Klinchuch *et al.* (1999)

(continued overleaf)

Table D.6   (*continued*)

Number	Location	Arsenic-bearing medium	Reference(s)
176	California: Owens Dry Lake	Groundwater	Ryu *et al.* (2002)
176	California: Tulare Lake Bed	Groundwater	Gao *et al.* (2006)
177	California-Nevada: Eastern Sierra Nevada	Geothermal	Johannesson *et al.* (1997)
178	California: Lassen County	Groundwater	Rahman, Sengupta and Chowdhury (2006)
179	California: Mono Lake	Geothermal and evaporation	Oremland *et al.* (2000)
180	California: Ruth Mine near Trona	Ore deposits and mining wastes	Foster *et al.* (1998)
181	California: San Diego County	Pegmatite	Clanin (2004)
182	California: Western Mojave Desert	Groundwater	Ball and Izbicki (2003)
183	Colorado: Upper Arkansas River	Ore deposits and mining wastes	Kimball, Callender and Axtmann (1995)
183	Colorado: Leadville mining district	Ore deposits and mining wastes	Kreps (1996)
183	Colorado: Near Idaho Springs	Ores deposits and mining	Bednar *et al.* (2005)
183	Colorado: Summit County: Blue River Basin	Ore deposits and mining wastes	Apodaca, Driver and Bails (2000)
184	Colorado: Sites near Ouray and Summitville	Ores deposits and mining	Bednar *et al.* (2005)
185	Colorado: Upper Colorado River Basin	Ore deposits and mining wastes	Deacon and Driver (1999)
186	Colorado-Northern New Mexico: Upper Rio Grande Basin	Ore deposits and mining wastes	Taylor *et al.* (2001)
187	Connecticut	Groundwater	Brown and Chute (2001); Ayotte *et al.* (2006)
188	Florida	Groundwater	Arthur, Cowart and Dabous (2000)
188	Florida: Tampa-Ft. Myers region: Sawannee Limestone	Groundwater	Price and Pichler (2006)
189	Idaho: Coeur d'Alene River and Lake	Ore deposits and mining wastes	La Force *et al.* (1998); Farag *et al.* (1998); Cummings *et al.* (2000)
190	Idaho: North of Soda Springs	Ores deposits and mining	Bednar *et al.* (2005)
191	Idaho: Southwest	Groundwater	Donato, Lamothe and Sanzolone (2004)
192	Illinois: Big Muddy and Saline River basins	Coal and coal mining	Toler (1982)
193	Illinois: Lower Illinois River Basin	Groundwater	Warner (2001); Kelly *et al.* (2005); Holm *et al.* (2004); Thomas (2003)
194	Indiana: Upper Illinois River Basin	Groundwater	Thomas (2003)

195	Iowa	Groundwater	Erickson and Barnes (2005)
196	Kentucky	Groundwater	Fisher (2002)
197	Kentucky: Clay County: Manchester coal bed	Coal	Sakulpitakphon et al. (2000)
197	Kentucky: Pond Creek and Fire Clay coal beds	Coal	Hower et al. (1997)
197	Kentucky: Whitley County: River Gem coal bed	Coal	Hower et al. (1996)
198	Louisiana: Southern	Groundwater	Hanor (1998a, 1998b)
199	Maine	Groundwater	Loiselle, Marvinnev and Smith (2001); Ayotte et al. (2006); Sidle, Wotten and Murphy (2001); Lipfert et al. (2006)
200	Massachusetts	Groundwater	Ayotte et al. (2006)
200	Massachusetts: Central: Landfills	Groundwater	Hon, Mayo and Brandon (2003)
200	Massachusetts: Woburn	Groundwater and sediment porewaters	Davis et al. (1994); Wilkin and Ford (2006)
201	Michigan: Eastern	Groundwater	Ryker (2001)
202	Michigan: Hell	Groundwater	McCall, Walter and Szramek (2002)
202	Michigan: Oakland County	Groundwater	Aichele (2000; 2004)
202	Michigan: Southeastern	Groundwater	Kim, Nriagu and Haack (2002; 2003); Thomas (2003)
203	Minnesota	Groundwater	Erickson and Barnes (2005); Minnesota Pollution Control Agency (1998); Kanivetsky (2000)
204	Missouri: Kansas City area	Groundwater	Shahnewaz et al. (2003)
205	Missouri: Ozarks	Ore deposits and mining wastes	Goldhaber, Lee and Hatch (2003)
206	Montana: Anaconda Smelter	Smelter	Walker and Griffin (1998)
206	Montana: Berkeley Pit Mine	Ore deposits and mining wastes	Abdo (1994)
206	Montana: Boulder River watershed	Ore deposits and mining wastes	Fey, Church and Finney (2000)
206	Montana: Powell County	Ore deposits and mining wastes	Milodragovich (2003)
206	Montana: Washoe Smelter	Smelter	Redente, Zadeh and Paschke (2002)
207	Montana: Clark Fork River	Ore deposits and mining wastes	Klarup (1997); Moore, Ficklin and Johns (1988)
207	Montana: Milltown	Ore deposits and mining wastes	Moore and Woessner (2003); Mickey and Moore (1997)
208	Montana: Madison River Valley	Groundwater	Sonderegger and Ohguchi (1988); Sonderegger and Sholes (1989)

(continued overleaf)

Table D.6 (continued)

Number	Location	Arsenic-bearing medium	Reference(s)
208	Montana-Wyoming: Madison and Missouri Rivers	Geothermal	Nimick et al. (1998); Nimick (1998)
177	Nevada: Carson Desert	Groundwater	Welch and Lico (1998)
177	Nevada-California: Carson River Basin	Groundwater	Welch et al. (1997)
177	Nevada: Fallon	Groundwater	Mandal and Suzuki (2002)
177	Nevada: Fernley area	Groundwater	Ghebremicael and Campana (2001)
209	Nevada: Betze-Post-Screamer deposits	Ore deposits and mining wastes	Kesler, Riciputi and Ye (2005)
209	Nevada: Carlin Trend	Ore deposits and mining wastes	Decker et al. (1999)
209	Nevada: Elko County	Ore deposits and mining wastes	Shallow (1999); Bednar et al. (2005); Earman and Hershey (2004)
210	Nevada: Getchell Mine Pit	Ore deposits and mining wastes	Davis et al. (2006); Bennett et al. (1998)
210	Nevada: Humboldt County: Twin Creeks	Ore deposits	Stenger et al. (1998)
210	Nevada: Lone Tree Mine	Groundwater	Nicholson et al. (1996)
210	Nevada: North-central	Ore deposits and mining wastes	Cline (2001)
200	New England	Groundwater	Ayotte et al. (2006)
200	New Hampshire	Groundwater	Peters et al. (2006); Ayotte et al. (2006); Montgomery et al. (2003)
186	New Mexico-Colorado: Upper Rio Grande Basin	Ore deposits and mining wastes	Taylor et al. (2001)
211	New Mexico: Albuquerque Basin: Santa Fe Group sediments	Groundwater	Stanton et al. (2001)
211	New Mexico: Middle Rio Grande Basin	Groundwater	Bexfield (2001); Bexfield and Plummer (2003)
211	New Mexico: Socorro Basin	Groundwater	Brandvold (2001)
212	New Mexico: Cleveland Mine	Ore deposits and mining wastes	Boulet and Larocque (1996)
213	New Mexico: Española Basin	Groundwater	Finch (2005)
214	New Mexico: McKinley and Cibola counties: Grants Mineral Belt	Ore deposits and mining wastes	Thomson, Longmire and Brookins (1986)
214	New Mexico: Valles Caldera and Jemez Mountains	Geothermal	Criaud and Fouillac (1989); Reid, Goff and Counce (2003)
215	North Dakota: North and east	Groundwater	Erickson and Barnes (2005)

No.	Location	Source	Reference
216	Ohio: Northeast	Groundwater	Eshete and Chyi (2003); Welch et al. (2000)
217	Ohio: Miami River Basins	Groundwater	Thomas (2003)
218	Oklahoma: Norman area	Groundwater	Smith (2005)
219	Oregon: Harney County: Alvord Basin	Geothermal	Koski and Wood (2003)
219	Oregon: Warner Valley: Anderson Lake	Geothermal and evaporation	Finkelstein et al. (2004)
220	Oregon: Lane County	Groundwater	Rahman, Sengupta and Chowdhury (2006)
220	Oregon: Western	Groundwater	Welch et al. (2000)
220	Oregon: Willamette Basin	Groundwater	Hinkle and Polette (1999)
221	Oregon: Malheur County	Groundwater	Gonthier and Collins (1979)
222	South Carolina: Simpsonville area	Groundwater	Price et al. (2004)
223	South Dakota	Groundwater	Erickson and Barnes (2005); Welch et al. (2000)
224	South Dakota: Black Hills	Ore deposits and mining wastes	May et al. (2001)
224	South Dakota: Black Hills: Spearfish Creek	Ore deposits and mining wastes	Driscoll and Hayes (1995)
224	South Dakota: Whitewood Creek	Ore deposits and mining wastes	Fuller and Davis (2003) 225
225	Tennessee-North Carolina: Great Smoky Mountains National Park	Ore deposits and mining wastes	Hammarstrom et al. (2003)
226	Texas: Nueces and San Antonio River watersheds	Groundwater	Lake, Herbert and Louchouarn (2001)
227	Texas: Panhandle: Ogallala Aquifer	Groundwater	Drake (2001)
227	Texas: Southern panhandle	Groundwater	Ryker (2001)
228	Texas: South	Groundwater	Ryker (2001)
229	Utah: North central	Groundwater	Ryker (2001)
229	Utah: Little Cottonwood Creek	Ore deposits and mining wastes	Gerner and Waddell (2003)
229	Utah: North Fork of American Fork River	Ore deposits and mining wastes	Burk, Lachmar and Kolesar (2004)
229	Utah: Salt Lake Valley	Groundwater	Thiros (2003)
229	Utah: West of Salt Lake City	Ores deposits and mining	Bednar et al. (2005)
230	Utah: Millard County	Groundwater	Rahman, Sengupta and Chowdhury (2006)
231	Utah: Roosevelt Hot Springs	Geothermal	Christensen, Capuano and Moore (1983)
232	Vermont	Groundwater	Ayotte et al. (2006)
233	Virginia: Floyd County	Ore deposits and mining wastes and groundwater	Brown (2005)

(continued overleaf)

Table D.6 *(continued)*

Number	Location	Arsenic-bearing medium	Reference(s)
234	Washington	Groundwater	Welch *et al.* (2000); Ryker (2001)
234	Washington: Snohomish County	Groundwater	Frost *et al.* (1993)
235	Washington: Pasco Basin	Groundwater	Petersen, Hoover and McKinley (1997)
236	Wisconsin: Fox River Valley	Groundwater	Burkel and Stoll (1999); Schreiber, Simo and Freiberg (2000); Schreiber *et al.* (2003); Thornburg and Sahai (2004)
237	Wisconsin: Near Lake Geneva	Groundwater	Root, Bahr and Gotkowitz (2003); Thomas (2003)
238	Wyoming: Central	Ore deposits and mining wastes	Hajj-Djafari, Antommaria and Crouse (1981)
239	Wyoming: Yellowstone National Park	Geothermal	Planer-Friedrich *et al.* (2006); Inskeep, Macur and Harrison (2004); Stauffer and Thompson (1984)

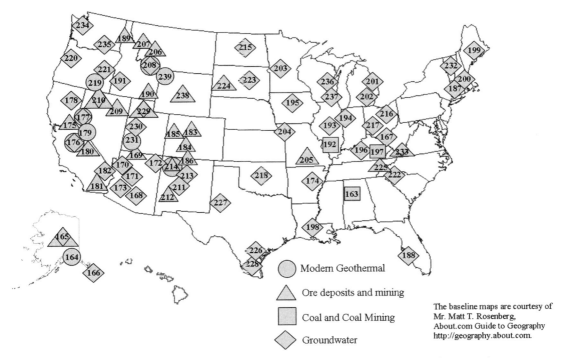

Figure D.6 *Map showing significant arsenic contamination problems in the United States of America. (Map reproduced by permission of Mr. Matt T. Rosenberg, http://geography.about.com.)*

Table D.7 *Locations of significant arsenic contamination problems in Central America and the Caribbean*

Number	Location	Arsenic-bearing medium	Reference(s)
240	Dominica	Geothermal	Criaud and Fouillac (1989)
241	Guatemala: Northwestern	Ore deposits and mining wastes	Guillemette (1991)
242	Mexico: Baja California Sur	Groundwater	Carrillo-Chávez, Drever and Martinez (2000)
242	Mexico: Baja California Sur: San Antonio-El Triunfó	Ore deposits and mining wastes	Carillo and Drever (1998)
243	Mexico: Baja California Sur: Bahia Concepción: Punta Santa Barbara	Geothermal	Melwani and Forrest (2003)
244	Mexico: Chihuahua	Groundwater	Cebrián, Albores and García-Vargas (1994)
245	Mexico: Guanajuato: Guanajuato mining district	Ore deposits and mining wastes	Morton, Carrillo-Chávez and Hernandez (2000)
245	Mexico: Guanajuato: Mineral de Pozos district	Ore deposits and mining wastes	Carrillo-Chávez, Gonzalez-Partida and Morton-Bermea (2002)
245	Mexico: Salamanca aquifer	Groundwater	Rodriguez et al. (2002)
245	Mexico: Guanajuato: San Nicolas-Mata Basin	Ore deposits and mining wastes	Solorzano and Carrillo-Chávez (2000)
246	Mexico: Guerrero: Taxco	Ore deposits and mining wastes	Mendoza et al. (2006)
246	Mexico: Morelos	Groundwater	Cebrián, Albores and García-Vargas (1994)
246	Mexico: Puebla	Groundwater	Cebrián, Albores and García-Vargas (1994)
247	Mexico: Hidalgo: Zimapán	Groundwater	Armienta et al. (2001)
247	Mexico: Rioverde Basin	Groundwater	Planer-Friedrich, Armienta and Merkel (2001)
248	Mexico: La Primavera Geothermal Field	Geothermal	Welch (1999)
249	Mexico: Lagunera region: Coahuila	Groundwater	Cebrián, Albores and García-Vargas (1994)
249	Mexico: Lagunera region: Durango	Groundwater	Cebrián, Albores and García-Vargas (1994)
250	Mexico: Lerma-Chapala Basin	Groundwater	Ortega-Guerrero (2003)
250	Mexico: Michoacán: Los Azufres	Geothermal	Birkle and Merkel (2000)
251	Mexico: Nuevo Léon	Groundwater	Cebrián, Albores and García-Vargas (1994)
252	Mexico: San Luis Potosí	Ore deposits, mining, and smelting	Razo et al. (2004); Cebrián, Albores and García-Vargas (1994)
253	Mexico: Tabasco: Activo Luna Oilfield	Formation waters	Birkle, Cid Vá´zquez and Fong Aguilar (2005)

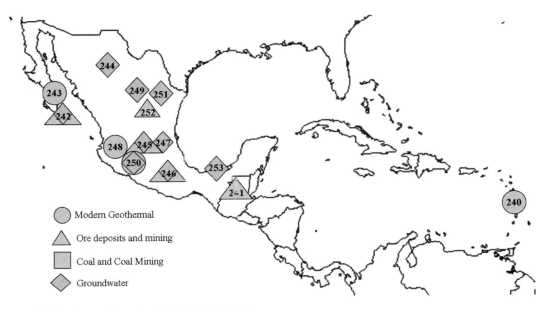

The baseline map is courtesy of Mr. Matt T. Rosenberg,
About.com Guide to Geography http://geography.about.com.

Figure D.7 *Map showing significant arsenic contamination problems in Central America and the Caribbean. (Map reproduced by permission of Mr. Matt T. Rosenberg, http://geography.about.com.)*

Table D.8 *Locations of significant arsenic contamination problems in South America*

Number	Location	Arsenic-bearing medium	Reference(s)
254	Argentina	Ore deposits and mining wastes	Williams (2001)
255	Argentina: Cordoba Province	Groundwater	Warren, Burgess and Garcia (2005)
256	Argentina: Jujay Province	Groundwater	Rahman, Sengupta and Chowdhury (2006)
256	Argentina: Salta Province	Groundwater	Mandal and Suzuki (2002)
256	Argentina: Salta Province: Lerma Valley	Ore deposits and mining wastes	Bundschuh and Ebhardt (1999)
256	Argentina: Tucumán Province	Groundwater	Warren, Burgess and Garcia (2005)
257	Argentina: Santiago del Estero Province	Groundwater	Grimolizzi and Martin (1989)
258	Argentina: La Pampa Province	Groundwater	Smedley *et al.* (2002); Warren, Burgess and Garcia (2005)
259	Bolivia: Rio Pilcomayo Basin	Ore deposits and mining wastes	Archer *et al.* (2005); Hudson-Edwards *et al.* (2001)
260	Brazil: Amazon: Amapá: Santana area	Ore deposits and mining wastes	De Figueiredo, Borba and Angélica (2007)
261	Brazil: Minas Gerais: Iron Quadrangle	Ore deposits and mining wastes	Borba, Figueiredo and Matschullat (2003); De Figueiredo, Borba and Angélica (2007)
261	Brazil: Minas Gerais: Iron Quadrangle: Ouro Preto region	Ore deposits and mining wastes	Pimentel, De Lena and Nalini (2003)
261	Brazil: Minas Gerais: Iron Quadrangle: Passagem and Morro Velho	Ore deposits and mining wastes	Williams (2001)
261	Brazil: Minas Gerais: Iron Quadrangle: Sao Bento deposit	Ore deposits and mining wastes	Márquez *et al.* (2006)
262	Brazil: Ribeira Valley	Ore deposits and mining wastes	De Figueiredo, Borba and Angélica (2007)
262	Brazil: São Paulo	Groundwater	Campos (2002)
263	Chile: Antofagasta	Groundwater	Rahman, Sengupta and Chowdhury (2006)
263	Chile: Rio Loa	Geothermal and evaporation	Romero *et al.* (2003)
264	Chile: Elqui River watershed	Groundwater and ore deposits and mining wastes	Oyarzun *et al.* (2006)
265	Chile: Lluta River Basin	Groundwater	Yamasaki and Hata (2000)
266	Ecuador	Ore deposits and mining wastes	Appleton *et al.* (2001); Williams (2001)
267	Peru: Huancavelica district	Ore deposits and mining wastes	Wise *et al.* (2004)

Figure D.8 *Map showing significant arsenic contamination problems in South America. (Map reproduced by permission of Mr. Matt T. Rosenberg, http://geography.about.com.)*

many countries in Africa, the Middle East, and Latin America (including the Caribbean). The baseline maps in this appendix are courtesy of Mr. Matthew T. Rosenberg, About.com Guide to Geography at http://geography.about.com.

References

Abdo, G.N. (1994) Baseline ground-water monitoring for documenting impact of abandoned mine flooding, Berkeley Pit, Butte, Montana. *Abstracts with Programs–Geological Society of America*, **26**(7), 412–13.

Abrecht, J. (1990) An As-rich manganiferous mineral assemblage from the Koednitz Valley (Eastern Alps, Austria): geology, mineralogy, genetic considerations, and implications for metamorphic Mn deposits. *Neues Jahrbuch für Mineralogie*, **8**, 363–75.

Aggett, J. and ÓBrien, G.A. (1985) Detailed model for the mobility of arsenic in lacustrine sediments based on measurements in Lake Ohakuri. *Environmental Science and Technology*, **19**(3), 231–38.

Aichele, S. (2000). *Ground-water Quality Atlas of Oakland County, Michigan*. Water-Resources Investigation Report 00–4120, U.S. Geological Survey Fact Sheet 135–98, p. 53.

Aichele, S. (2004) *Arsenic, Nitrate, and Chloride in Groundwater, Oakland County, Michigan*, U.S. Geological Survey, Water Resources Division, Lansing, MI 48911.

Aiuppa, A., Avino, R., Brusca, L. *et al.* (2006) Mineral control of arsenic content in thermal waters from volcano-hosted hydrothermal systems: insights from island of Ischia and Phlegrean fields (Campanian volcanic province, Italy). *Chemical Geology*, **229**(4), 313–30.

Apodaca, L.E., Driver, N.E. and Bails, J.B. (2000) Occurrence, transport, and fate of trace elements, blue river basin, Summit County, Colorado: an integrated approach. *Environmental Geology*, **39**(8), 901–13.

Appleton, J.D., Williams, T.M., Orbea, H. and Carrasco, M. (2001) Fluvial contamination associated with arsenical gold mining in the Ponce Enri'quez, Portovelo-Zaruma and Nambija areas, Ecuador. *Water Air and Soil Pollution*, **131**(1–4), 19–39.

Appleyard, S.J., Angeloni, J. and Watkins, R. (2006) Arsenic-rich groundwater in an urban area experiencing drought and increasing population density, Perth, Australia. *Applied Geochemistry*, **21**(1), 83–97.

Archer, J., Hudson-Edwards, K.A., Preston, D.A. *et al.* (2005) Aqueous exposure and uptake of arsenic by riverside communities affected by mining contamination in the Ri'o Pilcomayo basin, Bolivia. *Mineralogical Magazine*, **69**(5), 719–36.

Armienta, M.A., Villaseñor, G., Rodriguez, R. *et al.* (2001) The role of arsenic-bearing rocks in groundwater pollution at Zimapán valley, México. *Environmental Geology*, **40**(4–5), 571–81.

Arthur, J.D., Cowart, J.B. and Dabous, A.A. (2000) Arsenic and uranium mobilization during aquifer storage and recovery in the Floridan Aquifer system. *Abstracts with Programs–Geological Society of America*, **32**(7), 356.

Asghar, U., Perveen, F., Alvi, S.K. *et al.* (2006) Contamination of arsenic in public water supply schemes of Larkana and Mirpurkhas Districts of Sind. *Journal of the Chemical Society of Pakistan*, **28**(2), 130–35.

Ashley, P.M., Craw, D., Graham, B.P. and Chappell, D.A. (2003) Environmental mobility of antimony around mesothermal stibnite deposits, New South Wales, Australia and southern New Zealand. *Journal of Geochemical Exploration*, **77**(1), 1–14.

Ashley, P.M., Graham, B.P., Tighe, M.K. and Wolfenden, B.J. (2007) Antimony and arsenic dispersion in the Macleay River catchment, New South Wales: a study of the environmental geochemical consequences. *Australian Journal of Earth Sciences*, **54**(1), 83–103.

Ashley, R.P. and Ziarkowski, D.V. (1999) Arsenic in waters affected by mill tailings at the Lava Cap Mine, Nevada County, California. *Abstracts with Programs–Geological Society of America*, **31**(6), 35.

Atanassova, R. Kerestedjian, T. and Satchanska, G. (2003) Mineralogical, Geochemical and Microbiological Study of Metal Polluted Soils, Sediments and Dump Materials in the Vicinity of the KCM Pb-Zn Smelter. International Conference Modern Management of Mine Producing, Geology and Environmental Protection 3, Plovdiv region, Bulgaria, pp. 305–15.

Aucamp, P. (2003) Trace-element pollution of soils by abandoned goldmine tailings near Potchefstroom, South Africa. *Bulletin – Council for Geoscience of South Africa*, **130**, 69.

Ayora, C. and Casas, J.M. (1986) Strata-bound as-au mineralisation in Pre-Caradocian rocks from the Vall de Ribes, eastern Pyrenees, Spain. *Mineralium Deposita*, **21**(4), 278–87.

Ayotte, J.D., Nolan, B.T., Nuckols, J.R. *et al.* (2006) Modeling the probability of arsenic in groundwater in New England as a tool for exposure assessment. *Environmental Science and Technology*, **40**(11), 3578–85.

Azcue, J.M. and Nriagu, J.O. (1995) Impact of abandoned mine tailings on the arsenic concentrations in Moira Lake, Ontario. *Journal of Geochemical Exploration*, **52**(1–2), 81–89.

Azcue, J.M., Mudroch, A., Rosa, F. *et al.* (1995) Trace elements in water, sediments, porewater, and biota polluted by tailings from an abandoned gold mine in British Columbia, Canada. *Journal of Geochemical Exploration*, **52**(1–2), 25–34.

Baba, A. and Ármannsson, H. (2006) Environmental impact of the utilization of geothermal areas. *Energy Sources, Part B: Economics, Planning and Policy*, **1**(3), 267–78.

Baker, J.H., Papakonstantinou, A., van Os, B. *et al.* (1995) Natural sources of high As and U in groundwaters from the Mygdonia region of northern Greece. *Terra Abstracts*, **7**(1), 254.

Ball, J.W. and Izbicki, J.A. (2003) Geochemical relations among arsenic, chromium, and other trace elements in ground water underlying the Sheep Creek fan, western Mojave Desert, California. *Abstracts with Programs–Geological Society of America*, **35**(6), 565.

Baroni, F., Boscagli, A., Di Lella, L.A. *et al.* (2004) Arsenic in soil and vegetation of contaminated areas in southern Tuscany (Italy). *Journal of Geochemical Exploration*, **81**(1–3), 1–14.

Bashkin, V.N. and Wongyai, K. (2002) Environmental fluxes of arsenic from lignite mining and power generation in northern Thailand. *Environmental Geology*, **41**(8), 883–88.

Beauchemin, S. and Kwong, Y.T.J. (2006) Impact of redox conditions on arsenic mobilization from tailings in a wetland with neutral drainage. *Environmental Science and Technology*, **40**(20), 6297–303.

Becker, A., Klöck, W., Friese, K. *et al.* (2001) Lake Süßer See as a natural sink for heavy metals from copper mining. *Journal of Geochemical Exploration*, **74**(1–3), 205–17.

Bednar, A.J., Garbarino, J.R., Ranville, J.F. and Wildeman, T.R. (2005) Effects of iron on arsenic speciation and redox chemistry in acid mine water. *Journal of Geochemical Exploration*, **85**(2), 55–62.

Belzile, N. (1988) The fate of arsenic in sediments of the Laurentian Trough. *Geochimica et Cosmochimica Acta*, **52**(9), 2293–302.

Bennett, J.C., Tempel, R.N., Shevenell, L. *et al.* (1998) Arsenic variations in groundwaters at the Getchell Mine, Nevada. *Abstracts with Programs–Geological Society of America*, **30**(7), 59.

Berg, M., Giger, W., Tron, H.C. (2006) Extent and severity of arsenic pollution in Vietnam and Cambodia, in *Managing Arsenic in the Environment: From Soil to Human Health* (eds R. Naidu, E. Smith, G. Owens, P. Bhattacharya and P., Nadebaum), CSIRO Publishing, Collingwood, pp. 495–509.

Berg, M., Tran, H.C., Nguyen, T.C. *et al.* (2001) Arsenic contamination of groundwater and drinking water in Vietnam: a human health threat. *Environmental Science and Technology*, **35**(13), 2621–26.

Bexfield, L.M. (2001) *Occurrence and sources of arsenic in ground water of the middle Rio Grande Basin, Central New Mexico*. Master's Thesis. New Mexico Institute of Mining and Technology, Socorro, p. 183.

Bexfield, L.M. and Plummer, L.N. (2003) Occurrence of arsenic in ground water of the Middle Rio Grande Basin, central New Mexico, in *Arsenic in Ground Water* (eds A.H. Welch and K.G. Stollenwerk), Kluwer Academic Publishers, Boston, MA, pp. 295–327.

Birkle, P., Cid Va'zquez, A.L. and Fong Aguilar, J.L. (2005) Legal aspects and technical alternatives for the treatment of reservoir brines at the Activo Luna oilfield, Mexico. *Water Environment Research*, **77**(1), 68–77.

Birkle, P. and Merkel, B. (2000) Environmental impact by spill of geothermal fluids at the geothermal field of Los Azufres, Michoacán, Mexico. *Water Air and Soil Pollution*, **124**(3–4), 371–410.

Black, A. and Craw, D. (2001) Arsenic, copper and zinc occurrence at the Wangaloa coal mine, southeast Otago, New Zealand. *International Journal of Coal Geology*, **45**, 181–93.

Bojar, H.-P., Bojar, A.-V., Mogessie, A. *et al.* (2001) Evolution of veins and sub-economic ore at Strassegg, Paleozoic of Graz, eastern Alps, Austria: evidence for local fluid transport during metamorphism. *Chemical Geology*, **175**(3–4), 757–77.

Borba, R.P., Figueiredo, B.R. and Matschullat, J. (2003) Geochemical distribution of arsenic in waters, sediments and weathered gold mineralized rocks from Iron Quadrangle, Brazil. *Environmental Geology*, **44**(1), 39–52.

Bortnikova, S.B., Smolyakov, B.S., Sidenko, N.V. *et al.* (2001) Geochemical consequences of acid mine drainage into a natural reservoir: inorganic precipitation and effects on plankton activity. *Journal of Geochemical Exploration*, **74**, 127–39.

Bottomley, D.J. (1984) Origins of some arseniferous groundwaters in Nova Scotia and New Brunswick, Canada. *Journal of Hydrology*, **69**(1–4), 223–57.

Boulet, M.P. and Larocque, A.C.L. (1996) A mineralogical and geochemical study of the Cleveland Mine-tailings in New Mexico, U.S.A. *Abstracts with Programs–Geological Society of America*, **28**(7), 518.

Bouška, V. and Pešek, J. (1999) Quality parameters of lignite of the North Bohemian Basin in the Czech Republic in comparison with the world average lignite. *International Journal of Coal Geology*, **40**(2–3), 211–35.

Bowell, R.J., Morley, N.H. and Din, V.K. (1994) Arsenic speciation in soil porewaters from the Ashanti mine, Ghana. *Applied Geochemistry*, **9**(1), 15–22.

Bowell, R.J., Warren, A., Minjera, H.A. and Kimaro, N. (1995) Environmental impact of former gold mining on the Orangi River, Serengeti N.P., Tanzania. *Biogeochemistry*, **28**(3), 131–60.

Boyle, D.R., Turner, R.J.W. and Hall, G.E.M. (1998) Anomalous arsenic concentrations in groundwaters of an island community, Bowen Island, British Columbia. *Environmental Geochemistry and Health*, **20**(4), 199–212.

Brandvold, L. (2001) Arsenic in ground water in the Socorro Basin, New Mexico. *New Mexico Geology*, **23**(1), 2–8.

Breit, G.N. and Hiza, M.M. (2002) Variations in water composition in aquifers of the Hopi Buttes (Tsezhin Bii) area of the Navajo Nation, northeastern Arizona. *Abstracts with Program - Geological Society of America*, **34**(6), 395–96.

Brown, B.V. (2005) *Arsenic transport in groundwater, surface water, and the hyporheic zone of a mine-influenced stream-aquifer system*. Master's Thesis, Virginia Polytechnic Institute and State University, p. 44.

Brown, C.J. and Chute, S.K. (2001) Arsenic in bedrock wells in Connecticut, *USGS Workshop on Arsenic in the Environment*, February 21–22, US Geological Survey, Denver, CO.

Bundschuh, J. and Ebhardt, G. (1999) Umweltbeeinflussung durch Verarbeitung arsenhaltiger Borate: Fallbeispiel Lermatal, NW-Argentinien (Environmental effects from the processing of arsenic-containing borates: case study of the Lerma Valley, northwestern Argentina.). *Schriftenreihe der Deutschen Geologischen Gesellschaft*, **6**, 107–11.

Burgess, W. and Ahmed, K.M. (2006) Arsenic in aquifers of the Bengal Basin, in *Managing Arsenic in the Environment: From Soil to Human Health* (eds R. Naidu, E. Smith, G. Owens, P. Bhattacharya and P. Nadebaum), CSIRO Publishing, Collingwood, pp. 31–56.

Burgess, W.G. and Pinto, L. (2005) Preliminary observations on the release of arsenic to groundwater in the presence of hydrocarbon contaminants in UK aquifers. *Mineralogical Magazine*, **69**(5), 887–96.

Burk, N.I., Lachmar, T.E. and Kolesar, P.T. (2004) Geochemistry of metals in surface and ground water from an abandoned mine in the North Fork of the American Fork River, Utah. *Abstracts with Programs–Geological Society of America*, **36**(4), 29.

Burkel, R.S. and Stoll, R.C. (1999) Naturally occurring arsenic in sandstone aquifer water supply wells of north-eastern Wisconsin. *Ground Water Monitoring and Remediation*, **19**(2), 114–21.

Buschmann, J., Berg, M., Stengel, C. and Sampson, M.L. (2007) Arsenic and manganese contamination of drinking water resources of Cambodia: coincidence of risk areas with low relief topography. *Environmental Science and Technology*, **41**, 2146–52.

Camm, G.S., Butcher, A.R., Pirrie, D. *et al.* (2003) Secondary mineral phases associated with a historic arsenic calciner identified using automated scanning electron microscopy: a pilot study from Cornwall, UK. *Minerals Engineering*, **16**(Suppl 11), 1269–77.

Campos, V. (2002) Arsenic in groundwater affected by phosphate fertilizers at Saño Paulo, Brazil. *Environmental Geology*, **42**(1), 83–87.

Carillo, A. and Drever, J.I. (1998) Adsorption of arsenic by natural aquifer material in the San Antonio-El Triunfo mining area, Baja California, Mexico. *Environmental Geology*, **35**(4), 251–57.

Carlson, L., Bigham, J.M., Schwertmann, U. *et al.* (2002) Scavenging of as from acid mine drainage by schwertmannite and ferrihydrite: a comparison with synthetic analogues. *Environmental Science and Technology*, **36**(8), 1712–19.

Carrillo-Chávez, A., Drever, J.I. and Martinez, M. (2000) Arsenic content and groundwater geochemistry of the San Antonio-El Triunfo, Carrizal and Los Planes aquifers in southernmost Baja California, Mexico. *Environmental Geology*, **39**(11), 1295–303.

Carrillo-Chávez, A., Gonzalez-Partida, E. and Morton-Bermea, O. (2002) Heavy metals distribution in rock, sediments, groundwater, mine tailings and leaching experiments in Mineral de Pozos historical mining site, central Mexico. *Abstracts with Programs–Geological Society of America*, **34**(6), 415–16.

Cebrián, M.E., Albores, A. and García-Vargas, G. (1994) Chronic arsenic poisoning in humans: the case of Mexico, in *Arsenic in the Environment: Part II: Human Health and Ecosystem Effects* (ed J.O. Nriagu), John Wiley & Sons Ltd, New York, pp. 93–107.

Chakraborti, D., Rahman, M.M., Paul, K. *et al.* (2002) Arsenic calamity in the Indian subcontinent: what lessons have been learned? *Talanta*, **58**(1), 3–22.

Chen, S.-L., Dzeng, S.R., Yang, M.-H. *et al.* (1994) Arsenic species in groundwaters of the Blackfoot disease area, Taiwan. *Environmental Science and Technology*, **28**(5), 877–81.

Christensen, O.D., Capuano, R.M. and Moore, J.N. (1983) Trace-element distribution in an active hydrothermal system, Roosevelt Hot Springs Thermal Area, Utah. *Journal of Volcanology and Geothermal Research*, **16**(1–2), 99–129.

Clanin, R.J. (2004) The Cryo-Genie Pegmatite: a complexly zoned, LCT pegmatite in the Peninsular Ranges Batholith, San Diego County, California. *Abstracts with Programs–Geological Society of America*, **36**(5), 44–45.

Cleverley, J.S., Benning, L.G. and Mountain, B.W. (2003) Reaction path modelling in the As-S system: a case study for geothermal As transport. *Applied Geochemistry*, **18**(9), 1325–45.

Cline, J.S. (2001) Timing of gold and arsenic sulfide mineral deposition at the Getchell Carlin-type gold deposit, north-central Nevada. *Economic Geology*, **96**(1), 75–89.

Coetsiers, M. and Walraevens, K. (2006) Chemical characterization of the Neogene aquifer, Belgium. *Hydrogeology Journal*, **14**(8), 1556–68.

Çolak, M., Gemici, Ü. and Tarcan, G. (2003) The effects of colemanite deposits on the arsenic concentrations of soil and ground water in Igdeköy-Emet, Kütahya, Turkey. *Water Air and Soil Pollution*, **149**(1–4), 127–43.

Compton-ÓBrien, A.M., Foust, R.D., Jr., Ketterer, M.E. and Blinn, D.W. (2003) Total arsenic in a fishless desert spring, in *Biochemistry of Environmentally Important Trace Elements*, ACS Symposium Series 835 (eds Y. Cai and O.C. Braids), Montezuma Well, AZ, pp. 200–9.

Courtin-Nomade, A., Bril, H., Néel, C. and Lenain, J.-F. (2003) Arsenic in iron cements developed within tailings of a former metalliferous mine — Enguiale's, Aveyron, France. *Applied Geochemistry*, **18**(3), 395–408.

Craw, D., Chappell, D. and Reay, A. (2000a) Environmental mercury and arsenic sources in fossil hydrothermal systems, Northland, New Zealand. *Environmental Geology*, **39**(8), 875–87.

Craw, D., Falconer, D. and Youngson, J.H. (2003) Environmental arsenopyrite stability and dissolution: theory, experiment, and field observations. *Chemical Geology*, **199**(1–2), 71–82.

Craw, D., Koons, P.O. and Chappell, D.A. (2002b) Arsenic distribution during formation and capping of an oxidised sulphidic minesoil, Macraes Mine, New Zealand. *Journal of Geochemical Exploration*, **76**(1), 13–29.

Crespo, J.L., Moro, M.C., Fado'n, O. *et al.* (2000) The Salamon gold deposit (Leon, Spain). *Journal of Geochemical Exploration*, **71**(2), 191–208.

Criaud, A. and Fouillac, C. (1989) The distribution of arsenic (III) and arsenic (V) in geothermal waters: examples from the Massif Central of France, the island of Dominica in the Leeward Islands of the Caribbean, the Valles Caldera of New Mexico, USA, and southwest Bulgaria. *Chemical Geology*, **76**(3–4), 259–69.

Crock, J.G., Gough, L.P., Wanty, R.B. *et al.* (1999) Regional Geochemical Results from the Analyses of Rock, Water, Soil, Stream Sediment, and Vegetation Samples–Fortymile River Watershed, East-Central, Alaska, 1998 Sampling. Open-File Report 00–511, U.S. Geological Survey, Reston, Virginia.

Cummings, D.E., March, A.W., Bostick, B. *et al.* (2000) Evidence for microbial Fe(III) reduction in anoxic, mining-impacted lake sediments (Lake Coeur d'Alene, Idaho). *Applied and Environmental Microbiology*, **66**(1), 154–62.

Dai, S., Ren, D., Tang, Y. *et al.* (2005) Concentration and distribution of elements in Late Permian coals from western Guizhou Province, China. *International Journal of Coal Geology*, **61**(1–2), 119–137.

Dai, S., Zeng, R. and Sun, Y. (2006) Enrichment of arsenic, antimony, mercury, and thallium in a Late Permian anthracite from Xingren, Guizhou, southwest China. *International Journal of Coal Geology*, **66**(3), 217–26.

Daniele, L. (2004) Distribution of arsenic and other minor trace elements in the groundwater of Ischia Island (southern Italy). *Environmental Geology*, **46**(1), 96–103.

Davis, A., Bellehumeur, T., Hunter, P. *et al.* (2006) The nexus between groundwater modeling, pit lake chemogenesis and ecological risk from arsenic in the Getchell main pit, Nevada, U.S.A. *Chemical Geology*, **228**(1–3 Special Issue), 175–96.

Davis, A., Kempton, J.H., Nicholson, A. and Yare, B. (1994) Groundwater transport of arsenic and chromium at a historical tannery, Woburn, Massachusetts, USA. *Applied Geochemistry*, **9**(5), 569–82.

De Figueiredo, B.R., Borba, R.P. and Angélica, R.S. (2007) Arsenic occurrence in Brazil and human exposure. *Environmental Geochemistry and Health*, **29**(2), 109–18.

De Mora, S., Sheikholeslami, M.R., Wyse, E. *et al.* (2004) An assessment of metal contamination in coastal sediments of the Caspian Sea. *Marine Pollution Bulletin*, **48**(1–2), 61–77.

Deacon, J.R. and Driver, N.E. (1999) Distribution of trace elements in streambed sediment associated with mining activities in the Upper Colorado River Basin, Colorado, USA, 1995–96. *Archives of Environmental Contamination and Toxicology*, **37**(1), 7–18.

Deak, J., Liebe, P. and Deseo, E. (1998) Nation-wide characteristic of the quality of subsurface waters. *Vizugyi Kozlemenyek*, **80**(1), 67–83.

Decker, D.L., Tyler, S.W., Papelis, C. and Simunek, J. (1999) A reactive transport model for arsenic in unsaturated gold mine heap and waste rock structures. *Abstracts with Programs–Geological Society of America*, **31**(7), 70.

Diamond, M.L. (1995) Application of a mass balance model to assess in-place arsenic pollution. *Environmental Science and Technology*, **29**(1), 29–42.

Diehl, S.F., Goldhaber, M.B. and Hatch, J.R. (2004) Modes of occurrence of mercury and other trace elements in coals from the Warrior field, Black Warrior Basin, northwestern Alabama. *International Journal of Coal Geology*, **59**(3–4), 193–208.

Dogan, M. and Dogan, A.U. (2007) Arsenic mineralization, source, distribution, and abundance in the Kutahya region of the western Anatolia, Turkey. *Environmental Geochemistry and Health*, **29**(2), 119–29.

Donato, M.M., Lamothe, P.J. and Sanzolone, R.F. (2004) Geologic sources and geochemical mechanisms of arsenic contamination in southwestern Idaho ground water: preliminary results. *Abstracts with Programs–Geological Society of America*, **36**(4), 32.

Donkor, A.E., Bonzongo, J.-C.J., Nartey, V.K. and Adotey, D.K. (2005) Heavy metals in sediments of the gold mining impacted Pra River Basin, Ghana, West Africa. *Soil & Sediment Contamination*, **14**, 479–503.

Dowling, C.B., Poreda, R.J., Basu, A.R. *et al.* (2002) Geochemical study of arsenic release mechanisms in the Bengal Basin groundwater. *Water Resources Research*, **38**(9), 12-1–18.

Drake, J.F. (2001) Natural gas, uranium, helium, radon, arsenic and groundwater quality of the Ogallala Aquifer. *AAPG Bulletin*, **85**(9), 1692.

Driscoll, D.G. and Hayes, T.S. (1995) *Arsenic Loads in Spearfish Creek, Western South Dakota, Water Years 1989–91.* U.S. Geological Survey Water-Resources Investigations Report 95–4080, Rapid City, South Dakota.

Duker, A.A., Carranza, E.J.M. and Hale, M. (2005) Spatial relationship between arsenic in drinking water and Mycobacterium ulcerans infection in the Amansie West district, Ghana. *Mineralogical Magazine*, **69**(5), 707–17.

Earman, S. and Hershey, R.L. (2004) Water quality impacts from waste rock at a Carlin-type gold mine, Elko County, Nevada. *Environmental Geology*, **45**(8), 1043–53.

Erickson, M.L. and Barnes, R.J. (2005) Glacial sediment causing regional-scale elevated arsenic in drinking water. *Ground Water*, **43**(6), 796–805.

Eshete, D.W. and Chyi, L.L. (2003) Source identification and natural attenuation of arsenic contaminated groundwater in northeastern Ohio. *Abstracts with Programs–Geological Society of America*, **35**(6), 565.

Eskenazy, G.M. (1995) Geochemistry of arsenic and antimony in Bulgarian coals. *Chemical Geology*, **119**(1–4), 239–54.

Farag, A.M., Woodward, D.F., Goldstein, J.N. *et al.* (1998) Concentrations of metals associated with mining waste in sediments, biofilm, benthic macroinvertebrates, and fish from the Coeur d'Alene River Basin, Idaho. *Archives of Environmental Contamination and Toxicology*, **34**(2), 119–27.

Fengthong, T., Dethoudom, S. and Keosavanh, O. (2002) Drinking Water Quality in the Lao People's Democratic Republic. Seminar on the Environmental and Public Health Risks Due to Contamination of Soils, Crops, Surface and Groundwater from Urban, Industrial and Natural Sources in South East Asia, UN/ESCAP, Hanoi, Vietnam.

Fey, D.L., Church, S.E. and Finney, C.J. (2000) Analytical Results for Bullion Mine and Crystal Mine Waste Samples and Bed Sediments from a Small Tributary to Jack Creek and from Uncle Sam Gulch, Boulder River Watershed, Montana. Open-File Report 00–031, U.S. Geological Survey, p. 63.

Filippi, M., Goliáš, V. and Pertold, Z. (2004) Arsenic in contaminated soils and anthropogenic deposits at the Mokrsko, Roudný, and Kašperské Hory gold deposits, Bohemian Massif (CZ). *Environmental Geology*, **45**(5), 716–30.

Finch, S.T. (2005) Occurrence of elevated arsenic and fluoride concentrations in the Española Basin, in *Geologic and Hydrogeologic Framework of the Española Basin –Proceedings of the 4th Annual Española Basin Workshop, Santa Fe, New Mexico, March 1–2, 2005* (ed K.C. McKinney), U.S. Geological Survey, Open-File Report 2005–1130, p. 9.

Finkelstein, D.B., Munhall, A., Pratt, L.M. and Bauer, C.E. (2004) A baseline study of evaporative water chemistry and microbial mat diversity from alkaline lakes in Warner Valley, Oregon. *Abstracts with Programs–Geological Society of America*, **36**(5), 87.

Fisher, R.S. (2002). *Groundwater Quality in Kentucky: Arsenic*, Kentucky Geological Survey Information Circular 5, Series XII, Lexington, KY.

Foster, A.L., Brown, G.E., Tingle, T.N. and Parks, G.A. (1998) Quantitative arsenic speciation in mine tailings using X-ray absorption spectroscopy. *American Mineralogist*, **83**(5–6), 553–68.

Foust, R.D., Jr., Mohapatra, P., Compton-ÓBrien, A.-M. and Reifel, J. (2004) Groundwater arsenic in the Verde Valley in central Arizona, USA. *Applied Geochemistry*, **19**(2), 251–55.

Frau, F. and Ardau, C. (2003) Geochemical controls on arsenic distribution in the Baccu Locci stream catchment (Sardinia, Italy) affected by past mining. *Applied Geochemistry*, **18**(9), 1373–86.

Frentiu, T., Vlad, S.-N., Ponta, M. *et al.* (2007) Profile distribution of As(III) and As(V) species in soil and groundwater in Bozanta area. *Chemical Papers*, **61**(3), 186–93.

Frost, F., Frank, D., Pierson, K. *et al.* (1993) A seasonal study of arsenic in groundwater, Snohomish County, Washington, USA. *Environmental Geochemistry and Health*, **15**(4), 209–14.

Fuller, C.C. and Davis, J.A. (2003) *Section II. Evaluation of the Processes Controlling Dissolved Arsenic in Whitewood Creek, South Dakota. U.S. Geological Survey Professional Paper1681*, Reston, VA, pp. 27–48.

Gao, S., Goldberg, S., Herbel, M.J. *et al.* (2006) Sorption processes affecting arsenic solubility in oxidized surface sediments from Tulare Lake bed, California. *Chemical Geology*, **228**(1–3 Special Issue), 33–43.

García-Sánchez, A., Moyano, A. and Mayorga, P. (2005) High arsenic contents in groundwater of central Spain. *Environmental Geology*, **47**(6), 847–54.

Gayer, R.A., Rose, M., Dehmer, J. and Shao, L.-Y. (1999) Impact of sulphur and trace element geochemistry on the utilization of a marine-influenced coal-case study from the South Wales Variscan foreland basin. *International Journal of Coal Geology*, **40**(2–3), 151–74.

Gbadebo, A.M. (2005) Occurrence and fate of arsenic in the hydrogeological systems of Nigeria. *Abstracts with Programs–Geological Society of America*, **37**(7), 375.

Gelova, G.A. (1977) *Hydrogeochemistry of Ore Elements*, Nedra, Moscow.

Gentzis, T., Goodarzi, F., Koukouzas, C.N. and Foscolos, A.E. (1996) Petrology, mineralogy, and geochemistry of lignites from Crete, Greece. *International Journal of Coal Geology*, **30**(1–2), 131–50.

Gerner, S.J., Waddell, K.M. (2003) *Hydrology and Water Quality of an Urban Stream Reach in the Great Basin - Little Cottonwood Creek near Salt Lake City, Utah, Years 1999–2000*. U.S. Geological Survey Water-Resources Investigations Report 02–4276, p. 46.

Ghebremicael, S.T. and Campana, M. (2001) Geochemistry of arsenic in the ground water of Fernley, Nevada. *Abstracts with Programs–Geological Society of America*, **33**(6), 54.

Giere, R., Sidenko, N. and Lazareva, E.V. (2003) The role of secondary minerals in controlling the migration of arsenic and metals from high-sulfide wastes. *Applied Geochemistry*, **18**(9), 1347–59.

Goldhaber, M.B., Lee, R.C. and Hatch, J.R. (2003) Role of large scale fluid-flow in subsurface arsenic enrichment, in *Arsenic in Ground Water* (eds A.H. Welch and K.G., Stollenwerk), Kluwer Academic Publishers, Boston, MA, pp. 127–64.

Gonthier, J.B. and Collins, C.A. (1979) *High Arsenic Concentrations in Ground-Water Samples from Northern Malheur County*. U.S. Geological Survey Professional Paper, No. 1150, p. 132.

Goodarzi, F. and Swaine, D.J. (1993) Chalcophile elements in western Canadian coals. *International Journal of Coal Geology*, **24**(1–4), 281–92.

Govil, P.K., Reddy, G.L.N. and Gnaneswara Rao, T. (1999) Environmental pollution in India: heavy metals and radiogenic elements in Nacharam Lake. *Journal of Environmental Health*, **61**(8), 23–28.

Grimolizzi, O.M. and Martin, A.P. (1989) Concentraciones de arsenico en aguas subterraneas de la region semiarida de Santiago del Estero, Argentina (Arsenic concentrations in the ground waters of the semi-arid region of Santiago del Estero, Argentina). *Geofisica*, **45**(2), 157–68.

Guillemette, N. (1991) Geology and geochemistry of the Ixtahuacan Sb-W Deposits, Northwestern Guatemala. Master's Thesis, McGill University, Montreal, QC, p. 110.

Guo, H. and Wang, Y. (2005) Geochemical characteristics of shallow groundwater in Datong Basin, northwestern China. *Journal of Geochemical Exploration*, **87**(3), 109–20.

Guo, R., Yang, J. and Liu, Z. (2004) Thermal and chemical stabilities of arsenic in three Chinese coals. *Fuel Processing Technology*, **85**(8–10), 903–12.

Haji-Djafari, S., Antommaria, P.E. and Crouse, H.L. (1981) Attenuation of radionuclides and toxic elements by in situ soils at a uranium tailings pond in central Wyoming. *ASTM Special Technical Publication*, **746**, 221–42.

Halbert, B.E., Phillips, H.A., Fernandes, S.L. and Scharer, J.M. (2003) Risk assessment of arsenic trioxide management alternatives at the Giant Mine, NWTB, in *Proceedings — Assessment and Remediation of Contaminated Sites in Arctic and Cold Climates* (eds M. Nahir, K. Biggar and G., Cotta), University of Alberta, Edmonton, AB, Vol. **3**, pp. 29–39.

Hammarstrom, J.M., Seal, R.R., II, Meier, A.L. and Jackson, J.C. (2003) Weathering of sulfidic shale and copper mine waste: secondary minerals and metal cycling in Great Smoky Mountains National Park, Tennessee, and North Carolina, USA. *Environmental Geology*, **45**(1), 35–57.

Hanor, J.S. (1998a) Migration pathways of saline oil field wastes at a surface storage and disposal site in South Louisiana. *AAPG Bulletin*, **82**(9), 1782.

Hanor, J.S. (1998b) Pathways of shallow subsurface migration of saline oil field wastes at a commercial disposal site in south Louisiana. *Gulf Coast Association of Geological Societies Transactions*, **XLVIII**, 107–18.

Haque, I. (2005) *National Action Plan for Arsenic Mitigation, 2005–09 (Pakistan)*. Final draft report prepared for the Ministry of Science and Technology, Pakistan.

He, B., Liang, L. and Jiang, G. (2002) Distributions of arsenic and selenium in selected Chinese coal mines. *Science of the Total Environment*, **296**(1–3), 19–26.

Hedenquist, J.W. and Henley, R.W. (1985) Hydrothermal eruptions in the Waiotapu geothermal system, New Zealand: their origin, associated breccias, and relation to precious metal mineralization. *Economic Geology*, **80**(6), 1640–68.

Heinrichs, G. and Udluft, P. (1999) Natural arsenic in Triassic rocks: a source of drinking-water contamination in Bavaria, Germany. *Hydrogeology Journal*, **7**(5), 468–76.

Helvaci, C. and Alonso, R. (2000) Borate deposits of Turkey and Argentina: a summary and geological comparison. *Turkish Journal of Earth Sciences*, **9**(1), 1–27.

Hering, J.G. and Chiu, V.Q. (2000) Arsenic occurrence and speciation in municipal ground-water-based supply system. *Journal of Environmental Engineering*, **126**(5), 471–74.

Hering, J.G., Wilkie, J.A., Kneebone, P.E. and Salmassi, T. (1999) Redox cycling of arsenic in gothermally-influenced surface waters. *Abstracts with Programs–Geological Society of America*, **31**(7), 448.

Hernández-García, M.E. and Custodio, E. (2004) Natural baseline quality of Madrid tertiary detrital aquifer groundwater (Spain): a basis for aquifer management. *Environmental Geology*, **46**(2), 173–88.

Hinchey, J.G., O'Driscoll, C.F. and Wilton, D.H.C. (2000). *Breccia-Hosted Gold on the Northern Burin Peninsula, Newfoundland*. Geological Survey, Report 2000–1, Current Research Newfoundland Department of Mines and Energy, 299–309.

Hinkle, S.R. and Polette, D.J. (1999) *Arsenic in Ground Water of the Willamette Basin, Oregon. Water-Resources Investigations Report 98–4205*, U.S. Geological Survey, Portland, OR.

Hirner, A.V., Feldmann, J., Krupp, E. *et al.* (1993) Metal(loid)organic compounds in geothermal gases and waters. *Organic Geochemistry*, **29**(5–7 pt 2), 1765–78.

Hofmann, B.A., Helfer, M., Diamond, L.W. *et al.* (2004) Topography-driven hydrothermal breccia mineralization of Pliocene age at Grimsel Pass, Aar Massif, central Swiss Alps. *Schweizerische Mineralogische und Petrographische Mitteilungen*, **84**(3), 271–302.

Holm, T.R., Kelly, W.R., Wilson, S.D. *et al.* (2004) *Arsenic Geochemistry and Distribution in the Mahomet Aquifer, Illinois*. Waste Management and Research Center Report RR-107, Illinois State Water Survey and Illinois Waste Management and Research Center.

Hon, R., Mayo, M.J. and Brandon, W.C. (2003) Arsenic equilibria in ground water at landfills, central Massachusetts. *Abstracts with Programs–Geological Society of America*, **35**(6), 48.

Hossain, M.F. (2006) Arsenic contamination in Bangladesh –an overview. *Agriculture, Ecosystems & Environment*, **113**(1–4), 1–16.

Hower, J.C., Calder, J.H., Eble, C.F. *et al.* (2000) Metalliferous coals of the Westphalian a Joggins Formation, Cumberland Basin, Nova Scotia, Canada: petrology, geochemistry, and palynology. *International Journal of Coal Geology*, **42**(2–3), 185–206.

Hower, J.C., Robertson, J.D., Wong, A.S. *et al.* (1997) Arsenic and lead concentrations in the pond creek and fire clay coal beds, eastern Kentucky coal field. *Applied Geochemistry*, **12**(3), 281–89.

Hower, J.C., Ruppert, L.F., Eble, C.F. and Graham, U.M. (1996) Geochemical and palynological indicators of the paleoecology of the river gem coal bed, Whitley County, Kentucky. *International Journal of Coal Geology*, **31**(1–4), 135–49.

Hozhina, E.I., Khramov, A.A., Gerasimov, P.A. and Kumarkov, A.A. (2001) Uptake of heavy metals, arsenic, and antimony by aquatic plants in the vicinity of ore mining and processing industries. *Journal of Geochemical Exploration*, **74**, 153–62.

Hudson-Edwards, K.A. and Edwards, S.J. (2005) Mineralogical controls on storage of As, Cu, Pb and Zn at the abandoned Mathiatis massive sulphide mine, Cyprus. *Mineralogical Magazine*, **69**(5), 695–706.

Hudson-Edwards, K.A., Jamieson, H.E., Charnock, J.M. and Macklin, M.G. (2005) Arsenic speciation in waters and sediment of ephemeral floodplain pools, Ríos Agrio-Guadiamar, Aznalcóllar, Spain. *Chemical Geology*, **219**(1–4), 175–92.

Hudson-Edwards, K.A., Macklin, M.G., Miller, J.R. and Lechler, P.J. (2001) Sources, distribution and storage of heavy metals in the Ri'o Pilcomayo, Bolivia. *Journal of Geochemical Exploration*, **72**(3), 229–50.

Huisman, D.J., Vriend, S.P., Gunnink, J. and Kloosterman, F. (2002) Arsenic in the Netherlands: processes, spatial distribution and risk. *Geochimica et Cosmochimica Acta*, **66**(15A), 347.

Huntsman-Mapila, P., Mapila, T., Letshwenyo, M. *et al.* (2006) Characterization of arsenic occurrence in the water and sediments of the Okavango delta, NW Botswana. *Applied Geochemistry*, **21**(8), 1376–91.

Inskeep, W.P., Macur, R.E., Harrison, G. *et al.* (2004) Biomineralization of As(V)-hydrous ferric oxyhydroxide in microbial mats of an acid-sulfate-chloride geothermal spring, Yellowstone National Park. *Geochimica et Cosmochimica Acta*, **68**(15), 3141–55.

Ito, H., Masuda, H. and Kusakabe, M. (2003) Some factors controlling arsenic concentrations of groundwater in the northern part of Osaka Prefecture. *Journal of Groundwater Hydrology*, **45**(1), 3–18.

Jakubick, A.T., Jenk, U. and Kahnt, R. (2002) Modelling of mine flooding and consequences in the mine hydrogeological environment: flooding of the Koenigstein mine, Germany. *Environmental Geology*, **42**(2–3), 222–34.

Jdid, E.A., Blazy, P., Kamoun, S. *et al.* (1999) Environmental impact of mining activity on the pollution of the Medjerda River, north-west Tunisia. *Bulletin of Engineering Geology and the Environment*, **57**(3), 273–80.

Johannesson, K.H., Lyons, W.B., Huey, S. *et al.* (1997) Oxyanion concentrations in eastern Sierra Nevada rivers –2. Arsenic and phosphate. *Aquatic Geochemistry*, **3**(1), 61–97.

Jung, W., Knitzschke, G. and Gerlach, R. (1974) Zu den "Schadstoffkomponenten" Arsen, Antimon, Wismut, Tellur und Quecksilber im Kupferschiefer des Südostharzvorlands (The toxic compounds in the Kupferschiefer of the southeastern Harz forelands: arsenic, antimony, bismuth, tellurium, and mercury). *Zeitschrift für Angewandte Geologie*, **20**(5), 205–11.

Jung, M.C., Thornton, I. and Chon, H.-T. (2002) Arsenic, Sb and Bi contamination of soils, plants, waters and sediments in the vicinity of the Dalsung Cu-W mine in Korea. *Science of the Total Environment*, **295**(1–3), 81–89.

Juntunen, R., Vartiainen, S. and Pullinen, A. (2004) Arsenic in water from drilled bedrock wells in Pirkanmaa, in *Arseeni Suomen Luonnossa Ymparistovaikutukset Ja Riskit(Arsenic in Finland: Distribution, environmental impacts and risks)* (eds R. K., Loukola and P. Lahermo), Geologian Tutkimuskeskus, Espoo, pp. 111–22.

Kalbitz, K. and Wennrich, R. (1998) Mobilization of heavy metals and arsenic in polluted wetland soils and its dependence on dissolved organic matter. *Science of the Total Environment*, **209**(1), 27–39.

Kanivetsky, R. (2000) *Arsenic in Minnesota Ground Water; Hydrochemical Modeling of the Quaternary Buried Artesian Aquifer and Cretaceous Aquifer Systems*. Report of Investigations- Minnesota Geological Survey, No. 55, St. Paul, MN.

Karayigit, A.I., Spears, D.A. and Booth, C.A. (2000) Antimony and arsenic anomalies in the coal seams from the Gokler coalfield, Gediz, Turkey. *International Journal of Coal Geology*, **44**(1), 1–17.

Kelly, W.R., Holm, T.R., Wilson, S.D. and Roadcap, G.S. (2005) Arsenic in glacial aquifers: sources and geochemical controls. *Ground Water*, **43**(4), 500–10.

Kerr, A. (2004). *An Overview Of Sedimentary-Rock-Hosted Gold Mineralization in Western White Bay (NTS map area 12H/15)*. Current Research Newfoundland Department of Mines and Energy, Geological Survey, Report 04–1, pp. 23–42.

Kerswill, J.A., Kjarsgaard, B.A., Bretzlaff, R. *et al.* (1999) *Metallogeny and Geology of the Half Way Hills Area, Central Churchill Province, Northwest Territories (Nunavut)*. Current Research — Geological Survey of Canada, Current Research 1999-C, pp. 29–41.

Kesler, S.E., Riciputi, L.C. and Ye, Z. (2005) Evidence for a magmatic origin for Carlin-type gold deposits: isotopic composition of sulfur in the Betze-Post-Screamer Deposit, Nevada, USA. *Mineralium Deposita*, **40**, 127–36.

Kim, M.-J., Nriagu, J. and Haack, S. (2002) Arsenic species and chemistry in groundwater of southeast Michigan. *Environmental Pollution*, **120**(2), 379–90.

Kim, M.-J., Nriagu, J. and Haack, S. (2003) Arsenic behavior in newly drilled wells. *Chemosphere*, **52**(3), 623–33.

Kimball, B.A., Callender, E. and Axtmann, E.V. (1995) Effects of colloids on metal transport in a river receiving acid mine drainage, Upper Arkansas River, Colorado, U.S.A. *Applied Geochemistry*, **10**(3), 285–306.

Kizilshtein, L.Ya. and Kholodkov, Yu.I. (1999) Ecologically hazardous elements in coals of the Donets Basin. *International Journal of Coal Geology*, **40**(2–3), 189–97.

Klarup, D.G. (1997) The influence of oxalic acid on release rates of metals from contaminated river sediment. *Science of the Total Environment*, **204**(3), 223–31.

Klinchuch, L.A., Delfino, T.A., Jefferson, J.L. and Waldron, J.M. (1999) Does biodegradation of petroleum hydrocarbons affect the occurrence or mobility of dissolved arsenic in groundwater? *Environmental Geosciences*, **6**(1), 9–24.

Klukanová, A. and Rapant, S. (1999) Impact of mining activities upon the environment of the Slovak Republic: two case studies. *Journal of Geochemical Exploration*, **66**(1–2), 299–306.

Kneebone, P.E. and Hering, J.G. (2000) Behavior of arsenic and other redox-sensitive elements in Crowley Lake, CA: a reservoir in the Los Angeles aqueduct system. *Environmental Science and Technology*, **34**(20), 4307–12.

Komnitsas, K., Xenidis, A. and Adam, K. (1995) Oxidation of pyrite and arsenopyrite in sulphidic spoils in Lavrion. *Mining Engineering*, **8**(12), 1443–54.

Koski, A. and Wood, S. (2003) Aqueous geochemistry of thermal waters in the Alvord Basin, Oregon. *Abstracts with Programs–Geological Society of America*, **35**(6), 407.

Kouprianov, V.I., Bashkin, V.N., Towprayoon, S. *et al.* (2002) Emission of arsenic and gaseous pollutants from power generation in northern Thailand: impact on ecosystems and human health. *World Resources Review*, **14**(1), 99–116.

Kreps, J. 1996. *Impact of historic mining activities on vadose zone pore-water and soil geochemistry*. Master's Thesis, University of Colorado, Boulder, CO, p. 158.

Kubota, Y., Yokota, D. and Ishiyama, Y. (2001) Arsenic concentration in hot spring waters from the Niigata Plain and Shinji Lowland, Japan: source supply of arsenic in arsenic contaminated ground water problem: part 2. *Earth Science*, **55**(1), 11–22.

Kwong, Y.T.J. (2003) Characteristics of impounded tailings at Mount Nansen, Yukon Territory: implications for remediation, in *Proceedings–Assessment and Remediation of Contaminated Sites in Arctic and Cold Climates* (eds M. Nahir, K. Biggar and G. Cotta), University of Alberta, Edmonton, AB, pp. 57–62.

La Force, M.J., Fendorf, S.E., Li, G.C. *et al.* (1998) A laboratory evaluation of trace element mobility from flooding and nutrient loading of Coeur d'Alene River sediments. *Journal of Environmental Quality*, **27**(2), 318–28.

Lake, G.E., Herbert, B.E. and Louchouarn, P. (2001) Quantification of arsenic bioavailability in spatially varying geologic environments at the watershed scale using chelating resin. *Abstracts with Programs–Geological Society of America*, **33**(6), 53.

Langmuir, D., Mahoney, J., MacDonald, A. and Rowson, J. (1999) Predicting arsenic concentrations in the porewaters of buried uranium mill tailings. *Geochimica et Cosmochimica Acta*, **63**(19–20), 3379–94.

Langmuir, D., Mahoney, J. and Rowson, J.W. (2002) Arsenic releases from buried uranium mill tailings at McClean Lake: application of geochemical concepts and license approved by the Canadian government. *Abstracts with Programs–Geological Society of America*, **34**(6), 459.

Lawrence, L.J., Smith-Munro, V., Ramsden, A.R. *et al.* (1999) Geology and mineralogy of the Lorena gold mine, Cloncurry District, northwest Queensland. *Journal and Proceedings of the Royal Society of New South Wales*, **132**(1–2), 29–35.

Lee, C.H. and Lee, H.K. (2004) Environmental impact and geochemistry of old tailing pile from the Sanggok mine creek, Republic of Korea. *Environmental Geology*, **46**(6–7), 727–40.

Lee, C.H., Lee, H.K. and Lee, J.C. (2001) Hydrogeochemistry of mine, surface and groundwaters from the Sanggok Mine Creek in the Upper Chungju Lake, Republic of Korea. *Environmental Geology*, **40**(4–5), 482–94.

Lerouge, C., Bouchot, V. and Guerrot, C. (2000) Fluids and the W (+/-As,Au) ore deposits of the Enguialès-Leucamp District, La Châtaigneraie, French Massif Central. *Journal of Geochemical Exploration*, **69–70**, 343–47.

Lestinen, P., Kontas, E., Niskavaara, H. and Virtasalo, J. (1991) Till geochemistry of gold, arsenic and antimony in the Seinajoki district, western Finland. *Journal of Geochemical Exploration*, **39**(3), 343–61.

Lett, R.E., Bobrowsky, P., Cathro, M. and Yeow, A. (1998) Geochemical pathfinders for massive sulphide deposits in the southern Kootenay Terrane, *Geological Fieldwork 1997*, British Columbia Geological Division, Victoria, BC, 15.1–15.9.

Levy, D.B., Schramke, J.A., Esposito, K.J. *et al.* (1997) Geochemistry of Arsenic and Fluorine in Shallow Ground Water: Eastern Owens Lake, California. 4th International Symposium on Environmental Geochemistry: Proceedings, Vail, Colorado, USA. U.S. Geological Survey Open-File Report, No. 97–0496, p. 53.

Lianfang, W. and Jianzhong, H. (1994) Chronic arsenism from drinking water in some areas of Xinjiang, China, in *Arsenic in the Environment: Part I: Cycling and Characterization* (ed J.O. Nriagu), John Wiley & Sons Ltd, New York, pp. 159–72.

Lintern, M.J. and Butt, C.R.M. (1998) *The Distribution of Gold and Other Elements in Soils at the Granny Smith Gold Deposit, Western Australia.* CRC LEME Open File Report, CSIRO Division of Exploration Geoscience Report 385R, 1st impression, 1993.

Lipfert, G., Reeve, A.S., Sidle, W.C. and Marvinney, R. (2006) Geochemical patterns of arsenic-enriched ground water in fractured, crystalline bedrock, Northport, Maine, USA. *Applied Geochemistry*, **21**(3), 528–45.

Liu, G. (2000) Groundwater environment of Jinan City. *CCOP Technical Bulletin*, **28**, 107–21.

Liu, C.-W., Wang, S.-W., Jang, C.-S. and Lin, K.-H. (2006) Occurrence of arsenic in ground water in the Choushui River Alluvial Fan, Taiwan. *Journal of Environmental Quality*, **35**, 68–75.

Loiselle, M.C., Marvinney, R.G. and Smith, A.E. (2001) Spatial distribution of arsenic in groundwater in Maine. *Abstracts with Programs–Geological Society of America*, **33**(6), 54.

Loredo, J., Ordónez, A., Álvarez, R. and García Iglesias, J. (2004) The potential for arsenic mobilisation in the Caudal River catchment, north-west Spain. *Transactions of the Institution of Mining and Metallurgy, Section B: Applied Earth Science*, **113**(1), B65–75.

Loredo, J., Ordónez, A., Baldo, C. and García-Iglesias, J. (2003) Arsenic mobilization from waste piles of the El Terronal Mine, Asturias, Spain. *Geochemistry: Exploration, Environment, Analysis*, **3**(3), 229–37.

Loredo, J., Ordónez, A., Gallego, J.R. *et al.* (1999) Geochemical characterisation of mercury mining spoil heaps in the area of Mieres (Asturias, northern Spain). *Journal of Geochemical Exploration*, **67**(1–3), 377–90.

Loukola, R.K. and Lahermo, P. (eds) (2004) *Arseeni Suomen Luonnossa Ymparistovaikutukset Ja Riskit(Arsenic in Finland: Distribution, Environmental Impacts And Risks)*, Geologian Tutkimuskeskus, Espoo.

Loukola, R.K., Tanskanen, H. and Lahermo, P. (1999) Anomalously high arsenic concentrations in spring waters in Kittilä, Finnish Lapland. *Special Paper of the Geological Survey of Finland*, No. 27, 97–102.

34333

Lu, Z.M., Liu, Z.J., Zhang, L.H. *et al.* (1998) An investigation on endemic arsenism in Jilin province. *Chinese Journal of Control of Endemic Diseases*, **15**, 77–79.

Lu, X. and Zhang, X. (2005) Environmental geochemistry study of arsenic in western Hunan mining area, P.R. China. *Environmental Geochemistry and Health*, **27**(4), 313–20.

Lutz, T.M. (2001) *Arsenic in Recharged Central Arizona Project Water in Avra Valley*, American Water Resources Association Technical Publication Series TPS 01–1, Marana, AZ, American Water Resources Association: Middleburg, Virginia, pp. 163–68.

Malmström, M.E., Gleisner, M. and Herbert, R.B. (2006) Element discharge from pyritic mine tailings at limited oxygen availability in column experiments. *Applied Geochemistry*, **21**(1), 184–202.

Mandal, B.K. and Suzuki, K.T. (2002) Arsenic round the world: a review. *Talanta*, **58**(1), 201–235.

Márquez, M., Gaspar, J., Bessler, K.E. and Magela, G. (2006) Process mineralogy of bacterial oxidized gold ore in São Bento mine (Brasil). *Hydrometallurgy*, **83**(1–4), 114–23.

Marszalek, H. and Wasik, M. (2000) Influence of arsenic-bearing gold deposits on water quality in Zloty Stok mining area (SW Poland). *Environmental Geology*, **39**(8), 888–92.

Martin, C.J., Al, T.A. and Cabri, L.J. (1997) Surface analysis of particles in mine tailings by time-of-flight laser-ionization mass spectrometry (TOF-LIMS). *Environmental Geology*, **32**(2), 107–13.

Martin, A.J. and Pedersen, T.F. (2002) Seasonal and interannual mobility of arsenic in a lake impacted by metal mining. *Environmental Science and Technology*, **36**(7), 1516–23.

Martley, E., Gulson, B.L. and Pfeifer, H.-R. (2004) Metal concentrations in soils around the copper smelter and surrounding industrial complex of Port Kembla, NSW, Australia. *Science of the Total Environment*, **325**(1–3), 113–27.

Mascaro, I., Benvenuti, B., Corsini, F. *et al.* (2001) Mine wastes at the polymetallic deposit of Fenice Capanne (southern Tuscany, Italy). Mineralogy, geochemistry, and environmental impact. *Environmental Geology*, **41**(3–4), 417–29.

Masuda, H., Ibuki, Y. and Tonokai, K. (1999) Mechanism of natural arsenic pollution of shallow groundwater in the northern part of Osaka Prefecture, Japan. *Journal of Groundwater Hydrology*, **41**(3), 133–46.

Mattu, G.S., Clarkson, L. and Carmichael, V. (2000) An assessment of spatial and temporal variations of arsenic in groundwater in the Sunshine Coast and Powell River regions of British Columbia. *Abstracts with Programs–Geological Society of America*, **32**(6), 29.

May, T.W., Wiedmeyer, R.H., Gober, J. and Larson, S. (2001) Influence of mining-related activities on concentrations of metals in water and sediment from streams of the Black Hills, South Dakota. *Archives of Environmental Contamination and Toxicology*, **40**(1), 1–9.

McCall, P.J., Walter, L.M. and Szramek, K.J. (2002) Arsenic sources and sinks in a surface water/groundwater system: tracking recharge to discharge in glacial drift deposits (Hell, Michigan). *Abstracts with Programs–Geological Society of America*, **34**(6), 294–95.

McDonough, M.R. (1997) Structural controls and age constraints on sulphide mineralization, southern Taltson magmatic zone, northeastern Alberta. *Bulletin of the Geological Survey of Canada*, **500**, 13–29.

McGregor, R.G., Blowes, D.W., Jambor, J.L. and Robertson, W.D. (1998) The solid-phase controls on the mobility of heavy metals at the Copper Cliff tailings area, Sudbury, Ontario, Canada. *Journal of Contaminant Hydrology*, **33**(3–4), 247–71.

Meck, M., Love, D. and Mapani, B. (2006) Zimbabwean mine dumps and their impacts on river water quality –a reconnaissance study. *Physics and Chemistry of the Earth*, **31**(15–16), 797–803.

Mehrabi, B., Yardley, B.W.D. and Cann, J.R. (1999) Sediment-hosted disseminated gold mineralisation at Zarshuran, NW Iran. *Mineralium Deposita*, **34**(7), 673–96.

Meladiotis, I., Veranis, N. and Nikolaidis, N. (2002) Arsenic contamination in central Macedonia, northern Greece: extent of the problem and potential solutions, in *Protection and Restoration of the Environment: Proceedings of an International Conference,* Skiathos, Greece (eds A.G. Kungolos, A.B. Liakopoulos, G.P. Korfiatis *et al.*), Vol **6** (2), Stevens Institute of Technology: Hoboken, New Jersey, USA. 913–20.

Melwani, A.R. and Forrest, M.J. (2003) Patterns in macrofaunal abundance and diversity at a shallow-water hydrothermal vent near Punta Santa Barbara, Bahia Concepcion, BCS, Mexico. *Abstracts with Programs–Geological Society of America*, **35**(6), 492.

Mendoza, O.T., Herna'ndez, Ma.A.A., Abundis, J.G. and Mundo, N.F. (2006) Geochemistry of leachates from the El Fraile sulfide tailings piles in Taxco, Guerrero, southern Mexico. *Environmental Geochemistry and Health*, **28**(3), 243–55.

Mickey, J.W. and Moore, J.N. (1997) The effects of discharge variation on the dissolved concentrations of trace metals and arsenic through Milltown Reservoir, Montana. *Abstracts with Programs–Geological Society of America*, **29**(6), 149.

Migon, C. and Mori, C. (1999) Arsenic and antimony release from sediments in a Mediterranean estuary. *Hydrobiologia*, **392**(1), 81–88.

Milintawisamai, M., Boonchaleamkit, S., Fukuda, M. and TabucAnon, M.S. (1998) Application of isotope techniques to the study of groundwater pollution by arsenic in Nakorn Si Thammarat Province, Thailand, *Isotope Techniques in the Study of Environmental Change: Proceedings Series*, International Atomic Energy Agency, pp. 473–481.

Milodragovich, E.L. (2003) *Hydrogeochemistry of a natural wetland receiving acid mine drainage, Ontario Mine, Powell County, Montana*. Master's Thesis. University of Montana, Butte, MT, p. 113.

Minnesota Pollution Control Agency (1998) *Arsenic in Minnesota's Ground Water*, Ground Water Monitoring and Assessment Program, St. Paul, MN.

Mitsunobu, S., Harada, T. and Takahashi, Y. (2006) Comparison of antimony behavior with that of arsenic under various soil redox conditions. *Environmental Science and Technology*, **40**(23), 7270–76.

Moncur, M.C., Ptacek, C.J., Blowes, D.W. *et al.* (2002) Impact of mine drainage on a lake from an abandoned tailings impoundment. *Abstracts with Programs–Geological Society of America*, **34**(6), 51.

Montgomery, D.L., Ayotte, J.D., Carroll, P.R. and Hamlin, P. (2003) *Arsenic Concentrations in Private Bedrock Wells in Southeastern New Hampshire*, U.S. Geological Survey Fact Sheet, p. 6.

Moon, C.J. (2002) The spatial distribution of arsenic in East Cornwall: geological and anthropogenic signatures. *Proceedings of the Ussher Society*, **10**(3), 343–51.

Moore, J.N., Ficklin, W.H. and Johns, C. (1988) Partitioning of arsenic and metals in reducing sulfidic sediments. *Environmental Science and Technology*, **22**(4), 432–37.

Moore, J.N. and Woessner, W.W. (2003) Arsenic contamination in the water supply of Milltown, Montana, in *Arsenic in Ground Water* (eds A.H. Welch and K.G. Stollenwerk), Kluwer Academic Publishers, Boston, MA, pp. 329–50.

Morin, G., Juillot, F., Casiot, C. *et al.* (2003) Bacterial formation of tooeleite and mixed arsenic(III) or arsenic(V)–iron(III) gels in the Carnoule's acid mine drainage, France. A XANES, XRD, and SEM study. *Environmental Science and Technology*, **37**(9), 1705–12.

Morton, O., Carrillo-Chávez, A. and Hernandez, E. (2000) Geochemical characterization of historical mine tailings and groundwater in the Guanajuato mining district, central Mexico: environmental considerations. *Abstracts with Programs–Geological Society of America*, **32**(7), 488.

Motyka, R.J., Hawkins, D.B., Poreda, R.J. and Jeffries, A. (1986) *Geochemistry, Isotopic Composition, and the Origin of Fluids Emanating from Mud Volcanoes in the Copper River Basin, Alaska*, Alaska Division of Geological and Geophysical Surveys, Public-data File 86–34, p. 87.

Mroczek, E.K. (2005) Contributions of arsenic and chloride from the Kawerau geothermal field to the Tarawera River, New Zealand. *Geothermics*, **34**(2 Special Issue), 223–38.

Mudroch, A., Joshi, S.R., Sutherland, D. *et al.* (1989) Geochemistry of sediments in the Back Bay and Yellowknife Bay of Great Slave Lake. *Environmental Geology and Water Sciences*, **14**(1), 35–42.

Mueller, S.H., Goldfarb, R.J., Farmer, G.L. *et al.* (2001b) A seasonal study of the arsenic and groundwater geochemistry in Fairbanks, Alaska, in *Mineral Deposits at the Beginning of the 21st Century* (eds A. Piestrzynski *et al.*), Society for Geology Applied to Mineral Deposits, Swets & Zeitlinger Publishers, Lisse, pp. 1043–46.

Mueller, S., Goldfarb, R. and Verplanck, P. (2001a) *Ground-water Studies in Fairbanks, Alaska — A Better Understanding of Some of the United States' Highest Natural Arsenic Concentrations.*, U. S. Geological Survey Fact Sheet, FS–111–01.

Mukherjee, A.B., Bhattacharya, P., Jacks, G. (2006) Groundwater arsenic contamination in India, in *Managing Arsenic in the Environment: From Soil to Human Health* (eds R. Naidu, E. Smith, G. Owens, P. Bhattacharya and P. Nadebaum), CSIRO Publishing, Collingwood, pp. 553–93.

Mukherjee, S. and Srivastava, S.K. (2005) Trace elements in high-sulfur Assam coals from the Makum coalfield in the northeastern region of India. *Energy & Fuels*, **19**(3), 882–91.

Mukhopadhyay, P.K., Goodarzi, F., Crandlemire, A.L. *et al.* (1998) Comparison of coal composition and elemental distribution in selected seams of the Sydney and Stellarton Basins, Nova Scotia, eastern Canada. *International Journal of Coal Geology*, **37**(1–2), 113–41.

Muloin, T. and Dudas, M.J. (2005) Aqueous phase arsenic in weathered shale enriched in native arsenic. *Journal of Environmental Engineering and Science*, **4**(6), 461–68.

Naseem, M., Farooqi, A., Masih, D. and Anwar, M. (2001) Investigation of Toxic Elements in the Ground Water of Kalalanwala Area Near Lahore, Punjab, Pakistan. Proceedings of GEOSAS-III, Lahore, Pakistan.

Nasution, A.H., Muraoka, I. *et al.* (1999). Preliminary survey of Bejawa geothermal area, Ngada district, Flores, East Nusa Tenggara, Indonesia. *Transactions - Geothermal Resources Council*, **23**, 467–72.

Nath, B., Berner, Z., Mallik, S.B. *et al.* (2005) Characterization of aquifers conducting groundwaters with low and high arsenic concentrations: a comparative case study from West Bengal, India. *Mineralogical Magazine*, **69**(5), 841–54.

Néel, C., Bril, H., Courtin-Nomade, A. and Dutreuil, J.-P. (2003) Factors affecting natural development of soil on 35-year-old sulphide-rich mine tailings. *Geoderma*, **111**(1–2), 1–20.

Nicholson, A.D., Hanna, T., Mansanti, J. *et al.* (1996) Evolution of groundwater chemistry during dewatering: Lone Tree Mine, Nevada. *Abstracts with Programs–Geological Society of America*, **28**(7), 467.

Nickson, R.T., McArthur, J.M., Shrestha, B. *et al.* (2005) Arsenic and other drinking water quality issues, Muzaffargarh district, Pakistan. *Applied Geochemistry*, **20**(1), 55–68.

Nimick, D.A. (1998) Arsenic hydrogeochemistry in an irrigated river valley –a reevaluation. *Ground Water*, **36**(5), 743–53.

Nimick, D.A., Moore, J.N., Dalby, C.E. and Savka, M.W. (1998) The fate of geothermal arsenic in the Madison and Missouri Rivers, Montana and Wyoming. *Water Resources Research*, **34**(11), 3051–67.

Nishikawa, O., Okrugin, V., Belkova, N. *et al.* (2006) Crystal symmetry and chemical composition of yukonite: TEM study of specimens collected from Nalychevskie hot springs, Kamchatka, Russia and from Venus mine, Yukon Territory, Canada. *Mineralogical Magazine*, **70**(1), 73–81.

Niskavaara, H., Reimann, C. and Chekushin, V. (1996) Distribution and pathways of heavy metals and sulphur in the vicinity of the copper-nickel smelters in Nikel and Zapoljarnij, Kola Peninsula, Russia, as revealed by different sample media. *Applied Geochemistry*, **11**(1–2), 25–34.

Noguchi, K. and Nakagawa, R. (1994a) Arsenic in thermal waters and deposits of Manza Hot Springs, Gumma, Japan, *Geochemical Studies of Volcanoes, Hot Springs, Brine Waters of Oil Fields and Ground Waters*, Tokyo Metropolitan University, Tokyo, Vol. II, pp. 159–63.

Noguchi, K. and Nakagawa, R. (1994b) Arsenic content of the hot springs water in Narugo and Onikobe areas, Miyagi, Japan, *Geochemical Studies of Volcanoes, Hot Springs, Brine Waters of Oil Fields and Ground Waters*, Tokyo Metropolitan University, Tokyo, Vol II, pp. 33–34.

Noguchi, K. and Nakagawa, R. (1994c) Arsenic in waters and deposits of Osoreyama Hot Springs, Aomori Prefecture, *Geochemical Studies of Volcanoes, Hot Springs, Brine Waters of Oil Fields and Ground Waters*, Tokyo Metropolitan University, Tokyo, Vol II, pp. 3–8.

O'Brien, S.J., Dubé, B. and O'Driscoll, C.F. (1999) *High-Sulphidation, Epithermal-Style Hydrothermal Systems in Late Neoproterozoic Avalonian Rocks on the Burin Peninsula, Newfoundland: Implications for Gold Exploration*. Geological Survey Report 99–1, Current Research Newfoundland Department of Mines and Energy, 275–96.

Ódor, L., Wanty, R.B., Horváth, I. and Fügedi, U. (1998) Mobilization and attenuation of metals downstream from a base-metal mining site in the Matra Mountains, northeastern Hungary. *Journal of Geochemical Exploration*, **65**(1 pt 2), 47–60.

Öhlander, B., Müller, B., Axelsson, M. and Alakangas, L. (2007) An attempt to use LA-ICP-SMS to quantify enrichment of trace elements on pyrite surfaces in oxidizing mine tailings. *Journal of Geochemical Exploration*, **92**(1), 1–12.

Ohnuki, T., Sakamoto, F., Kozai, N. *et al.* (2004) Mechanisms of arsenic immobilization in a biomat from mine discharge water. *Chemical Geology*, **212**(3–4 Special Issue), 279–90.

Ollivier, P., Radakovitch, O. and Hamelin, B. (2006) Unusual variations of dissolved As, Sb, and Ni in the Rhône River during flood events. *Journal of Geochemical Exploration*, **88**, 394–98.

Ollson, C.A., Reimer, K.J., Koch, I. and Cullen, W.R. (2003) Contaminant bioavailability and its consequences for ecological and human health risk assessment case study: arsenic in Yellowknife, NWT, in *Proceedings — Assessment and Remediation of Contaminated Sites in Arctic and Cold Climates* (eds M. Nahir, K. Biggar and G., Cotta) University of Alberta, Edmonton, AB, Vol. **3**, pp. 129–38.

Oremland, R.S., Dowdle, P.R., Hoeft, S. *et al.* (2000) Bacterial dissimilatory reduction of arsenate and sulfate in meromictic Mono Lake, California. *Geochimica et Cosmochimica Acta*, **64**(18), 3073–84.

Ortega-Guerrero, A. (2003) Arsenic in groundwater at the southernmost end of the Cordilleran. *Abstracts with Programs–Geological Society of America*, **35**(4), 24.

ÓShea, B., Jankowski, J. and Sammut, J. (2007) The source of naturally occurring arsenic in a coastal sand aquifer of eastern Australia. *Science of the Total Environment*, **379**(2–3), 151–66.

Oyarzun, R., Guevara, S., Oyarzu'n, J. *et al.* (2006) The As-contaminated Elqui River Basin: a long lasting perspective (1975–1995) covering the initiation and development of Au-Cu-As mining in the high Andes of northern Chile. *Environmental Geochemistry and Health*, **28**(5), 431–43.

Pandey, P.K., Sharma, R., Roy, M. *et al.* (2006) Arsenic contamination in the Kanker district of central-east India: geology and health effects. *Environmental Geochemistry and Health*, **28**(5), 409–20.

Panov, B.S., Dudik, A.M., Shevchenko, O.A. and Matlak, E.S. (1999) On pollution of the biosphere in industrial areas: the example of the Donets coal Basin. *International Journal of Coal Geology*, **40**(2–3), 199–210.

Panthi, S.R., Sharma, S. and Mishra, A.K. (2006) Recent status of arsenic contamination in groundwater of Nepal — a review. *Kathmandu University –Journal of Science, Engineering and Technology*, **2**, 1–11.

Park, J.M., Lee, J.S., Lee, J.-U. *et al.* (2006) Microbial effects on geochemical behavior of arsenic in As-contaminated sediments. *Journal of Geochemical Exploration*, **88**, 134–38.

Patel, K.S., Shrivas, K., Brandt, R. *et al.* (2005) Arsenic contamination in water, soil, sediment and rice of central India. *Environmental Geochemistry and Health*, **27**(2), 131–45.

Penczak, R.S. and Mason, R. (1997) Metamorphosed Archean epithermal Au-As-Sb-Zn-(Hg) vein mineralization at the Campbell mine, northwestern Ontario. *Economic Geology*, **92**(6), 696–719.

Pentari, D., Foscolos, A.E. and Perdikatsis, V. (2004) Trace element contents in the Domeniko lignite deposit, Elassona Basin, central Greece. *International Journal of Coal Geology*, **58**(4), 261–68.

Pentcheva, E.N., Velitchkova, N. and Karadjov M. (2003) Occurrence and Evolution of Heavy Metals and as Pollution in Waters and Vegetables in Plovdiv-Assenovgrad Area. International Conference Modern Management of Mine Producing, Geology and Environmental Protection 3, June 2003, Varna city, Bulgaria, pp. 295–304.

Peters, S.C., Blum, J.D., Karagas, M.R. *et al.* (2006) Sources and exposure of the New Hampshire population to arsenic in public and private drinking water supplies. *Chemical Geology*, **228**, 72–84.

Petersen, S.W., Hoover, J.D. and McKinley, J.P. (1997) Determination and use of groundwater background chemistry in the Pasco Basin, south-central Washington. *Abstracts with Programs–Geological Society of America*, **29**(6), 435.

Petrunic, B.M. and Al, T.A. (2005) Mineral/water interactions in tailings from a tungsten mine, Mount Pleasant, New Brunswick. *Geochimica et Cosmochimica Acta*, **69**(10), 2469–83.

Pfeifer, H.-R., Gueye-Girardet, A., Reymond, D. *et al.* (2004) Dispersion of natural arsenic in the Malcantone watershed, southern Switzerland: field evidence for repeated sorption-desorption and oxidation-reduction processes. *Geoderma*, **122**(2–4 Special Issue), 205–34.

Pfeifer, H.-R. and Zobrist, J. (2002) Arsenic in deep groundwater of Switzerland and their environmental impact and health risk. *Geochimica et Cosmochimica Acta*, **66**(15A), 597.

Phoenix, V.R., Renaut, R.W., Jones, B. and Ferris, F.G. (2005) Bacterial S-layer preservation and rare arsenic-antimony-sulphide bioimmobilization in siliceous sediments from Champagne pool hot spring, Waiotapu, New Zealand. *Journal of the Geological Society* **162**(2), 323–31.

Pichler, T., Hendry, M.J. and Hall, G.E.M. (2001) The mineralogy of arsenic in uranium mine tailings at the Rabbit Lake in-pit facility, northern Saskatchewan, Canada. *Environmental Geology*, **40**(4–5), 495–506.

Pimentel, H.S., De Lena, J.C. and Nalini, H.A., Jr. (2003) Studies of water quality in the Ouro Preto region, Minas Gerais, Brazil: the release of arsenic to the hydrological system. *Environmental Geology*, **43**(6), 725–30.

Pirrie, D., Power, M.R., Wheeler, P.D. *et al.* (2002) Geochemical signature of historical mining: Fowey Estuary, Cornwall, UK. *Journal of Geochemical Exploration*, **76**(1), 31–43.

Planer-Friedrich, B., Armienta, M.A. and Merkel, B.J. (2001) Origin of arsenic in the groundwater of the Rioverde Basin, Mexico. *Environmental Geology*, **40**(10), 1290–98.

Planer-Friedrich, B., Lehr, C., Matschullat, J. *et al.* (2006) Speciation of volatile arsenic at geothermal features in Yellowstone National Park. *Geochimica et Cosmochimica Acta*, **70**(10), 2480–91.

Polya, D.A., Gault, A.G., Diebe, N. *et al.* (2005) Arsenic hazard in shallow Cambodian groundwaters. *Mineralogical Magazine*, **69**(5), 807–23.

Power, M.R., Pirrie, D., Jedwab, J. and Stanley, C.J. (2004) Platinum-group element mineralization in an as-rich magmatic sulphide system, Talnotry, southwest Scotland. *Mineralogical Magazine*, **68**(2), 395–411.

Price, R.E. and Pichler, T. (2005) Distribution, speciation and bioavailability of arsenic in a shallow-water submarine hydrothermal system, Tutum Bay, Ambitle Island, PNG. *Chemical Geology*, **224**(1–3), 122–35.

Price, R.E. and Pichler, T. (2006) Abundance and mineralogical association of arsenic in the Suwannee Limestone (Florida): implications for arsenic release during water-rock interaction. *Chemical Geology*, **228**(1–3 Special Issue), 44–56.

Price, V., Temples, T., McCary, L.T. *et al.* (2004) Investigations of high uranium contents in domestic water wells, South Carolina Piedmont. *Abstracts with Programs–Geological Society of America*, **36**(5), 243.

Pršek, J., Mikuš, T., Makovicky, E. and Chovan, M. (2005) Cuprobismutite, kupčíkite, hodrushite and associated sulfosalts from the black shale hosted Ni-Bi-As mineralization at Cierna Lehota, Slovakia. *European Journal of Mineralogy*, **17**(1), 155–62.

Querol, X., Whateley, M.K.G., Fernández-Turiel, J.L. and Tuncali, E. (1997) Geological controls on the mineralogy and geochemistry of the Beypazari lignite, central Anatolia, Turkey. *International Journal of Coal Geology*, **33**(3), 255–71.

Rahman, M.M., Sengupta, M.K., Chowdhury, U.K. (2006) Arsenic contamination incidents around the world, in *Managing Arsenic in the Environment: From Soil to Human Health* (eds R. Naidu, E. Smith, G. Owens, P. Bhattacharya and P., Nadebaum), CSIRO Publishing, Collingwood, pp. 3–30.

Rapant, S. and Krčmová, K. (2007) Health risk assessment maps for arsenic groundwater content: application of national geochemical databases. *Environmental Geochemistry and Health*, **29**(2), 131–41.

Rasilainen, K., Nurmi, P.A. and Bornhorst, T.J. (1993) Rock geochemical implications for gold exploration in the late Archean Hattu Schist belt, Ilomantsi, eastern Finland. *Special Paper of the Geological Survey of Finland*, **17**, 353–62.

Ravenscroft, P., Burgess, W.G., Ahmed, K.M. *et al.* (2005) Arsenic in groundwater of the Bengal Basin, Bangladesh: distribution, field relations, and hydrogeological setting. *Hydrogeology Journal*, **13**(5–6), 727–51.

Razo, I., Carrizales, L., Castro, J. *et al.* (2004) Arsenic and heavy metal pollution of soil, water and sediments in a semi-arid climate mining area in Mexico. *Water Air and Soil Pollution*, **152**(1–4), 129–52.

Redente, E.F., Zadeh, H. and Paschke, M.W. (2002) Phytoxicity of smelter-impacted soils in southwest Montana, USA. *Environmental Toxicology and Chemistry*, **21**(2), 269–74.

Reid, K.D., Goff, F. and Counce, D.A. (2003) Arsenic concentration and mass flow rate in natural waters of the Valles Caldera and Jemez Mountains region, New Mexico. *New Mexico Geology*, **25**(3), 75–81.

Reith, F., McPhail, D.C. and Christy, A.G. (2005) Bacillus cereus, gold and associated elements in soil and other regolith samples from Tomakin Park gold mine in southeastern New South Wales, Australia. *Journal of Geochemical Exploration*, **85**(2), 81–98.

Riemann, U. (1999) Impacts of urban growth on surface water and groundwater quality in the city of Dessau, Germany, in *Impacts of Urban Growth on Surface Water and Groundwater Quality: Proceedings of an International Symposium Held during IUGG 99* (eds E.J. Bryan), IAHS-AISH Publication, Vol. **259**, pp. 307–14.

Robb, L.J., Freeman, L.A. and Armstrong, R.A. (2001) Nature and longevity of hydrothermal fluid flow and mineralisation in granites of the Bushveld Complex, South Africa, in *Fourth Hutton Symposium: The origin of granites and related rocks: Proceedings* (eds B. Barbarin, W. E. Stephens, B. Bonin *et al.*) Geological Society of America, Special Paper, Vol. **350**, pp. 269–281.

Rodriguez, R., Armienta, A., Berlin, J. and Mejia, J.A. (2002) Arsenic and lead pollution of the Salamanca Aquifer, Mexico: origin, mobilization and restoration alternatives, in *Groundwater Quality: Natural and Enhanced Restoration of Groundwater Pollution* (eds S.F. Thornton and S.E., Oswald), IAHS-AISH Publication, Vol. **275**, pp. 561–65.

Romero, L., Alonso, H., Campano, P. *et al.* (2003) Arsenic enrichment in waters and sediments of the Rio Loa (Second Region, Chile). *Applied Geochemistry*, **18**(9), 1399–416.

Root, T.L., Bahr, J. and Gotkowitz, M.G. (2003) Arsenic in groundwater in southeastern Wisconsin: sources of arsenic and mechanisms controlling arsenic mobility. *Abstracts with Programs–Geological Society of America*, **35**(6), 575.

Rösner, U. (1998) Effects of historical mining activities on surface water and groundwater – An example from northwest Arizona. *Environmental Geology*, **33**(4), 224–30.

Routh, J., Bhattacharya, A., Saraswathy, A. *et al.* (2007) Arsenic remobilization from sediments contaminated with mine tailings near the Adak mine in Västerbotten District (northern Sweden). *Journal of Geochemical Exploration*, **92**(1), 43–54.

Rubinos, D., Barral, M.T., Ruíz, B. *et al.* (2003) Phosphate and arsenate retention in sediments of the Anllóns River (northwest Spain). *Water Science and Technology*, **48**(10), 159–66.

Ruiz-Chancho, M.J., López-Sánchez, J.F. and Rubio, R. (2007) Analytical speciation as a tool to assess arsenic behaviour in soils polluted by mining. *Analytical and Bioanalytical Chemistry*, **387**(2), 627–35.

Ryker, S. (2001) Mapping arsenic in groundwater. *Geotimes*, **46**(11), 34–36.

Ryu, J.-H., Gao, S., Dahlgren, R.A. and Zierenberg, R.A. (2002) Arsenic distribution, speciation and solubility in shallow groundwater of Owens Dry Lake, California. *Geochimica et Cosmochimica Acta*, **66**(17), 2981–94.

Sadiq, M., Locke, A., Spiers, G. and Pearson, D.A.B. (2002) Geochemical behavior of arsenic in Kelly Lake, Ontario. *Water Air and Soil Pollution*, **141**(1–4), 299–312.

Sahoo, N.R. and Pandalai, H.S. (2000) Secondary geochemical dispersion in the Precambrian auriferous Hutti-Maski schist belt, Raichur district, Karnataka, India. Part II. Application of factorial design in the analysis of secondary dispersion of As. *Journal of Geochemical Exploration*, **71**(3), 291–303.

Sakulpitakphon, T., Hower, J.C., Schram, W.H. and Ward, C.R. (2000) Geochemistry of the Manchester Coal Bed, Clay County, Kentucky. *Abstract with Programs–Geological Society of America*, **32**(2), 70–71.

Salzsauler, K.A., Sidenko, N.V. and Sherriff, B.L. (2005) Arsenic mobility in alteration products of sulfide-rich, arsenopyrite-bearing mine wastes, Snow Lake, Manitoba, Canada. *Applied Geochemistry*, **20**(12), 2303–14.

Samecka-Cymerman, A. and Kempers, A.J. (1994) Aquatic bryophytes as bioindicators of arsenic mineralization in Polish and Czech Sudety Mountains. *Journal of Geochemical Exploration*, **51**(3), 291–97.

Sanderson, M., Karrow, P.F., Greenhouse, J.P. *et al.* (1994) Susceptibility of groundwater to contamination in Kitchener-Waterloo: a case study with policy implications. *Program with Abstracts –Geological Association of Canada, Mineralogical Association of Canada, and Canadian Geophysical Union*, **19** 98.

Savage, K.S., Ashley, R. and Bird, D. (2001) Mineral contributions to pit lake arsenic at the Jamestown Mine, California. *Abstracts with Programs–Geological Society of America*, **33**(6), 117.

Savage, K.S., Bird, D.K. and Ashley, R.P. (2000a) Legacy of the California gold rush: environmental geochemistry of arsenic in the southern mother lode gold district. *International Geology Review*, **42**(5), 385–415.

Savage, K.S., Tingle, T.N., O'Day, P.A. *et al.* (2000b) Arsenic speciation in pyrite and secondary weathering phases, Mother Lode Gold District, Tuolumne County, California. *Applied Geochemistry*, **15**(8), 1219–44.

Schreiber, M.E., Gotkowitz, M.B., Simo, J.A. and Freiberg, P.G. (2003) Mechanisms of arsenic release to water from naturally occurring sources, eastern Wisconsin, in *Arsenic in Ground Water* (eds A.H. Welch and K.G. Stollenwerk), Kluwer Academic Publishers, Boston, MA, pp. 259–80.

Schreiber, M.E., Simo, J.A. and Freiberg, P.G. (2000) Stratigraphic and geochemical controls on naturally occurring arsenic in groundwater, eastern Wisconsin, USA. *Hydrogeology Journal*, **8**(2), 161–76.

Scott, K.M. (1986) Sulphide geochemistry and wall rock alteration as a guide to mineralization, Mammoth Area, NW Queensland, Australia. *Journal of Geochemical Exploration*, **25**, 283–308.

Seki, Y. (2000) Hydrothermal alteration of the Sunagohara formation in the Okuaizu geothermal area, Japan: alteration of lacustrine sediments formed by the present geothermal activity. *Bulletin of the Geological Survey of Japan*, **51**(8), 329–67.

Serfor-Armah, Y., Nyarko, B.J.B., Adotey, D.K. *et al.* (2006) Levels of arsenic and antimony in water and sediment from Prestea, a gold mining town in Ghana and its environs. *Water Air and Soil Pollution*, **175**(1–4), 181–92.

Shahnewaz, M., Saunders, J.A., Lee, M.K. *et al.* (2003) Naturally occurring arsenic in Holocene alluvial aquifer and is implications for biogeochemical linkage among arsenic, iron and sulfur form source to sink. *Abstracts with Programs–Geological Society of America*, **35**(6), 247.

Shallow, L.J. (1999) *Refractory ores at the Rain Mine, Elko County, Nevada: structural controls, wallrock alteration, petrography and geochemistry*. Master's Thesis, University of Nevada, Reno, NV.

Shiber, J.G. (2005) Arsenic in domestic well water and health in central Appalachia, USA. *Water Air and Soil Pollution*, **160**, 327–41.

Shih, C.-J. and Lin, C.-F. (2003) Arsenic contaminated site at an abandoned copper smelter plant: waste characterization and solidification/stabilization treatment. *Chemosphere*, **53**(7), 691–703.

Sidle, W.C., Wotten, B. and Murphy, E. (2001) Provenance of geogenic arsenic in the Goose River Basin, Maine, USA. *Environmental Geology*, **41**(1–2), 62–73.

Sjöblom, A., Håkansson, K. and Allard, B. (2004) River water metal speciation in a mining region –the influence of wetlands, liming, tributaries, and groundwater. *Water Air and Soil Pollution*, **152**(1–4), 173–94.

Skanes, M., Kerr, A. and Sylvester, P.J. (2004) *The VBE-2 Gold Prospect, Northern Labrador: Geology, Petrology and Mineral Geochemistry*. Geological Survey, Report 04–1, Current Research Newfoundland Department of Mines and Energy, pp. 43–61.

Šlejkovec, Z. and Kanduč, T. (2005) Unexpected arsenic compounds in low-rank coals. *Environmental Science and Technology*, **39**(10), 3450–54.

Smedley, P.L. (1996) Arsenic in rural groundwater in Ghana. *Journal of African Earth Sciences*, **22**(4), 459–70.

Smedley, P.L. (2003) Arsenic in groundwater –south and east Asia, in *Arsenic in Ground Water* (eds A.H. Welch and K.G. Stollenwerk), Kluwer Academic Publishers, Boston, pp. 179–209.

Smedley, P. (2005) *Arsenic Occurrence in Groundwater in South and East Asia — Scale, Causes and Mitigation.Towards a More Effective Operational Response: Arsenic Contamination of Groundwater in South and East Asian Countries II*. Technical Report, World Bank Report No. 31303.

Smedley, P.L. and Kinniburgh, D.G. (2002) A review of the source, behaviour and distribution of arsenic in natural waters. *Applied Geochemistry*, **17**(5), 517–68.

Smedley, P.L., Nicolli, H.B., Macdonald, D.M.J. *et al.* (2002) Hydrogeochemistry of arsenic and other inorganic constituents in groundwaters from La Pampa, Argentina. *Applied Geochemistry*, **17**(3), 259–84.

Smedley, P.L., Zhang, M., Zhang, G. and Luo, Z. (2003) Mobilisation of arsenic and other trace elements in fluvio-lacustrine aquifers of the Huhhot Basin, Inner Mongolia. *Applied Geochemistry*, **18**(9), 1453–77.

Smith S.J. (2005). *Naturally Occurring Arsenic in Ground Water, Norman, Oklahoma, 2004, and Remediation Options for Produced Water*, U. S. Geological Survey, Fact Sheet 2005–3111.

Smith, J.V.S., Jankowski, J., Sammut, J. (2003) Vertical distribution of As(III) and As(V) in a coastal sandy aquifer: factors controlling the concentration and speciation of arsenic in the Stuarts Point groundwater system, northern New South Wales, Australia. *Applied Geochemistry*, **18**(9), 1479–96.

Smith, J.V.S., Jankowski, J. and Sammut, J. (2006) Natural occurrences of inorganic arsenic in the Australian coastal groundwater environment, in *Managing Arsenic in the Environment: From Soil to Human Health* (eds R. Naidu, E. Smith, G., Owens *et al.*), CSIRO Publishing, Collingwood, pp. 129–53.

Solorzano, H. Carrillo-Chávez, A. (2000) Experimental and modeled adsorption of Cd, Pb, Cu, As, and Zn onto natural sediments from the San Nicolas-Mata Dam Basin, Guanajuato: natural control of heavy metals in the environment. *Abstracts with Programs–Geological Society of America*, **32**(7), 488.

Sonderegger, J.L. Ohguchi, T. (1988) Irrigation related arsenic contamination of a thin, alluvial aquifer, Madison River valley, Montana. *Environmental Geology and Water Sciences*, **11**(2), 153–61.

Sonderegger, J.L. Sholes, B.R. (1989) Arsenic contamination of aquifers caused by irrigation with diluted geothermal water. *Abstracts with Programs–Geological Society of America*, **21**(5), 147

Sriwana, T., Van Bergen, M.J., Sumarti, S. *et al.* (1998) Volcanogenic pollution by acid water discharges along Ciwidey River, West Java (Indonesia). *Journal of Geochemical Exploration*, **62**(1–3), 161–82

Stanton, M.R., Sanzolone, R.E., Sutley, S.J. *et al.* (2001) Abundance, residence, and mobility of arsenic in Santa Fe Group sediments, Albuquerque Basin, New Mexico. *Abstracts with Programs–Geological Society of America*, **33**(5), 2

Stauffer, R.E. Thompson, J.M. (1984) Arsenic and antimony in geothermal waters of Yellowstone National Park, Wyoming, U.S.A. *Geochimica et Cosmochimica Acta*, **48**(12), 2547–61

Stenger, D.P., Kesler, S.E., Peltonen, D.R. Tapper, C.J. (1998) Deposition of gold in Carlin-type deposits: the role of sulfidation and decarbonation at Twin Creeks, Nevada. *Economic Geology and the Bulletin of the Society of Economic Geologists*, **93**(2), 201–15

Sterckeman, T., Douay, F., Proix, N. *et al.* (2002) Assessment of the contamination of cultivated soils by eighteen trace elements around smelters in the north of France. *Water Air and Soil Pollution*, **135**(1–4), 173–94.

Subba, R.C. and Subba, R.N.V. (1999) The impact of urban industrial growth on groundwater quality in Visakhapatnam, India, in *Impacts of Urban Growth on Surface Water and Groundwater Quality: Proceedings of an International Symposium Held during IUGG 99* (ed E. J. Bryan), IAHS-AISH Publication, Vol. **259**, pp. 203–9.

Tandukar, N., Bhattacharya, P., Neku, A. and Mukherjee, A.B. (2006) Extent and severity of arsenic occurrence in groundwater of Nepal, in *Managing Arsenic in the Environment: From Soil to Human Health* (eds R. Naidu, E. Smith, G. Owens, P., Bhattacharya and P., Nadebaum), CSIRO Publishing, Collingwood, pp. 541–52.

Tanskanen, H., Lahermo, P., Loukola, R. K. *et al.* (eds) (2004) Arseeni Kittilan pohjavesissa keski-lapissa (Arsenic in ground water in Kittila, Finnish Lapland), in *Arseeni Suomen Luonnossa Ymparistovaikutukset Ja Riskit (Arsenic in Finland: Distribution, Environmental Impacts and Risks)*, Geologian Tutkimuskeskus, Espoo, pp. 123–34.

Taylor, H.E., Antweiler, R.C., Roth, D.A. *et al.* (2001) The occurrence and distribution of selected trace elements in the Upper Rio Grande and tributaries in Colorado and northern New Mexico. *Archives of Environmental Contamination and Toxicology*, **41**, 410–26.

Tazaki, K., Okrugin, V., Okuno, M. *et al.* (2003) Heavy metallic concentration in microbial mats found at hydrothermal systems, Kamchatka, Russia. *Science Reports of the Kanazawa University*, **47**(1–2), 1–48.

Temgoua, E. and Pfeifer, H.-R. (2002) Arsenic in spring waters and soils in southern Switzerland: evidence of complex weathering and redeposition processes. *Geochimica et Cosmochimica Acta*, **66**(15A), 768.

Thiros, S.A. (2003) *Quality and Sources of Shallow Ground Water in Areas of Recent Residential Development in Salt Lake Valley, Salt Lake County, Utah*. Water-Resources Investigations Report 03–4028, U. S. Geological Survey, Salt Lake City, UT.

Thomas, M.A. (2003) *Arsenic in Midwestern Glacial Deposits–Occurrence and Relation to Selected Hydrogeologic and Geochemical Factors*. U. S. Geological Survey Water-Resources Investigations Report 03–4228, Columbus, OH.

Thompson, T.S., Le, M.D., Kasick, A.R. and Macaulay, T.J. (1999) Arsenic in well water supplies in Saskatchewan. *Bulletin of Environmental Contamination and Toxicology*, **63**(4), 478–83.

Thomson, B.M., Longmire, P.A. and Brookins, D.G. (1986) Geochemical constraints on underground disposal of uranium mill tailings. *Applied Geochemistry*, **1**, 335–343.

Thornburg, K. and Sahai, N. (2004) Arsenic occurrence, mobility, and retardation in sandstone and dolomite formations of the Fox River valley, eastern Wisconsin. *Environmental Science and Technology*, **38**(19), 5087–94.

Toler, L.G. (1982) *Some Chemical Characteristics of Mine Drainage in Illinois*, U. S. Geological Survey Water-Supply Paper, Reston, VA, pp. 47.

Turner, S.J., Flindell, P.A., Hendri, D. *et al.* (1994) Sediment-hosted gold mineralisation in the Ratatotok District, North Sulawesi, Indonesia. *Journal of Geochemical Exploration*, **50**(1–3), 317–36.

US Environmental Protection Agency (US EPA) (2002) *Arsenic Treatment Technologies for Soil, Waste, and Water*, EPA-542-R-02–004. Office of Solid Wastes and Emergency (5102G).

Vallance, J., Cathelineau, M., Boiron, M.C. *et al.* (2003) Fluid-rock interactions and the role of late Hercynian aplite intrusion in the genesis of the Castromil gold deposit, northern Portugal. *Chemical Geology*, **194**(1–3), 201–24.

Varnavas, S.P. and Cronan, D.S. (1988) Arsenic, antimony and bismuth in sediments and waters from the Santorini hydrothermal field, Greece. *Chemical Geology*, **67**, 295–305.

Varsányi, I. and Kovács, L.O. (2006) Arsenic, iron and organic matter in sediments and groundwater in the Pannonian Basin, Hungary. *Applied Geochemistry*, **21**(6), 949–63.

Verplanck, P.L., Mueller, S.H., Youcha, E.K. *et al.* (2003) *Chemical Analyses of Ground and Surface Waters, Ester Dome, Central Alaska, 2000–2001*. U. S. Geological Survey Open-File Report 03–244, Boulder, CO.

Walker, S. and Griffin, S. (1998) Site-specific data confirm arsenic exposure predicted by the US Environmental Protection Agency. *Environmental Health Perspectives*, **106**(3), 133–39.

Walker, S.R., Jamieson, H.E., Lanzirotti, A. *et al.* (2005) The speciation of arsenic in iron oxides in mine wastes from the giant gold mine, N.W.T.: application of synchrotron micro-XRD and micro-XANES at the grain scale. *Canadian Mineralogist*, **43**(4), 1205–24.

Wang, J., Lunshan, Z. and Wu, Y. (1998) Environmental geochemical study on arsenic in arseniasis areas in Shanyin and Yingxian, Shanxi China. *Geoscience (Xiandai Dizhi)*, **12**(2), 243–48.

Wang, X.C. and Zhang, Z.R. (2001) Geology of sedimentary rock-hosted disseminated gold deposits in northwestern Sichuan, China. *International Geology Review*, **43**(1), 69–90.

Warner, K.L. (2001) Arsenic in glacial drift aquifers and the implication for drinking water — Lower Illinois River Basin. *Ground Water*, **39**(3), 433–42.

Warren, C., Burgess, W.G. and Garcia, M.G. (2005) Hydrochemical associations and depth profiles of arsenic and fluoride in Quaternary loess aquifers of northern Argentina. *Mineralogical Magazine*, **69**(5), 877–86.

Welch, D. (1999) *Arsenic geochemistry of stream sediments associated with geothermal waters at La Primavera Geothermal Field, Mexico* Master's Thesis. New Mexico Institute of Mining and Technology, Socorro, p. 91.

Welch, A.H., Lawrence, S.J., Lico, M.S. *et al.* (1997) *Ground-Water Quality Assessment of the Carson River Basin, Nevada and California: Results of Investigations*, U.S. Geological Survey Water-Supply Paper, p. 931 sheet.

Welch, A.H. and Lico, M.S. (1998) Factors controlling As and U in shallow ground water, southern Carson Desert, Nevada. *Applied Geochemistry*, **13**(4), 521–39.

Welch, A.H., Westjohn, D.B., Helsel, D.R. and Wanty, R.B. (2000) Arsenic in ground water of the United States: occurrence and geochemistry. *Ground Water*, **38**(4), 589–604.

Wennrich, R., Mattusch, J., Morgenstern, P. *et al.* (2004) Characterization of sediments in an abandoned mining area; a case study of Mansfeld region, Germany. *Environmental Geology*, **45**(6), 818–33.

Wenrich, K.J. and Silberman, M.L. (1984) Potential precious and strategic metals as by-products of uranium mineralized breccia pipes in northern Arizona. *AAPG Bulletin*, **68**(7), 954.

Widerlund, A. and Ingri, J. (1995) Early diagenesis of arsenic in sediments of the Kalix River estuary, northern Sweden. *Chemical Geology*, **125**(3–4), 185–96.

Wilkin, R.T. and Ford, R.G. (2006) Arsenic solid-phase partitioning in reducing sediments of a contaminated wetland. *Chemical Geology*, **228**(1–3 Special Issue), 156–74.

Williams, M. (2001) Arsenic in mine waters: international study. *Environmental Geology*, **40**(3), 267–78.

Williams, M., Fordyce, F., Paijitprapapon, A. and Charoenchaisri, P. (1996) Arsenic contamination in surface drainage and groundwater in part of the Southeast Asian tin belt, Nakhon Si Thammarat Province, southern Thailand. *Environmental Geology*, **27**(1), 16–33.

Williams, T.M., Rawlins, B.G., Smith, B. and Breward, N. (1998) In-vitro determination of arsenic bioavailability in contaminated soil and mineral beneficiation waste from Ron Phibun, southern Thailand: a basis for improved human risk assessment. *Environmental Geochemistry and Health*, **20**(4), 169–77.

Williamson, B.J., Udachin, V., Purvis, O.W. *et al.* (2004) Characterization of airborne particulate pollution in the Cu smelter and former mining Town of Karabash, South Ural Mountains of Russia. *Environmental Monitoring and Assessment*, **98**, 235–59.

Willis-Richards, J. and Jackson, N.J. (1989) Evolution of the Cornubian Ore Field, Southwest England: part I. Batholith modeling and ore distribution. *Economic Geology*, **84**, 1078–1100.

Wise, J.M., Noble, D.C., Vidal, C.E. and Gustafson, L.B. (2004) Geology and structural control of the Huancavelica mercury district, central Peru. *Abstracts with Programs–Geological Society of America*, **36**(4), 22.

Wong, H.K.T., Gauthier, A., Beauchamp, S. and Tordon, R. (2002) Impact of toxic metals and metalloids from the Caribou gold-mining areas in Nova Scotia, Canada. *Geochemistry: Exploration, Environment, Analysis*, **2**(3), 235–41.

Woo, N.C. and Choi, M.C. (2001) Arsenic and metal contamination of water resources from mining wastes in Korea. *Environmental Geology*, **40**(3), 305–11.

Xia, Y. and Liu, J. (2004) An overview on chronic arsenism via drinking water in PR China. *Toxicology*, **198**(1–3), 25–29.

Xie, Z.M., Zhang, Y.M. and Naidu, R. (2006) Extent and severity of arsenic poisoning in China, in *Managing Arsenic in the Environment: From Soil to Human Health* (eds R. Naidu, E. Smith, G. Owens, P. Bhattacharya and P. Nadebaum), CSIRO Publishing, Collingwood, pp. 541–52.

Yamasaki, Y. and Hata, Y. (2000) Changes and their factors of concentrations of arsenic and boron in the process of groundwater recharge in the lower Lluta River Basin, Chile. *Journal of Groundwater Hydrology*, **42**(4), 341–53.

Yang, S.X. and Blum, N. (1999) A fossil hydrothermal system or a source-bed in the Madiyi formation near the Xiangxi Au-Sb-W deposit, NW Hunan, PR China? *Chemical Geology*, **155**(1–2), 151–69.

Yokoyama, T., Takahashi, Y. and Tarutani, T. (1993) Simultaneous determination of arsenic and arsenious acids in geothermal water. *Chemical Geology*, **103**(1–4), 103–11.

Yudovich, Ya.E. and Ketris, M.P. (2005) Arsenic in coal: a review. *International Journal of Coal Geology*, **61**(3–4), 141–96.

Zachariáš, J., Frýda, J., Paterova, B. and Mihaljevič, M. (2004) Arsenopyrite and As-bearing pyrite from the Roudný Deposit, Bohemian Massif. *Mineralogical Magazine*, **68**(1), 31–46.

Zänker, H., Moll, H., Richter, W. *et al.* (2002) The colloid chemistry of acid rock drainage solution from an abandoned Zn-Pb-Ag mine. *Applied Geochemistry*, **17**(5), 633–48.

Zeng, S. (1994) Geological characteristics and origin of Xinzhou gold deposit of Qingyuan County, Guangdong. *Guangdong Geology*, **9**(1), 1–11.

Zhang, H. (2004) Heavy-metal pollution and arseniasis in Hetao region, China. *Ambio*, **33**(3), 138–40.

Zhang, J., Ren, D., Zhu, Y. *et al.* (2004) Mineral matter and potentially hazardous trace elements in coals from Qianxi Fault Depression Area in southwestern Guizhou, China. *International Journal of Coal Geology*, **57**, 49–61.

Zhang, X.C., Spiro, B., Hall, C. *et al.* (2003) Sediment-hosted disseminated gold deposits in Southwest Guizhou, PRC: their geological setting and origin in relation to mineralogical, fluid inclusion, and stable-isotope characteristics. *International Geology Review*, **45**(5), 407–70.

Zhou, Y. and Ren, Y. (1992) Distribution of arsenic in coals of Yunnan province, China, and its controlling factors. *International Journal of Coal Geology*, **20**(1–2), 85–98.

Appendix E

Regulation of Arsenic: A Brief Survey and Bibliography

E.1 Introduction

This appendix provides a brief survey of some of the readily available English-language government documents, websites, and other key references on the regulation of arsenic in various nations. Regulations related to arsenic in domestic water, solid and liquid wastes, food, commercial products, and emissions to the atmosphere are often very complex (sometimes industry- and site-specific), vary from nation to nation, and frequently change over time. Considering the complexity and the number of regulations and how often modifications occur, a comprehensive and up-to-date summary of the regulations is not possible in this appendix. Readers that desire more current or additional regulations should access the websites listed in this appendix or search the Internet for additional information. Individuals should also recognize that some state, provincial, and municipal governments have more stringent regulations than those enforced by their federal agencies. For example, since 2006, the American state of New Jersey has enforced an arsenic drinking water standard or *maximum contaminant level* (MCL) of $5 \mu g \, l^{-1}$ instead of the federal MCL of $10 \mu g \, l^{-1}$ (New *Jersey Administrative Code* 7 : 10–5.2). Italicized terms in the text of this appendix or words derived from the italicized terms are defined in the glossary of Appendix B.

E.2 Regulation of arsenic in water

To protect human health, many nations have established regulatory standards for contaminants in drinking water. In some cases, nations also regulate contaminants in: (1) *groundwater*, (2) various aqueous wastes, including wastewaters that may be discharged into the environment, (3) irrigation and other agricultural waters, and/or (4) surface waters to protect aquatic life and other organisms.

Arsenic   Edited by Kevin R. Henke
© 2009 John Wiley & Sons, Ltd

E.2.1   Drinking water

E.2.1.1   *World Health Organization (WHO) Guideline*

World Health Organization (WHO) (1998) *Guidelines for Drinking-water Quality*, 2nd edn, World Health Organization, Geneva.

After considering the cancer and other health risks in humans associated with arsenic exposure, the World Health Organization (WHO) lowered their recommended drinking water criterion from 50 to $10 \, \mu g \, l^{-1}$ in 1993 (Smith and Smith, 2004). Although WHO is not a regulatory agency with the power to set standards and enforce them, its arsenic recommendation has been adopted by many regulatory agencies in industrialized nations, including the United States, Canada, and the European Union (Table E.1). In contrast, many developing nations continue to retain a $50 \, \mu g \, l^{-1}$ drinking water standard for arsenic (Table 5.1). These countries often lack sampling programs, analytical equipment, and funding to effectively enforce lower standards (Smith and Smith, 2004). Smith, A.H. and Smith, M.M.H. (2004) Arsenic drinking water regulations in developing countries with extensive exposure. *Toxicology*, **198**(1-3), 39–44.

E.2.1.2   *Australia*

National Health and Medical Research Council (2004) *Australian Drinking Water Guidelines 6*, Natural Resource Management Ministerial Council, Australian Government.

When compared with other nations, Australia has a very stringent arsenic drinking water standard of $7 \, \mu g \, l^{-1}$ (National Health and Medical Research Council, 2004; Table 5.1).

E.2.1.3   *Bangladesh*

Ministry of Environment and Forest, Bangladesh http://www.moef.gov.bd/. Last accessed October 14, 2008. Department of Environment (Bangladesh) (1997) *The Environmental Conservation Rules, 1997*, Ministry of Environment and Forests, ECR '97, pp. 197–227. See also: http://www.moef.gov.bd/html/laws/env_law/178-189.pdf. Last accessed October 14, 2008.

The Ministry of Environment and Forest enforces drinking water regulations in Bangladesh. The arsenic drinking water standard for Bangladesh is $50 \, \mu g \, l^{-1}$ (Department of Environment (Bangladesh), 1997; Table 5.1). Updates are available at the Ministry of Environment and Forest website (http://www.moef.gov.bd/). See also the website of the Bangladesh Environmental Management Force, http://bemfbd.tripod.com/.

E.2.1.4   *Canada*

Federal-Provincial-Territorial Committee on Drinking Water of the Federal-Provincial-Territorial Committee on Health and the Environment (2007) *Guidelines for Canadian Drinking Water Quality: Summary Table*, Health Canada, Ottawa. See also: www.healthcanada.gc.ca/waterquality.

Canada recently lowered their arsenic drinking water standard from 25 to $10 \, \mu g \, l^{-1}$ (Federal-Provincial-Territorial Committee on Drinking Water of the Federal-Provincial-Territorial Committee on Health and the Environment, 2007; Table 5.1).

E.2.1.5   *China*

Ministry of Environmental Protection (China) http://english.mep.gov.cn/.

Table E.1   *National and other drinking water standards for arsenic. Most information taken from recent and readily available English-language references.*

Nation: state	Drinking water standard ($\mu g\,l^{-1}$)	References
Argentina	50	Codigo Alimentario Argentino (1994)
Australia	7	National Health and Medical Research Council (2004)
Bangladesh	50	Bangladesh Rural Advancement Committee (2000); Department of Environment (Bangladesh) (1997)
Canada	10	Federal-Provincial-Territorial Committee on Drinking Water of the Federal-Provincial-Territorial Committee on Health and the Environment (2007)
Chile	50	Caceres *et al.* (2005)
China	50	Guo and Wang (2005)
Croatia	50	Kutle *et al.* (2004)
Ecuador	50	Appleton *et al.* (2001)
European union	10	European Union (1998)
Ghana	50	Duker *et al.* (2005)
India	50	Indian Bureau of Standards 10500 (1991); Rahman *et al.* (2005)
Japan	10	Takanashi *et al.* (2004); Japanese Ministry of the Environment (2008)
Mexico	35	Mendoza *et al.* (2006); Mexican Guidelines for Industrial Discharge Waters (1996)
Nepal	50	Shrestha *et al.* (2003)
Taiwan	10	Republic of China Environmental Protection Administration (1998)
Thailand	50	Zhang *et al.* (2003)
US	10	Code of Federal Regulations (2006)
US: New Jersey	5	New Jersey Administrative Code (2004)
Vietnam	50	Berg *et al.* (2001)
World Health Organization	10	World Health Organization (1998)

The English-language website of the Chinese Ministry of Environmental Protection (http://english.mep. gov.cn/) would have any updates of the Chinese arsenic drinking water standard. As of 2005, the Chinese drinking water standard for arsenic was $50\,\mu g\,l^{-1}$ (Guo and Wang, 2005; Table 5.1).

E.2.1.6   *European Union*

European Union (1998) Council directive 98/83/EC on the quality of water intended for human consumption. *Official Journal of the European Communities*, **L 330/32**, 32–54.

The European Union has an arsenic drinking water standard of $10\,\mu g\,l^{-1}$ (European Union, 1998; Table 5.1).

E.2.1.7   Japan

Ministry of the Environment (Japan): Water Standards http://www.env.go.jp/en/water/wq/wp.html.

Ministry of the Environment (Japan): Groundwater Standards http:// www.env.go.jp/en/water/gw/gwp.html.

Japan has an arsenic standard of $10\,\mu g\,l^{-1}$ for both drinking water (Table 5.1) and all groundwater. Any updates are at the Ministry of the Environment English-language website: http://www.env.go.jp/en/standards/.

E.2.1.8   Russia

Ministry of Natural Resources of the Russian Federation, http://enc.ex.ru/cgi-bin/n1firm.pl?lang = 2&f = 1245.

The Ministry of Natural Resources regulates environmental protection in Russia. Few English-language documents are available.

E.2.1.9   United States of America

Code of Federal Regulations (CFR), US Government Printing Office, Superintendent of Documents, Washington, DC, 2006, 2007.

New Jersey Administrative Code, State of New Jersey Department of Environmental Protection, Bureau of Safe Drinking Water, Safe Drinking Water Act Regulations, 7:10-5.2. See also: New Jersey Department of Environmental Protection: http://www.state.nj.us/dep/

American federal regulations on arsenic in drinking water and most other arsenic-related environmental issues are in Title (volume) 40 of the *Code of Federal Regulations* (CFR). The United States Environmental Protection Agency (US EPA) recently lowered the MCL from 50 to 10 $\mu g\,l^{-1}$ (40 CFR 141.62; Smith *et al.*, 2002; Table E.1). Han *et al.* (2003) review the cost effects on treatment technologies from lowering the arsenic US MCL. They found that with appropriate filtering, a Fe(III) dose of $6\,mg\,l^{-1}$ at a *pH* of 6.8 can reduce arsenic concentrations in water to below $2\,\mu g\,l^{-1}$. Similarly, Chwirka *et al.* (2004) endorsed the use of *coagulation/microfiltration* processes to meet the lower MCL (also see Chapter 7). The US EPA also has a recommended, but unenforceable, *maximum contaminant level goal* (MCLG) of $0\,\mu g\,l^{-1}$ for arsenic (40 CFR 141.51). Furthermore, US States have the option to adopt even lower MCLs. As an example, the state of New Jersey has an arsenic MCL of $5\,\mu g\,l^{-1}$ (*New Jersey Administrative Code* 7:10-5.2; Table E.1).

In the United States, bottled water is considered a beverage and is regulated by the United States Food and Drug Administration (US FDA) rather than the US EPA. The arsenic standard for bottled water is the same as the US MCL, $10\,\mu g\,l^{-1}$ (21 CFR 165.110).

Chwirka, J.D., Colvin, C., Gomez, J.D. and Mueller, P.A. (2004) Arsenic removal from drinking water using the coagulation/microfiltration process. *Journal of American Water Works Association*, **96**(3), 106–14.

Han, B., Zimbron, J., Runnells, T.R. *et al.* (2003) New arsenic standard spurs search for cost-effective removal techniques. *Journal of American Water Works Association*, **95**(10), 109–18.

Smith, A.H., Lopipero, P.A., Bates, M.N. and Steinmaus, C.M. (2002) Arsenic epidemiology and drinking water standards. *Science*, **296**, 2145–46.

E.2.2 Arsenic standards of natural surface waters and groundwaters

E.2.2.1 Australia

National Environment Protection Council (1999) *Schedule B(1) Guideline on the Investigation Levels for Soil and Groundwater*, National Environment Protection (Assessment of Site Contamination) Measure, Adelaide.

Marine and fresh waters in Australia should not contain more than $50 \mu g l^{-1}$ of arsenic (National Environment Protection Council, 1999, 10). Australian irrigation and livestock water should not contain more than 100 and $500 \mu g l^{-1}$ of arsenic, respectively (National Environment Protection Council, 1999, 10).

E.2.2.2 Japan

Ministry of the Environment (Japan): Groundwater Standards, http://www.env.go.jp/en/water/gw/gwp.html.

Japan has an arsenic standard of $10 \mu g l^{-}$ for all groundwaters. Any updates are at the Ministry of the Environment English-language website: http://www.env.go.jp/en/standards/.

E.2.2.3 United States of America: California surface water standards

Code of Federal Regulations (CFR), US Government Printing Office, Superintendent of Documents, Washington, DC.

To protect aquatic organisms from arsenic and other contaminants, the US state of California has established *criteria maximum concentrations* (CMC) and *criteria continuous concentrations* (CCC) for most inland surface waters, bays, and *estuaries* (40 CFR 131.38). The CMC is the highest concentration of a contaminant in a surface water that will not show any deleterious effects to aquatic life after exposure for a brief period of time (an average of 1 hour) (40 CFR 131.38, 131.36). The CCC is the highest concentration of a contaminant that aquatic life will not show any deleterious effects after exposure for four days (an 'extended period of time') (40 CFR 131.38). For freshwater, the dissolved arsenic CMC is $340 \mu g l^{-1}$. The CMC for saltwater is $69 \mu g l^{-1}$. The dissolved arsenic CCCs for freshwater and saltwater are 150 and $36 \mu g l^{-1}$, respectively (40 CFR 131.38).

E.3 Regulation of arsenic in solid and liquid wastes

E.3.1 Bangladesh

Ministry of Environment and Forest, Bangladesh, http://www.moef.gov.bd/.

Department of Environment (Bangladesh) (1997) *The Environmental Conservation Rules, 1997*, Ministry of Environment and Forests, ECR '97, 197–227. See also: http://www.moef.gov.bd/html/laws/env_law/178-189.pdf.

The Bangladesh Department of Environment in the Ministry of Environment and Forest prohibits liquid wastes with more than $200 \mu g l^{-1}$ of arsenic from being discharged into inland surface waters or used to irrigate land (Department of Environment (Bangladesh), 1997). Updates are available at the Ministry of Environment and Forest website (http://www.moef.gov.bd/). Also see the website of the Bangladesh Environmental Management Force, http://bemfbd.tripod.com/.

E.3.2   European Union (EU)

According to Ottosen *et al.* (2004, 296) Denmark, an EU member, has banned chromated copper arsenate (CCA) wood preservatives.

Ottosen, L., Pedersen, A. and Christensen, I. (2004) Characterization of residues from thermal treatment of CCA impregnated wood. Chemical and electrochemical extraction, in *Environmental Impacts of Preservative-Treated Wood*, Florida Center for Environmental Solutions, Conference, February 8–11, Gainesville, Orlando, FL, pp. 295–311.

E.3.3   Japan

For discharging wastewaters into Japanese environments, Itakura *et al.* (2006) states that the Japanese National Effluent Standard for arsenic is $100\,\mu g\,l^{-1}$.

Itakura, T., Sasai, R. and Itoh, H. (2006) Arsenic recovery from water containing arsenic ions by hydrothermal mineralization. *Chemistry Letters*, **35**(11), 1270–71.

E.3.4   Norway

Like Denmark, Norway has reportedly banned chromated copper arsenate (CCA) wood preservatives (Ottosen *et al.* 2004, 296).

Ottosen, L., Pedersen, A. and Christensen, I. (2004) Characterization of residues from thermal treatment of CCA impregnated wood. Chemical and electrochemical extraction, in *Environmental Impacts of Preservative-Treated Wood*, Florida Center for Environmental Solutions, Conference, February 8–11, Gainesville, Orlando, FL, pp. 295–311.

E.3.5   Taiwan

Huang *et al.* (2007, 292–293) report that the arsenic discharge limit for industrial wastewaters in Taiwan is $500\,\mu g\,l^{-1}$.

Huang, C., Pan, J. R., Lee, M. and Yen, S. (2007) Treatment of high-level arsenic-containing wastewater by fluidized bed crystallization process. *Journal of Chemical Technology and Biotechnology* **82**(3), 289–94.

E.3.6   United States of America

E.3.6.1   *Identification of arsenic-bearing hazardous wastes under US regulations*

Code of Federal Regulations (CFR), US Government Printing Office, Superintendent of Documents, Washington, DC.

Unofficial *Code of California Regulations* (2006) California Department of Toxic Substances Control, Sacramento, http://www.dtsc.ca.gov/LawsRegsPolicies/Title22/index.cfm.

US Environmental Protection Agency (US EPA) (2007) *Test Methods for Evaluating Solid Waste, Physical/Chemical Methods*, SW-486, 6th Revision, Office of Solid Waste, National Technical Information Service, Springfield, VA.

Hazardous wastes are gases, liquids, solids, or mixtures of liquids and solids that are considered threats to human health and life. The US EPA maintains lists of specific solid and liquid wastes that by definition are considered hazardous. These listed wastes are grouped into F, K, P, and U categories. F-listed wastes are from 'nonspecific sources', which are generated by a variety of manufacturing processes (40 CFR 261.31). For example, F035 listed wastes contain arsenic, chromium, and lead. K-listed wastes are 'source-specific' and are generated by specific industries, such as pesticide manufacturers, drug companies, or the petroleum industry (40 CFR 261.32). For example, K084 listed wastes refer to wastewater treatment sludges generated by the production of veterinary pharmaceuticals containing arsenic compounds. P- and U-listed wastes include unused chemicals or other unwanted commercial products, such as old barrels of unused arsenical pesticides (40 CFR 261.33).

Some liquid wastes undergo disposal by injection into the deep subsurface. Underground injection disposal is not allowed for liquid hazardous wastes and sludges that contain 500 mg l^{-1} or more of arsenic (40 CFR 148.12). The underground injection of many K and other listed arsenic wastes is also not permitted (40 CFR 148.14–148.15).

The US EPA also identifies hazardous wastes by evaluating them on the basis of four recognized characteristics of hazardousness: ignitability, reactivity, corrosivity, and *toxicity* (40 CFR 261.3, 261.20–261.24). Although some arsenic wastes are specifically listed as hazardous (e.g. 40 CFR 261.31–261.33), most arsenic-bearing solid and liquid wastes would need to be evaluated for hazardousness and, for the vast majority of them, the *toxicity characteristic* is most relevant. Under current US federal regulations, the toxicity characteristic of hazardousness for solid and liquid wastes is evaluated with the *toxicity characteristic leaching procedure* (TCLP) (40 CFR 261.24; US EPA, 2007, Method 1311). The TCLP, which is a laboratory *batch leaching method*, is designed to simulate conditions in an unlined municipal waste landfill if acetic acid solutions from rotting garbage were to leach contaminants from buried solid wastes followed by the movement of the contaminated *leachates* into surrounding groundwaters (Greene, 1991). Local and state regulations in the United States also may require that the TCLP be supplemented with other leaching tests, such as the *California waste extraction test* (CWET) (Unofficial Code of California Regulations, 2006, Title 22, Division 4.5, Chapter 11, Appendix 2), before the materials can be legally declared nonhazardous and possibly landfilled.

For the TCLP, liquid wastes containing less than 0.5% dry solids are filtered through 0.6–0.8 μm glass fiber filters. The filtrates are TCLP leachates. For an arsenic-bearing nonvolatile solid waste, the TCLP uses either a pH 2.88 ± 0.05 acetic acid leaching solution or a pH 4.93 ± 0.05 acetic acid and sodium hydroxide leaching solution, depending upon the pH of an aqueous slurry of the waste. At least 100 g of the solid waste is mixed with 20 times the mass of the leaching solution (i.e. 2.0 kg of solution for 100 g of waste). The mixture is then tumbled for 18 ± 2 hours at 23 ± 1 °C on an end-over-end stirrer. After tumbling, the liquid is separated from the sample with a 0.6–0.8 μm glass fiber filter. The filtrate is a TCLP leachate. If a TCLP leachate contains more than 5 mg l^{-1} of arsenic, the liquid or solid waste has the toxic characteristic of hazardousness and the waste is identified as hazardous under US federal regulations.

Some wastes, such as CCA-treated wood or *coal flyashes*, are exempt from US federal hazardous waste regulations (40 CFR 261.4) even though individual samples may produce TCLP leachates that exceed the maximum concentration of arsenic for the toxicity characteristic (i.e. 5 mg l^{-1}) (40 CFR 261.24; Townsend *et al.*, 2004, 171-172; Stook *et al.*, 2005; Appendix B; Chapter 5). Although the US federal government may exempt certain wastes from federal hazardous waste regulations, state and local laws and regulations may still prohibit the disposal of these wastes in municipal, construction debris, and other landfills that are designed for nonhazardous wastes. In particular, the US state of Minnesota has adopted stricter regulations for the disposal of CCA-treated wood and individual landfill operators in the United States may not accept the material (Oskoui, 2004, 241; Chapter 7).

Greene, G.C. (1991) TCLP: Where are we and what lies ahead? *American Environmental Laboratory*, December, 9–14.

Oskoui, K. (2004) Recovery and reuse of the wood and chromated copper arsenate (CCA) from CCA-treated wood - A technical paper, in *Environmental Impacts of Preservative-Treated Wood*, Florida Center for Environmental Solutions, Conference, February 8–11, Gainesville, Orlando, FL, pp. 238–44.

Stook, K., Tolaymat, T., Ward, M. *et al.* (2005) Relative leaching and aquatic toxicity of pressure-treated wood products using batch leaching tests. *Environmental Science and Technology*, **39**(1), 155–63.

Townsend, T., Dubey, B. and Solo-Gabriele, H. (2004) Assessing potential waste disposal impact from preservative-treated wood products, in *Environmental Impacts of Preservative-Treated Wood*, Florida Center for Environmental Solutions, Conference, February 8–11, Gainesville, Orlando, FL, pp. 169–88.

E.3.6.2 The US National Priority List and Superfund sites

After lead, arsenic is the most common contaminant at US *Superfund National Priorities List* (NPL) sites (EPA, 2002a, 2). As of 2002, arsenic was a contaminant of concern at 568 or 47% of 1209 NPL sites with records of decision (RODs) (EPA, 2002a, 2). Among the arsenic-contaminated sites, 380 of them have contaminated groundwater, a total of 86 have arsenic-contaminated surface water, and the number of sites with arsenic-contaminated *soils* and *sediments* are 372 and 154, respectively (EPA, 2002b, 2.2). Appendix B in US EPA (2002b) lists the locations of Superfund sites where arsenic is a contaminant of concern.

US Environmental Protection Agency (US EPA) (2002a) *Proven Alternatives for Aboveground Treatment of Arsenic in Groundwater*, EPA-542-S-02-002, Office of Solid Wastes and Emergency (5102G).

US Environmental Protection Agency (US EPA) (2002b) *Arsenic Treatment Technologies for Soil, Waste, and Water*, EPA-542-R-02-004, Office of Solid Wastes and Emergency (5102G).

E.3.6.3 Best (demonstrated) available technologies

Code of Federal Regulations (CFR), US Government Printing Office, Superintendent of Documents, Washington, DC.

The US EPA has designated certain technologies as the 'best available technology' (BAT) or the '*best demonstrated available technology*' (BDAT) for treating arsenic in liquid and solid wastes. For removing *arsenate* (inorganic As(V)) from water, the US EPA recommends as BATs treatment with activated *alumina, ion exchange, reverse osmosis, electrodialysis*, or, under certain conditions, coagulation/filtration and *precipitation* with lime (40 CFR 141.62; Chapter 7). The US EPA also recommends that any inorganic As(III) (*arsenite*) in wastewaters be oxidized to arsenate and then treated with the As(V) BATs (40 CFR 141.62). For arsenic-contaminated soils and sediments, the US EPA considers *vitrification* to be the BDAT (US EPA, 1999, C.1).

US Environmental Protection Agency (US EPA) (1999) *Presumptive Remedy for Metals-in-Soils Sites*, EPA-540-F-98-054, Office of Solid Wastes and Emergency (5102G).

E.4   Sediment and soil guidelines and standards for arsenic

E.4.1   Introduction

Several countries have established methodologies, standards, guidelines, or goals for *remediating* arsenic in soils and sediments. Rather than establishing one set of national standards, many nations have preliminary criteria that may be modified depending on local conditions. In other cases, regulatory agencies in consultation with the site owners establish procedures that can be used to calculate site-specific cleanup goals for arsenic in contaminated soils and sediments.

In many situations, soil and sediment guidelines and standards will vary depending on how the remediated land will be used; namely, industrial or residential applications. Although at least the American criteria are no longer current (http://www.epa.gov/region09/waste/sfund/prg/index.html), Provoost *et al.* (2006) reviews soil cleanup guidelines and standards for the United States (Region 9 Superfund sites in Arizona, California, Hawaii, Nevada, US Pacific Islands, and Tribal Nations) and several European nations. For areas that are planned for residential use, the standards and guidelines range from 2 (Norway) to 110 (Flemish Belgium) mg kg^{-1} (dry mass) of arsenic in the soil. If the areas will be utilized by industries, the standards are usually higher (ranging from 12 mg kg^{-1}, dry mass for Canada to 500 mg kg^{-1}, dry mass for the United Kingdom). For Sweden, the soil standard for industrial applications is 15 mg kg^{-1} (dry mass) if the groundwater at the site will be utilized and 40 mg kg^{-1} (dry mass) if not (Provoost *et al.*, 2006, 177).

Provoost, J., Cornelis, C. and Swartjes, F. (2006) Comparison of soil cleanup standards for trace elements between countries: Why do they differ? *Journal of Soils and Sediments*, **6**(3), 173–81.

E.4.2   Australia

National Environment Protection Council (1999) *Schedule B(1) Guideline on the Investigation Levels for Soil and Groundwater*, National Environment Protection (Assessment of Site Contamination) Measure, Adelaide.

Australia has established 'investigation levels' for soils. Investigation levels are not cleanup or response levels (National Environment Protection Council, 1999, 4). If the levels are exceeded at a site, additional investigations or evaluations are required to ensure that the local ecology and human population are protected (National Environment Protection Council, 1999). An interim ecological investigation level of 20 mg kg^{-1} of arsenic has been established for Australian urban areas to protect plant life (National Environment Protection Council, 1999, 9).

E.4.3   Canada

Canadian Sediment Quality Guidelines, http://www.ec.gc.ca/ceqg-rcqe/English/ceqg/sediment/default.cfm.

Canadian Soil Quality Guidelines, http://www.ec.gc.ca/ceqg-rcqe/English/ceqg/soil/default.cfm.

Data on soil and sediment quality and associated information on the websites allow individuals to derive site-specific guidelines for protecting people and the environment from contaminants in soils and sediments.

E.4.4   European Union

European Union Environment: Soil Protection, http://europa.eu/scadplus/leg/en/s15010.htm.

Regulations for Protecting European soils from Contamination are at http://europa.eu/scadplus/leg/en/s15010.htm.

E.4.5   Italy

Spadoni *et al.* (2005, 88) mention that Italy has limits for arsenic in urban and residential soils of 20 and 50 mg kg^{-1}, respectively.

Spadoni, M., Voltaggio, M. and Cavarretta G. (2005) Recognition of areas of anomalous concentration of potentially hazardous elements by means of a subcatchment-based discriminant analysis of stream sediments. *Journal of Geochemical Exploration*, **87**(3), 83–91.

E.4.6   Japan

Ministry of the Environment (Japan): Standards for Soil Pollution, http://www.env.go.jp/en/water/soil/sp.html.

According to Japanese regulations, the arsenic standard for soil waters is 10 μg l^{-1} and the soils of paddy fields must not contain more than 15 mg kg^{-1} of arsenic (http://www.env.go.jp/en/water/soil/sp.html).

E.4.7   Korea (South)

Lee *et al.* (2007, 321) state that the (South) Korean Soil Pollution Warning Limit for arsenic is 6 mg kg^{-1}.

Lee, M., Paik, I.S., Do, W. *et al.* (2007) Soil washing of As-contaminated stream sediments in the vicinity of an abandoned mine in Korea. *Environmental Geochemistry and Health*, **29**(4), 319–329.

E.4.8   Thailand

Pollution Control Department (Thailand), http://www.pcd.go.th/Info_serv/en_reg_std_soil01.html.

Thailand has an arsenic soil quality standard of 3.9 mg kg^{-1} for habitable and agricultural sites and a standard of 27 mg kg^{-1} for other areas (http://www.pcd.go.th/Info_serv/en_reg_std_soil01.html).

E.4.9   United States of America

Florida Administrative Code, https://www.flrules.org/.

Site Remediation Program (New Jersey), http://www.nj.gov/dep/srp/guidance/scc/.

Special Updates or Reports on Arsenic and Other Contaminants of Concern (Florida), http://www.dep.state.fl.us/waste/categories/csf/pages/SpecialUpdatesorReports.htm.

Superfund Region 9: Preliminary Remediation Goals, http://www.epa.gov/region09/waste/sfund/prg/index.html.

Until recently (http://www.epa.gov/region09/waste/sfund/prg/index.html), US EPA Region 9 (Arizona, California, Hawaii, Nevada, US Pacific Islands, and Tribal Nations) had preliminary remediation goals

(PRGs) of 22 mg kg^{-1} for arsenic in residential soils and 260 mg kg^{-1} for industrial soils at Superfund sites in the region. The updated regulations are at Superfund Region 9: Preliminary Remediation Goals: http://www.epa.gov/region09/waste/sfund/prg/index.html.

A number of US States have guidelines or standards for soils and sediments. For example, New Jersey has a cleanup criterion of 20 mg kg^{-1} for arsenic in residential and nonresidential soils (http://www.nj.gov/dep/srp/guidance/scc/). Like many regulations, the New Jersey soil criteria may be modified for local conditions. The arsenic soil cleanup target levels in Florida are 2.1 mg kg^{-1} for residential and 12 mg kg^{-1} for commercial and industrial sites (Florida Administrative Code 62-777.170).

E.5   Regulation of arsenic in food and drugs

E.5.1   Australia and New Zealand

Food Standards Australia New Zealand, http://www.foodstandards.gov.au/.

Unless specified differently, most foods in Australia and New Zealand may not contain more than 1 mg kg^{-1} (dry mass) of arsenic (Standard 1.3.4.-). Salt used in food may not contain more than 0.5 mg kg^{-1} of arsenic (Standard 2.10.2).

E.5.2   Canada

Canadian Food Inspection Agency, http://www.inspection.gc.ca/english/toce.shtml.

The Canadian Food Inspection Agency issues alerts on excessive arsenic in commercial food products.

E.5.3   United States of America

Code of Federal Regulations (CFR), US Government Printing Office, Superintendent of Documents, Washington, DC.

US Food and Drug Administration Fact Sheet on Trisenox, http://www.fda.gov/cder/consumer-info/druginfo/Trisenox.HTM.

In the United States, the United States Food and Drug Administration (US FDA) has the primary role of regulating the amount of arsenic in food and drugs. For chicken and turkey meat and eggs, the US FDA limits the amount of arsenic to 0.5 mg kg^{-1} in uncooked muscle tissues, 2 mg kg^{-1} in uncooked edible byproducts (the organs), and 0.5 mg kg^{-1} in eggs (21 CFR 556.60). Chickens are also not to be fed arsenic supplements within five days of slaughter (21 CFR 558.35, 558.55). In swine, arsenic limits are 2 mg kg^{-1} in uncooked liver and kidneys, and 0.5 mg kg^{-1} in uncooked muscle tissues and byproducts other than liver and kidneys (21 CFR 556.60). The US FDA also sets limits for the amount of arsenic in various food additives (21 CFR 73.30–73.3110), including 1 mg kg^{-1} of arsenic in titanium dioxide (21 CFR 73.575) and 3 mg kg^{-1} in synthetic iron oxides (21 CFR 73.200). Although arsenical drugs are generally prohibited in the United States (Chapter 5), the US FDA allows the use of As_2O_3 in the treatment of leukemia (Trisenox®   http://www.fda.gov/cder/consumerinfo/druginfo/Trisenox.HTM). The US EPA also establishes limits on the amounts of pesticide and herbicide residues in fruits and vegetables, which includes a limit of 0.35 mg kg^{-1} as As_2O_3 in citrus fruits from herbicides containing sodium salts of methanoarsonic acid ($CH_3As(O)(OH)_2$; Table 4.1) (40 CFR 180.289).

E.6 Regulation of arsenic in air

E.6.1 European Union

European Union: Air Pollution, http://europa.eu/scadplus/leg/en/s15004.htm.

European Union (2004) Directive 2004/107/EC of the European Parliament and of the Council of 15 December 2004 relating to arsenic, cadmium, mercury, nickel, and polycyclic aromatic hydrocarbons in ambient air. *Official Journal of the European Communities*, **L23/3**, 3–15.

The European Union has an arsenic target value for ambient air (PM_{10}) of 6 ng m^{-3}, where PM_{10} refers to air that passes through an inlet with a 50% efficiency cutoff of 10 μm *aerodynamic diameter*. Beginning on 31 December 2012, the target value will be enforced in the member states (European Union, 2004).

E.6.2 United States of America

Code of Federal Regulations (CFR), US Government Printing Office, Superintendent of Documents, Washington, DC.

Arsenic is on the US EPA list of 'high risk' air pollutants (40 CFR 63.74). The US EPA specifically regulates arsenic in air emissions from a number of industries, including: glass manufacturing plants that use commercial arsenic (40 CFR 61.160–61.165), copper ore smelters (40 CFR 61.170–61.177), phosphoric acid manufacturing plants (40 CFR 63.601), secondary aluminum production plants (40 CFR 63.1500–63.1503), plywood and composite wood manufacturing facilities (40 CFR 63.2292), wood pulp mills (40 CFR 63.860–63.868), boilers and process heaters (40 CFR 63.7480–63.7575), tire manufacturing plants (40 CFR 63.6015), and facilities that produce elemental arsenic or arsenic trioxide (As_2O_3) (40 CFR 61.180–61.186).

The US Occupational Safety and Health Administration (OSHA) limits exposure to *organoarsenicals* in air. The *permissible exposure limit* (PEL) for organoarsenicals is 0.5 mg m^{-3} as a time-weighted average over an 8-hour period (29 CFR 1910.1000). The PEL (time-weighted average for 8 hours) for arsine (AsH_3) gas is 0.05 parts of arsine per million parts of contaminated air (ppm) by volume at 25 °C and 760 torr (Appendix A), or about 0.2 mg m^{-3} (29 CFR 1910.1000). The PEL for inorganic arsenic is 10 μg m^{-3} of air as a time-weighted average over an 8-hour period (29 CFR 1910.1018). The *action level* for inorganic arsenic in air is 5 μg m^{-3} as a time-weighted average over an 8-hour period (29 CFR 1910.1018).

References

Appleton, J.D., Williams, T.M., Orbea, H. and Carrasco, M. (2001) Fluvial contamination associated with arsenical gold mining in the Ponce Enríquez, Portovelo-Zaruma and Nambija areas, Ecuador. *Water, Air, and Soil Pollution*, **131**(1-4), 19–39.

Bangladesh Rural Advancement Committee (BRAC) (2000) *Combating a Deadly Menace: Early Experiences with a Community-Based Arsenic Mitigation Project in Bangladesh*, Research Monograph Series No. 16, BRAC, Dhaka.

Berg, M., Tran, H.C., Nguyen, T.C. *et al.* (2001) Arsenic contamination of groundwater and drinking water in Vietnam: A human health threat. *Environmental Science and Technology*, **35**(13), 2621–26.

Caceres, D.D., Pino, P., Montesinos, N. *et al.* (2005) Exposure to inorganic arsenic in drinking water and total urinary arsenic concentration in a Chilean population. *Environmental Research*, **98**(2), 151–59.

Code of Federal Regulations (CFR). US Government Printing Office, Superintendent of Documents, Washington, DC.

Codigo Alimentario Argentino (CAA) (1994) Art. 1 Res. MS y AS N8 494. Ley 18284. Dec. Reglamentario 2126. Anexo I y II, *Marzocchi*, Buenos, Aires.

Department of Environment (Bangladesh) (1997) *The Environmental Conservation Rules, 1997*, Ministry of Environment and Forests, ECR '97, pp. 197–227. Also: http://www.moef.gov.bd/html/laws/env_law/178-189.pdf. Last accessed October 14, 2008.

Duker, A.A., Carranza, E.J.M. and Hale. M. (2005) Spatial relationship between arsenic in drinking water and *Mycobacterium ulcerans* infection in the Amansie West district, Ghana. *Mineralogical Magazine*, **69**(5), 707–17.

European Union (1998) Council directive 98/83/EC on the quality of water intended for human consumption. *Official Journal of the European Communities*, **L 330/32**, 32–54.

Federal-Provincial-Territorial Committee on Drinking Water of the Federal-Provincial-Territorial Committee on Health and the Environment (2007) *Guidelines for Canadian Drinking Water Quality: Summary Table*, Health Canada.

Guo, H. and Wang, Y. (2005) Geochemical characteristics of shallow groundwater in Datong Basin, northwestern China. *Journal of Geochemical Exploration*, **87**(3), 109–20.

Indian Bureau of Standards (1991) *Specifications for Drinking Water*, IS 10500, New Delhi.

Japanese Ministry of the Environment, http://www.env.go.jp/en/standards/. Last accessed October 14, 2008.

Kutle, A., Oreščanin, V., Obhodaš, J. and Valkov.c, V. (2004) Trace element distribution in geochemical environment of the island Krk and its influence on the local population. *Journal of Radioanalytical and Nuclear Chemistry*, **259**(2), 271–76.

Mendoza, O.T., Hernández, Ma.A.A., Abuncis, J.G. and Mundo, N.F. (2006) Geochemistry of leachates from the El Fraile sulfide tailings piles in Taxco, Guerrero southern Mexico. *Environmental Geochemistry and Health*, **28**(3), 243–55.

Ministry of the Environment, Natural Resources, and Fisheries of Mexico, *Mexican Guidelines for Industrial Discharge Waters*. 1996, NOM-002-ECOL-1996.

National Health and Medical Research Council (2004) *Australian Drinking Water Guidelines 6*, Natural Resource Management Ministerial Council, Australian Government.

New Jersey Administrative Code. State of New Jersey Department of Environmental Protection, Bureau of Safe Drinking Water, Safe Drinking Water Act Regulations, 7:10-5.2. Also see: New Jersey Department of Environmental Protection: http://www.state.nj.us/dep/.

Rahman, M.M., Sengupta, M.K., Ahamed, S. *et al.* (2005) The magnitude of arsenic contamination in groundwater and its health effects to the inhabitants of the Jalangi - one of the 85 arsenic affected blocks in west Bengal, India. *Science of the Total Environment*, **338**(3), 189–200.

Republic of China Environmental Protection Administration (1998) *National Drinking Water Quality Standards*, Environmental Protection Administration, Taiwan, Republic of China.

Shrestha, R.R., Shrestha, M.P., Upadhyay, N.P. *et al.* (2003) Groundwater arsenic contamination, its health impact and mitigation program in Nepal. *Journal of Environmental Science and Health - Part A Toxic/Hazardous Substances and Environmental Engineering*, **38**(1), 185–200.

Takanashi, H., Tanaka, A., Nakajima, T. and Ohki, A. (2004) Arsenic removal from groundwater by a newly developed adsorbent. *Water Science and Technology*, **50**(8), 23–32.

World Health Organization (WHO). (1998) *Guidelines for Drinking-Water Quality*, 2nd edn, Geneva, World Health Organization.

Zhang, J.J., Parkpian, P. and Wu, L.P. (2003) Arsenic risk assessment in Ronphibun District, Thailand. *Toxicology*, **191**(1), 62.

Index

Page numbers in *italic* refer to tables or figures.

Readers seeking definitions of terms are directed to the comprehensive glossary on pages 437–471

Arsenic   Edited by Kevin R. Henke
© 2009 John Wiley & Sons, Ltd